Lenses are used in a great variety of applications, ranging from photography to fiberoptic communications systems. This book is a comprehensive introduction to lens design, covering the fundamental physical principles and key engineering issues. It describes clearly how to carry out the design of a lens, from the initial layout to the final analysis and tolerance evaluation. In illustrating this process, several practical examples of modern computer-aided lens design are worked out in detail from start to finish.

The basic theory and results of geometrical and physical optics are presented early on in the book, along with a discussion of optical materials. Aberrations, and their correction, and image analysis are then covered in great detail. Subsequent chapters deal with design optimization and tolerance analysis. Several design examples are then given, beginning with basic lens design forms, progressing to zoom and aspheric lenses, and to advanced systems, such as gradient index and diffractive optical components.

In covering all aspects of optical design, from the fundamental physical principles through to the use of modern lens design software, this book will be invaluable to students of optical engineering as well as to anyone engaged in optical design at any stage.

THE ART AND SCIENCE OF OPTICAL DESIGN

THE ART AND SCIENCE OF OPTICAL DESIGN

ROBERT R. SHANNON
University of Arizona

PUBLISHED BY THE PRESS SYNDICATE OF THE UNIVERSITY OF CAMBRIDGE
The Pitt Building, Trumpington Street, Cambridge CB2 1RP, United Kingdom

CAMBRIDGE UNIVERSITY PRESS
The Edinburgh Building, Cambridge CB2 2RU, United Kingdom
40 West 20th Street, New York, NY 10011-4211, USA
10 Stamford Road, Oakleigh, Melbourne 3166, Australia

First published 1997

Printed in the United States of America

Typeset in 11/14pt Times New Roman

Library of Congress Cataloging-in-Publication Data

Shannon, Robert Rennie, 1932–
The art and science of optical design / Robert R. Shannon.
p. cm.
ISBN 0-521-45414-X (hc)
1. Lenses–Design and construction. I. Title.
QC385.2.D47S53 1997
681′.423–dc20 96-2867
CIP

A catalog record for this book is available from the British Library

ISBN 0 521 45414 X hardback
ISBN 0 521 58868 5 paperback

To Helen,
the light of my life,
and
the focus of my endeavors.

CONTENTS

PREFACE

The purpose of this book is to provide an introduction to the practice of lens design. As the title suggests, successful design will require the application of individual creativity as well as artful manipulation and thorough comprehension of the numeric tools available in lens design programs. The technology, user connection, and breadth of the commercial lens design programs have reached a very high level. The availability of inexpensive, high-speed personal computers and user-friendly operating systems has brought the computational tools within economic reach of any individual.

This book covers the basics of image formation, system layout, and image evaluation, and contains a number of examples of lens designs. There are several excellent books in existence that are principally a compilation of the results of the design of several types of lenses. In this book, it is my intention to describe the process rather than the results. The explanations of the basics are provided here in a practical manner and in a level of detail sufficient to provide an understanding of the principles. The selected examples of designs include a narrative of the thinking and approach toward the decisions that need to be made by the designer when carrying out the work. The principles are, of course, independent of the software used. Each example shown does provide the opportunity to exploit different avenues of approach to the design.

Several different lens design programs have been used to provide the majority of the illustrations in this book. All of the programs can handle most of the problems posed in this book, but of course in somewhat different ways. The capabilities of the programs do, of course, differ, and some programs will be easier to use for a specific purpose than others. The selection of the programs for a given purpose was almost random, but knowledge of the program properties did play some role. The reader is not to presume that any program is better in certain operations or different from the others except as may specifically be mentioned in the text. In general one or another program is used consistently for a specific design example, although in some cases it is useful to show the same or similar data as it appears from different programs. In order to give credit to the software supplier, the source of the information used in each figure will be identified with the program. In several cases, to improve clarity within the page space limitations of this book, the graphical output of the program has been rearranged to retain just the significant information.

It is important to note the following. CODE V® is a trademark of Optical Research Associates Inc.; OSLO® is a trademark of Sinclair Optics, Inc.; ZEMAX® is a trademark of Focus Software, Inc.; Excel® is a trademark of Microsoft Inc. The programs used were in versions current in 1995.

There are so many possible optical design problems in existence today that only a few examples can be covered in a single book. The exposition of the principles and approaches discussed here will serve as a basis for a designer exploring other regions of design space. I hope that this book is helpful in conveying the reasons why lens design is done the way it is, and that it provides insight helpful for any designer or design student to understand how to go about the process of lens design.

Tucson, Arizona; January 1996

ACKNOWLEDGMENTS

Any individual depends more on the knowledge and encouragement that he or she gains from those around him or her than often seems the case. I am no exception in this regard. My first introduction to lens design occurred at the Institute of Optics at the University of Rochester in the 1950s. Lens design was just moving from tedious hand calculation toward the broad and interesting field that computers have made possible. I was fortunate to have the opportunity to learn the practical aspects of design from Professor Robert Hopkins, who properly saw the possibilities engendered in the new resources of digital computers. His influence can be seen in many of the ideas and explanations that appear in this book. The basic theory of aberrations was at that time taught by Professor Rudolf Kingslake. The solidity of his organization of the subject served as a basis for understanding much of what goes on inside a lens.

Following education at Rochester, I spent a fascinating decade at Itek Corporation in Lexington, Massachusetts. During that time, the applications of lenses to many topics, including high altitude photographic reconnaissance, began to explode. I am grateful to such people as F. Dow Smith and Richard W. Philbrick for encouraging and permitting me to spend my career in industry in the many interesting projects that provided much practical experience in the design and fabrication of optical systems. During that time I benefited greatly from association with such now well-known colleagues in the business as Edward O'Neill, Berge Tatian, Robert Hilbert, Richard Forkey, Richard Barakat, and many others who provided an opportunity to understand the different aspects of the total system problem.

A special note needs to be made of the contributions of the late H. H. Hopkins, who during this time taught me much about the basic theory of image formation. Conversations with Dave Grey and Warren Smith added much to my understanding of the theoretical and practical basis of lens design.

A move, in 1969, to the then new and growing Optical Sciences Center at the University of Arizona provided a new opportunity. Not only did I have to work on design problems, I had to learn to explain to students how to do design. This provided many new, and frequently unexpected, challenges. Since beginning at Arizona, more than 300 students have been through the two-course sequence of design offered at the Optical Science Center. I have also been the supervisor to forty-five PhD and MS students. All of these students contributed to my education in the subject as I did to theirs. I have to especially thank my faculty colleagues at the Center for discussions and insight provided over the years. In particular, Aden Meinel, James Wyant, Roland Shack, William Wolfe, George Lawrence, and John Greivenkamp are to be particularly thanked.

Within this book, several different lens design programs are used to describe the process of design. The program suppliers have been extremely generous in permitting me to have access to their product, and have made the programs freely available over the years to the students in the design courses at the Optical Sciences Center. I wish to express extreme gratitude to Robert Hilbert, Tom Harris, Chuck Rimmer, Rick Juergens, and others at Optical Research Associates for access to CODE V. Doug Sinclair at Sinclair Optics has provided access to successive versions of the OSLO series of programs. Ken Moore of Focus Software has made available up-to-date versions of the ZEMAX program. In addition, Michael Kidger provided access to his Sigma program, and Don Dilworth to his SYNOPSYS program. These companies and individuals constitute the state of the art in the optical design software business, and have been instrumental in providing the opportunity for a wide variety of engineers and scientists to participate in the process of lens design. These programs are continuously changing, and the versions next year will undoubtedly be even better than those of today.

Credit for getting me interested in writing a book goes to Beatrice Shube, a good friend for many years. At Cambridge University Press, Simon Capelin and Philip Meyler provided the opportunity to convert thoughts into printed material. The excellent work by Richard Cook and others at Keyword Publishing Services, Ltd is gratefully acknowledged.

Finally, I must express the deepest thanks to my family. Betsy, Barbara, Jennifer, Amy, John, and Rob provided the stimulus to find a productive and profitable career. But beyond that, they made, and still make, life fun. Without the constant encouragement, friendly criticism, and willing partnership of my wife, Helen Lang Shannon, none of this would have been possible.

1 INTRODUCTION: THE ART AND SCIENCE OF OPTICAL DESIGN

1.1 Science and Art in Optical Design

The process of optical design is both an art and a science. There is no closed algorithm that creates a lens, nor is there any computer program that will create useful lens designs without general guidance from an optical designer. The mechanics of computation are available within a computer program, but the inspiration and guidance for a useful solution to a customer's problems come from the lens designer. A successful lens must be based upon technically sound principles. The most successful designs include a blend of techniques and technologies that best meet the goals of the customer. This final blending is guided by the judgment of the designer.

Let us start by looking at a lens design. Figure 1.1 shows the layout of a photographic type of lens, showing some of the ray paths through the lens. The object is located a long distance (100,000,000 mm) to the left. This is what the computer program considers equivalent to an infinite object distance. The bundles of rays from each object point enter the lens as parallel bundles of rays. Each ray bundle passes through the lens and is focused toward an image point. On the lens shown, the field covered is 21° half width, which defines the size of the object that will be imaged by the lens.

The diameter of the bundles of light rays entering the lens determines the brightness of the image, and is established by the aperture stop of the lens. The aperture stop is represented by a physical aperture placed on a surface within the lens. Because the object is at infinity, the image will be formed in the focal plane of the lens. The focal length of the lens provides a scaling factor between the angular coordinates on the object and the linear coordinates on the image plane of the lens. The lens shown here has a focal length of 50.0 mm, and accepts an incoming bundle in the center of the field which is 26 mm in diameter. Using standard optical designations, this lens is a 50 mm

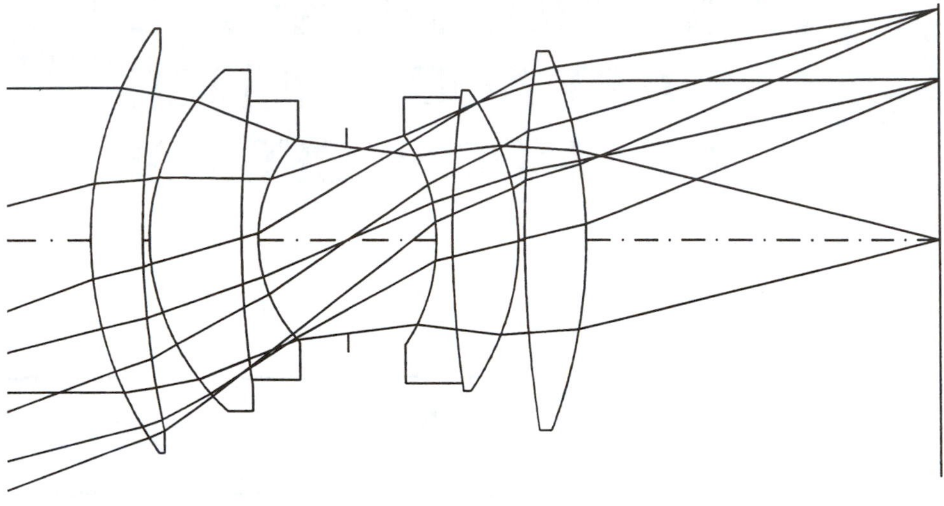

FIGURE 1.1 Layout plot of a double Gauss lens, indicating paths of on- and off-axis ray bundles (CODE V).

effective focal length lens, F/1.92, covering a 21° half field. This is approximately a "normal" lens for a 35 mm camera.

The parameters describing the field size, axial aperture, and image size are first order properties of the lens. In a perfect lens, these properties would be all that is needed to describe the image formation, as each point on the object would be represented as a perfect point in the image. This is as far as simple optics needs to go. If image formation were aberration free once the selection of the image scale, location, and brightness were stated, there would be no difficulty in lens design. Also this book would not have a reason to exist.

The reality is that the laws of geometrical and physical optics do not permit the formation of a perfect image except in a very small number of simple cases. In the lens being considered here, the image formation is not, and cannot be, entirely perfect. The lens shown contains a number of individual elements, chosen by the designer and optimized on the computer to obtain the best possible solution to the customer's requests, within a set of practical boundary conditions.

Image formation will be limited by the aberrations intrinsic to the passage of light through the lens. One measure of the aberration is the size of the blur of rays surrounding the central ray through the aperture stop. Another is the extent to which the optical paths along each ray in a bundle through the lens differs from the optical path along the central ray from the object. The aberrations are also dependent upon the color or wavelength of the light within each ray bundle, and by the distortion in the image, or the extent to which the central ray of each bundle fails to intersect at each image point determined by the first-order optical description of the image.

This lens consists of several elements; the shapes and locations of which are determined by the lens designer to provide the best possible match between the customer's needs and the physical limitations upon image formation. The laws of geometrical optics determine the passage of rays through a lens system. The laws of physical optics determine how the light within each bundle combines to form an image of each object point. The description of a lens in terms of aberrations and image quality can be calculated to any desired accuracy. All of the descriptions are based upon numerical computation, and are only approximately represented by analytic equations. A closed-form analytic description of the image forming process does not exist for a practical lens. This is due to the nonlinearity relating the angles of incidence and refraction of each ray, and because of the complexity in computing the physical image due to diffraction of light.

The lens designer is working with a very complicated system of physical components that can be described in numerical detail. The complexity is such that there are possibly hundreds or even thousands of closely equivalent solutions for each set of parameters provided by the customer for the lens. Thus, although the scientific or technical description of the lens can be expressed to any accuracy by using a computer, the artistry of the lens designer is required to guide the design process and select the best solution from the myriad possible "close fits" to the required lens parameters.

There are explicit parameters, such as the required focal length, but there are also somewhat "hidden" parameters that drive the path of the design. Vignetting, which is the deliberate reduction of irradiance off axis by proper selection of element diameters, is an extremely important tool in eliminating some of the worst parts of off-axis aberrations. This is a tool used frequently by designers in finding the best possible and most economic solution to design problems. This vignetting effect changes with the setting of the working aperture of the lens, and must be agreed to by the designer and the customer as part of the design goals.

Description of image formation requires tracing about five rays at each field, to produce about sixty different aberrations to correct. This amounts to 180 ray surfaces to be computed for each examination of the aberrations in this sample lens. Each step in the optimization requires that each parameter be changed and a finite differential formed of the change in each of the aberrations. This will require 6,480 ray surfaces for each step in the computation. The six-element lens shown contains at least thirty-six possible parameters. Even though a computer can trace rays at a fantastic rate, about 50,000 to 200,000 ray surfaces per second on a high-level personal computer, it is not possible to explore all possible alternatives for a practical lens. If it is assumed that each parameter can take any of 10,000 distinct values, the complete evaluation would require about 10^{108} possible combinations be evaluated. At a rate of about 0.1 seconds per evaluation this would take approximately 10^{99} years. There obviously is a better way to proceed.

The artistic part of design extends beyond the selection of the image quality to include the mechanical layout, selection of materials and minimization of problems with tolerance required for fabricating a successful lens. A high-speed computer permits optimization to proceed very rapidly. In the usual case, hundreds of possible designs can be evaluated in an hour, and the optimum selection of these made by the use of a computer program. Few of these are desirable or acceptable solutions. Only in a very small number of cases can the design be completed without application of the judgment of a designer to guide the outcome of the computer program.

It should be evident by now that successful design is more than computer program manipulation. Conversely, an understanding of how to creatively use the process of computerized optimization is essential to successful designing. The art of the designer begins with definition of the starting point. No matter how effective the optimization process, an inappropriate selection of starting point can lead to a failure in the design.

As the design proceeds, alteration of the requested goals is important. The designer needs to learn from the steps taken by the optimization process what the limitations are for the lens, and how the parameters and the merit function describing the goals can best be altered.

Understanding of the basis of geometrical optics and aberration content in lenses is essential to knowing when to stop. Continuing to attempt to optimize a lens whose image forming capabilities cannot be improved is inefficient, costly, and a bit foolish. Completing the design with a lens that cannot be fabricated is equally foolish.

By now it is obvious that successful optical design is more than mere computer operation and interpretation of ray traces. The successful designer must be aware of many properties of lenses that will affect the eventual outcome of the design. Figure 1.2 shows a short list of many of the items that should be considered by a designer approaching a lens design. It is likely that others can add (but certainly not eliminate) many items to this list.

In this list, the items in bold type are those that are covered in some detail in this book. The items in italics are mentioned or discussed in somewhat less detail and completeness. The ordinary type face lists some of the topics that a designer needs to be aware of that are not discussed in any detail in this book.

Focal length
Field angle or field size
F/number
Numerical aperture
Wavelength and spectral range
Magnification
Magnification range
Type of lens
Back focus
Front focus
Pupil locations
Illumination
Irradiance uniformity
  vignetting
  transmission
Ghost images
Distortion
Variation with conjugates
Variation with spectral region
Size and configuration
Folding components
Interference with optical path
Zoom range
Zoom mechanization
Focus mechanization
Image quality
  Aberrations
  Resolution
  OTF
  MTF
  Energy concentration
  Effect of aperture stop
Scattered light
Polarization
Veiling glare
Light baffling
Off-axis rejection
Field stop definition
Diffraction effects
Tolerances
Depth of focus
Interface with variable aperture
Interface with autofocus system
Image quality at various apertures
Cost of design
Cost of prototype
Cost of production
Schedule and delivery time
Optical interfacing with instrument
Materials
  Availability
  Cost
  Continued supply
  Suitability for processing
  Environmental considerations
  Hazardous materials
Environment
  Temperature range
  Storage conditions
  Atmospheric pressure
  Humidity
  Vibration and shock
Availability of subcontractors
Level of technology
Coatings
  Transmission
  Reflectivity
  Absorption
  Availability
  Risk
  Environmental effects
Weight
Moment about mounting
Producibility
Manufacturability
Manufacturing processes
Mounting procedures
Mounting interfaces
Mechanical interface with instrument
Detector
  Photographic
  Sampling array
  Signal to noise
Surface finish, cosmetics
Beam parameters
Radiation damage
Irradiance damage
Prior experience
Track record
Prior art
Patentability
Patent conflict situation
Competitive situation
Marketability
Interface to other products
Lifetime of product
Rate of production
Environmental hazards
Liability issues
Delay to market
Timing of disclosure
Integration with products
Customer view of product
Styling
Financial viability
Investment requirements
Investment risk
Access to funding

FIGURE 1.2 Some of the important topics that a successful lens designer needs to address during the process of designing a lens.

1.2 Starting a Design

The starting point is data supplied by a customer for the lens. The goals for the lens need to be stated in a form that can be translated by a designer into the initial selection of parameters for the lens. Frequently the specifications will be unclear, sometimes redundant or conflicting with physical reality. Sometimes the customer or intended user is not well versed in optics and image formation, leading to some innocently expensive or physically impossible requirements.

The designer has the responsibility of sorting through the specifications and constructing a realistic set of goals for the design. Since the specifications usually include cost and delivery, the refining of lens specifications is not entirely a technical activity.

Figure 1.2 includes most of the properties of a lens that need to be included in a complete set of specifications. The first-order optical specifications, near the beginning of the list, are required to establish the paraxial base set of coordinates in which the image will be evaluated. These quantities should be familiar to anyone who has ever taken an optics course. The image quality requirements describe how faithfully the image reproduces the object. Although radiometry would seem to be a property of the first-order requirements, additional effects such as losses by transmission through the lens and the vignetting that is used to control image quality off axis in some lenses are image quality considerations. In Chapter 2 on basic optics, the fundamental relationship between paraxial coordinates, radiometry of images, and change of image quality across the field are discussed.

Mechanical and fabrication requirements deal with the need to be able to produce a lens if it is to be of any use to the customer. The tolerances break into three parts. The first are the requirements on construction parameters to ensure that the image falls in the proper location and contains aberrations within an acceptable degradation from the base system. The second deals with the need to use the lens in a defined environment. The third specifies the acceptable irregularity and randomness that can be allowed on the surfaces of the lens to control both aberrations and scattered light. Specification of these tolerances requires running an emulation of the design of the lens, in which the computer perturbs the state of the lens according to specific algorithms and calculates the likelihood that any lens assembled within the tolerances will meet the requirements.

The final set of specifications deals with the "other" things about lenses that are important. Significant here are cost and schedule for delivery. Success in meeting these specifications actually is very closely tied to the choice of the parameters and tolerances that are needed for the lens.

The type of lens that is selected needs to be responsive to these requirements. There are many hidden considerations, such as the focal length, weight, spectral range, and actual required image quality that will affect the choice of lens type. The number of elements permitted also influences the type of lens that is used, as available space and cost are two of the perennial difficult-to-express limits that are used in design. The discussion in Chapter 7 on the design of specific types of lenses will indicate how this selection process is developed, and will permit modification of the starting configuration by the designer.

In many cases, the starting point can be obtained from a basic optical layout, beginning with a first-order calculation. In Chapter 7, the discussion of doublets and triplets will be carried out using this approach. For more complicated lenses, such as Double Gauss types, wide-angle and some types of zoom lenses, the best starting point is usually prior art. In most cases this will be from a lens data library or a lens patent.

1.2.1 DETAILED DESCRIPTION OF A LENS

A lens can usually be described in terms of only a few parameters, any of which may become variables in the subsequent optimization process. The basic lens consists of an ordered set of spherical surfaces, separations, or thicknesses, and refractive or reflective materials. The surfaces are stored in sequential numbered order with the curvature, thickness to the next surface, and the index of refraction of the medium after the surface attached to each surface number. Additional information about the surface shape, orientation, and dimension may also be attached to the surface number.

In most designs, the goal is to produce a lens which provides uniform imagery over a two-dimensional image plane. Therefore, most lenses of interest will consist of rotationally symmetric surfaces. Data for the lens shown in the previous section is listed in Figure 1.3. In addition to the surface data, certain other information describing the optical operating conditions of the lens is also listed. Although each lens design program will have its own set of acronyms and format for the lens data, the goal for all is the same.

The specific lens data actually put into a computer will usually appear in a different form, as in Figure 1.4, which is a data input file for the CODE V program. There is a natural flow to this form of input, and each program will have its own set of acronyms and abbreviations that make the job of communication simpler. Most programs also permit interactive input of the lens data in a spreadsheet format as in Figure 1.5. Here the user is prompted for the data necessary to complete the lens data set. In most cases the most likely

```
Sample Lens
                RDY              THI              GLA               CCY   THC
> OBJ:       INFINITY         INFINITY                               100   100
     1:      31.79843         4.009367         LAF2_SCHOTT             0   100
     2:      81.76294         0.500000                                 0   100
     3:      19.00541         7.206837         LAF2_SCHOTT             0   100
     4:      90.39788         1.303044         SF3_SCHOTT              0   100
     5:      12.41339         7.000000                                 0   100
   STO:      INFINITY         7.000000                               100   100
     7:     -15.50210         1.303044         SF64A_SCHOTT            0   100
     8:      91.47280         5.242247         LAF2_SCHOTT             0   100
     9:     -21.41355         0.500000                                 0   100
    10:     116.05450         4.800000         LAF2_SCHOTT             0   100
    11:     -43.97434        27.980887                                 0   PIM
   IMG:      INFINITY        -0.066048                               100     0

 SPECIFICATION DATA
    EPD        25.00000
    DIM              MM
    WL           656.30     587.60     486.10
    REF               2
    WTW               1          1          1
    XAN         0.00000     0.00000      0.00000
    YAN         0.00000    14.70000     21.00000

 APERTURE DATA/EDGE DEFINITIONS
    CA
    CIR S1                  16.004532
    CIR S2                  15.526973
    CIR S3                  12.769472
    CIR S4                  10.368361
    CIR S5                   8.307168
    CIR S6                   7.578557
    CIR S7                   8.530869
    CIR S8                  10.655071
    CIR S9                  11.220801
    CIR S10                 13.938192
    CIR S11                 14.269285

 REFRACTIVE INDICES
     GLASS CODE                      656.30       587.60       486.10
    SF3_SCHOTT                     1.732416     1.739997     1.758671
    SF64A_SCHOTT                   1.699097     1.705846     1.722403
    LAF2_SCHOTT                    1.739046     1.743999     1.755690

NFINITE CONJUGATES
    EFL        49.9998
    BFL        27.9809
    FFL       -19.4621
    FNO         2.0000
    IMG DIS    27.9148
    OAL        38.8645
    PARAXIAL  IMAGE
     HT        19.1931
    ANG        21.0000
    ENTRANCE  PUPIL
     DIA       25.0000
     THI       27.1596
    EXIT  PUPIL
     DIA       26.8115
     THI      -25.6419
```

FIGURE 1.3 A sample listing of the data describing a lens (CODE V).

```
RDM;LEN
TITLE 'Sample Lens'
EPD   25.0
NFO   7
FFO   -0.2
IFO   0.05
DIM   M
WL    656.3 587.6 486.1
REF   2
WTW   1 1 1
INI   'RRS'
XAN   0.0 0.0 0.0
YAN   0.0 14.7 21.0
VUX   0.00910997119141 0.0115763173828 0.0994343730469
VLX   0.00910997119141 0.0115763173828 0.0994343730469
VUY   0.00910989619054 0.0461619962916 0.349698303521
VLY   0.00910989619054 0.302554428909 0.497028541314
SO    0.0 5011708836836.3
S     31.7984318367 4.00936706947 LAF2_SCHOTT
  CCY 0
  CIR 16.0045318604
S     81.7629405148 0.5
  CCY 0
  CIR 15.5269727707
S     19.0054101873 7.20683730737 LAF2_SCHOTT
  CCY 0
  CIR 12.7694721222
S     90.3978818762 1.30304429758 SF3_SCHOTT
  CCY 0
  CIR 10.3683605194
S     12.4133854383 7.0
  CCY 0
  CIR 8.3071680069
S     0.0 7.0
  STO
  CIR 7.5785574913
S     -15.5021009259 1.30304429758 SF64A_SCHOTT
  CCY 0
  CIR 8.53086853027
S     91.4728040296 5.24224744333 LAF2_SCHOTT
  CCY 0
  CIR 10.6550712585
S     -21.4135490344 0.5
  CCY 0
  CIR 11.2208013535
S     116.054499482 4.8 LAF2_SCHOTT
  CCY 0
  CIR 13.9381923676
S     -43.974336366 27.9808869798
  CCY 0
  CIR 14.269285202
  PIM
SI    0.0 -0.0660477586985
  THC 0
GO
```

FIGURE 1.4 The same data as in Figure 1.3, but expressed in the serial form used in entering data into a computer program (CODE V).

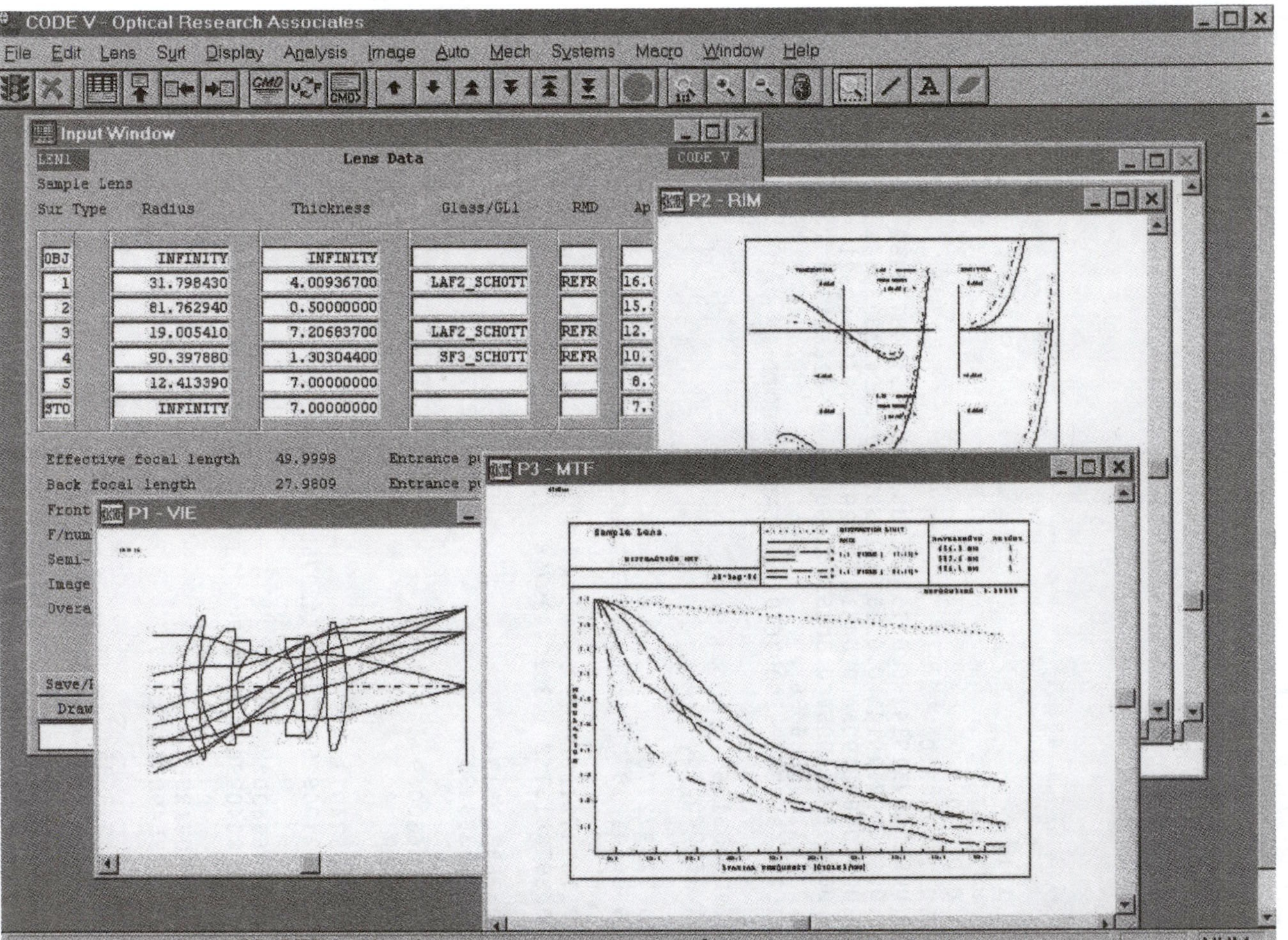

(a)

FIGURE 1.5 Examples of spreadsheet form of lens data input for three different commercially available lens design programs: (a) CODE V; (b) ZEMAX; (c) OSLO.

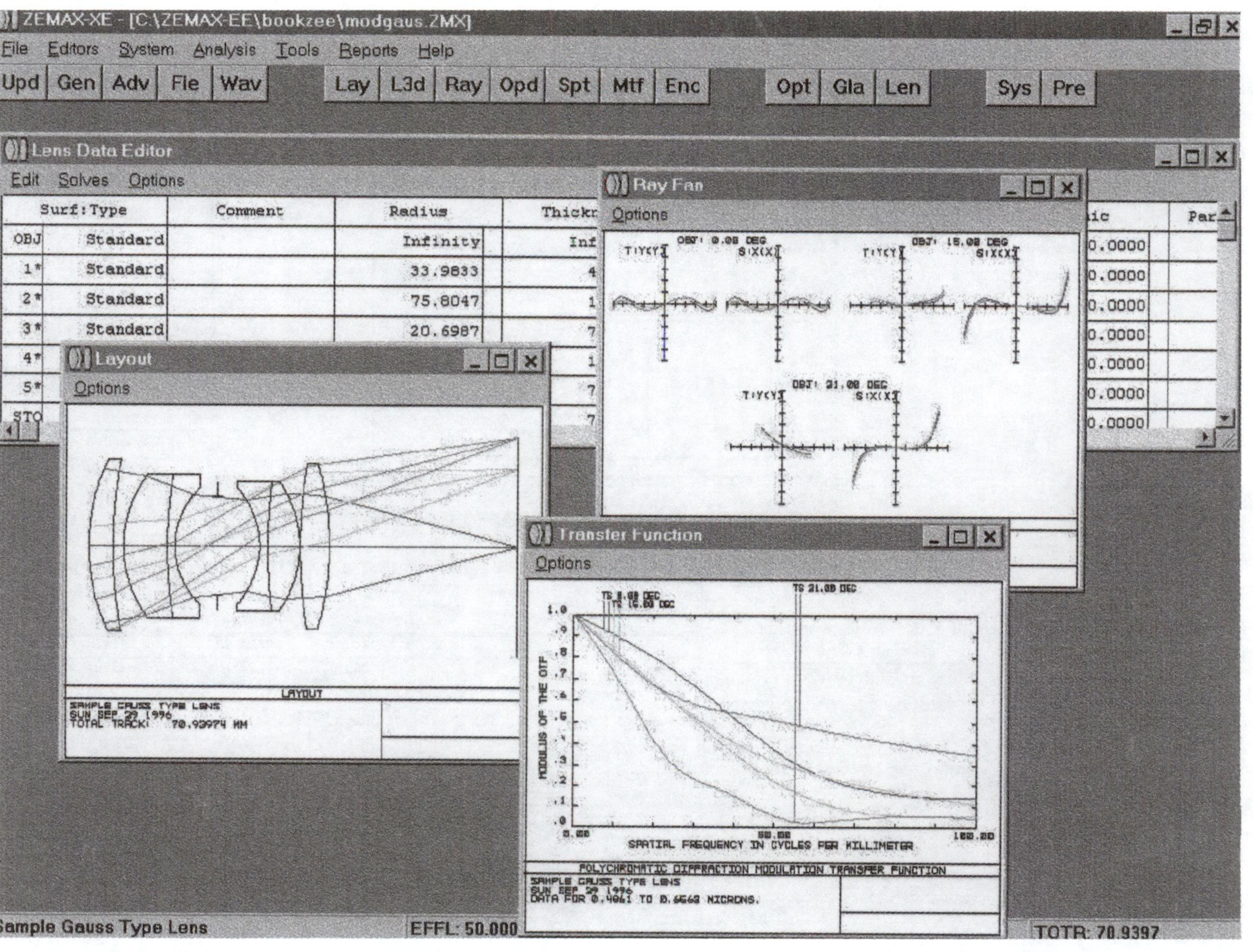
ZEMAX-XE - [C:\ZEMAX-EE\bookzee\modgaus.ZMX]
File Editors System Analysis Tools Reports Help
Upd Gen Adv Fie Wav Lay L3d Ray Opd Spt Mtf Enc Opt Gla Len Sys Pre
Lens Data Editor
Edit Solves Options
Surf:Type Comment Radius Thickn
OBJ Standard Infinity Inf
1* Standard 33.9833
2* Standard 75.8047
3* Standard 20.6987
Layout
Options
LAYOUT
TOTAL TRACK: 70.93974 MM
Ray Fan
Options
Transfer Function
Options
MODULUS OF THE OTF
SPATIAL FREQUENCY IN CYCLES PER MILLIMETER
POLYCHROMATIC DIFFRACTION MODULATION TRANSFER FUNCTION
Sample Gauss Type Lens
EFFL: 50.000
TOTR: 70.9397

(b)

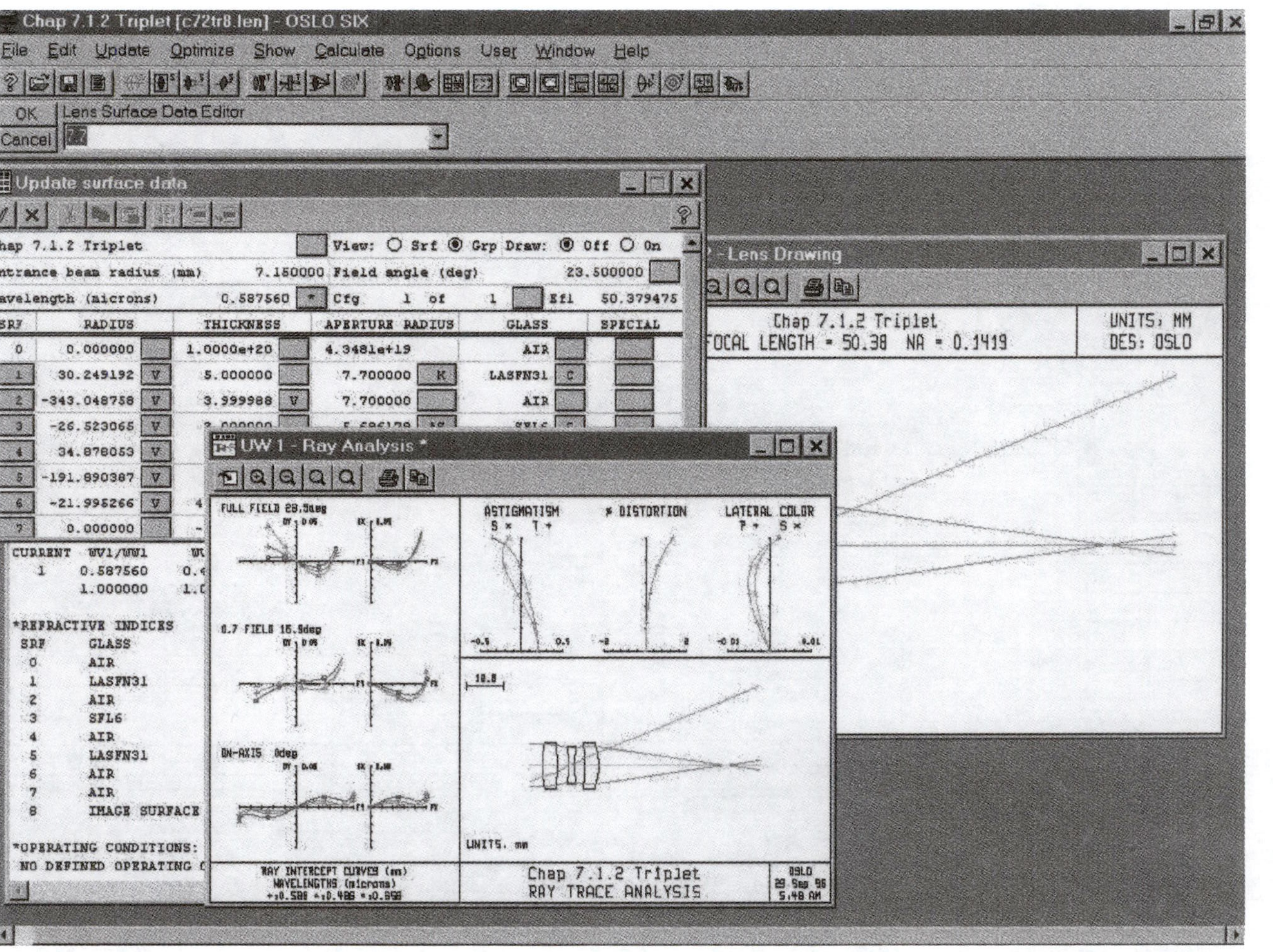

Chap 7.1.2 Triplet [c72tr8.len] - OSLO SIX
File Edit Update Optimize Show Calculate Options User Window Help
OK
Cancel
Lens Surface Data Editor
Update surface data
Chap 7.1.2 Triplet
View: Srf Grp Draw: Off On
Entrance beam radius (mm) 7.150000 Field angle (deg) 23.500000
Wavelength (microns) 0.587560 Cfg 1 of 1 Efl 50.379475
SRF RADIUS THICKNESS APERTURE RADIUS GLASS SPECIAL
0 0.000000 1.0000e+20 4.3481e+19 AIR
1 30.249192 V 5.000000 7.700000 K LASFN31 C
2 -343.048758 V 3.999988 V 7.700000 AIR
3 -26.523065 V
4 34.878053 V
5 -191.890387 V
6 -21.995266 V
7 0.000000
CURRENT WV1/WW1
1 0.587560
1.000000
*REFRACTIVE INDICES
SRF GLASS
0 AIR
1 LASFN31
2 AIR
3 SFL6
4 AIR
5 LASFN31
6 AIR
7 AIR
8 IMAGE SURFACE
*OPERATING CONDITIONS:
NO DEFINED OPERATING
- Lens Drawing
Chap 7.1.2 Triplet
FOCAL LENGTH = 50.38 NA = 0.1419
UNITS: MM
DES: OSLO
UW 1 - Ray Analysis *
FULL FIELD 23.5deg
0.7 FIELD 16.5deg
ON-AXIS 0deg
ASTIGMATISM
S × T +
% DISTORTION
LATERAL COLOR
-0.5
0.5
-2
2
-0.01
0.01
10.8
UNITS, mm
RAY INTERCEPT CURVES (mm)
WAVELENGTHS (microns)
Chap 7.1.2 Triplet
RAY TRACE ANALYSIS
OSLO
29 Sep 96
5:48 AM

(c)

default values are initially placed in the data locations by the program, and the user is permitted to alter these to make a correct input table.

As graphical user interfaces develop and become more common, many new computer programs will provide alternate methods for communication between the designer and the program. In all cases, the basic data describing the lens is the data set upon which all image computation depends. This data set will be altered as the optimization of the lens proceeds.

Understanding of the input system is best acquired by reference to the operating manual of the specific lens design program being used. In most cases, the manuals accompanying software are not examples of enduring literature. Partly because of this literary reputation, the reading of lens design program manuals is not a favorite occupation for most program users, but the effort can be well worth the time invested. In this book, specific procedures for the lens data input will not be examined in detail, but the important lens parameters and the approach to optimization of the lens will be treated in a somewhat universal manner. The details as to how to enter requests for this information will need to be obtained from the specific program manuals.

1.3 Optimizing a Lens

The construction parameters describing the lens are obtained from the starting configuration. A set of these is selected as the variables that may be used in the design. These are the curvatures, spacings, and sometimes the optical materials and aspheric surface shapes. In some cases, tilts and decenters may become variables. In the case of zoom lens design, certain variables are reserved as variable separations to permit the change in magnification during use.

Once the starting parameters are determined, the designer can calculate the initial aberration content of the lens. The designer selects a set of target values for the final aberrations in the lens. The selected combinations form a merit function that describes the state of correction of the lens in a single summary number. In most cases the designer will select a merit function that is suggested, at least in part, by the design program that is being used. Such "default merit functions" have evolved over a number of years, and have reached a high degree of sophistication. As will be seen in Chapter 5, it is essential that the designer understand the reasons for the choice of components of the merit function, as it will often be necessary to adjust the targets and the components of the merit function during the optimization process.

The heaviest use of computer programs in design is in the optimization of lens parameters against a specified merit function. The goal is to minimize the distance in aberration space between the present state of correction of the lens and the desired state of correction. This minimization of residual aberrations needs to be carried out within specified boundaries, such as finite thickness of lens elements and exact correction of specified optical parameters of the lens.

1.4 Evaluating a Design

The aberrations that are used in setting up the merit function are a carefully selected sampling of the state of aberration correction of the lens. The customer or user of the lens is usually interested in how the image quality affects the intended purpose of the lens. This requires computation of the light distribution in the image, including the effect of diffraction.

A rapid idea of the aberration content of a lens can be obtained from examination of the ray intercept plots for a lens. In Figure 1.6, the ray plots for the Double Gauss lens discussed earlier are presented. The method of analyzing these curves is covered in Chapter 3, with significant detail in the design examples (Chapter 7). At this point, it is necessary only to observe that the ordinates are the extent of ray displacement on the image surface, and that the abscissae are coordinates in the pupil of the lens. The shape of the curves is a clue to the nature of the aberration content and the order of the aberration. The vignetting which limits the dimensions of the pupil off axis is also clearly evident in these plots.

A more detailed representation of the ray errors in the lens can be obtained by plotting spot diagrams, or ray scatter diagrams. These are shown in Figure 1.7. The symmetry of the aberrations is somewhat evident in these plots. These rapid assessments of the aberrations are of the geometrical images formed by rays passing through the lens.

A complete assessment of the image formation requires the inclusion of physical optics, or diffraction, in the calculations. Figure 1.8 shows a representation of the light distribution in the diffraction image. This function is usually referred to as the Point Spread Function. Some similarity to the geometrical spot images is expected, and can be observed. This intensity distribution is what would be expected if the lens were made perfectly and the image of point objects measured with a scanning radiometer.

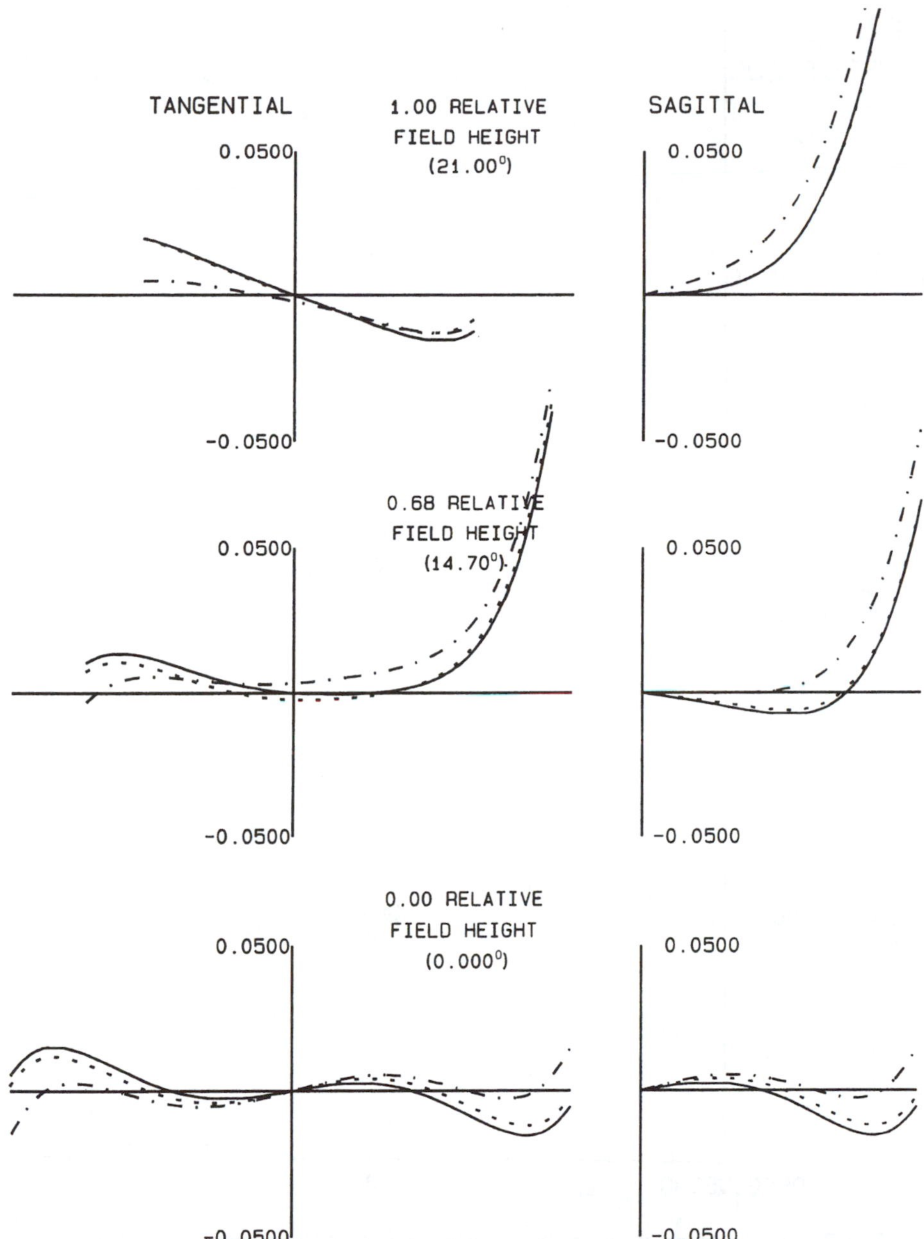

FIGURE 1.6 Ray intercept plots for the sample lens; these provide a measure of the local spread of the rays from several object points across the image surface. . . . . . . , 656.3 nm; ———, 587.6 nm; –·–·–, 486.1 nm (CODE V).

A description of the image in terms of functions that can be combined with objects and detectors would be of more use to system engineers. It has become common to describe the image quality in terms of Optical Transfer Functions. These provide a functional representation of the spatial frequency

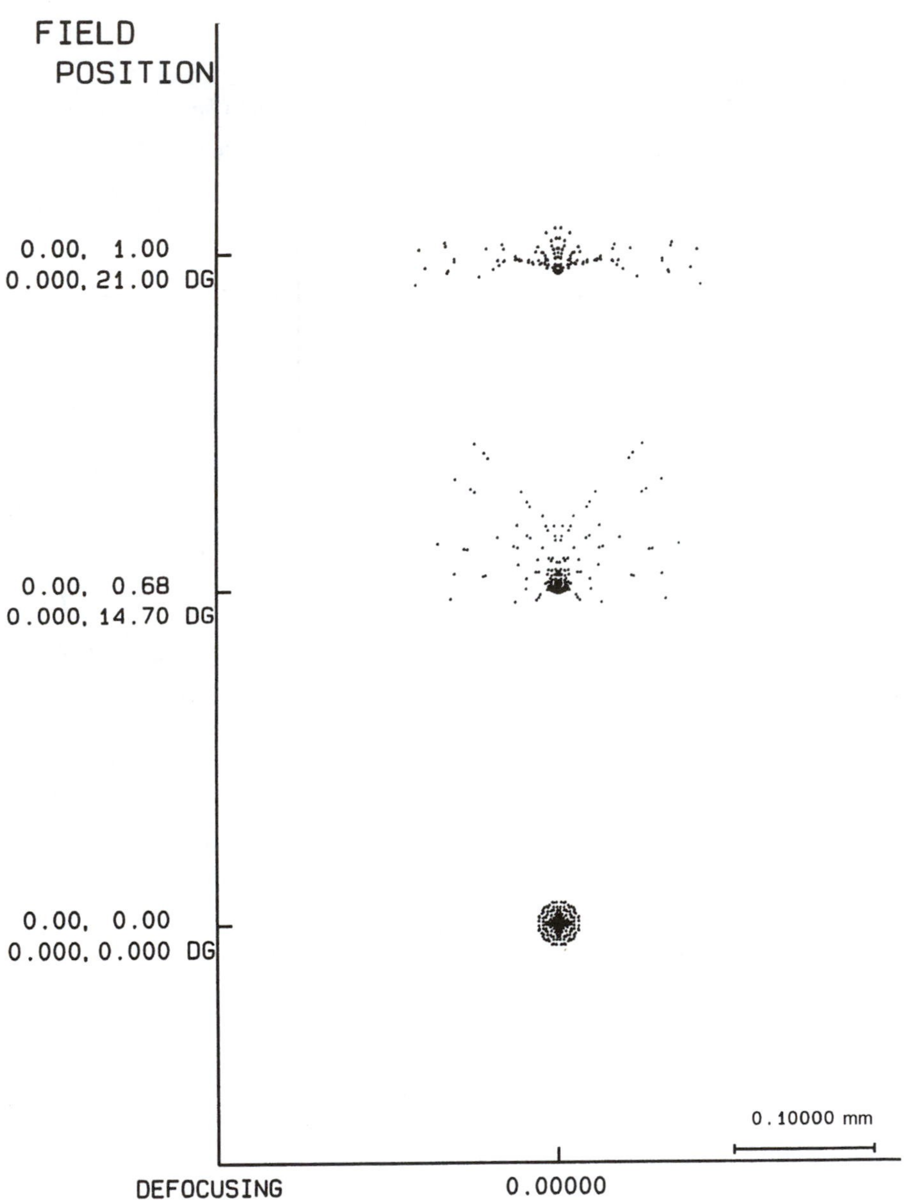

FIGURE 1.7 Spot diagrams produced by the sample lens at the same field positions as the plots in Figure 1.6 (CODE V).

response of the lens. Basically, the OTF is a Fourier transform of the point spread function. In practical applications, the OTF describes the contrast in images of sinusoidal objects of specified spatial frequency. Using these functions to describe the lens permits image analysis similar to the familiar frequency bandpass analysis used in communication systems. In general, the

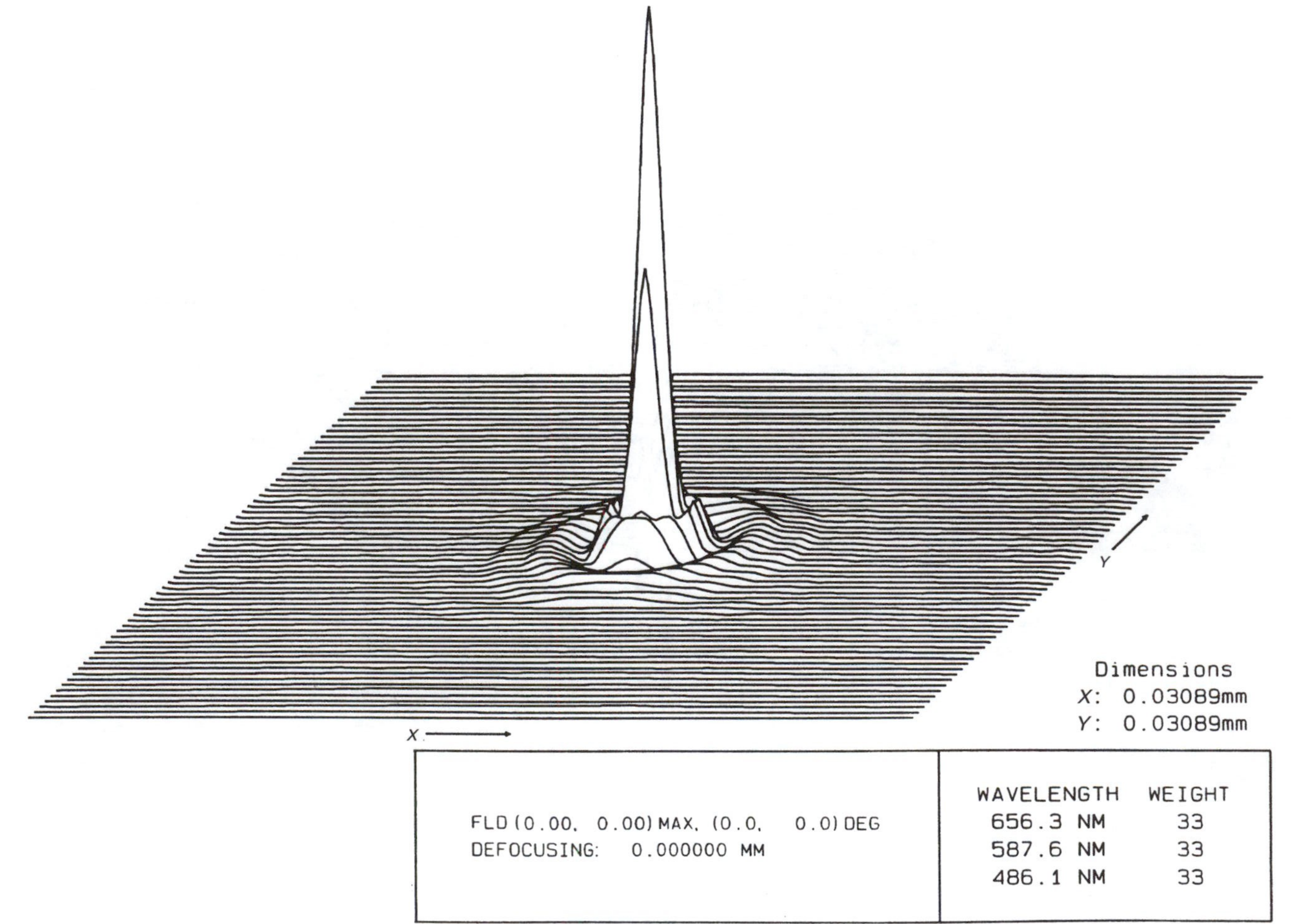

(a)

FIGURE 1.8 Point Spread Function for the images produced by the sample lens. The scale and sample density is changed at each field angle. Discussed in more detail in Chapter 2 (CODE V).

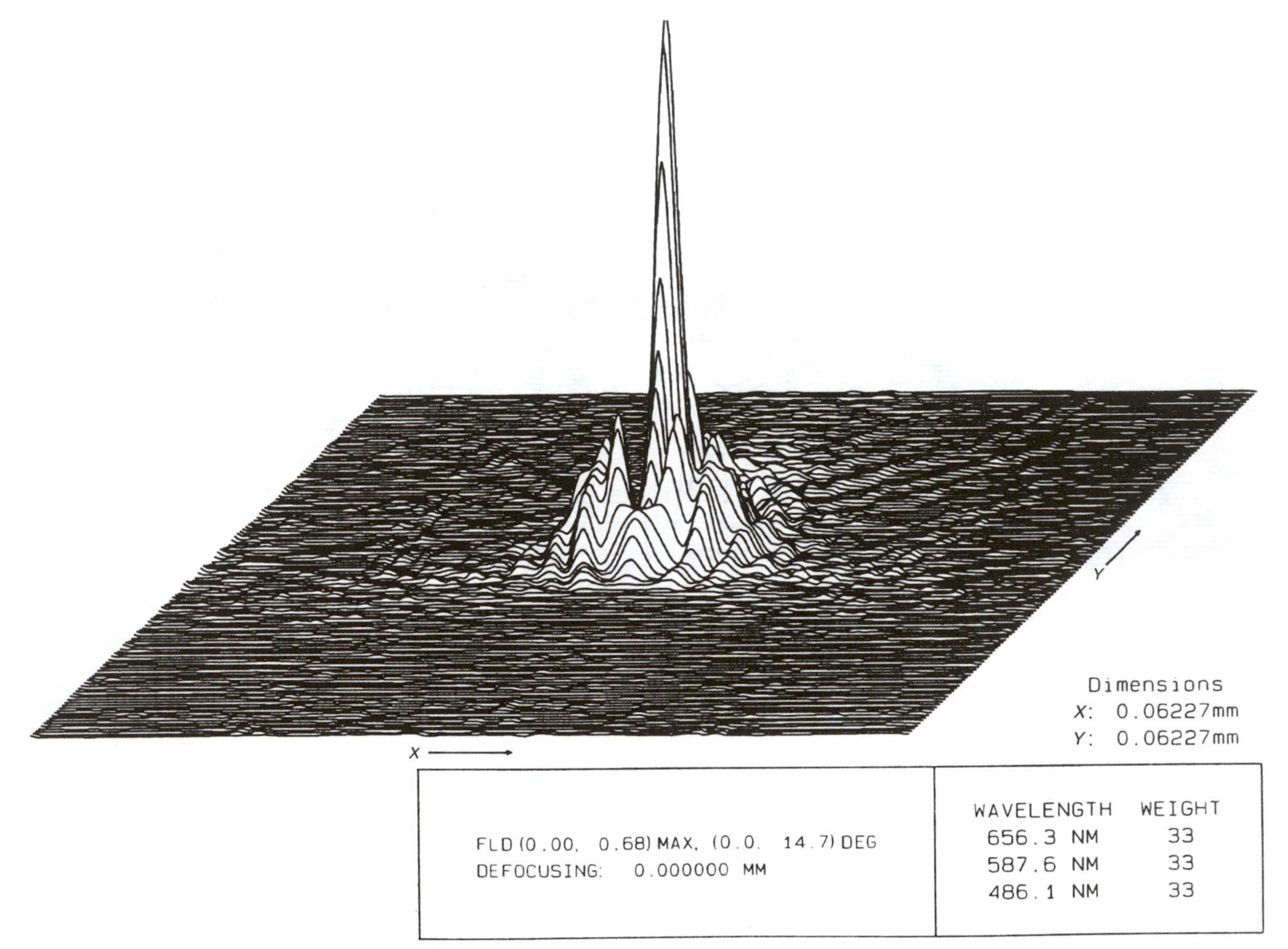
Dimensions
X: 0.06227mm
Y: 0.06227mm
X
Y
FLD (0.00, 0.68) MAX. (0.0, 14.7) DEG
DEFOCUSING: 0.000000 MM
WAVELENGTH WEIGHT
656.3 NM 33
587.6 NM 33
486.1 NM 33

(b)

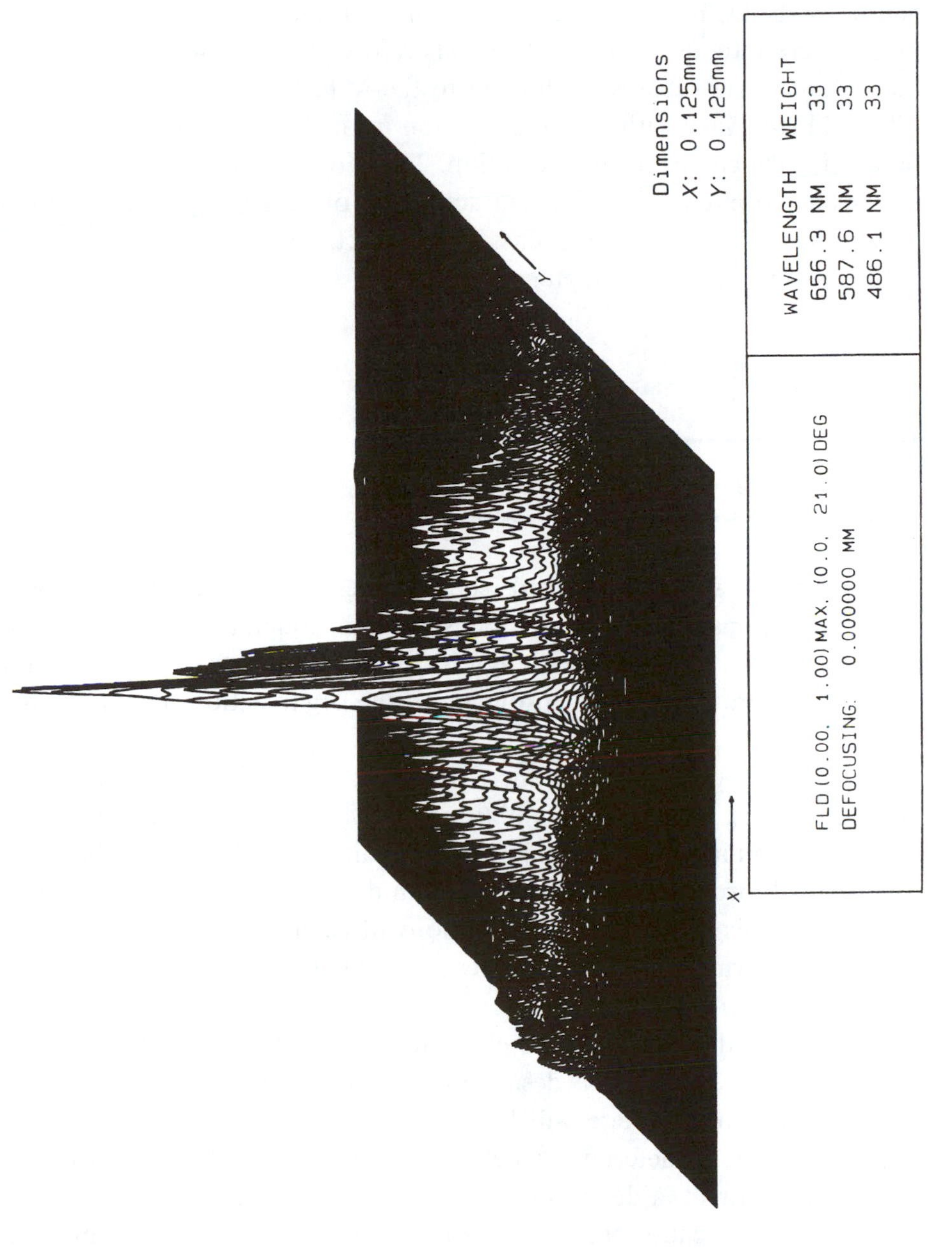

Dimensions
X: 0.125mm
Y: 0.125mm
X
Y
FLD(0.00, 1.00)MAX, (0.0, 21.0)DEG
DEFOCUSING: 0.000000 MM
WAVELENGTH WEIGHT
656.3 NM 33
587.6 NM 33
486.1 NM 33

(c)

modulus of the OTF, the Modulation Transfer Function (MTF) is the function of interest to users of the lens. Therefore most presentations of the transfer functions of lenses are limited to the MTF.

The OTF will vary with the field location in the lens, and will also change rapidly with the choice of focal position. Therefore, there are many transfer functions describing the imaging characteristics of a lens. Figure 1.9 shows a family of MTFs for the sample lens. These functions and their computation are discussed in detail in Chapter 4.

1.5 Completing a Lens Design

The design process can be stopped when the design goals are met. Usually, this is when the image produced by a lens is good enough to meet the needs of the customer. Just stopping the design is not sufficient to complete the task. Fabrication tolerances must be provided to the organization that is to build the lens.

Establishing appropriate tolerances is often more difficult than the actual design of the lens. Tolerances are obtained by varying the parameters of the lens and determining the changes in image quality. The allowable fabrication errors that will provide image quality within desired levels are found from a statistical analysis of possible combinations of each individual error. These are stated in terms that may be measured by the shop fabricating and assembling the lens components.

A drawing of the lens and lens elements that contain these tolerances is delivered as the result of the design process. If the designer is interested, and fortunate enough, there will be an opportunity to be included in the testing of the lens to determine whether the design goals have been met. If the goals are not met, a determination of whether the design or the fabrication process is the cause needs to be made. Modern lens analysis methods provide a sufficient basis for predicting performance of a lens built as specified. There is an art as well as a science to identifying the sources of problems and failures. The designer needs to be able to relate the knowledge obtained during the design to the practical difficulties involved in manufacturing a lens.

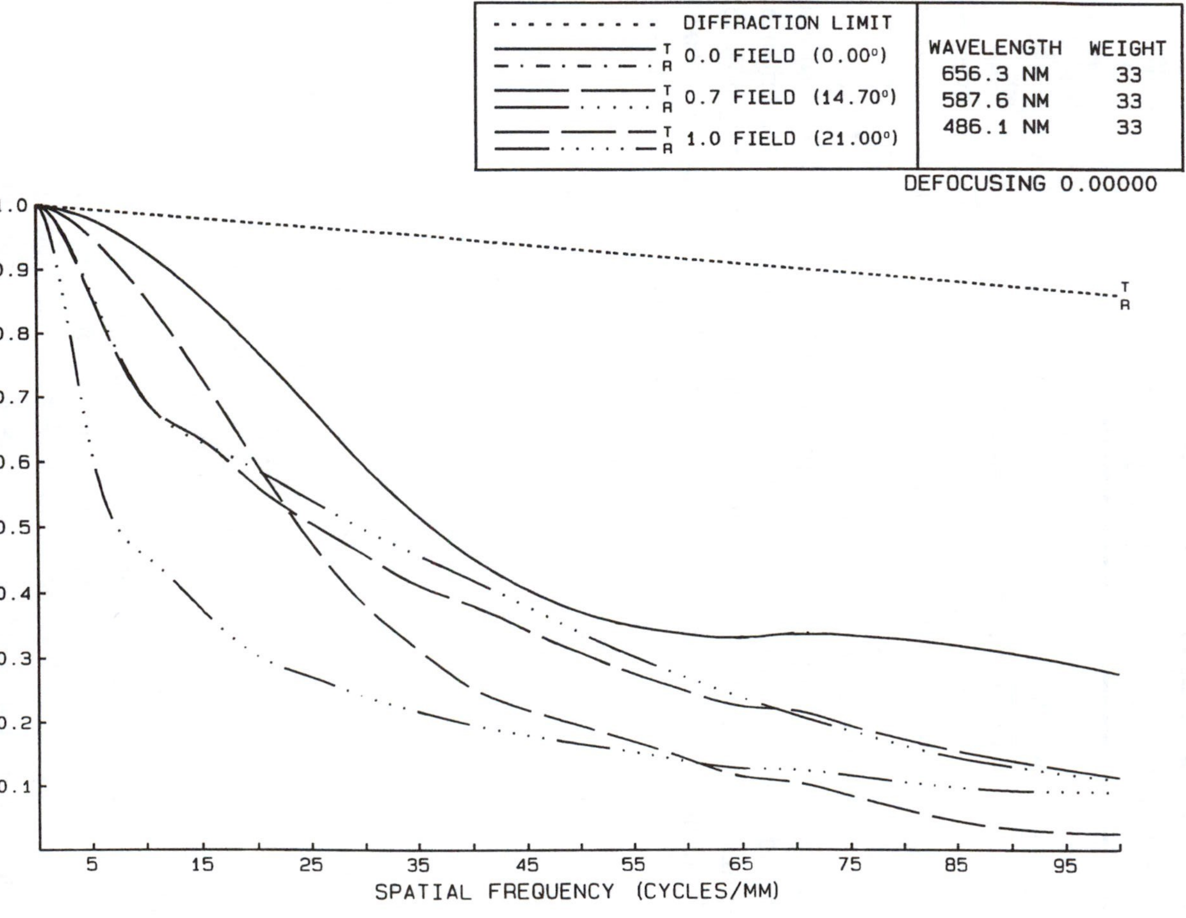

FIGURE 1.9 Modulation Transfer Function (MTF) plots for the sample lens (CODE V).

1.6 This Book

This book is intended to provide a guide to the successful design of a lens. Just enough of the optical theory is included to provide a basis for understanding the numerical procedures that are used in lens design. An investigation of the process of varying the parameters of several typical lens designs is included to permit the reader to understand the process and decisions necessary to direct the course of a highly numerical procedure. Methods for determining the quality of the image produced by a lens, and the methods for deciding upon the allowable tolerances that may be applied in fabricating the lens are also explored. Some effort is placed in demonstrating how the limits to image quality that can be achieved in a lens of specified complexity can be determined. The purpose is to show the designer how to blend judgment and technology in producing an acceptable result.

This book covers the practical aspects of lens design. Chapters 2, 3, and 4 develop some of the basic optical concepts and ideas necessary for understanding how lenses operate. Chapter 5 sketches the process of design using computer optimization. Chapter 6 describes methods of obtaining tolerances for a lens design.

Practical design examples for several types of lenses are worked out in Chapter 7, with the options that must be exercised by the designer explained at each stage. These examples serve as a sample of paths that lead to a final acceptable design. The successful designer will blend some of the basic knowledge in the first chapters with the techniques described in Chapter 7 to develop a personal approach to optical design.

Since design is a practical engineering subject, the approach used in this book is a pragmatic one. Understanding of the operation of lenses and their uses is the priority rather than concentration on a rigorous derivation of optical theory. It will become obvious to the design student that numerical evaluation and understanding of lenses is far more important than extensive theoretical discussion. However, an understanding and appreciation of the sources of aberrations that can limit the performance of lenses is extremely important.

Obviously, lens design involves the use of one or more of several computer programs. With almost no exceptions, the examples to be described in this book can be carried out on any of the leading lens design programs. The input of lens design data, setting of the optimization commands and presenting the evaluation information will differ. Some programs are easier to use for certain classes of design than others. Each program can provide output that provides a good interpretation of the state of correction of the lens, and can provide very misleading information when not wisely used. The designer must be aware of this, and must understand the basic physics and engineering of the design being carried out to avoid these problems.

The basic goals for the design of any lens are somewhat universal. Each designer will naturally develop his or her own understanding and techniques and will approach each lens design task differently. The principles and procedures contained in this book should provide sufficient guidance to make lens design achievable by any interested designer using any computer program.

BASIC OPTICS FOR DESIGN

Design and analysis of lens systems uses numerical calculations based upon geometrical optics. The calculation of the form of the image requires interpretation of these geometrical results by the use of physical optics. Closed-form mathematical solutions are available only in a few relatively simple cases so that all realistic lens design is based upon manipulation of computed evaluations of the imagery produced by a lens.

The geometrical optical model is sufficient to define the properties of image formation by a lens and to relate them to the construction parameters of the lens. The geometrical ray-based model permits determination of image location and aberrations, and enables calculation of the pupil function describing the wavefront. Evaluation of the image quality requires the introduction of concepts of physical optics and wave propagation. The intensity distribution in the image is calculated by applying a diffraction integral to the pupil function.

The basic optics of image location and size is established by the application of paraxial or first-order optics. Ray tracing is the basic tool used in optical design. Aberrations are defined as the deviation of the ray path for real rays from the paraxial basis coordinates. Physical optics and beam propagation extend the geometrical model to include diffraction and interference. Investigation of these models permits determination of the limits on image formation.

Optical glasses are the most commonly used materials, although many other materials have become important in modern optical design. A knowledge of the nature and behavior of optical, physical, and mechanical properties of these materials is important to the designer. The last section of this chapter provides basic information that is needed by a designer to make decisions on the choice of materials for a lens during the design process.

2.1 Geometrical Optics

Geometrical optics is based upon the fundamental assumption that light propagates along rays. Each ray through an optical system follows the path of the shortest time through the system from points in the object space to points located in the image space. Rays in a homogeneous medium follow straight lines.

The velocity of propagation of a light ray in a homogeneous isotropic medium is given by

$$v = \frac{c}{n_{med}} \tag{2.1}$$

where c is the velocity of light in vacuum and n_{med} is the index of refraction of the medium. (In this book both n and N will be used on different occasions to denote the refractive index. The choice will be made to avoid confusion with other quantities in each individual equation.) The length of time for a ray to propagate from point P in object space to point P' in image space is

$$t = \frac{1}{c}\int ns\, ds \tag{2.2}$$

where s is a length along all the ray vectors from the origin, P_1, to the conclusion point, P_2, of the ray. The integral on the right side is called the Optical Path Length along the ray. It is the actual geometrical length along the ray multiplied by the index of refraction at each location along the path of the ray. Fermat's principle, known as the principle of the shortest optical path states that *the optical path length of an actual ray between two points P_1 and P_2 is shorter than the optical length of any other curve which joins these points and which lies in a certain regular neighborhood of the curve describing the ray.* This principle can be extended to state that the curve describing the actual path of a ray between two points satisfies a stationary value for the optical path length. See Born and Wolf (1959, chapter 3) for a mathematical discussion of this point.

Homogeneous, isotropic materials are usually employed in optical design. In such materials the path of light rays between interfaces is as straight lines with refraction or reflection producing a change in the ray direction at interfaces between media. Nonhomogeneous, or index gradient, materials will cause rays to pass along curved lines. Anisotropic materials will cause the direction of the rays to depend upon the polarization of the light being represented by the rays.

Optical calculations use a right-handed coordinate system, with the positive z axis oriented in the initial light propagation direction. If the optical

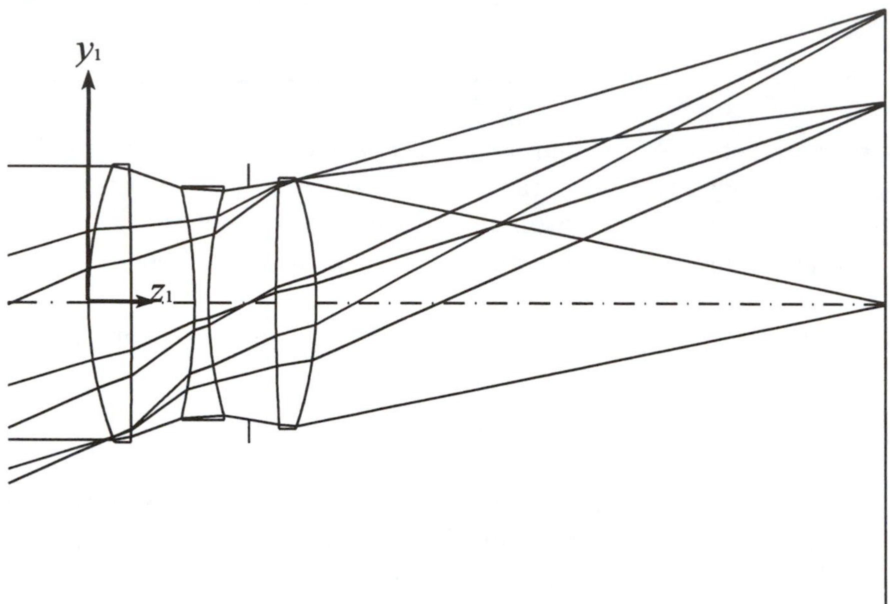

FIGURE 2.1 Layout of a triplet lens, indicating the local coordinates for the surfaces.

system is rotationally symmetric, the z axis corresponds to the axis of the system. Although the surfaces of a lens can be described in a global coordinate system, with the origin of coordinates arbitrarily established, it is conventional to describe each surface in local coordinates attached to the center of symmetry of each surface, as is illustrated in Figure 2.1. The distance between local coordinates is the separation between successive surfaces.

A lens system is defined by a number of parameters stating the surfaces, separations and materials used in constructing the lens, as indicated in Figure 2.2. There will always be at least one limiting aperture that determines the size of the bundle of rays that can pass from an object point to the image surface. The surface on which this is located is called the aperture stop for the lens system. The entrance pupil for a lens is usually described as the image of the aperture stop in the lens as viewed from the object side. The exit pupil is the image of the aperture stop as viewed from the image side. The pupils are both images of the same object, and are thus images of each other. Such a relationship is indicated in Figure 2.3. Each individual ray in the possible bundle of rays passing through the lens can be uniquely identified by coordinates of the ray in the pupils.

Use of the lens with any detector requires that sufficient light per unit area is conveyed to the image so that the noise limits of the detector are exceeded. The overall irradiance on the image surface will be determined by the aperture size and the image scale. The fundamental parameter limiting the irra-

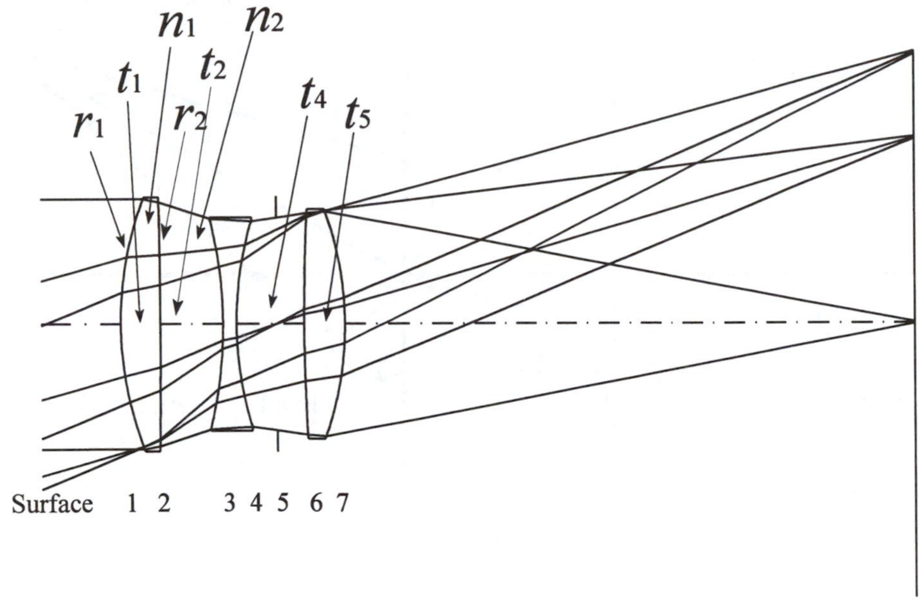

FIGURE 2.2 Triplet lens layout indicating the sequential surfaces for the lens.

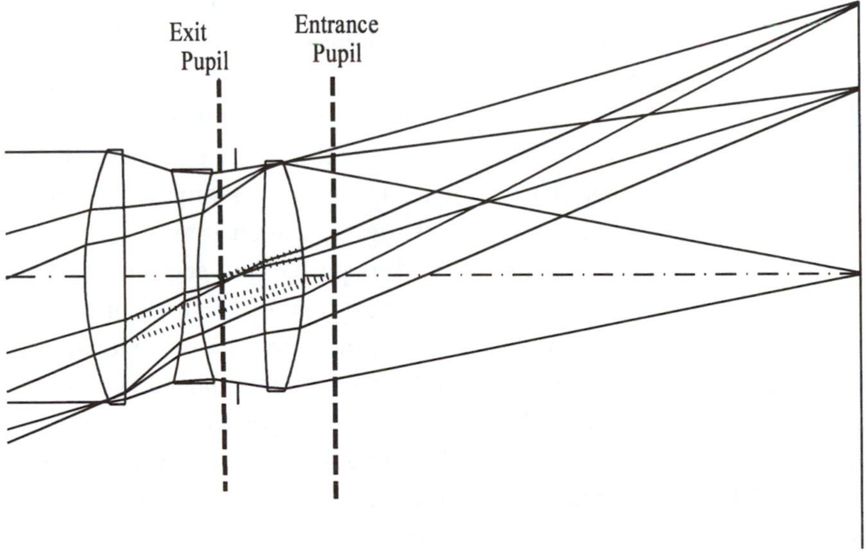

FIGURE 2.3 Location of the effective entrance and exit pupil locations for the lens. These are defined by the intersection of the actual chief ray at each field with the optical axis.

diance on the image surface will be the solid angle subtended by the pupil from each point in the image. The irradiance so computed will provide the average radiance level in the image, and will be correct for large-scale object structure. For fine-scale structure, the irradiance profile in the image must be

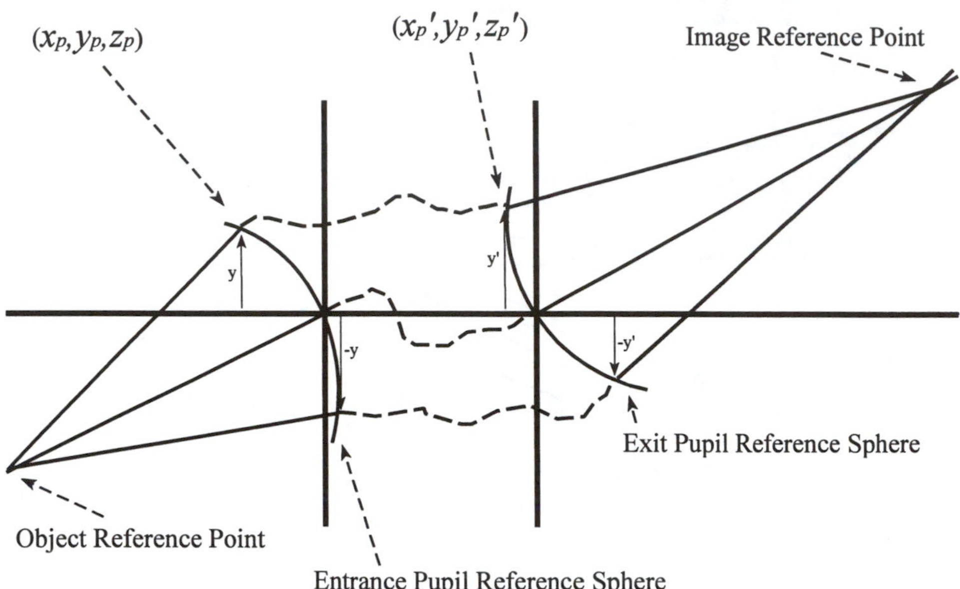

FIGURE 2.4 Entrance and exit pupil reference spheres for a lens system.

computed for specific objects, and the contrast will be reduced as a result of aberrations and diffraction during image formation.

A geometrically perfect image would be an exact replica of the light distribution of the object, modified by the irradiance limitations of the aperture size. Diffraction imposed by physical optics will provide a diffraction limit on the minimum dimensions of the image of a point. Geometrically, a perfect image of a point object is a point of negligible dimensions.

If P_1 and P_2 define an object point and its perfect image point then all rays leaving the object point and passing through the aperture of the optical system would meet at the perfect image point, for all ray coordinates are in the pupils. A consequence of Fermat's principle is that all rays satisfy the stationary path condition and must therefore have the same optical path between these two conjugate points. Figure 2.4 shows a diagrammatic representation of the entrance and exit pupil locations which are used to describe the imagery through a lens.

When a spherical reference surface is drawn about a selected object point, the coordinates of any ray from the object point can be described by the coordinates on the object reference sphere. Normally, the coordinates of this reference sphere are established at the center of the entrance pupil of the system. This reference surface is called the entrance pupil reference sphere.

A similar reference sphere can also be established in the image space following the optical system. This image space reference sphere is centered at the desired image point, and is located at the exit pupil of the system. This

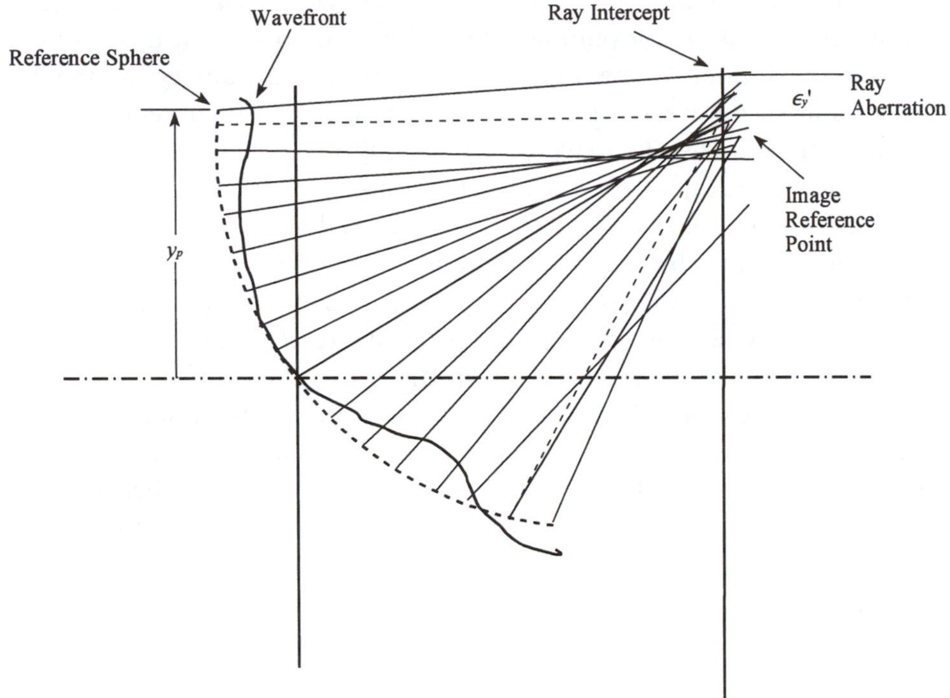

FIGURE 2.5 Relationship between the wavefront, the optical path difference (OPD), and the ray direction and intercept.

is called the exit pupil reference sphere. Since both pupil spheres are located in images of the aperture stop, the selection of a ray coordinate on the entrance pupil will lead to a matching coordinate in the exit pupil. Later, it will be demonstrated that stable imagery requires that a simple scale relationship exist between the ray coordinates in the entrance and exit pupils. The ray and wavefront aberrations can be described by the behavior of rays passing through the pupils, as indicated in Figure 2.5.

The usual choice of coordinates for an optical system is a right-handed set of coordinates, with light initially proceeding in the direction of the positive z axis. For the most prevalent system, a rotationally symmetric optical system, the axis of symmetry will be taken as the z axis. The x and y axes form a right-handed pair with respect to the axis of symmetry. The off-axis objects for a symmetric system will be located in the y–z plane, with the object located in the negative y direction.

This decision regarding coordinates permits great simplification in the entering of lens data into a ray trace. It also allows simplification of the derivation of the ray trace. Such a coordinate choice is not essential, and alternate approaches as well as nonsequential general ray tracing can be used for applications that require such treatment.

Although a global coordinate system could be used for describing the lens system, the usual convention is to assign a number of local coordinates with origins located along the rotational axis of symmetry of the lens system, and successive local coordinates placed at the intersection of each surface with the axis. Thus we describe a ray in a set of successive local coordinates (x_j, y_j, z_j) where $j = 0$ for the object coordinate plane, $j = 1$ for the first surface, $j = k$ for the final surface and $j = k + 1$ for the image plane coordinates. (In some conventions, 1 is reserved for the entrance pupil and $k + 1$ for the exit pupil; and other combinations are possible. The user of any lens design program will have to refer to the documentation for the details of the lens setup in that program.)

The description of a lens system as an ordered series of surfaces is not essential to the discussion of image formation. In most practical cases, the restriction to an ordered set is not limiting. A special case, albeit the most common case, is that of a centered optical system, where all surfaces are related to a single unique axis that defines the system. For the time being, presume that the optical system being discussed is a rotationally symmetric system. The cases where this is not true will be discussed later in the book.

The basic optical surface is a spherical surface. It is represented by a single number which is either the radius of curvature, r, or the surface curvature, c. The diameter of the surface is given by a clear aperture, d, or sometimes by the semiaperture, y_m, which is obviously half of the diameter. The surface is described in a tangent plane set of coordinates, with the z axis coinciding with the axis of symmetry of the lens system. The sag is the departure of the surface from the tangent plane. A simple functional relationship is used to describe the spherical surface sag:

$$z = \frac{c(x_s^2 + y_s^2)}{1 + \sqrt{1 - c^2(x_s^2 + y_s^2)}} \tag{2.3}$$

where x_s and y_s are the coordinates of points lying on the spherical surface. While it looks complicated, this is actually a very simple formula. The sag values z have the same sign as the sign of c. Therefore, a negative c represents a spherical surface with the radius to the left of the coordinate origin, and a positive c indicates the center of curvature lies to the right.

Surfaces that depart from a sphere are called aspheric surfaces. They will be discussed later in this book. The spherical surface is important because it permits rays from an object point to be directed toward a focus position. The failure of all rays to intersect at a common focal point is aberration. Under most imaging conditions, spherical aberration is intrinsic to image formation by a spherical surface.

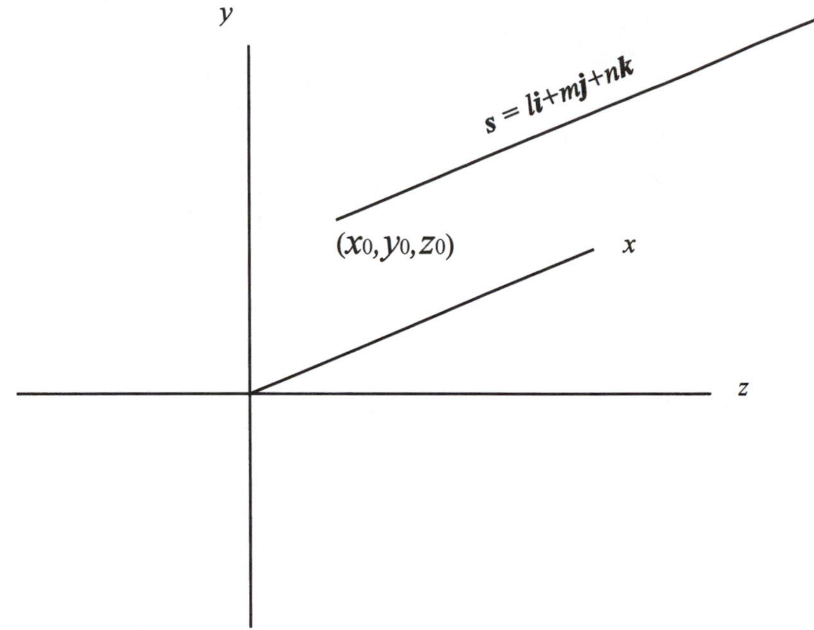

FIGURE 2.6 A single ray vector.

A general ray in a coordinate set is defined by the origin of the ray in the (x, y, z) coordinate set and direction cosines (l, m, n) relative to the axes. In general, a ray is defined as

$$\boldsymbol{s} = l\boldsymbol{i} + m\boldsymbol{j} + n\boldsymbol{k} \tag{2.4}$$

where $(\boldsymbol{i}, \boldsymbol{j}, \boldsymbol{k})$ are the unit vectors along the (x, y, z) directions. If A is the length along the ray, then the coordinates along the ray in any given space with constant refractive index are given by

$$\begin{aligned} x &= x_0 + Al \\ y &= y_0 + Am \\ z &= z_0 + An \end{aligned} \tag{2.5}$$

where (x_0, y_0, z_0) are the origin coordinates for the ray. A sample vector ray is indicated in Figure 2.6.

Tracing a ray along a path in a homogeneous medium is easily done by projecting along the ray vector. For example, Figure 2.7 shows the separation between two planes, the ray origination plane and the next tangent plane in the lens. Here the distance along the ray is easily found from

$$\begin{aligned} A &= \frac{t_j}{n} \\ x_{j+1} &= x_j + Al \\ y_{j+1} &= y_j + Am \end{aligned} \tag{2.6}$$

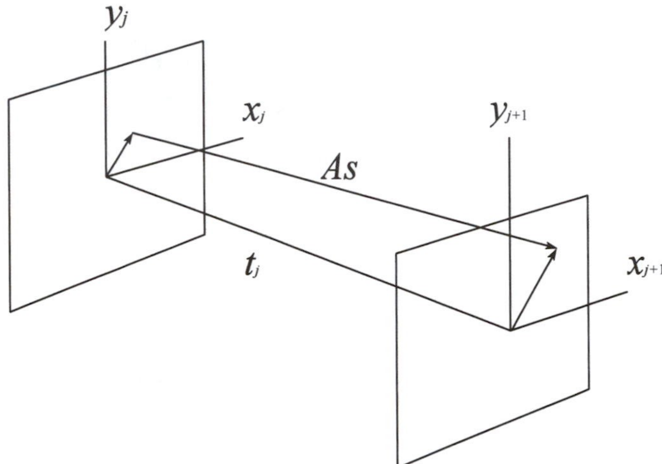

FIGURE 2.7 Coordinates for tracing a ray between two reference planes.

where A is a scalar which is equal to the length along the ray between the planes. The optical path along the ray is the index of refraction times the value of A. The optical path is the index of refraction times the path length and thus is obtained as a part of the process of tracing the rays.

The optical path from the object point to the entrance pupil reference sphere is constant for all rays. Likewise, the optical path from the exit pupil reference sphere to the image reference point would be the same for all rays. Therefore, we can conclude that the optical path from the point where any ray passes through the entrance pupil reference sphere to the point where that ray passes through the exit pupil reference sphere will also be constant for all rays originating at the selected object point.

Aberrations or errors in image formation can be measured in two ways. The failure of all rays to intersect at the final desired image point represents a ray displacement aberration for the lens. The coordinate offset of the rays in the image surface can be used to measure the amount of geometrical light blur that can be expected about the image of a point. Aberrations can also be expressed by the failure of the path from entrance pupil coordinate to exit pupil coordinate to be the same for all rays. The calculation of the Optical Path is obtained by tracing rays and computing the path length from pupil to pupil. These two expressions of aberration are obviously not independent, the ray displacement error being a derivative of the ray displacement error with respect to pupil coordinates.

The process of computing the ray trajectories for an optical system will be discussed in a later section in this chapter. At this point, all that is necessary is to understand that the computation can be carried out. The path of any ray through the system can be defined in terms of the coordinates of origin of the ray, and the coordinates in the entrance pupil for the specific ray.

All of the rays originating at an object point will not usually pass through the image reference point. The distribution of rays about the reference point is defined as the aberration associated with each ray. If the image reference point has coordinates (x_0', y_0') on the desired image surface, then the ray aberration for each ray is shown as in Figure 2.5 to be

$$(\epsilon_x', \epsilon_y') = (x' - x_0', y' - y_0') \tag{2.7}$$

where the aberrations can be expressed as a function of the pupil coordinates which define the path through the pupil for each ray. These values can be computed by tracing exact rays through the system. In most cases, the aberration will be expressed in terms of the distance from the intersection of the chief ray with the image surface. In some cases the desired image location is determined by the paraxial image reference locations.

This concept provides a useful method for measuring the degree to which an optical system meets perfection. Rays from the object may be traced through the system between the reference spheres. Then, a value called the Optical Path Difference (OPD), is computed for each ray:

$$\text{OPD} = [\text{Reference ray path}] - [\text{Ray path}] \tag{2.8}$$

The reference ray is usually taken as a chief ray, the real ray from the object point to the center of the entrance pupil, which consequently passes through the center of the exit pupil. Since each ray is described by its coordinates in either the entrance or exit pupil, the OPD represents a path difference as a function of the pupil coordinates, or is generally called the pupil function for the optical system at the indicated field position:

$$\text{Pupil function} = \text{OPD}(x_p, y_p) \tag{2.9}$$

where x_p and y_p are coordinates in either the entrance or exit pupil.

Light propagates as a wave. This does not directly enter into the geometrical optics model except as a measure of the propagation length along the ray. The color of the light can be described by the wavelength of the light, λ. It is conventional to describe the pupil function in terms of the wavefront function

$$W(x_p, y_p) = \frac{\text{OPD}(x_p, y_p)}{\lambda} \tag{2.10}$$

The values of the wavefront function are referred to as the wavefront aberration. As we will see later, there is a well-defined relationship between the ray aberration and the wavefront aberration. This leads to a determination of the phase of the wavefront on the exit pupil as

$$\phi(x_p, y_p) = 2\pi W(x_p, y_p) \tag{2.11}$$

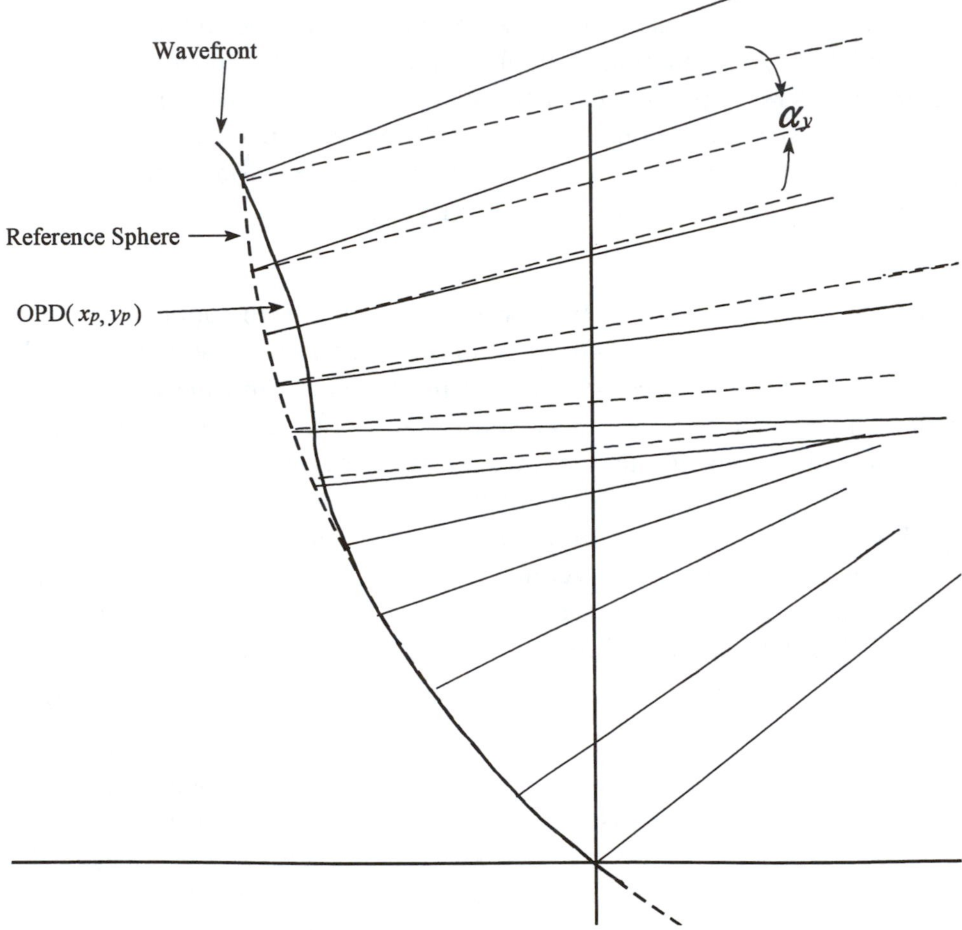

FIGURE 2.8 Relationship between the OPD and the ray direction. $\alpha_y = d(\mathrm{OPD})/dy_p$.

where the phase is in radians. The propagation of a ray is normal to the wavefront, and is related to the wavefront error by a derivative as shown in Figure 2.8.

The failure of the wavefront function to be identically zero over the pupil is evidence of aberration. Aberrations can be produced as a consequence of the passage of a bundle of rays from an object point through the system, by failure of the components of the system to meet the designated form, or by perturbations produced by the environment through which the rays propagate.

The wavefront function is used extensively to describe the state of correction for an optical system. It is not possible to obtain a pupil function in analytic form for any but the simplest of systems. In general the pupil function will be obtained numerically by tracing rays from the entrance pupil

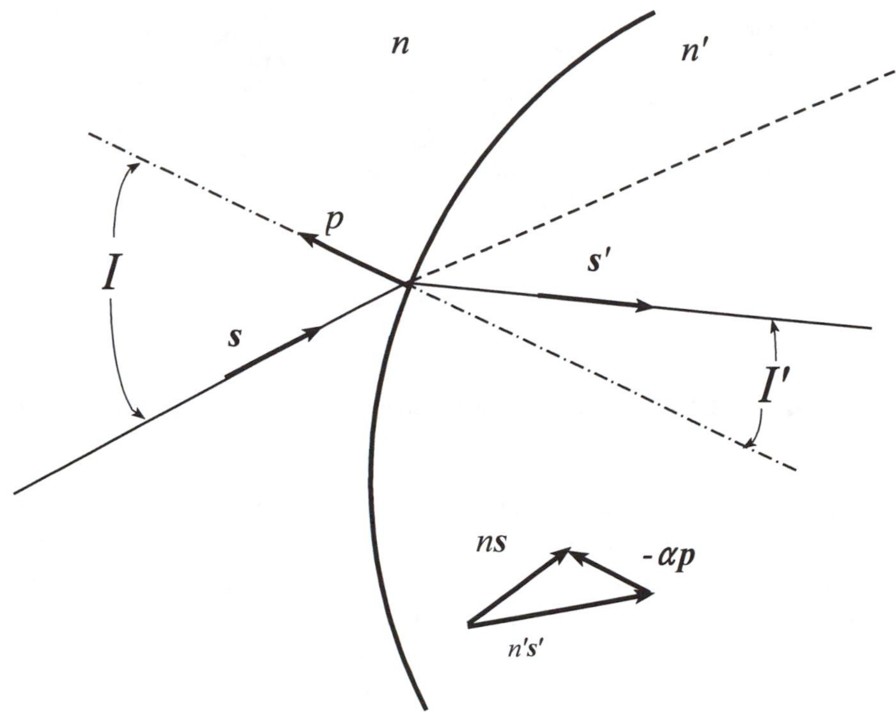

FIGURE 2.9 Vector rules for refraction at a surface.

reference sphere to the exit pupil reference sphere and logging the optical path along each ray.

There is a basic rule for the refraction of rays at an interface or surface between two isotropic media. This is the so-called Snell's law for refraction at a surface between media of index n and n':

$$n' \sin I' = n \sin I \tag{2.12}$$

where I and I' are angles of incidence between the ray for the incident and refracted ray, respectively. There is the important additional condition that *the vectors describing both the incident ray and the refracted ray are coplanar with the local normal to the surface.*

This expression can be converted into a convenient vector relationship

$$n'\boldsymbol{s}' - n\boldsymbol{s} = -\alpha\boldsymbol{p} \tag{2.13}$$

where $\boldsymbol{s}$ and $\boldsymbol{s}'$ are the vectors describing the incident and refracted ray, and $\boldsymbol{p}$ is the unit vector describing the outward normal of the surface. The vector relationship between these rays and the surface normal at a location on the interface between two media is shown in Figure 2.9. The scalar α is given by

$$\alpha = n' \cos I' - n \cos I \tag{2.14}$$

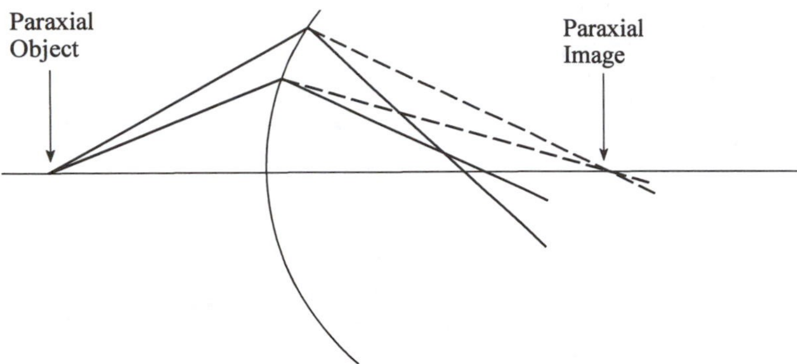

FIGURE 2.10 Comparison of paraxial and real rays at a single surface.

where the cosines of the angle of incidence will be seen later to be obtained as a byproduct of the ray tracing process. The direction and optical path for each real ray is a complicated, nonlinear function of the system parameters, and must be obtained by numerical ray tracing.

The result of reflection at a surface can be evaluated by setting the index following the surface to the negative of that preceding the surface, or $n' = -n$. Then the ray trace equation reduces to

$$\mathbf{s}' - \mathbf{s} = -2\mathbf{p} \tag{2.15}$$

and reference to the sine expression of Snell's law shows that the angles of incidence and reflection are equal in magnitude, but lie on opposite sides of the normal to the surface. The use of this simple vector relation to find the refracted ray direction cosines requires calculation of the intersection point on the surface, as well as the direction cosines of the incident ray and the outward surface normal. For spherical and conic surfaces, the surface intersection and the normal cosines can be obtained from a formula. For general aspherics, an iterative method is employed to locate the intersection point.

When the aperture and angles of the rays are small, a very important approximation can be used in which the actual values for a real ray are replaced by a linear calculation called the paraxial ray trace. The coordinates used are the base set of coordinates for the lens system. The paraxial ray trace serves to establish the location and magnification of images for an ideal aberration-free optical system. The paraxial imaging conditions serve as the base set of object and image coordinates.

For a spherical surface, the direction of the surface normal is found by a linear relation, y/r, the values i and i' are small and the direction cosines for the incident and refracted paraxial ray are u and u'. Thus, in the paraxial approximation

$$n'u'_y - nu_y = -y(n' - n)c$$
$$n'u'_x - nu_x = -x(n' - n)c \tag{2.16}$$

for paraxial rays in the y or x direction. For a rotationally symmetric situation, these will be identical, and by tradition the y, z plane is used to describe the optical system. In the event that the system is not rotationally symmetric, the values of c_x and c_y may be different, indicating cylindric or toric power in the lens system. In this case both the x and y direction paraxial rays are used to describe the image position and sizes in the system. Tracing of paraxial rays determines the locations and size of images in the system as determined by the spherical surface power component of each surface in the lens.

2.1.1 PARAXIAL OPTICS

Paraxial optics is used to determine the location and size of images and pupils in the optical system. Sometimes, this is referred to as first-order optics or Gaussian optics. The paraxial quantities provide information about ideal image formation in the selected set of coordinates. It will be seen later that the paraxial quantities serve as a basis for the actual ray paths through the lens.

The lens is defined by the parameters of the curvature of the surfaces, the separation of surfaces and the index of refraction of the medium between the surfaces. If the medium is dispersive, the index of refraction will vary with wavelength, and the dispersive characteristics of the material need to be stated.

Paraxial variables are angles and ray coordinates that describe the passage of a paraxial ray through the lens. These angles may be selected in object space to correspond to the sines or tangents of the real ray angles that will pass through the lens. Because the equations for paraxial ray tracing are linear, this choice is somewhat arbitrary. This choice is extremely useful in that the paraxial ray heights and angle on each surface of the lens will bear a direct relation to the actual real ray heights and angles to be expected within the lens.

In much of the literature, paraxial optics is considered to describe the passage of rays for infinitesimal field and aperture. In reality, paraxial optics has a considerably greater significance in describing the ideal imaging conditions for a lens. When properly interpreted, paraxial optics is seen to describe the ray path through the lens for real rays in an ideal aberration-free imaging situation. The relation between paraxial constant and the real apertures of an isoplanatic imaging system will be discussed in section 2.1.4. For the present, it may be considered that paraxial optics is a linearized description of the lens system.

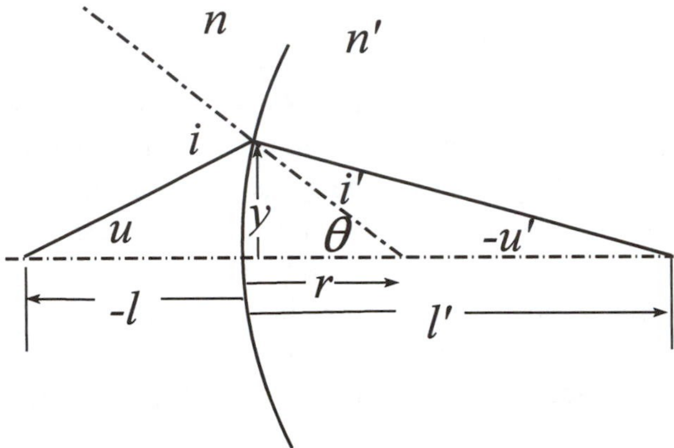

FIGURE 2.11 Variables used in deriving the paraxial ray equations.

The image positions and apertures defined by paraxial ray tracing are important in establishing the base set of coordinates for analyzing the properties of the optical system. The paraxial coordinates establish ideal object and image coordinates, as well as entrance and exit pupil coordinates. In the usual imaging case, the paraxial image coordinates are used as the image reference point locations.

A paraxial ray is defined by the origin of the ray (x, y) and a slope paraxial angle (u, v). By general convention, for a centered optical system, the axis of symmetry is the z axis, and the off-axis object is located in the negative y direction. Once the angle of incidence of the ray on a surface is calculated, the refraction law for paraxial rays is

$$n'i' = ni \tag{2.17}$$

which is a linearized form of Snell's law stated above. The relation between the intersection height of a ray and the slope angle of a ray is given by

$$\begin{aligned} y &= -lu \\ y' &= -l'u' \end{aligned} \tag{2.18}$$

where l or l' is the distance of the intersection of the ray with the axis from the surface upon which the intersection height is being computed. (The minus signs that appear in these equations indicate that the quantities are taken with respect to a local right-hand coordinate system for each surface.) In general, the rule for propagation of a paraxial ray between two successive planes in a rotationally symmetric optical system is

$$y_{j+1} = y_j + t_j u_{j+1} \tag{2.19}$$

The angle of incidence of a paraxial ray on a spherical surface can be found from examination of Figure 2.11 to be

$$\theta + u + (\pi - u) = \pi$$
$$i = \theta + u$$
$$i = \left(\frac{y}{r}\right) + u = yc + u \qquad (2.20)$$

as the angle θ is obtained by dividing y by the radius of the surface (again, taking account of the local coordinate origin).

The paraxial rules for tracing a ray through a centered optical system are obtained by using the equations above for the paraxial angle of incidence and substituting these into the paraxial refraction rule. A small bit of algebra then gives

$$n'u' = nu - y(n' - n)c \qquad (2.21)$$

for the refraction at the jth surface. Use of the relation between the axial intercept of a ray and the slope angle provides a method for calculating the intercept of the ray on the next surface:

$$y' = y + \frac{t}{n'} n'u' \qquad (2.22)$$

In these equations, c is the curvature of the local surface with n and n' as indices of refraction before and after the surface. The thickness t follows the surface. In the usual case wherein surfaces follow in succession, the object is taken as the zeroth surface, the first surface of the lens as 1, the last surface as k and a general surface as j. The image surface is the $(k+1)$th surface.

One of the surfaces must be designated as the aperture stop for the lens. The diameter of this stop is either used as stated or computed by the program to meet a specific requirement. The imaging characteristics of the lens can then be described by two paraxial rays. The marginal ray is directed from the axial object point with a paraxial angle in the object space chosen so it passes through the edge of the aperture stop. The paraxial chief ray originates at the edge-of-field object point for the lens and is directed with such a paraxial chief ray angle in object space that the ray passes through the center of the aperture stop. The definition of these terms is noted in Figure 2.12.

The tracing of this paraxial chief ray through the lens then uses the same formulae for paraxial tracing,

$$\begin{aligned} n'\bar{u}' &= n\bar{u} + \bar{y}(n' - n)c \\ \bar{y}' &= \bar{y} + t\bar{u}' \end{aligned} \qquad (2.23)$$

where the barred quantities refer to the chief ray intersections and angles. The process of paraxial ray tracing is easily carried out by following a format table such as that shown in Figure 2.13.

The image of the aperture stop, as viewed from the object side, forms the paraxial entrance pupil for the lens through all of the surfaces of the lens preceding the stop. Since this entrance pupil optically resides in object space,

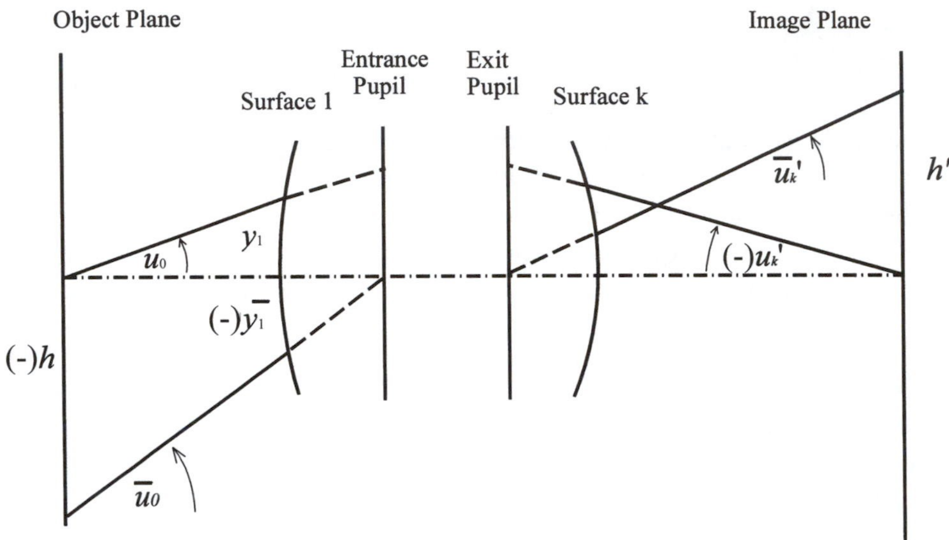

FIGURE 2.12 Location of important features for marginal and chief ray traces.

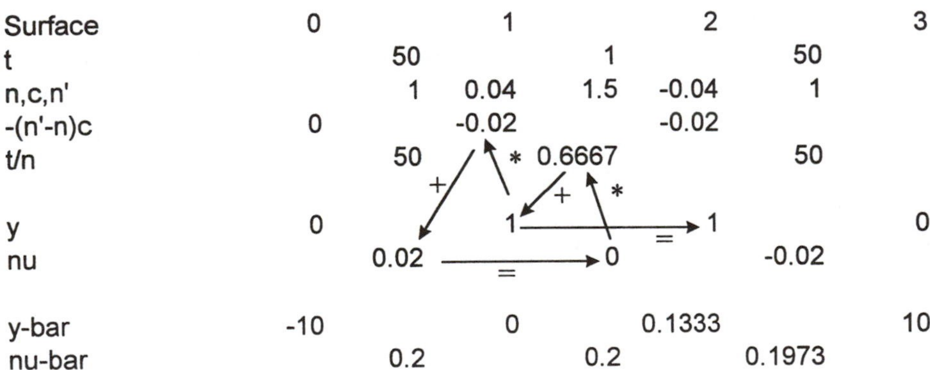
Paraxial Ray Trace Format

Surface	0		1		2		3
t		50		1		50	
n,c,n'		1	0.04	1.5	-0.04	1	
-(n'-n)c	0		-0.02		-0.02		
t/n		50		0.6667		50	
y	0		1		1		0
nu		0.02		0		-0.02	
y-bar	-10		0		0.1333		10
nu-bar		0.2		0.2		0.1973	

FIGURE 2.13 Tabular function for tracing of paraxial rays (Excel).

and is a conjugate paraxial image of the stop, any ray directed from the object to a point in the entrance pupil will pass through the image of that point in the stop.

Likewise, the exit pupil is found as the image of the aperture stop formed by all surfaces of the lens following the stop, and is viewed from the image side of the lens. Therefore, any ray passing through the stop should pass through a conjugate image point in the exit pupil. Since the entrance and exit pupils are paraxial images of the same surface, the aperture stop, it is evident that each point in the entrance pupil has a conjugate image point in the exit pupil. Any paraxial ray traced to a point in the entrance pupil will appear to emerge from a scaled point in the exit pupil.

Paraxial bundles that pass through the lens are defined by the object and image locations and the pupils. The ray from the object point to the center of the entrance pupil is called the chief ray. That ray will pass through the center of the exit pupil and on to the image point. The marginal paraxial ray is directed to the edge of the entrance pupil, and leaves from the edge of the exit pupil to the image point.

An important concept relates to the relative values of any two rays passing through the lens. The two rays used are generally the axial marginal paraxial ray, which passes from the axial object point to the edge of the entrance pupil, and the chief ray for a specified field point. The optical system is common between the chief ray and the marginal ray, so that rewriting the equations for the ray trace, and eliminating the common value $(n' - n)c$ between them leads to

$$n'(\bar{y}u' - \bar{u}'y) = n(\bar{y}u - \bar{u}y) \tag{2.24}$$

which applies to all spaces in the lens system. This value is constant throughout the lens, from object to image. Then, we can define an invariant

$$H = n(\bar{y}u - \bar{u}y) \tag{2.25}$$

where the H is the optical invariant, sometimes called the Lagrange invariant for the optical system.

When taken between the image and object planes, the chief ray heights are the object and image heights and the invariant leads to

$$H = nhu = n'h'u' \tag{2.26}$$

The magnification is the ratio of the object and image height:

$$m = \frac{h'}{h} = \frac{nu}{n'u'} \tag{2.27}$$

producing an important relation between the numerical apertures of the bundles of rays on the object and image side of the lens.

If the invariant is evaluated between the entrance and exit pupils, the value of $\bar{y}$ will be zero and the y value is the pupil radius h_p so that

$$H = n'h'_p\bar{u}' = nh_p\bar{u} \tag{2.28}$$

and the magnification of the pupils is

$$m_p = \frac{h'_p}{h_p} = \frac{n\bar{u}}{n'\bar{u}'} \tag{2.29}$$

which relates the linear pupil size to the chief ray angles.

The choice of the chief ray was determined by the location of the physical aperture stop of the system. This is representative of what happens in a real

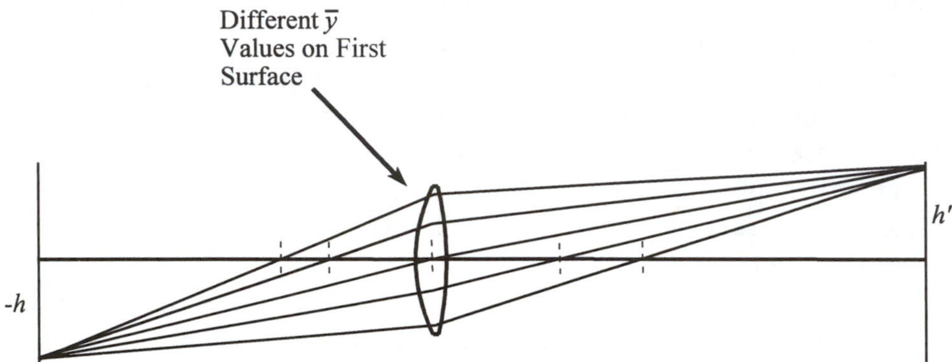

FIGURE 2.14 Selection of specific chief ray from the possible set of chief rays by location of stop position at the point where the desired chief ray crosses the axis.

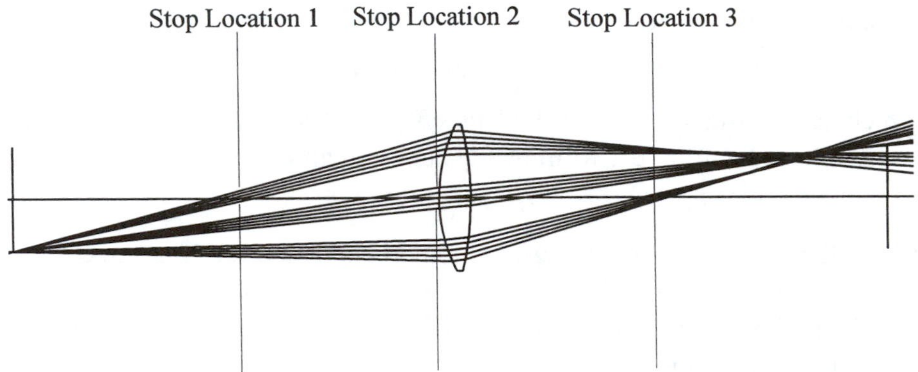

FIGURE 2.15 Demonstration of the effect of moving the stop to a new location in the passage of ray bundles through the lens.

lens system. There are some additional points or locations describing the first-order properties of a lens. To define these cardinal points, it is useful to extend the concept of invariant to special sets of chief rays and effective pupil locations. These can be used to determine the Gaussian properties of the lens, which relate to the locations of images and objects.

The meaning of the stop location and the chief ray is illustrated by examination of Figures 2.14 and 2.15. In Figure 2.14, the tracing of a number of rays from the object to the image point is shown. As far as the lens is concerned, these are all legitimate rays. In Figure 2.15, a stop is located at three different positions. This stop limits the ray bundle passing through the lens to a size required to meet the aperture requirements. Moving the stop to different locations selects a different portion of the lens for the off-axis bundle of rays to pass through the lens. All of these different ray passages indicate the same invariant, and are simply selections of different portions of the lens for image formation at the off-axis images. The diameter of the lens must necessarily be large enough to pass the selected bundle of imaging rays. In paraxial optics, any stop location provides similar imagery. The selection of one stop position or another affects the aberration content of the imagery, and is thus a very powerful tool in lens design.

Since the invariant establishes a relation between the parameters of the marginal and chief rays on each surface, the knowledge of any three parameters defining this pair of rays leads directly to finding the remaining parameter for the rays. There is an important relation in which the parameters of any desired third ray can be determined as a linear combination of the parameters of the first two rays. If the third ray is designated by a double bar above the ray parameters, the rule is

$$\begin{aligned}\bar{\bar{u}} &= Au + B\bar{u}\\ \bar{\bar{y}} &= Ay + B\bar{y}\end{aligned} \tag{2.30}$$

The value of this concept is that it may be used to obtain information about any desired third ray from a trace of an arbitrary pair of paraxial marginal and chief rays. The process is to trace the first two rays. The coefficients A and B are found by knowing the desired characteristics of the third ray on any surface in the lens. The starting or finishing values of the ray trace for the third ray can then be determined by applying the pair of equations to the known paraxial ray trace data at the appropriate surfaces. One example of the application of this technique is the determination of appropriate starting values for a chief ray if the stop location is changed. Another application is the identification of the principal or nodal points for a complicated lens system simply by knowing the path of any two paraxial rays through the system.

This concept can be used to derive useful paraxial data describing the lens. For example, consider finding the focal length of a complex lens system, given the availability of paraxial ray data from a ray trace. The data for two paraxial rays are known in object and image space. The third ray of interest is a ray from infinity that will provide the focal length of the lens by finding

$$\phi = \frac{-u'_\infty}{y_1} \tag{2.31}$$

These values can be found by setting up two sets of equations in object and image space:

$$\begin{aligned}Y &= Ay_1 + B\bar{y}_1 \qquad & Y_{k'} &= Ay_{k'} + B\bar{y}_{k'}\\ 0 &= Au_1 + B\bar{u}_1 \qquad & u'_\infty &= Au_{k'} + B\bar{u}_{k'}\end{aligned} \tag{2.32}$$

which may be solved for A and B in object space, and then evaluated for u'_∞ in image space to obtain

$$\phi = \frac{-Y}{u'_\infty} = \frac{n(\bar{u}'u - \bar{u}u')}{H} \tag{2.33}$$

with H being the optical invariant, and n the index of refraction in image space. This allows direct computation of the power within any portion of a complex system.

Most of the discussion in this book will deal with optical systems with a finite power and finite image or object distance. There is a special case of optical systems that may not form images at a finite distance. These systems, called afocal systems, have very special optical properties that make them invaluable for certain purposes. There is an excellent summary of this topic by Wetherell (1992). All of the paraxial rules apply to afocal systems, but the interpretation of the imaging behavior may be somewhat different, and occasionally surprising.

First-order imagery is defined by the focal length of the lens, and the distances from the locations named as cardinal points. Any complex lens or imaging system can be described in terms of these cardinal points, which permits replacing any lens system by its equivalent set of cardinal points. Paraxial ray trace rules can be used to locate the cardinal points.

The first of these special or cardinal points is the location of the principal points for a lens system. A special set of pupil locations can be found for which the magnification of the pupils is unity. That is m_p is $+1$. Under those conditions, a special chief ray is found such that $y' = y$ and then $n'\bar{u}' = n\bar{u}$. The locations of these special "pupils" are called the principal points for the lens. When the imagery is measured from these points, some new relations develop.

Application of algebra yields

$$m = \frac{h'}{h} = \frac{n}{n'}\frac{l'}{l} \tag{2.34}$$

where l and l' are measured from the locations of the principal points to the object and the image. If the paraxial rays are traced to the principal planes, then the ray trace equation can be written in the form

$$n'u' = nu - y\phi \tag{2.35}$$

where y is the height on the principal planes and the value of ϕ is found by letting the object go to infinity, or $nu = 0$. Then

$$\phi = \frac{-n'u'_\infty}{y} = \frac{n'}{f} \tag{2.36}$$

where f is the distance from the rear principal point to the image when the object is at infinity. This f is called the focal length of the system. Once the principal plane locations and the focal length are determined, the locations of the object and the image are found by paraxial ray tracing using the principal planes as equivalent refracting surfaces for the lens.

Similarly, if the object is placed at the front focal point, such that the value of the marginal ray leaving the lens is zero, then $\phi = n/f$, where f is the front focal distance. In this concept the constant ϕ is called the power of the lens, and is related to the inverse of the focal length.

Another set of points of importance are the nodal points, defined by locating a special set of "pupil" planes located such that the angles are preserved in image and object space. Then the magnification of this set of effective pupils is

$$m_{np} = \frac{nu_{np}}{n'u'_{np}} = \frac{n}{n'} \tag{2.37}$$

which says that the nodal points are located at a different magnification than the principal point planes unless the indexes of refraction of image and object space coincide. This definition of principal and nodal points leads to a method for computing the location of these points in any system by finding the location of the axis crossing of the rays that correspond to the required conditions. This approach is very useful in that it always provides the correct location of the required cardinal points in unusual circumstances.

The chief ray angles, u and $\bar{u}$ provide directions of rays between the pupil locations and the object and image heights. For a given lens, with fixed magnification and object size, moving the stop will alter the position of the pupils, but will not change the basic field coverage required of the lens. Therefore, the chief ray angles are not necessarily a valid measure of the field angle requirements for the lens. This is indicated in Figure 2.15, but a more dramatic example of this is a telecentric system, in which a finite field is imaged, but the chief ray angles in either image or object space are zero across the field. The nodal rays must be used as a measure of the functional field angle required for the lens. For the case of unity nodal point magnification in air, the most usual case, this leads to a simple concept for an unambiguous field angle. If the values of the marginal and chief rays is known on the principal planes, then the nodal angle, or field angle requirements will be

$$\theta = \frac{-(\bar{y}u - \bar{u}y)}{y} = \frac{-\text{Invariant}}{y} \tag{2.38}$$

This concept is clear, compression of the aperture height on the principal planes indicates a corresponding increase in the required field angle coverage required for a lens, with the invariant determining the field and aperture requirement.

The paraxial imaging properties of any complex lens are defined by the locations of object and image relative to the principal points of the lens. The power of a single element lens can be derived using the paraxial ray trace by carrying out a paraxial trace in which the parameters are inserted algebraically rather than numerically. The result for a lens of finite thickness, index of refraction n and located in air is

$$\phi = (n-1)(c_1 - c_2) - t\,\frac{(n-1)^2}{n}\,c_1 c_2 \tag{2.39}$$

Other properties of the single element can be obtained such as

$$\begin{aligned} \text{bfl} &= f\left[1 - t\frac{n-1}{n}c_1\right] \\ \text{ffl} &= f\left[1 + t\frac{n-1}{n}c_2\right] \\ Z &= t(n-1)\left[\frac{(r_1 - r_2) - 1}{n(r_1 - r_2) - t(n-1)}\right] \end{aligned} \tag{2.40}$$

where bfl is the back focal length, ffl the front focal length and Z is the separation between the principal points.

There is an important case of a "thin lens" which has a zero thickness. Since most of the action in a lens occurs at the surfaces, the aberrations and imaging properties of a thin lens are a reasonable approximation of a lens with finite thickness, but the calculation of the properties is considerably easier. For example, the power for a "thin" lens is simply

$$\phi = (n-1)(c_1 - c_2) \tag{2.41}$$

and the front and back focal lengths are identical, with both of the principal points coinciding at the location of the lens.

A convenient ray trace applies to calculating the paraxial path through a series of thin lenses, or a series of thick lenses whose principal point locations and focal lengths are defined. The paraxial trace for lenses in air is

$$\begin{aligned} u'_j &= u_j - y_j\phi_j \\ y'_{j+1} &= y_j + t_j u'_j \end{aligned} \tag{2.42}$$

In the case of thin lenses the separation between lenses is counted from lens to lens. For equivalent thick lenses, the separations, t, are distances between the successive principal points of the lenses. Additional space must be added to account for the principal plane separations. In this way, paraxial tracing mixing actual lens data and equivalent lens data can be carried out in the same paraxial ray trace.

When tracing paraxial rays, the starting conditions include the description of the optical system along with the specification of some surface in the system as an aperture stop location. The starting values for the paraxial rays will include the object location and height, as well as the values for the rays on the first surface of the lens. It is necessary to calculate the proper entering height for the marginal and chief rays on the first surface so that these rays will each pass through the edge or the center of the stop. In some cases, the size of the aperture stop must be adjusted to achieve a specific entrance pupil diameter or final numerical aperture (or F/number). Methods for achieving this can be derived from the paraxial ray trace.

The procedure is to select a nominal set of parameters for the paraxial rays on the first surface, and then carry out the trace to the surface upon which the definition of the ray must be made. The difference between the computed values for the paraxial rays that have been traced and the desired values are used to obtain an adjustment for the entering values. The paraxial rays can then be retraced using the correct entry values, and the paraxial parameters on each surface obtained. (It is possible to use a formula to adjust the height of these rays on each surface, but the speed of tracing rays on modern computers often makes a complete retrace more efficient.)

The adjustment of entering marginal ray height on the first surface is easily accomplished by scaling the entering paraxial ray height by the error found in the quantity of interest. Since the paraxial model is a linear model, this always works.

Once the correct entering values for the rays are obtained, the location and size of the pupils can readily be calculated. The location of the pupils is directly provided by the height and angle of the chief ray on the first and last surfaces of the lens. The size of the pupils is determined by tracing the marginal rays to the pupil locations.

After this has been done, the values of paraxial aperture and paraxial chief ray intersection can be obtained at each surface. The required clear aperture for each surface to pass the bundles of rays from the object through the aperture stop can then be calculated by

$$\text{Diameter of clear aperture}_j = D_j = 2(|y_j| + |\bar{y}_j|) \tag{2.43}$$

which will provide a set of paraxial clear apertures for the lens. Since these are paraxial values, they may not exactly match the requirements on lens apertures for real rays to pass through the apertures.

The field angle associated with this component can be defined as the angle of the chief ray which passes though the center of the thin lens, that is, the nodal ray from the maximum field height. This is a definition of field angle that is independent of the position of the stop, and provides an indication of the difficulty of designing the lens. This can be found from

$$\theta = \frac{-H}{y} \tag{2.44}$$

where H is the invariant applicable to the system. This will be the tangent of the actual field angle, and will also apply to tracing of rays to the principal planes of a thick lens or lens assembly.

Paraxial ray tracing also permits several other functions to be performed during setup of the lens. It is usually necessary to know the location of the paraxial image. This can be found by permitting the program to solve for the location of the surface following the last surface of the lens by calculating the intersection point of the marginal ray with the axis. In most programs, it is possible to carry this out by specifying a paraxial thickness solve for the thickness attached to a specific surface.

In many cases, it will be desirable to maintain a specified convergence angle or power following the last surface of the lens. This is accomplished by a paraxial marginal ray angle solve placed on the last surface using

$$c_j = \frac{-(n_j' u_j' - n_j u_j)}{y(n' - n)} \tag{2.45}$$

where some caution is obviously required if the height of the marginal ray on the surface should be close to zero. This is a very common way of maintaining the focal length for a system when the object is at infinity, or of maintaining the magnification for a finite conjugate imaging situation. In some cases, the angle solve is assigned to a surface within the lens, and an adjustment of the specified surface is used to maintain the magnification or focal length.

If a surface at or near an image location is selected for an angle solve, the major effect is to control the angle of the chief ray, or the location and magnification of the pupils following the surface. A solve on a field lens can be used to place the exit pupil of the lens in a desired location.

The thickness or spacing following the surface may also be defined in terms of an action on a ray. For example, a spacing solve can be introduced to make the distance following a surface correspond to the distance to the paraxial image location. This would be accomplished by solving for the thickness on surface k, so that the height of the marginal ray on surface $k+1$ is zero. Subsequent calculations of ray intercepts or other system parameters will then automatically be referred to the paraxial image location.

Clearly, there are many combinations of solving for curvature, spacing, or aperture that can be placed on a surface. Some can be incompatible, or can provide ridiculous results. Solves are an important part of a design process, because they add special conditions to the process of solving for a lens meeting certain conditions.

There is an important role that paraxial variables have in calculating the aberration content of a lens. If the aberrations can be considered to arise intrinsically on a specific surface, one can imagine the errors producing a specific distribution of errors on the image plane following the surface. It does not matter whether the image is actually physically formed behind each surface, as only one object and image space associated with the surface is being considered. The function of the remainder of the lens can be considered as transferring that image to the final image plane. The Lagrange rule states that the relative dimensions of the error in the two focal planes are given by

$$n_k' u_k' \epsilon_k' = n_j' u_j' \epsilon_j' \tag{2.46}$$

Therefore, if we can calculate the contribution of the error, ϵ_k', then we can sum up the contributions from each surface to the final aberration of that type on the image plane as

$$\epsilon_{k_{total}}' = \frac{-1}{n_k' u_k'} \sum_{j=1}^{k} (-n_j' u_j' \epsilon_j') \tag{2.47}$$

The minus signs that have been introduced are a recognition that the paraxial marginal angle following the last surface is usually a convergent bundle and is negative as a consequence of the sign convention that has been used.

The paraxial optics that defines the base coordinates for the lens system is also useful in determining the radiometry of the images formed by the system. The paraxial marginal variable is equal to the numerical aperture (NA) of the lens. (More detail on this is given in section 2.1.4 on the generalized sine condition.) Then

$$\begin{aligned} \mathrm{NA} &= n' \sin U' = -n'u' \\ \mathrm{F/number} &= \frac{1}{2NA} = \frac{-1}{2n'u'} \end{aligned} \tag{2.48}$$

provides parameters important in determining the image irradiance. The irradiance incident on an image surface is given by

$$E' = \pi t L_0 u'^2 = \pi t L_0 \frac{1}{4(\mathrm{F/number})^2} \tag{2.49}$$

where L_0 is the object irradiance in units such as watts per square centimeter per steradian and E is the image plane irradiance in watts per square centimeter. The t is the transmission factor for the lens. The total flux collected by the lens is E' times the area h^2. Thus the throughput of the lens system is

$$\Lambda = \pi t u'^2 h^2 = \mathrm{Constant(Invariant)}^2 \tag{2.50}$$

which determines the total energy that can be passed through a lens. The square of the invariant is sometimes called the etendue or throughput of the system. It is assumed that the lens meets the sine condition or isoplanatism rule (described in section 2.1.4). The paraxial description of the irradiance on the focal plane must be modified by the actual geometry of the off-axis imaging ray bundles, as well as adjusted for losses due to vignetting within the lens.

An understanding of paraxial rules is important in setting up the initial layout of a lens system. The process involves setting goals for the first-order

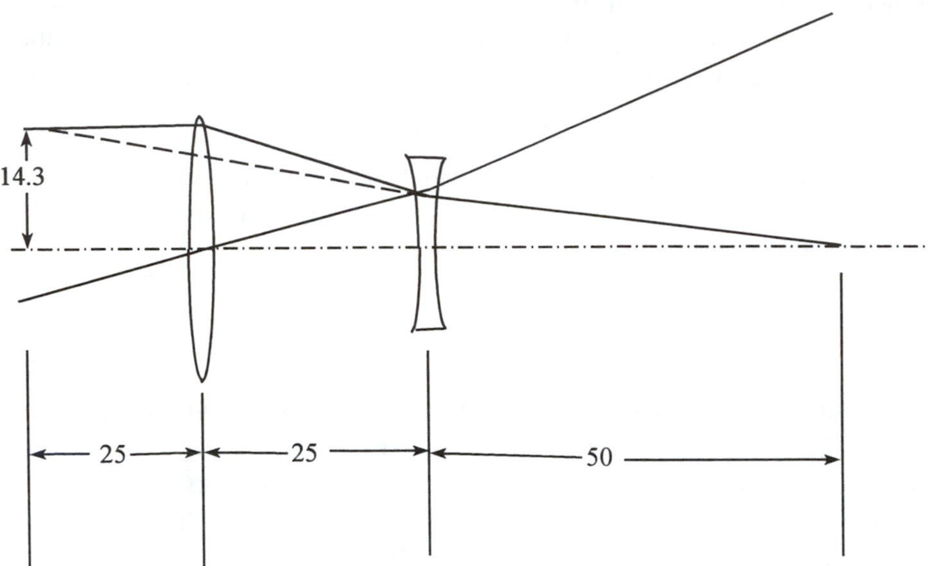

FIGURE 2.16 Setup for a telephoto lens (example 1). Dimensions in millimeters.

parameters for the lens system, and then using the paraxial equations to determine the characteristics of the lenses that will be required. Some examples will now be considered that illustrate some of the important concepts involved in the use of paraxial optics formulae.

EXAMPLE 1

A starting point for an F/3.5 telephoto lens is desired which will have a focal length of 100 mm, a back focal length of 50 mm and an overall length of 75 mm from the front of the lens to the focal plane. Find the powers and locations of two thin lenses that will accomplish this imaging condition. Find the clear apertures required for the lenses if a field angle of 10° is required.

A sketch is a good starting point, and is supplied as Figure 2.16. The ray paths are readily defined as shown. The marginal ray heights are easily computed from the geometry. The ray angle between the lenses is

$$u'_a = \frac{y_b - y_a}{t} = -0.286 \tag{2.51}$$

Having all of the angles permits computation of the individual powers:

$$\phi_a = \frac{u'_a - u_a}{-y_a} = 0.02$$
$$\phi_b = \frac{u'_b - u_b}{-y_b} = -0.02 \tag{2.52}$$

which shows that the powers of the individual elements are twice that of the total power of the lens.

Finding the clear apertures requires the tracing of a chief ray with paraxial angle of tan $10^\circ = 0.1763$ with a zero height at the first lens. This leads to a chief ray height of 4.4082 mm on the second lens. The paraxial clear aperture is the sum of the absolute values of the marginal and chief ray heights on that lens or 11.552 mm. The diameter of that element is 23.11 mm. The diameter of the first lens is simply the diameter of the entering bundle, 28.6 mm, since the stop is located at the lens.

If the lenses were actual simple thin lenses, the curvatures could be computed, presuming some desired bending shape for the lenses. This data could then be entered into the lens design program as starting values for the lens design. Acting on additional knowledge, the designer would know that additional achromatizing elements would be required, and some modification of this data would be made prior to starting the detailed design process.

EXAMPLE 2

A lens is needed for use in a projector. The object is a 35 mm slide, dimensions of 24 mm by 36 mm. The slide is to be projected on a screen at ten times magnification. The screen is located one meter away from the object. The working F/number on the object side is to be F/3.5. Find the focal length and F/no$_\infty$ for the required lens, as well as the field angle required.

The requirement for an object side F/number of 3.5 indicates that $u_0 = 0.143$. A 10 times magnification with a real image will require $m = -10$. On the image side, $u' = -0.0143$. Construction of a simple ray diagram shows that if there is no separation between the principal planes, the rays will intersect at a location 90.909 mm from the object, with an object distance for the lens of -90.909 as measured from the lens. The height of the marginal ray on the principal plane is then $y = -lu = 13.0$ mm.

The focal length can be found by applying the equation for ray tracing through a lens,

$$u' = u - y\phi$$
$$\phi = \frac{u' - u}{-y} \tag{2.53}$$

so that the focal length is 82.6446 mm. The F/no.$_\infty$ required is 3.1786. The paraxial field angle is

$$\bar{u} = \frac{-h}{-l} = \frac{0.5\sqrt{24^2 + 36^2}}{90.909} = 0.2380 \tag{2.54}$$

which at this point can be taken as the entering chief ray angle for the lens, since no assumptions have yet been made about the location of the stop other than the location of the principal planes. This is about 13.38° semifield, if the field is presumed to be flat and the tangent of the paraxial angle is taken.

In any real lens, the principal planes may be separated by some amount, perhaps even crossed. This can be taken into account by introducing a known separation into the diagram for the lens setup.

This problem could be approached by using a lens design program. It also can be carried out manually, or by setting up a spreadsheet computation. Modern spreadsheets, such as the Excel program, permit very great flexibility in carrying out the computation. Figure 2.17 is an example of such a calculation. The desired lens has been solved for. The apparent added places for lenses with zero power were inserted in anticipation of the need to solve the next problem. By incorporating the paraxial ray trace into the spreadsheet, such items as field angles, lens diameters and F/numbers are easily obtained.

EXAMPLE 3

The customer for the lens in example 1 has accepted delivery of the completed viewer, but now has changed his mind and requests that lenses be added between the lens and the screen that would change the magnification to twenty times. There is a desire to keep the costs as low as possible, so neither the existing lens nor the screen may be altered in any way. The customer further requests that the change be done with only two lenses, if possible.

There are two conditions to be met, a magnification and an image position. With two lenses, a solution for these two conditions should be possible. The starting point is again a sketch of the system, with a possible starting point indicated. It is evident that a number of possible solutions exist. Selection of the best possible solution will require that the apertures needed be computed and the powers required be determined. Since the initial lens is not to be changed, the stop is fixed in location at that lens. The chief ray will have a significant height at both of the added lenses.

A quick consideration of the layout will show that the two lenses must be chosen to provide a final marginal paraxial ray angle of half that from the basic lens, with a focus located at the original position. A combination of a separated positive and negative lens will be required to accomplish this. For every choice of locations of these lenses, there will be one choice of powers of the two lenses to accomplish the imaging. Though not stated explicitly, the orientation of the image should not be altered, thus the final paraxial angle must have the same sign.

This problem can be solved using analytic approaches, but the best approach is to use the paraxial ray trace to determine the conditions on the two lenses to be added. The use of a spreadsheet for this purpose is extremely useful, as the designer can see what is happening at each step of the solution. Figure 2.18 is a view of an Excel spreadsheet which accomplishes this task. The paraxial ray

Spreadsheet for thin lens trace										
		invariant	-3.09356				Lens Data			
								a	b	c
surf num	phi	thi	y	u	ybar	ubar	phi	0.0121	0	0
							efl	82.64463	#DIV/0!	#DIV/0!
0			0		-21.6333		dia	26	31.23416	48.36664
		90.90909		0.143		0.237966	Fno inf	3.17864	#DIV/0!	#DIV/0!
1	*0.0121*		13		0		Field Ang	13.38551	13.74984	14.13415
		25		-0.0143		0.237966				
2	0		12.6425		5.94916					
		25		-0.0143		0.237966				
3	0		12.285		11.89832					
		859.0909		*-0.0143*		0.237966				
4			*-1.3E-08*		216.3331					
5										
Drawing of the rays										
anmorph factor		1								
0	0	-21.6333								
90.90909	13	0								
115.9091	12.6425	5.94916								
140.9091	12.285	11.89832								
1000	-1.3E-08	216.3331								

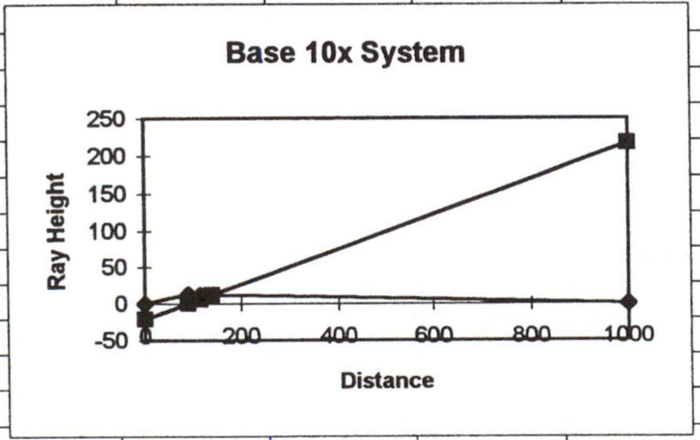

FIGURE 2.17 Spreadsheet setup for example 2, indicating the required 10× magnification (Excel).

Spreadsheet for thin lens trace										
		invariant	-3.09356				Lens Data			
								a	b	c
surf num	phi	thi	y	u	ybar	ubar	phi	0.0121	0.009382	-0.02132
							efl	82.64463	106.5868	-46.9101
0			0		-21.6333		dia	26	46.93664	70.73328
		90.90909		0.143		0.237966	Fno inf	3.17864	2.270866	-0.6632
1	0.0121		13		0		Field Ang	13.38551	14.96946	29.68225
		100		-0.0143		0.237966				
2	0.009382		11.57		23.79664					
		50		-0.12285		0.014706				
3	-0.02132		5.4275		24.53193					
		759.0909		-0.00715		0.537662				
4			-3.5E-05		432.666					
5										
Drawing of the rays										
anmorph factor		1								
0	0	-21.6333								
90.90909	13	0								
190.9091	11.57	23.79664								
240.9091	5.4275	24.53193								
1000	-3.5E-05	432.666								

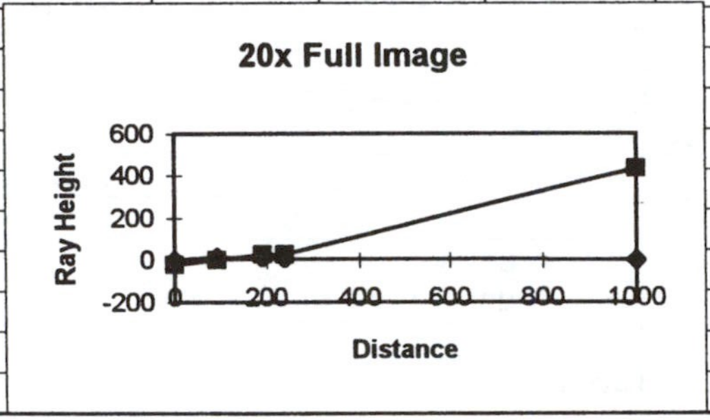

FIGURE 2.18 Spreadsheet for the insertion of two lenses to change the magnification (Excel).

trace is set up in the upper left portion of the sheet. The requirements are that the magnification and the image position be constrained. The designer selected arbitrarily the separations of 100 and 50 for the two added lenses.

There is an additional trick needed here, as the fixed screen size requires that the image overfill the screen for magnifications greater than ten, thus the object height is set to be -10.8167, because of the field limitation by the screen. The two variables are the powers of the lenses. These are found by using the solver portion of the spreadsheet program to simultaneously find a zero image height for the marginal ray, and the specified image height for the chief ray. The spreadsheet in Figure 2.19 includes this screen size limitation.

The powers and diameters of the added lenses can be computed from the solved powers and the y and $\bar{y}$ obtained from the paraxial ray trace. The F/no.$_\infty$ is also computed. The indicated solution does not seem to be completely unreasonable as a starting point for detailed design. This is only one of a large number of possible solutions. The spreadsheet form permits the designer to change the initial starting assumptions of the separations and rapidly examine the nature of a number of possible solutions.

EXAMPLE 4

The customer for the viewer has once again changed his mind, and now desires the introduction of a moving pair of lenses between the fixed lens and the screen so that the slide magnification varies continuously from five times to ten times magnification. As before, no changes in the existing layout are permitted, and the customer has also specified that only two moving components may be inserted.

This is now a more complicated problem. The use of a spreadsheet and paraxial ray trace approach permits locating solutions for the five times magnification in the same manner as done in example 3. The difficulty now is that the same lens powers must be used to locate separations that would permit finding magnifications that cover the entire range. Again a spreadsheet approach is used, the difference now being that three different magnifications are picked for the ray trace. Now the variables are the two lens powers, as well as the three sets of zoom variables, the separations of the lenses at the three magnifications chosen. This setup is shown in Figure 2.20. The solving condition is set to be that the mean square sum of the marginal ray heights at the image planes for the three zoom positions is chosen to be set to zero. To this is added the boundary conditions that the magnifications at the three zoom positions be set to specified values. The designer also thinks ahead and decides that all separations must of course be positive, and furthermore for practical reasons the separations shall never be less than ten.

In this case, the paraxial marginal angle for the five times magnification needs to be twice that of the base system. Therefore, a sketch of the situation suggests

Spreadsheet for thin lens trace										
		invariant	-1.54678				Lens Data			
								a	b	c
surf num	phi	thi	y	u	ybar	ubar	phi	0.0121	0.009382	-0.02132
							efl	82.64463	106.5868	-46.9101
0			0		-10.8167		dia	26	35.03832	46.93664
		90.90909		0.143		0.118983	Fno inf	3.17864	3.042007	-0.99944
1	0.0121		13		0		Field Ang	6.785335	7.614664	15.90701
		100		-0.0143		0.118983				
2	0.009382		11.57		11.89832					
		50		-0.12285		0.007353				
3	-0.02132		5.4275		12.26596					
		759.0909		-0.00715		0.268831				
4			-3.5E-05		216.333					
5										

Drawing of the rays		
anmorph factor		1
0	0	-10.8167
90.90909	13	0
190.9091	11.57	11.89832
240.9091	5.4275	12.26596
1000	-3.5E-05	216.333

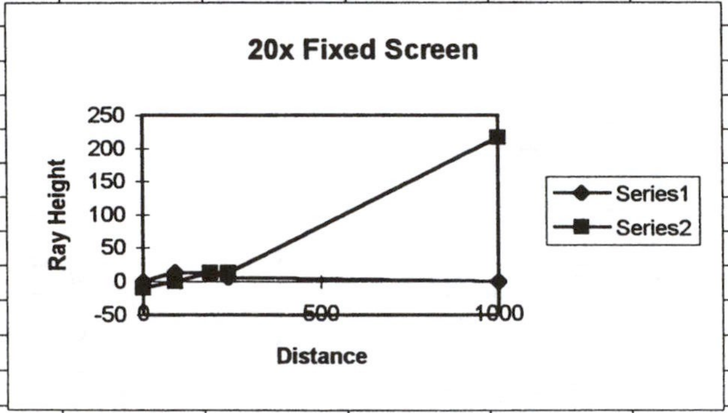

FIGURE 2.19 Adjustment of ray trace to include fixed screen size (Excel).

Spreadsheet for thin lens trace

Zoom pos 1						
surf num	phi	thi	y	u	ybar	ubar
0			0		-21.6333	
		90.90909		0.143		0.237966
1	0.0121		13		0	
		93.85614		-0.0143		0.237966
2	-0.00303		11.65786		22.33461	
		234.8972		0.020998		0.305592
3	0.002989		16.59031		94.11742	
		580.3375		-0.02859		0.024295
4			0.000424		108.2168	
5						

Zoom pos 2						
surf num	phi	thi	y	u	ybar	ubar
0			0		-21.6333	
		90.90909		0.143		0.237966
1	0.0121		13		0	
		152.1457		-0.0143		0.237966
2	-0.00303		10.82432		36.20556	
		124.4603		0.018475		0.347592
3	0.002989		13.12366		79.46694	
		632.4849		-0.02075		0.110082
4			8.87E-08		149.0919	

Zoom pos 3						
surf num	phi	thi	y	u	ybar	ubar
0			0		-21.6333	
		90.90909		0.143		0.237966
1	0.0121		13		0	
		10		-0.0143		0.237966
2	-0.00303		12.857		2.379664	
		17.0408		0.024629		0.245172
3	0.002989		13.2767		6.557584	
		882.0501		-0.01505		0.225572
4			3.61E-07		205.5238	

FALSE	8	FALSE	FALSE
FALSE	100		

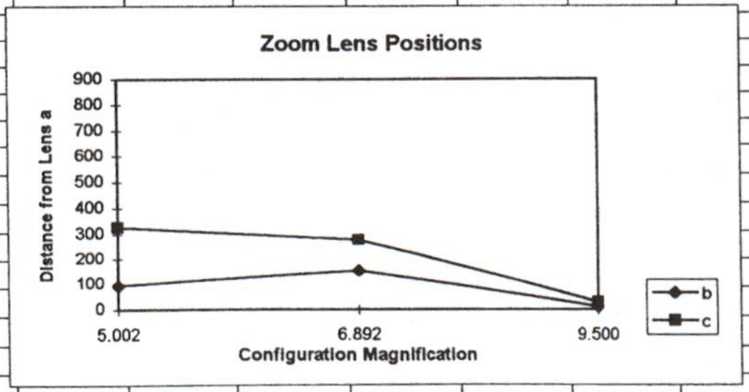

Lens Data

	a	b	c
phi	0.0121	-0.00303	0.002989
efl	82.64463	-330.266	334.5835
dia 1	26	67.98493	221.4155
dia 2	26	94.05976	185.1812
dia 3	26	30.47333	39.66857
min Fnum	3.17864	-3.51124	1.511112

mag1	5.002324	ht1	0.000424
mag 1 targ	5		
mag 1 res	0.002324		
mag 2	6.891777	ht2	8.87E-08
mag 2 targ	6.892024		
mag 2 res	0.000247		
mag 3	9.500339	ht3	3.61E-07
mag 3 tar	9.5		
mag 3 res	0.000339		
mag sq	5.58E-06	ht sq	1.8E-07
merit	5.76E-06		

Lens locations

z	b	c
5.002	93.85614	328.7534
6.892	152.1457	276.606
9.500	10	27.0408

Zoom Lens Positions

Distance from Lens a

900
800
700
600
500
400
300
200
100
0

5.002
6.892
9.500

Configuration Magnification

b
c

FIGURE 2.20 Solution of variable magnification system (Excel).

that the two lenses will be a negative lens followed by a positive lens. The starting choice is provided by the designer as a guess at the likely powers. The spreadsheet solver is set to work, and no useful solution with lens parameters that are even remotely buildable results. After some frustration the designer decides to think about what is happening, and realizes that there is a severe hidden boundary condition in the problem. In order that the magnification of exactly ten times be achieved, the net effect of the two lenses must be zero. This greatly constrains the solution for two components to where they can become neutral when in contact, but contact is not a permissible boundary condition.

Realizing this, the designer changes the upper limit to the magnification to something less, such as nine times, and a solution results. Extending the zoom range and finding a practical solution can be carried out by successive tries while extending the target for the upper magnification limit once a solution has been found. Any solver is likely to be more satisfied with locating an acceptable solution which lies in the neighborhood of a known existing solution. One solution for a zoom range of five times to 9.5 times is indicated in Figure 2.20. The focal lengths, diameters, and F/numbers of the added components are reasonable for a starting point to design a functional system. This is obviously only one of many possible solutions. Carrying out this problem using a spreadsheet is extremely instructive in teaching the beginning designer how to consider boundary conditions and approaches to locating a reasonable solution, as all aspects of the problem are visible to the designer.

Understanding the layout process is essential to gaining a knowledge of how optical systems work. The proposed solutions suggested in these examples serve as the starting point for final designs. Use of simple paraxial ray diagrams, manual computations, and spreadsheets helps to provide insight, and exploration of many possible solutions prior to launching into the use of a full-scale lens design program. The designer also obtains some new ideas, such as understanding that coverage of the entire zoom magnification range can only be accomplished by adding a third stationary lens that adjusts the basic magnification and image position to a new location. Then the moving components can cover a different range of magnifications, and do not have to be constrained to have no net power at one end of the range.

2.1.2 RAY TRACING

Ray tracing is the numerical method used for determining the behavior of lenses. Data obtained from ray tracing provides the geometrical path of light through the lens, and defines the aberration content of the image. Optical path values are obtained from the ray trace data and used for computation of the physical image. All of the required information about clear apertures, tolerances and image quality is obtained from computations based upon ray tracing.

In this section, the basic information and algorithms required for tracing rays through an optical system will be described. The depth of this discussion will only be sufficient for understanding the process and determining the links to presentation of data from ray tracing. Anyone interested in developing design software will find that reference to the original literature would be advisable. The actual algorithms used in each of the available design programs will vary somewhat in detail from the approach described here. Because the ray trace process is so crucial to the optical design process, the user of design programs must be sensitive to possible errors or misinterpretations of data that can arise. The principles remain constant, and can be used to evaluate the accuracy and completeness of this critical process, which is at the heart of the design approach.

The coordinates most frequently used in ray tracing are local coordinates, associated with each surface in the lens. The surfaces are expressed in the local coordinates, and the rays are traced between the coordinate systems. In some cases, the ray will be traced or evaluated using global coordinates that are referenced to a specified location in the optical system. The convention today is to use a right-handed coordinate system, and that the positive direction for light propagation is from left to right.

In most cases, the surfaces describing a lens are placed into the computer program in sequence, and the tracing of a ray proceeds from the object to the image surface, encountering each surface in sequence. (For some problems, this is not adequate, and rays proceed through the system in a nonsequential order. Such nonsequential ray tracing uses the principles discussed here, but the implementation and algorithms are quite a bit more complex.)

The concept of a ray vector was introduced in the previous section. A ray is described by the coordinates of origin or intersection in the coordinate set being used, with a direction given by direction cosines measured against the coordinate axes. The ray is a unit vector

$$\boldsymbol{s} = L\boldsymbol{i} + M\boldsymbol{j} + N\boldsymbol{k} \tag{2.55}$$

(As a point of notation, the upper case letters are used for the direction cosines here in an attempt to avoid confusion with the lower case n, which is generally used to note the index of refraction. In some texts, the upper case direction cosines are used to represent "optical" direction cosines, which are the geometrical cosines multiplied by the index of refraction. In all cases, the user must determine the intent of any formula in any reference. There can be traps for the unwary!)

The coordinates of a point along the ray are found by using the coordinate origin of the ray (x_0, y_0, z_0) and the path length, p, along the ray

$$\boldsymbol{R} = \boldsymbol{R}_0 + A\boldsymbol{s} \tag{2.56}$$

the coordinates of the end of the vector $\boldsymbol{R}$ are

$$\begin{aligned} x &= AL + x_0 \\ y &= AM + y_0 \\ z &= AN + z_0 \end{aligned} \tag{2.57}$$

These describe the point along the ray that is a distance A from the coordinate start.

As an example of using sequential local coordinates for the ray, consider the diagram shown in Figure 2.7. The z axis is coincident with the axis of the rotationally symmetric lens system. The coordinate origin is set at the axial position of the first plane, which might be the object coordinate, or any local coordinate in the system. For generality, call this the coordinate set for surface j. The next surface is located a distance t_j away along the positive z axis. If a ray intersects the j-th coordinate at (x_j, y_j, z_j) the intersection of the ray with coordinate set $j + 1$ is

$$\begin{aligned} A &= \frac{t_j - z_j}{N} \\ x_{j+1} &= AL + x_j \\ y_{j+1} &= AM + y_j \\ z_{j+1} &= 0 \end{aligned} \tag{2.58}$$

and the ray can be seen to have been traced from one surface coordinate set to the tangent plane of the surface in the subsequent coordinate set.

The most common surface in lenses is a spherical surface. As noted previously, a spherical base curvature is required to form an image. Figure 2.21 illustrates the process of tracing the ray to an intersection with the surface. Since the direction cosines of the ray are known, it is necessary only to find the length along the ray to the surface. This can be done for a spherical surface (and for general conic surfaces) by use of a formula rather than an iterative process.

The ray has been traced to the coordinate plane associated with the surface. The distance along the ray to the surface must be found. The approach is to transfer the ray to the spherical surface by using the unknown transfer distance A, and then solving for A by confining the surface intercept coordinates to lie on the form of the surface in question. This approach is generally applicable to the tracing of rays to any surface, the example worked out here for a spherical surface is a basic approach. Later sections on aspherics will

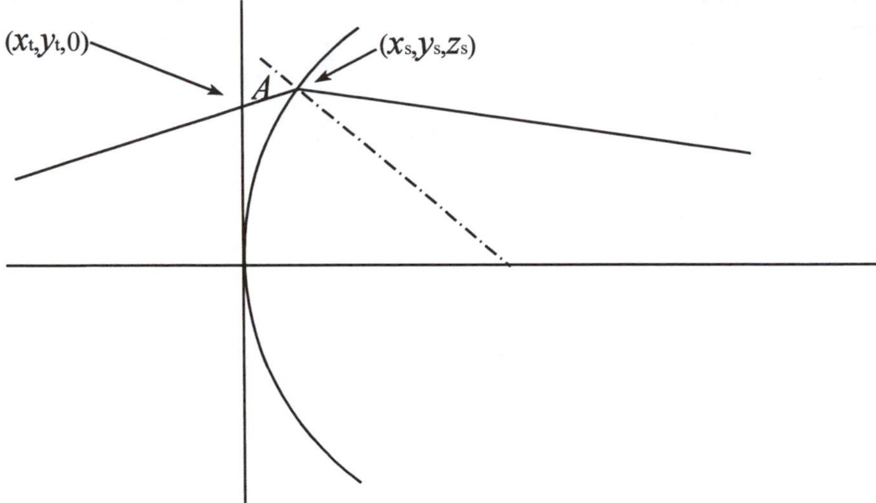

FIGURE 2.21 Coordinates for tracing a ray to and through a surface.

deal with more complex surfaces, including those that cannot be analytically defined.

Continuing with the spherical surface transfer, the transfer equations are

$$\begin{aligned} x_s &= x_t + AL \\ y_s &= y_t + AM \\ z_s &= AN \end{aligned} \tag{2.59}$$

for the transfer from $(x_t, y_t, 0)$ to (x_s, y_s, z_s) on the surface. To confine the intercept to the surface, the relation

$$z_s = \frac{c}{2}(x_s^2 + y_s^2 + z_s^2) \tag{2.60}$$

describing the spherical surface must be satisfied. The ray intercept coordinate values are written in terms of the known values of the intercept on the tangent plane as well as the unknown transfer length A.

$$AN = \frac{c}{2}((x_t + AL)^2 + (y_t + AM)^2 + (AN)^2) \tag{2.61}$$

which can be rearranged to give

$$\frac{c}{2}(x_t^2 + y_t^2) + (cx_tAL + cy_tAM - AN) + \frac{c}{2}A^2(L^2 + M^2 + N^2) = 0 \tag{2.62}$$

The sum of the squares of the direction cosines is unity, and rearranging leads to

$$A^2\frac{c}{2} + A(cx_tL + cy_tM - N) + \frac{c}{2}(x_t^2 + y_t^2) = 0 \tag{2.63}$$

which is now in the form of a quadratic equation. To save a bit of writing, define

$$F = c(x_t^2 + y_t^2)$$
$$G = N - c(x_t L + y_t M) \tag{2.64}$$

so that the quadratic equation becomes

$$cA^2 - 2GA + F = 0 \tag{2.65}$$

The solution of this quadratic equation is then

$$A = \frac{2G \pm \sqrt{4G^2 - 4cF}}{2c} \tag{2.66}$$

which is written as

$$cA = G \pm \sqrt{G^2 - cF} \tag{2.67}$$

The sign can be found from logical considerations. Since the origin of the coordinates of the tangent plane coincides with the pole of the surface, the transfer distance must go to zero for a ray intersecting at coordinates (0,0) on the tangent plane. Since F is proportional to the square of the distance from the origin of the tangent plane, F becomes zero under those conditions, and A must likewise be zero. Therefore, the negative sign for the root should be chosen.

The positive sign for the root can be a legitimate choice under some conditions. In that case, the ray is transferred to the "back side" or alternate surface intercept for the surface described by the equation defining the surface form (Cheatham 1980). In the vast majority of cases, this positive root is not the proper choice.

The equation for A is correct as it stands, but is subject to some inaccuracy when finite word length is used in numerical calculations. The value of G is usually of the order of unity in most calculations, while F can be small. It is not advisable to use the number resulting from the difference of what could be two relatively large numbers, as there can be error buildups due to successive application of these equations when tracing a ray through the system. Embarrassment can also occur when c is zero, a common value for a plano surface.

Multiplying and dividing the equation by the same value allows a rearrangement

$$cA = G - \sqrt{G^2 - cF}\left(\frac{G + \sqrt{G^2 - cF}}{G + \sqrt{G^2 - cF}}\right) = \frac{cF}{G + \sqrt{G^2 - cF}} \tag{2.68}$$

leading to

$$A = \frac{F}{G + \sqrt{G^2 - cF}} \tag{2.69}$$

which will not offer significant problems in numerical calculation.

Finding the intercept on the surface is now straightforward by applying the transfer equations

$$\begin{aligned} x_s &= x_t + AL \\ y_s &= y_t + AM \\ z_s &= AN \end{aligned} \tag{2.70}$$

where all of the quantities on the right side are now known. The ray has intercepted the surface, and the next step is to carry out refraction of the ray at the surface and find the direction cosines (l', m', n') in terms of the incident direction cosines (l, m, n).

It is necessary to find the angle of incidence, or at least the cosine of the angle of incidence to permit application of the vector refraction equation. Since the incident ray vector, $\boldsymbol{s}$, and the normal to the surface, $\boldsymbol{p}$, are known, the cosine of the angle of incidence can be found by taking a scalar product of these two vectors. The normal to the surface at coordinates (x_s, y_s, z_s) for a sphere is

$$\boldsymbol{p} = -cx_s\hat{i} - cy_s\hat{j} + (1 - cz_s)\hat{k} \tag{2.71}$$

and the scalar product leads to

$$\begin{aligned} \cos I &= \boldsymbol{sp} \\ &= -Lcx_s - Mcy_s + N - Ncz_s \\ &= N - c(Lx_s + My_s + Nz_s) \end{aligned} \tag{2.72}$$

which is more conveniently rewritten by substituting in the tangent plane intercepts to get

$$\begin{aligned} \cos I &= N - c(L(x_t + AL) + M(y_t + AM) + N(NA)) \\ &= N - c(Lx_t + My_t + A(L^2 + M^2 + N^2)) \\ &= N - cA - c(Lx_t + My_t) \\ &= -cA + (N - c(Lx_t + My_t)) \end{aligned} \tag{2.73}$$

This can then be written in terms of the F and G that are calculated as a by-product of transferring the ray to the surface

$$\begin{aligned} \cos I &= -cA + G \\ &= -(G - \sqrt{G^2 - cF}) + G \\ &= \sqrt{G^2 - cF} \end{aligned} \tag{2.74}$$

and all of the important information comes almost for free as a part of the computation.

The refraction equation derived earlier can be applied to finding the direction cosines of the refracted ray. However, $\cos I'$ is needed in the computation.

The cosine of the angle of refraction is found by rewriting the familiar Snell's law of angles to yield

$$n' \cos I' = \sqrt{n'^2 - n^2(1 - \cos^2 I)} \tag{2.75}$$

then the factor α is

$$\alpha = n' \cos I' - n \cos I \tag{2.76}$$

and the refracted ray direction cosines are

$$\begin{aligned} n'L' &= nL - \alpha c x_s \\ n'M' &= nM - \alpha c y_s \\ n'N' &= nN + (1 - c z_s)\alpha \end{aligned} \tag{2.77}$$

and will have no special cases, since reflection can be obtained by setting $n' = -n$ and suitably laying out the optical system.

Transfer of the ray to the following tangent plane can then be carried out by use of the original ray transfer equations, because the coordinates of the ray on the surface are known. The sequence of application of the equations is: transfer to tangent plane, transfer to surface, refract (or reflect), and then continue the process until the final image surface is encountered.

There are some problems that can arise. A ray can encounter total internal reflection at a surface. Examination of the Snell's law relation indicates that this will happen when the value for the angle of refraction becomes imaginary, which occurs when the argument of the square root taken to obtain $\cos I$ is negative. Therefore a test should be made of the sign of this quantity. If it is found to be zero or negative, the ray can be identified as totally internally reflected and the tracing of the specific ray interrupted. A warning is given to the user, and the ray tracing proceeds to start the next ray. In some cases, the identification of total internal reflection can cause the ray tracing to take an alternate path. For example, in the case of a nonsequential ray trace, the chosen option could be to treat the surface as reflecting, and proceed along with the ray, because there is not necessarily a way for the ray trace to know whether the designer intended the surface to act as a reflective surface under this condition.

There is another square root taken in determining the intercept. That square root is used in the solution of the quadratic equation computing the transfer length along the ray to the surface. In the case of a negative discriminant, there is no real solution to the equation, and the ray consequently misses the surface entirely. Thus a check for a negative value for $G^2 - cF$ indicates a ray-missed surface condition and the ray trace is terminated for the specific ray. Again, other options could be taken for nonsequential ray tracing, or special applications.

There are other optional methods of handling the total internal reflection and missed surface conditions. The purpose of the ray trace may be more general than a simple sequence of surfaces, and the required correction may be to change to reflection at the surface for the specific ray, and continue following the ray through the surface. A missed surface may just be ignored and another surface sought for intersection with the ray in a nonsequential ray tracing method.

After determination of the surface coordinates and direction of the ray, the transfer equations are used to carry the ray to the next tangent plane. The ray trace equations are applied sequentially for each surface in the lens. When the coordinates and refracted ray direction cosines are found on the last surface, the ray is transferred to the specified image surface. In most cases, this will be a plane surface, but a curved field system is occasionally encountered. The ray intercept coordinates are then available for use in describing the imagery.

Other information is available from the ray trace. Since the distance along the ray to the surface is obtained as a byproduct of carrying out the transfer of rays, the path length along the ray can be easily obtained. When this is compared with the path length for a reference ray from a selected field point, the Optical Path Difference (OPD) can be obtained for each ray that is traced. The OPD can be obtained at each surface or coordinate plane by subtracting the A value from that of the reference ray, and maintaining a continuing sum of path differences as the ray is traced through the system. In addition to the path difference lengths along the ray, information about the chromatic variation of optical path can be obtained by calculating the difference in path length for two different colors (Kingslake 1978). Additional information regarding the behavior of rays in the close vicinity of the pupil coordinates of the ray can be obtained by carrying out a close skew ray trace, similar to that discussed later in this chapter. These added pieces of information do not significantly slow down the ray tracing speed.

Starting and ending the rays are not a particular problem, once a decision has been made as to the reference surface to be used. In general, the ray is defined by an object point position, and an entrance pupil coordinate which is a fraction of the maximum pupil dimension. The ray is transferred to pupil coordinate. Transfer to the tangent plane of the first surface follows directly. Closing out the ray trace is accomplished by transferring the ray to the exit pupil surface, and then to the image surface. The absolute position on the image surface is available, but usually the data of interest is the aberration, or difference between the intercept of the specific ray and the chief ray.

The ray trace equations are designed to operate with any entering ray. In order to be useful, there are rules regarding the selection of rays. The lens

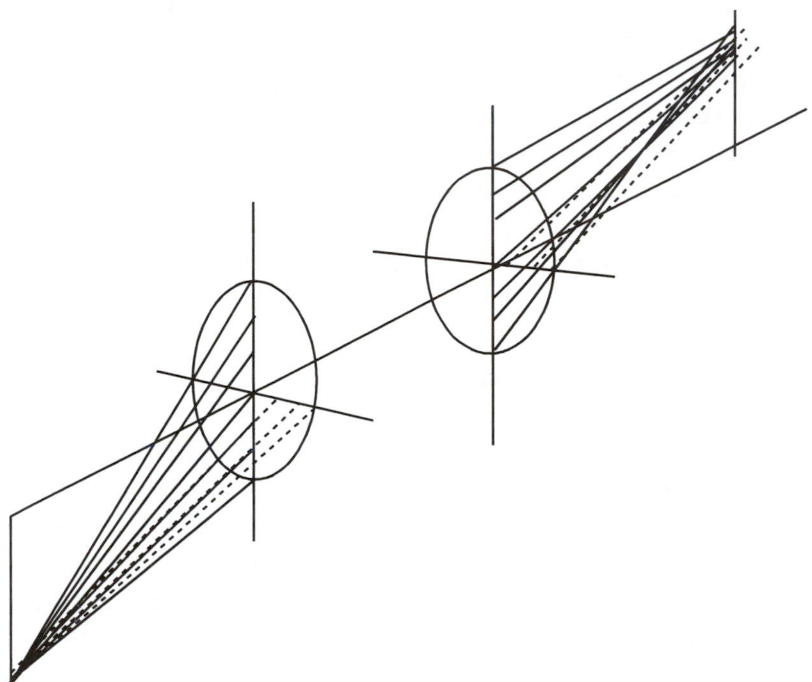

FIGURE 2.22 Definition of sagittal (– – –) and tangential (——) ray fans.

system consists of a specification of the object and image surfaces as well as the surfaces of the lens. The stored lens will also include information about the field size, either linear or angular, and the dimensions of the aperture that defines the size of the bundle of rays that can get through the lens. In most programs, the selection of a ray will be made by picking an object point as a fraction of the stated maximum field, and the coordinates of the ray in the aperture as a fraction of the maximum aperture.

The precise definition of the aperture of the lens is obtained by checking at each surface to see if the ray falls within the specified clear aperture of the surface. This will yield an accurate picture of the vignetted pupil for off-axis images. Conversely, certain rays at specified fractions of the full pupil height can be named as aperture defining rays at each field, and used to determine what clear aperture is required on each surface to obtain a specified amount of vignetting. This is a more accurate method of defining the required clear apertures than the use of paraxial aperture definitions described earlier. The paraxial aperture definitions serve as a good starting point for this calculation.

A rotationally symmetric lens will have an additional symmetry that is of importance in understanding the ray paths. Figure 2.22 indicates the definition of a meridional plane that contains the object point and the axis of the lens. For a rotationally symmetric system, the tangential plane is redundant with any plane around the axis, and the imagery in the lens will have a symmetry about the lens axis. Any ray from the object point which is contained within the meridional plane stays within that plane. The meridional plane also defines the tangential plane for the lens. Rays contained within this plane are also called tangential rays, for reasons that will become clear in the discussion of astigmatism (section 2.1.5). Any ray launched to a pupil location not in the meridional plane is called a skew ray. The special set of rays that are traced to the pupil with a zero y coordinate in the pupil are called sagittal rays. Again, this will become clearer in the discussion of astigmatism.

If the lens does not have a single axis of symmetry, the definitions of tangential and sagittal are more complex. A typical application of this is evaluation of errors introduced as a result of misalignments of elements in a lens. Evaluation of ray tracing for tilted and decentered surfaces is required for analyzing the assembly tolerances to be specified in the lens. In general, deliberate use of tilted and decentered surfaces is required for folded systems, prisms and unobscured aperture reflective systems. The ray tracing approach is identical to that described above, except that when a displaced surface is encountered, the ray vector is transformed into the new displaced set of surface coordinates using well-known rotation and coordinate shift matrices. These techniques are available on most design programs. The major difficulty in use of these tilt and decenter ray traces is keeping track of the coordinates during setup of the lens data.

The starting point for the ray selection is the paraxial definition of the field and pupils. Just as in the paraxial ray case, the ray heights can be specified in terms of a fraction of the pupil radius and the field height. In the presence of perfect imagery, all rays from a given object point would pass through the conjugate image point. In real systems, aberrations cause individual rays to intersect the image surface at a point different from the chief ray position. In real lenses, the pupils that must be used to define the ray selection are not the paraxial pupils, but real pupils defined by tracing real rays through the lens system. In some cases these may differ significantly from the paraxial pupils. Correct assessment of the imaging conditions require that the ray coordinate selection be modified to include pupil aberrations and possible large distortion of the chief ray.

The chief ray is traced first. This defines an image height as the ray intercept for the real chief ray. (In some cases, a surface within the lens is defined as a reference surface, and the chief ray coordinates are referenced to that surface, using an iteration of starting coordinates for the ray.) The aberrations relative to the chief ray intercept are computed by tracing a specified set of rays at different relative heights in the entrance pupil. (There are some legitimate cases of systems with obscuration in which the apparent chief ray

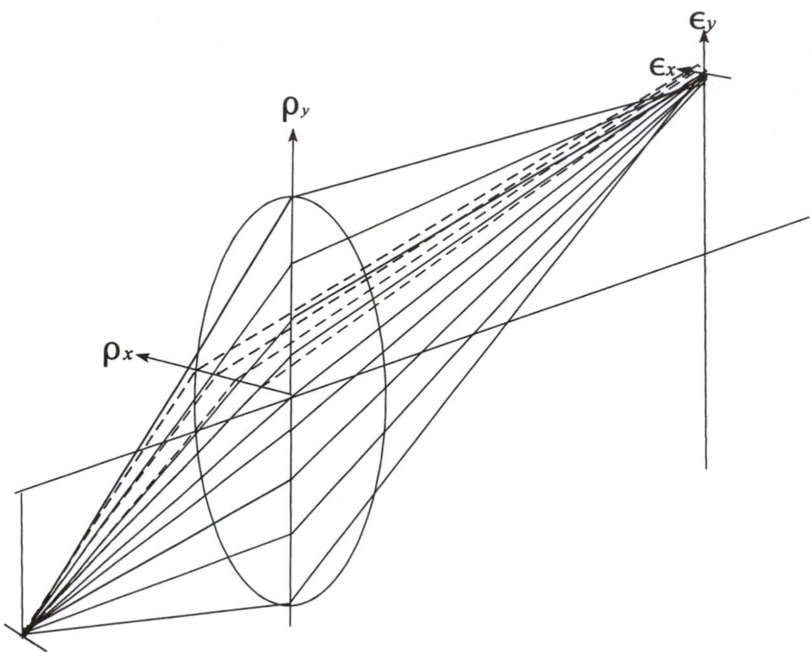

FIGURE 2.23 Selection of rays for plotting ray intercept plots.

does not pass through the system. The method of handling such an event differs between programs.) A table of ray intercept errors can be obtained for each field point and wavelength specified in the lens. These numbers can be used in many ways to obtain image error information. There are two common methods of graphical presentation of the ray intercept data. The most informative method for designers is in the form of ray intercept curves or rim ray curves obtained by tracing rays in the sagittal and tangential directions in the pupil. The second most common method is a spot diagram, which is the two-dimensional graphical presentation of the ray intercepts for a bundle of rays sampling the pupil.

The ray intercept plot is the basic tool used by the designer in analyzing the effect of ray tracing. It is generated as shown in Figure 2.23. Two fans of rays are traced through the lens at specified field angles. The first ray being traced in each fan is the chief ray for the lens. Usually this is the ray that passes through the center of the paraxial entrance pupil. When pupil aberration exists, or when there is vignetting in the lens, the selection of the appropriate chief ray should be the ray which is midway between the maximum upper and lower tangential ray coordinates. The intercept location of this ray on the image surface is obtained and kept as a reference for all of the other rays to be traced. (The distance of this chief ray intercept from the paraxial image reference point will provide the distortion associated with that ray.)

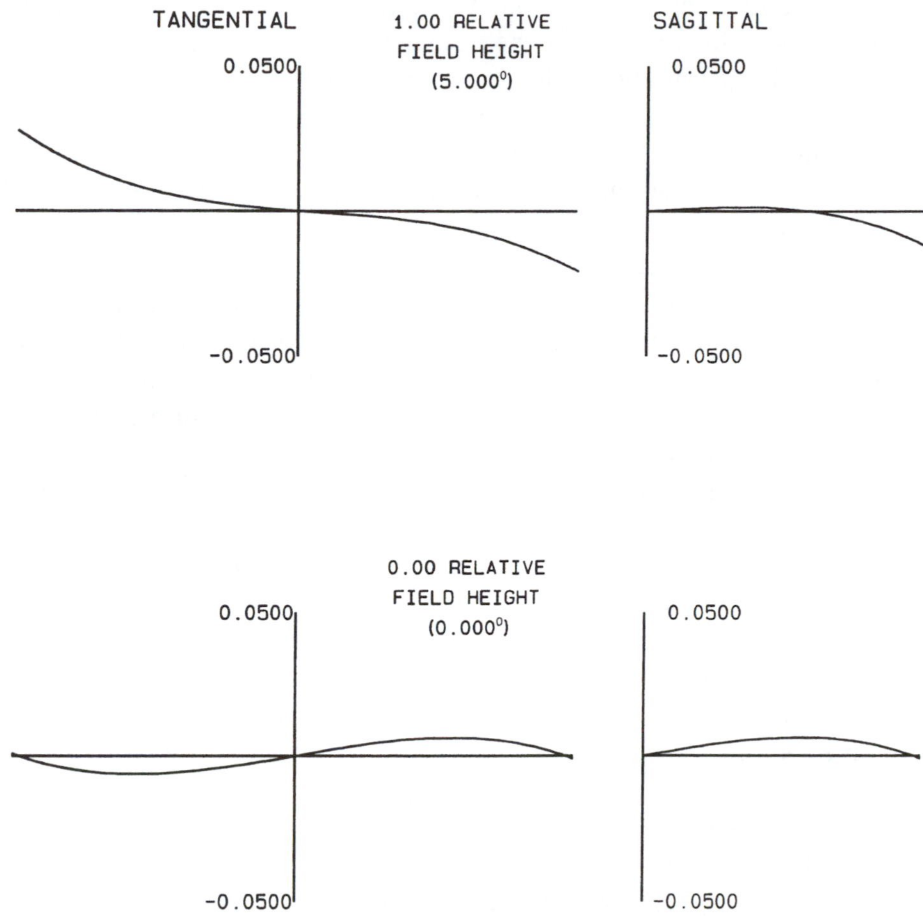

FIGURE 2.24 Sample ray intercept plot of a lens; wavelength 632.8 nm (CODE V).

A fan of tangential rays is then traced to provide a set of ray intercepts that describe the aberration present in each ray. The selection of the pupil coordinates for each ray is used to provide a fan of rays that lie within the vignetted aperture. (This is not an essential, as the user may be interested in the aberration information about all rays that could get through the lens in the absence of vignetting.) Similarly, a set of sagittal rays is traced which will provide additional aberration information.

For a rotationally symmetric lens, the tangential ray fan direction is the y direction, and the sagittal is the x direction. The ray intercept plot of interest is made by plotting the y coordinate of the ray intercept relative to the chief ray against the relative y height of the ray in the pupil. The sagittal aberration plot is found by plotting the x coordinate for the sagittal fan against the relative x pupil coordinate. The result is a pair of plots such as those in Figure 2.24. Should there be vignetting, then the plotted ray fan will not fill the full relative pupil dimension. Vignetting is indicated by the unfilled pupil dimension. In some cases, the designer may elect to examine the ray fans that would get through if there were no vignetting.

These ray intercept plots are extremely useful in diagnosing the aberration content of a lens. Later discussions of aberrations will demonstrate that the shape of these curves are related to specific aberrations of each order. The amount of variation of the ray error in the x and y directions is the sum of all of the aberration contributions. For a rotationally symmetric lens, all of the tangential ray data is included in these plots. The sagittal ray fans indicate only the sagittal component of the ray error. Since the sagittal rays are skew rays, they may actually have an appreciable amount of y direction component in the ray error. Usually this information is available as an option for the lens design program. In most cases, it is of considerably less interest than the sagittal data which is plotted. (Note however, that the y component of the sagittal ray error is fundamentally related to the coma produced by a nonisoplanatic lens. This is discussed in some detail in section 2.1.4.)

Another view of the ray intercept aberration can be found by tracing a bundle of rays with orderly sets of x and y coordinates that eventually fill the pupil. This two-dimensional bundle of rays is traced to the image plane, and the intercept coordinates relative to the chief ray coordinate are plotted. This provides a spot diagram as shown in Figure 2.25. The ray intercept plots provide information about the aberration in the principal sagittal and tangential portions of this image. While the spot diagram provides a visual impression of the light distribution about the chief ray coordinate, the ray intercept curves provide information as to the nature of the aberrations that are present. The shape of these curves permits the designer to deduce the aberration content of the image, as well as the magnitude of the aberration blur. In many cases, the result of a combination of aberrations will be so complicated that visual judgments about aberration content are not easy to obtain.

The ray intercept curves must be obtained for all field positions of interest. For systems that are rotationally symmetric, the convention is to trace ray fans at 0, 0.7, and 1.0 of the maximum field height. The choice of these field heights is determined by the normal behavior of aberrations that change principally linearly and quadratically with the field. There are, of course, higher-order field dependencies in many cases. The choice of these heights permits a quick assessment of the rate of change of aberrations with field.

The Optical Path Difference describing the wavefront can be presented in a similar manner. The plot of tangential and sagittal section of the wavefront error can be used to provide important aberration information about a lens.

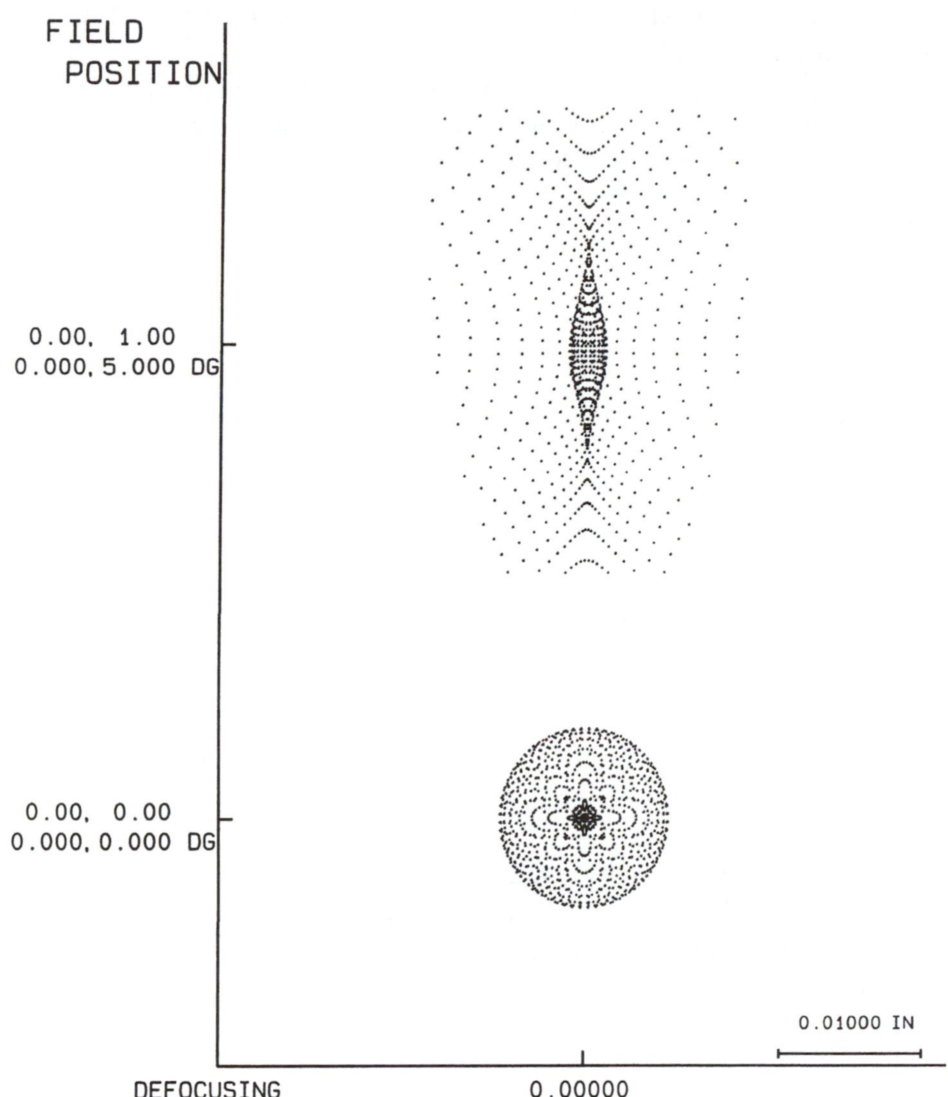

FIGURE 2.25 Spot diagrams for the ray intercept plots in Figure 2.24 (CODE V).

The shape of these curves is related to the individual aberrations, and is often dissected by fitting to well-known aberration polynomials to obtain numerical descriptions of the aberrations. In many cases, where the aberrations are comparable to or smaller than a few wavelengths, the appropriate design goal may be to minimize the wavefront error. The OPD plot is usually made in wavelength units, and permits the designer to compare the amount of aberration with the diffraction limit goal of a quarter wavelength.

The ray intercept or OPD data is the basis for many other calculations of the nature of the image. These include the Point Spread Function, Optical

Transfer Function, Encircled Energy, and many others. These will be discussed later in this book.

The ray trace described above was derived for successive spherical surfaces aligned along a specified optical axis. Simple modifications of the formulae apply to conic surfaces of revolution. The process of derivation follows the same course, except that the formula defining the surface is that of a rotational conic or conicoid. That is, the surface is a conic section rotated about the axis of the system.

$$z_{conic} = \frac{c(x^2 + y^2)}{1 + \sqrt{1 - (1 + \kappa)c^2(x^2 + y^2)}} \tag{2.78}$$

where the meaning of κ is

$\kappa = 0$	Sphere
$\kappa = -1$	Paraboloid
$-1 < \kappa < 0$	Ellipsoid (prolate spheroid)
$\kappa < -1$	Hyperboloid
$\kappa > 0$	Oblate spheroid

Tracing the ray to a conicoid surface requires that a quadratic equation similar to equation 2.61 is then solved and a set of ray trace equations are obtained. The process of ray tracing is quite the same as described above, except that the intercept coordinates now lie on a rotationally symmetric conic.

Similarly, the definition of a cylinder or toroid, which can also be a conic in one dimension, leads to a set of ray trace equations that can be carried out similarly to the spherical ray trace.

For other surfaces, the closed-form equations of the ray trace are not applicable. As an example, the rotationally symmetric high-order aspheric is defined by the equation

$$z = z_{conic} + (ad)(x^2 + y^2)^2 + (ae)(x^2 + y^2)^3 + (af)(x^2 + y^2)^4 + (ag)(x^2 + y^2)^5 \tag{2.79}$$

This surface form cannot be expressed in a set of equations as is the spherical or conic surface. The surface sag and normal to the surface can be computed. Ray tracing is done iteratively using Newton's method to find the intersection point of the incoming ray with the surface to a satisfactory accuracy. The type of surfaces that can be ray traced by iterative methods are even more general than this polynomial form. (The iterative ray trace can, of course, be used on any type of surface when the maximum accuracy is required for tracing of rays.)

The iterative ray tracing equations require the ability to compute the sag and the normal to the surface at all points on the surface. For a general surface $f(x, y, z) = 0$ which meets this differentiability condition, the ray is traced to the tangent plane or to the conic or spherical base surface. The equation defining the surface is evaluated at that set of coordinates and the difference between the actual surface and the intersection along the ray is obtained. This difference is used as an adjusting value and the ray intersection evaluated again. This is carried out repeatedly until the coordinates of the surface and the coordinates along the ray agree to within some stated tolerance. The iteration on the path length to the surface is carried out until the intersection point is located to an accuracy comparable to the least significant bit in the computer, or to something like 0.000001 wavelengths. This is actually the most accurate method of tracing rays on a computer since the iterative process ensures that no constant errors are propagated along with the ray. The necessity of taking several iterative steps at a surface slows the ray tracing speed to below that of the formula trace used on spheres and conics.

The equation describing the surface may take many different forms. The surface may not be describable by a single equation, but perhaps only by a two-dimensional array of sags. Spline interpolation can be used between the specifying points, which not only forms a smooth local fit to the surface, but meets the condition of being continuous and differentiable. This spline-fit surface can mimic any arbitrary surface, but handling the sag data defining the splines can become quite difficult and extensive. Limited types of splines can be used, such as a rotational spline, which permit some limited matching to otherwise seemingly intractable surfaces.

The surface may be the sum of a set of differentiable functions, such as two-dimensional power series, or a set of Zernike polynomials. These may be used in the design process, or may be used to fit the errors measured on surfaces during fabrication. Once this is done, the effects of fabrication residual errors can be incorporated into the design and evaluation process.

The surface may be represented as a locally periodic grating lying on the surface. In that case, the traced ray will be refracted by the index of refraction change at the point of intersection and then diffracted by using the grating equation. The diffraction order needs to be specified along with the specification of the grating on the surface. Since only the period and direction is used, the shape of the grooves does not enter into the calculation, and the relative efficiency of the diffraction of the ray cannot be determined. It is obviously possible to add some mathematical modeling which permits computation of the efficiency, and indeed some programs enable the computation of the actual wave propagating from a local section of the grating.

The choice of grating can vary considerably. A simple linear grating can be used, or a rotational grating such as a Fresnel zone plate. A holographic grating can be described either by an equation that gives the grating spacing and direction for each point on the surface, or it may be described by parameters that establish the location of the sources that produce the interference pattern generating the grating.

The most general diffractive optical component is the computer-generated grating, which may, or may not, be a binary or stepped binary surface. This permits the integrated design of diffractive components and reflective or refractive components. Such diffractive components can be manipulated by computer generation into a potentially very broad family of optical components. All diffractive surfaces are strongly wavelength dependent. In some cases, this may be used to compensate for the chromatic dispersion of optical materials.

Determination of the clear apertures that are required on each surface can be initially made from the paraxial ray trace as described in the previous section. This will provide reasonable guidance. In some cases, the paraxial path will not be a sufficiently good representation of the real ray path that realistic clear apertures result. The determination of clear apertures by real rays is therefore indicated. The clear apertures are defined by the required entrance pupil diameter on axis, and by the vignetted pupil dimensions off axis.

The actual clear aperture required on the surfaces of a lens is determined by tracing the maximum upper and lower rays that are expected to pass through the system without being blocked. This is accomplished by specifying fractional upper and lower coordinates in the entrance pupil for each field of the lens that must be passed. These provide vignetting factors that can generate the appropriate rays for checking for clear apertures. In most lens design programs, this is a relatively simple specification. However, just specifying these values for the ray bundles does not necessarily mean that there is a physical combination of apertures that will achieve the requested vignetting at all field heights. The designer must evaluate a drawing of the lens with the critical rays, as well as the clear apertures plotted that can verify that the requested vignetting is achievable.

Ray tracing is usually very accurate in any program. This is a consequence of the basic importance of the ray trace as the cornerstone of all lens design computation. Therefore, the programmer spends considerable time verifying and optimizing the ray tracing subroutines. There are cases where erroneous results can occur. Some errors are due to the truncation resulting from the finite word length of any computer. Therefore, for very large-diameter systems, the great disparity between the lateral intercept of a ray and the small size of the direction cosines may produce some round-off errors. This can be found by careful observation of the numeric output and comparison with expected results. Should this be suspected, the surfaces can be renamed as aspherics, and iterative ray tracing takes place. In this case all quantities are computed to the precision asked in the aspheric ray trace at each surface and there will be no cumulative error, as can be experienced by successive application of a formula ray trace.

Other apparent errors are not as easy to identify. Sometimes, in the presence of significant pupil aberration, the rays are not mapped properly in the pupil, and an erroneous placement of aberration in the pupil coordinates can be obtained. This can be avoided by examining the ray intercepts on critical surfaces to ensure that the proper pupils are actually being used in tracing. Apparent errors can also occur if the program plots entrance pupil coordinates instead of exit pupil coordinates in carrying out ray trace plots.

In most cases, ray trace errors occur in systems with significant aberrations, and are not the result of simple computational deficiencies. The designer should always question any unusual-appearing ray plot errors and be sure that the source of the unusual-appearing ray plot is fully understood. Errors that seem to be appearing on more complex calculations, such as MTF plots, should always be examined first by ensuring that the ray trace data are correct.

Ray tracing through cylinders or torics can also be handled by a formula ray trace. The paraxial setup must be considered with respect to the two principal orientations in the lens. The surface will be described by a sag formula on one azimuth, which is rotated about a cylindric axis in the other azimuth.

Other forms of nonspheric surfaces are not as easy to handle in ray tracing. General aspherics, which range from addition of a radially symmetric polynomial in aperture height to the basic conic sag to surfaces which can only be described by a spline function, require a different approach to ray tracing. Numeric methods are used to iterate to find the intersection of a ray with the surface. The process used is Newton's method, in which the difference between coordinates along a ray and the formula describing the surface is minimized below a designated tolerance. Once the intersection with the surface is obtained, the surface slope and the normal can be computed and refraction or reflection takes place as before. This type of ray tracing is obviously slower than a single evaluation of a formula, but has the advantage of always converging to below some required tolerance at each surface.

The conditions for total internal reflection are the same as before. The condition for missing a surface is a bit more complicated and may be surface type specific.

So far, the coordinate sets for all surfaces have been coincident with the optical axis, or z axis. Tilts and decenters can occur. The approach used to include this in ray traces is to trace a ray to the location in which the base surface should be located. The ray coordinates are then transformed by the usual means to the new set of surface coordinates in which the tilted and decentered surface resides. Ray tracing can then proceed as usual.

After passing the surface, a second inverse transform may be called for in order to return the coordinate reference to the original coordinate system, or the coordinate system may remain changed. In some cases, the desire is to tilt an entire component or lens assembly that may have several surfaces. The options in this are those of the designer who is attempting to set up a lens with some specified set of surfaces in a known coordinate location. Each of the lens design programs uses basically the same ray trace procedure, but differs in the methods of describing the lens. In some cases, the interpretation of the rotation angles and their relation to the coordinates may be different. The designer needs to refer to the program documentation which may relieve the confusion. There are many complications in designing nonrotationally symmetric systems, and a designer must frequently develop graphics of the location of components in order to ensure that the intended arrangement is being accomplished.

A major use for tilt and decenter ray traces is in the evaluation of tolerances for assembly of a lens system. Here small changes are introduced into the location of surfaces and components and the change in the merit function or other image quality function computed. The use of this procedure in tolerancing is discussed in Chapter 6.

The coordinates for tracing rays are attached to the surface. An alternate approach exists in which the ray intercept coordinates and direction cosines are interpreted with a global coordinate set. The designer can choose an origin of coordinates, and all ray data will be presented in terms of that coordinate set. This can be very useful when dealing with a complicated nonrotationally symmetric system. The global coordinate set is also very useful when interfacing a design to a mechanical arrangement in which the relative position of beams passing through the lens, as well as the lens components, need to be referenced to some summary drawing. Uses of this in multifocal systems and scanning systems are obvious. Once again, the specific procedure for calling for this differs between lens design programs, and reference to the program documentation is required.

Some optical systems are defined as multiconfiguration systems. Zoom lenses are one example of this type of system. Each ray that is traced is requested to be for a specified member of the set, perhaps for a specific zoom setting. Each surface has attached alternate data which will be used during the ray tracing, depending upon the zoom number attached to the ray. Other examples of multiconfiguration systems include spectrometers, where the light may take a different direction at some surfaces, dependent upon the wavelength. Each member of the multiconfiguration lens is a sequential set of surfaces, but the action taken at critical surfaces differs according to which member of the set is being evaluated.

The most general optical system is one in which the designer may not have a clue as to which surface will be encountered next by a specific ray. These systems, called nonsequential surface systems, are specified in a global coordinate system, as is the source or entrance pupil. Rays are traced and each surface is examined to see if it is the next intersecting surface. In each case, when the surface is encountered, the decision is made as to whether the ray transmits through or is reflected from the surface. In some cases, such as a beamsplitter, both operations may occur.

These systems differ from the multiconfiguration systems mentioned above in that the light does not pass through an ordered set of surfaces. Each ray direction is defined, and the ray trace program determines which is the first surface to be encountered. One example is a complex prism, in which the path through the prism depends upon the entry conditions.

There are many other options that may be added to a ray trace. Surfaces may be described in terms of interferometer measured data, and the ensuing phase differences added to the wavefront errors at a given surface. In order to be correct, the alteration in ray direction must also be considered in such computations. The index of refraction of the medium may vary with location, requiring the ray to be traced along a curved path incrementally through the medium.

In each of these cases, the lens design program must contain options for describing the optical system and selecting the ray set to be traced.

2.1.3 WAVEFRONT DETERMINATION

The optical path along a ray is the integrated value of geometrical length along each ray multiplied by the index of refraction of the medium the light is passing through. For the case of homogeneous material, the optical path is found by multiplying the length of the ray within each medium by the index of the medium, and then adding the paths. The units of the optical path are length, since the index of refraction is dimensionless. Since the index of refraction is the ratio of the velocity of light in vacuum (or in air at standard temperature and pressure, STP) divided by the velocity in the medium, the Optical Path Length is proportional to the time necessary for the ray to propagate through the medium relative to that in the reference medium (vacuum or air). Thus a surface of constant Optical Path Length is a surface of constant time from the origin of the ray, and corresponds to a wavefront associated with the ray. The measure of deviation of a wavefront from perfection is the Optical Path Difference (OPD).

The Optical Path Difference is a measure of the aberration content of a lens. The basic definition of the OPD is

$$\mathrm{OPD}(x_p, y_p) = [\text{Path along reference ray}] - [\text{Path along ray}] \quad (2.80)$$

The paths are computed from the ray intersection with the entrance pupil reference sphere to the intersection of this ray with the exit pupil reference sphere. Here x_p and y_p are coordinates of the ray intersection in the exit pupil sphere for the lens. Usually, the chief ray is used as the reference ray, and will have (0,0) coordinates in the pupil. For a perfect image the OPD will be zero for all pupil locations. This indicates that the wavefront coincides with the pupil reference sphere. The value of the OPD is a measure of the time lag or phase difference between the reference ray and the arbitrary ray at coordinates (x_p, y_p). If the OPD is divided by the wavelength, then the number-of-waves error can be obtained. This is the standard way of referring to the OPD for most optical systems.

The wavefront error, in wavelength units, is then

$$W(x_p, y_p) = \frac{\mathrm{OPD}(x_p, y_p)}{\lambda} \quad (2.81)$$

An error of $W = 0.25$ means a quarter-wave error at the pupil coordinate selected. A value of $W = 1$ is, of course, a full wavelength of error.

The optical path is calculated as a byproduct of carrying out the ray trace. The optical path between the surface and reference plane is found as a part of the ray trace. The path to the next reference surface is found in tracing the ray between surfaces. The specific definition of importance is the OPD calculated for the ray path length between the entrance and exit pupil reference spheres. Usually the object-side reference sphere will be established in the entrance pupil and centered upon the object point for the chief ray at the chosen field location. The exit pupil reference sphere will be located in the exit pupil and will be centered upon the intersection point of the chief ray and the image surface. The optical paths between the surfaces of the lens and the actual reference sphere can be calculated as part of the opening and closing parts of the ray trace. For added accuracy, the optical path can be obtained as a cumulative difference between the paths along the rays at each surface.

The conventional choice of reference is the entrance and exit pupil locations. For various purposes, the designer may choose to select the actual total path along the ray to be computed, or to project the pupils to an infinite or flat reference surface, and compute the optical path at that location. It will be seen in section 2.2.2 that the pupil spheres are a natural location for computing the OPD, and in section 2.1.4 that the pupil aberrations are closely related to isoplanatic image formation.

The relation between the ray intercept error and the wavefront error is shown in section 2.1.4 to be

$$\epsilon'_x = -\frac{R'}{n'}\frac{\partial \text{OPD}}{\partial x'_p}$$
$$\epsilon'_y = -\frac{R'}{n'}\frac{\partial \text{OPD}}{\partial y'_p} \quad (2.82)$$

Here the ϵ values are the ray aberrations on the image surface measured with respect to a chosen reference location. R is the distance from the center of the pupil to the image reference location and x_p and y_p are coordinates in the exit pupil. These directly provide the interpretation in terms of lens dimensions. Understanding and usefulness is improved by conversion into a more convenient form, scaling the off-axis values of R' and x'_p and y'_p by recognizing that

$$R' = \left(\frac{R_0}{\cos \bar{U}'}\right)$$
$$x'_p = \rho_x y'_{0m}$$
$$y'_p = \rho_y y'_{0m} \cos \bar{U}'$$
$$u'_0 = -\frac{y'_{0m}}{R_0} \quad (2.83)$$

In this equation, R_0 is the axial exit pupil distance, and $\bar{U}'$ is the chief ray angle associated with the field position being evaluated. The ρ values are fractional values in the pupil and y_{0m} is the maximum pupil height for the axial field position. The obliquity in the y direction produces the cosine relation between the axial and off-axis maximum pupil height. These are, of course, paraxial values which provide the proper scaling of the image across the field. Their relationship is explained in section 2.1.4. The axial–paraxial ray slope angle is u'_0.

With all of these "simplifications" introduced, the result is a relationship in terms of relative heights in the exit pupil:

$$\epsilon'_x = \frac{\lambda}{n'u'_0}\frac{\partial W}{\partial \rho'_x}\frac{1}{(\cos \bar{U}')}$$
$$\epsilon'_y = \frac{\lambda}{n'u'_0}\frac{\partial W}{\partial \rho'_y}\frac{1}{(\cos^2 \bar{U}')} \quad (2.84)$$

Here the λ has been inserted to permit the function W to be expressed in wavelength units. In the interpretation of this, remember that in most cases the paraxial slope angle will be negative. The inverse cosine scaling takes into account the intrinsic anamorphism of the pupil. The presence of vignetting will limit the pupil for the calculation of diffraction images but will not change the scaling onto the focal surface.

This is a rather simple relationship between the wavefront errors and the ray errors. The presence of any aberration in the pupil will be scaled to the ray intercept on the image plane by this relationship. In several references this

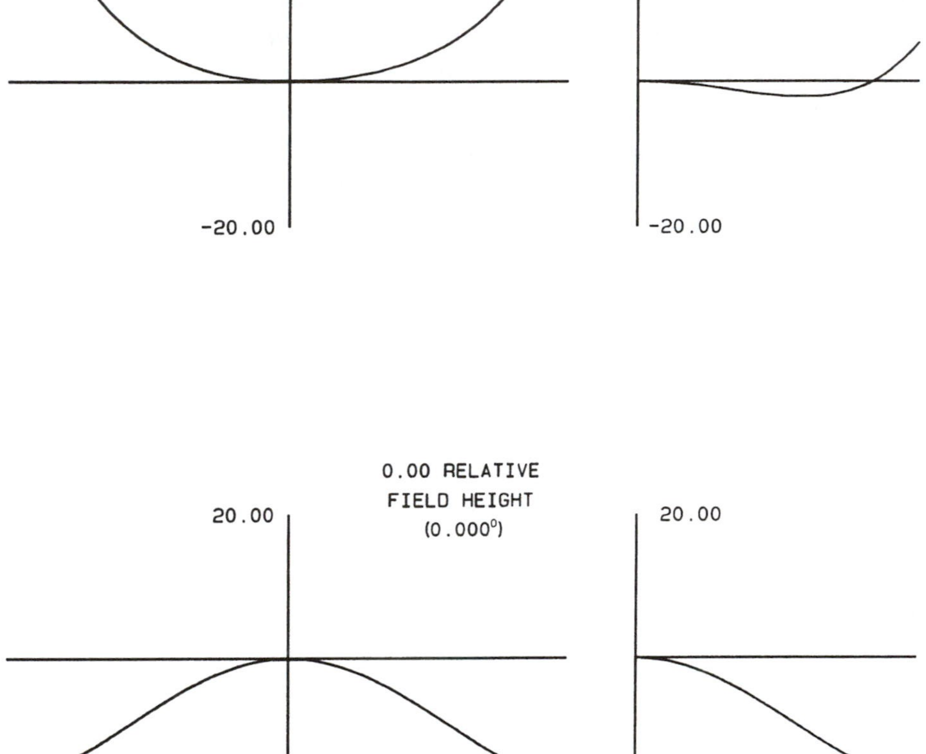

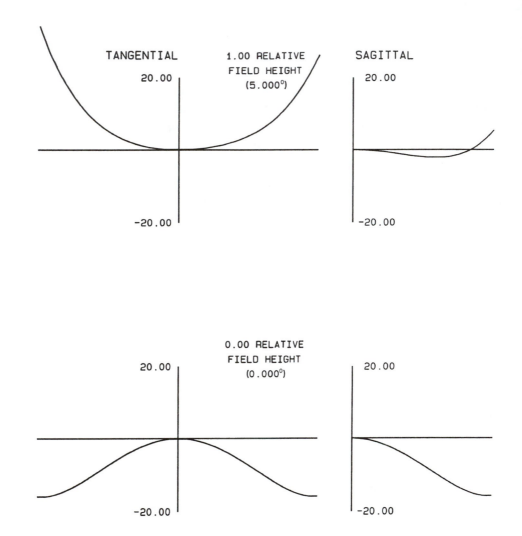

FIGURE 2.26 OPD plots for the system shown in the last two figures (CODE V).

form of the aberration expression is related to the Spherical Point Characteristic Function (Buchdahl 1970).

A comparison of the ray intercept and wavefront error plots for one example of a lens are shown in Figures 2.24 and 2.26. The ray plots provide a measure of the blur or spread of light on the image surface. The OPD plots provide a good measure of the relative correction of the aberrations against the wavefront criterion. The slope of these curves will correlate to the ray aberration plots. Visual comparison of these figures indicates that a zero ray aberration is accompanied by a zero slope in the OPD plots.

The calculation of the OPD is accomplished as a byproduct of ray tracing. The calculation of the transfer distance A in section 2.1.2 provides the path length along the ray. Adding this up as the ray is traced from the entrance pupil to the exit pupil will provide the total path through the lens. Tracing each ray from the entrance pupil to the exit pupil reference sphere will provide a value for the optical path along the ray. Then, subtracting this from the reference ray path provides the Optical Path Difference. When this is carried out for a large number of rays distributed in the pupil, a numerical matrix describing the wavefront error is produced.

Two different ways of plotting the OPD are shown in Figures 2.27 and 2.28. The contour map in Figure 2.27 is useful in assessing the symmetry of the wavefront error. The perspective plot in Figure 2.28 provides an assessment of the magnitude and irregularity of the OPD values. These plots are for the central wavelength only. This OPD data is obtained as a two-dimensional matrix of values of phase error across the pupil, and is used as the basic information in computations of the diffraction image.

The accuracy of the OPD is determined by the numerical precision of the computation of the ray trace. For a 32-bit computer, the error is one part in 4×10^{10}. If the total length of the lens being evaluated is about a meter (quite long for most lenses) then the possible error from simple subtraction of the two paths is of the order of about 0.0002 wavelengths, in the visible. This is usually an acceptable level of numerical error. This accuracy can be increased if required by treating each surface as an iteratively traced aspheric surface and by accumulating the OPD as each surface is traced. In most cases, the accuracy of the calculation of the OPD is not an issue. The designer should be aware that questions of accuracy can become an issue in some cases, especially with very large optical systems.

The OPD describes the passage of the wavefront across the pupil reference sphere. Construction of the actual wavefront would require adding the OPD increment to the spherical pupil surface at the position and in the direction of the ray passing through the pupil. The OPD is a measure of the aberration at a single position in the field. Section 2.1.4 discusses the rate of change of OPD with field position.

The sign of the OPD is significant. The convention of subtracting the ray path from the reference path is used. Then a positive value for the OPD indicates that the ray will have propagated from the entrance pupil to the exit pupil in a shorter time than the reference ray. This would indicate that at the time the wavefront associated with the reference ray has reached the exit pupil, the wavefront portion corresponding to the ray of interest will have proceeded beyond the pupil. Normal undercorrected spherical aberration, usually found in a single-element lens, would result in a positive OPD at the edge of the aperture, when the exit pupil sphere is centered at the paraxial focus position.

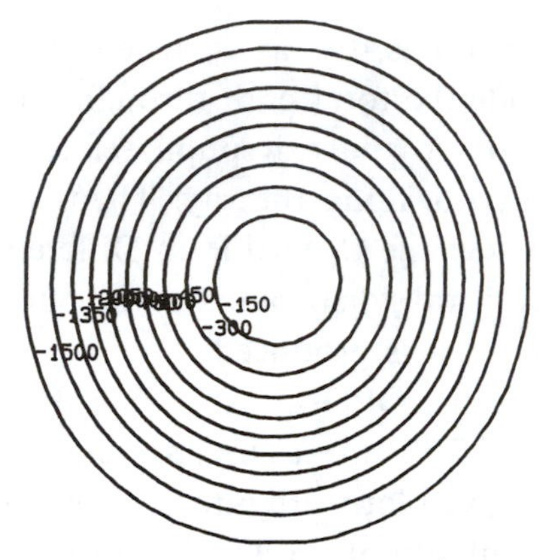

(a)

WAVELENGTH: 632.8 NM

FLD (0.00, 0.00) MAX, (0.0, 0.0) DEG

DEFOCUSING: 0.000000 IN.

CONTOUR INTERVAL: 1.50 WAVES

MIN/MAX: -15.85 / 0.00 WAVES

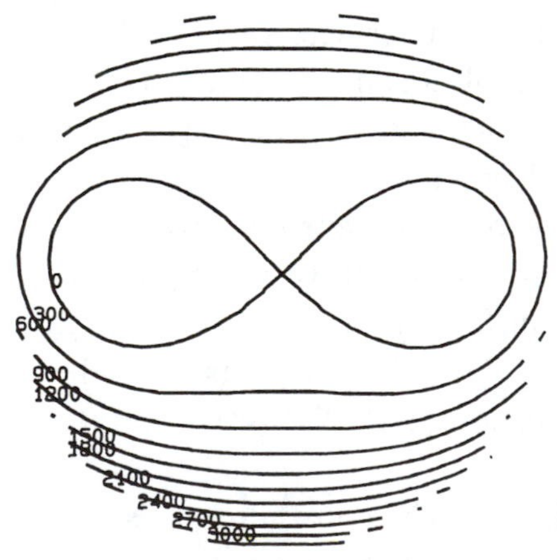

(b)

WAVELENGTH: 632.8 NM

FLD (0.00, 1.00) MAX, (0.0, 5.0) DEG

DEFOCUSING: 0.000000 IN.

CONTOUR INTERVAL: 3.00 WAVES

MIN/MAX: -2.10 / 32.76 WAVES

FIGURE 2.27 Contour maps for the OPD in the lens: (a) on axis; (b) at edge of field (CODE V).

The OPD can be plotted as a representation of the aberration, just as the ray intercept data is plotted. Other alternate representations are obtained by fitting the OPD to a polynomial representing the class of aberration expected in the wavefront. This is discussed in detail in sections 3.1.1 and 3.1.2. Other representations using orthogonal polynomials are described in section 3.2.

The importance of the OPD in computing the physical image is discussed in section 2.2.2. The OPD provides the phase errors in the pupil that define the diffraction that takes place in image formation. Calculation of the OPD in the design data permits computation of all of the image quality functions associated with the lens alone. Additional information about the lens can be gained by adding measured interferometric test data from shop tests of components or the entire assembled lens.

Addition of shop test data is accomplished by collecting an interferogram of the wavefront passing through the lens under known conditions. The interferometric data is collected by a CCD detector, converted to phase information and stored as an interferometer file in a specified format. Almost all lens design programs have an interface that accepts data from interferometers made by the principal manufacturers, such as Zygo, Wyko, or Phase Shift Technologies. Use of this interface is not difficult, but care is required in editing the data to match the actual clear apertures in the lens. In addition, most interferometers do not permit operation at a number of wavelengths, so that the use of measured wavefront data to add test data to a lens design may require some assumptions as to the source of residual errors.

2.1.4 EXTENDED SINE CONDITION – ISOPLANATISM

The design of a lens to cover a wide field of view requires that the imaging properties of the lens change as little as possible over the field. To understand the conditions necessary for accomplishing this isoplanatism, it is necessary to begin with some fundamental considerations from geometrical optics.

The most general function defining the aberrations in a lens is the characteristic function, first defined by Hamilton, and later used in modified form by Bruns as the eikonal. To define the characteristic function first select two points, P and P' in the object space and the image space, respectively. Let $\boldsymbol{a}$ and $\boldsymbol{a}'$ be the vectors describing the location of these points in appropriate coordinate systems. The optical path between these points will be defined by $V(\boldsymbol{a}, \boldsymbol{a}')$ in which it is implicitly assumed that only one ray passes between the points for a specified field position. (This restriction will be overcome later by defining the interesting function in the pupils, rather than the image locations.) The function V describes a geometrical ray field; knowledge of this

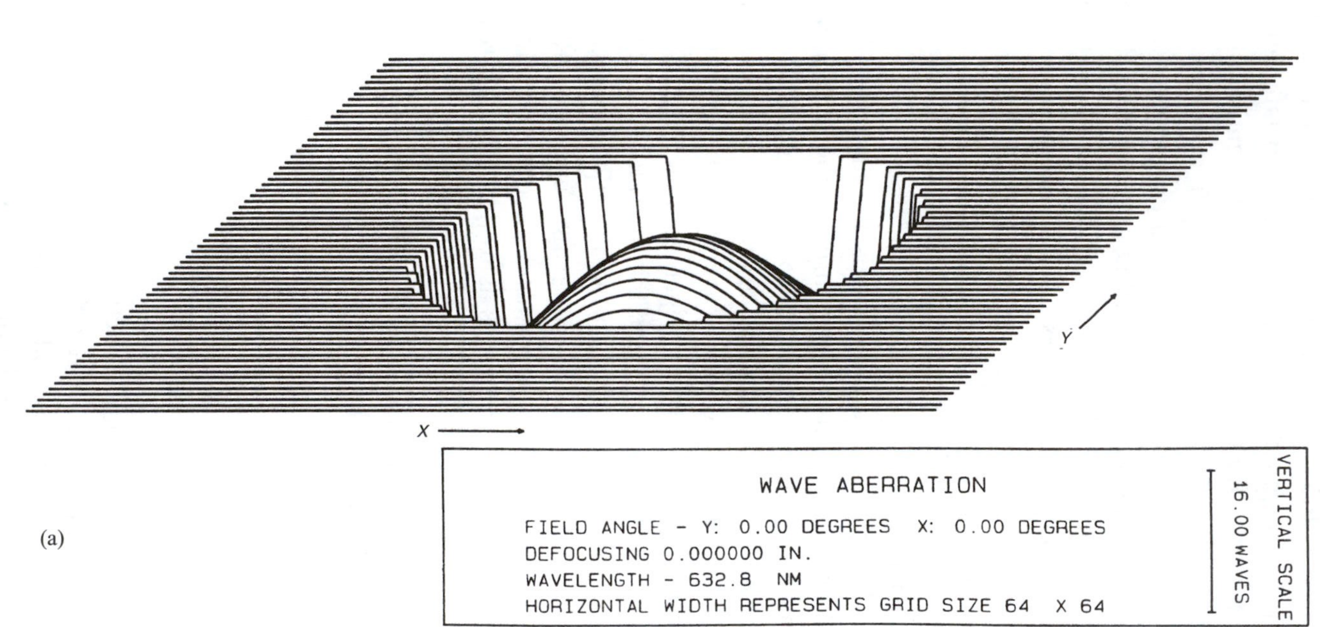
X
Y
WAVE ABERRATION
FIELD ANGLE - Y: 0.00 DEGREES X: 0.00 DEGREES
DEFOCUSING 0.000000 IN.
WAVELENGTH - 632.8 NM
HORIZONTAL WIDTH REPRESENTS GRID SIZE 64 X 64
VERTICAL SCALE
16.00 WAVES

(a)

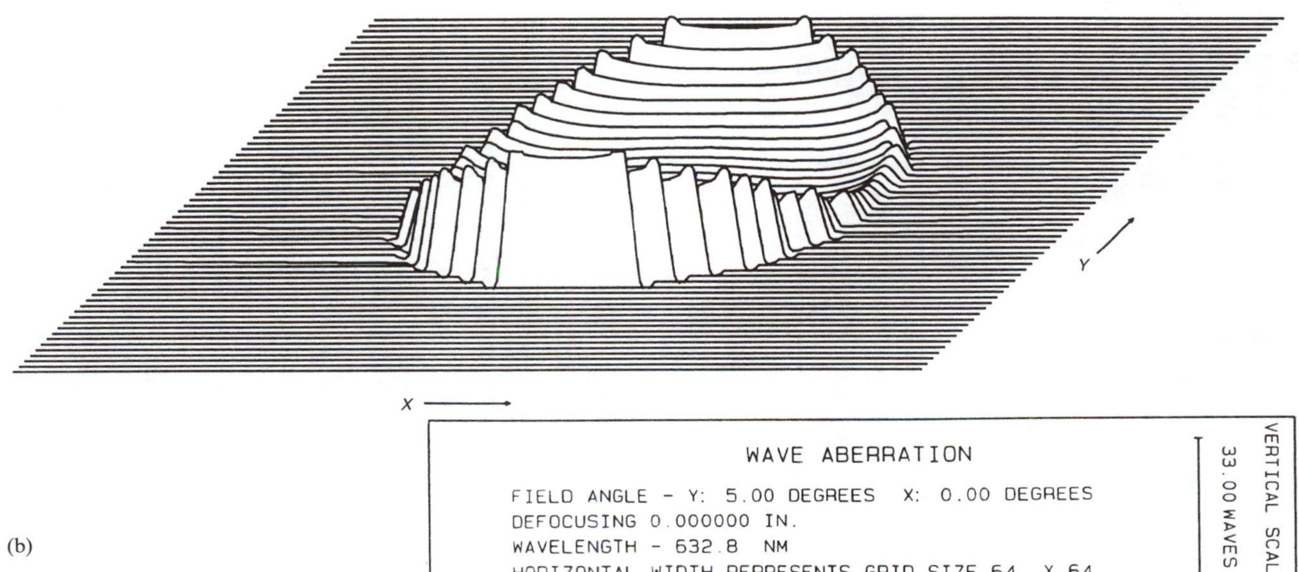

(b)

FIGURE 2.28 Surface plot of the OPD (a) on axis, (b) at edge of field (CODE V).

function, if it could be obtained, permits definition of the complete optical properties of the lens. Because there is no way to obtain this function in the general case, it is almost always numerically determined. However, understanding of the basics of geometrical optics can permit development of an understanding of the properties and behavior of the function. In particular, the properties of this function can be used to determine the rate of change of imagery with changes in the field location.

By Fermat's principle, the path between the points is an extremum, usually a minimum. This means that the second derivative of the function with respect to the rays defined by the function is zero. If a very small perturbation in the coordinates of the ray along the path is taken, then there will be a first-order change in the optical path function, but not a second-order change. A small change of location is made from point p' to point p''. The small vector element $\delta \boldsymbol{a}'$ represents the change, where this change vector is presumed to be small. Because $V(\boldsymbol{a}, \boldsymbol{a}')$ represents a ray field, then a small change in the function will be given by

$$V(\boldsymbol{a}' + \delta \boldsymbol{a}', \boldsymbol{a}) - V(\boldsymbol{a}', \boldsymbol{a}) = \delta \boldsymbol{a}' \nabla V \tag{2.85}$$

where the path being evaluated is a change in the path along the ray described by r'. The vector describing the change, $\delta \boldsymbol{a}'$, is arbitrary in direction, but very small in length. The left side of the equation is just the optical path along the ray for a perturbation in the ray position between p to p''. Therefore

$$n' \delta \boldsymbol{a}' r' = \delta \boldsymbol{a}' \nabla V \tag{2.86}$$

which can actually be expressed as three scalar equations in the three coordinates. Because the choice of $\delta \boldsymbol{a}'$ is arbitrary,

$$n' \boldsymbol{r}' = \nabla V \tag{2.87}$$

and the three scalar equations describing the ray are

$$\begin{aligned} n'L' &= \frac{\partial V}{\partial x'} \\ n'M' &= \frac{\partial V}{\partial y'} \\ n'N' &= \frac{\partial V}{\partial z'} \end{aligned} \tag{2.88}$$

where L', M', N' are direction cosines along the ray. A similar relationship exists in object space:

$$nL = \frac{\partial V}{\partial x}$$

$$nM = \frac{\partial V}{\partial y}$$

$$nN = \frac{\partial V}{\partial z} \tag{2.89}$$

where the unprimed quantities represent coordinates in the object space. These functions enable determination of the most general changes in imagery that take place in the lens system.

In order to examine in a practical manner the effect of object and image location changes on the characteristic function, it is necessary to develop a new characteristic function. This function, called the Spherical Point Characteristic Function, or the eikonal, includes a choice of specific coordinate sets in object and image space. To develop this function, we add optical paths related to changes in coordinate reference position:

$$E(\boldsymbol{r}, \boldsymbol{r}') = V(\boldsymbol{a}, \boldsymbol{a}') - n'\boldsymbol{a}'\boldsymbol{r}' + n\boldsymbol{a}\boldsymbol{r} \tag{2.90}$$

Here the added terms are the optical paths introduced by selecting a specific set of coordinates for expressing the characteristic function. We would now like to know the conditions upon the coordinate sets describing the imagery in object and image space that will make the first derivative of the aberration with respect to a small reference coordinate change equal to zero. This will produce imagery that is locally isoplanatic.

Differentiating the function with respect to L' and M' leads to the relations

$$\frac{\partial E}{\partial L'} = \frac{\partial V}{\partial x'}\frac{\partial x'}{\partial L'} + \frac{\partial V}{\partial y'}\frac{\partial y'}{\partial L'} - n'x' - n'L'\frac{\partial x'}{\partial L'} - n'M'\frac{\partial y'}{\partial L'}$$

$$\frac{\partial E}{\partial M'} = \frac{\partial V}{\partial x'}\frac{\partial x'}{\partial M'} + \frac{\partial V}{\partial y'}\frac{\partial y'}{\partial M'} - n'x' - n'L'\frac{\partial x'}{\partial M'} - n'M'\frac{\partial y'}{\partial M'} \tag{2.91}$$

which can be simplified by using parts of the equation 2.88 to yield

$$\frac{\partial E}{\partial L'} = -n'x'$$

$$\frac{\partial E}{\partial M'} = -n'y' \tag{2.92}$$

which express a relatively simple relation of the change in characteristic function to the change in image position. A similar set applies to the object space:

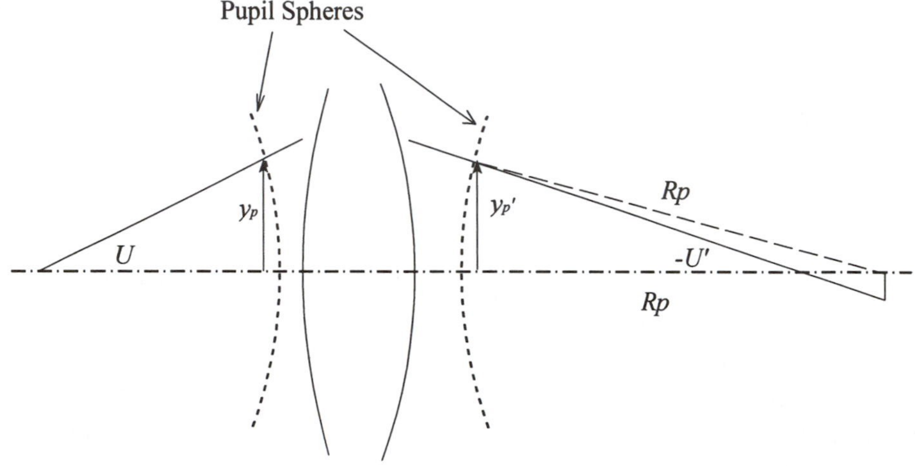

FIGURE 2.29 Reference sphere coordinate definition, on axis.

$$\frac{\partial E}{\partial L} = -nx$$

$$\frac{\partial E}{\partial M} = -ny \tag{2.93}$$

This symmetric relationship must be maintained in both the object and image coordinates in order to maintain a constant imaging condition. These equations indicate the relation between the derivative of the wavefront error for a characteristic function defined on a pupil surface with respect to the coordinates of the ray in the pupil and the location of the ray intercept on the image surface.

The wavefront aberration on a spherical pupil surface linked to the choice of object and image reference coordinates can be defined in the following manner. The optical path difference is defined on the reference pupil surfaces, as shown in Figure 2.29.

$$W(x'_p, y'_p) = V(P', Q') - V(P', Q'_0) \tag{2.94}$$

where the function V is a characteristic function defined on the exit pupil surface. This is the OPD or Optical Path Difference function defined earlier in this chapter. The OPD will be confined to points on the spherical pupil reference surface. From the reference point, which is at the center of the reference surface, the coordinates on the surface are x_p and y_p obtained in terms of the direction cosines by

$$x'_p = R'(L' - L'_0)$$

$$y'_p = R'(M' - M'_0)$$

$$z'_p = \frac{x'^2_p + y'^2_p + z'^2_p - 2y'_0 y'_p}{-2l'} \tag{2.95}$$

the last equation stating that the z_p pupil coordinate is confined to the spherical exit pupil surface. This then yields

$$\frac{\partial W}{\partial x'_p} = -\frac{n'}{R'_i}(x' - x'_0) = -\frac{n'}{R'_i}\epsilon'_x$$
$$\frac{\partial W}{\partial y'_p} = -\frac{n'}{R'_i}(y' - y'_0) = -\frac{n'}{R'_i}\epsilon'_y \tag{2.96}$$

where the x' and y' are coordinates on the image surface and the values R'_i are the actual distance from the current x'_p, y'_p coordinates in the pupil and the actual intersection points x', y' on the image plane. The ϵ'_x and ϵ'_y are ray aberration errors in the x and y directions on the image surface, measured from the center of the exit pupil reference sphere.

In almost every imaging case, the aberrations are small compared to the distance from the pupil, and for all practical purposes R is constant over the region of interest on the image surface. In that case, the value of R'_i can be replaced with the distance along the chief ray to the image plane. Then

$$\epsilon'_x = -\frac{R'}{n'}\frac{\partial W}{\partial x'_p}$$
$$\epsilon'_y = -\frac{R'}{n'}\frac{\partial W}{\partial y'_p} \tag{2.97}$$

where a small approximation is that the ray distribution is being computed on a surface of radius R' rather than on the actual image surface. In ray tracing, the exact calculations are of course evaluated, placing the ray distributions on the actual focal surface. This minor approximation makes it easier to relate the ray aberrations to the wavefront aberrations when discussing the nature of image formation.

The coordinate sets being used are established by the pupil coordinate sets. These in turn are usually established by the axis of a centered optical system, with the z direction along the axis and the x and y directions orthogonal to each other and the z axis. If the optical system is not centered, then appropriate coordinate sets can be assigned to the object and image space so that the aberration calculations are interpreted in terms of the correct local coordinate system.

The paraxial base set provides a convenient starting point for the pupil coordinates. The interpretation of the paraxial parameters describing the optical system can be used in defining the extent to which the lens meets the isoplanatism condition. The conditions for stability of the image with respect to coordinate location change can be evaluated by expressing the pupil coordinates and the wavefront function in the paraxial base set, and

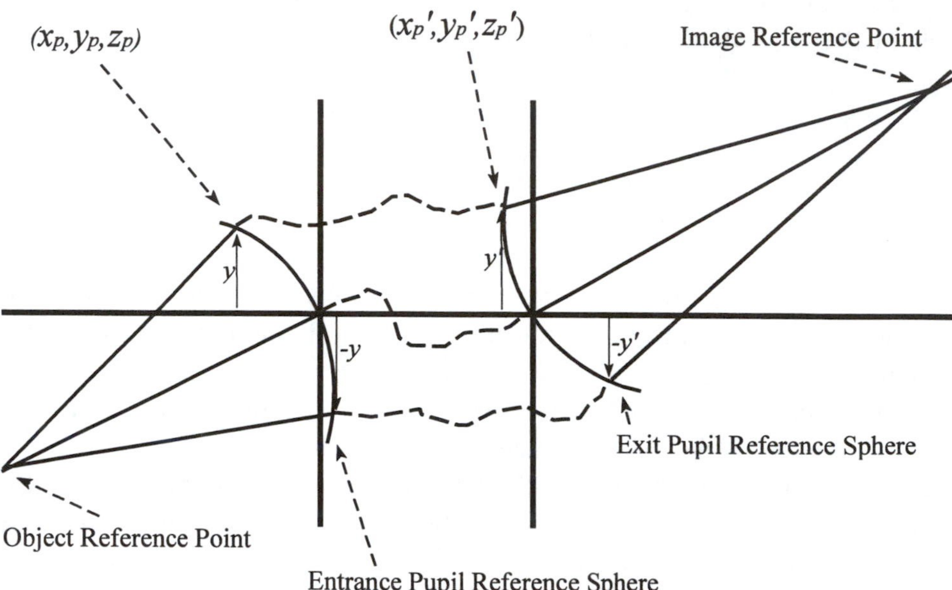

FIGURE 2.30 Reference sphere coordinate definition, general.

then investigating the change in the aberrations with a change in reference point location.

In order to determine the change in W for a shift of the reference point, let the image reference point in Figure 2.30 move from P'_0 to P'_1. The coordinates in the image surface are

$$(\delta x', y'_0 + \delta y', \delta z') \tag{2.98}$$

which determine the equation for the change in the reference sphere centered at the center of the exit pupil. Differentiating the equation of the exit pupil reference sphere yields

$$\begin{aligned}(x'_pL' + y'_pM' + z'_pN')\,\frac{\delta W}{n} &= -\,x'_p\delta x' - y'_p\delta y' + y'_0M'\,\frac{\delta W}{n}\\ &\quad - z'_p\delta z' + z'_0N'\,\frac{\delta W}{n}\end{aligned} \tag{2.99}$$

where n is the refractive index and δW is the change in the value of the optical path difference on the reference sphere attributed to the shift in reference point location. Dividing this out leads to

$$\delta W = \frac{n(x'_p\delta x' + y'_p\delta y' + z'_p\delta z')}{L'x'_p + M'(y'_p - y'_0) + N'(z'_p - z'_0)} \tag{2.100}$$

which reduces to

$$\delta W = \frac{n'}{R'}\,(x'_p\delta x' + y'_p\delta y' + z'_p\delta z') \tag{2.101}$$

for the change in the OPD resulting from a shift in the reference point location. All of the quantities are, so far, derived for the image space. A similar set of relations exist, unprimed, for the object space.

The equation for the reference sphere is

$$x_p^2 + y_p^2 + z_p^2 - 2y_0'y_p' - 2z_0'z_p' = 0 \tag{2.102}$$

which can be solved for the redundant value z_p'. A simple case is for the axial image, with y_0' equal to zero. The change in OPD for a shift in the z coordinate is

$$\delta W = \frac{n'}{R'} \frac{(x_p^2 + y_p^2 + z_p^2)}{2z_0'} \delta z' \tag{2.103}$$

and since the distance z_0' is the radius of the reference sphere,

$$\delta W = \frac{n'}{2} \frac{(x_p'^2 + y_p'^2 + z_p'^2)}{R'^2} \delta z' \cong \frac{n'}{2} \frac{(x_p'^2 + y_p'^2)}{R'^2} \delta z' \tag{2.104}$$

The latter equation is an excellent approximation for most cases of moderate relative aperture.

The off-axis imaging situation is more complex, as the issue of focusing in a direction parallel to the z axis or along the chief ray enters.

Defining the relative ray height in the pupil will be extremely useful in demonstrating the relation of real ray effects to the paraxial base set. In this section, to avoid confusion regarding paraxial and real values, we will define the paraxial values

$$\begin{aligned} \eta &= \text{Paraxial aperture radius in the entrance pupil} \\ \eta' &= \text{Paraxial aperture radius in the exit pupil} \end{aligned} \tag{2.105}$$

These are, of course, equal to the y and y' heights in the entrance and exit pupils as used in the section on paraxial optics.

The location of the pupil coordinates for an on-axis image is indicated in Figure 2.29. The pupil heights are related to the paraxial values indicating that the path of real rays will match that of the paraxial rays when interpreted in this manner. Off axis, there is an anamorphism of the pupils, as demonstrated in Figure 2.30. The scaled pupil coordinates are fixed in the base coordinates and include the necessary scaling to place an image of the right magnification on the image surface.

Using this definition, the actual ray coordinates in the pupils can be scaled by

$$\rho_x' = \frac{x_p'}{\eta'} \qquad \rho_y' = \frac{y_p'}{\eta'} \tag{2.106}$$

where x_p' and y_p' are actual ray heights in the exit pupil and η' is the paraxial parameter for the pupil height calculated at the exit pupil. There are some useful alternate forms for the aberration introduced by a focus shift adjustment.

$$\begin{aligned}\delta W &= \frac{n'}{2}\frac{\eta'^2}{R'^2}(\rho_x^2+\rho_y^2)\delta z' \\ &= n'\frac{D^2}{8R^2}(\rho_x^2+\rho_y^2)\delta z' \\ &= \frac{n'}{8(\mathrm{F/No.})^2}(\rho_x^2+\rho_y^2)\delta z' \end{aligned} \tag{2.107}$$

These equations are the familiar quadratic form of the dependence of the Optical Path Difference upon the shift of focus. The interpretation is that these errors are an additional optical path introduced by selecting a different image reference point. The difference between the original reference sphere and the new reference sphere is then given to satisfactory accuracy by these formulae.

As a practical matter, it is usually best to recalculate the actual optical path errors on a new reference sphere, as the above formulae presume that there is a linear mapping from the entrance pupil to the exit pupil.

The lateral shift of the reference sphere introduces a different form of Optical Path Difference error. The change due to the lateral shift is

$$\delta W = \frac{n'}{R'}(x_p'\delta x' + y_p'\delta y') \tag{2.108}$$

Making a similar replacement as above, the Optical Path Difference introduced in the exit pupil is

$$\delta W = \frac{n'}{R'}\eta'(\rho_x\delta x' + \rho_y\delta y') \tag{2.109}$$

The literal interpretation of this equation is that the Optical Path Difference added as a consequence of a lateral shift is a linear "tilt" in the wavefront. Any aberrations present in the wavefront will be altered by the addition of the tilt term. Under most conditions, a tilt in the wavefront can also be compensated by an appropriate reference coordinate shift. This option is used to compensate or balance the effect of lateral higher-order aberrations such as coma.

The effect of a small lateral shift of the reference field position in the object will introduce a linear or tilt component to the aberration entering the exit pupil. This will appear as an added aberration at the exit pupil. If the aberration change at the exit pupil is accurately linear, a compensating shift of the image reference point will cancel the aberration, and the imagery can be said to be unchanged under a small change in field.

The conditions under which this is true will now be examined. There are two equations of importance in the object and image space:

$$\text{Object space, entrance pupil } \delta W = \frac{n}{R}\eta(\rho\delta x + \rho\delta y)$$

$$\text{Image space, exit pupil } \delta W' = \frac{n'}{R'}\eta'(\rho_x'\delta x' + \rho_y'\delta y') \tag{2.110}$$

The symmetry of these equations is evident. The addition of the prime to the wavefront in the exit pupil is an indication that this is the compensating wavefront error introduced in the pupil by the lateral shift of the object. For imagery to be constant between the two object locations, the values of the wavefront error in both the entrance and exit pupil need to exactly compensate.

The aberration difference between the entrance and exit pupils after a small shift of image and object position is

$$\Delta W = \delta W' - \delta W \tag{2.111}$$

where the full expression is

$$\Delta W = \frac{n'}{R'}\eta'(\rho_x'\delta x' + \rho_y'\delta y') - \frac{n}{R}\eta(\rho_x\delta x + \rho_y\delta y) \tag{2.112}$$

The paraxial magnifications lead to a set of relations between the object and image plane,

$$m = \frac{nu}{n'u'}$$
$$\delta x' = m\delta x$$
$$\delta y' = m\delta y \tag{2.113}$$

for the shift in object and the compensating shift in the image position. The invariant and the definition of paraxial angles lead to two more relationships,

$$\eta = \eta'\frac{n'u'}{nu}$$
$$\eta' = -R'u' \qquad \eta = -Ru \tag{2.114}$$

which are inserted into the equation for ΔW to get

$$\Delta W = -n'u'[(\rho_x'\delta x' + \rho_y'\delta y') - (\rho_x\delta x + \rho_y\delta y)] \tag{2.115}$$

To see the effect more clearly, take the usual case that the off-axis direction is in the y direction. Then

$$\frac{\Delta W}{\delta y'} = -n'u'(\rho_y' - \rho_y) \tag{2.116}$$

Finally, noting that the change in the invariant due to a lateral shift in the object and a corresponding shift in the image is

$$\Delta I = n'u'\delta y' \tag{2.117}$$

the requirement for ΔW is

$$\frac{\Delta W}{\Delta I} = -(\rho' - \rho) \tag{2.118}$$

which is the most fundamental statement that can be made about the requirements for isoplanatism. The most direct interpretation of this relation is that as long as the entrance and exit pupil coordinates are scaled representations of each other, the slope or first derivative of the change of aberration with field angle will be zero.

Since the aberration which varies linearly with field is coma, this condition is equivalent to stating that the lens will be free of linear coma if the pupil scaling relationships are met. Using the knowledge that the real ray coordinates are on the reference pupil spheres, another relationship is demonstrated:

$$\rho = \frac{y_p}{\eta} = \frac{R \sin \Theta}{Ru} = \frac{\sin \Theta}{u} \tag{2.119}$$

In this equation, Θ is the angle subtended by the real ray height at the location of the reference point. The condition for isoplanatism is then

$$\frac{\sin \Theta}{u} = \frac{\sin \Theta'}{u'} \tag{2.120}$$

for the correspondence of entrance and exit pupils. This is the well known sine condition. In short, the condition requires that the mapping between entrance and exit pupil spheres must be free of distortion in order that the imagery be locally isoplanatic.

In the usual textbook treatment, the real ray slope angles are frequently used. This causes some confusion when there is spherical aberration present, as the angles of importance are related to the subtend of the pupil and not to the actual ray direction. In classical discussions of this property, a condition known as the Stahele–Lihotsky principle is applied, which is actually a correction for the spherical aberration present in the lens.

This discussion of the local stable imaging properties of a lens leads naturally to the question of image radiometry. Presume a small but finite-size, Lambertian, uniform source is the object. If the sine condition is satisfied, all points within this source will be imaged on the image surface in a dimension given by the magnification of the optical system. The Lagrange rules described above show that a conservation of the object dimension times the sine of the numerical aperture will occur. Then

$$hn \sin U = h'n' \sin U' \tag{2.121}$$

If a circular aperture exists in the lens, both sides of the equation can be squared to obtain

$$h^2 n^2 \sin^2 U = h'^2 n'^2 \sin^2 U' \tag{2.122}$$

in which the h^2 and h'^2 are the areas of the object and image segments and the solid angle subtended by the aperture at the object or image respectively is given by

$$\Omega = \tfrac{1}{2}\sin^2 U \tag{2.123}$$

indicating a relationship between the solid angle of the pupil viewed at the object and image and the square of the invariant relation.

In radiometry, the irradiance of a Lambertian source is given by the flux per unit area per unit solid angle. The units generally used are watts per square centimeter. For any detector application, the value must be spectrally weighted by the spectral sensitivity of the source. In photometry, the flux is spectrally weighted by the sensitivity of the human eye to obtain values that are proportional to the visual brightness of the source or image. The flux from a Lambertian source passing through the entrance pupil of the lens will be the product of the irradiance times the area and solid angle. The amount of flux incident upon the exit pupil will be the entering flux multiplied by the transmission factor of the lens. The entirety of the flux passing through the pupil will be incident upon an image area given by the sine relation stated above. The most usual case of irradiance computation is for a Lambertian object surface, in which the irradiance appears to be constant with viewing angle. Thus the irradiance of the image of this Lambertian object will be given by

$$E' = \pi t L_0 \sin^2 U \tag{2.124}$$

where E' is the irradiance in watts per centimeter on the image surface and L_0 is the irradiance of the object in watts per square centimeter per steradian. The angle U is the angle subtended by the edge of the pupil at the image location. The transmission factor for the lens is given by t.

Using concepts discussed above regarding paraxial variables, this can be written as

$$E' = \pi t L_0 \sin^2 U' = \pi t L_0 u'^2 = \pi t L_0 \frac{1}{4(F/\text{number})^2} \tag{2.125}$$

where the F/number is defined by

$$\text{F/number} = \frac{-1}{2u'} = \frac{-1}{2\sin U'} \tag{2.126}$$

which has to be interpreted as applying to imagery satisfying the sine condition. The angle is signed, but the F/number is used in radiometric computations, and is usually defined as a positive number, so a minus sign is attached to the numerator of the equation.

There are several F/numbers to be considered in a lens application. If the object is located at infinity, the F/number may be easily found by dividing the diameter of the entering beam aperture into the focal length. This will be called the F/Number for an object at infinity and is usually stated in describing the relative aperture of a lens:

$$(\text{F/number})_\infty = \frac{f}{D} \tag{2.127}$$

When the same lens is used to image an object at finite conjugates, the aperture of the lens is fixed, but the conjugates change so that the paraxial angle changes, and consequently the effective F/number will be different:

$$(\text{F/number})_{\text{effective}} = \frac{1}{2u'} = (\text{F/number})_\infty (1 - m) \tag{2.128}$$

Here m is the magnification, equal to zero for an object at infinity and equal to -1 for an object being imaged at unit conjugates. It can be seen that the F/number will change by a factor of two in shifting from infinite conjugates to unit magnification. A lens will thus produce an image with one-quarter the irradiance as a consequence of the shift of conjugates. A fallout from this consideration is the definition of the focal length of the lens, which is consistent with the paraxial value of the focal length. Then, for an aperture height Y for a lens with object at infinity,

$$Y = f \sin\theta = F \sin U' \tag{2.129}$$

Use of this definition simplifies the calculation of the radiometry of the lens at various conjugate positions.

There is an additional concept sometimes used to refine the radiometric nature of the F/number. This is the definition of a "T/number" which includes the effect of the lens transmission factor in the computations. This is defined by

$$\text{T/number} = \frac{\text{F/number}}{\sqrt{t}} \tag{2.130}$$

and is an attempt to include all of the factors regarding image irradiance into one number. This concept has not found universal acceptance.

The radiometric concepts so far discussed apply only to the axial image for a lens. If there is no vignetting, then three effects contribute to a change in the irradiance off axis for a lens forming an image on a flat image surface. The first effect is the geometry of the off-axis image distance, which changes by the secant of the off-axis angle. Using the previously stated method of calculating the solid angle, this introduces a $\cos^2\theta$ factor to the effective numerical aperture for a field angle θ. The pupil itself will be viewed obliquely, and will appear to shrink in the off-axis direction by a factor of $\cos\theta$. The light will be obliquely incident upon the image surface introducing another factor of $\cos\theta$. The net result will be that the theoretical irradiance off axis for a lens with a flat field will be governed by a $\cos^4$ law:

$$E_\theta = E_0 \cos^4 \theta \tag{2.131}$$

This law is one consequence of the sine condition required to form an isoplanatic image. It would seem that forming an image with uniform irradiance across the image surface is completely impossible. However, this law is frequently broken, both increasing and decreasing the relative irradiance across the field.

The sources of lawbreaking are deliberate or accidental modification of each of the geometrical factors discussed above. The aperture size can change with obliquity if there is sufficient pupil aberration that modifies the apparent shape of the stop as viewed from the object or image side, changing this factor from a simple cos relationship. The $\cos^2$ factor due to the flat field can be changed by using a curved object or image surface. The cos factor due to incidence on the image surface can be controlled by designing a lens with telecentricity of the chief ray so that the obliquity of the image-forming bundle is reduced.

There is an additional effect on the image irradiance. If the lens contains distortion, the local magnification will be changed. Large amounts of negative distortion, for example, can decrease the area into which the illuminance from successive object segments falls. The radiometry of extremely wide-angle lenses can be considerably affected by this geometrical factor.

In addition to the geometrical effects, there are two other sources of irradiance change with field angle. The most significant is vignetting, in which the shape of the pupil is deliberately reduced by clipping of the ray bundles by surfaces located away from the stop. Vignetting is often used as a control on image quality by eliminating portions of the ray bundle that have large amounts of aberration. This will lead to the introduction of a vignetting factor V_θ which is the relative area of the stop that is used at each field angle. This choice of description of vignetting is not universal, and some users describe vignetting by the fraction of light that is removed by blockage at a specific point in the field. Vignetting of 15% at a point in the field implies a vignetting factor of 85%, for example. Vignetting may also occur as a result of the introduction of a deliberate obstruction, such as a folding mirror or secondary mirror, that blocks light from getting through some portion of the aperture.

The precise calculation of the irradiance on an image surface is straightforward, but requires careful inclusion of all of the effects occurring in a lens. The actual numerical apertures and distribution of rays in the entrance and exit pupils need to be included. Magnification errors produced by the presence of distortion will also influence the local irradiance pattern. The angular distribution of light from the source will also weight the pupil distribution, as well as the irradiance with field position.

In addition to the geometrical effects, there are practical problems in modeling an actual lens. Any real lens will contain surfaces, usually with thin film coatings, whose reflectivity varies with the angle of incidence of each ray bundle on each surface, so that the basic lens transmission can change with field angle. In some cases, a change in total glass path that may occur at different field angles may increase the bulk absorption of the light with field angle. There is, in addition, loss and redistribution of light by scattering and ghost image reflections within the lens.

When all of these effects are considered, it is seen that the accurate computation of the image plane irradiance is somewhat more complex than usually described in a textbook. In this book, there will be no further discussion of the details of irradiance computation. The designer is encouraged to think about the points made in the last few sections, and use caution when evaluating the expected irradiance produced by an object through a lens. In most cases, the simple geometrical considerations will be sufficient. Precise knowledge of the irradiance profile requires precise understanding of all of the miscellaneous factors existing within a lens.

2.1.5 ASTIGMATISM AND FIELD CURVATURE

The discussion of paraxial optics was based upon the establishment of a base coordinate set for an optical system. The image positions and sizes were determined in this set of coordinates. The last section indicated how the paraxial and real ray passages from object to image were linked by imaging conditions that linked the paraxial optics and the real ray optics if isoplanatic images are to be formed. If the concept of paraxial optics is linked to the tracing of real rays through a lens, additional information about the imagery can be obtained. The concept of close skew rays consists of tracing paraxial rays that ride along the coordinates of each real ray. The tracing of the ray provides information on the optical path difference and the ray direction, or first derivative, of the optical path difference. The use of close skew rays permits determination of the local focusing properties around each real ray, and adds information regarding the second derivative of the optical path difference in the vicinity of the ray.

The paraxial optics indicates that a second-order, or spherical, surface shape is necessary to obtain optical power and form an image. The base paraxial focal locations are obtained by considering the focal power of the components along the axis of symmetry. The focal power of second-order surfaces for a nonaxial imaging bundle will vary with angle of incidence, leading to a change in the second-order power, or field curvature, and to differing power in two meridians. This will introduce astigmatism into the image.

The local power at each surface can be computed for a real ray traced through the surface. In ray tracing, a summation of the path length along the ray provides the optical path. The calculation of the direction of the ray provides the first derivative of the ray path, and permits computation of the ray intercept position on the image surface. The calculation of the local focal power of the surface along the ray provides a value for the second derivative of the wavefront, and permits computation of the field focal properties of a lens for each aperture coordinate being evaluated. Calculation of these values for chief rays provides the field focus characteristics of the lens. The additional information that is obtained from these calculations provides great insight into the process of image formation for a finite imaging field.

The technique used is to apply a formula for a pair of close skew rays that are traced along with the real ray, using equations that are similar to paraxial ray formulae. There are several methods of deriving the appropriate equations, and several ways of writing the equations. The most insight is obtained by using a semigraphical mode of derivation.

The basic mechanism involved in the generation of astigmatism and field curvature at a second order surface is that the surface "appears" to have a changed local curvature when viewed from a point along the incoming ray. Increasing the angle of incidence of a ray bundle increases the apparent local power of surface, and the obliquity causes the apparent power to be dependent upon the azimuth direction of the ray bundle. Symmetry indicates that there will be two principal meridians for which the changed power will be the maximum and minimum. From this argument, it can be seen that astigmatic imagery is intrinsic to optical surfaces used at oblique incidence. The obliquity effect is determined by the cosine of the angle of incidence, indicating that the effective change in power is a second-order effect with change in angle of incidence.

The easiest development of the appropriate formulae is obtained by considering the principal meridians separately. The principal meridians are the tangential direction, which lies in the plane of the angle of incidence and the sagittal, which is the direction perpendicular to the plane of incidence. Figure 2.31 shows the concepts involved. A spherical surface is shown, with a real ray traced to and through the surface. Indicated along the ray are close skew rays originating from a focal point on the ray. The parameter h_t in the figure represents a small distance along the surface from the real ray, or a paraxial ray height measured along the ray. Because the coordinates for this derivation travel along with the ray, the surface intersect can be considered as an effective pupil location for the family of paraxial rays being traced along with the real ray.

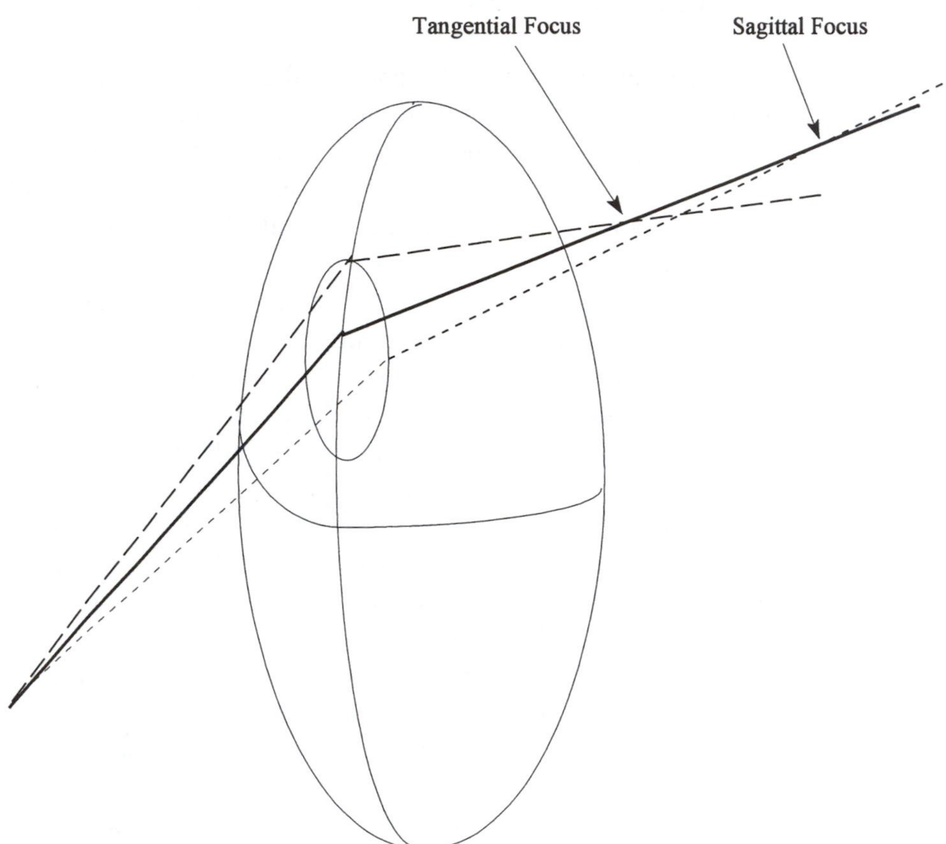

FIGURE 2.31 Geometry for the close skew ray trace: ——, ray; — —, close tangential ray; - - - -, close sagittal ray.

The derivation used here presumes a spherical surface for a centered optical system, and further considers an arbitrary meridional or chief ray passing through the lens. Because the close skew rays are related to the local ray coordinates, the geometry is not restricted to centered optical systems. This becomes very important in understanding of lens systems with tilted or displaced components. More extensive derivations of the theory for this rule will be found in other texts. Most theoretical discussions fail to provide an adequate physical evaluation of the consequences of the close skew ray error. The treatment here is directed toward assessing the use of this concept in lens design.

First consider the tangential direction. Figure 2.32 shows an exaggeration of the imaging situation for a real ray, r, and another ray, r_1 close to the real ray in the tangential direction. For the purpose of the derivation, the object and image locations are both shown on the same side of the surface. Differentiating the law of refraction to find the small angle change gives

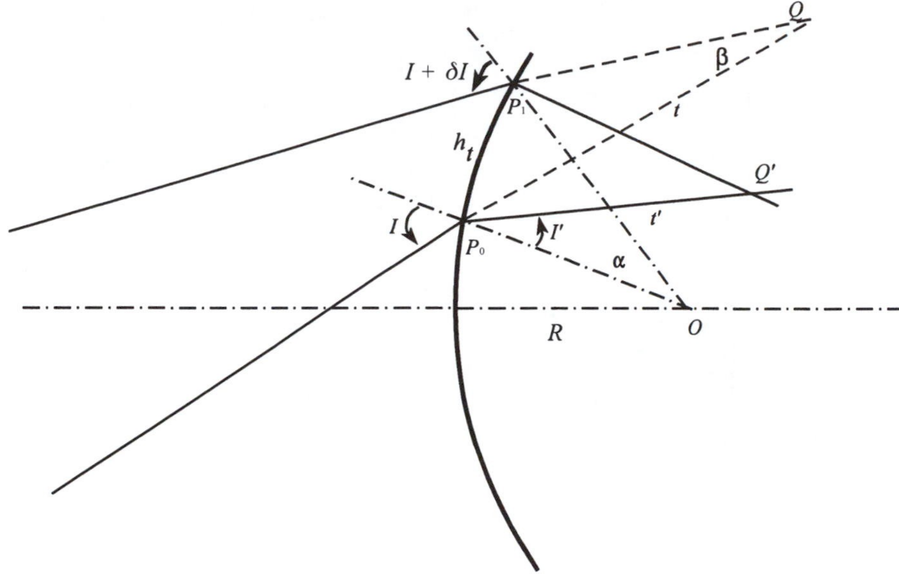

FIGURE 2.32 Geometry of the tangential ray intercept.

$$n \cos I \delta I = n' \cos I' \delta I' \tag{2.132}$$

The r_1 ray is acting as a close or paraxial ray relative to the real ray with h_t as the parameter describing the distance from the real ray along the surface. This is a sort of paraxial marginal ray height for the adjacent ray in the tangential direction,

$$\delta I = \alpha - \beta \qquad \beta = \frac{h_t \cos I}{t} \qquad \alpha = \frac{h_t}{R} \tag{2.133}$$

leading after some algebra to

$$n \cos I \left(\frac{1}{R} - \frac{\cos I}{t} \right) = n' \cos I' \left(\frac{1}{R} - \frac{\cos I'}{t'} \right) \tag{2.134}$$

which is rearranged to

$$\frac{n' \cos^2 I'}{t'} = \frac{n \cos^2 I}{t} + \frac{n' \cos I' - n \cos I}{R} \tag{2.135}$$

where the portion of the equation on the right can be considered as an "oblique power" for the surface.

$$\phi_I = \frac{n' \cos I' - n \cos I}{R} \tag{2.136}$$

The calculation of the sagittal close skew rays could follow a similar pattern, but a different diagram is useful. Figure 2.33 shows a spherical surface with an auxiliary axis passing through the center of the surface and

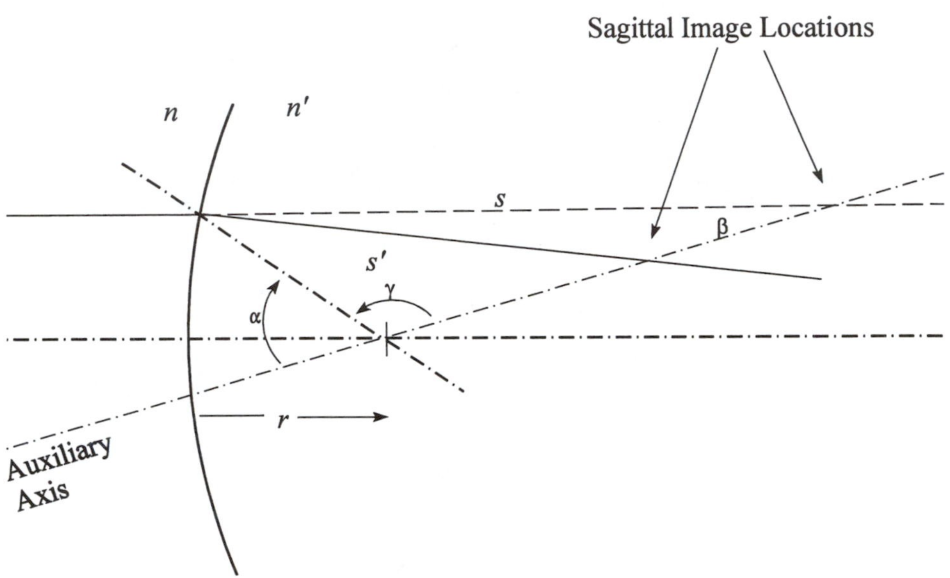

FIGURE 2.33 Geometry of the sagittal ray intercept.

intersecting with the real ray before and after refraction. The symmetry of the diagram is used to indicate that the paraxial ray is in the sagittal direction, which has a paraxial parameter h_s lying in the local surface coordinates and perpendicular to the direction of the tangential paraxial parameter direction.

Using the paraxial nature of the angles,

$$\beta + \gamma + I = \pi \qquad \alpha + \gamma = \pi \qquad \beta = \alpha - I \tag{2.137}$$

which leads to

$$\frac{R}{\sin(\alpha - I)} = \frac{s}{\sin\alpha} \qquad \frac{R}{\sin(\alpha - I')} = \frac{s'}{\sin\alpha} \tag{2.138}$$

for the sagittal focal positions along the ray before and after refraction. Using trigonometric relations for the sum of angles, and including Snell's law leads to

$$\frac{s\sin I}{1 - \frac{s}{R}\cos I} = \frac{s'\left(\frac{n}{n'}\right)\sin I}{1 - \frac{s}{R}\cos I'} \tag{2.139}$$

and rearranging gives a similar formula to the tangential refraction equation:

$$\frac{n'}{s'} = \frac{n}{s} + \frac{n'\cos I' - n\cos I}{R} \tag{2.140}$$

where the oblique power again appears in the equation.

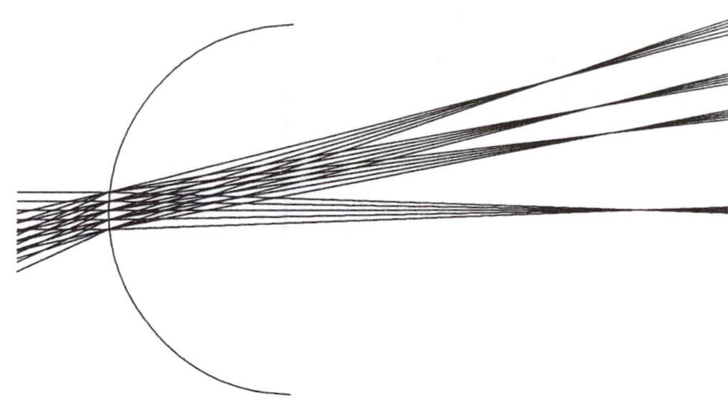

FIGURE 2.34 Demonstration of tangential focal location with stop at the surface (OSLO).

These two equations together provide the location of the sagittal and tangential focus along any real ray traced through the lens. The rays of most interest are the chief rays for any field angle. The geometry of the lens can be used to construct the locations of the sagittal and tangential field locations for the lens.

The angle of incidence of the real ray is not limited to small values. The calculation for the paraxial ray intercept along the ray is, of course, correct only for a ray vanishingly close in aperture to the real ray and provides a base set of coordinates locating image position and magnification along a real ray bundle. The astigmatic fields obtained by application of this rule will include high-order astigmatic field errors as a result of the interaction of the angle of incidence with the surface.

The astigmatism generated by a single surface will depend upon the angle of incidence of the real ray on the surface. Figure 2.34 shows several ray bundles incident at various angles on a single surface with the stop located at the surface. The change in distance of the tangential focus from the surface intersection is visually evident from examination of the figure. Figure 2.35 is a plot of the location of the tangential and the sagittal foci relative to a flat image surface. Both these surfaces are inward curving from the flat surface.

If the stop is moved to the center of curvature of the single spherical surface, then the angle of incidence of any real chief ray on the surface will be zero. By equations 2.135 and 2.140, the location of the sagittal and tangential foci should be the same distance from the surface when measured along the ray. Visual examination of Figure 2.36 indicates that this is true. A plot of the astigmatic fields relative to the flat image surface now appears as in Figure 2.37. Both astigmatic foci correspond, but there is a curvature of

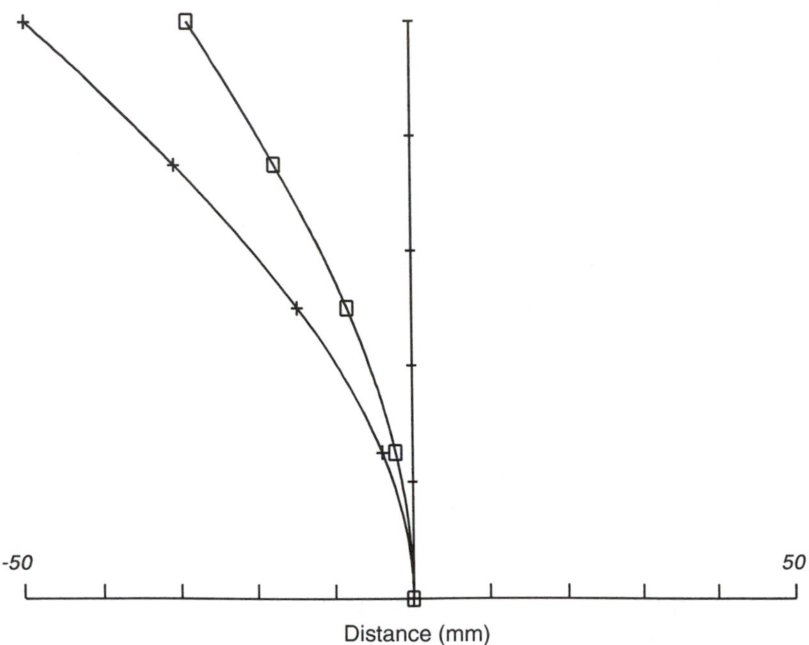

FIGURE 2.35 Field curves with stop at the surface +, tangential; □, sagittal. The abscissa is in lens units (OSLO).

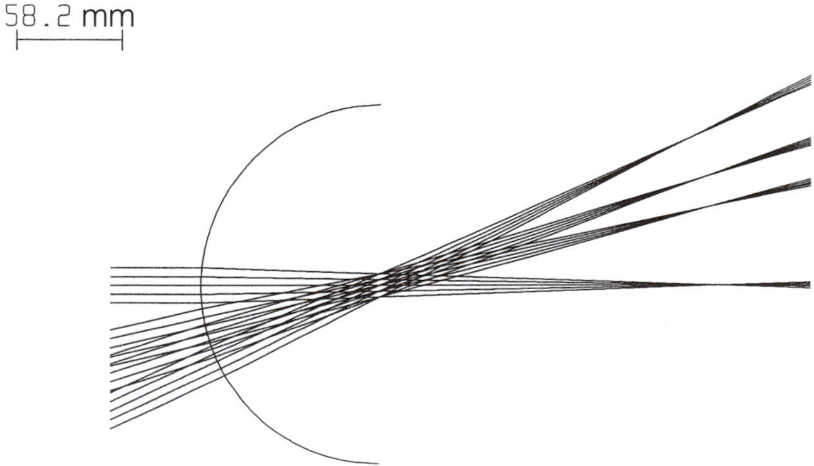

FIGURE 2.36 Demonstration of tangential focus with stop at center of curvature of the surface (OSLO).

this common focal surface relative to a flat focal surface. This curvature is intrinsic to the spherical focusing surface, and indicates that there is a well-defined surface upon which the astigmatic image will be formed. This is the Petzval surface that will be discussed in Chapter 3.

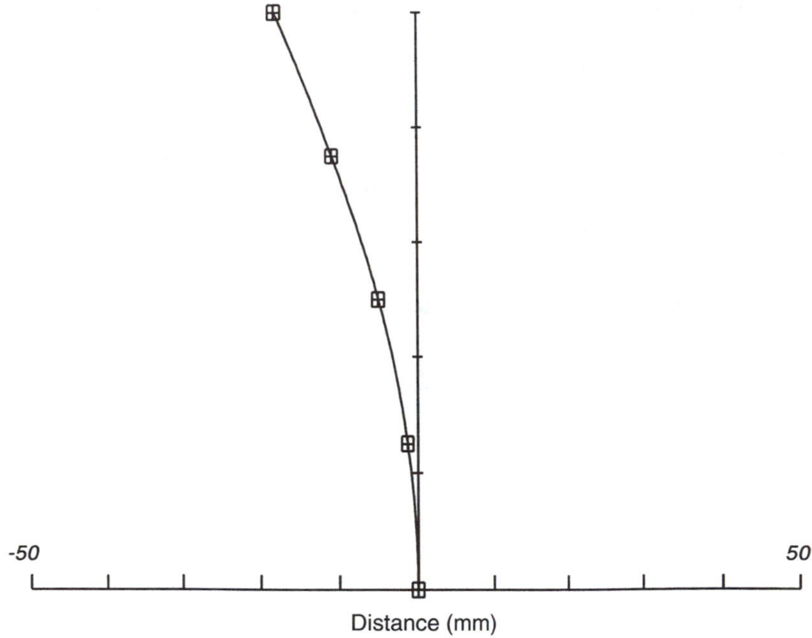

FIGURE 2.37 Field curves with stop at center of curvature (OSLO).

The selection of the sagittal and tangential directions is clear in the examples and derivation used. The equations have a more general application in that they are not limited to spherical surfaces, nor are they limited to chief rays in symmetric systems. For any optical surface, the local curvature can be computed, and its value may depend upon the direction on the surface. Simple examples are cylindric or toric optical components, which are defined by a principal pair of curvatures across the surface. Aspheric surfaces, including rotationally symmetric aspherics, will have a change in the local surface curvature that varies with location and direction on the surface. A conic surface, for example, will have a specific variation of surface curvature with aperture, imparting a specific toric shape to the surface which varies with location on the surface. The direction of the principal surface curvatures will be obvious for a rotationally symmetric aspheric surface.

There are alternate ways of writing the close skew ray equations that can be applied to a wide variety of optical problems. These are based upon recognizing that the close skew rays are paraxially related to the real base ray, and provide a pair of paraxial values for the sagittal and tangential directions. Define as sagittal parameters (y_t, u_t) and (x_s, u_s) in the tangential and sagittal directions, respectively. Then, the paraxial relations

$$\begin{aligned} y_t &= -tu_t = -t'u_t' \\ x_s &= -su_s = -s'u_s' \end{aligned} \tag{2.141}$$

provide close skew ray paraxial trace equations

$$n'u_t' \cos^2 I' = nu_t \cos^2 I - y_t \phi_I$$
$$n'u_s' = nu_s - y_s \phi_I \tag{2.142}$$

for a spherical surface. Recognizing that the focusing property defining the paraxial focal locations is also interpretable as a second-order curvature of the wavefront, these equations can be rewritten as a relationship between the second-order wavefronts before and after refraction and the local surface curvature, expressed in waves of surface sag over the paraxial aperture. Using this form leads to

$$W_{2_t} = \frac{y_t^2}{2t}$$
$$W_{2_s} = \frac{x_s^2}{2s}$$
$$W_{2_{surf}} = (n' \cos I' - n \cos I) \frac{y^2}{2R} \tag{2.143}$$

describing the process as a pair of defocus terms in the principal meridians.

2.2 Physical Optics

Geometrical optics describes the passage of rays through a lens. The path along the ray provides the phase of the light passing through the lens. The wavefront associated with the ray passage can be obtained as a phase function evaluated in the exit pupil of the lens. The wave propagation through a lens can be obtained by successive transforms applied to the wavefront function from the pupil to the image surface. While the geometrical optical model serves as the framework on which the aberration properties of the lens are based, physical optics is required to construct the actual light intensity that will be observed as a result of propagation of a wavefront to the image surface.

Full consideration of physical image formation requires the inclusion of the entire imaging process. This process starts with a light source from which rays (or waves) are propagated to the object. The transmission, reflection, or scattering at each point on the object produces radiation that propagates to the entrance pupil of the lens system. The previous section on isoplanatic imagery demonstrates that the light propagated to the entrance pupil is then propagated to the exit pupil through the lens, and emerges from the exit pupil modified by the phase errors and transmission limitations imposed by the lens. The radiation emerging from the exit pupil then propagates to the image surface.

The characteristics of the light source determine the coherence relations between adjacent portions of the object. In the most frequent case of incoherent imagery, there is a negligible coherence relation between adjacent object points. The imaging process can be characterized by evaluating the propagation of a point object to the image surface to determine a point spread function. Intensity superposition rules are used to construct the image from the point spread function, weighted by the object irradiance distribution.

For imaging situations in which the coherence of the radiation on the object is significant, a more complex imaging situation occurs. In this text, most of the discussion will be on the case of incoherent imagery and the cases of partial coherence and coherent beam propagation will be treated only as required. The diffraction, or physical optics, imaging can then be characterized by the use of an intensity point spread function and its Fourier transform, the Optical Transfer Function.

Reasonable quality imagery is assumed to be required for most applications. The considerations for isoplanatic imagery apply, and it is presumed that the coordinates in the entrance pupil and exit pupil are related sufficiently well by a simple scaling factor. This greatly simplifies the computations required, and permits eventual application of the Fourier techniques in image formation analysis, since stationarity and linearity are required for application of these techniques.

The limitation of the size of the light bundle passing through the lens by stops and other blocking apertures can be described well by geometrical optics as long as the aperture dimension exceeds many wavelengths. Incorporation of the effect of aperture limitation or transmission variations in the aperture requires a summation of the contributions from all portions of the wavefront. Propagation of light through a lens can be calculated by integrating the contributions from each portion of the wavefront passing through the lens aperture. This computation is required in order to examine the effect of aperture limitation and interaction of light with a detector.

For most imaging problems, polarization is not a significant issue. The application of scalar diffraction theory for the computation of point spread functions is adequate. When this is not true, a vector computation of the diffraction through the lens is required. This more complicated, but more general, case will be discussed later.

There are many textbooks describing physical optics, so that it is necessary only to discuss those properties of the process that are unique or important in design and image analysis (Born and Wolf 1959; Klein 1970; Hecht and Zajac 1974).

2.2.1 DIFFRACTION IMAGES

Diffraction is a consequence of the interference of light propagated from the exit pupil and summed or integrated with respect to a stated image reference position. In most cases, diffraction images can be calculated to sufficient accuracy by using scalar diffraction theory, in which the polarization of the light is ignored. Scalar diffraction theory is used for almost all of the imaging situations encountered in lens design, and will be considered here.

The basis for scalar diffraction theory has been covered extremely well in many references, so there is no need to develop it from fundamental origins here (Born and Wolf 1959). The basic relationship is that the amplitude in the image plane can be obtained by summing or integrating over the entirety of the wavefront leaving the exit pupil, taking into account the propagation distance to the point of interest in the image. By including the phase and amplitude of the wavefront at each point in the pupil the effect of aberrations is included in the computation. The geometry associated with a realistic imaging system leads to some significant simplifications in most cases of importance.

The key process that must be considered here is derived from the diffraction integral obtained by using the Huygens–Fresnel approach, which leads to a straightforward diffraction integral:

$$U(P) = -\frac{i}{\lambda}\frac{Ae^{-ikR}}{R}\int\int\frac{e^{ik(\Phi+s)}}{s}\,dx_p\,dy_p \tag{2.144}$$

In this integral, the amplitude of the wave disturbance U at the point P is given by an integration over a reference sphere that fills the exit pupil with coordinates $(x_{pj}y_p)$. Choice of the reference sphere produces a zero phase base for distances from the image reference point for calculation of the phase errors due to aberration in the lens. The phase error in the pupil is Φ, which is a function of x_p and y_p, the coordinates in the pupil. The distance s is the distance from the pupil coordinates to the location on the image surface where the amplitude is being computed. The distance R is the radius of the reference sphere from the pupil to the image reference location. The use of a complex exponential is merely a mathematical notation shortcut leading to the amplitude at the image point as a complex number, denoting both amplitude and phase at the location of interest.

Detection of the light present in the image at optical frequencies is by a square law or intensity detector, which is sensitive to the intensity, or complex square of the light. The calculation represented by the integral in equation (2.144) appears to be quite simple, and indeed it is if the appropriate coordinates are used. The equation provides a complex value, which includes the phase and the amplitude of the wave at the point of interest. Because the intensity distribution in the image is required,

$$|U(P)|^2 = \frac{1}{\lambda^2}\frac{A^2}{R^2}\left|\iint \frac{e^{ik(\Phi+s)}}{s}\, dx_p\, dy_p\right|^2 \tag{2.145}$$

Here the important quantity is the integral within the absolute value lines. The factor preceding the integral establishes the absolute intensity level of the diffraction image that is being computed. In most cases of imaging the absolute phase and intensity is of less interest than the contour of the light distribution in the image.

Historically, the solution of this integral is divided into two regions, the Frauenhofer region and the Fresnel region. The distinction between these regions is not sharp, but is determined by the rate of change of phase over the aperture with distance of the computation point from the reference point. Frauenhofer diffraction is considered "far field" diffraction, in that a linear approximation to the relative phase change with separation between points on the image surface is adequate. In Fresnel diffraction, the geometry of the diffracting aperture relative to the point of observation is such that a higher-order change, at least quadratic, must be considered.

There is a straightforward method of calculating the intensity in the region of the reference point on the focal surface under the conditions that the aberration is "small," the numerical aperture is reasonable and the distance from the pupil to the focal plane is much greater than a few wavelengths. These restrictions are not significant in the vast majority of useful imaging cases, and some numerical limits upon these restrictions will be discussed later.

Figure 2.38 shows the geometry of the pupil reference sphere with the image reference location at the center of the reference sphere located a distance h_y off axis in the y coordinate direction. (The limitation to circular symmetry of the imaging system is not required for this calculation, but the derivation of the important equations is much clearer. This apparent restriction will be dealt with later.) Section 2.1.4 on isoplanatism defines the wavefront aberration to be located on the pupil reference spheres. The phase representing the aberration present in the imaging system will be calculated in coordinates established in the pupil–image coordinate sets.

The diffraction image at coordinates (ϵ_x, ϵ_y) in the image plane relative to the reference point located at $(0, h_y)$ will be found from the integral

$$U(\epsilon_x, \epsilon_y) = \mathrm{K}\iint e^{i\frac{2\pi}{\lambda}[(B-A)+P(x_p,y_p)]}\, dx_p\, dy_p \tag{2.146}$$

where K is a constant absorbing the absolute phase and intensity scaling, P is the OPD of the wavefront calculated in the pupil, and A and B are as in

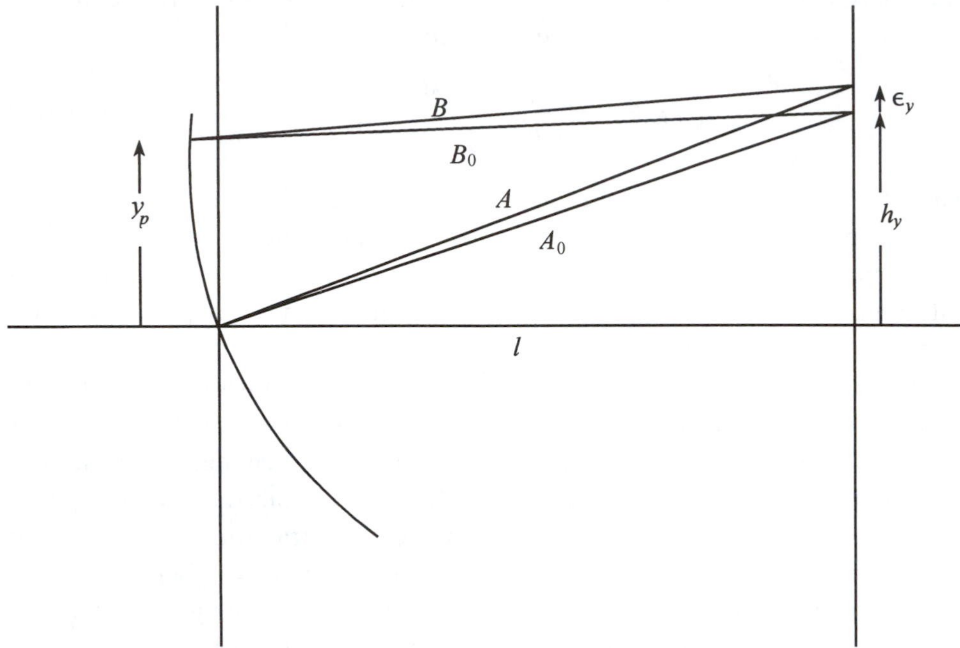

FIGURE 2.38 Geometry of the change in phase due to lateral displacement of the reference location.

Figure 2.39. Expressed in the pupil coordinate set, the coordinates of the point on the pupil reference sphere are (x_p, y_p, z_p), the coordinates of the reference point location are $(0, h_y, l)$, and the coordinates of the location where the amplitude is to be calculated are $(\epsilon_x, \epsilon_y + h_y, l)$ relative to the pupil coordinates. The radius of the pupil reference sphere is r, and the distance from the pupil to the image plane is also $l = r - z_p$.

The distances A and B are

$$\begin{aligned} B &= \sqrt{(\epsilon_x - x_p)^2 + (\epsilon_y + h_y - y_p)^2 + (x_0 - z_p)^2} \\ A &= \sqrt{(\epsilon_x)^2 + (\epsilon_y + h_y)^2 + (z_0)^2} \end{aligned} \tag{2.147}$$

which contain some difficult square roots. In order to make the equation simple use

$$(B - A)(B + A) = B^2 - A^2 = -2\epsilon_{x'}x_p - 2\epsilon_y y_p - 2h_y y_p - 2z_p z_0 + x_p^2 + y_p^2 + z_p^2 \tag{2.148}$$

which can be simplified by noting from Figure 2.39 that

$$\begin{aligned} r^2 &= x_p^2 + (y_p - h_y)^2 + (z_p - z_0)^2 \\ &= h_y^2 + z_0^2 \end{aligned} \tag{2.149}$$

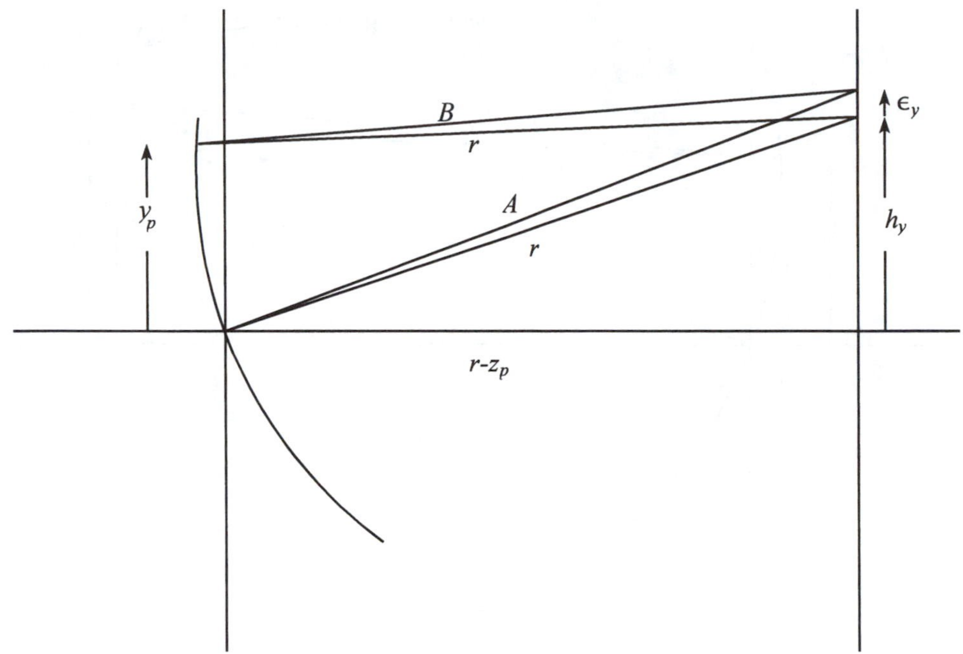

FIGURE 2.39 Substitution for reference surface radius.

which shows that

$$x_p^2 + y_p^2 + z_p^2 - 2z_p z_0 - 2h_y y_p = 0 \tag{2.150}$$

and the very simple result is

$$B - A = \frac{-2(\epsilon_x x_p + \epsilon_y y_p)}{A + B} \tag{2.151}$$

The sum of the distances is very closely

$$B + A \cong 2r \tag{2.152}$$

so that the needed value of the phase introduced by shifting from the reference coordinate origin to the coordinate (ϵ_x, ϵ_y) is

$$B - A = \frac{-(\epsilon_x x_p + \epsilon_y y_p)}{r} \tag{2.153}$$

The interpretation of this approximation is that the coordinates where the amplitude is being calculated actually lie on the surface shown in Figure 2.40. This is a surface with radius of curvature r, tangent to the image plane at the reference point. Since the extent of a diffraction image with a reasonably small amount of aberration is small compared to r, this is an acceptable approximation for almost all cases which are encountered.

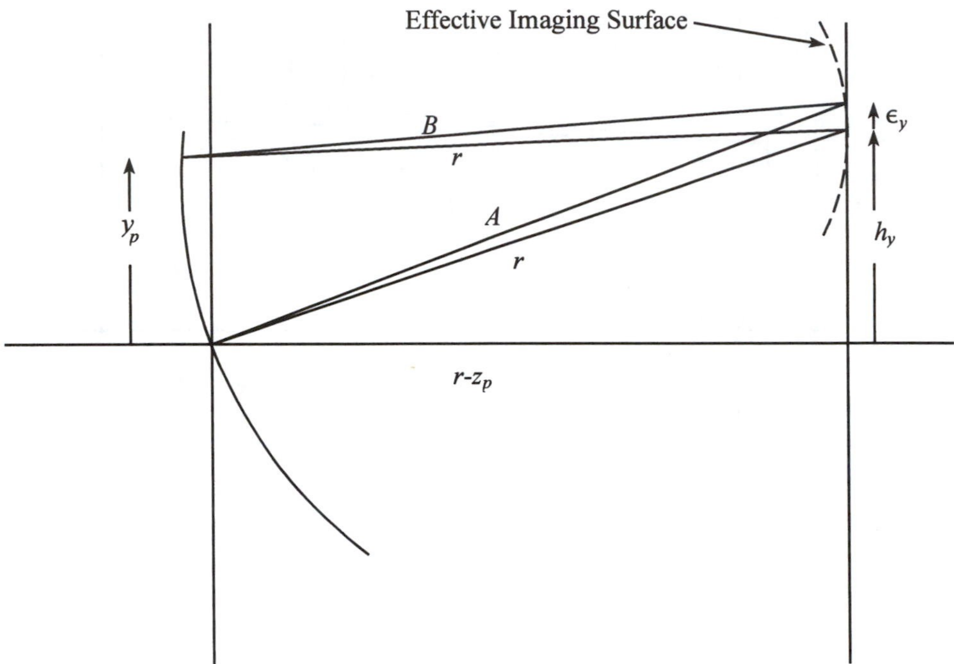

FIGURE 2.40 Interpretation of constant phase surface at the focus surface.

As an example, let $r = 50$ mm. If the aberration causes a blur ten times the diffraction dimension, or Airy disc diameter, then the sag error for an F/2 lens will be 0.0000018 mm, which is quite acceptably small and can be considered to lie on the image plane.

The new form for the integral is now

$$U(\epsilon_x, \epsilon_y) = \mathrm{K} \iint e^{i\frac{2\pi}{\lambda}\left[\left(\frac{-(\epsilon_x x_p + \epsilon_y y_p)}{r}\right) + P(x_p, y_p)\right]} dx_p\, dy_p \tag{2.154}$$

which can be rewritten in the form of a Fourier transform

$$U(\epsilon_x, \epsilon_y) = \mathrm{K} \iint e^{-i\frac{2\pi}{\lambda}\left[\frac{(\epsilon_x x_p + \epsilon_y y_p)}{r}\right]} e^{i\frac{2\pi}{\lambda}P(x_p, y_p)} dx_p\, dy_p \tag{2.155}$$

in which the pupil function, the complex phase introduced by $P(x_p, y_p)$, is transformed to obtain the amplitude distribution on the image surface. The intensity of the pattern, which will be the point spread function for the lens at the specific field and focal position is

$$f(\epsilon_x, \epsilon_y) = |U(\epsilon_x, \epsilon_y)|^2 \tag{2.156}$$

The coordinates in the pupil are the actual values of the pupil coordinates at the specific field for the lens. There is a transformation of coordinates that is useful in applying this integral to lens analysis problems. The off-axis image coordinates are related to the axial image coordinates by two scaling factors,

$$\left(\frac{x_{p_m}}{r}\right)_{\bar{U}} = \cos(\bar{U})\left(\frac{x_{0_m}}{r}\right)_0$$
$$\left(\frac{y_{p_m}}{r}\right)_{\bar{U}} = \cos^2(\bar{U})\left(\frac{y_{0_m}}{r}\right)_0 \tag{2.157}$$

as previously noted in the discussion on the spherical pupil coordinates in the last chapter. If the axial numerical apertures NA_0 are used to "normalize" the coordinates the integral can be rewritten using a temporary notation as (q_x, q_y)

$$U(\epsilon_x, \epsilon_y) = \mathrm{K}' \iint e^{-i2\pi[q_x\rho_x + q_y\rho_y]} e^{i2\pi W(\rho_x,\rho_y)} d\rho_x \, d\rho_y \tag{2.158}$$

with K' being a new proportionality coefficient and where the wavefront error function $W(\rho_x, \rho_y)$ is the wavefront error in wavelength units. The real coordinates on the image surface are

$$\epsilon_x = \frac{\lambda}{NA_0 \cos \bar{U}} q_x$$
$$\epsilon_x = \frac{\lambda}{NA_0 \cos^2 \bar{U}} q_y \tag{2.159}$$

The transform integral is now independent of the specific optical properties of the lens, and can be evaluated by standard mathematical techniques, such as the fast Fourier transform. The region of integration is the normalized aperture of the lens. The Point Spread Function on the image surface is asymmetric off axis, dependent upon the obliquity of the chief ray, $\bar{U}$. This compensates for the natural anamorphism of the pupils as well as the oblique distance from the pupil to the image off axis.

The computation can be carried out using a standard fast Fourier transform computation method with the pupil mapped into a symmetric set of values and the scaling in the x and y directions unwrapped in printing the data. There is an additional effect, that the pupil may also be vignetted. In general, the vignetting reduces the extent of the phase function in the matrix representing the exit pupil. If desired, the computation can be carried out with an additional scaling due to approximating the pupil with a two-dimensional set of points, and an additional anamorphic scaling used in unwrapping the intensity on the image surface following the computation.

There is an additional important aspect to the modeling of the lens system using the pupil spherical coordinates. If the lens is locally isoplanatic, or close to being so, the tracing of rays into a uniform grid in the set of coordinates on the entrance pupil will result in a uniform grid on the exit pupil sphere. In addition, the computation will apply to reasonable accuracy for a region of

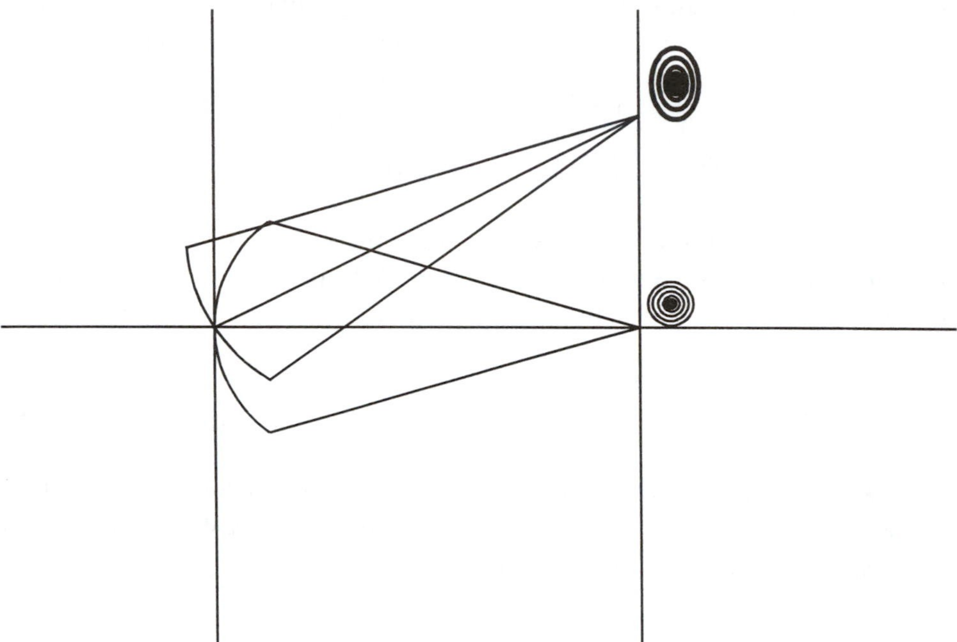

FIGURE 2.41 Demonstration of scaling of "perfect" image point spread function due to geometry of imaging situation.

the image surface, and can be used later in evaluating the image quality that will be produced using the Optical Transfer Function.

Figure 2.41 indicates in a general form the changes that can occur to an otherwise aberration-free diffraction image as a function of field angle. The scaling of the diffraction image is changed off axis according to the rules described above. The increased distance from the pupils decreases the numerical aperture off axis, increasing the scale of the diffraction image. The anamorphisn of the pupil caused by obliquity causes a different scale in the x and y directions.

There are some additional considerations. The discussion so far has covered reasonable values of the numerical aperture and field angle. The treatment of diffraction also has ignored the polarization aspects of the diffraction problem. In a real system, the effect of polarization on the diffraction image is usually quite small, as long as there is no analyzer or polarization-sensitive detector involved. If the numerical aperture increases, approaching unity, there is a likely effect on image shape due to the polarization, and a vector diffraction calculation is required in order to obtain a correct value for the diffraction pattern intensity. The reason is that the vector associated with the polarized light converging toward the image plane will mix states of polarization.

A second region where the diffraction problem must be carefully handled is in the case of extremely low numerical apertures. The diffraction integral is preceded by a propagation factor involving the inverse square of the distance to the image surface. For very small numerical apertures, any change of focus can produce a significant change in the distance to the pupil, even though the phase change in the aberration due to focus change may be of reasonable size. This will produce an asymmetry in the diffraction pattern with focus position that needs to be considered in carrying out the calculation.

The derivation above was based on the assumption that the distance from the pupil to the image surface is large compared to the wavelength. If the image is shifted to a location close to the pupil, the diffraction problem enters the Fresnel region. The assumption that the distance from the pupil to all points of interest on the image surface is constant becomes invalid, and the shift of phase with lateral position on the image surface is no longer linear, but contains high-order terms, at least quadratic, in the pupil dimension. The calculation under these conditions can be carried out by directly using the actual distance from the pupil to the image surface for each point in the integration. While this is tedious and lengthy, it is not really very complicated for a high-speed computer.

In general, wave propagation through the lens should be carried out by a successive propagation of the wavefront from surface to surface. In some programs, techniques for doing this are made available. In general, for imaging systems, the mapping between the entrance and exit pupil is quite exact, and the only propagation required is from the exit pupil to the image surface. Effects of transmission from the object to the entrance pupil, when effects such as turbulence are present, can be carried out by calculating the phase and amplitude effects upon the entrance pupil, and simply adding these to the effects in the exit pupil. This works satisfactorily as long as the entering phase perturbations are reasonably small.

There are some basic, and generally quite familiar, concepts regarding diffraction image formation that it is useful to review here. The most frequently used pupil is a circular pupil. The diffraction image resulting from such a pupil is the Airy disc or Airy function

$$f(\epsilon_x, \epsilon_y) = \left[2J_1 \left(\frac{\pi\sqrt{\epsilon_x^2 + \epsilon_y^2}}{\lambda(\mathrm{F/No.})} \right) \bigg/ \left(\frac{\pi\sqrt{\epsilon_x^2 + \epsilon_y^2}}{\lambda(\mathrm{F/No.})} \right) \right]^2 \tag{2.160}$$

with the well-known fact that the function first reaches zero at

$$\sqrt{\epsilon_x^2 + \epsilon_y^2} = 1.22\lambda(\mathrm{F/No.}) \tag{2.161}$$

which corresponds to a root of the Bessel function. Successive minima correspond to the successive zeros of the Bessel function.

Figure 2.42 is an isometric plot of this function. About 84% of the total energy through the pupil is contained within the central maximum. The remaining 16% is spread out over the image surface, with the height of the first ring, or maximum being less than 5% of the peak intensity in the image. A contour plot of the intensity distribution in this perfect point spread function is shown in Figure 2.43. The above images were calculated on the axis of a perfect optical system. Off axis, at 45° for a perfect lens, with the exiting chief ray having a 45° angle, the same intensity distribution exists, but the obliquity of the pupil and the image surface produces an elliptical image form, as shown in Figures 2.44 and 2.45. The intensity distribution in this image is the standard against which the perfection of the diffraction image is compared.

The radius of this pattern has come to be known as the Rayleigh diffraction limit. This is based upon an assumption that the images of two stars can be considered to be just resolved visually when the separation of the images of those stars coincides with this distance. At that separation, the peak of one image lies on the first minimum of the second image. Obviously, as can be seen from Figure 2.46, the presence of more than one star can be deduced by the asymmetric shape of the joint diffraction image of the two stars, so that this is really a simplified criterion for resolution. Nevertheless, this has become a rule-of-thumb value for the resolving power of a perfect optical system.

2.2.2 COMPUTATION OF DIFFRACTION IMAGES

The diffraction integral derived above has come to be used generally for the computation of diffraction images using computers. The Fourier transform nature of equation 2.155 is admirably suited for calculation using the discrete fast Fourier transform introduced by Cooley and Tukey in the 1960s. This approach requires that the pupil function to be transformed be represented as a two-dimensional matrix of points. The phase at each point in the pupil is obtained by tracing a ray to the exit pupil reference sphere and computing the optical path difference relative to the chief ray. The matrix of points to be transformed is the real and imaginary parts of the complex pupil function. The fastest algorithms require that the matrix be a power of two on each side, with 64×64 points a commonly used size.

Beam propagation can be carried out by considering the transforms between successive matrix planes. The input is a phase and amplitude pupil function.

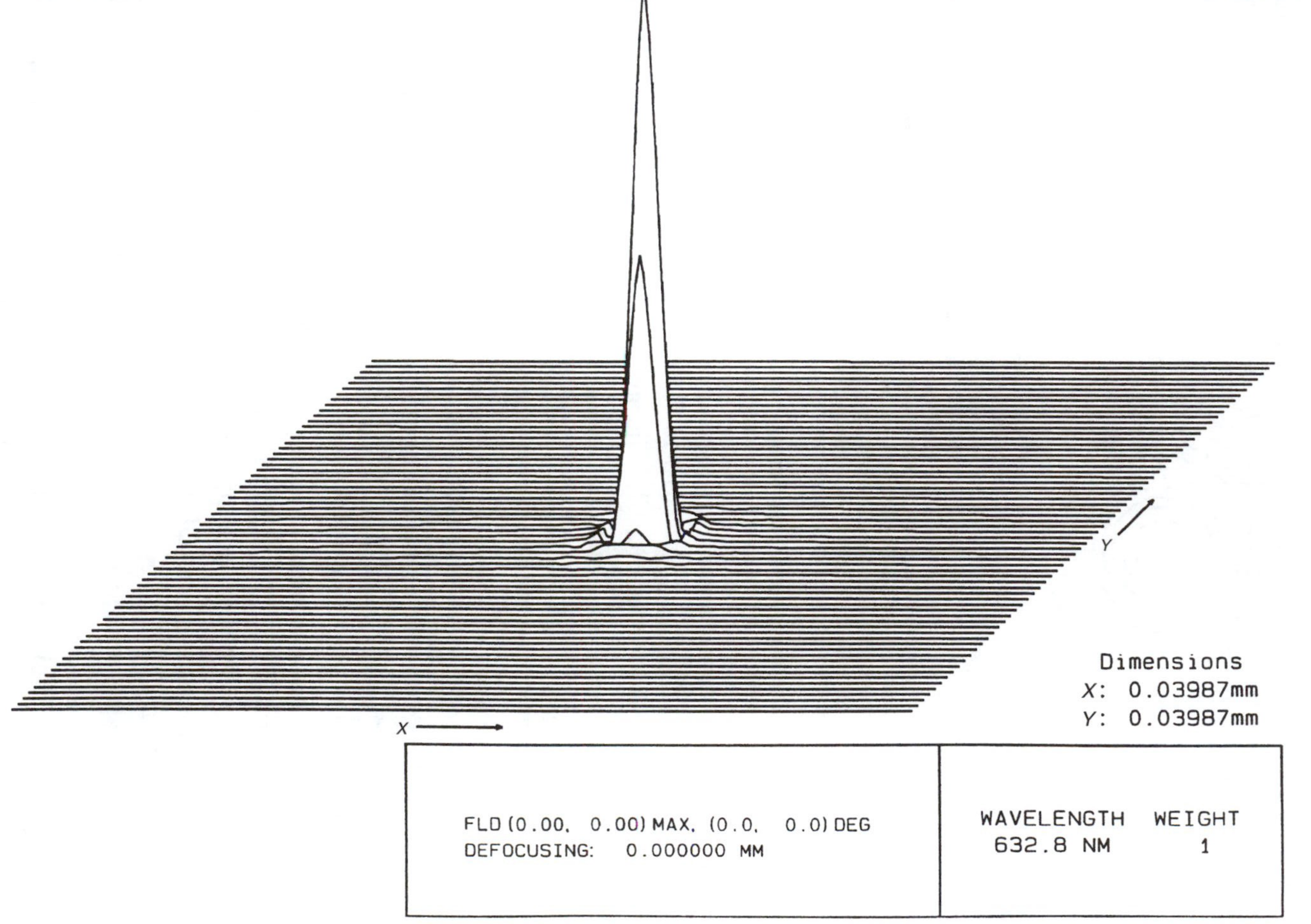

FIGURE 2.42 Surface plot of Point Spread Function on axis for F/2 system (CODE V).

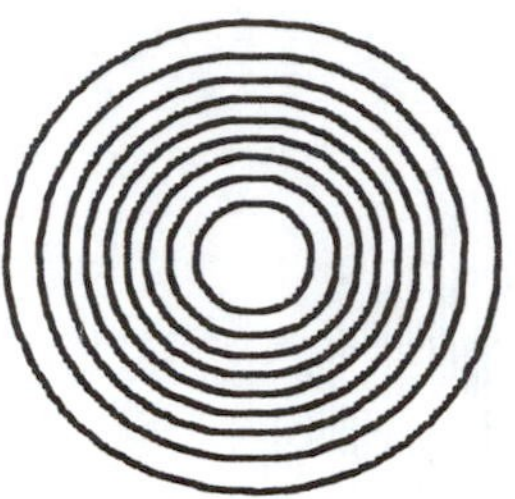

FIGURE 2.43 Contour plot of Point Spread Function on axis (CODE V).

The computation method is a discrete transform. The calculation may be a direct numerical integration or may be the application of a fast Fourier transform. The details of this calculation method will be discussed in Chapter 4 on image analysis. At this point, it is only important to note that the wavefront is represented as a matrix. The Fourier transform implementation result is contained within an area on the image surface determined by the number of sample points in the matrix. The width of the image area is inversely proportional to the numerical aperture of the pupil and directly proportional to the number of rays, or sample points of the pupil function, within the region of integration over the pupil.

The calculation is more accurate as a larger number of points is taken in the pupil. It is important that the separation of sampling points in the pupil should not leave any more than one quarter wavelength of phase difference between adjacent samples of the OPD function in the pupil. Using more points increases the time of computation, so that there is a judgment call to be used in selecting the fineness of the sampling mesh for the rays in the pupil.

Some examples are shown in Figures 2.47 to 2.49. If the number of points within the pupil is increased, the transform region on the image surface increases in diameter. Figure 2.47 shows the result of doubling the number of rays across the pupil representing the pupil function. The transform region on the focal surface doubles in size, but the size of the point spread function is, of course, limited by the actual fixed dimensions of the pupil. There is little new information provided for the aberration free case. This is even more striking when an aberration is present. Figure 2.48 is a plot of the point

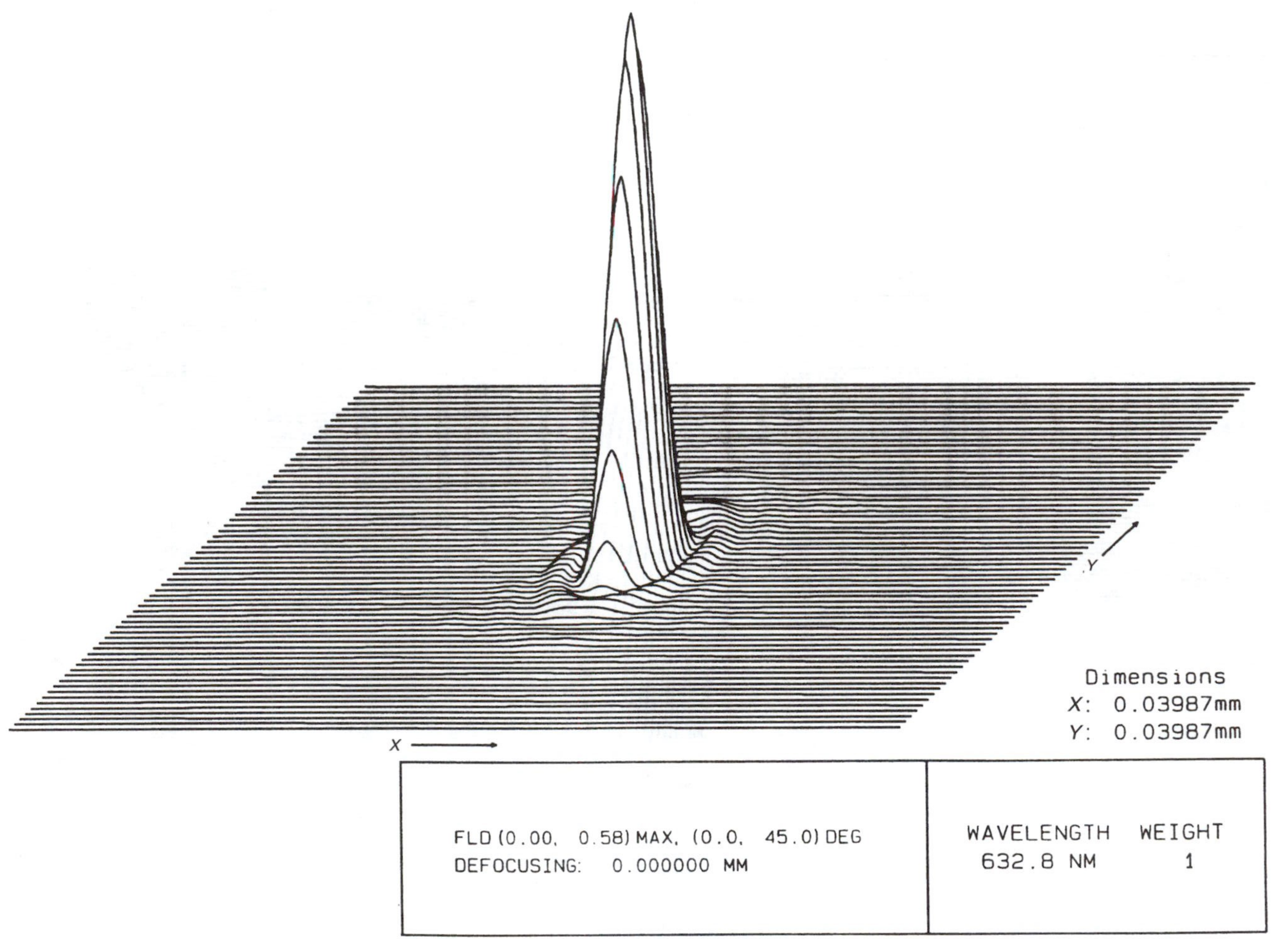

FIGURE 2.44 Surface plot of perfect Point Spread Function for system at 45° field (CODE V).

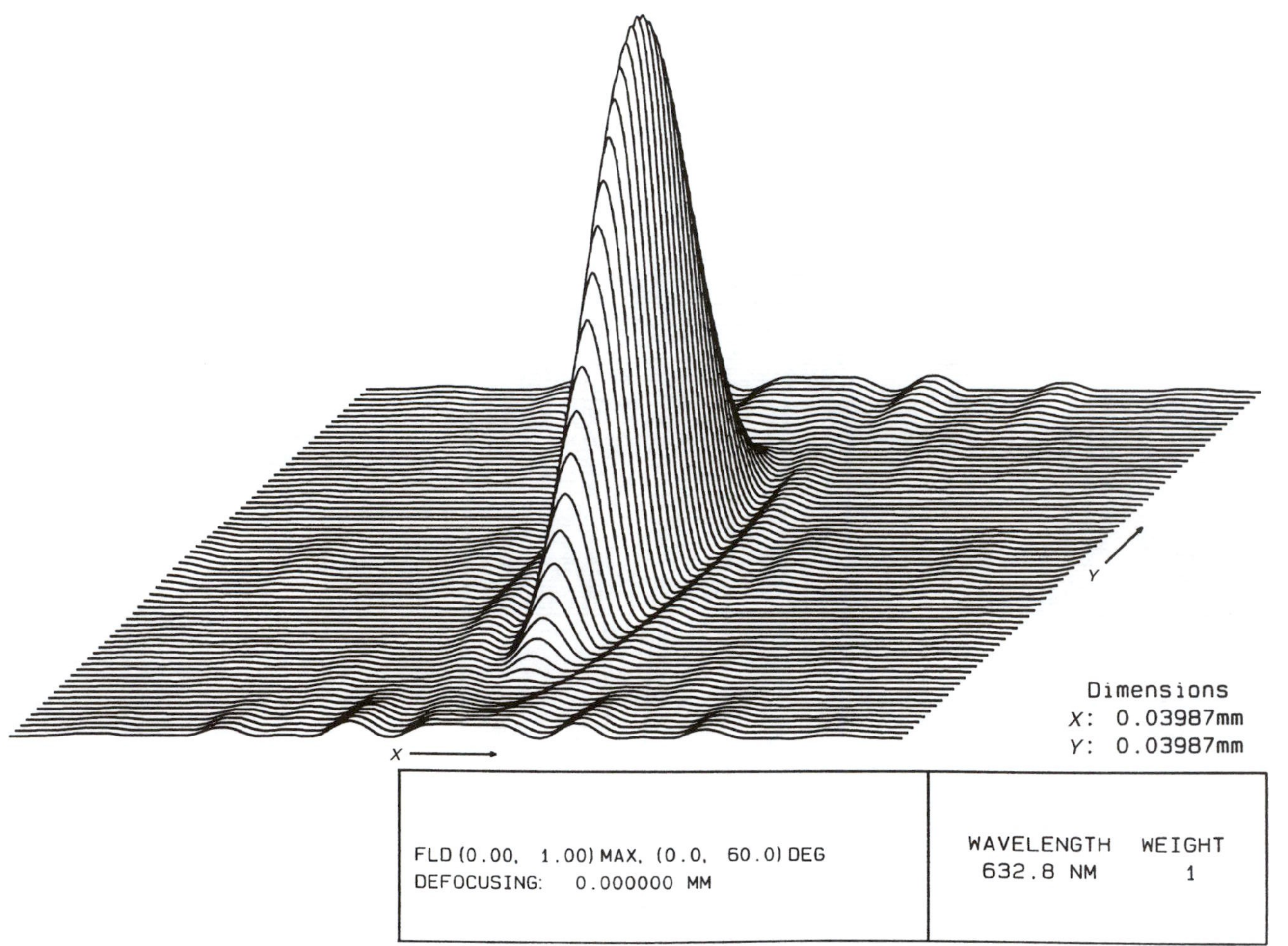

FIGURE 2.45 Surface plot of Point Spread Function for 60° field (CODE V).

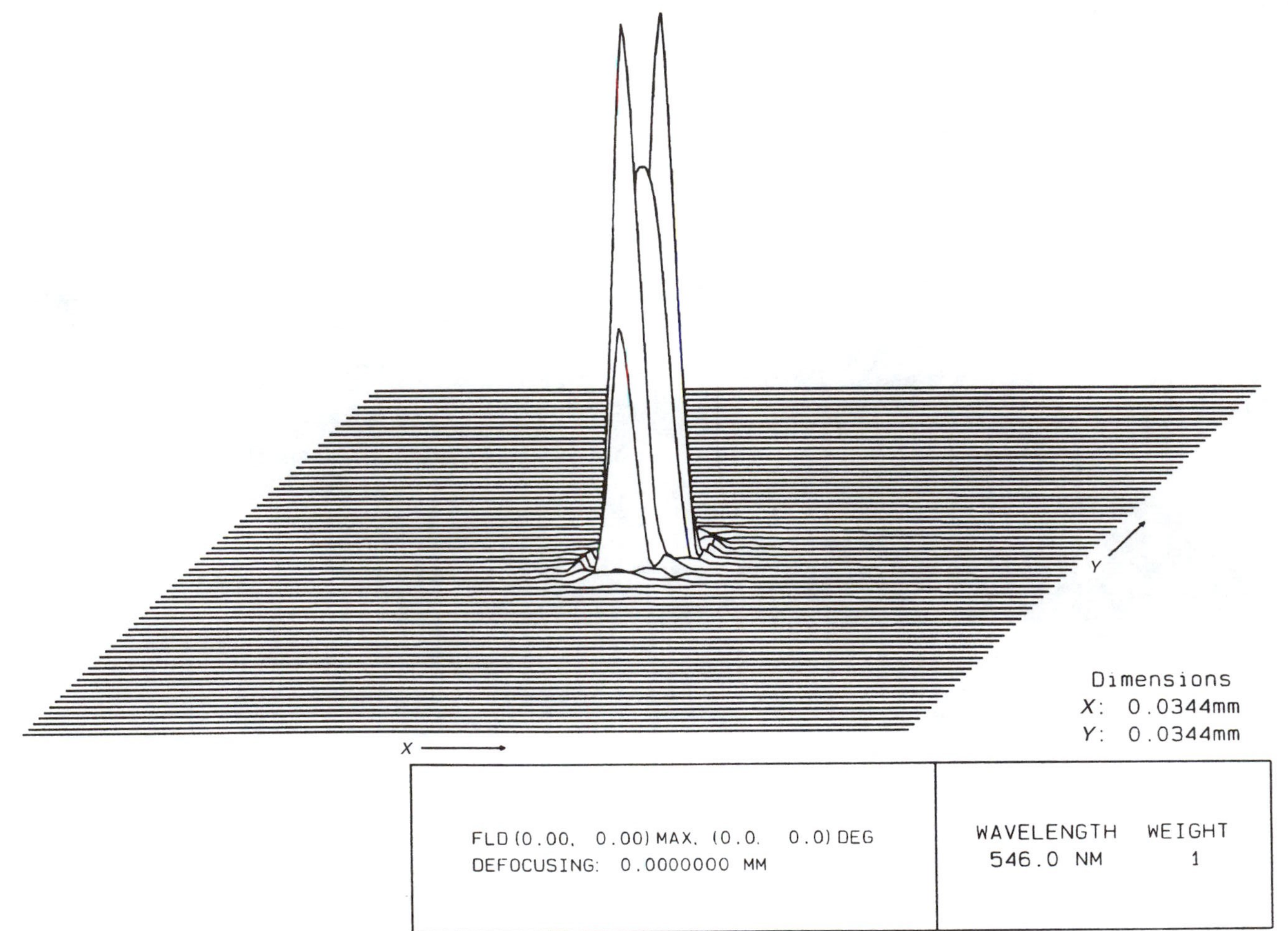

FIGURE 2.46 Image plot for two point sources separated by the Rayleigh distance (CODE V).

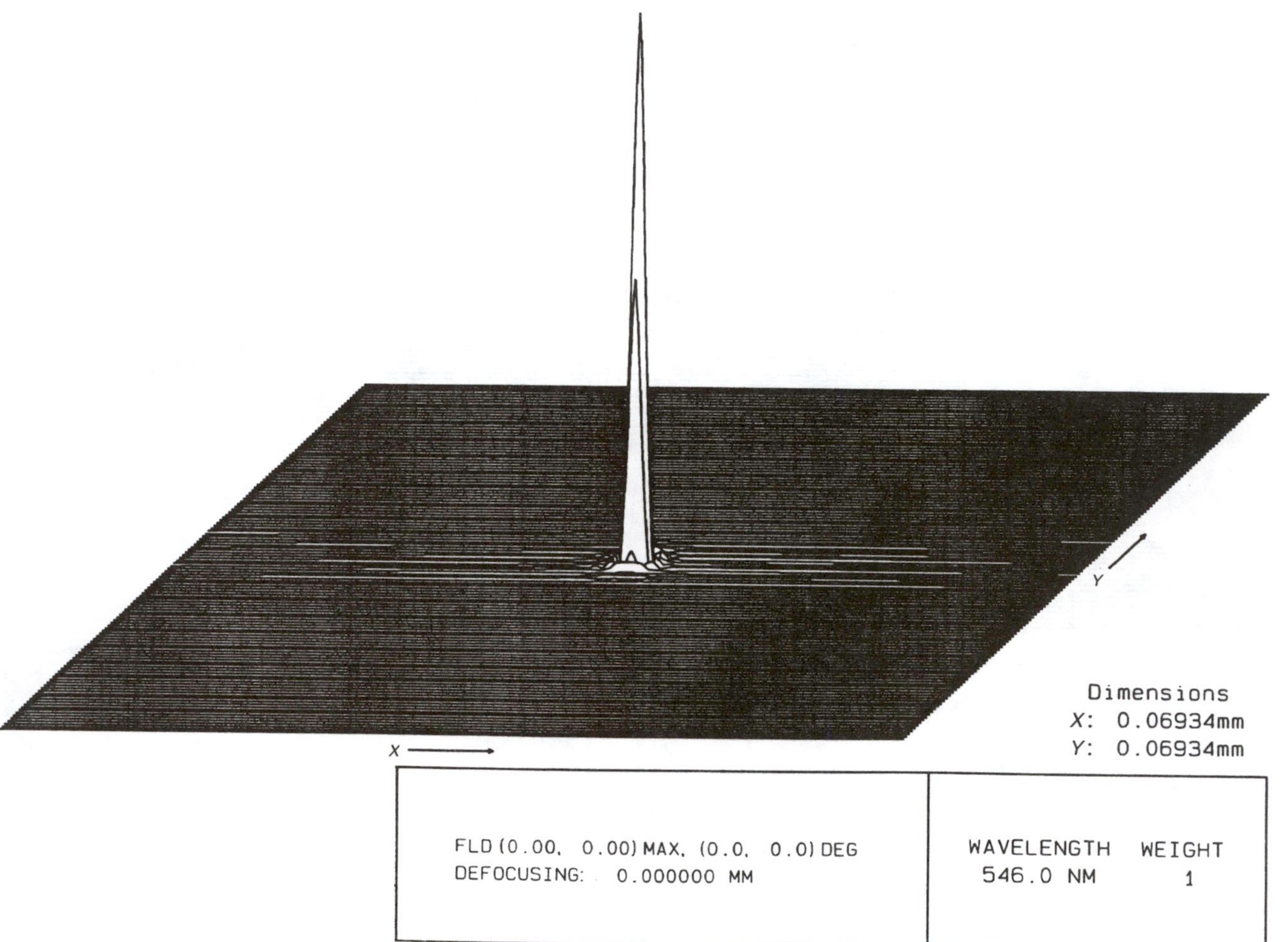

FIGURE 2.47 Point Spread Function computed with double the number of samples across the pupil (CODE V).

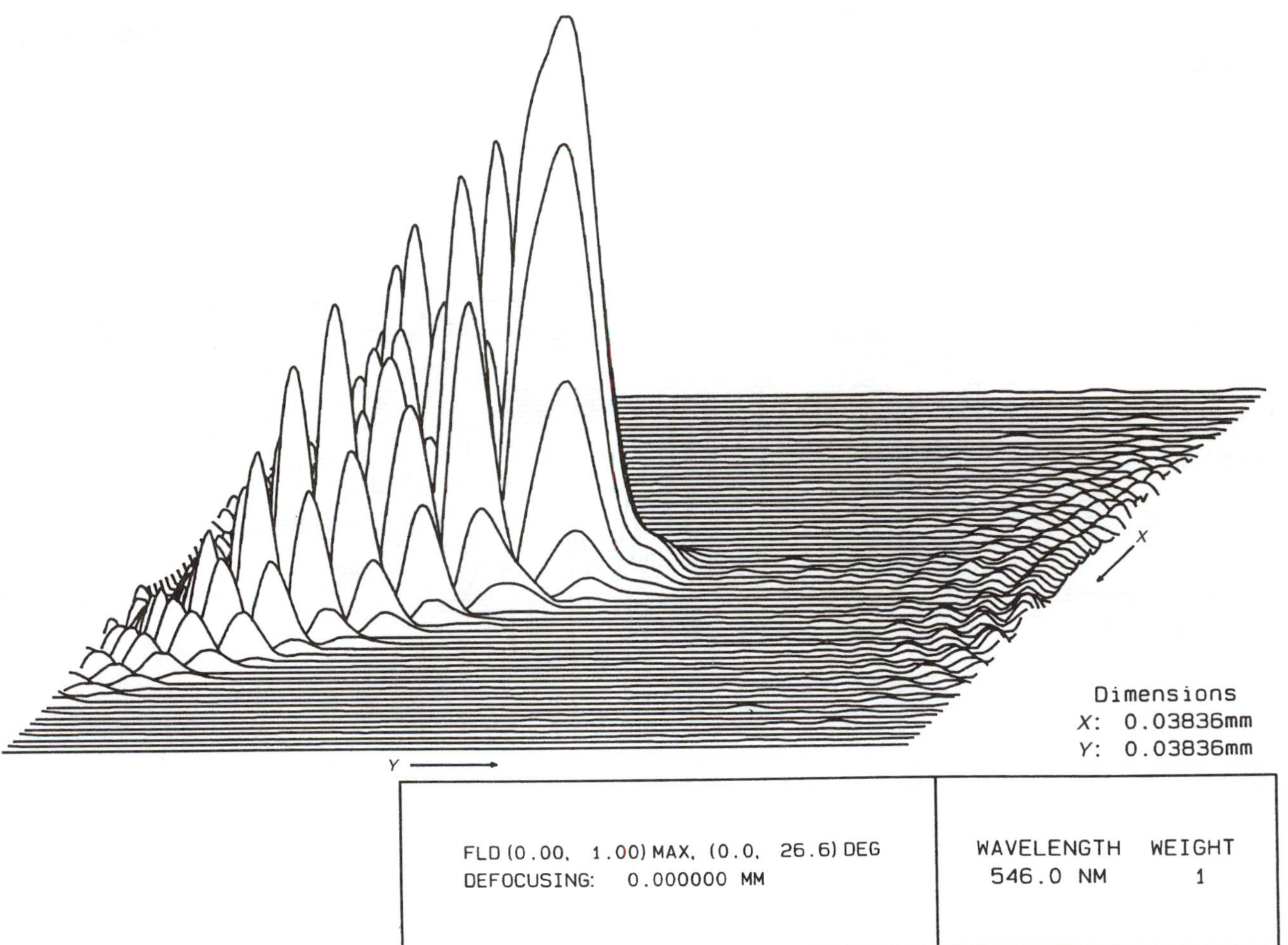

FIGURE 2.48 Point Spread Function for a lens with coma, illustrating the aliasing of the information in the function to the adjacent computational cell (CODE V).

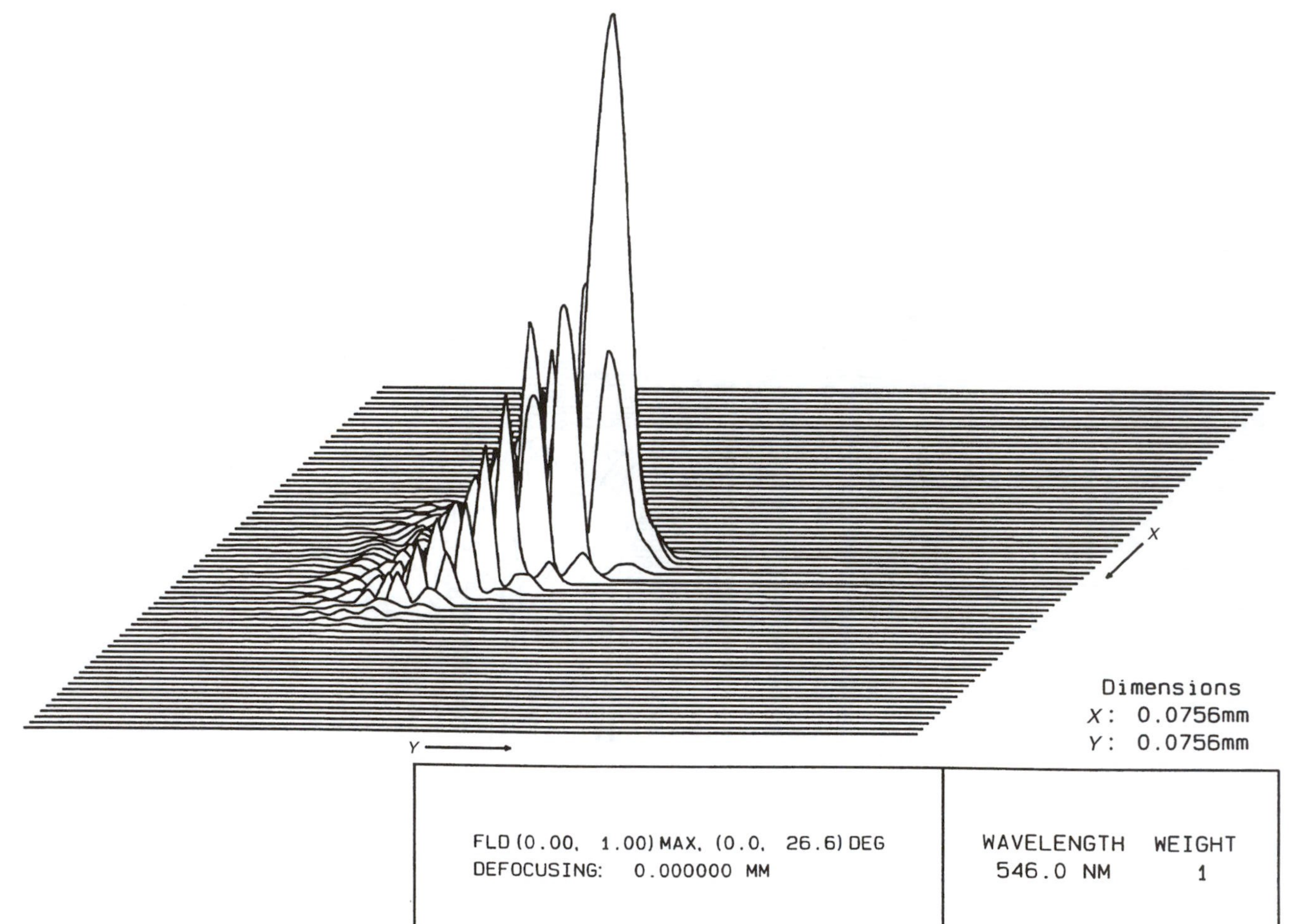

FIGURE 2.49 A repeat of the computation in Figure 2.48, but with the sampling changed to provide a larger sampling patch on the image surface (CODE V).

spread function for a significant amount of coma. The spread function almost fills the width of the transform region, and actually appears split into two parts by the selection of the origin of coordinates as being at the paraxial location on the image. Doubling the points across the pupil provides Figure 2.49, which shows the same point spread function within a larger region on the image surface.

The normalization of the point spread function is important. The amount of energy entering the pupil of the lens will be conserved (less any losses due to transmission through the lens). Therefore, the total energy within the point spread function is constant, and the peak intensity level will be reduced as the spread function spreads out over a larger and larger region of the image surface. The intensity at the center of the point spread function under this constant energy normalization will be reduced with changes in focus position or the introduction of aberrations. The ratio of this peak intensity to that of a perfect system with the same aperture is known as the Strehl ratio. This single value is frequently used as a measure of the image quality for lens systems that have small aberration content. The application of the Strehl ratio will be discussed in some detail under the section on image quality.

An example of the reduction in peak intensity is the effect of focal position. Figures 2.50 and 2.51 demonstrate the same set of focal positions for an F/2 perfect lens. The two sets of plots differ in that the ones in Figure 2.50 are normalized to constant peak level of intensity, while the plots in Figure 2.51 are normalized to a constant total energy content. The latter is representative of the appearance of an image in the real world and is what would be seen by a detector collecting an image from a point source of a given total energy. The plots in Figure 2.50 provide a better view of the details of the image, but fail to show the significant loss in peak intensity as the point spread function blurs out across the focal surface.

The weighting of the intensity in the pupil will also affect the appearance of the point spread function. A common intensity distribution from a laser is a Gaussian-weighted beam pattern. Figure 2.52 is a repeat of the focus changes in Figure 2.50, but the data has a Gaussian beam with intensity level of 0.00001 at the edge of the pupil for the plots on the left, and 0.01 for the plots on the right. The transition from a Gaussian beam to a fully filled aperture is demonstrated by the plots. All of these plots are normalized to unity at the central peak.

All of the discussion of physical optics imagery in this chapter is based upon scalar diffraction and the application of Fourier methods to point spread function calculation. The vast majority of lenses are satisfactorily represented by scalar calculations based upon Frauenhofer diffraction. For cases in which very large aberrations are present, or in which the amount of

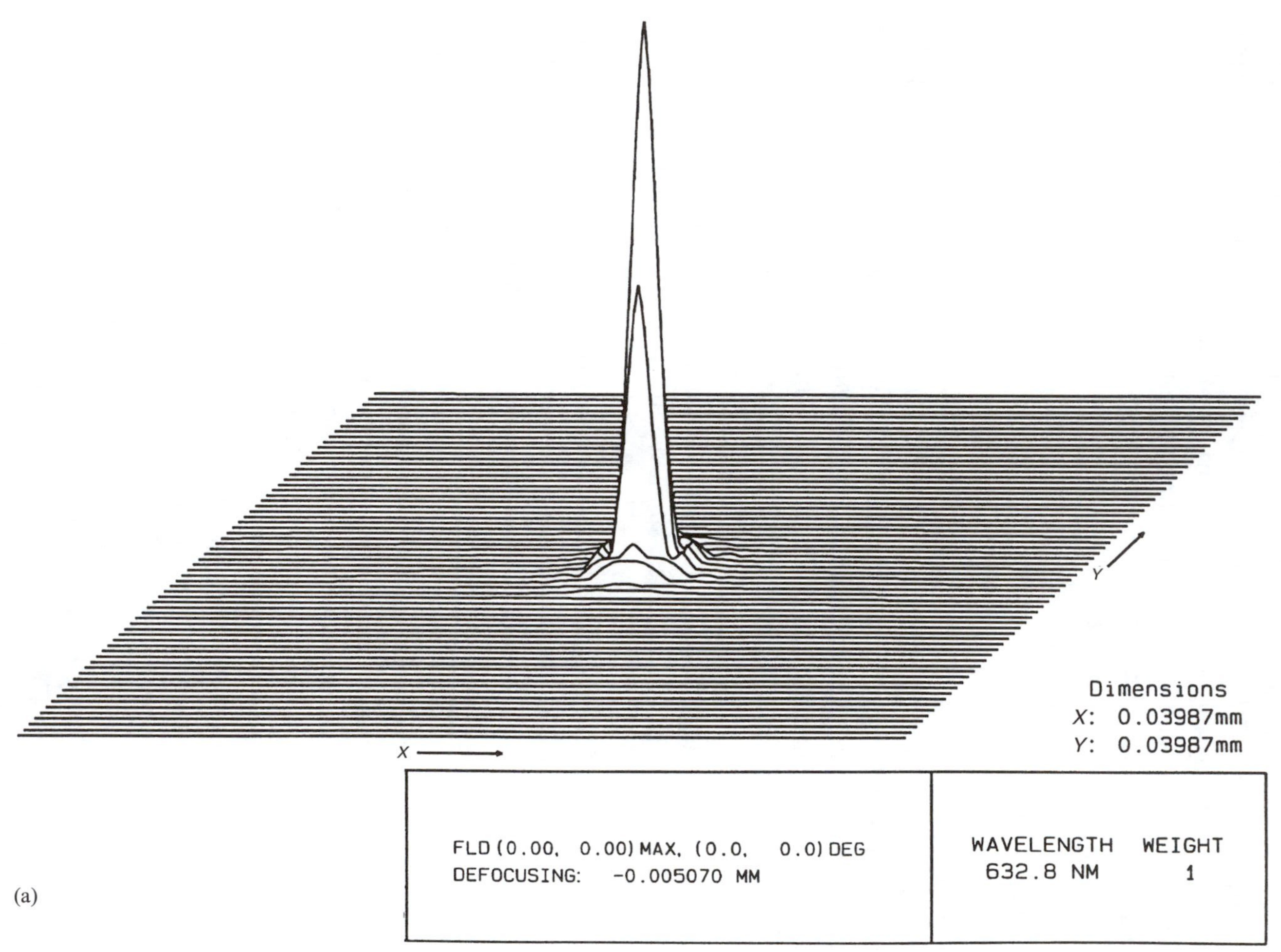
Y
X
Dimensions
X: 0.03987mm
Y: 0.03987mm
FLD (0.00, 0.00) MAX, (0.0, 0.0) DEG
DEFOCUSING: -0.005070 MM
WAVELENGTH WEIGHT
632.8 NM 1

(a)

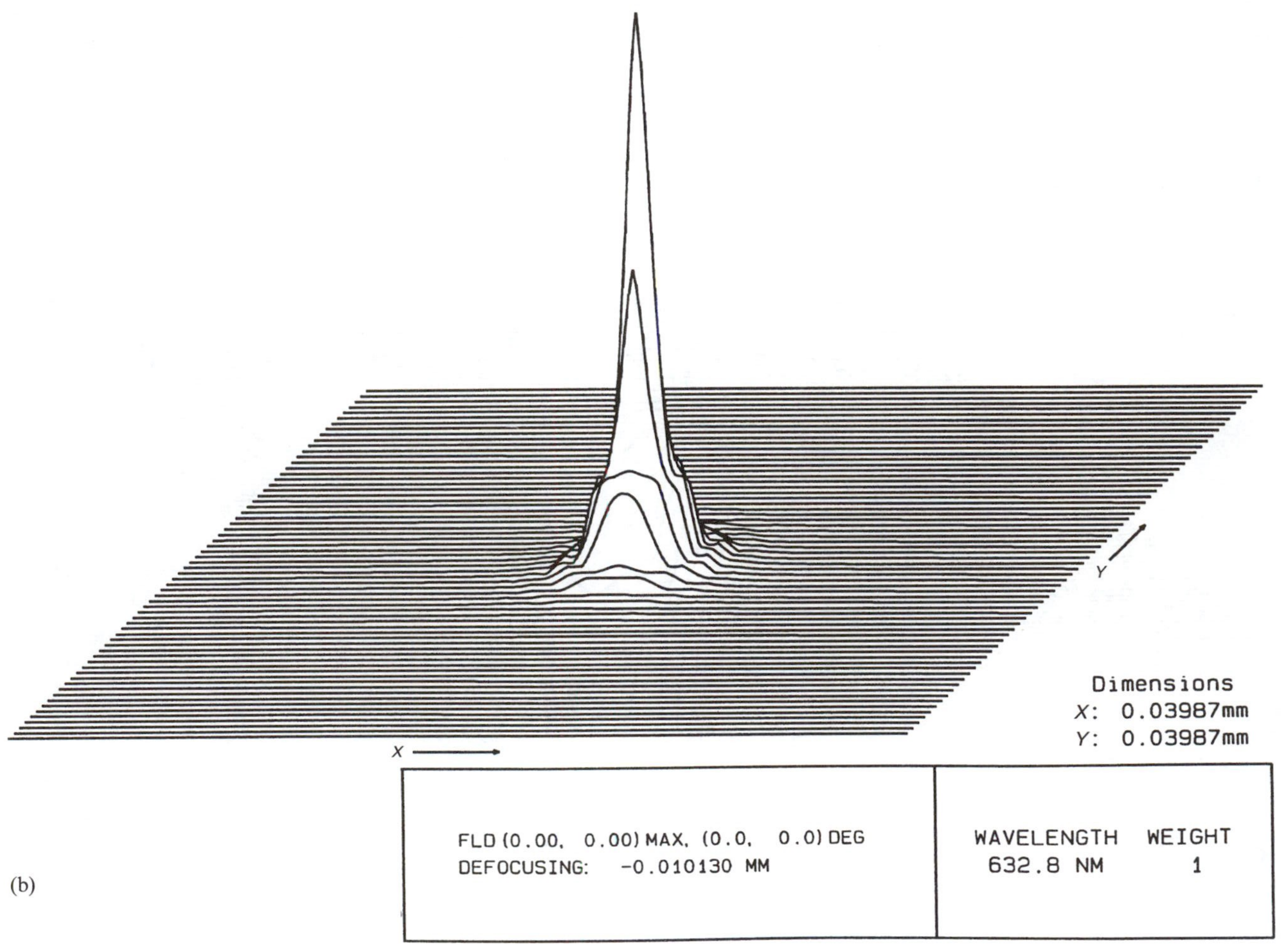
Y
Dimensions
X: 0.03987mm
Y: 0.03987mm
X
FLD (0.00, 0.00) MAX, (0.0, 0.0) DEG
DEFOCUSING: -0.010130 MM
WAVELENGTH WEIGHT
632.8 NM 1

(b)

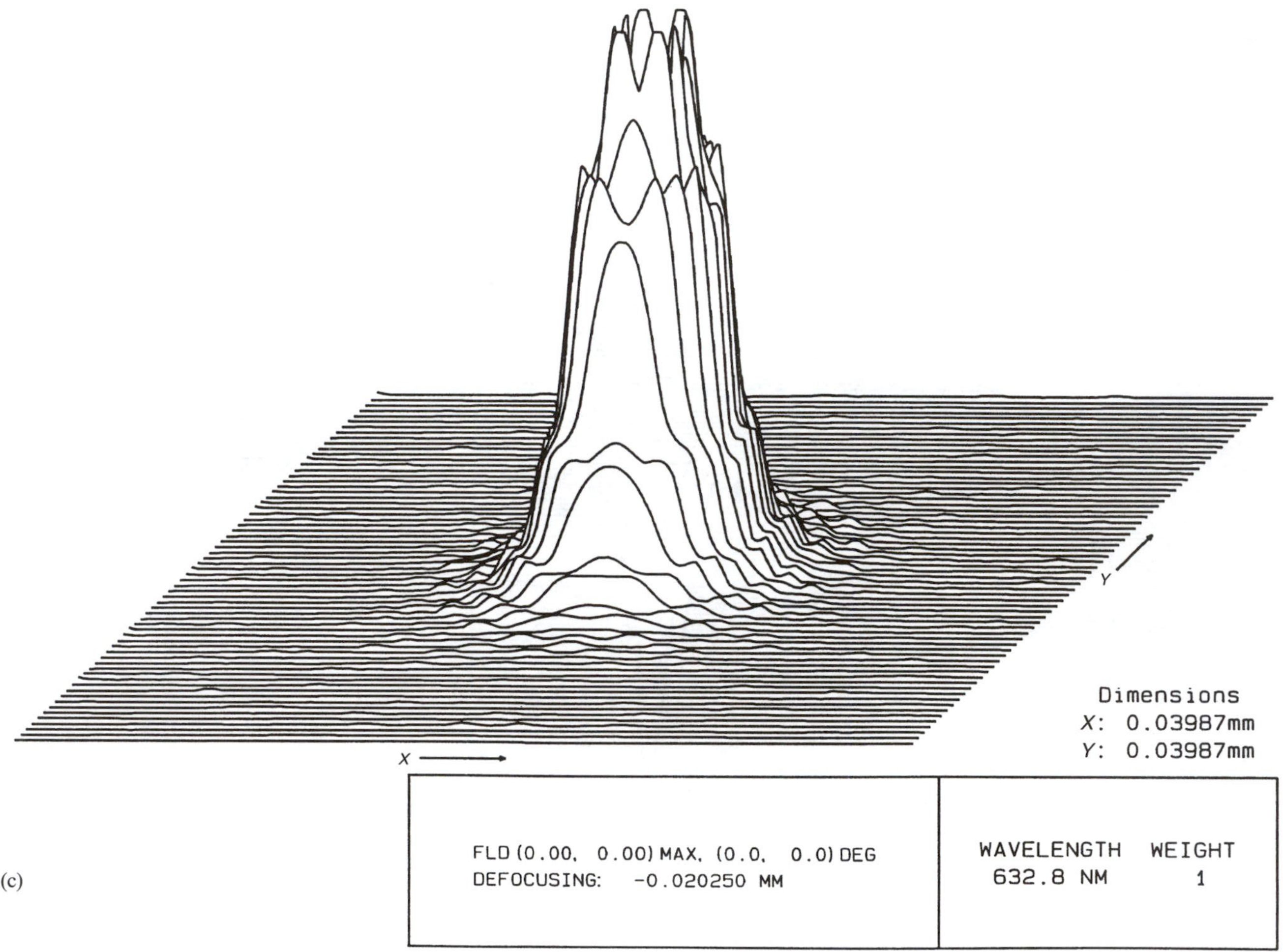

FIGURE 2.50 Point Spread Function for three focal positions, with the peak of the point spread function normalized to unity at all focal positions (CODE V).

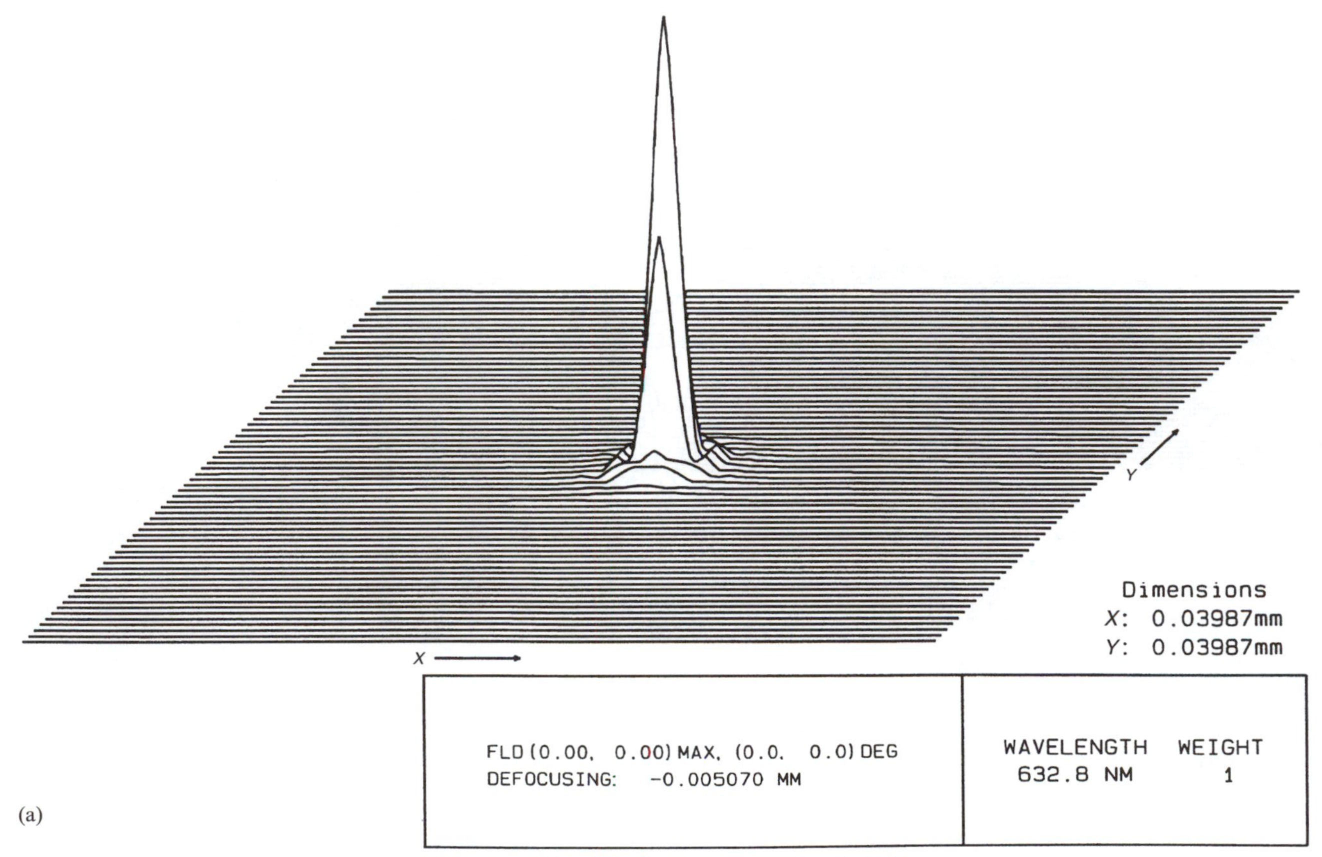
Y
Dimensions
X: 0.03987mm
Y: 0.03987mm
X
FLD (0.00, 0.00) MAX, (0.0, 0.0) DEG
DEFOCUSING: -0.005070 MM
WAVELENGTH WEIGHT
632.8 NM 1

(a)

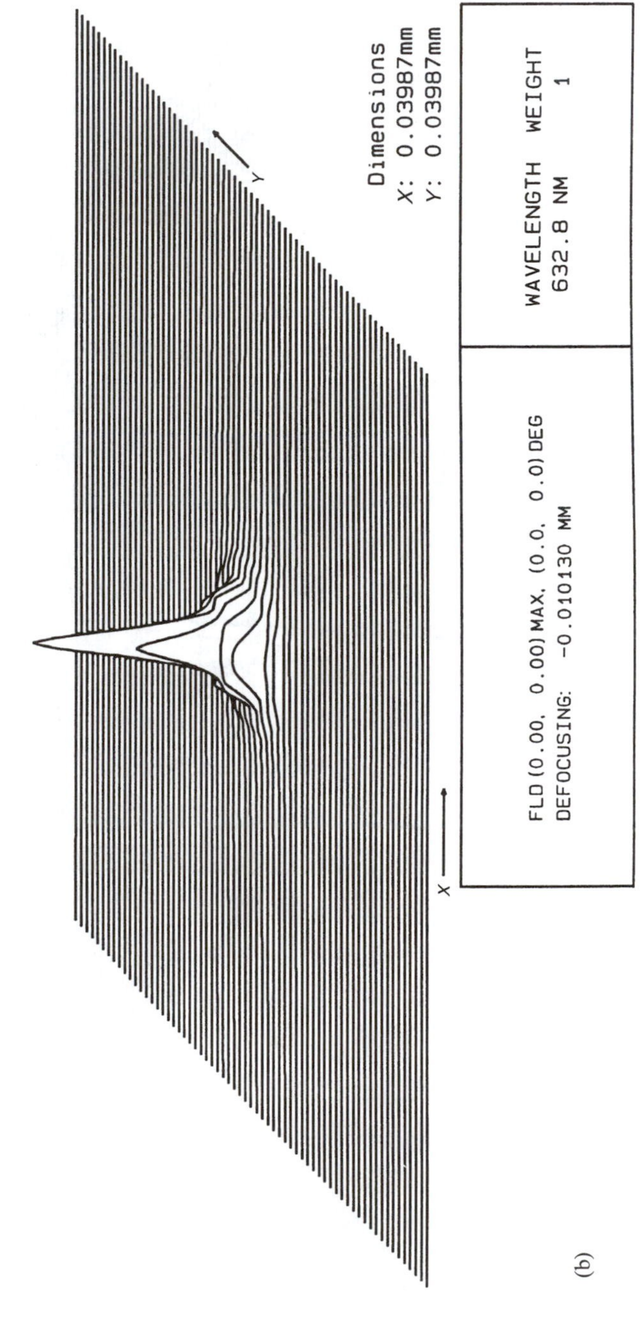
Dimensions
X: 0.03987mm
Y: 0.03987mm
Y
X
WAVELENGTH WEIGHT
632.8 NM 1
FLD (0.00, 0.00) MAX, (0.0, 0.0) DEG
DEFOCUSING: -0.010130 MM

(b)

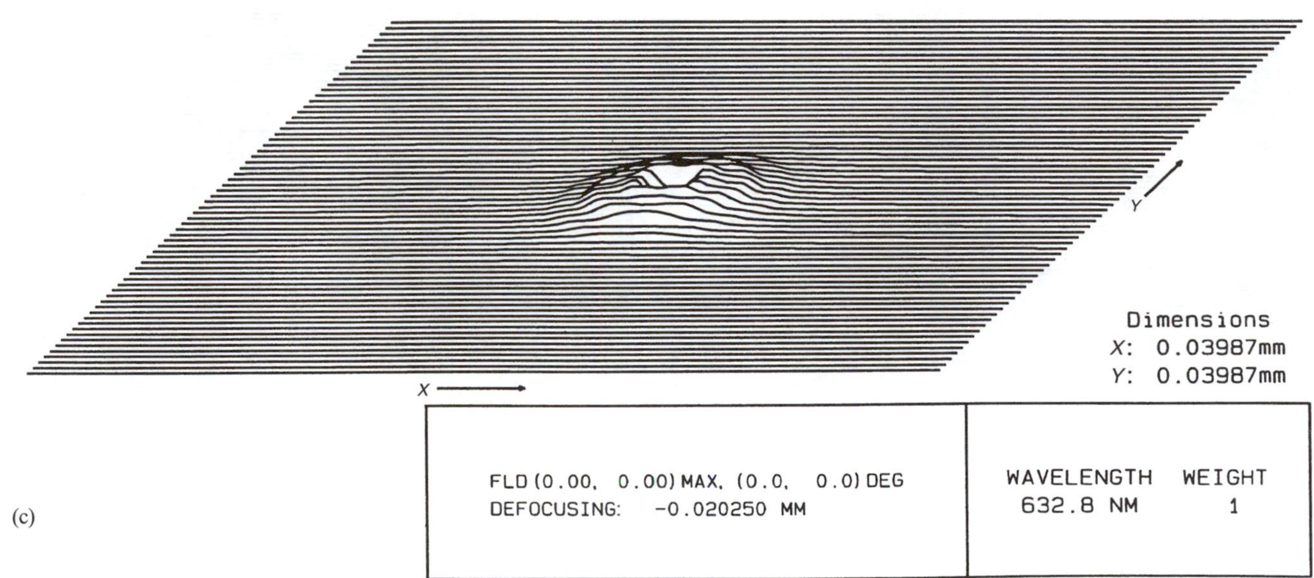

FIGURE 2.51 Point Spread Function at the same three focal positions; the normalization is now for a constant total amount of energy in the image. The height of the point spread function is then proportional to what would be observed in an actual situation (CODE V).

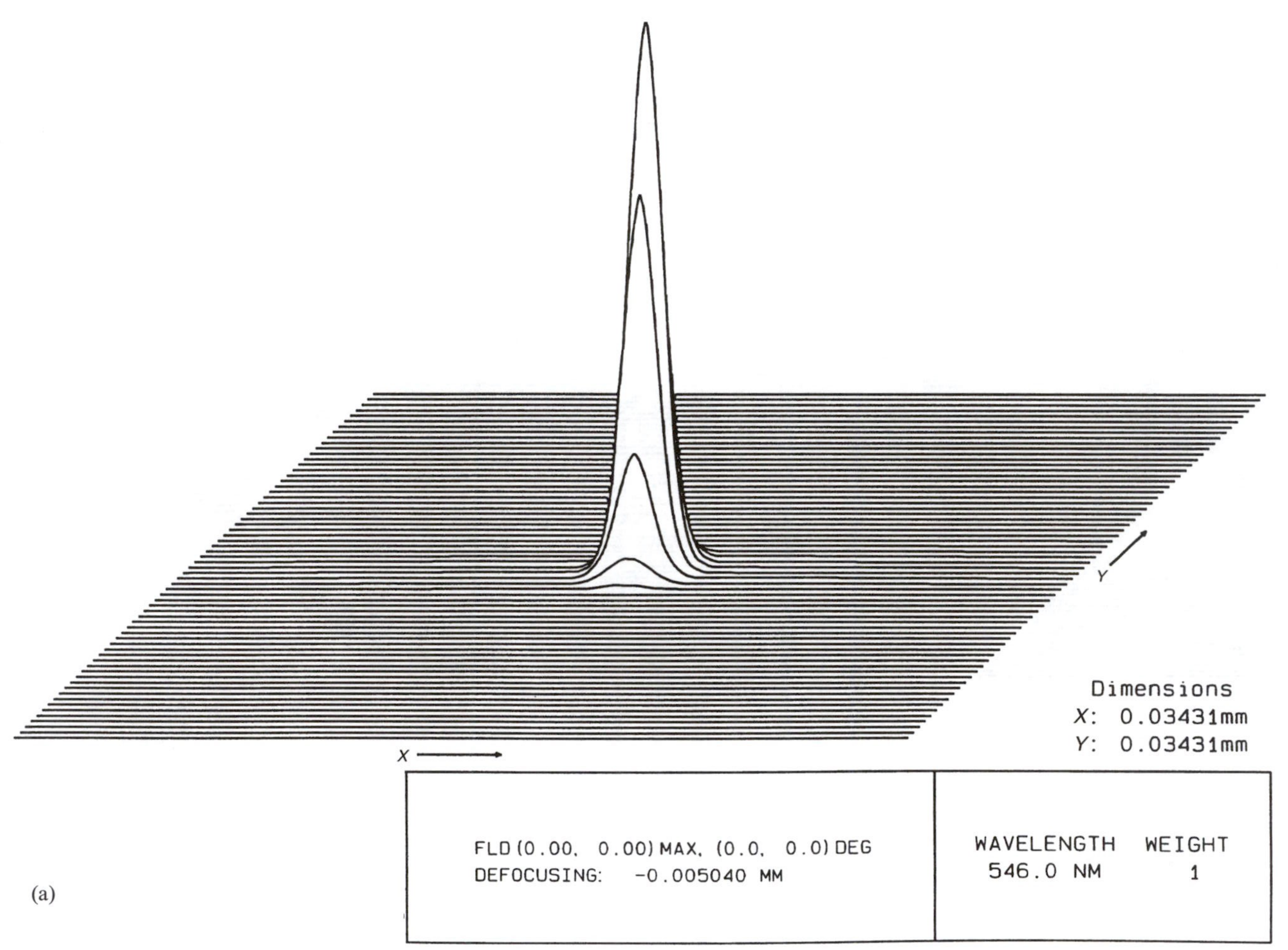

Y
X
Dimensions
X: 0.03431mm
Y: 0.03431mm
FLD (0.00, 0.00) MAX, (0.0, 0.0) DEG
DEFOCUSING: -0.005040 MM
WAVELENGTH WEIGHT
546.0 NM 1

(a)

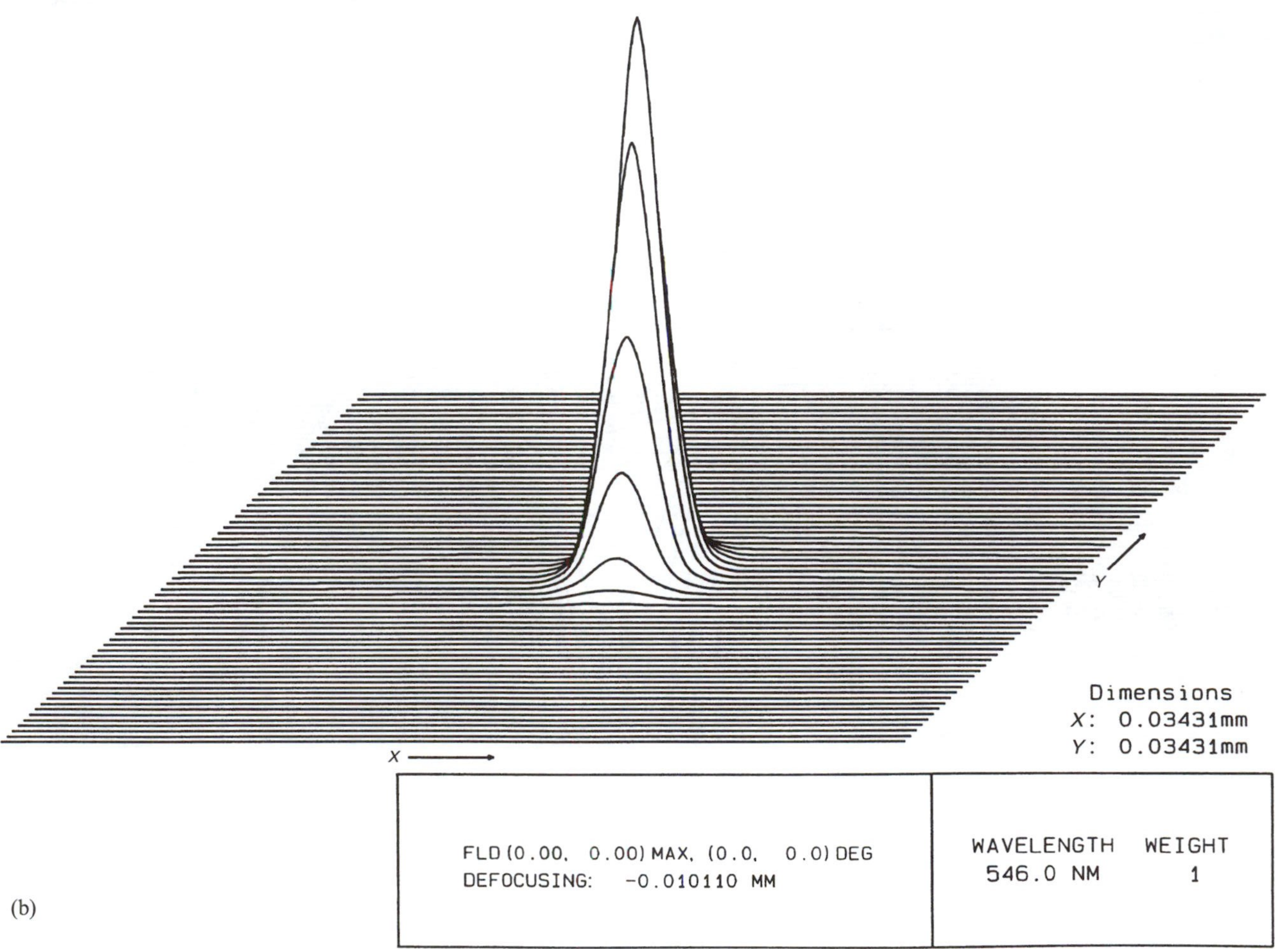
Y
X
Dimensions
X: 0.03431mm
Y: 0.03431mm
FLD (0.00, 0.00) MAX, (0.0, 0.0) DEG
DEFOCUSING: -0.010110 MM
WAVELENGTH WEIGHT
546.0 NM 1
(b)

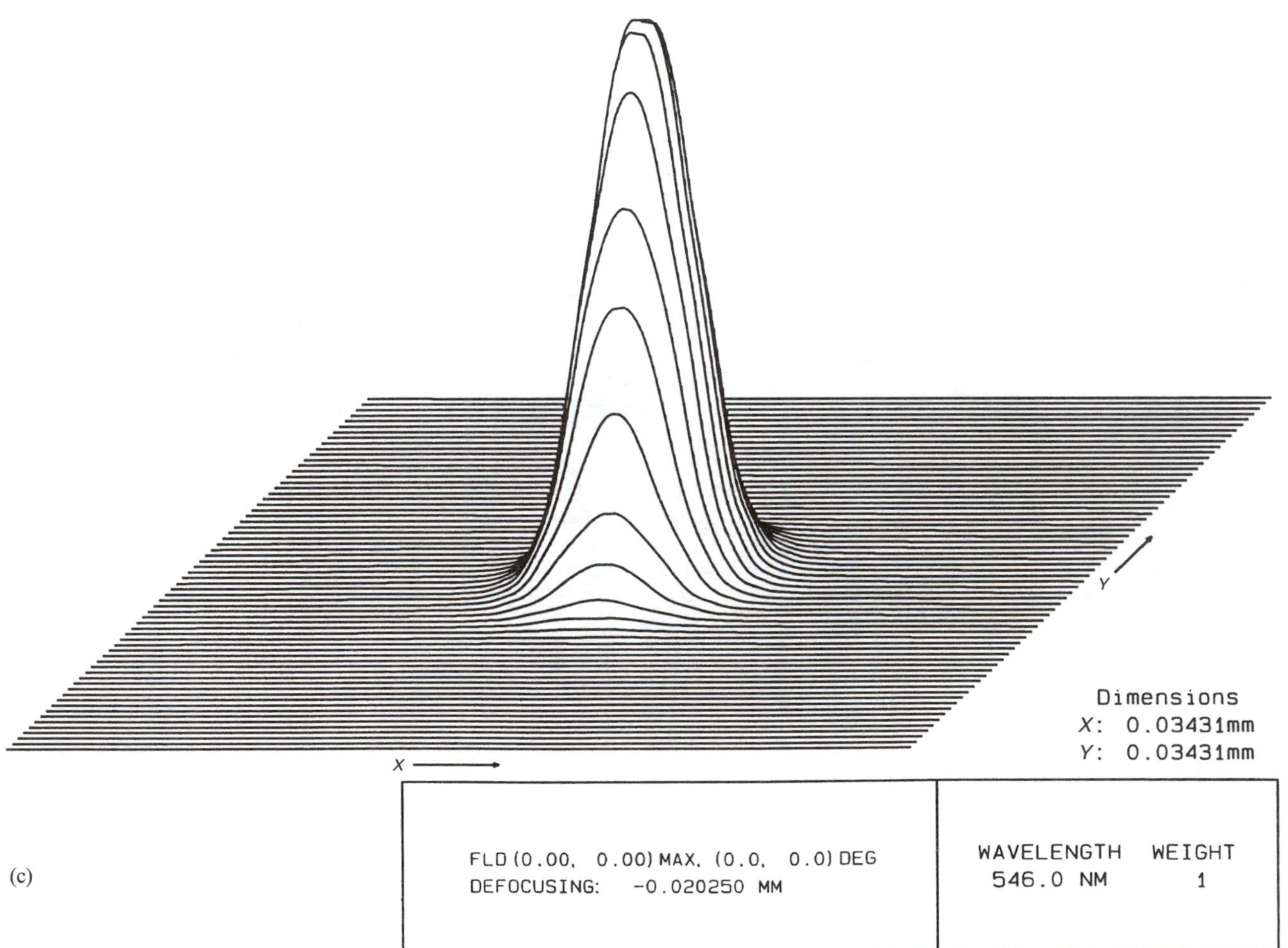

FIGURE 2.52 Point Spread Function at the same focal positions as in Figure 2.50. The aperture is weighted with a Gaussian having a 1% intensity level at the edge of the pupil (CODE V).

focus shift is very large, Fresnel computations may be required. Polarization effects for incoming unpolarized light are not significant in most lenses of reasonable aperture, usually F/numbers greater than 1.0.

Additional assumptions need to be made when the illumination is partially coherent, or is not randomly polarized.

2.3 Optical materials

Most lenses are made of optical glass. Other materials, including crystals, plastics, and reflective substrate materials are important for some applications. Selection of a material for use in a lens design initially requires consideration of the optical properties of the material. Success in the use of this material in a particular environment requires additional consideration of the mechanical, physical, and chemical properties of the material. Availability of the material is dependent upon the cost, forms of supply, and required working and finishing procedures for the material. A successful optical designer will consider all of the parameters when selecting or recommending materials for a lens.

The most important optical property of materials used in lens elements is the refractive index. The refractive index is defined as

$$n = \frac{v_{med}}{v} = \frac{\text{Velocity of light in the surrounding medium}}{\text{Velocity of light in the material}} \qquad (2.162)$$

where, most often, the surrounding medium is air. Thus the refractive index is usually stated relative to air. Most optical glass catalogs use this definition of refractive index, selecting standard conditions of temperature and air pressure for defining the listed values. Absolute refractive index is referred to vacuum as the surrounding medium, and can be obtained by multiplying the value for the index as defined above by the index of refraction of air at standard temperature and pressure. The refractive property of the material is related to the dielectric constant of the material at the frequency of the electromagnetic radiation being transmitted through the material, and is dependent upon the wavelength or color of the light (Born and Wolf 1959; Fanderlik 1983; Izumitani 1985).

In the majority of lens designs, materials are required to be homogeneous and isotropic, contribute negligible scattering, and have high transmittance over the desired spectral region. In specialist applications, variable index of refraction, nonlinear behavior of the refractive index, and birefringence are important properties.

Optical glass is one of the most homogeneous materials produced in large quantity, with physical and optical properties which can have better than one part per million homogeneity throughout each glass blank. Glass is a vitreous material with only short-order crystalline properties, and is therefore isotropic unless some physical change, such as strain, is introduced to change this property. The available glasses are produced by dissolving specific inorganic materials in a silica-based melt of sand with a material that reduces the melting temperature. The limited range of materials that can be used successfully in forming glasses define the optical properties of the glass, and a limited region of possible optical properties. The vitreous, noncrystalline, character of glass causes the material to soften and eventually become liquid as the temperature rises, rather than liquefy or condense at a specific temperature. This transition range permits extremely fine annealing of the glass to attain the required homogeneous properties.

Newer developments in optical engineering have required the use of a variety of additional materials for transmitting optics, including plastics, crystals, and microcrystalline materials. These materials have specific transmission, refractive, or dispersive properties that make them useful. Plastics, because of their low density and moldable properties, are attractive as lightweight inexpensive lens components, but are not as stable or homogeneous as glass. Most crystals have problems in fabrication or in maintaining constant birefringence properties in useful sizes. Crystals also will generally exhibit directionally dependent mechanical and thermal properties.

A new approach to lens design incorporates the use of material that intrinsically has a well-defined index of refraction gradient. These materials, in combination with traditional materials, permit the solution of some specific optical problems.

Special materials for forming holographic optical components and other types of diffracting components have become important. Components made from these materials have complementary properties to refractive materials, and are widely used in laser systems. A newer approach is to approximate the continuity of phase steps in holographic optics as a set of binary steps. Such binary optical components have fabrication and physical properties that have found use in many specialized optical designs, especially for monochromatic-imaging situations.

A wide variety of possible materials for use as substrates in reflecting components have been developed over the past two decades. These materials are only indirectly included in the lens design process, as light does not pass through them. The environmental properties of these reflective substrate materials, and the material-dependent ability to form them into desired aspheric components with sufficient thermal and mechanical stability is of importance to the designer in making decisions.

TABLE 2.1 Standard wavelengths used in defining glasses

Wavelength (nm)	Designation	Spectral source
2325.42		Mercury
1970.09		Mercury
1529.58		Mercury
1060.00		Neodymium laser
1013.98	t	Mercury
852.11	s	Cesium
706.52	r	Helium
656.27	C	Hydrogen
643.85	C'	Cadmium
632.80		He–Ne laser
589.29	D	Sodium (Doublet)
587.56	d	Helium
546.07	e	Mercury
486.13	F	Hydrogen
479.99	F'	Cadmium
435.83	g	Mercury
404.66	h	Mercury
365.01	l	Mercury
334.15		Mercury
312.59		Mercury

2.3.1 OPTICAL GLASS

Index of refraction varies with wavelength. This means that the speed of light in the glass varies with color and the medium is dispersive. In almost every spectral region of interest, glasses follow a dispersion curve of the form shown in Figure 2.53, in which the refractive index of the glass is highest for shorter wavelengths. There are an immense number of optical glasses. These are only a small sample. Table 2.1 lists a number of wavelengths conventionally used in defining the dispersive properties of an optical glass. In order to characterize an optical glass in a simple manner, three wavelengths are selected, and the refractive index measured for those wavelengths. By convention, for glasses to be used in the visible spectral region, the wavelengths d, F, and C are used. These are the spectral lines for helium at 587.6 nm and those of hydrogen at 486.1 and 656.3 nm, respectively. In most of the examples used in this book, these wavelengths are used, by

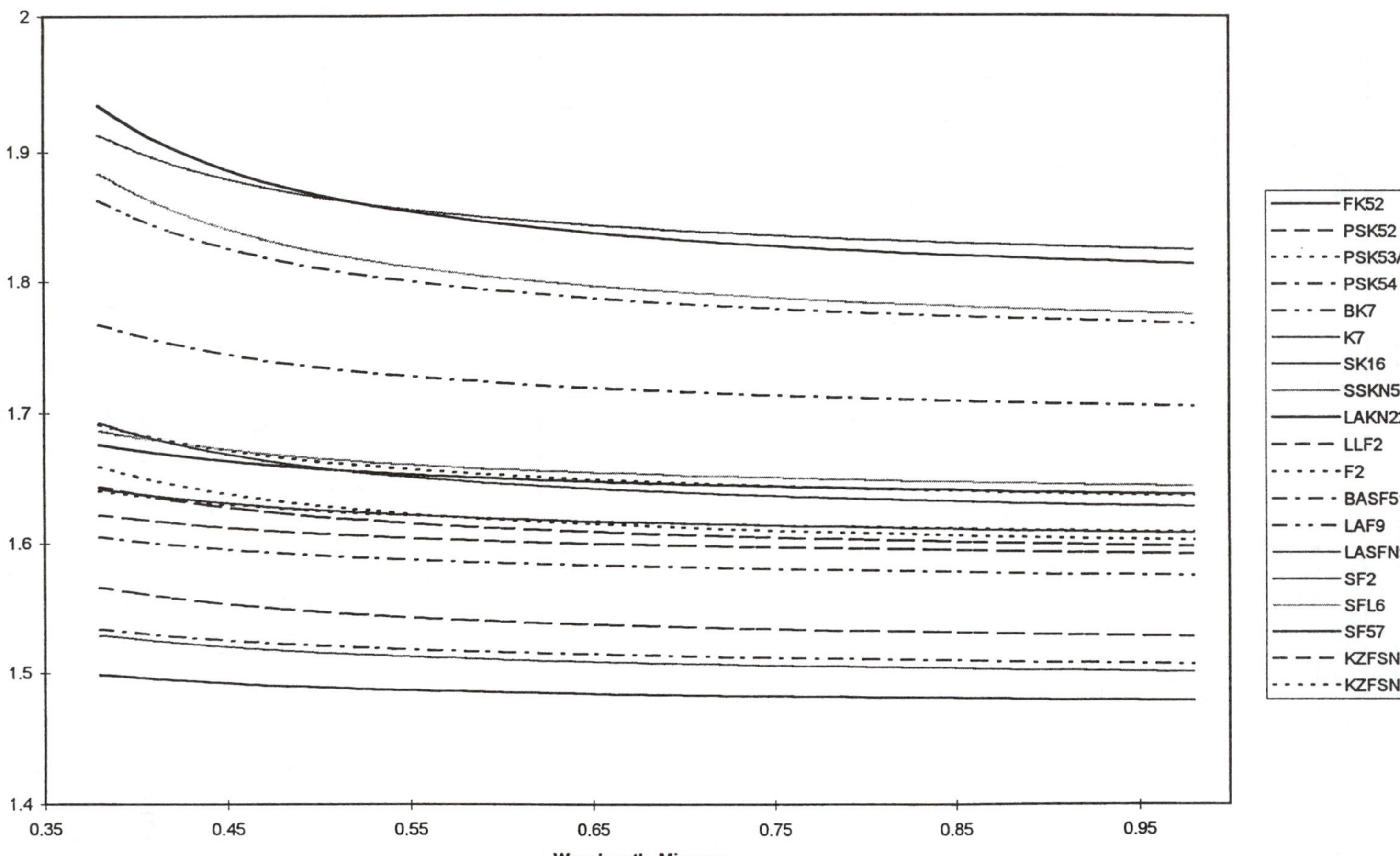

FIGURE 2.53 Plot of refractive index versus wavelength for a number of optical glasses.

TABLE 2.2 Optical properties for some glasses for the *F–d–C* spectral range

Type	Designation	n_d	V_d	P_{dC}
BK7	517642	1.51680	64.1	0.3077
F2	620364	1.62004	36.4	0.2938
SF2	648339	1.64769	33.9	0.2922
FK52	486818	1.48605	81.8	0.3047
KzFSN4	613443	1.61340	44.3	0.3004
SF6	805254	1.80518	25.4	0.2871

default, to define the operational wavelength range. In any specific application the designer will choose wavelength sets appropriate for the problem.

An optical glass will usually be defined by a triad of numbers for index, Abbe number (reciprocal dispersion or V-number) and partial dispersion defined by

$$\begin{aligned} n_d &= n_d = \text{Index} \\ V_d &= \frac{n_d - 1}{n_F - n_C} = \text{Abbe number} \\ P_{dC} &= \frac{n_d - n_C}{n_F - n_C} = \text{Partial dispersion} \end{aligned} \tag{2.163}$$

These numbers are conventionally stated for the nominal visible range by using the *F–d–C* spectral lines. For any other spectral range, similar quantities can be defined using an appropriate set of wavelengths which denote the center, upper bound, and lower bound of the spectral region of interest. In glass catalogs, the glasses are conventionally referred to using the standard wavelengths mentioned above. Table 2.2 provides some sample numbers for several typical optical glasses. In this table the designation of the glass type is a standard six digit international manufacturer's code. The glass type designations used by Schott Glass Corporation are also listed. Since the refractive properties of the glass are primarily determined by the chemical constituents in the glass, the naming of a glass type is meaningful, as all purchases of one type of glass will be of glass that is a member of the same chemical and physical family.

The optical meaning of these glass parameters is the following. The refractive index at the center of the spectral band is a measure of the degree of bending of a ray of light by the glass. It was shown earlier in this chapter that the power of a thin lens is proportional to the index of refraction of the glass minus the index of refraction of the surrounding material times the total curvature of the lens. The chromatic variation of power over the selected wavelength range for this same lens is the power divided by the Abbe number:

$$\frac{-\delta f}{f} = \frac{\delta\phi}{\phi} = \frac{(n_F - 1)c - (n_C - 1)c}{(n_d - 1)c} = \frac{\phi}{V_d} \tag{2.164}$$

Thus, the smaller the Abbe number, the greater will be the chromatic aberration contribution of a lens fabricated from the glass. Use of two optical elements of opposite sign and different Abbe number permits compensation for the chromatic aberration. The partial dispersion describes the fraction of the total chromatic dispersion that resides to one side of the wavelength defining the nominal center of the wavelength band. This glass parameter is useful in determining the residual or secondary chromatic aberration introduced by a lens element.

Figure 2.54 is a plot of the glass map using the glass types included in the previous figure. This plot shows the glass presented with respect to two of the optical parameters, the index and the Abbe number. Figure 2.55 shows a plot of these same glasses for the partial dispersion versus the Abbe number.

The glass map in Figure 2.54 is clearly two-dimensional, with glasses filling most of the plotted region. The glasses presented in this map are a result of about 150 years of development of glass types. Each glass represents a specific glass, and glasses are not produced that lie between the plotted glasses. There are some obvious limitations in the coverage of the map.

By convention, glasses with an Abbe number greater than fifty are known as *crown glasses* and glasses with Abbe numbers less than fifty are known as *flint glasses*. This terminology is historical in nature. The earliest optical glasses were all soda-lime glasses, in which sodium and calcium salts were added to the silica sand in order to produce a good glass melt. This meant that there was only one type of optical glass, which led Isaac Newton to decide that there was no hope of reducing the chromatic aberration of refractive lenses. In about 1750, John Dollond of England discovered that adding flint minerals, or lead oxides, to the sand produced a new type of glass with greater dispersive power. Thus was born the terminology still in use.

The other designations on the glass map describe general types of glasses obtained by adding different materials to the sand in the melt. By the end of the nineteenth century, the types of glasses generally followed the curved line defining the right side of the glass map. Around the turn of the twentieth century, the addition of heavy metals to the melt produced glasses that showed higher index and low dispersion, the SK or dense crown materials. The addition of rare earths to the melt permitted development of very-high-index, low-dispersion glasses in both crowns and flints, denoted by the LaK and LaF regions on the glass map. For all practical purposes, this glass availability map has remained basically unchanged for the past 50 years,

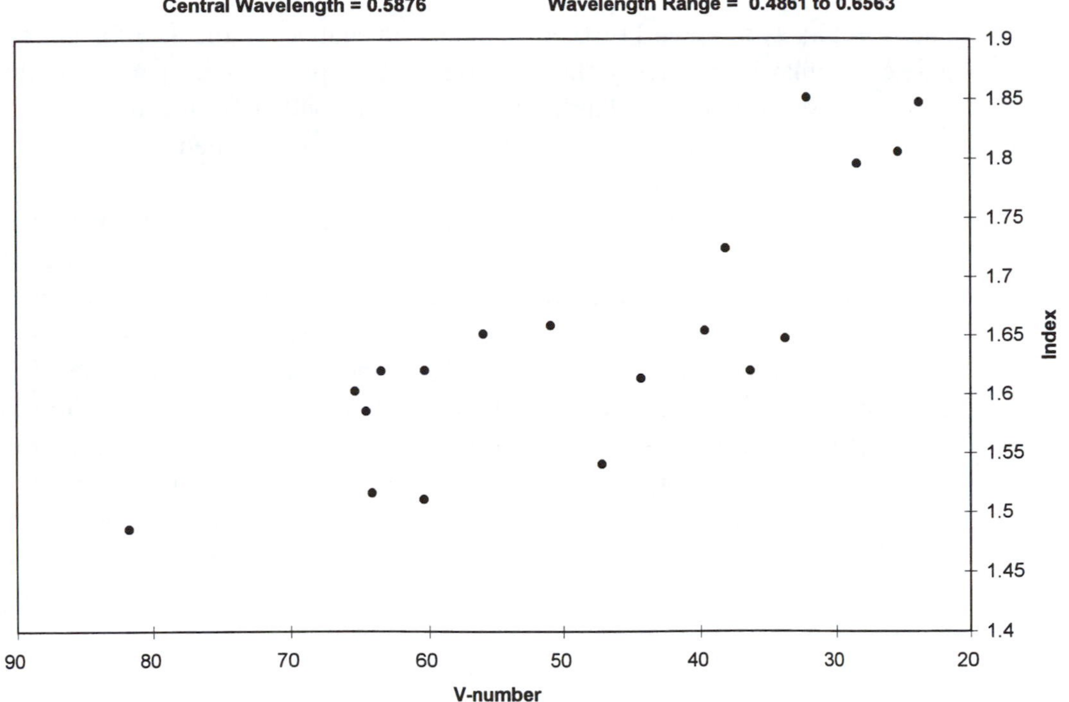

FIGURE 2.54 The index versus V-number plot for glasses in Figure 2.53.

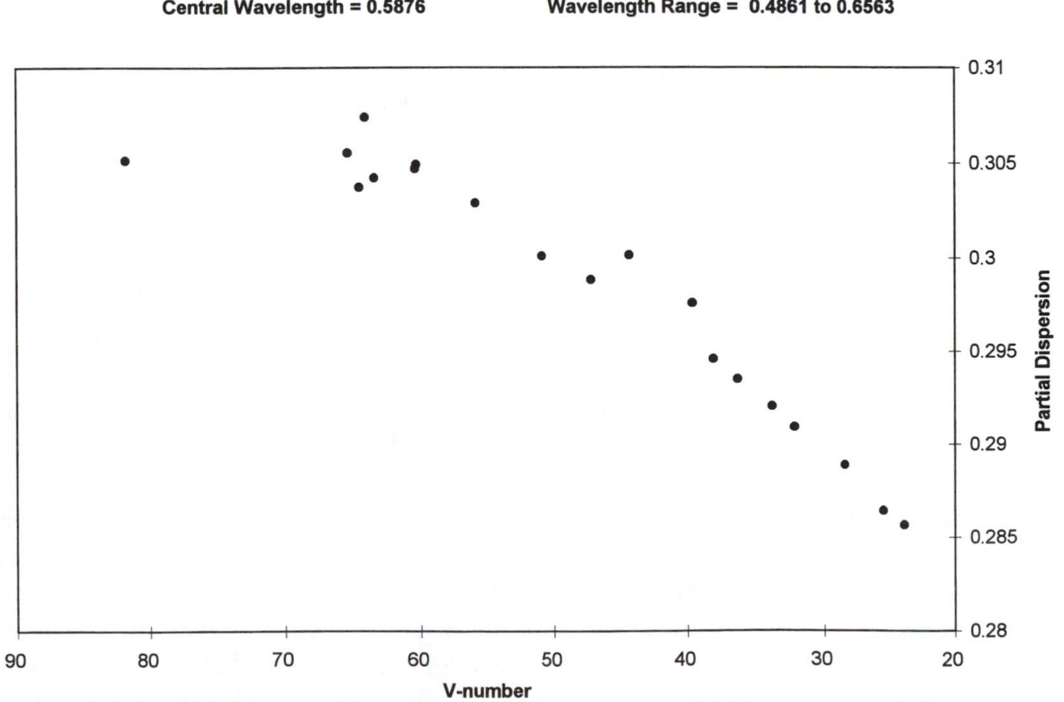

FIGURE 2.55 Partial dispersion versus V-number for the glasses represented in Figure 2.53.

and is not likely to spread further. Incremental changes in the list of available glasses, primarily involving the improvement of physical properties of the glasses, have been made. The limitations are provided by the difficulty of producing practical, economic, homogeneous, stress-free melts of glass outside the region of the map as shown.

The large number of glasses shown are not needed today, as the availability of good lens design programs has permitted the substitution of glasses in most cases. The present large number of glasses is a consequence of a tradition in which some designers used certain glasses in an attempt to obtain a patent on a lens that might otherwise infringe upon another design. There is a trend toward reducing the number of glasses that are listed in the catalogs in an effort to bring more economy into the glass supply business. In addition, some glass types have been discontinued in response to new environmental protection laws. There is now an opportunity, offered by using global optimization methods, to explore the possibilities in reducing the number of glasses that are used to a small set of stable, economically viable glass types (Zhang and Shannon 1995).

Another picture of the glass map is obtained by looking at Figure 2.55, the partial dispersion plot. Most of the glasses cluster about a line of "normal dispersion" glasses on the plot, with a few departing from this line, indicating special abnormal dispersive behavior. These special glasses, generally, but not always, indicated by a Kz prefix, are valuable for the reduction of secondary residual chromatic aberration in a lens. The fluorocrown glasses also show a considerable deviation from the normal glass line. A practical apochromat, a lens with reduced residual color, requires the use of one of these abnormal types of glass to be successful. The special glasses are thus of great importance to the optical designer for solving certain problems when designing lenses with very large diameters or large numerical apertures, or both, and those requiring color correction over an extended spectral range. The slope of the normal glass line for the F–C spectral range is approximately 1/2400.

These considerations have so far been for the specific wavelength region between the C and F wavelengths. The basic rules apply to any spectral range. Figures 2.56 to 2.58 are a set of data in which the central wavelength of the chosen spectral region is varied, with the upper and lower wavelengths having the same ratio to the central wavelength as in the standard C–F range. All of the Schott glasses are included in these plots. The distribution of points representing the glasses varies as the spectral region is changed. As the central wavelength becomes longer, with the spectral region moving toward the near infrared, the plots show a progressive change. In a general sense, the points on the index versus V-number plot cluster together and change their pattern, with most of the flint glasses becoming more like crown glasses and vice versa for the crowns. The fluorocrown glasses remain clustered as a group in the low-dispersion region of the plot. (The transmission of the glass is not considered in making these plots, and some of the glasses will not be applicable for the extended wavelength ranges.)

The behavior of the partial dispersion plots also changes, with the "normal line" being less well defined. When a designer is selecting glasses for a spectral region quite different from the conventional spectral region, the roles of the various glasses are different, and the spectral range of interest needs to be examined. The dispersive behavior is only one of the parameters of importance in selecting an optical glass for use in a lens design, however. The transmission has not been considered in the plots just presented. Nor have the mechanical or chemical properties been discussed. Although the optical properties are most important in choosing a glass, the other properties need to be considered in order to be sure that the best possible choice of material has been made.

Figures 2.59, 2.60, and 2.61 are three examples of data sheets from the Schott Glass catalog. The designer should understand the significance of all of the numbers and designations on these pages. The data at the top of the page provides the glass type designation as well as the index and Abbe number for the glass. An international six-digit code is also stated. For the BK7 example it is 517642, the first three digits standing for the index minus one and the last three for the Abbe number expressed without a decimal point. This international glass designation is of use in searching for equivalent materials in various glass catalogs. The partial dispersion is not referred to in this designation.

The left columns of the glass data sheets provide the index of refraction of the glass for a number of standard spectral wavelengths. In many cases, these are not sufficient for the complete analysis of a lens design. Therefore, the index of refraction is also available by use of a dispersion formula whose constants, evaluated for the specific glass, are given in the data sheet. It is these dispersion formula constants that are stored by lens design programs along with the name of the type of glass, so that the designer merely has to designate the glass type and the desired wavelengths, and these are computed by the program and inserted into the data being used for computation on the lens.

The two most commonly used dispersion formulae for representing index of refraction data in the glass catalog are the Schott formula and the Sellmeier formula. The Schott dispersion formula:

$$n^2 - 1 = A_0 + A_1\lambda^2 + A_2\lambda^{-2} + A_3\lambda^{-4} + A_4\lambda^{-6} + A_5\lambda^{-8} \tag{2.165}$$

is an empirical formula, linear in the coefficients, which is easily fitted to a set of measured index of refraction data using least-squares methods. It was introduced by Schott in the late 1950s and has became a de facto standard

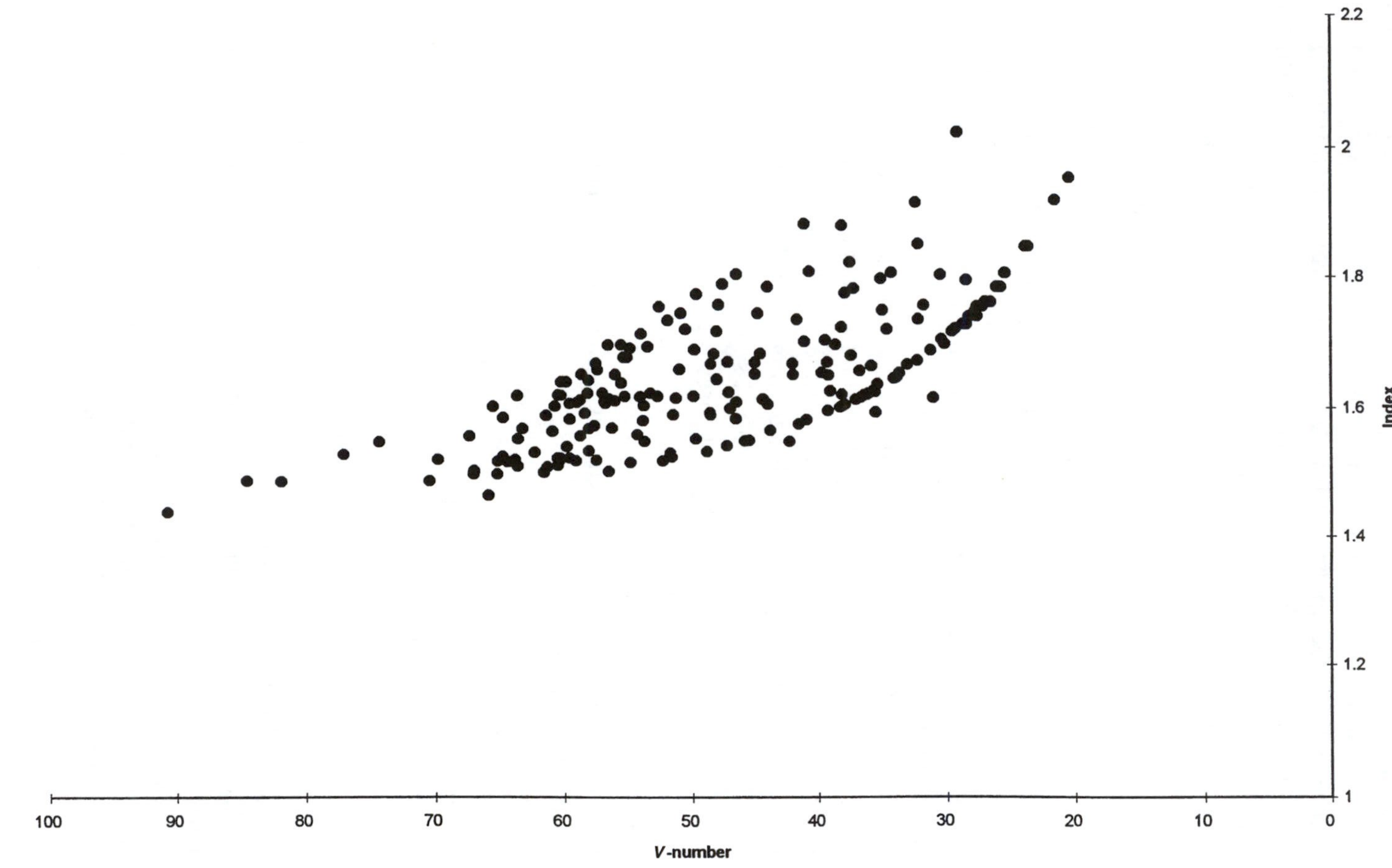

(a)

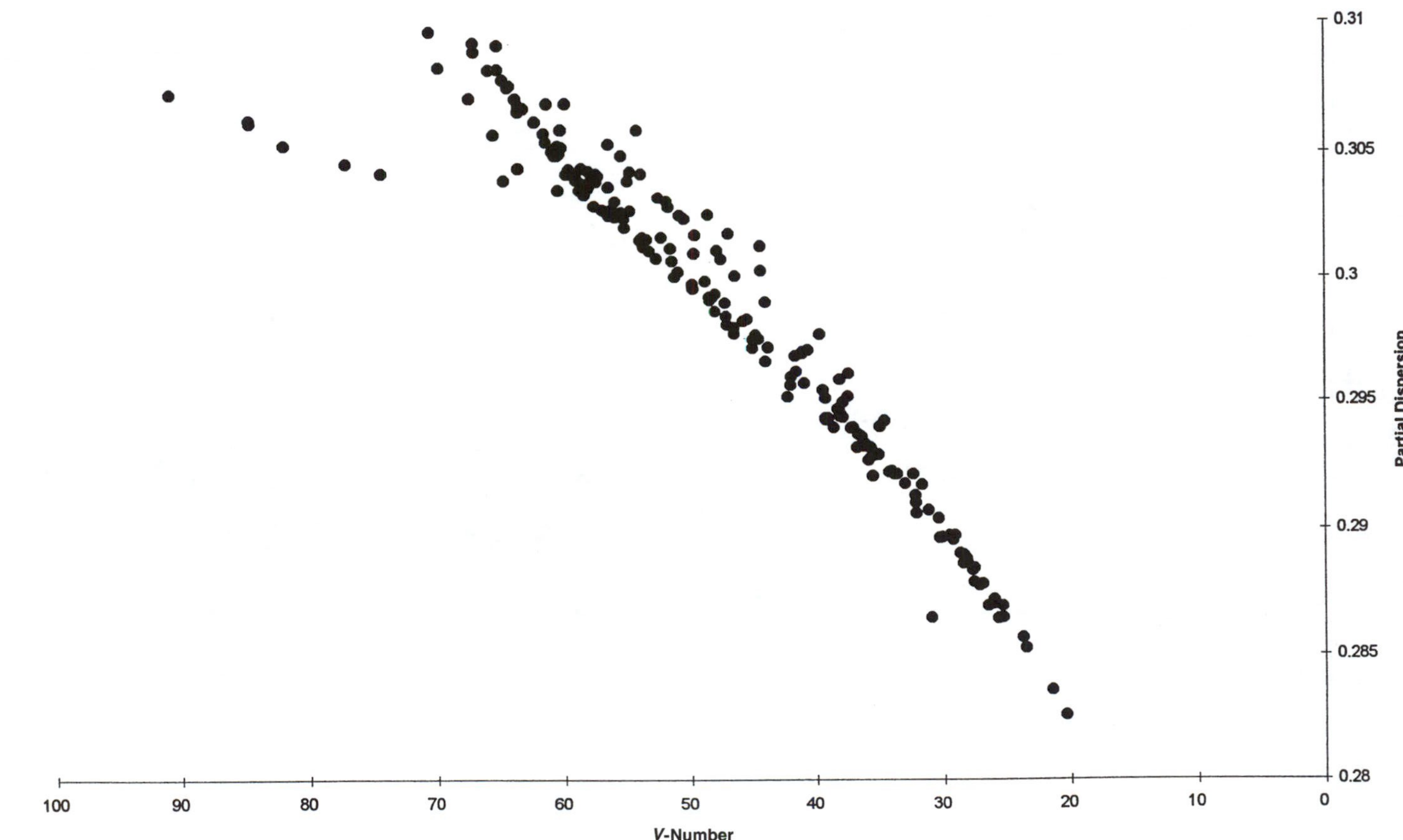

FIGURE 2.56 Glass map for the F–d–C spectral region (486.1–587.6–656.3 nm): (a) index versus V-number; (b) partial dispersion versus V-number.

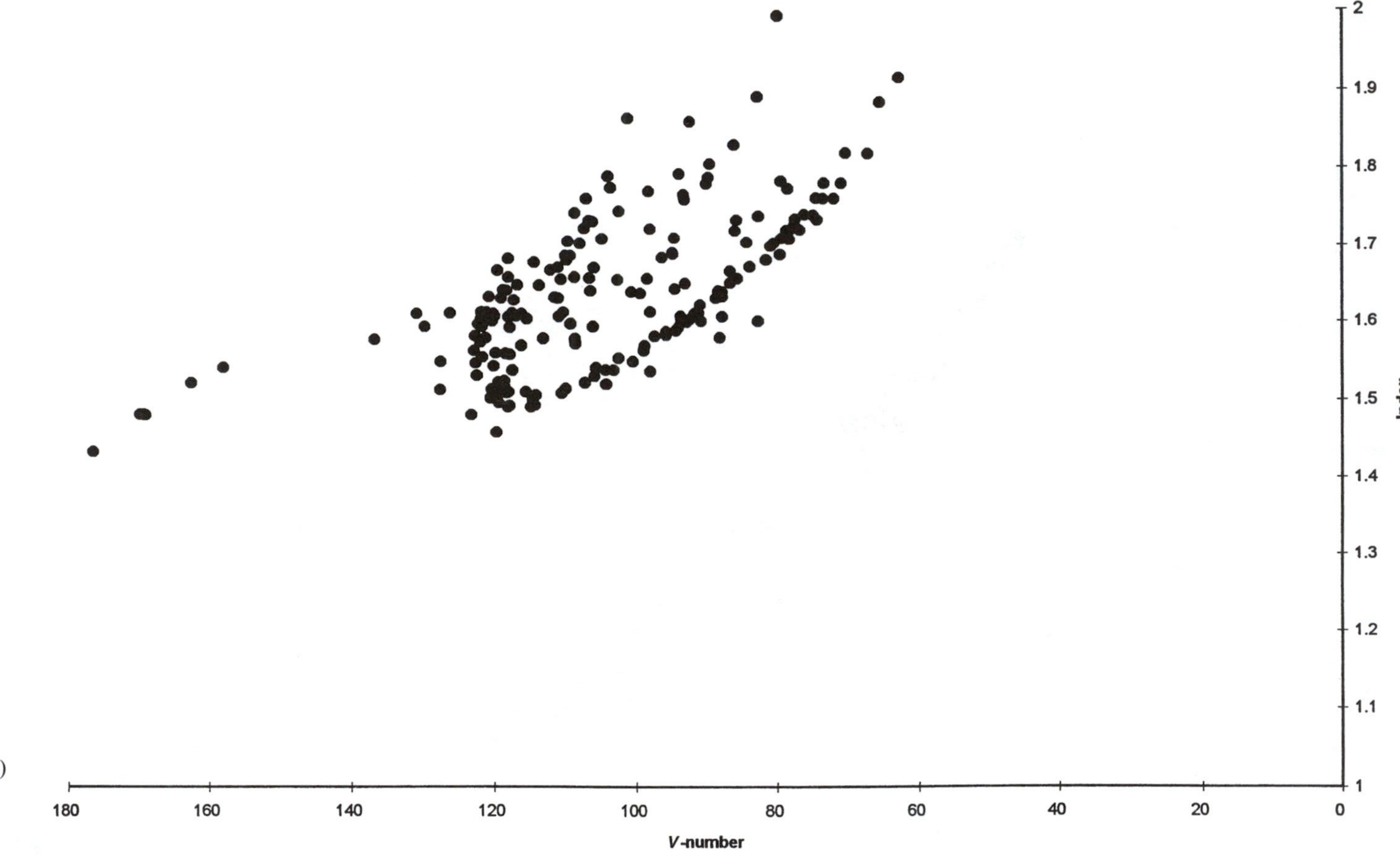
Index
2
1.9
1.8
1.7
1.6
1.5
1.4
1.3
1.2
1.1
1
180
160
140
120
100
80
60
40
20
0
V-number

(a)

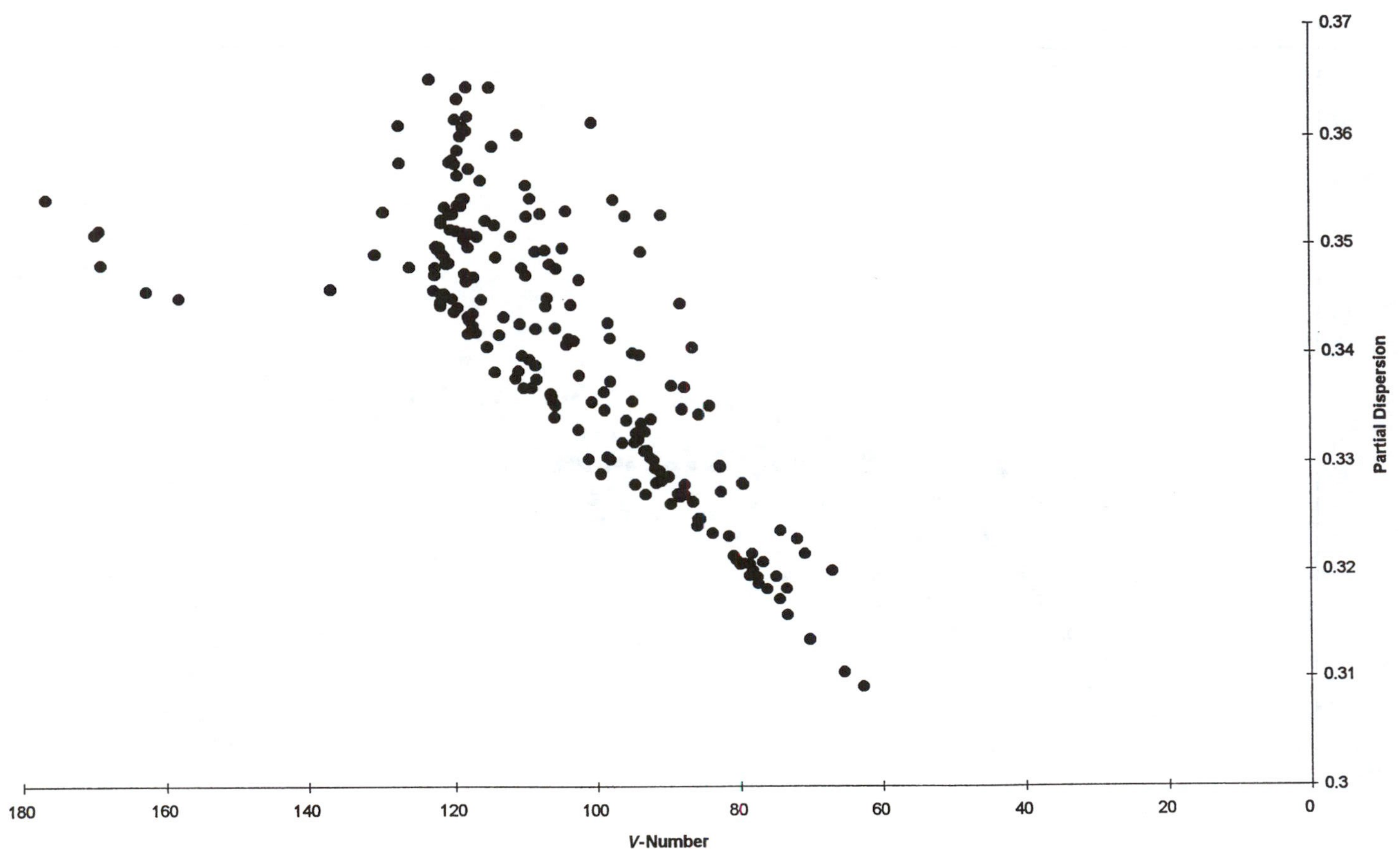

FIGURE 2.57 Glass map for the 827.3–1,000–1,117 nm region: (a) index versus V-number; (b) partial dispersion versus V-number.

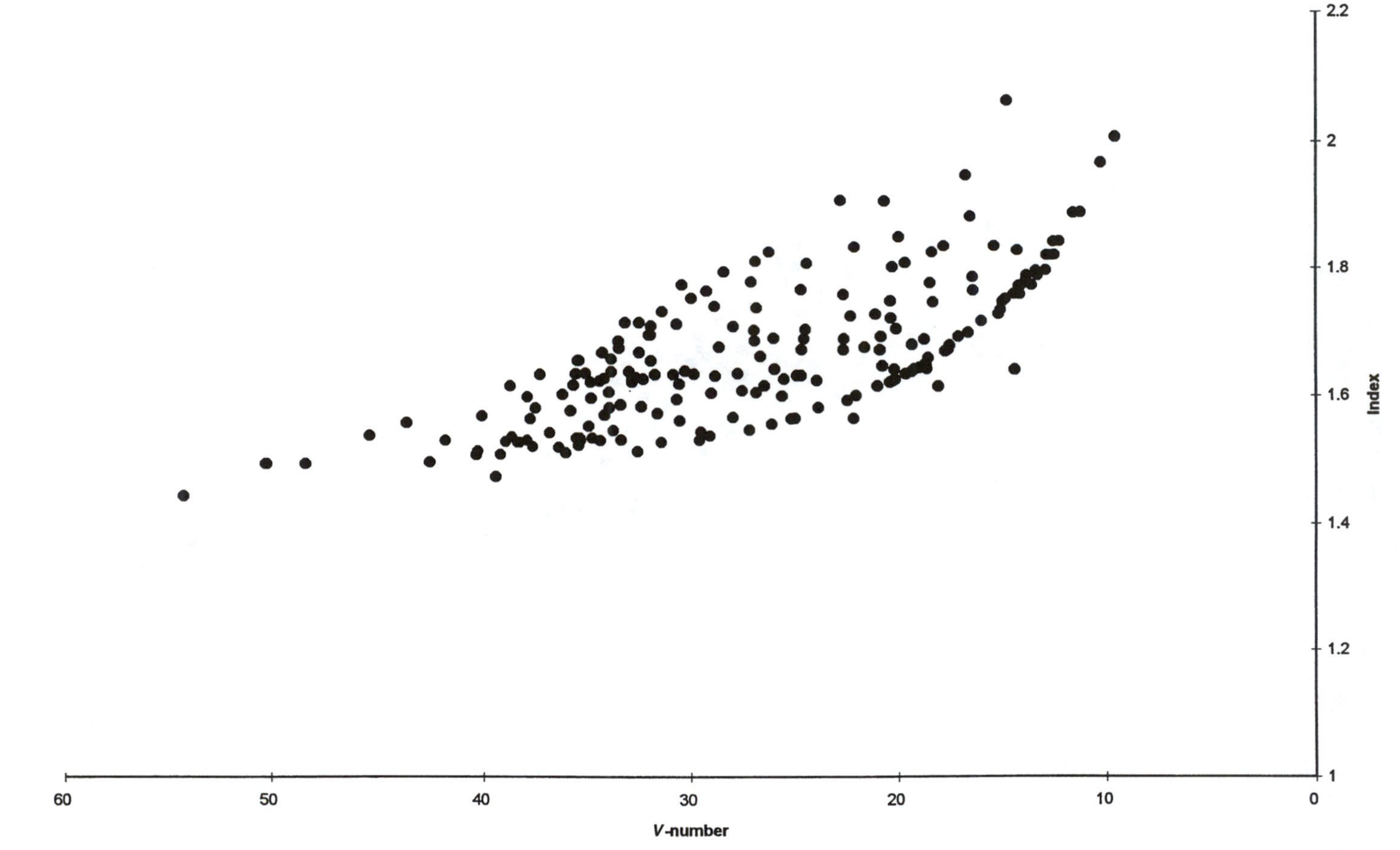

(a)

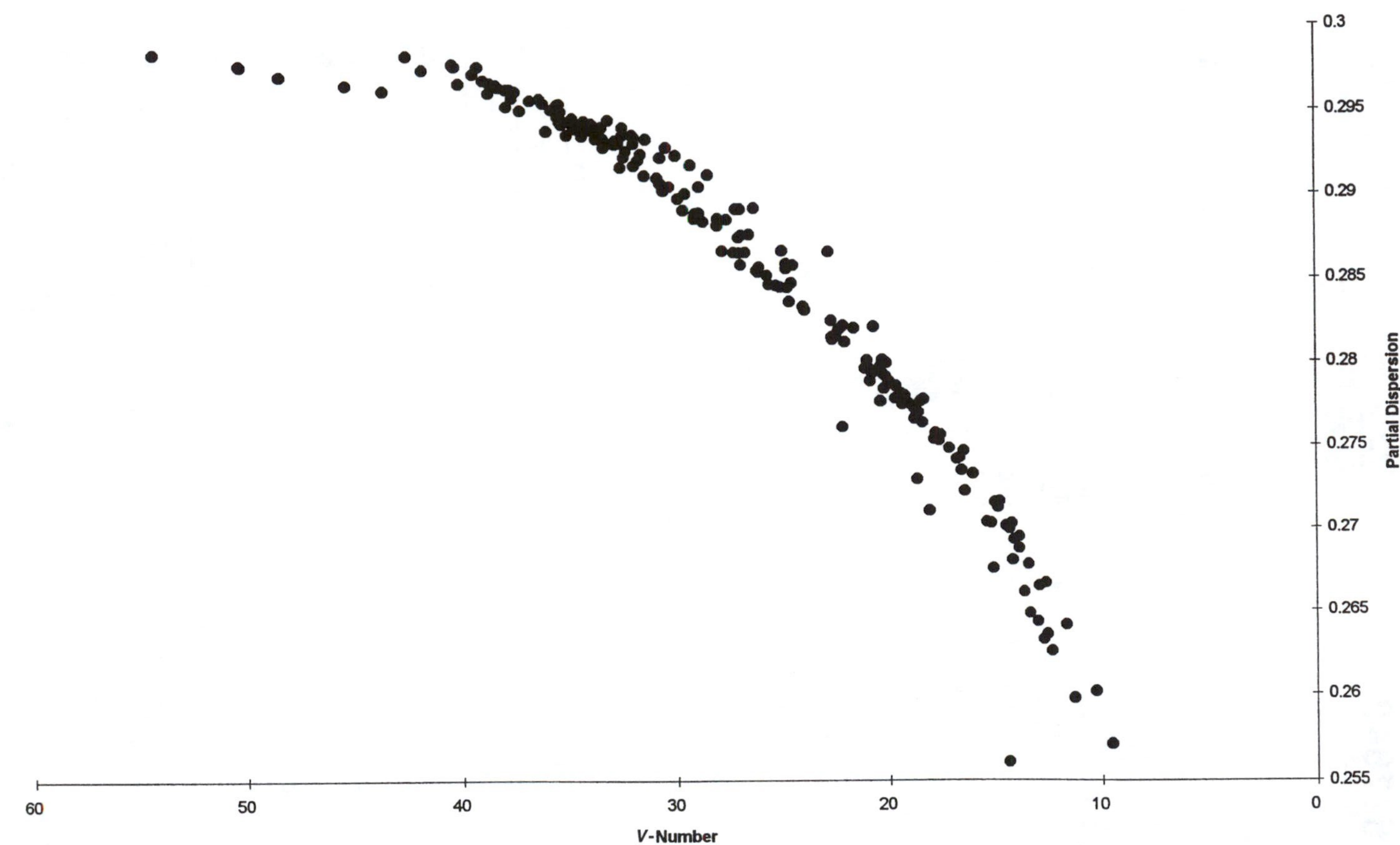

FIGURE 2.58 Glass map for the 372.3–450–502.6 nm spectral region: (a) index versus *V*-number; (b) partial dispersion versus *V*-number.

BK7	**517642**	n_d = 1.51680	ν_d = 64.17	$n_F - n_C$ = 0.008054
		n_e = 1.51872	ν_e = 63.96	$n_{F'} - n_{C'}$ = 0.008110

Refractive Indices		
	λ [nm]	
$n_{2325.4}$	2325.4	1.48921
$n_{1970.1}$	1970.1	1.49495
$n_{1529.6}$	1529.6	1.50091
$n_{1060.0}$	1060.0	1.50669
n_t	1014.0	1.50731
n_s	852.1	1.50980
n_r	706.5	1.51289
n_C	656.3	1.51432
$n_{C'}$	643.8	1.51472
$n_{632.8}$	632.8	1.51509
n_D	589.3	1.51673
n_d	587.6	1.51680
n_e	546.1	1.51872
n_F	486.1	1.52238
$n_{F'}$	480.0	1.52283
n_g	435.8	1.52668
n_h	404.7	1.53024
n_i	365.0	1.53627
$n_{334.1}$	334.1	1.54272
$n_{312.6}$	312.6	1.5486_2
$n_{296.7}$	296.7	
$n_{280.4}$	280.4	
$n_{248.3}$	248.3	

Constants of Dispersion Formula	
B_1	1.03961212
B_2	$2.31792344 \cdot 10^{-1}$
B_3	1.01046945
C_1	$6.00069867 \cdot 10^{-3}$
C_2	$2.00179144 \cdot 10^{-2}$
C_3	$1.03560653 \cdot 10^{2}$

Constants of Formula for dn/dT	
D_0	$1.86 \cdot 10^{-6}$
D_1	$1.31 \cdot 10^{-8}$
D_2	$-1.37 \cdot 10^{-11}$
E_0	$4.34 \cdot 10^{-7}$
E_1	$6.27 \cdot 10^{-10}$
λ_{TK} [μm]	0.170

Internal Transmittance τ_i		
λ [nm]	τ_i (5 mm)	τ_i (25 mm)
2500.0		
2325.4	0.89	0.57
1970.1	0.968	0.85
1529.6	0.997	0.985
1060.0	0.999	0.998
700	0.999	0.998
660	0.999	0.997
620	0.999	0.997
580	0.999	0.996
546.1	0.999	0.996
500	0.999	0.996
460	0.999	0.994
435.8	0.999	0.994
420	0.998	0.993
404.7	0.998	0.993
400	0.998	0.991
390	0.998	0.989
380	0.996	0.980
370	0.995	0.974
365.0	0.994	0.969
350	0.986	0.93
334.1	0.950	0.77
320	0.81	0.35
310	0.59	0.07
300	0.26	
290		
280		
270		
260		
250		

Color Code	
λ_{80}/λ_5	33/30

Remarks	

Relative Partial Dispersion	
$P_{s,t}$	0.3098
$P_{C,s}$	0.5612
$P_{d,C}$	0.3076
$P_{e,d}$	0.2386
$P_{g,F}$	0.5349
$P_{i,h}$	0.7483
$P'_{s,t}$	0.3076
$P'_{C',s}$	0.6062
$P'_{d,C'}$	0.2566
$P'_{e,d}$	0.2370
$P'_{g,F'}$	0.4754
$P'_{i,h}$	0.7432

Deviation of Relative Partial Dispersions ΔP from the "Normal Line"	
$\Delta P_{C,t}$	0.0216
$\Delta P_{C,s}$	0.0087
$\Delta P_{F,e}$	−0.0009
$\Delta P_{g,F}$	−0.0009
$\Delta P_{i,g}$	0.0036

Other Properties	
$\alpha_{-30/+70°C}$ [10^{-6}/K]	7.1
$\alpha_{20/300°C}$ [10^{-6}/K]	8.3
Tg [°C]	557
$T_{10^{13.0}}$ [°C]	557
$T_{10^{7.6}}$ [°C]	719
c_p [J/(g·K)]	0.858
λ [W/(m·K)]	1.114
ρ [g/cm³]	2.51
E [10^3 N/mm²]	82
μ	0.206
K [10^{-6} mm²/N]	2.77
$HK_{0.1/20}$	610
B	0
CR	2
FR	0
SR	1
AR	2.0
PR	2.3

Temperature Coefficients of Refractive Index						
	$\Delta n_{rel}/\Delta T$ [10^{-6}/K]			$\Delta n_{abs}/\Delta T$ [10^{-6}/K]		
[°C]	1060.0	e	g	1060.0	e	g
−40/−20	2.4	2.9	3.3	0.3	0.8	1.2
+20/+40	2.4	3.0	3.5	1.1	1.6	2.1
+60/+80	2.5	3.1	3.7	1.5	2.1	2.7

SCHOTT Optical Glass Nr. 10 000 9/92

FIGURE 2.59 Data sheet for BK7 glass from the Schott Glass Catalog no. 10000 9/92.

SF 2	648339	n_d = 1.64769	ν_d = 33.85	$n_F - n_C$ = 0.019135
		n_e = 1.65222	ν_e = 33.60	$n_{F'} - n_{C'}$ = 0.019412

Refractive Indices		
	λ [nm]	
$n_{2325.4}$	2325.4	1.61003
$n_{1970.1}$	1970.1	1.61494
$n_{1529.6}$	1529.6	1.62055
$n_{1060.0}$	1060.0	1.62766
n_t	1014.0	1.62861
n_s	852.1	1.63289
n_r	706.5	1.63902
n_C	656.3	1.64210
$n_{C'}$	643.8	1.64297
$n_{632.8}$	632.8	1.64379
n_D	589.3	1.64752
n_d	587.6	1.64769
n_e	546.1	1.65222
n_F	486.1	1.66123
$n_{F'}$	480.0	1.66238
n_g	435.8	1.67249
n_h	404.7	1.68233
n_i	365.0	1.70027
$n_{334.1}$	334.1	
$n_{312.6}$	312.6	
$n_{296.7}$	296.7	
$n_{280.4}$	280.4	
$n_{248.3}$	248.3	

Constants of Dispersion Formula	
B_1	1.40301821
B_2	$2.31767504 \cdot 10^{-1}$
B_3	$9.39056586 \cdot 10^{-1}$
C_1	$1.05795466 \cdot 10^{-2}$
C_2	$4.93226978 \cdot 10^{-2}$
C_3	$1.12405955 \cdot 10^{2}$

Constants of Formula for dn/dT	
D_0	$1.10 \cdot 10^{-6}$
D_1	$1.75 \cdot 10^{-8}$
D_2	$-1.29 \cdot 10^{-11}$
E_0	$1.08 \cdot 10^{-6}$
E_1	$1.03 \cdot 10^{-9}$
λ_{TK} [μm]	0.249

Internal Transmittance τ_i		
λ [nm]	τ_i (5 mm)	τ_i (25 mm)
2500.0	.	
2325.4	0.94	0.74
1970.1	0.979	0.90
1529.6	0.999	0.994
1060.0	0.999	0.998
700	0.999	0.999
660	0.999	0.998
620	0.999	0.998
580	0.999	0.998
546.1	0.999	0.998
500	0.999	0.997
460	0.998	0.992
435.8	0.997	0.986
420	0.997	0.983
404.7	0.995	0.973
400	0.994	0.970
390	0.989	0.95
380	0.979	0.90
370	0.965	0.84
365.0	0.954	0.79
350	0.87	0.50
334.1	0.49	0.03
320	0.01	
310		
300		
290		
280		
270		
260		
250		

Color Code	
λ_{80}/λ_5	37/33

Remarks

Relative Partial Dispersion	
$P_{s,t}$	0.2233
$P_{C,s}$	0.4813
$P_{d,C}$	0.2923
$P_{e,d}$	0.2367
$P_{g,F}$	0.5886
$P_{i,h}$	0.9376
$P'_{s,t}$	0.2201
$P'_{C',s}$	0.5196
$P'_{d,C'}$	0.2430
$P'_{e,d}$	0.2334
$P'_{g,F'}$	0.5209
$P'_{i,h}$	0.9242

Deviation of Relative Partial Dispersions ΔP from the "Normal Line"	
$\Delta P_{C,t}$	−0.0010
$\Delta P_{C,s}$	−0.0005
$\Delta P_{F,e}$	0.0004
$\Delta P_{g,F}$	0.0017
$\Delta P_{i,g}$	0.0112

Other Properties	
$\alpha_{-30/+70°C}$ [10^{-6}/K]	8.4
$\alpha_{20/300°C}$ [10^{-6}/K]	9.2
Tg [°C]	446
$T_{10^{13.0}}$ [°C]	428
$T_{10^{7.6}}$ [°C]	600
c_p [J/(g·K)]	0.498
λ [W/(m·K)]	0.735
ρ [g/cm³]	3.86
E [10^3 N/mm²]	55
μ	0.227
K [10^{-6} mm²/N]	2.62
$HK_{0.1/20}$	410
B	1
CR	1
FR	0
SR	2
AR	2.3
PR	2.0

Temperature Coefficients of Refractive Index						
	$\Delta n_{rel}/\Delta T$ [10^{-6}/K]			$\Delta n_{abs}/\Delta T$ [10^{-6}/K]		
[°C]	1060.0	e	g	1060.0	e	g
−40/−20	2.3	4.0	6.0	0.1	1.8	3.7
+20/+40	2.7	4.6	6.9	1.3	3.2	5.4
+60/+80	3.1	5.2	7.6	2.0	4.1	6.4

SCHOTT Optical Glass Nr. 10 000 9/92

FIGURE 2.60 Data sheet for Schott SF2 glass.

FK 52 486818 n_d = 1.48605 ν_d = 81.80 $n_F - n_C$ = 0.005942
n_e = 1.48747 ν_e = 81.40 $n_{F'} - n_{C'}$ = 0.005989

Refractive Indices

	λ [nm]	
$n_{2325.4}$	2325.4	1.46929
$n_{1970.1}$	1970.1	1.47229
$n_{1529.6}$	1529.6	1.47552
$n_{1060.0}$	1060.0	1.47896
n_t	1014.0	1.47936
n_s	852.1	1.48102
n_r	706.5	1.48320
n_C	656.3	1.48424
$n_{C'}$	643.8	1.48453
$n_{632.8}$	632.8	1.48480
n_D	589.3	1.48600
n_d	587.6	1.48605
n_e	546.1	1.48747
n_F	486.1	1.49018
$n_{F'}$	480.0	1.49052
n_g	435.8	1.49338
n_h	404.7	1.49601
n_i	365.0	1.50046
$n_{334.1}$	334.1	1.50519
$n_{312.6}$	312.6	1.5094_8
$n_{296.7}$	296.7	
$n_{280.4}$	280.4	
$n_{248.3}$	248.3	

Constants of Dispersion Formula

B_1	1.01571114
B_2	$1.69998431 \cdot 10^{-1}$
B_3	$8.05158532 \cdot 10^{-1}$
C_1	$5.15884949 \cdot 10^{-3}$
C_2	$1.73380788 \cdot 10^{-2}$
C_3	$1.58603308 \cdot 10^{2}$

Constants of Formula for dn/dT

D_0	$-2.03 \cdot 10^{-5}$
D_1	$-8.12 \cdot 10^{-9}$
D_2	$1.12 \cdot 10^{-11}$
E_0	$4.09 \cdot 10^{-7}$
E_1	$4.22 \cdot 10^{-10}$
λ_{TK} [μm]	0.139

Internal Transmittance τ_i

λ [nm]	τ_i (5 mm)	τ_i (25 mm)
2500.0	0.972	0.87
2325.4	0.981	0.91
1970.1	0.991	0.958
1529.6	0.996	0.981
1060.0	0.997	0.985
700	0.997	0.985
660	0.997	0.985
620	0.997	0.985
580	0.997	0.984
546.1	0.997	0.984
500	0.997	0.983
460	0.996	0.982
435.8	0.996	0.982
420	0.996	0.981
404.7	0.996	0.981
400	0.996	0.981
390	0.996	0.981
380	0.996	0.979
370	0.995	0.973
365.0	0.993	0.967
350	0.977	0.89
334.1	0.950	0.75
320	0.80	0.33
310	0.70	0.17
300	0.58	0.07
290	0.47	0.02
280		
270		
260		
250		

Color Code

λ_{80}/λ_5	33/29

Remarks

	"0"

Relative Partial Dispersion

$P_{s,t}$	0.280_8
$P_{C,s}$	0.540_8
$P_{d,C}$	0.305_4
$P_{e,d}$	0.238_7
$P_{g,F}$	0.538_2
$P_{i,h}$	0.748_5
$P'_{s,t}$	0.278_6
$P'_{C',s}$	0.584_8
$P'_{d,C'}$	0.254_7
$P'_{e,d}$	0.236_9
$P'_{g,F'}$	0.477_8
$P'_{i,h}$	0.742_6

Deviation of Relative Partial Dispersions ΔP from the "Normal Line"

$\Delta P_{C,t}$	-0.111_4
$\Delta P_{C,s}$	-0.052_8
$\Delta P_{F,e}$	0.010_5
$\Delta P_{g,F}$	0.031_9
$\Delta P_{i,g}$	0.153_4

Other Properties

$\alpha_{-30/+70°C}$ [10^{-6}/K]	14.4
$\alpha_{20/300°C}$ [10^{-6}/K]	16.0
Tg [°C]	437
$T_{10^{13.0}}$ [°C]	424
$T_{10^{7.6}}$ [°C]	
c_p [J/(g·K)]	0.716
λ [W/(m·K)]	0.861
ρ [g/cm³]	3.64
E [10^3 N/mm²]	78
μ	0.291
K [10^{-6} mm²/N]	0.78
$HK_{0.1/20}$	400
B	1
CR	2
FR	0
SR	52.3
AR	2.2
PR	4.3

Temperature Coefficients of Refractive Index

	$\Delta n_{rel}/\Delta T$ [10^{-6}/K]			$\Delta n_{abs}/\Delta T$ [10^{-6}/K]		
[°C]	1060.0	e	g	1060.0	e	g
−40/−20	−5.7	−5.3	−5.0	−7.7	−7.4	−7.1
+20/+40	−6.8	−6.4	−6.1	−8.1	−7.7	−7.4
+60/+80	−7.3	−6.9	−6.5	−8.3	−7.9	−7.6

SCHOTT Optical Glass Nr. 10 000 9/92

FIGURE 2.61 Data sheet for Schott FK52 glass.

throughout the industry. In general, this formula is capable of representing the index of refraction to within five in the sixth place following the decimal point over the spectral range 0.386 to 0.750 μm. Outside this range the formula deviates to a larger extent.

The other popular approach to data fitting is based upon a common formula describing the dispersion of uncoupled molecules. This is the Sellmeier formula:

$$n^2 - 1 = \sum_{j=1}^{3} \frac{a_j \lambda^2}{\lambda^2 - b_j} \tag{2.166}$$

which can be mean-square fitted to measured index of refraction data. When three terms, or six constants are used, this formula provides index values that are within 0.5 in the sixth place over the wide spectral range 0.36 to 1.5 μm. The form of the terms of the equation has a physical basis in dispersion theory, and thus fits the glass behavior better over a wider spectral range. Because this contains coefficients that are nonlinear, the fitting is somewhat more complicated, but is very easily achieved with present-day computer power. Recently, Schott has adopted the Sellmeier six-constant formula as the standard for expressing the dispersion curve of their materials (Schott 1992).

There are several other dispersion formulae that have been used historically in an effort to understand the nature of the dispersion of optical materials. The following dispersion representations have mostly fallen into disuse because of the ready availability of the fitted formulae discussed above. These include the Cauchy formula, the Hartmann formula, the Herzberger formula, and a glass description formula by Buchdahl. Each one of these is based upon different presumptions about the dispersive properties of glass, and each has different ranges of validity.

The usual textbook formula is the Cauchy formula:

$$n - 1 = A\left(1 + \frac{B}{\lambda^2}\right) \tag{2.167}$$

which is simple, but of limited accuracy, generally representing the dispersion of glass over a limited range to ±0.005 in the visible. The constants A and B are usually found by curve fitting. Another approach is the empirical Hartmann formula:

$$n = D + \frac{E}{\lambda - F} \tag{2.168}$$

where, again, the different constants are found by curve fitting the data for each glass.

In an attempt to generalize the fitting of index of refraction by a formula, two unusual formulae were developed by Herzberger and Buchdahl. The Herzberger formula,

$$n = a + b\lambda^2 + \frac{c}{\lambda^2 - 0.028} + \frac{d}{(\lambda^2 - 0.028)^2} \tag{2.169}$$

is an attempt to develop the index of refraction in the form of four universal functions that may be summed when weighted by the values of the index of refraction at specific wavelengths to obtain the index at any other wavelength.

Buchdahl examined the range of glass values and developed a fitting function that provided the difference of the index of refraction from a base value as a function of the difference in wavelength from the base value wavelength:

$$\delta N = a_1\omega + a_2\omega^2 + a_3\omega^3$$

where

$$\omega = \frac{\delta\lambda}{1 + \frac{5}{2}\delta\lambda} \tag{2.170}$$

This function can be evaluated by finding coefficients at appropriate wavelengths that sum to the required values. Buchdahl claims a good fit to this formula well into the short- and long-wavelength regions (Buchdahl 1968). Recent work has used this formula as a basis for locating sets of glasses for secondary color correction (Mercado 1992).

The glass data sheets also provide partial dispersion values for a number of spectral regions, as well as a section describing deviations from the normal partial dispersion line. While these numbers are interesting, the designer will find that a graphical representation of the partial dispersion data provides a better basis for selecting likely glasses to work with.

The optical behavior of glasses is different in various spectral regions. The familiar *F–d–C* spectral region plot of glass types for index versus *V*-number and partial dispersion versus *V*-number is shown in Figure 2.56. If the spectral range is changed, with a fixed ratio maintained for the new wavelength replacing the *F*, *d* and *C*, some different relations between glasses emerge. Figure 2.57 shows index and partial dispersion plots for all of the glasses for spectral regions moving through the red region into the near-infrared region. It can be seen that the relation between the glasses changes considerably, and decisions about glass choice for lenses will be changed. Not as evident from these plots is that because all glasses change their relative positions on the plot, some glasses that were "crowns" are now "flints."

Figure 2.58 is a plot for a spectral region toward the blue end of the spectrum. Here there is less perceived change in the index plot, but the partial dispersion plot is now changed. From these plots the designer can note that the basis for the choice of a glass may be different in spectral regions at either end of the visual spectrum. It is obvious that the transmission properties of the glass need to be reviewed as well as the dispersion properties.

Additional information about the refractive properties of the glass is given in the table of changes in refractive index with temperature (Figures 2.59–2.61). Recently, Schott scientists showed that a formula could be developed for predicting the change with temperature for each glass. Two cases are important, the relative index change with temperature, in which the glass is immersed in air, and the absolute index change with temperature, in which the glass is surrounded by vacuum. Note that the effect of temperature changes on the refractive characteristics of a lens must include the effect of expansion on changing the dimensions and surface radii of the lens. Prediction of thermal properties is further complicated if a thermal gradient exists through the lens.

The listing of optical data is completed by a listing of the internal transmission of the glass for several wavelengths. The internal transmission is usually given for glass thicknesses of 10 and 25 mm, and is the bulk transmission after correction for the Fresnel reflection losses at the surfaces. All glasses exhibit a near ultraviolet absorption peak, which is associated with the rapid rise of the index of refraction in the ultraviolet end of the transmitted range. Flint glasses will contain absorptions that extend into the blue end of the spectrum, which can cause some dense flints to appear yellowish. At the infrared end, most glasses transmit, with absorption beginning to become important for wavelengths beyond 1.75 to 2.0 μm.

Other data on the glass data sheets discusses chemical, mechanical, and thermal properties of the glass. These are of importance to optical designers as their lens may eventually be fabricated and used in the real world. Failure to consider some of these properties can lead to significant losses in performance of the lens.

Some of the data on glass is best expressed in various numeric forms. The expansion coefficient, α, is stated in units of inverse degrees, and is generally given as two or more values that cover different temperature ranges. Glasses with a high expansion coefficient must be approached with caution in several regards. Subjecting a high expansion glass to a thermal shock, as often occurs in production process, can cause the glass to fracture. Greatly mismatched expansion coefficients between the lens and its mounting material can cause strain and even breakage under temperature changes. The optical effect of a change in temperature will be induced by both the change of index with temperature and the change in shape of the lens due to temperature gradients. Special attention must be paid to elements that are to be cemented together.

The transformation temperature (T_g) and the temperature at which the glass achieves some stated viscosity are usually given, and provide an idea of the temperature range over which the glass will retain its form without slumping. While this may not seem to be of much importance to a designer, it must be remembered that heating of a completed optical element to receive a dielectric antireflection coating is a normal procedure. Use of a low T_g glass can lead to coating problems.

Other quantities that are stated numerically are specific heat, thermal conductivity, Young's modulus, Poisson's ratio, density, and Knoop hardness. These numbers are usually of interest to people other than the optical designer, but can sometimes be critical in determining the suitability of a glass for an application.

Some optical and chemical properties are not readily specified as numbers. The manufacturers designate classes for these properties and denote these classes in the catalog data. In general, the custom is to use number categories in which the lower the number, the less the property will be a problem. The presence of a high number does not necessarily mean that the selected glass will be inappropriate for any specific application, but it does mean that some discussion or investigation as to the suitability of the glass for the application or environment must be considered.

Bubble class is stated to lie from zero to three, which is a statement of the number of bubbles likely to be found in a cubic centimeter of glass. The value zero means few bubbles, and three corresponds to numerous bubbles. The designer should check this number to be sure that a glass with many bubbles is not specified for a critical application. In many cases the presence of a bubble will be of cosmetic interest only, but in laser systems and elements located close to a focal plane, any bubbles can be disturbing. Bubbles and inclusions are sources of scattering as well as sources of cosmetic concern in product appearance.

The available qualities and forms of supply are of interest. The process of manufacture of optical glass today consists of producing a large volume of glass in a continuous melter or glass tank. The raw materials are dumped in at the top of the back end of the tank and are melted together. The melted glass flows into an accumulator tank where it is stirred and mixed into a homogeneous liquid. From the accumulator the molten glass flows toward the front end of the tank in a temperature controlled stream. The molten glass is tapped off from the bottom of the tank and allowed to flow into molds. For specific precision blanks of glass, a metal or ceramic mold approximating the size of the desired final blank is filled. In other cases, the glass will be allowed to run into a continuous strip, or be cut off into gobs which are allowed to flow into a pressing mold where an approximate final lens shape is produced. The choice of supply is usually made by the fabrication engineer, but consultation with the optical designer is usually beneficial.

The glass is then allowed to cool to a solid state at a fairly rapid rate, which places the glass in a coarse annealed state. The glass blank is then reheated to a point just above the transition temperature and allowed to cool slowly under a specific annealing schedule. The result is a fine annealed blank, which has been cooled sufficiently slowly that the residual stress within the glass is extremely small. For volume production of glass blanks, the molds are rapidly chilled to below the critical temperature and the glass blank ejected from the mold. The blank is dropped onto a continuous belt running through a lehr in which the temperature is reduced along the length of the lehr. The blank is annealed during transit through the lehr, and is air cooled at the exit, inspected and inserted automatically into a shipping container.

The optical parameters are obtained at several stages in the production. An occasional sample of the glass is poured, cooled, and measured for index. For precision blanks, a sample prism will be poured and allowed to proceed through all of the anneal steps along with the blanks to be delivered. The prism is then assumed to be representative of the melt of glass and is measured for refractive index. Usually the prism is polished and measured on a prism spectrometer. In some cases an immersion method of index measurement is used, or the glass may be checked on a commercial refractometer. The prism spectrometer provides the most precise measurements of index, usually with errors of less than ± 0.000005 at several wavelengths. Since the annealing schedule effect on the glass is usually well known, a prediction of the final fine-anneal index from the coarse-anneal stage is possible. For critical high-precision lenses, measurement of the sample that has been through the entire manufacturing process is recommended.

Each batch or production melting of the glass constitutes a melt. It can be presumed that all glass blanks having the same melt number are identical in their properties. Variation in the index of refraction within a given blank is called inhomogeneity, and is minimized by careful mixing of the materials in the melting stage, by the use of well-characterized raw materials, and by careful attention in the cooling and annealing stages. Variations of index of refraction of less than 0.000001 in a glass blank are common. Some glass types are intrinsically capable of better homogeneity than others. For critical applications, reference to the catalog for a list of these glass types is necessary.

One important modification of the glass blank production process is the direct precision molding of finished glass elements. A precision mold made of silicon carbide or silicon nitride is used to form the material from the gob

TABLE 2.3 Optical properties of some plastics

Name	N_d	V_d	dN/dT per °C
Polymethylmethacrylate	1.4918	57.4	−0.000105
Polystyrene	1.5905	30.9	−0.000140
Polycarbonate	1.5855	29.9	−0.000107
Styrene acrylonitrile	1.5674	34.8	−0.000110

state. The mold and glass are cooled below the transition temperature and the glass element then annealed. This is an efficient process for producing elements up to 20 mm diameter or so that require no further processing to be used in several commercial-grade applications. The initial cost of the precision molds and development of the process requires that a very large production volume is required in order to make the cost per element economical. As yet, the process has been only partially successful because of the high entry cost to starting a production line for a specific element.

2.3.2 OTHER TRANSMITTING MATERIALS

Other transmitting materials are used in optics. Crystals and plastics are the most common. Liquids are sometimes used, but they are not of major importance.

Plastic lenses have the great advantage of low density and the potential for inexpensive production by an injection molding process. Relative to glass, plastics have the disadvantage of surface softness, sensitivity to humidity and temperature, and worse homogeneity and scattering. The choice of a plastic element in a lens should not be made lightly, as the startup costs for serial production of molded elements is very significant. The cost advantages of plastic in very high volume production are extremely great. There are only a few useful optical plastics. These are listed in Table 2.3. In addition to the sole use of plastics, hybrid lenses of glass and plastic elements can be designed which achieve some of the environmental stability of lenses that use only glass elements.

The use of plastics in high-quality applications is limited by the achievable parameters in optical, mechanical, and thermal behavior of plastics. The table of plastics shows that only a small number of materials are available with high transmission in the visible spectral region. The materials also will show light scattering much greater than most glasses. Injection molding is a severe environment, so that the index of refraction homogeneity is also considerably poorer than glasses. Plastics show a high variation of index of refraction with temperature and humidity changes, and mechanical instabilities can occur due to the absorption of water from the environment by most optical plastics. The surface of plastic materials, such as polymethylmethacrylate, is quite soft and subject to abrasion. Protection of plastic optics to avoid contact with abrasive materials is quite important.

With all of these problems, the great cost advantages, low density, and design flexibility of plastic optics encourages increasing use of these materials. The possibilities inherent in the production of extreme aspherics and possibly diffractive components in the injection molding process are very high, and plastics will undoubtedly continue to increase in popularity for many applications.

Plastic components can be produced in a wide variety of forms. Once the molds have been made and tested, volume production is extremely economical. Aspheric surfaces in plastics are very common, if the volume justifies the cost of producing the original molds. The addition of a cast plastic aspheric to a spherical glass substrate has been used in volume production of aspheric components for zoom lenses.

Crystals are important for special design applications. A crystal has long-range molecular order that provides the opportunity of the material to exhibit many special characteristics such as birefringence, electrooptical effects, or special environmental interactions. These properties can also be a disadvantage, as birefringence can be a problem in imaging systems.

Some crystals provide transmission over very wide wavelength regions. In particular calcium fluoride (fluorite) provides excellent transmission from 0.135 to 9.4 μm wavelength, along with very low dispersion. Fluorite components are used to provide apochromatic correction in the short-wavelength region.

2.3.3 REFLECTIVE MATERIALS

Reflective materials are, in principle, not of importance in the design process. The function of a reflective material is to hold in place a thin reflective coating that actually serves as the optical element. In practice, the nature of the reflective material becomes very important in setting the tolerances as well as the basic configuration of the system.

Reflective substrates fall into three classes: so-called zero-expansion ceramics, higher-expansion ceramics such as glass and crystal materials, and metallic materials. In most cases the reflective substrate can be opaque to optical wavelengths. For those materials that are transparent, the optical properties are usually poor, with the material being quite inhomogeneous, but accessible to optical measurement of some properties, such as birefrin-

TABLE 2.4 Properties of some reflective substrate materials

Material[a]	Density (g/cm^3)	Expansion ($°C^{-1} \times 10^{-6}$)	Stiffness[b]	Thermal diffusivity[c] (relative to fused silica)
Fused silica	2.2	0.55	1.0	1.0
Pyrex	2.35	3.2	0.91	0.1
ULE	2.21	0.02	1.39	26.8
Zerodur	2.50	0.01	1.68	59.7
Beryllium	1.82	12.4	4.84	1.86
Aluminum	2.70	23.9	0.80	3.14
Silicon carbide	3.1	3.5	3.69	4.32
Super Invar	8.13	0.1	0.53	68.0

[a] ULE is a product of Corning Inc.; Zerodur is a product of Schott Glaswerke. Other materials are generic samples of data.
[b] Relative stiffness = (Young's modulus)/(Density) compared to fused silica.
[c] Thermal diffusivity = (Conductivity)/(Heat capacity × Expansion). Relative values are obtained by calculating a figure of merit; larger is better.

gence, to measure internal stresses. Table 2.4 is a list of some reflective substrate properties.

The frequently used low-expansion materials are ultralow-expansion fused silica (ULE) from Corning Glass and low-expansion ceramic (Zerodur) from Schott Glass. These materials have similar properties by having a temperature range over which the expansion coefficient is substantially zero. These materials can nominally be described by an expansion coefficient of $0.0 \pm 0.03 \times 10^{-6}/°C$ over the temperature range of $-10°C$ to $+40°C$. The materials differ in mode of manufacture and forms in which mirror blanks can be made from the material. The designer is cautioned that the use of these materials in optomechanical design can be complicated by the forms of supply.

Glassy materials with a higher coefficient of expansion are fused silica with a nominal expansion coefficient of $0.63 \times 10^{-6}/°C$ for normal operating temperatures. Fused silica has a significant variation of properties with temperature, and actually shows a zero expansion coefficient at cryogenic temperatures. Fused silica can be obtained with extremely good homogeneity, but the designer is warned that it exhibits an extremely high net thermal index of refraction change with temperature. Glasses with higher expansion coefficients that are commonly used as mirror substrates include Pyrex or

Duran50 low-expansion glasses, with a nominal expansion coefficient of $3.2 \times 10^{-6}/°C$.

2.3.4 DIFFRACTIVE AND HOLOGRAPHIC MATERIALS

In most cases, the diffractive properties of importance are restricted to the surfaces of lens components. The basic diffractive component is a diffraction grating whose period and direction are controlled from point to point on the surface. A ray incident upon the grating will be changed in direction by diffraction according to the basic grating formula

$$nd(\sin I' - \sin I) = m\lambda \qquad (2.171)$$

where n is the refractive index of the medium in which the diffraction takes place, d is the grating period and m is the diffraction order. There is an additional implicit condition regarding the direction of the grating and the direction of the diffracted ray. Detailed discussion of the use of this relation can be found in section 7.2.6, on diffractive optics.

Diffraction is a physical optics, not a geometrical optics, phenomenon. A ray alone does not diffract, but has a direction determined by geometrical optics. The equation above describes the change in direction of an infinitesimal section of a wavefront. The direction of diffraction is determined by the local period. The efficiency of diffraction, or the fraction of light diffracted into a specific order, depends upon the contour of the phase or amplitude variation of the periodic grating. The change in ray direction, or wavefront modification, and the efficiency of the diffractive effect are usually separated in the design process.

The requirement upon any material to be used in diffractive applications is that it support the recording of phase or transmission information with a detail size that is comparable to, or smaller than, the wavelength transmitted through the diffracting structure. Holograms are a common example of a diffractive optical component made using a high-resolution photographic material. Computer-generated holograms are obtained by directly impressing a calculated structure on a recording medium. The recording of steps on the surface produces phase shifts in the reflected or transmitted light. There are many possible methods of encoding the step or phase information on the surface. Some of the most flexible and effective approaches are based upon lithographic techniques used to manufacture microcircuits. Approximation of the desired phase steps by a sequence of binary steps leads to binary optical surfaces, a special case of the possible range of diffractive components.

Recording of the surface relief is possible by photography, in various types of photographic emulsions. Application of photoresist to the surface, with recording of interference patterns from carefully located sources, or by writing the desired phase steps through photomasking permits addition of diffractive phase steps to a reflecting or refracting substrate. Etching of the substrate through a processed resist permits embedding the phase steps on the surface of the substrate. In some cases, diffractive structures in depth, using the Bragg condition to enhance efficiency in a particular diffractive order, are also possible in photosensitive or photorefractive media.

For volume applications the most likely approach is the molding of diffractive surfaces. In this case many replicas of a mold can produce inexpensive hybrids of diffractive and refractive elements. The techniques are similar to the manufacture of compact discs, except that the structures will have large-scale order and may not lie on a flat substrate surface. The efficiency is controlled by continuous or binary shaping of the phase steps within the period of the diffractive structure. Structural dimensions of less than a micron seem to be readily possible in such molding processes. Injection-molded plastics, usually similar to polymethylmethacrylate, seem to be the most likely candidates for these applications.

REFERENCES

Born and Wolf 1959 — Born, M. and Wolf, E. *Principles of Optics*, Pergamon Press, London

Buchdahl 1968 — Buchdahl, H. A. *Optical Aberration Coefficients*, Dover Publications, New York

Buchdahl 1970 — Buchdahl, H. A. *An Introduction to Hamiltonian Optics*, Cambridge University Press

Cheatham 1980 — Cheatham, P. S. Alternate surface intersection points, *Proc. SPIE* 237, 142–7

Fanderlik 1983 — Fanderlik, I. *Optical Properties of Glass*, Elsevier, Amsterdam

Hecht and Zajac 1974 — Hecht, E. and Zajac, A. *Optics*, Addison-Wesley, Reading, MA

Izumitani 1985 — Izumitani, T. S. *Optical Glass*, translated by Livermore, Livermore National Laboratory, Livermore, CA

Kingslake 1978 — Kingslake, R. *Lens Design Fundamentals*, Academic Press, New York

Klein 1970 — Klein, V. *Optics*, Wiley, New York

Mercado 1992 — Mercado, R. I. Design of apochromats and superachromats, *Proc. SPIE* CR41, 270

Schott 1992 — *Schott Optical Glass Catalog*, Schott Glass Technologies, Duryea, PA

Wetherell, 1992	Wetherell, W. B. Afocal optics, Chapter 3 in *Applied Optics and Optical Engineering*, vol 10, Academic Press, New York, pp. 109–92
Zhang and Shannon 1995	Zhang, S. and Shannon, R. Lens design using a minimum number of glasses, *Optical Engineering*, v34, 12, 3536–44

3 ABERRATIONS

3.1 Definition

Aberrations are deviations from the perfect geometrical imaging case. An understanding of the influence and correction of aberrations obviously requires that somewhat more detail be developed.

Ideal image formation requires that the relation between object and image would follow paraxial rules, and all rays from each object point would pass through its paraxial conjugate image point, with all rays having the same optical path length from object to image. The quantitative measure of the aberration at any field location is a spread of ray intercepts on the image plane or an associated optical path difference error evaluated on the exit pupil. The paraxial point is usually taken as the image reference point, and the image errors with real rays measured with respect to this point. Perfect imagery also can be defined as zero wavefront error in which the exiting wavefront coincides with the exit pupil reference sphere. (Additional considerations relate to the distribution of rays in the pupil, as pupil aberrations may indicate a change of aberration with field position, even though the optical path errors are zero at some specified field location.)

The aberration at a given field point produces a distribution of rays about the image reference point, the symmetry of which is determined by the magnitude and combination of aberrations that are present in the lens. In the case of low-order aberrations, such as lateral displacement or a focus position error, the choice of a new reference position can negate the aberration. In the case of lenses containing higher-order aberrations, the introduction of reference point shifts can be used to minimize or balance the effects of the aberration.

The efficiency of optimization against aberration targets is orders of magnitude greater than against direct image quality functions, so that an understanding of these basic contributors to image formation is necessary. In addition, the aberration contributions from individual surfaces can be computed, permitting evaluation of correction possibilities from the various components.

Paraxial image formation requires spherical surfaces to form images. Aberrations result from the failure of optical surfaces to produce a perfect wavefront following the surface. The refraction rule, $n \sin I = n' \sin I'$, is a nonlinear relation. The shape of a spherical surface obviously does not match the default reference coordinates of the image and object surfaces. These two effects provide a nonlinear relation between the incoming and refracted wavefront shapes.

There are several methods of classifying and describing aberrations. The most familiar method is by expressing the aberrations as coefficients of a series of terms of successively higher order in the pupil coordinates. The first-order aberrations are deviations of the paraxial imaging conditions from the perfect base imagery as determined by tracing rays at a central wavelength. The third, fifth, and so on, orders are aberrations which express a deviation from the paraxially determined imaging conditions. The third-order aberrations will be discussed in some detail, with a summary of the properties of the higher-order aberrations. In modern lens design it is most convenient to rely upon ray tracing to provide the majority of the information about the aberration content of the lens, and to use calculated aberrations as a guide to the initial design of the lens.

Classification of the aberrations permits some general statements to be made regarding choices of symmetries in the parameters of the lens and the effect of the aberrations upon image quality. The origin of the aberrations needs to be known to a lens designer, but the computational details required to obtain the aberration are of considerably less interest.

This chapter will discuss the types of aberrations that are commonly encountered in optical systems, classify the symmetries, and establish the basis for setting up merit functions constructed from weighted sums of aberrations to serve as optimization targets for the design. The discussion will concentrate upon rotationally symmetric optical systems. Optical systems that have other symmetries will generally follow the principles of classifications indicated here, but the details will obviously be different.

The aberration present in the intercept of a ray with the image plane can be defined by the coordinate pair

$$\begin{aligned} \epsilon'_x &= x'_k - x'_{k_0} \\ \epsilon'_y &= y'_k - y'_{k_0} \end{aligned} \tag{3.1}$$

where the values of x_k and y_k are ray intercepts for an arbitrary ray, and the zero subscripted coordinates represent the location of the chosen image reference point in the image plane. The image reference point is usually taken to be the paraxial reference location for any field point. The aberrations can be written as functions of the selection of the ray in the pupil for any image height as

$$\begin{aligned} \epsilon'_x &= \epsilon'_x(\rho_x, \rho_y; \eta) \\ \epsilon'_y &= \epsilon'_y(\rho_x, \rho_y; \eta) \end{aligned} \tag{3.2}$$

where the values of ρ and η are scaled representations of the actual height of the ray in the pupil and in the field. The relation to the actual values is

$$\begin{aligned} \rho'_x &= \frac{x'_p}{y'_{parax}} \\ \rho'_y &= \frac{y'_p}{y'_{parax} \cos \bar{U}'} \\ \eta &= \frac{h'}{h'_{parax}} \end{aligned} \tag{3.3}$$

for a rotationally symmetric optical system with the field direction in the y axis direction. The added factor of the cosine of the actual chief ray in the y coordinate corrects the normal anamorphic off-axis pupil dimension to a circular axial reference pupil. The values of ρ are normalized to unity, as is the relative field position η.

The next step will be to examine the nature of aberrations that can be represented in a lens system, to various orders of accuracy.

There are many thorough discussions in the literature of the symmetries inherent in aberrations for a lens system. Especially recommended are analyses by Walter Welford (Welford 1986), H. H. Hopkins (Hopkins 1950), and Hans Buchdahl (Buchdahl 1968). Other books of interest are by G. G. Slyusarev (Slyusarev 1984), Arthur Cox (Cox 1964), and Max Herzberger (Herzberger 1958). In modern optical design, the direct use of ray intercept errors has largely usurped the description of aberrations in terms of aberration coefficients, but a knowledge of aberration symmetry is important in understanding the way in which lenses form images.

Aberrations are introduced by interaction of rays with the surfaces of an optical system. The symmetry of the aberrations is related to the symmetry of the surfaces within the system. In order to examine the basic properties of aberrations, the usual first step is to expand the aberration function as a power series in the aperture coordinates.

The most general aberration polynomial would be of the form

$$W(\rho_x, \rho_y, \eta) = \sum a_{\alpha\beta\gamma} \rho_x^{\alpha} \rho_y^{\beta} \eta^{\gamma} \tag{3.4}$$

as a general power series in the aperture and field. The symmetry of the surfaces in the lens system reduces the number of possible terms by restricting the possible combinations that can exist in the sum. The symmetry case most often encountered is the rotationally symmetric lens system. In this choice of system symmetry, all surfaces are individually rotationally symmetric, spherical, conic, or aspheric, and are aligned with an optical axis, with all poles of the surfaces and all centers of curvature along the axis.

This symmetry imposes restrictions upon the combination of terms in the power series. If the field is zero, then a rotation of the coordinates of the pupil about the axis must leave the aberration unchanged. Thus terms in $\rho^2 = \rho_x^2 + \rho_y^2$ raised to any power are allowed. Similarly, if the aberration of the chief ray, for $\rho = 0$, is considered, then a symmetry must exist for all directions of the field, thus terms in η^2 are allowable. Combinations or products of terms in ρ^2 and η^2 are allowed.

The third allowable symmetry for a rotationally symmetric lens is a bit more difficult to visualize. The symmetry of off-axis images is established by choice of a meridional plane. Conventionally, this is chosen as the y direction. Under this condition, the wavefront error must be symmetric for plus and minus x directions. In addition, if the off-axis object direction is reversed from plus to minus, the wavefront error for a given x and y pupil coordinate must reverse. This is the same as saying that a 180° rotation of the image position about the axis must result in a 180° rotation of the wavefront aberrations in the pupil. The symmetry that permits this to hold is the product $\eta\rho_y$. Changing the sign of both the field position and the y height in the pupil maintains the same sign for the wavefront error.

To make this easier to use, the common procedure is to express the pupil coordinates in polar coordinates. By convention the choice is different from the usual choice for analytic geometry, in that the y axis, or meridional plane, is chosen as the base direction. Thus

$$\begin{aligned} \rho_x &= \rho \sin \psi \\ \rho_y &= \rho \cos \psi \end{aligned} \tag{3.5}$$

which can the be introduced into the power series, and after rearrangement and inclusion of the rotationally symmetric limitations discussed above yields

$$W = \sum a_{\alpha\beta\gamma}\eta^{2\alpha}\rho^{2\beta}(\eta\rho\cos\psi)^{\gamma} \tag{3.6}$$

The indices, temporarily noted here as α, β, and γ, can take on all values from $0, 1, 2, 3, \ldots$, and so on, to include all components in the aberration set. However, the summation indices are only symbolic, and can be combined and rewritten as

$$W(\eta, \rho, \psi) = \sum a_{i,j,k}\eta^i\rho^j \cos^k \psi \tag{3.7}$$

with the conditions that the order $n = i + j$, which defines the net power of the length components can only be $0, 2, 4, 6, \ldots$, as a result of the rotational symmetry conditions. Because the $\cos\psi$ factor is dimensionless, this does not count in determining the order of the aberration. These terms are each referred to as second, fourth, sixth, and so on, orders of wavefront error. Because the ray displacement aberrations are given by the derivative of the wavefront function, the ray aberrations are one order lower, or first-, third-, fifth-, seventh-, and so on, order aberrations. Some traditions refer to the first-order aberration as paraxial or Gaussian, the third-order aberrations as primary aberrations, the fifth-order as secondary aberrations, the seventh as tertiary, and so on, in an obvious pattern.

The units in which the aberrations are to be expressed are usually those of the lens being evaluated. Another convention is commonly used in optical design in which the ray displacement aberrations are referred to in length units that apply to the lens, usually in millimeters, but the wavefront aberrations are stated in units of wavelength. This leads to the definition of the wavefront aberration in wavelength units,

$$W(\eta, \rho, \psi) = \frac{W_g(\eta, \rho, \psi)}{\lambda} = \frac{\mathrm{OPD}(\eta, \rho, \psi)}{\lambda} \tag{3.8}$$

without the subscript denoting a geometrical value using the units of the lens. This introduces the wavelength into the definition of geometrical aberrations in a natural way. Thus, following section 2.1.4,

$$\begin{aligned} \epsilon'_x &= \frac{-\lambda}{n'u' \cos \bar{U}'} \frac{-\partial W(\eta, \rho, \psi)}{\partial \rho'_x} \\ \epsilon'_y &= \frac{-\lambda}{n'u' \cos^2 \bar{U}'} \frac{-\partial W(\eta, \rho, \psi)}{\partial \rho'_y} \end{aligned} \tag{3.9}$$

This set of equations relates the aberration polynomial evaluated in the exit pupil to the ray displacement on the image surface. The investigation of the symmetries in wavefront and ray aberrations can be carried out without regard to the actual values, which are obtained by including the numerical aperture, $n'u'$, and the obliquity and necessary pupil anamorphism. The actual values for the lens can then be obtained from this scaling.

The aberrations in the polynomial for each order of interest can be obtained by writing out the first few terms of the series. Reality in describing the aberrations of lenses leads to requiring an investigation of orders up to six in the wavefront polynomial. This will, of course, include fifth-order ray aberrations.

For order $n = 0$,

$$W = a_{000} \tag{3.10}$$

which is obviously a constant added to the wavefront. This adds a constant path error across the field of the lens, and produces no ray displacement error. Usually, this term is ignored, but can be an issue for coherent systems.

For order $n = 2$,

$$W = a_{020}\rho^2 + a_{111}\eta\rho\cos\psi + a_{200}\eta^2$$
$$= a_{020}(\rho_x^2 + \rho_y^2) + a_{111}\eta\rho_y + a_{200}\eta^2 \quad (3.11)$$

The last term, in η^2, contains no pupil coordinates, and thus produces no ray displacement error, only a quadratic base phase as a function of the square of the field. For normal incoherent imagery this term can be ignored. The first term is a focus shift, and the second indicates an error in magnification, or a tilt in the wavefront which varies linearly with field. This indicates a lateral shift which varies with field height.

For order $n = 4$,

$$w = a_{040}\rho^4 + a_{131}\eta\rho^3\cos\psi + a_{220}\eta^2\rho^2$$
$$+ a_{222}\eta^2\rho^2\cos^2\psi + a_{311}\eta^3\rho\cos\psi + a_{400}\eta^4 \quad (3.12)$$

This was obtained by writing out all products of the allowable symmetry terms that have length powers that sum to four. It is presumed that these aberrations are referred to the paraxial image reference points. These are the aberrations that would be perceived if the lower-order terms are zero.

For order $n = 6$, the same symmetries are used, with the assumption that all lower-order aberrations are zero. This leads to the polynomial

$$W = a_{060}\rho^6 + a_{151}\eta\rho^5\cos\psi$$
$$+ a_{240}\eta^2\rho^4 + a_{242}\eta^2\rho^4\cos^2\psi + a_{331}\eta^3\rho^3\cos\psi + a_{333}\eta^3\rho^3\cos^3\psi$$
$$+ a_{420}\eta^4\rho^2 + a_{422}\eta^4\rho^2\cos^2\psi + a_{511}\eta^5\rho\cos\psi + a_{600}\eta^6 \quad (3.13)$$

This equation was obtained by writing out the allowable terms to order six, and then grouping the terms in order of decreasing field dependence. As before, these terms will appear as shown in the image formed by a lens only if all of the lower-order aberration terms are zero.

Each successively higher-order aberration polynomial and set of coefficients can be developed by applying the symmetry rules. At each order of aberration, symmetries that are the same as the lower order will be obtained, with a few new higher-order types of aberration. An understanding of the source of aberrations through the sixth order is generally adequate for understanding most lens systems. The actual appearance of the ray aberrations will change due to the pupil scaling and anamorphism, the obliquity of the imaging ray bundle against the image surface and the possibility of aberrations of the pupil which redistribute the ray positions in the exit pupil. Discussions in section 2.1.4 indicate that these effects will be reasonably small for any realistic imaging system.

For lenses having nonrotational symmetry, such as those with cylinders or toroids, or that have noncentered components, additional terms will appear within each order of aberration. These terms can be dominant or small, depending upon the strength of asymmetry introduced. For example, the small tilts and decenters of elements in an assembled lens will add in some aberrations of the types included above, but with field dependencies and orientations that are related to the asymmetry that has been introduced into the lens. Strong asymmetries, such as cylindric components will add new types of aberrations. In most cases, these aberrations will appear somewhat similar to those about to be discussed, but some very different forms may emerge in specific conditions.

When designing a lens, the goal is minimization of a merit function which includes knowledge of the likely aberration content as well as the image quality. Understanding of the aberrations in both ray and wavefront symmetries will permit a logical choice for the components of the merit function.

3.1.1 LOW-ORDER ABERRATIONS

3.1.1.1 First-Order Aberrations

For order $n = 2$, the aberration polynomial is

$$\begin{aligned} W &= a_{020}\rho^2 + a_{111}\eta\rho\cos\psi + a_{200}\eta^2 \\ &= a_{020}(\rho_x^2 + \rho_y^2) + a_{111}\eta\rho_y + a_{200}\eta^2 \end{aligned} \tag{3.14}$$

The first term is a focus shift, and the second indicates an error in magnification, or a tilt in the wavefront which varies linearly with field. This indicates a lateral shift which varies with field height.

The last term, in η^2, contains no pupil coordinates, and thus produces no ray displacement error, only a quadratic base phase as a function of the square of the field. For systems such as data processors that use coherent illumination across the field, this quadratic phase term can be important. Because the interest at this point is in aberrations that produce a ray error, this coefficient can be ignored.

3.1.1.1.1 FOCUS AND LATERAL SHIFT

The computation of the ray displacement aberration is carried out by differentiation of the wavefront error. Ignoring the pupil anamorphism and obliquity scale factors discussed above, the ray displacement error is

$$\begin{aligned} \epsilon_x &= \frac{\lambda}{n'u'} 2a_{020}\rho_x \\ \epsilon_y &= \frac{\lambda}{n'u'} 2a_{020}\rho_y + \frac{\lambda}{n'u'} a_{111}\eta \end{aligned} \tag{3.15}$$

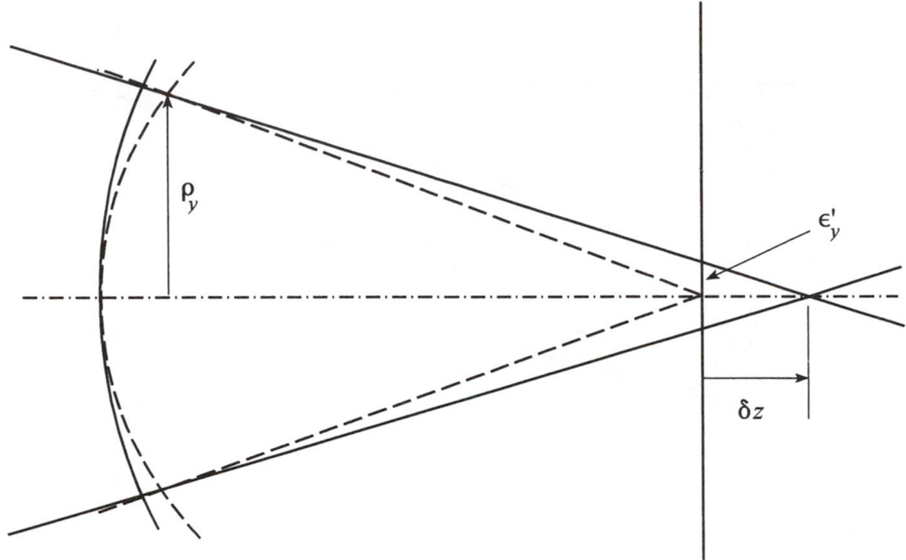

FIGURE 3.1 Important parameters for relating the OPD to the focus error.

Consider first the terms in a_{020}. In the region of the focal location, the relation between the axial intercept distance and the ray aberration for a specific ray, say the ith ray, is given by

$$\epsilon_i' = -\delta_{z_i} u_i' = -\delta_{z_i} \rho_i u' \tag{3.16}$$

where u' is the paraxial marginal ray angle from the edge of the pupil. Since both equations are linear in ρ, the result is

$$\frac{\lambda}{n'u'} 2a_{020} = -\delta_z u' \tag{3.17}$$

which can be rearranged into the more convenient forms

$$\delta_z = -\frac{1}{u'} \frac{\lambda}{n'u'} 2a_{020} = \frac{-8\lambda a_{020}(\mathrm{F/number})^2}{n'} \tag{3.18}$$

relating the wavefront defocus aberration and the amount of shift of focal position. By this formula, a negative OPD leads to a positive δ. Figure 3.1 shows this effect. Assurance that the sign is correct can be found by returning to the original definition of OPD as the reference ray path minus the path along the ray of interest. The diagram shows a case in which the wavefront phase front lags the reference sphere, and thus focuses to the right or positive direction from the paraxial reference location.

A similar argument applies to the second term in equation 3.15. The lateral shift wavefront error term indicates a lateral position error of the image as shown in Figure 3.2. Here the presence of a lateral direction error in the

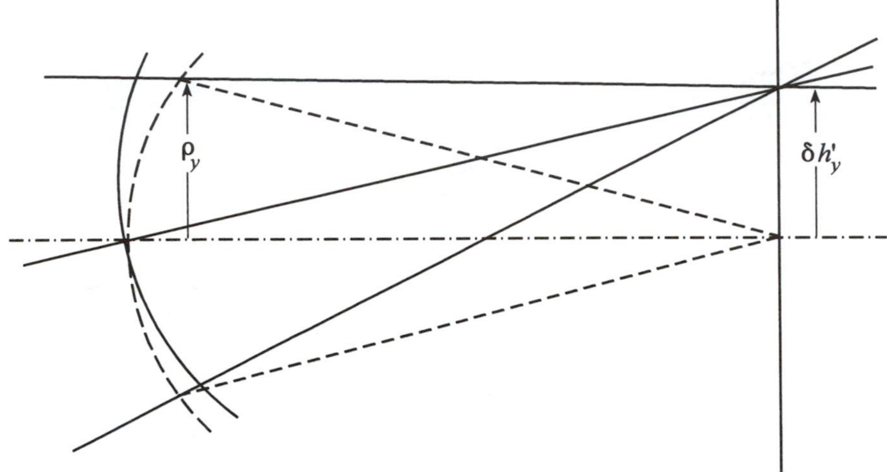

FIGURE 3.2 Parameters used in relating the OPD to a lateral shift from the reference position.

central or chief ray produces a positive lateral shift of the position of the image. The amount of shift for $\eta = 1$, or the edge of the field, is given by

$$\delta h_y = \frac{-\lambda}{n'u'}(-\alpha_{111}\eta) \tag{3.19}$$

In Figure 3.2, the wavefront will lag the reference sphere at the top of the pupil, and lead the reference sphere at the bottom of the pupil. This will be a negative OPD value at the top of the pupil. In the imaging case shown the sign of u' is negative, so that the sign of the shift of the image position is positive. The amount of the shift from the paraxial image location varies linearly with the image height. This indicates that the result of this first-order error is a change in magnification, or image scale, from the paraxially determined image scale.

It is necessary to introduce the concept of compensating aberrations for these first-order ray aberrations. Defocus adjustments a_d and lateral shift adjustments a_l can be introduced, which provide aberrations of

$$\begin{aligned} W_d &= a_d\rho^2 \\ W_l &= a_l\rho\cos\psi \end{aligned} \tag{3.20}$$

It is readily apparent that introduction of these compensating aberrations will cancel the effect of the first-order aberrations at any specified field point. These compensating aberrations amount to a relocation of the focus position and a lateral shift of the image reference point. The appearance of the aberration and the effect of the aberration upon the image quality will be altered

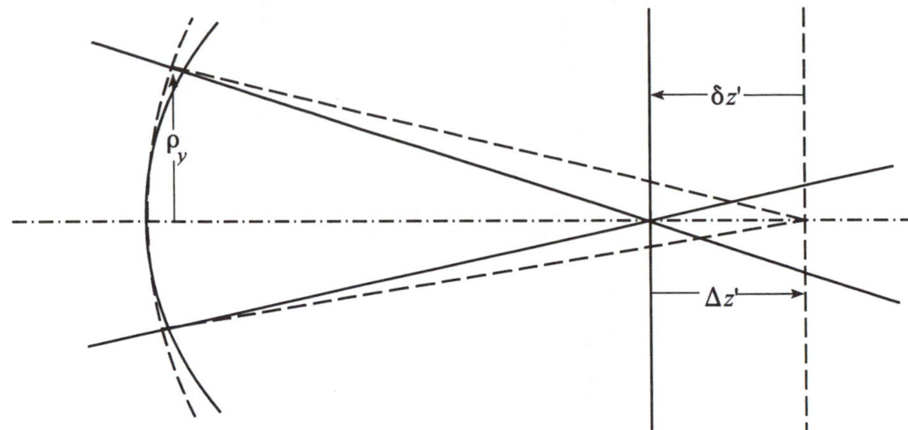

FIGURE 3.3 Effect on the OPD of introducing a reference position shift to compensate a focal error.

by these shifts. The basic nature of the aberration content of the lens is not altered, just viewed with respect to a new reference location.

Relating these quantities to real world values requires application of the scaling values from equation 3.15. Ignoring the effect of the obliquity of the chief ray, but including the paraxial magnification to the image plane,

$$a_d = \frac{n'\delta_z}{8\lambda(\text{F/number})^2}$$

$$a_l = \frac{n'\delta_y}{2\lambda(\text{F/number})} \tag{3.21}$$

The signs of these values can be understood by reference to Figures 3.2 and 3.3. If the original reference point is selected at the paraxial image location, then the introduction of a reference position shift introduces a compensating aberration. For example, if the value of a_{020} is calculated, and is found to be negative, the wavefront aberration added by shifting the reference point to the compensating location should be positive.

Table 3.1 provides a set of sample values for the coordinate shift in wavelength units for different F/number values. The index of refraction of the medium is presumed to be unity; conversion for an immersed image plane is obvious. Sometimes the signs of the relations between the wavefront aberrations and the ray aberrations are not obvious. When a question arises, it is usually beneficial to verify the signs by use of simple diagrams, such as Figure 3.2.

The presence of both a focus and a lateral shift will produce image position changes as indicated in Figure 3.4. Adjustment of focus position and image reference location can compensate the effect of these aberrations. As this

TABLE 3.1 Focus change and lateral shift for one-wavelength error

	Focus change (mm) for wavelength (μm) of				Lateral shift (mm) for wavelength (μm) of			
F/no.	**0.4**	**0.5876**	**0.6328**	**1.0**	**0.4**	**0.5876**	**0.6328**	**1.0**
10	0.3200	0.4701	0.5062	0.8000	0.0080	0.0118	0.0127	0.0200
5	0.0800	0.1175	0.1266	0.2000	0.0040	0.0059	0.0063	0.0100
2.5	0.0200	0.0294	0.0316	0.0500	0.0020	0.0029	0.0032	0.0050
2	0.0128	0.0188	0.0202	0.0320	0.0016	0.0024	0.0025	0.0040
1.5	0.0072	0.0106	0.0114	0.0180	0.0012	0.0018	0.0019	0.0030
1	0.0032	0.0047	0.0051	0.0080	0.0008	0.0012	0.0013	0.0020

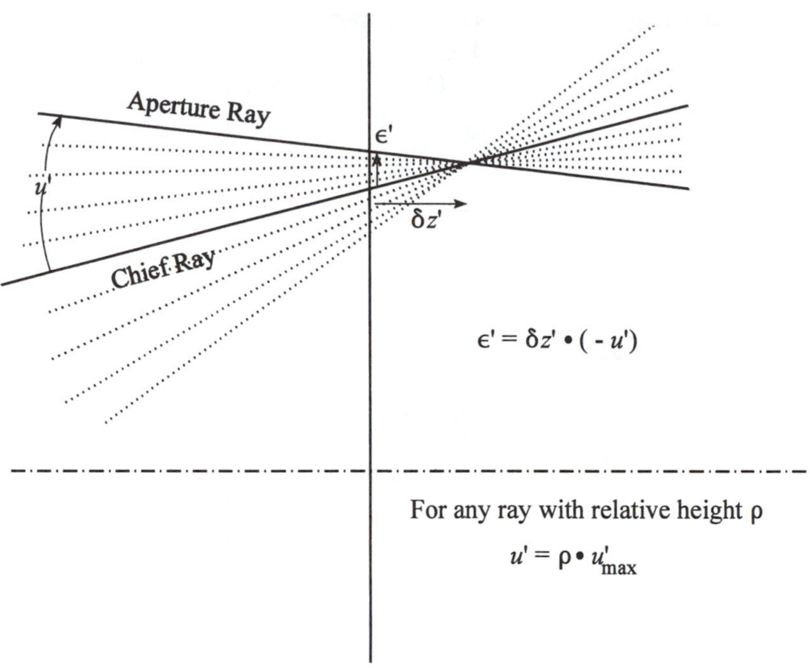

FIGURE 3.4 Combined lateral and longitudinal reference position shift.

diagram shows, focus adjustment may also change the lateral location of the chief ray intercept on the image surface, and can introduce a shift in apparent image scale which depends upon the chief ray angle. The interpretation of focus shift effects on image location needs to be considered carefully for each case of interest.

The presence of focus or lateral image position errors in a lens can be assessed from the ray intercept plots that are computed when analyzing a lens. Figure 3.5 shows a sample ray intercept plot for a lens in which the focal plane is shifted from the reference focus location. The plots are for axis, 0.7, and 1.0 (full field). The slopes of the ray intercept plots are similar over the field, indicating that the amount of focus error is constant with field. Figure 3.6 is a plot of the same focus shift but in this case the OPD has been plotted. The curves indicate the quadratic shape due to a focus wavefront error. Here a positive focus OPD is related to a negative shift of the image from the image reference location. Compensation of the focal shift would require introduction of a negative focus OPD. The dual nature of a focus shift as an aberration or a balancing change in reference position should be clear from these plots.

Figure 3.7 is a plot showing a magnification error in setting the paraxial reference height. The ray intercept plots are horizontal, indicating no aberra-

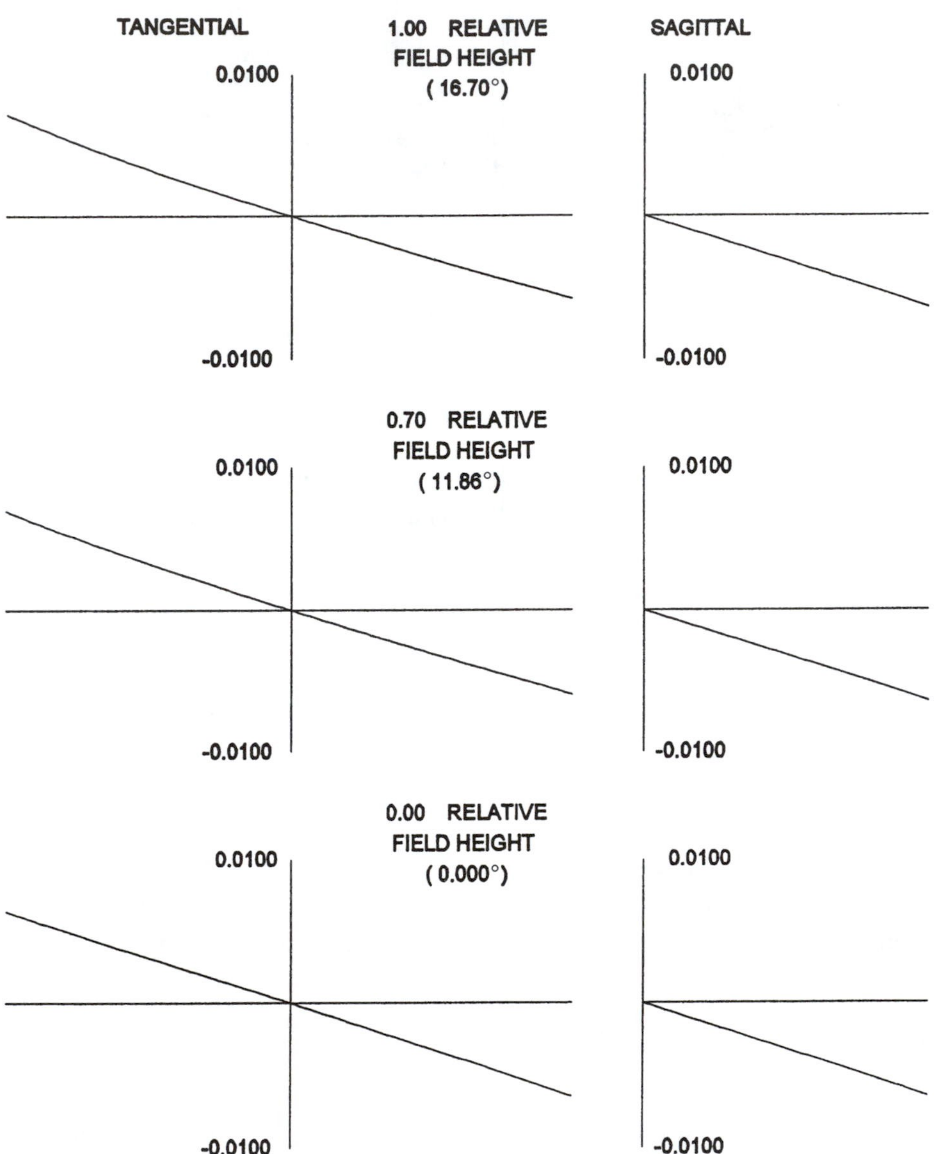

FIGURE 3.5 Appearance of the ray intercept plot due to a focus shift of the reference image surface from an otherwise perfect imaging situation (wavelength 546 nm).

tion blur, but a displacement from the reference location. The amount of the separation from the reference location varies linearly with height. Elimination of this particular aberration requires a lateral shift of the reference position at each image height, easily seen in this case of a magnification error. The reader should note that most lens design programs use the intercept height of the real chief ray in the reference wavelength as the reference

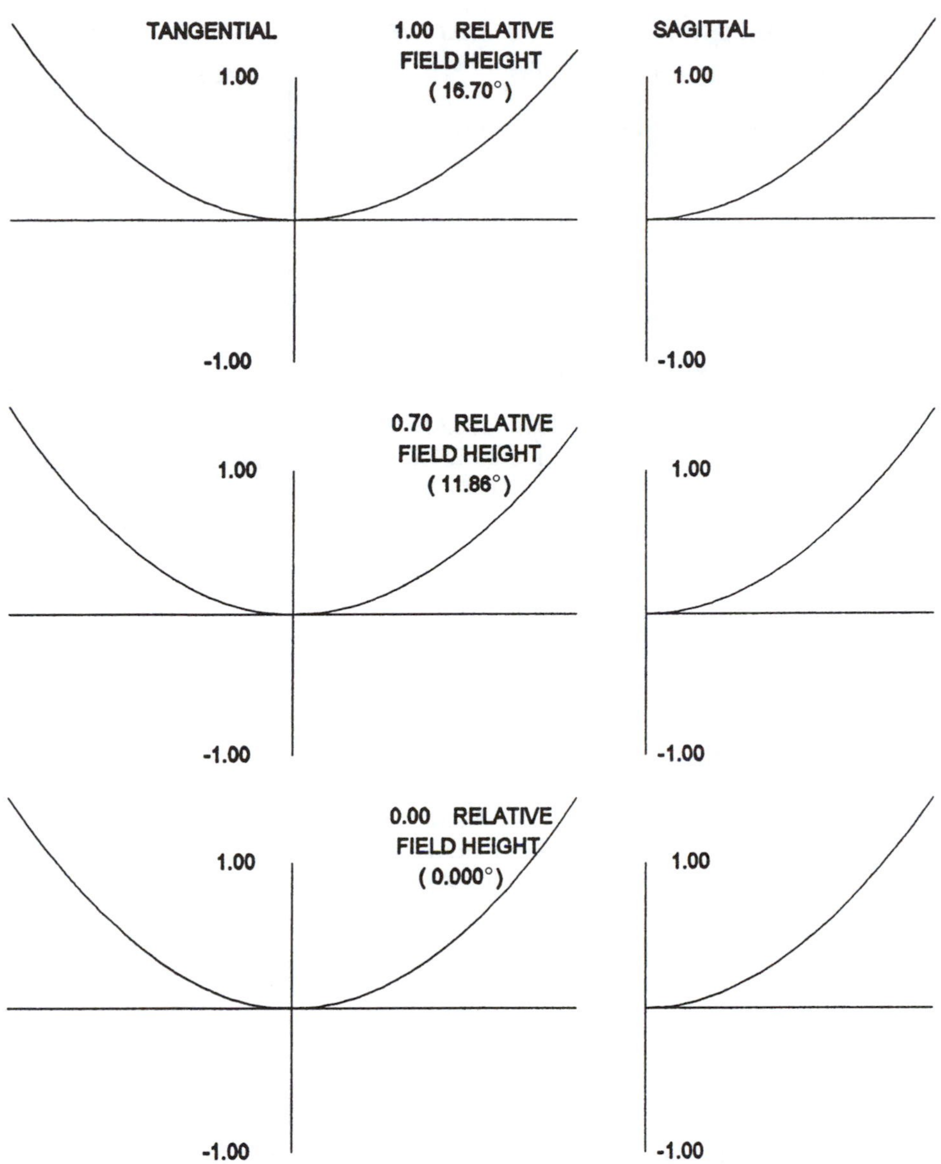

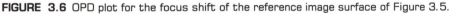

FIGURE 3.6 OPD plot for the focus shift of the reference image surface of Figure 3.5.

location for the ray intercept plot. Therefore in most cases this particular error would not show up on the default ray intercept plot.

If other aberrations are present, addition of these focus and displacement terms to the aberration equation permits computation of the appearance of the image including the effects of reference position shifts. It is not possible to cancel or eliminate the effect of aberrations beyond the first order by intro-

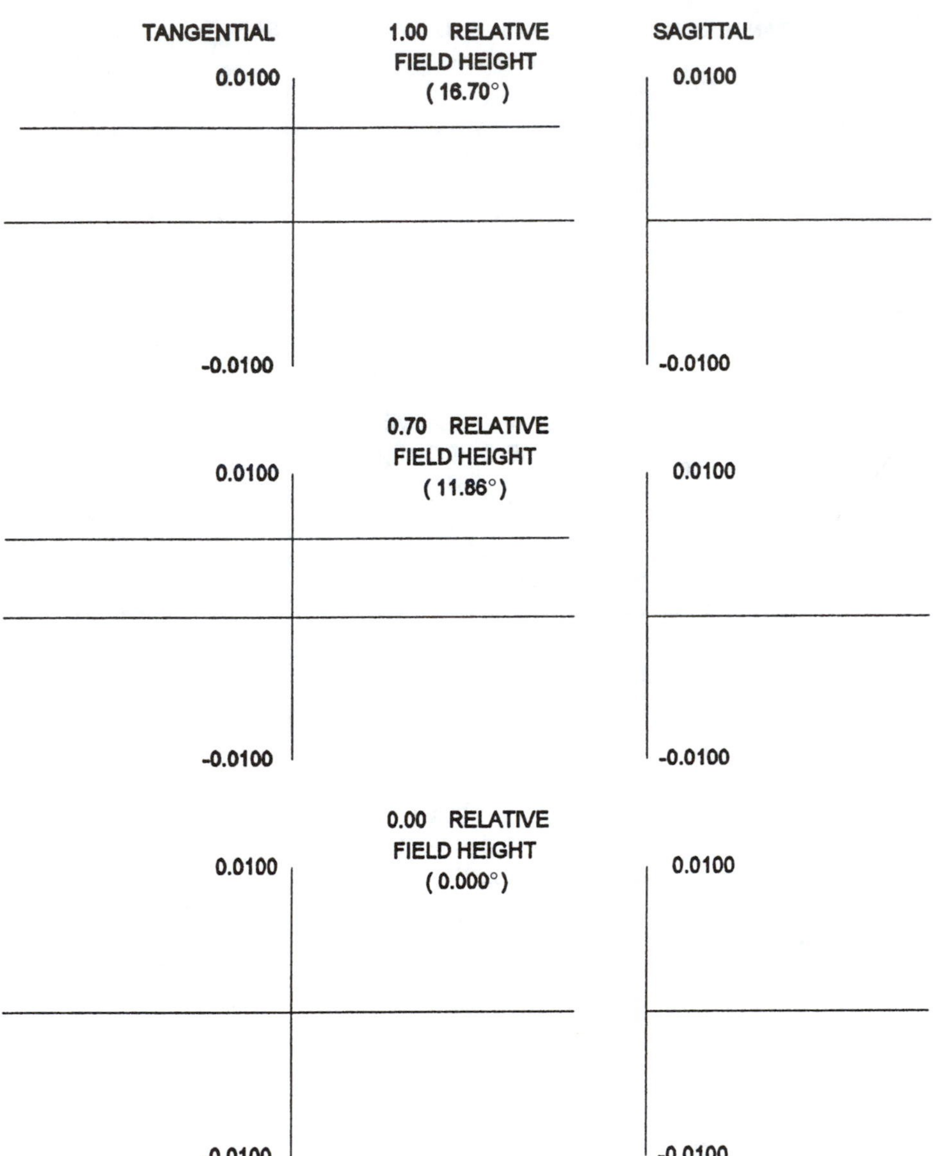

FIGURE 3.7 Ray intercept plot for a magnification error in establishing the reference image height. The amount of reference location error is linearly related to the image height (wavelength 546 nm).

ducing these shifts. However, the spread function dimensions can usually be reduced, providing better imagery.

3.1.1.1.2 CHROMATIC ABERRATIONS

Individually, these first-order aberrations state the shift of the image from the stated image reference points. In the usual case, the reference points are established by the paraxial optics. In design, it is necessary to select a specific wavelength to serve as the base wavelength for the optical system. The reference points are established by application of the paraxial ray trace equations using indices of refraction at the specified base wavelength. Because the index of refraction varies with wavelength, the paraxial image locations will vary with wavelength. It is conventional to represent the chromatic changes in terms of first-order aberrations. These two aberrations then will be dependent upon the chosen wavelength.

It is now appropriate to derive the relation for the first-order chromatic aberrations. A paraxial marginal ray is traced through the lens. If the space following an arbitrary surface is examined, the location of the paraxial image following the jth surface can be obtained from

$$l'_j = \frac{-y_j}{u'_j} \tag{3.22}$$

Then if the wavelength is changed from the base wavelength, the slope angle of the ray following the surface, in this case surface j, is changed by the change in index at the surface, and can be found by differentiating the ray trace equation, while holding the y height and the entering slope angle u constant,

$$n'du' + u'dn' = u\,dn - y(dn' - dn)c \tag{3.23}$$

which is rearranged into

$$\begin{aligned} n'du' &= (u + yc)dn - (u' + yc)dn' \\ &= i\,dn - i'dn' \\ &= i\,dn - \frac{n}{n'}\,i\,dn' \end{aligned} \tag{3.24}$$

so that

$$du' = i\left(\frac{n'}{n}\right)\left(\frac{dn}{n} - \frac{dn'}{n'}\right) \tag{3.25}$$

Multiplying by the distance from the surface to the image plane will give the spread between the paraxial marginal rays in the two wavelengths defining the differential in index of refraction

$$\begin{aligned} \epsilon'_j &= l'_j\,du'_j = \frac{-y_j}{u'_j}\,du'_j \\ &= \frac{-y}{u'}\left(\frac{n}{n'}\right)i\left(\frac{dn}{n} - \frac{dn'}{n'}\right) \end{aligned} \tag{3.26}$$

The contribution of the chromatic aberration at surface j to the final aberration is obtained by using the Optical Invariant to transfer the aberration from the local image plane to the final image plane:

$$\epsilon'_{k_j} = \frac{-1}{n'_k u'_k}(-n'_j u'_j \epsilon'_j)$$
$$= \frac{-1}{n'_k u'_k}\left[-yni\left(\frac{dn}{n} - \frac{dn'}{n'}\right)_j\right] \qquad (3.27)$$

where the primary axial chromatic aberration coefficient for surface j can be defined as

$$a_j = -yni\left(\frac{dn'}{n'} - \frac{dn}{n}\right) \qquad (3.28)$$

Since the calculation assumed that the aberration was individually generated at each surface, and that no aberration was transferred to the surface, the effect in the final focal surface can be obtained by adding the independent contributions from each surface. This yields the transverse primary axial chromatic aberration,

$$\mathrm{TA}'_{ch} = \frac{-1}{n'_k u'_k}\sum_{j=1}^{k} a_j \qquad (3.29)$$

A similar set of equations can be derived for the displacement of the chief ray,

$$T'_{ch} = \frac{-1}{n'_k u'_k}\sum_{j=1}^{k} b_j \qquad (3.30)$$

where b is the coefficient of primary lateral chromatic aberration and

$$b_j = -yn\bar{i}\left(\frac{dn'}{n'} - \frac{dn}{n}\right) = \left(\frac{\bar{i}}{i}\right)a_j \qquad (3.31)$$

provides a simple method of relating the aberrations. It would appear that problems can occur if the angle of incidence, i, is zero, but then the individual equations will provide a correct result. The last relationship involving the ratio of incidence angles is primarily an indication of the relative effect of elements on producing the aberration.

The interpretation of these aberrations is shown graphically in Figures 3.8 and 3.9. The aberration value indicated is for the separation of ray intercepts at two specific wavelengths. In almost all real cases, a broad spectrum of wavelengths is present, so that the actual appearance of the image of a point object will be the summation of all of the individual rays traced in many wavelengths. The chosen wavelengths for the aberration calculation are chosen to provide an estimate of the maximum spread of the rays in the spectral region of interest. Figure 3.10 indicates the result of the combination of these aberrations at some point in the field.

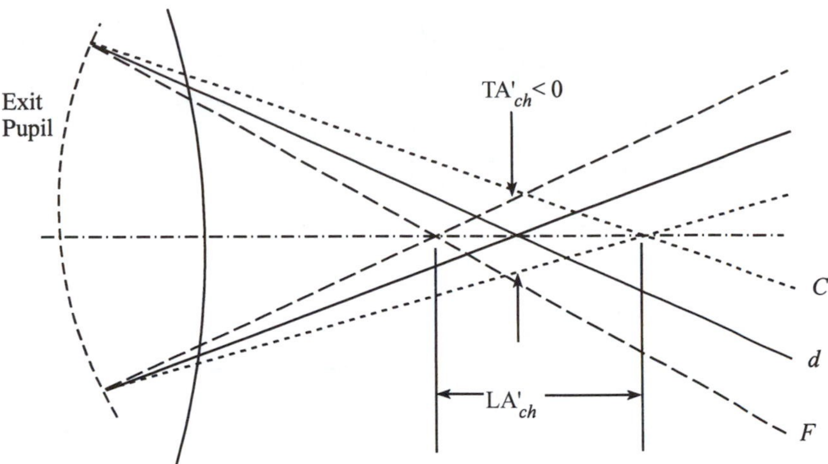

FIGURE 3.8 The effect of longitudinal chromatic aberration.

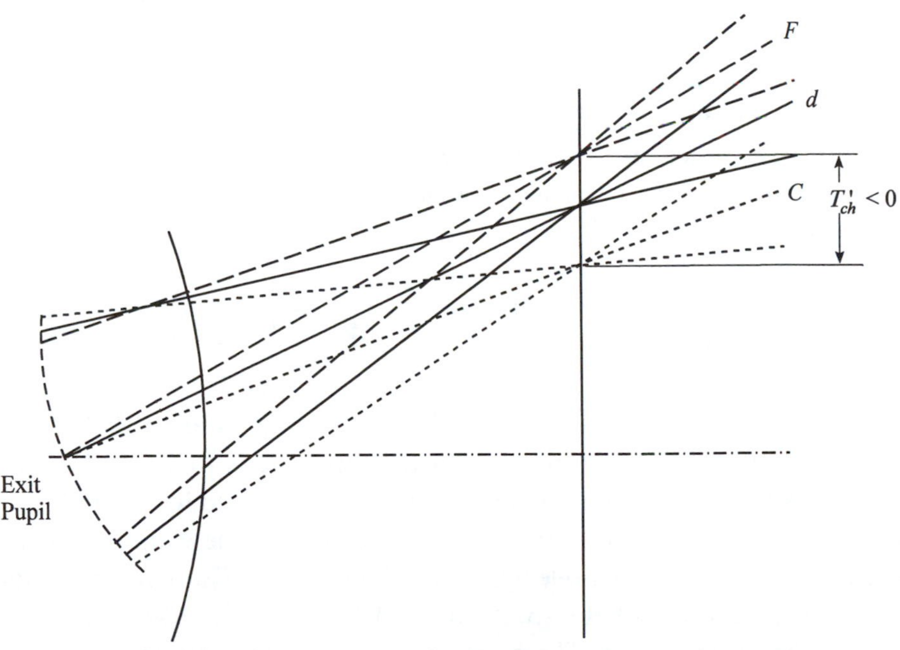

FIGURE 3.9 The effect of lateral chromatic aberration.

Ray intercept plots permit easy identification of the presence of these aberrations. Figure 3.11 shows a lens with some longitudinal, but no lateral, chromatic aberration. The pattern of focus shift with wavelength is evident, and the lack of lateral color is seen from the matching of the image heights at

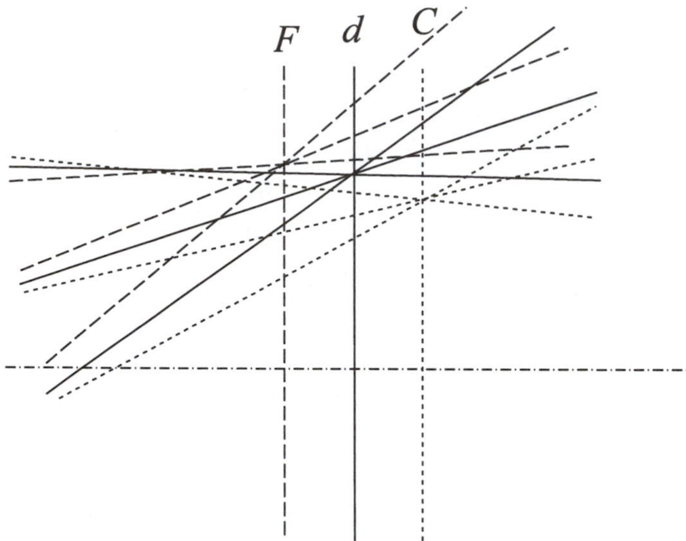

FIGURE 3.10 The effect at an off-axis image point of the combined longitudinal and lateral chromatic aberration.

the chief ray. Figure 3.12 is an example of a lens with lateral chromatic aberration. The chief ray locations now show a spread with wavelength.

Because the TA'_{ch} is a transverse aberration, the longitudinal aberration corresponding to this will be

$$\mathrm{LA}'_{ch} = \frac{-1}{u'_k}\mathrm{TA}'_{ch} = \frac{1}{n'_k u'^2_k}\sum_{j=1}^{k} a_j \tag{3.32}$$

The aberration formulae include a differential representing the index of refraction change. The interpretation of the formulae is that the aberration calculated is the difference in ray intercept for two selected wavelengths. The default values usually selected for the visible region are the *d* line (0.5876 μm) for the central or base wavelength, the *F* line (0.4861 μm) for the short-wavelength (blue) end of the spectrum, and the *C* line (0.6563 μm) for the long-wavelength (red) end. These are conventional choices; any actual lens design will be carried out with wavelengths selected to cover the operational spectral region.

The basic lens element is a single thin lens. For a lens with zero thickness, the contributions from the two surfaces can be written out algebraically and added together to obtain

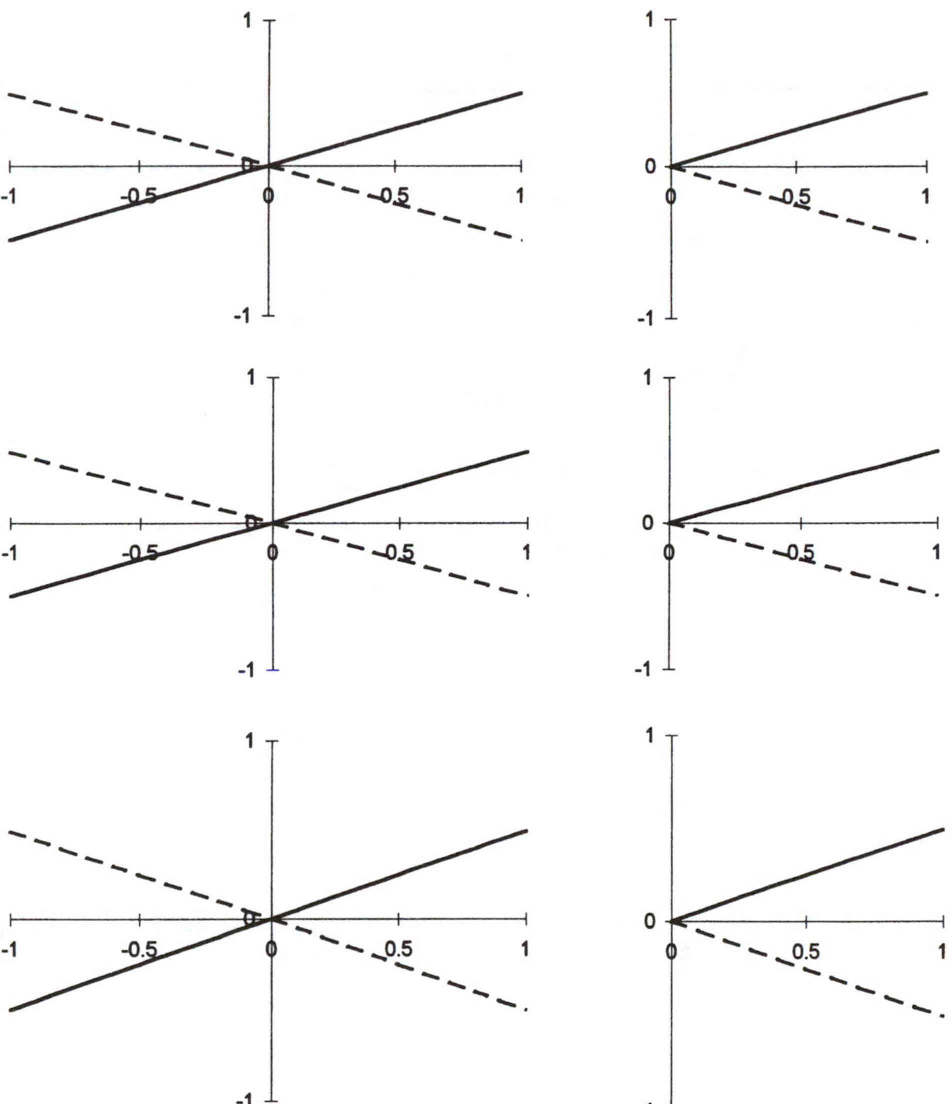

FIGURE 3.11 Ray intercept plot for longitudinal chromatic aberration.

$$a_j = -y_j^2 \frac{\phi_j}{V_j}$$
$$b_j = -y_j \bar{y}_j \frac{\phi_j}{V_j} \tag{3.33}$$

the sum of contributions from a sequence of components is then the sum of these terms:

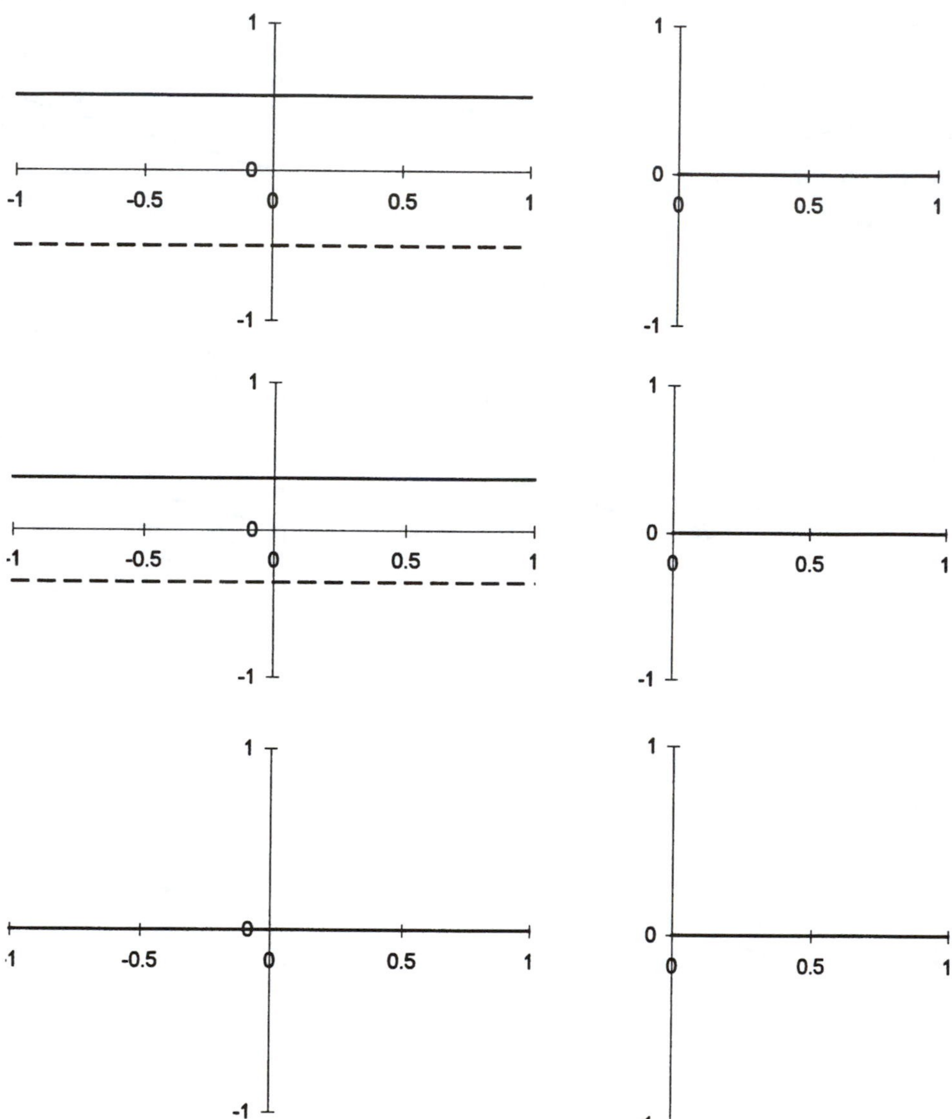

FIGURE 3.12 Ray intercept plot for lateral chromatic aberration.

$$\mathrm{TA}'_{ch} = \frac{-1}{n'_k u'_k} \sum_{j=1}^{k} -y_j^2 \frac{\phi_j}{V_j}$$

$$T'_{ch} = \frac{-1}{n'_k u'_k} \sum_{j=1}^{k} -y_j \bar{y}_j \frac{\phi_j}{V_j} \tag{3.34}$$

and are obtained using the paraxial ray trace data for the successive thin lenses in the system.

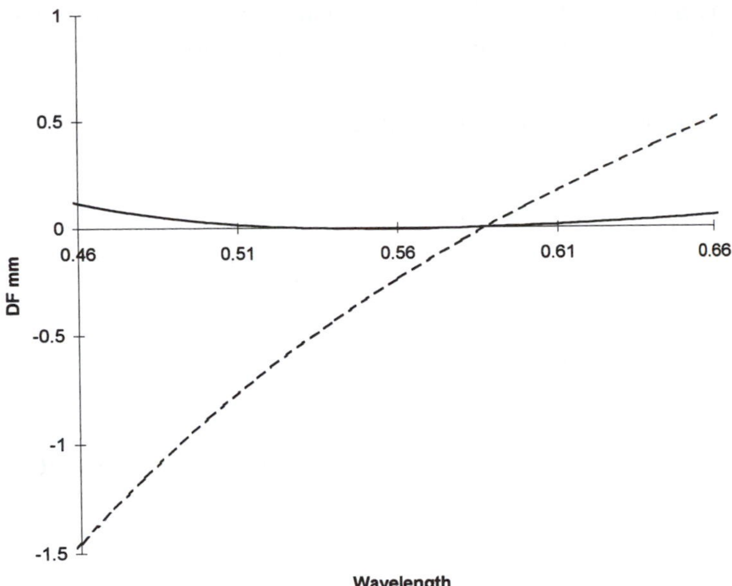

FIGURE 3.13 Comparison of focus position shift (DF) with wavelength for a singlet lens (– – –) and a doublet (——) of the same focal length.

For a single thin lens these equations simplify to

$$\mathrm{TA}'_{ch} = \frac{1}{n'u'}\, y^2\, \frac{\phi}{V} = \frac{-1}{n'}\, yl'\, \frac{\phi}{V}$$
$$T'_{ch} = \frac{1}{n'u'}\, y\bar{y}\, \frac{\phi}{V} = \frac{-1}{n'}\, \bar{y}l'\, \frac{\phi}{V} \tag{3.35}$$

or for the longitudinal value for the chromatic variation of focus the relation

$$L'_{ch} = \frac{-1}{n'}\, l'^2\, \frac{\phi}{V} \tag{3.36}$$

which for an object at infinity reduces to

$$\frac{L'_{ch}}{f} = \frac{-1}{V} \tag{3.37}$$

This states that the ratio of the longitudinal chromatic aberration to the focal length of a single-element lens is inversely equal to the V-number. For example, a single lens made of crown glass will have a longitudinal chromatic aberration of about one sixtieth of the focal length. A lens made of flint glass would have about twice the relative amount of longitudinal chromatic aberration.

The focus shift with wavelength of a 100 mm focal length, F/2.5 singlet made of Schott Bk7 glass is shown in Figure 3.13. This plot of the focus position with wavelength shows an almost linear change of focus of 1.57 mm. Selecting the best focus to be at the center of the range, at the d wavelength, the focus error is about ±0.78 mm. Using the optical path calculation formula, this is a focus error of about ±26 wavelengths. This is obviously significant.

The case for two thin lenses in contact is simple:

$$\mathrm{TA}'_{ch} = \frac{-1}{n'_k u'_k} - y^2\left(\frac{\phi_1}{V_1} + \frac{\phi_2}{V_2}\right)$$
$$T'_{ch} = \frac{-1}{n'_k u'_k} - y\bar{y}\left(\frac{\phi_1}{V_1} + \frac{\phi_2}{V_2}\right) \quad (3.38)$$

and the total power of the two components is

$$\phi = \phi_1 + \phi_2 \quad (3.39)$$

which can be used to solve the set of equations for a power distribution that will provide zero longitudinal and lateral color:

$$\phi_1 = \phi\left(\frac{V_1}{V_1 - V_2}\right)$$
$$\phi_2 = -\phi\left(\frac{V_2}{V_1 - V_2}\right) \quad (3.40)$$

These are the basic equations for color correction in a single component. If these are satisfied, the chromatic error between the colors represented by the extremes of the δn range will have the same intercept on the focal plane.

The chromatic error at other wavelengths can be found by using the partial dispersion ratio for the desired wavelengths. If the partial dispersion is defined by

$$P_{\lambda\lambda_0} = \frac{n_\lambda - n_{\lambda_0}}{n_F - n_C} \quad (3.41)$$

then the effective V-number for the pairs of wavelengths stated is

$$V_{\lambda\lambda_0} = \frac{n_d - 1}{n_\lambda - n_{\lambda_0}} = V_d P_{\lambda\lambda_0} \quad (3.42)$$

A secondary primary color coefficient can be defined by substituting the appropriate V-numbers (the notation gets a bit complex here in any attempt to retain uniformity)

$$\mathrm{TA}'_{chII_{\lambda\lambda_\eta}} = \frac{-1}{n'_k u'_k} \sum_{j=1}^{k} -y_j^2 \frac{\phi_j}{V_{\lambda\lambda_{0_j}}}$$
$$T'_{chII_{\lambda\lambda_\eta}} = \frac{-1}{n'_k u'_k} \sum_{j=1}^{k} -y_j\bar{y}_j \frac{\phi_j}{V_{\lambda\lambda_{0_j}}} \quad (3.43)$$

which can now be written in terms of partial dispersion values as

$$\mathrm{TA}'_{chII_{\lambda\lambda_0}} = \frac{-1}{n'_k u'_k} \sum_{j=1}^{k} -y_j^2 \phi_j \frac{P_{\lambda\lambda_{0_j}}}{V_d}$$

$$T'_{chII_{\lambda\lambda_0}} = \frac{-1}{n'_k u'_k} \sum_{j=1}^{k} -y_j \bar{y}_j \phi_j \frac{P_{\lambda\lambda_{0_j}}}{V_d} \qquad (3.44)$$

which looks rather complicated, but when applied to the case of two contacted elements leads to

$$\mathrm{TA}_{chII_{\lambda\lambda_0}} = \frac{-1}{n'u'} y^2 \left(\frac{\phi_1 P_1}{V_1} + \frac{\phi_2 P_2}{V_2} \right) \qquad (3.45)$$

where the subscripts 1 and 2 relate to the element number, and all of the other subscripts regarding wavelength are assumed to be present.

For the case of the color-corrected doublet, corrected for the choice of F and C wavelengths, the first-order chromatic aberration is zero, and a simplification leads to

$$\mathrm{TA}_{chII_{\lambda\lambda_0}} = \frac{-1}{n'u'} y^2 \left(\frac{\Delta P}{\Delta V} \right) \qquad (3.46)$$

where the large deltas represent the difference between the glass parameters for the two elements. It is obvious that for all other colors to be focused at the same location, the ratio of the differences would have to be zero. In section 2.3.1 on optical glasses, it was noted that this is generally not true. For most glasses, in the F–d–C region

$$\frac{\Delta P}{\Delta V} \cong \frac{1}{2{,}400} \qquad (3.47)$$

The longitudinal chromatic aberration for a doublet made of these normal glasses would be

$$\frac{l'_{ch}}{f} = \frac{\Delta P}{\Delta V} \cong \frac{1}{2{,}400} \qquad (3.48)$$

where this represents the distance from the combined C and F focus to the d light focus. Thus, using normal glasses, the effect of the chromatic aberration is not completely negated or canceled by designing a doublet, but is reduced by a factor of about forty. This residual chromatic error is generally referred to in the literature as Secondary Chromatic Aberration or sometimes "Secondary Spectrum." Note that this is actually a residual variation of the first-order chromatic aberration, or a variation in the paraxial focal position with wavelength.

Although only three wavelengths are stated in defining the amount of aberration, the image formed under these conditions is a continuous summation of the individual images at each wavelength within the passband of the system. The effect of the image at each wavelength will be weighted by the sensitivity of the detector and the chromatic content of the object and illumination. Although this discussion has generally considered the visual region of the spectrum, the same considerations obviously apply to any spectral range, from infrared through ultraviolet.

An explanation of why this type of correction occurs can be obtained by examining the details of the aberration correction by the two elements of a doublet. The crown, or convergent, lens forms an image of a point source whose focal position with wavelength is determined by the dispersion of the material. The flint, or diverging, component reimages this line of chromatically defocused images to the final focal position. The powers of the components have been selected such that two wavelengths fall at the same focal position. The intrinsic shape of the focal position with wavelength curves would match exactly if the partial dispersions were matched, so that all wavelengths were mapped into a single point.

The starting point for designing a color-corrected lens is the determination of an appropriate power distribution for a pair, or more, of elements. The simplest color-corrected lens is the achromatic doublet. The first-order layout of the doublet begins with the selection of two glasses for the lens elements. The formula for the first-order power distribution indicates that two glasses having a significant difference in dispersion, or V-number, will be required. Examination of the glass map in Chapter 2 indicates many possibilities. Because this step of the design determines only the paraxial imaging condition, any two glasses can be selected.

The design goal is a 100 mm focal length F/2.5 doublet for the C–F spectral region. The desired final power is $0.01\,\text{mm}^{-1}$, and the diameter of the entrance pupil, and the lenses, 40 mm. The required half field is nominally set at 5°. As a start, select Schott Bk7 for a crown, or low-dispersion, glass, and F2 as the high-dispersion glass. Entering this into the first-order formula leads to $\phi_1 = 0.02308$ and $\phi_2 = -0.01308$. The solution thus requires 2.308 times the power of a single Bk7 element, with −1.308 times the single-element power in the F2 element. The chromatic correction reduces the resultant chromatic variation of focus, but at the cost of significantly increased power for the individual elements. For the required F/number of F/2.5, the required F/number for the Bk7 element will be F/1.08. This results in a significant addition to the aperture-dependent aberrations.

Using the dispersion formula for the indices permits computation of the shift of the paraxial focus with wavelength. This is illustrated in Figure 3.14, in which the zero axial position is set to be at the d wavelength. It is evident that the focal positions of the C and F wavelengths are the same, but differ from the d light focus by about 0.050 mm. This secondary color defocus is the

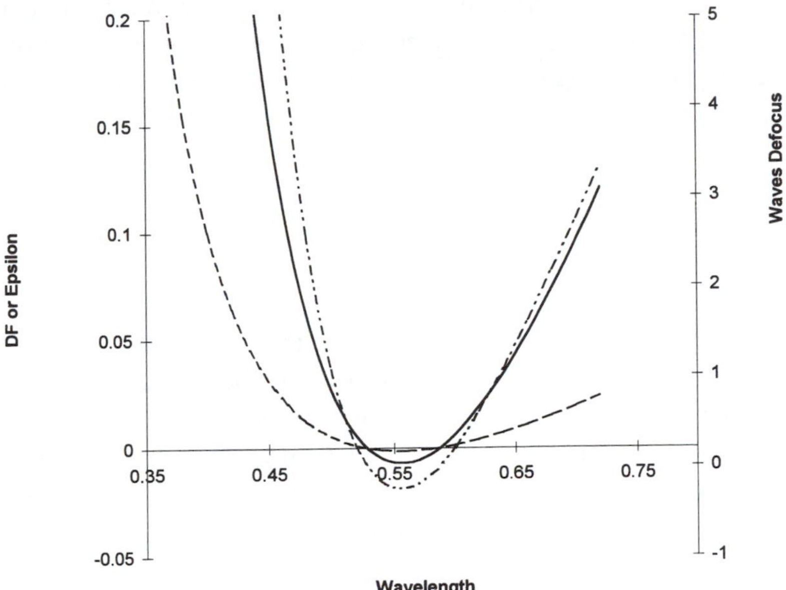

FIGURE 3.14 Plots of the transverse aberration error (Epsilon, ———), focus shift (DF, - - -) and wavelengths of focus shift (Waves defocus, -··-) for an achromatic doublet.

focal length divided by 2,000, or approximately the slope of the partial dispersion line, as expected.

The amount of the focus shift is approximately 1.7 wavelengths. This is significant relative to the desire to have the focus error less than 0.25 wavelengths for a perfect lens. However, the total image is a result of a continuous integration of the images over the wavelength region, so that the net effect on the image quality is likely to be less that indicated by the extreme wavelength defocus. The amount of the aberration depends upon the specific wavelength region used. The optimum focal position, which is chosen to minimize the effect of the residual chromatic aberration, will not be at the minimum of the curve but at a location that averages the effect of defocus in different colors.

Since the stop is located at the pair of contacted thin lenses, the lateral chromatic aberration will be zero. At this point the first-order aberrations are corrected as far as possible, leaving the residual secondary chromatic aberration. This is a starting point. The aberration correction has just begun, and the lens is still mathematically a thin lens. The detailed design of the doublet will be continued in section 7.1.1.

Correction for the residual first-order chromatic aberration can be accomplished by an appropriate choice of glass. The use of three different glasses should permit the correction of longitudinal chromatic aberration for three wavelengths, rather than two. Mathematically, the requirement is to satisfy three equations for the power, the primary chromatic aberration, and the secondary chromatic aberration at three specified wavelengths. The lens elements will be in contact, so the marginal ray height is the same on all the elements. The powers of the components must thus satisfy the equations

$$\begin{aligned} \phi &= \phi_a + \phi_b + \phi_c \\ 0 &= \frac{\phi_a}{V_a} + \frac{\phi_b}{V_b} + \frac{\phi_c}{V_c} \\ 0 &= \frac{\phi_a P_a}{V_a} + \frac{\phi_b P_b}{V_b} + \frac{\phi_c P_c}{V_c} \end{aligned} \tag{3.49}$$

Here, the glass characteristics are known and the powers must be found. The condition that these equations have a solution is that the discriminant of the equations be nonzero. The discriminant will be zero if the third row is a linear combination of the first two rows. If all glasses are normal (lying on the normal line) then this linear combination will happen, and no solution exists. Thus, at least one of the glasses must have a partial dispersion deviating from the normal glass line.

It also can be shown that the discriminant of this set of equations is proportional to the area encompassed on the partial dispersion versus V-number plot for the three glasses. Thus the larger the area enclosed by the three glasses, the larger will be the discriminant, and the smaller and more well-behaved will be the powers of the lens elements that are required to meet the solution. Conversely, if the three glasses fall on a straight line, the area is zero and no solution exists. Therefore, in order to find a paraxial or first-order solution in which three wavelengths have zero focus error simultaneously, three glasses must be chosen where one lies off a line between the other two. The farther off the line, the lower the powers of the three elements.

The required powers for any choice of glasses can be found by solving the set of equations above. An alternate approach is to choose the glasses, and then let a design program solve for the powers required in order that both the primary and the secondary axial chromatic aberration be zero.

As an example of a first-order apochromatic correction, three applicable glasses will be selected to initiate the design. Examination of the partial dispersion plots suggests that FK52 and F2 would serve as two glasses, with KzFSN5 defining the third corner of the triangle. These glasses were used in a design optimization of the first-order focus error at selected wavelengths. Several possible design starting points were obtained, with the resulting wavelength-dependent focus errors shown in Figure 3.15. For one of the solutions the required powers for the three elements are 0.022188, −0.02820, and 0.016007. Summing the powers of the convergent components shows that the excess convergent power is 3.82 times the power of a single equivalent element. This is a significantly greater excess power than for the achromat.

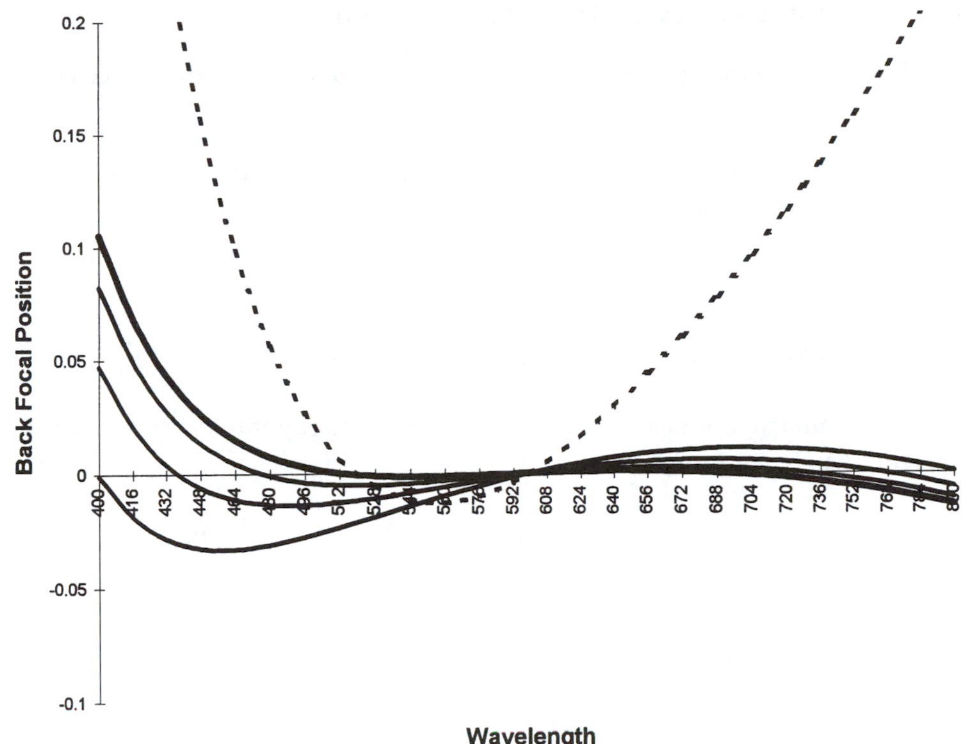

FIGURE 3.15 Comparison of the first-order focus shift versus wavelength for a doublet (– – –) and several possible apochromatic corrections (——).

In all cases the residual longitudinal chromatic aberration is much less than the achromat made of two normal glasses. The several focus curves for the apochromat are a consequence of using different wavelengths for the zero crossings. These curves also indicate that the ability to obtain good apochromatic correction (at least with this choice of glasses) depends upon the wavelength region being considered. The glasses for this example were chosen rather arbitrarily for their large separation on the partial dispersion versus V-number plot. Better choices for glasses can likely be found for spectral regions that are towards the ends of the spectrum.

The solution for both the achromat and the apochromat so far is only for first-order focus aberrations, and is independent of the order and shape of the elements. Aberration correction and balancing, the next stage of these designs, will be covered in the section on doublet design in Chapter 7.

3.1.1.2 Third-Order Aberrations

The next order of aberration is the third-order ray aberration which is related to the fourth-order wavefront aberration. The set of wavefront aberration terms associated with expansion to the fourth order are

$$W = a_{400}\eta^4 + a_{131}\eta\rho^3 \cos\psi + a_{220}\eta^2\rho^2 + a_{222}\eta^2\rho^2\cos^2\psi + a_{311}\eta^3\rho\cos\psi \tag{3.50}$$

The ray intercept aberrations are obtained by differentiating these terms with respect to the pupil coordinates

$$\begin{aligned} \epsilon_x &= \frac{\lambda}{u'}[4a_{040}\rho^2\rho_x + 2a_{131}\eta\rho_x\rho_y + 2a_{220}\eta^2\rho_x] \\ \epsilon_y &= \frac{\lambda}{u'}[4a_{040}\rho^2\rho_y + a_{131}\eta(\rho_x^2 + 3\rho_y^2) + 2(a_{220} + a_{222})\eta^2\rho_y + a_{311}\eta^3] \end{aligned} \tag{3.51}$$

Use of some algebra leads to another form of the ray intercepts in terms of aperture height and angle:

$$\begin{aligned} \epsilon_x &= \frac{\lambda}{u'}[4a_{040}\rho^3\sin\psi + a_{131}\eta\rho^2\sin 2\psi + 2a_{220}\eta^2\rho\sin\psi] \\ \epsilon_y &= \frac{\lambda}{u'}[4a_{040}\rho^3\cos\psi + a_{131}\eta\rho^2(2+\cos 2\psi) \\ &\quad + 2(a_{220} + a_{222})\eta^2\rho\cos\psi + a_{311}\eta^3] \end{aligned} \tag{3.52}$$

3.1.1.2.1 CLASSIFICATION AND INTERPRETATION

Each of these terms constitutes a member of the set of third-order aberrations. The coefficients of the terms, in order, are

a_{040} = Third-order spherical aberration
a_{131} = Third-order coma
a_{220} = Third-order sagittal field curvature
a_{222} = Third-order astigmatism
a_{311} = Third-order distortion

Third-order spherical aberration is a rotationally symmetric aberration in which there is a progressive change in the focal position of adjacent zones in the pupil. This aberration is constant with field, and is the only third-order aberration present on axis. Third-order coma varies linearly with field and is a consequence of the magnification changing with the zone in the pupil. Field curvature and astigmatism vary quadratically with field, and are a consequence of the wavefront having two principal curvatures in orthogonal directions in the pupil. Distortion represents a shift of the image position from the paraxially determined location, the shift varying with the cube of the field. The effect is that of a local nonlinearly varying magnification that produces a change in the metric of the image.

It is rare that only one of these aberrations is present; there is usually a mixture. In order to understand the nature of the aberrations, it is appro-

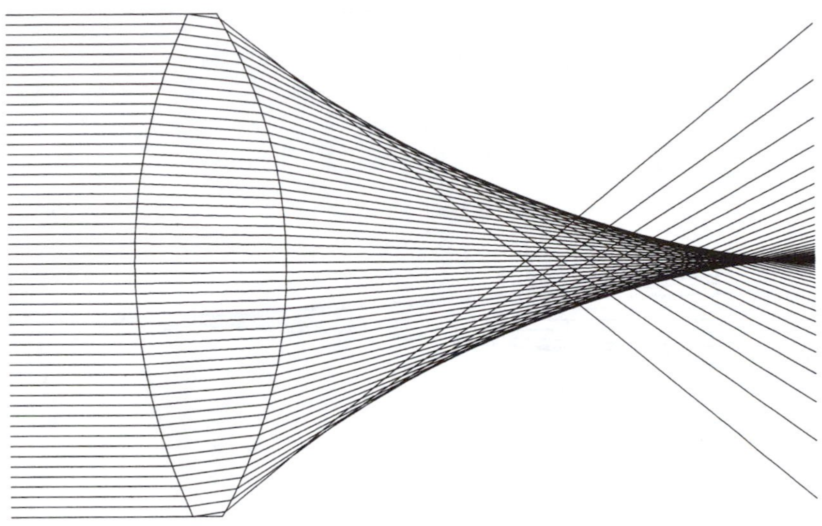

FIGURE 3.16 Ray diagram for spherical aberration for a single lens (OSLO).

priate to consider them each separately, in order, and later examine combinations of aberrations.

The conventional method of examining the behavior of lenses with these aberrations is to examine a theoretical model of the image. There are many descriptions of the aberrations in the literature (e.g. Kingslake 1978; Smith 1990). In this book, we will discuss each aberration in the detail necessary to understand how the aberration is recognized in the process of design. An example of a lens that contains primarily this aberration will be used to determine how the aberration will appear in the output of a lens design program.

Third-order spherical aberration is one of the most important aberrations in a lens system. Figure 3.16 is an extreme example of spherical aberration from a single lens operating at about F/1. Figure 3.17 is the spherical aberration in the region of focus for a lens with a more modest aperture, F/5, where the third-order spherical aberration is very dominant. The appearance of the image of a point object, as represented in a spot diagram, is indicated in a number of spot diagrams in Figure 3.18. In these diagrams, the spherical aberration is constant, but the focal position has been changed. It is obvious that shifting focus does not eliminate the aberration, but merely reduces its effect upon the image by reducing the size of the blur circle.

In Figures 3.19 and 3.20, ray intercept plots and OPD plots are presented as would be observed if the lens were set at the paraxial focal location. The characteristic cubic law of the spherical aberration curves permits the designer to identify the presence of the aberration. The aberration is, in

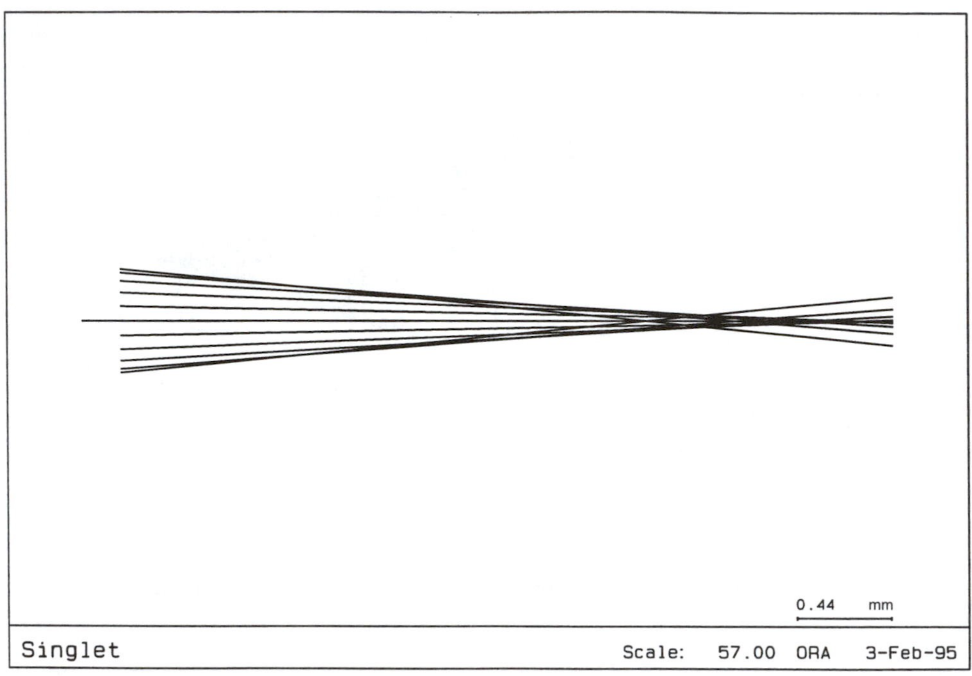

FIGURE 3.17 Ray pattern in the region of focus for an F/5 singlet lens (CODE V).

principle, constant with field position. Thus a lens with only spherical aberration should show a constant set of ray intercept curves across the field. However, only rare imaging cases will show third-order spherical aberration alone. Because the best focus is not at the paraxial location, an improvement in image quality is obtained by shifting to the best focus. As can be seen from the ray intercept and OPD plots in Figures 3.21 and 3.22, the best focus location corresponds approximately to the focal position of the ray at about 0.7 of the full aperture.

The wavefront error when including only spherical aberration and defocus is

$$\begin{aligned} W &= a_4\rho^4 + a_d\rho^2 \\ \epsilon &= \frac{\lambda}{u'}(4a_4\rho^3 + 2a_d\rho) \end{aligned} \tag{3.53}$$

where advantage is taken of the fact that the aberration is rotationally symmetric. In order to set the aberration to zero at 0.7, or $1/\sqrt{2}$, of the image height, it is necessary that $a_d = -a_4$. This also leads to the fact that the OPD at the edge of the aperture is zero. This is shown in Figures 3.21 and 3.22.

A single number suffices to define third-order spherical aberration. This is usually taken to be the radius of the blur defined by the marginal ray at the

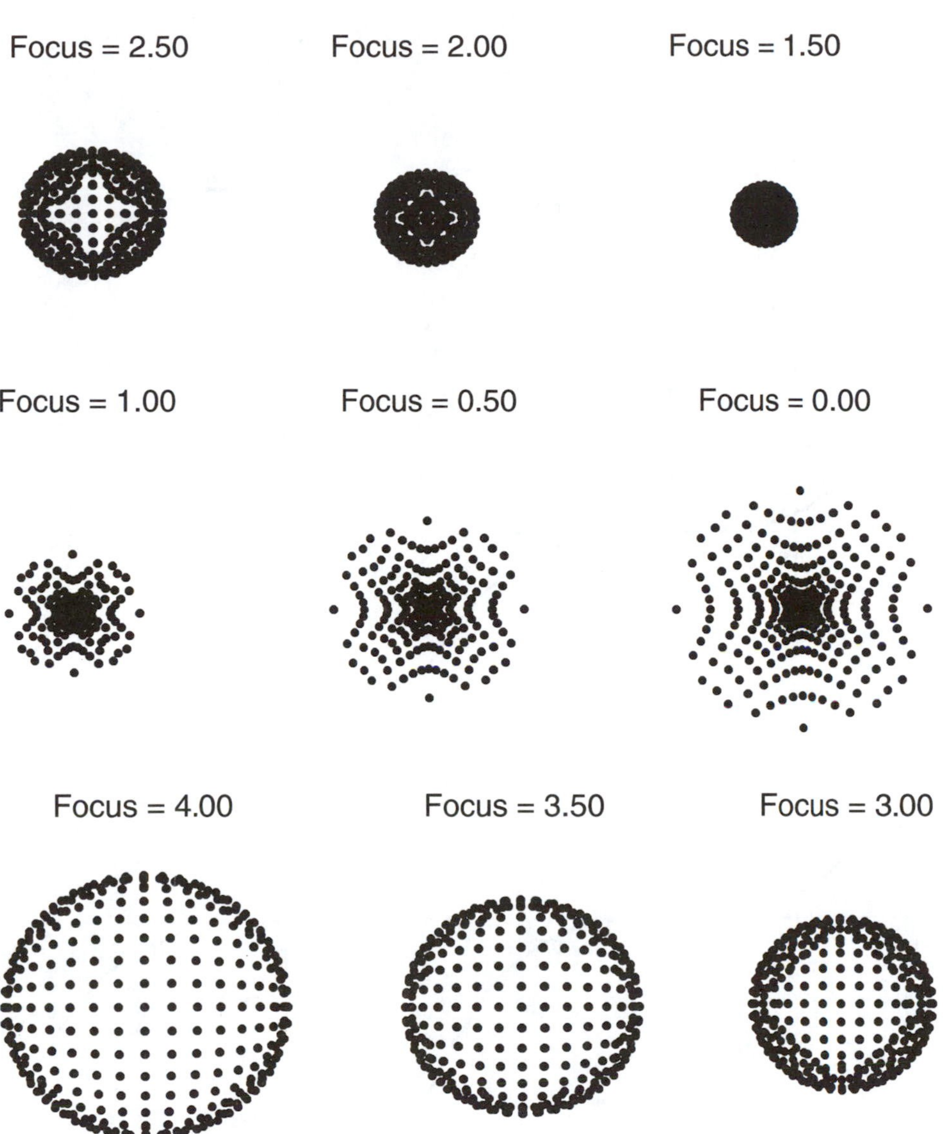

FIGURE 3.18 Spot diagrams for several focal positions from a lens with third-order spherical aberration.

paraxial focal position. The effect of changing focal position is equivalent to compensating or balancing the spherical aberration with a first-order focus shift aberration. The diameter of the effective point image blur is a minimum at a focal position midway between the paraxial and marginal foci. The size of this minimum blur is one-fourth the characteristic blur in the paraxial focal position.

Another interpretation of the spherical aberration can be obtained by plotting the focal position of the rays at each zone against the pupil height

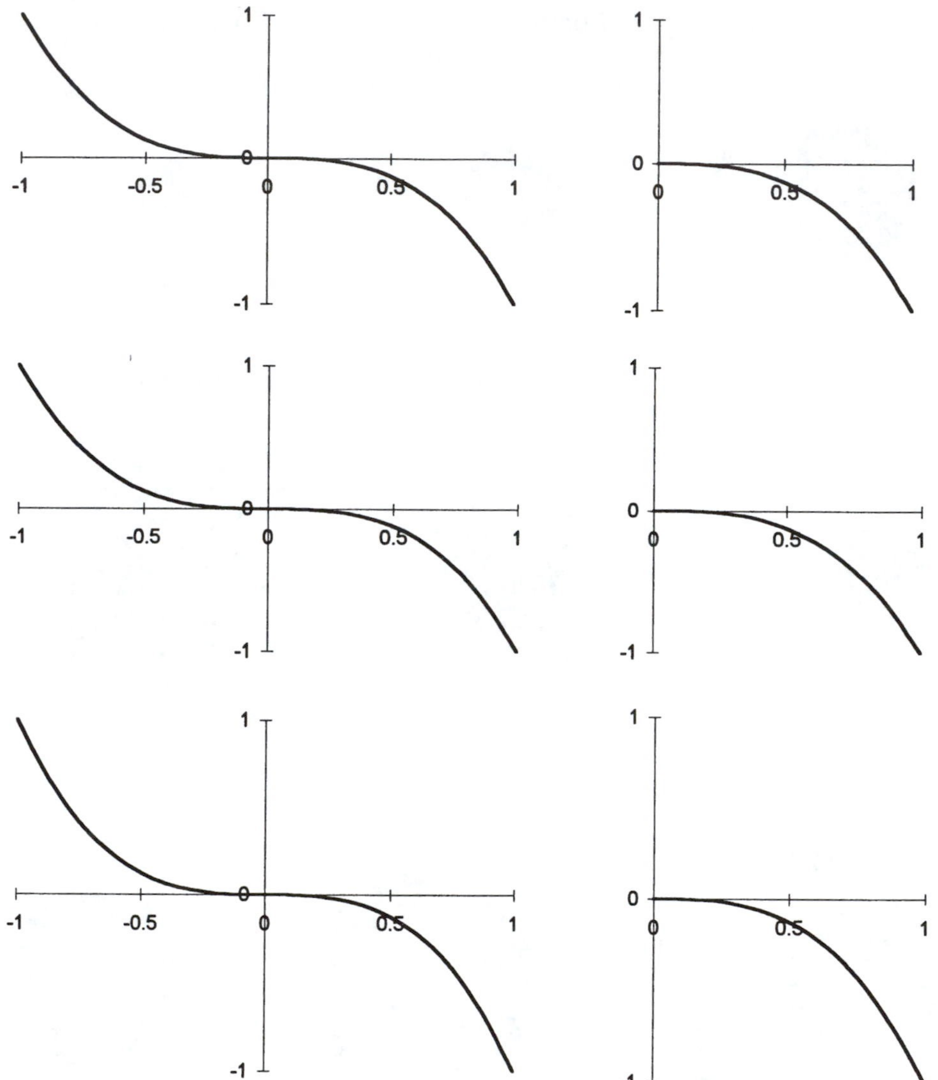

FIGURE 3.19 Ray intercept plots for third-order spherical aberration.

for the zone. The relation between the longitudinal and transverse aberration is dependent upon the F/number of the imaging cone. The longitudinal plot provides a direct view of the approximately quadratic change of focal position with aperture height that is characteristic of this aberration.

It is appropriate to digress for a moment and note that spherical aberration exists in various orders. A series expansion of the wavefront aberration to various orders would produce

$$W(\rho_x, \rho_y) = \sum_{j=2}^{\infty} a_{0,j,0}(\rho_x^2 + \rho_y^2)^j \tag{3.54}$$

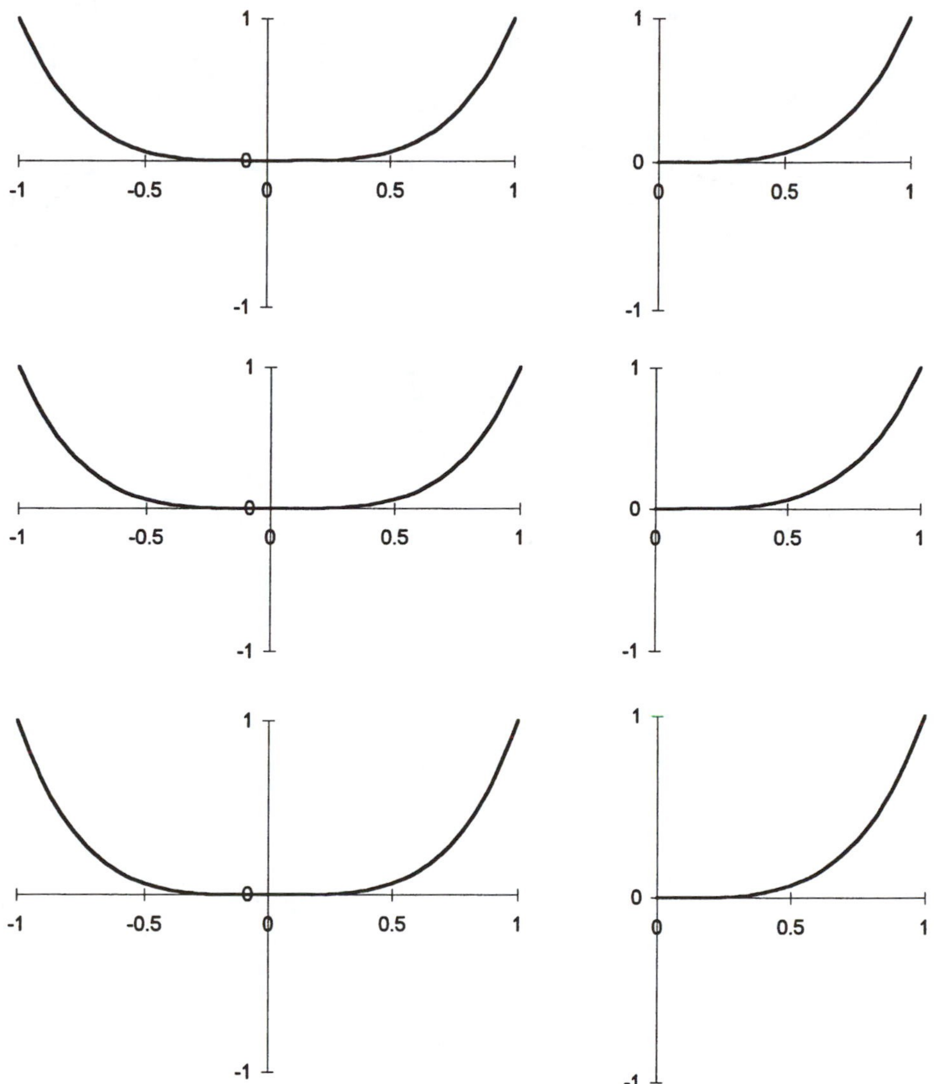

FIGURE 3.20 OPD plots for third-order spherical aberration.

where the order of the wavefront aberration is $n = 2j$. This can be written in terms of a ray intercept sum:

$$\begin{aligned}\epsilon_x &= \mathrm{SA3}\,\rho_x(\rho_x^2+\rho_y^2) + \mathrm{SA5}\,\rho_x(\rho_x^2+\rho_y^2)^2 + \mathrm{SA7}\,\rho_x(\rho_x^2+\rho_y^2)^3 + \cdots \\ \epsilon_y &= \mathrm{SA3}\,\rho_y(\rho_x^2+\rho_y^2) + \mathrm{SA5}\,\rho_y(\rho_x^2+\rho_y^2)^2 + \mathrm{SA7}\,\rho_y(\rho_x^2+\rho_y^2)^3 + \cdots \end{aligned} \tag{3.55}$$

or by using the wavefront error coefficients, and using the rotational symmetry for simplicity:

$$\epsilon = \frac{\lambda}{u'}\,(4a_4\rho^3 + 6a_6\rho^5 + 8a_8\rho^7 + \cdots) \tag{3.56}$$

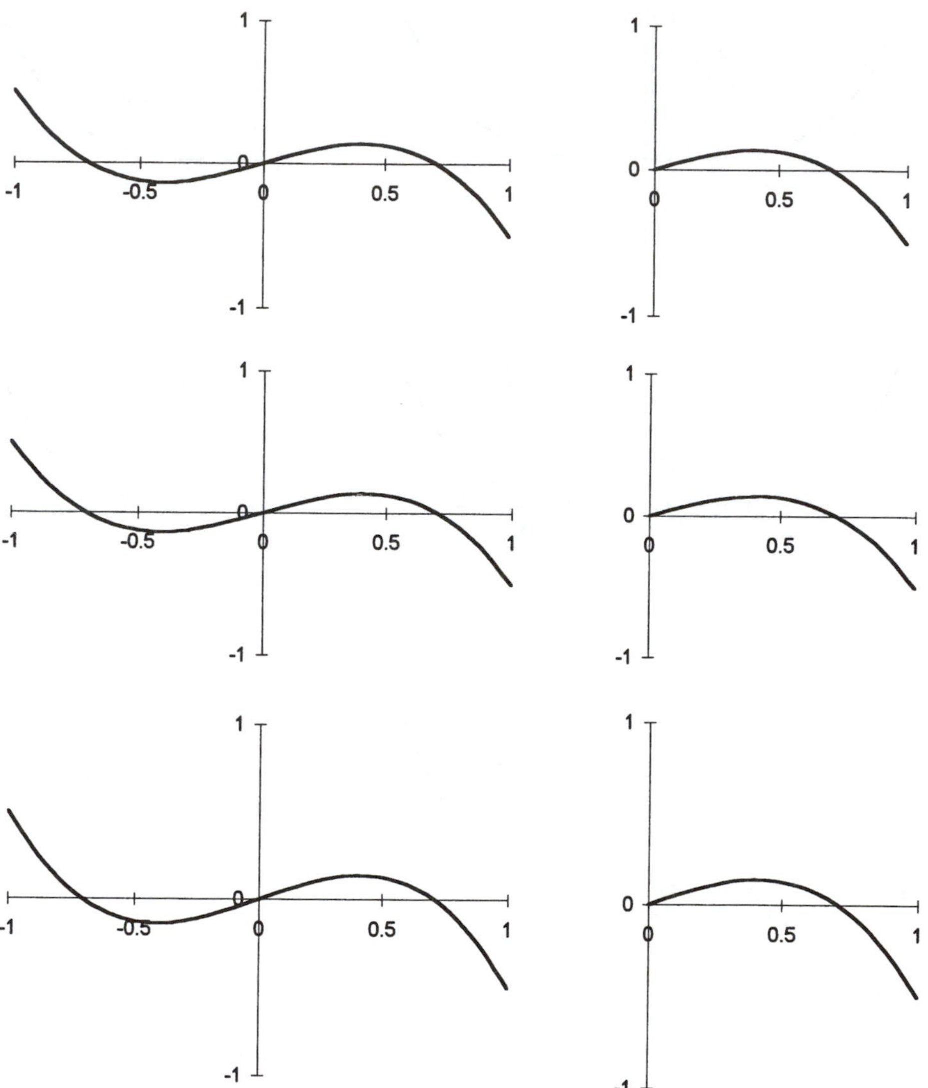

FIGURE 3.21 Ray intercept plot for the reference position shifted to the optimum focal position.

The magnitude and signs of the various orders of aberration are usually controllable by changes in the parameters of the lens design. A very common choice for the design is for the sum of the ray aberrations at the edge of the aperture to be zero. (The process for accomplishing this will be discussed later.) In most, but not all, cases of lenses in which the third-order spherical aberration can be controlled, the sign of the ray aberration for the higher-order spherical aberration is positive, or overcorrected, and is compensated by the deliberate introduction of a negative, or undercorrected, third-order spherical aberration. If only third- and fifth-order spherical aberration is

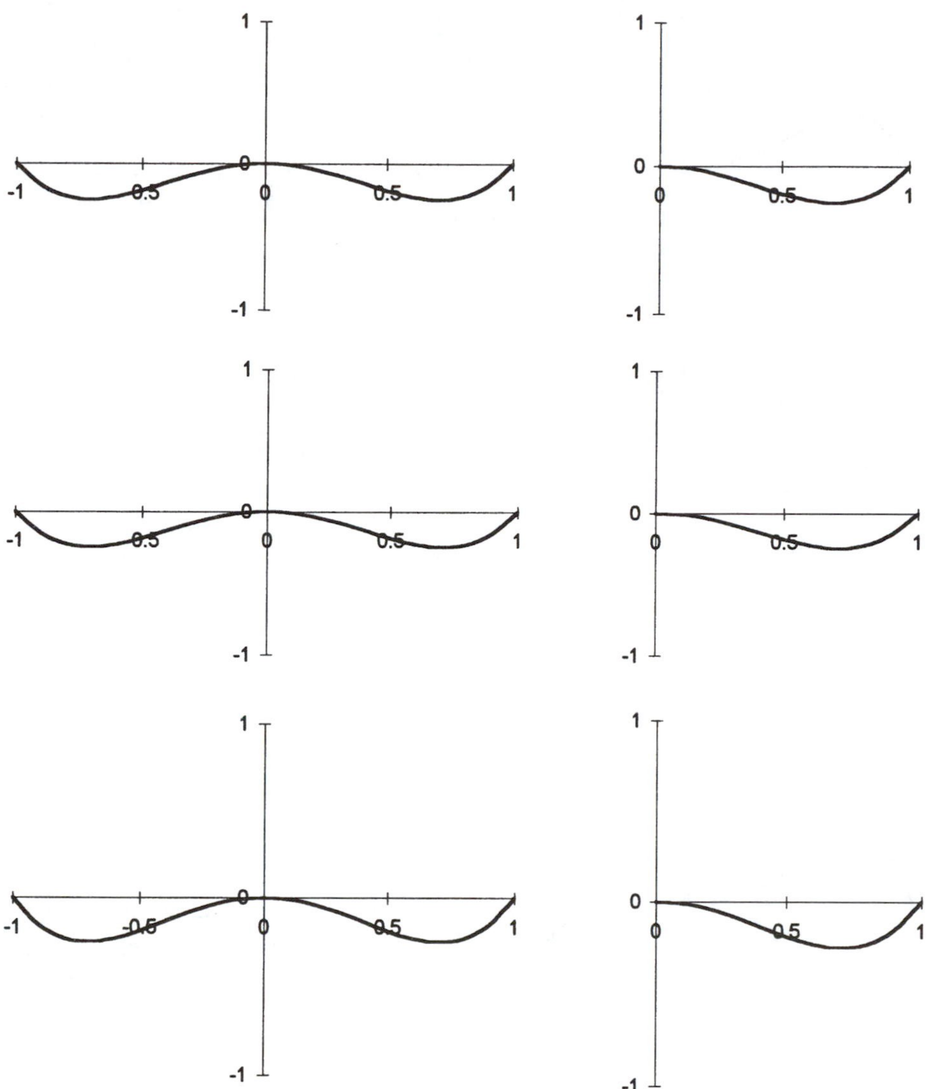

FIGURE 3.22 OPD Plot for third-order spherical aberration at optimum focus.

present the condition for this balancing is that $a_4 = -\frac{3}{2}a_6$. This conventional balancing of aberrations produces a set of ray intercept plots shown as transverse ray intercept plots in Figure 3.23 and as OPD plots in Figure 3.24.

The corresponding ray diagrams are shown in Figure 3.25, and spot diagrams are shown in Figure 3.26. These diagrams show a much more complicated set of forms of the aberration with respect to the choice of focal position. This type of aberration balancing is required as it is usually not possible to set the aberration of all orders to exactly zero. This type of

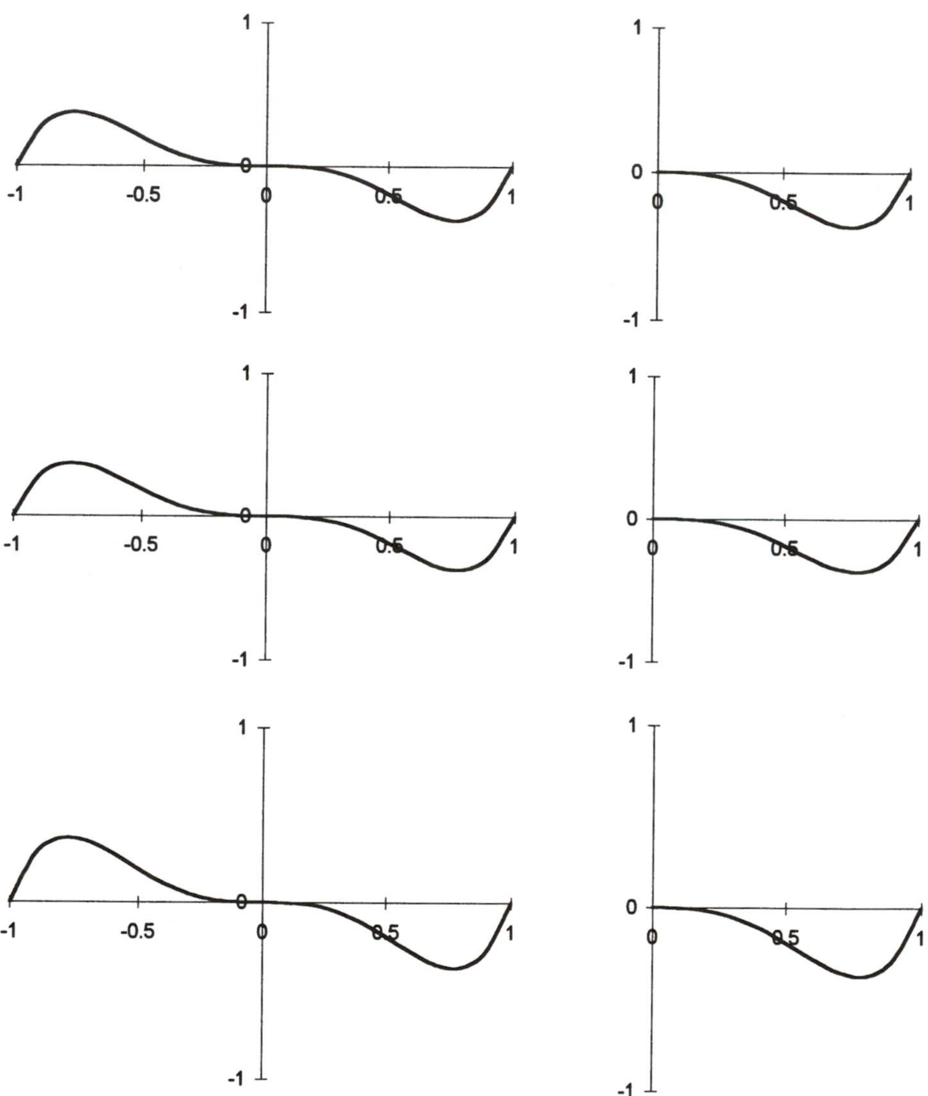

FIGURE 3.23 Ray intercept plots for edge-balanced spherical aberration.

balancing is usually encountered when spherical surfaces are used in the design.

It will be found later in this chapter, during the discussion of wavefront aberrations, that the edge correction is an optimum balancing for minimizing the root-mean-square wavefront aberration. In cases of small aberration, this is usually the best choice. For larger amounts of residual aberration, the determination of optimum balancing is somewhat more difficult, and the detailed purpose for application of the lens must be considered.

In many optical systems, there is a variable diameter aperture stop. In this case, edge correction may be optimum for one choice of aperture, but use of

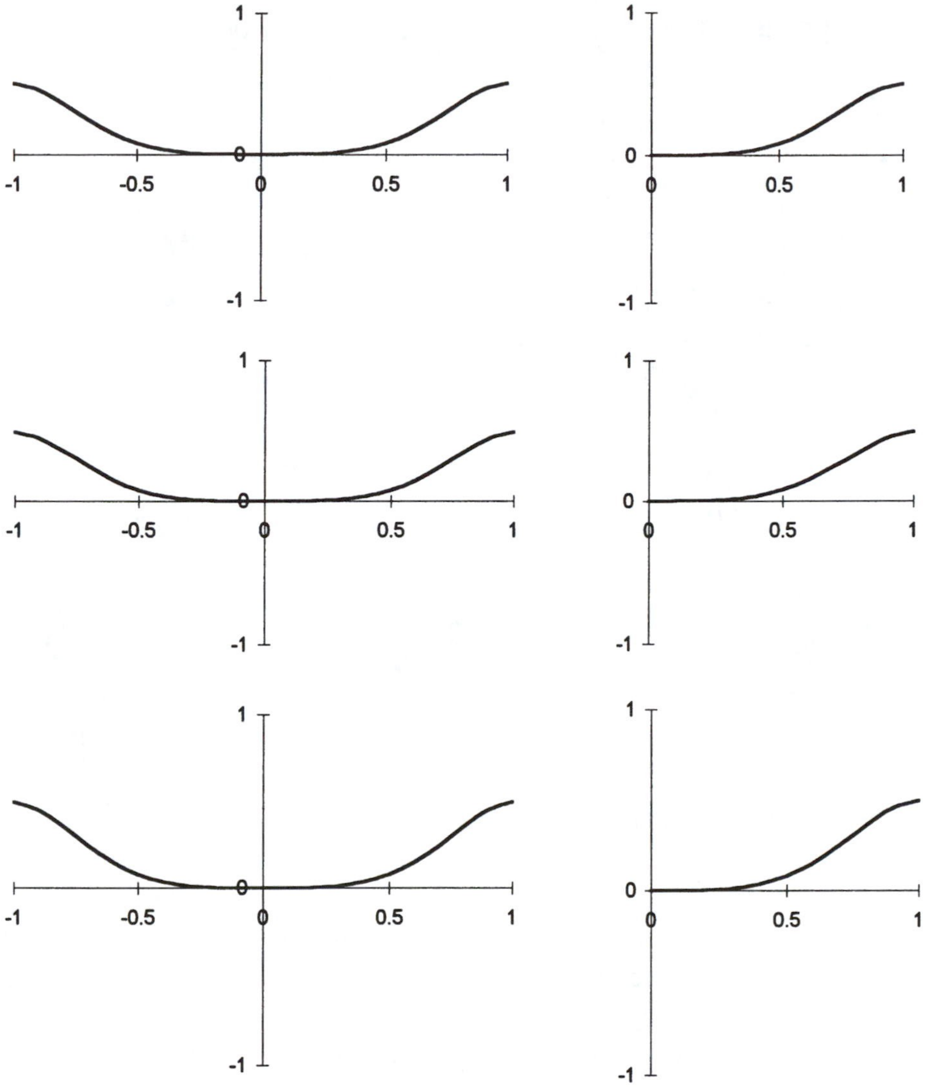

FIGURE 3.24 OPD plot for edge-balanced spherical aberration.

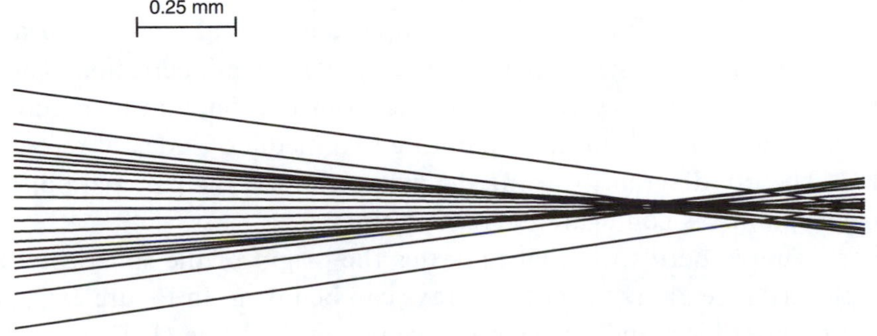

FIGURE 3.25 Ray diagram for balanced spherical aberration.

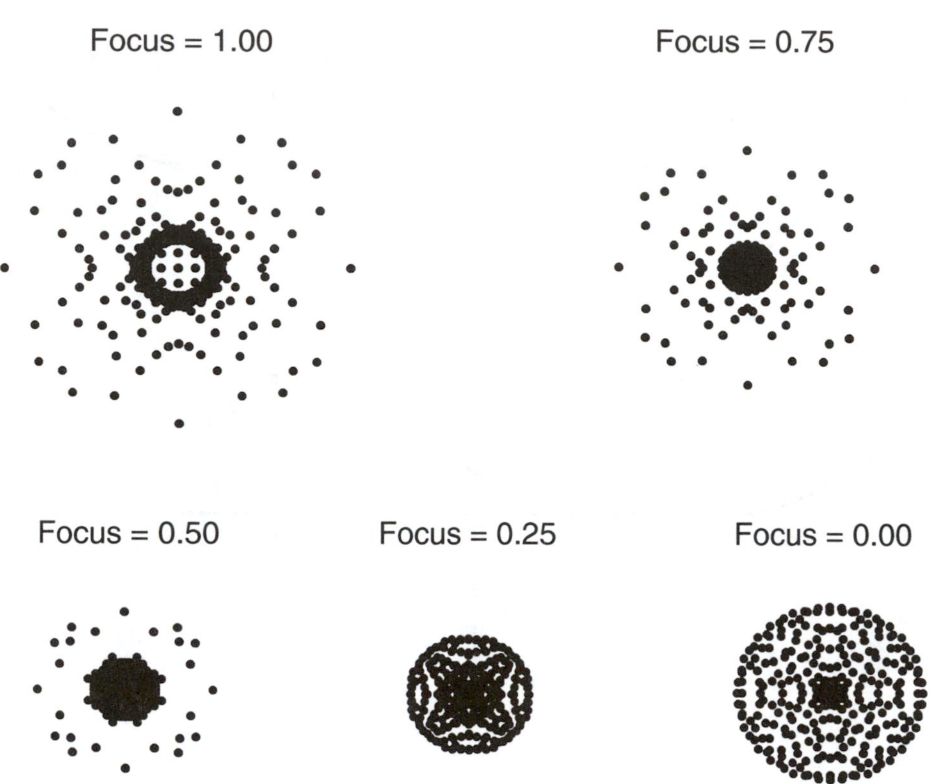

FIGURE 3.26 Spot diagrams for balanced spherical aberration at several focal positions.

the lens at other apertures will result in a nonoptimum correction for the aperture used. This acceptance of edge balancing as optimum also does not include a wide wavelength range, with the attendant likely variation of the aberrations with wavelength. Balancing of aberrations requires some careful considerations. There is no simple "best" choice of aberration balance that is optimum for all cases of imagery.

Spherical aberration is produced as a consequence of the geometry and the nonlinearity of the refraction rule. A simple geometrical model of axial imagery is useful in understanding the source of spherical aberration. The derivation of the spherical aberration begins from the diagram in Figure 3.27. This shows a spherical surface separating two media of index of refraction n and n'. The ray diagrams starting the derivation use the paraxial object and image locations as conjugate reference points.

For a simple derivation that indicates the origin of the spherical aberration, the distance along the effective rays can be taken. In Figure 3.27, points Q and Q' are object and image reference points. The line QPQ' represents a possible path between these points, which could be a ray passing from the

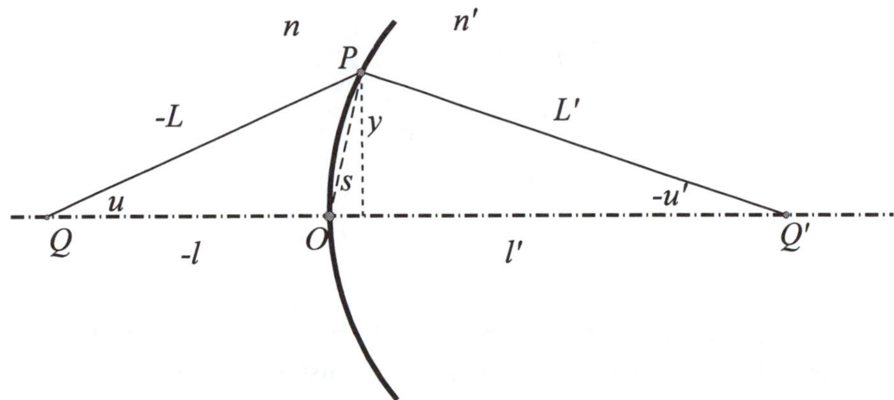

FIGURE 3.27 Parameters used in deriving the spherical aberration contribution.

object to the image reference point, through the point P on the surface. Point P is located at coordinates (x, y, z) on the spherical surface. Because the diagram is rotationally symmetric, we can take $x = 0$ without any loss of generality. The calculation is now to compute the OPD of the path QPQ' in terms of the height in the pupil.

Since the pupils do not reside on the surface, the parameter s will be used to select the aperture height. Later we will discuss the relation between this surface height and the pupil coordinates. This parameter s has the convenient feature that for a spherical surface, with the center of coordinates at the pole of the surface

$$x^2 + y^2 + (r - z)^2 = r^2 \tag{3.57}$$

with (x, y, z) confined to the surface. Rearranging gives

$$z = \frac{x^2 + y^2 + z^2}{2r} = \frac{s^2}{2r} \tag{3.58}$$

where s is the height parameter on the surface defined above.

To calculate the OPD along the path QPQ',

$$\begin{aligned} \text{OPD} &= [\text{Reference path}] - [\text{Ray path}] \\ &= [(-nl) + (n'l')] - [(-nL) + (n'L')] \\ &= n(L - l) - n'(L' - l') \end{aligned} \tag{3.59}$$

where the reference path is taken for a reference ray along the axis of the lens.

The values of the marginal ray path minus the axial path need to be found. Consider first the image space. The length along the marginal path will be

$$\begin{aligned}L'^2 &= (l' - z)^2 + x^2 + y^2 \\ &= l'^2 - 2zl' + z^2 + x^2 + y^2 \\ &= l'^2 - l'\frac{s^2}{r} + s^2 \\ &= l'^2\left[1 - \frac{s^2}{l'}\left(\frac{1}{r} - \frac{1}{l'}\right)\right]\end{aligned} \tag{3.60}$$

In general, the magnitude of s will be much less than the magnitude of l', or of r. In order to take the difference between the marginal length and the axial length, the square root can be expanded in the usual manner:

$$L' - l' = -s^2\left(\frac{1}{r} - \frac{1}{l'}\right) + \frac{1}{8}\frac{s^4}{l'}\left(\frac{1}{r} - \frac{1}{l'}\right)^2 - \frac{1}{16}\frac{s^6}{l'^2}\left(\frac{1}{r} - \frac{1}{l'}\right)^3 + \cdots \tag{3.61}$$

and the corresponding difference on the object side of the surface obtained by replacing the primed by the unprimed quantities

$$L - l = -s^2\left(\frac{1}{r} - \frac{1}{l}\right) + \frac{1}{8}\frac{s^4}{l}\left(\frac{1}{r} - \frac{1}{l}\right)^2 - \frac{1}{16}\frac{s^6}{l^2}\left(\frac{1}{r} - \frac{1}{l}\right)^3 + \cdots \tag{3.62}$$

The OPD is then found by taking the difference multiplied by the refractive indices of the spaces containing the differences:

$$\begin{aligned}\mathrm{OPD} = &- ns^2\left(\frac{1}{r} - \frac{1}{l}\right) + n's^2\left(\frac{1}{r} - \frac{1}{l'}\right) \\ &+ \frac{n}{8}\frac{s^4}{l}\left(\frac{1}{r} - \frac{1}{l}\right)^2 - \frac{n'}{8}\frac{s^4}{l'}\left(\frac{1}{r} - \frac{1}{l'}\right)^2 \\ &- \frac{n}{16}\frac{s^6}{l^2}\left(\frac{1}{r} - \frac{1}{l}\right)^3 - \frac{n'}{16}\frac{s^6}{l'^2}\left(\frac{1}{r} - \frac{1}{l'}\right)^3 + \cdots\end{aligned} \tag{3.63}$$

The series can be reorganized to bring out front the various powers of the aperture parameter, s:

$$\begin{aligned}\mathrm{OPD} = &\, s^2\left[\frac{n'}{r} - \frac{n'}{l'} - \frac{n}{r} + \frac{n}{l}\right] \\ &+ \frac{s^4}{8}\left[\frac{n}{l}\left(\frac{1}{r} - \frac{1}{l}\right)^2 - \frac{n'}{l'}\left(\frac{1}{r} - \frac{1}{l'}\right)^2\right] \\ &- \frac{s^6}{16}\left[\frac{n}{l^2}\left(\frac{1}{r} - \frac{1}{l}\right)^3 - \frac{n'}{l'^2}\left(\frac{1}{r} - \frac{1}{l'}\right)^3\right] + \cdots\end{aligned} \tag{3.64}$$

This is the aberration produced at the surface for the geometry described in Figure 3.27. It is desirable to express the aberration in terms of the pupil height in the entrance and exit pupils. Figure 3.28 includes in a diagrammatic form the pupils related to the surface. If the ray paths followed paraxial rules,

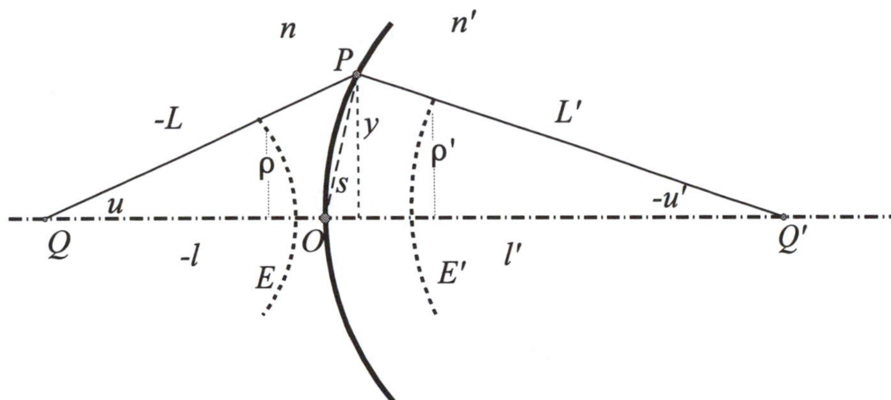

FIGURE 3.28 Diagram indicating the relation between the passage of rays through a surface and the coordinates on the effective pupil reference spheres.

there would be a proportional scaling between the surface height parameter s and the heights ρ and ρ' in the pupils. The relationships for small apertures would be

$$s \cong \rho y_p \cong \rho' y_p \tag{3.65}$$

where y_p is used here to denote the paraxial marginal ray height on the surface. The larger the aperture, the greater the aberration, or the higher the order of the aberration, the less accurate this approximation will be.

The precise relationship between the surface coordinates and the pupil coordinates is quite complicated. It is sufficient to note that because the third-order ray aberration error is an error about the first-order, or paraxial ray path, this substitution will be acceptable for the first term of the series. Application of the substitution for the fifth-order ray aberration term is more complicated, as the pupil coordinate relationship relative to the surface must contain the third-order ray path. Therefore, for the third-order term, the OPD determined in terms of the paraxial ray path is correct, and the aberration arises at the surface. For higher-order aberrations, there will be two terms, an intrinsic aberration arising at the surface, plus a transferred aberration resulting from including the third-order aberration in the relation between the surface coordinates and the pupil coordinates. The situation can become quite complicated if the pupils are located, optically, a long distance from the surface.

The first term of the OPD expansion contains the second-order wavefront aberration, or the first-order ray aberration. If the selection of l and l' is made such that the factor multiplying the s^2 term is zero, this implies that there is no focus aberration relative to the chosen reference positions. Examination of the factor shows that this is true for

$$\frac{n'}{r} - \frac{n'}{l'} - \frac{n}{r} + \frac{n}{l} = 0 \tag{3.66}$$

which is the standard paraxial ray trace relation. Therefore, choosing the paraxial conjugate points as reference points will eliminate the focus aberration, leaving only the spherical aberration terms.

Some general statements can be made about the spherical aberration arising at a surface. Each of the orders of spherical aberration contains factors which can individually go to zero, assuring that for certain cases there will be no spherical aberration introduced by the surface. The three conditions for zero spherical aberration arising at a surface are:

$$s = 0$$
$$l = l = l'$$
$$\frac{l'}{l} = \frac{n}{n'} \tag{3.67}$$

The first of these calls for a zero marginal ray height at the surface, which implies that the object and the image reside at the surface. Since the aperture ray height is zero, there is no spherical aberration contribution. The second states that the object and the image for the surface reside at the center of curvature of the surface. Since the angles of incidence and refraction are zero, no optical effect exists at the surface. The last is the so-called aplanatic condition, in which the magnification between the object and the image is given by the ratios of the refractive indices.

It is important to remember the conditions imposed by the derivation of these rules. The object and image ray path relations must be exact. If there is any aberration associated with the incoming ray bundle, then the conditions are not exactly met for all rays, and some spherical aberration will be introduced by the surface. As a general rule, the aberration will certainly be small for imaging circumstances that come close to the stated conditions. For example, placing a lens in an image plane to act as a field lens will have little effect upon the spherical aberration.

The next aberration to be considered is third-order coma. The functional form of this aberration is

$$W(\rho, \psi) = a_{131}\eta\rho^3 \cos\psi$$
$$\epsilon_x = \frac{-\lambda}{n'u'} 2a_{131}\eta\rho_x\rho_y$$
$$\epsilon_y = \frac{-\lambda}{n'u'} a_{131}\eta(\rho_x^2 + 3\rho_y^2) \tag{3.68}$$

for the wavefront and ray aberrations, respectively. The wavefront and ray plots for this aberration are shown in Figures 3.29 and 3.30. The appearance of the ray intercept plots indicates that there should be no sagittal aberration

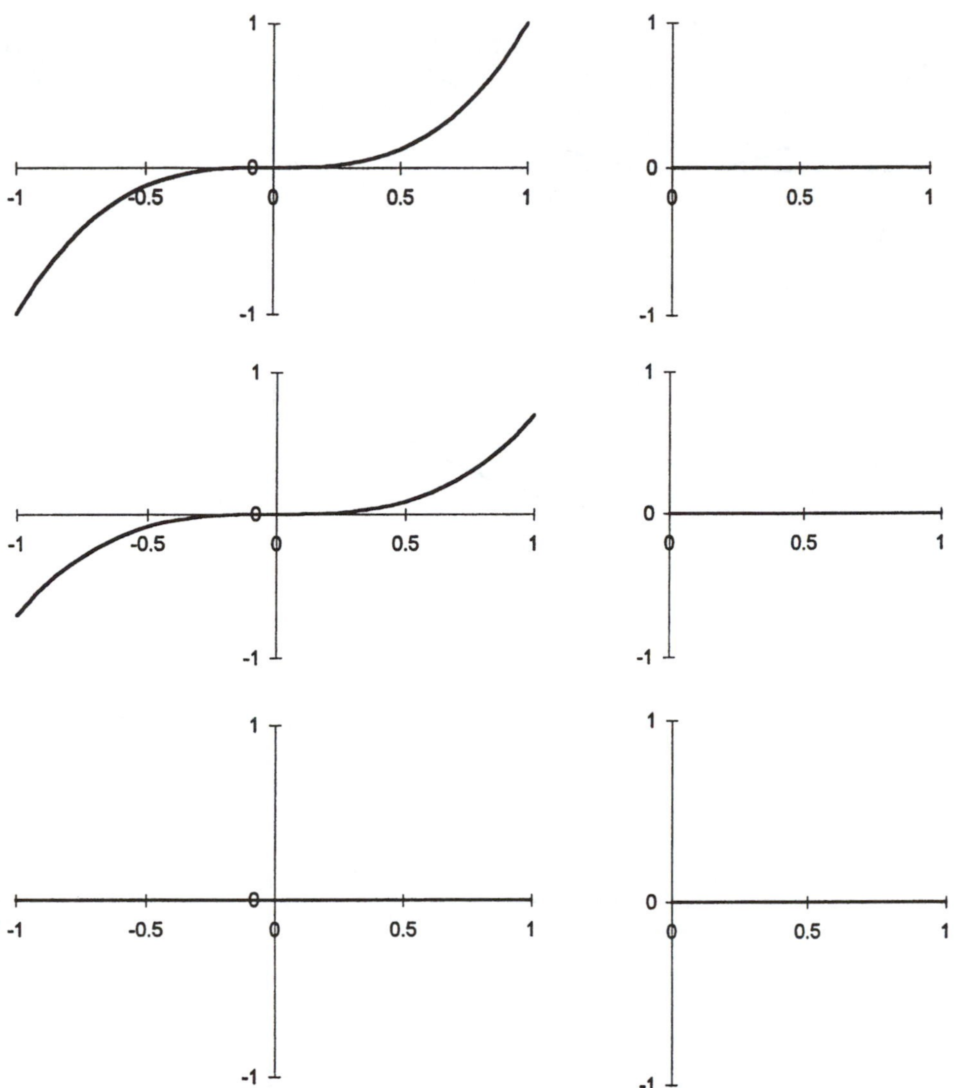

FIGURE 3.29 OPD plot for third-order coma.

from this aberration. This is true, but it must be remembered that the sagittal plot is for a ray fan in the $y = 0$ or sagittal plane. The actual x-direction extent of the image will be determined by the magnitude of the aberration and the symmetry of the aberration. If the ray intercept equations are rewritten in polar coordinates,

$$\epsilon_x = \frac{-\lambda}{n'u'} a_{131}\eta\rho^2 \sin(2\psi)$$

$$\epsilon_y = \frac{-\lambda}{n'u'} a_{131}\eta\rho^2(2 + \cos(2\psi)) \tag{3.69}$$

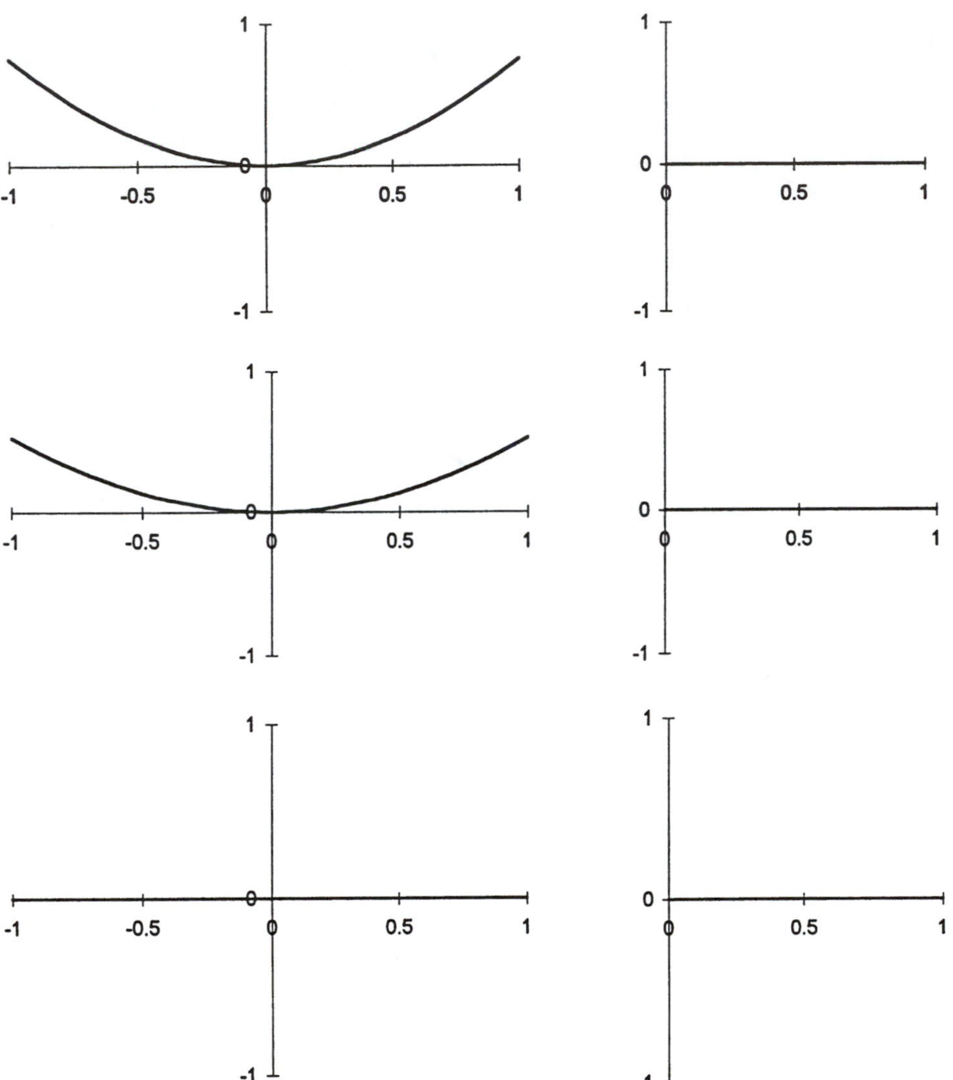

FIGURE 3.30 Ray intercept plot for third-order coma.

indicating a twice-around symmetry of the aberration with respect to pupil coordinates. The effect is quite complicated, with the geometrical effect shown in Figure 3.31. Each zone of the pupil generates a circular region on the image plane, traversed twice for a complete circuit of the pupil zone. The symmetry of the image that provides a 60° pair of asymptotes is established by the equations and indicates a three-to-one ratio of tangential to sagittal extent of the geometrical point image. Note that the rays that contribute to the maximum sagittal extent of the image arise at positions 2, 4, 6, and 8 in Figure 3.31.

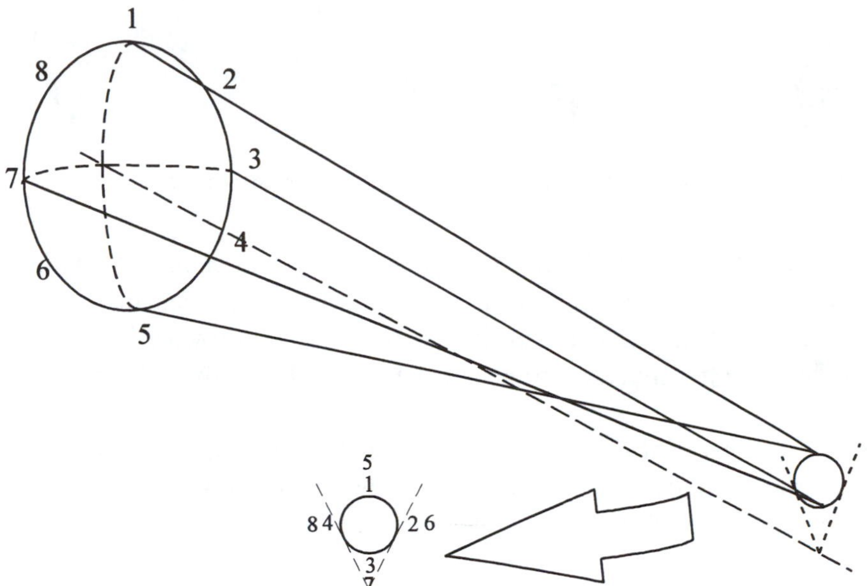

FIGURE 3.31 Ray behavior in the presence of coma.

The linear field dependence of this aberration makes it quite destructive to the appearance of the image. Since the amount of the aberration increases linearly from the axis, and since the image of each point forms a blur that is oriented towards the axis, objects will appear to be "flying out of the field."

There is an important relationship between this aberration and the fundamental optics of the lens as shown in section 2.1.4. In that section, the first-order variation of the aberration with field was shown to be zero if the sine condition is satisfied. Thus third-order coma aberration is the lowest order aberration generated by a failure of the sine condition in the lens. For this reason, it is required that the coma be made zero or at least be kept very small. It is a general practice in optical design to require that the linear coma be set to be zero.

Figure 3.30 shows a ray intercept plot for one wavelength of coma for an F/2 lens. The characteristic cubic wavefront error and the quadratic appearance of the ray aberration are indicators of the presence of coma. Note that the sagittal width appears to be zero, for the reason described above. A designer would also note that the magnitude of the aberration varies linearly with field position.

Spot diagrams showing the third-order coma pattern are provided in Figure 3.32. The effect of focus change is to rearrange the symmetry in the image. Because coma is an odd-power wavefront aberration, focus cannot be used to balance the effect of the error. The centroid of the coma image is not located at the chief ray location. Thus a lateral displacement of the image

FIGURE 3.32 Spot diagrams for third-order coma for several focal positions.

will appear with a coma aberration. Because coma varies linearly with field, there will be an apparent practical scale change introduced in the case of large amounts of coma. The blur increasing consistently with field also provides a disturbing visual appearance of the image flying away from the center of the field.

The next aberrations of importance are field curvature and astigmatism. These aberrations are usually treated together because both deal with a focal position change which varies quadratically with field. These aberrations are described by

$$\begin{aligned} W(x,y) &= a_{220}\eta^2\rho^2 + a_{222}\eta^2\rho^2\cos^2\psi \\ \epsilon_x &= \frac{-\lambda}{n'u'}\,2a_{220}\eta^2\rho_x \\ \epsilon_y &= \frac{-\lambda}{n'u'}\,2(a_{220}+a_{222})\eta^2\rho_y \end{aligned} \tag{3.70}$$

The terms in ρ^2 indicate that one circuit around a zone in the pupil leads to one circuit around a zone in the image. The different coefficients in the x and y directions indicate that in the general case the geometrical image will appear to be elliptical.

Understanding the implications of this aberration is best accomplished by adding a controlled focal shift to the aberration. The amount of this focal shift can be varied to shift to focal locations of interest:

$$\begin{aligned} \epsilon_x &= \frac{-\lambda}{n'u'}\,2a_{220}\eta^2\rho_x + 2a_d\rho_x \\ \epsilon_y &= \frac{-\lambda}{n'u'}\,2(a_{220}+a_{222})\eta^2\rho_y + 2a_d\rho_y \end{aligned} \tag{3.71}$$

Because the aberration varies quadratically with field, the best approach is to add a focal shift which varies with the second power of image height. The focus shift is then to surfaces which represent a specific characteristic of the aberration. These equations can then be rearranged to give

$$\begin{aligned} \epsilon_x &= \frac{-\lambda}{n'u'}\,2(a_{220}+a_d)\eta^2\rho_x \\ \epsilon_y &= \frac{-\lambda}{n'u'}\,2(a_{220}+a_{222}+a_d)\eta^2\rho_y \end{aligned} \tag{3.72}$$

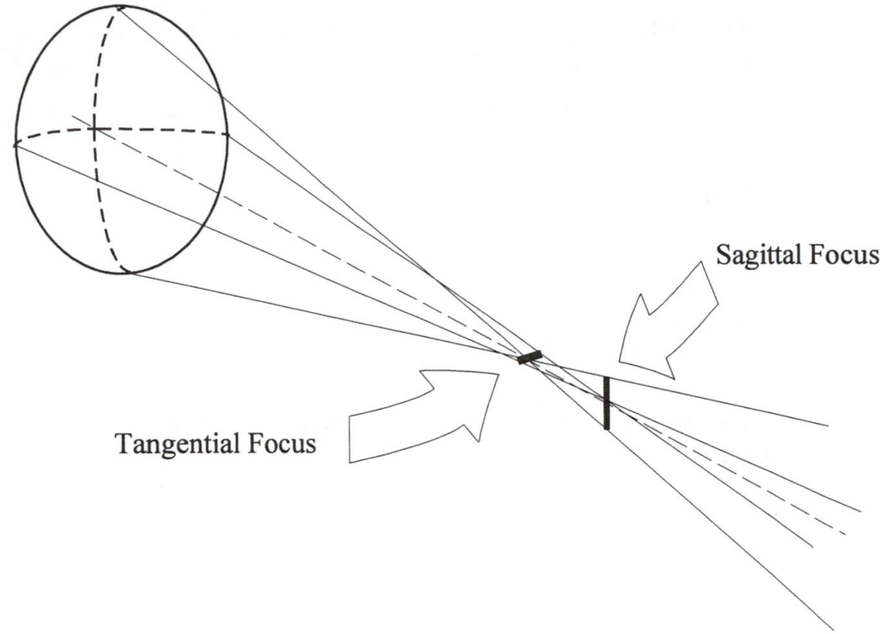

FIGURE 3.33 Ray behavior in the presence of astigmatism.

There are several obvious focus choices reached by setting the a_d term:

$$\begin{aligned} \text{Sagittal focus, } a_d &= -a_{220} \\ \text{Tangential focus, } a_d &= -(a_{220} + a_{222}) \\ \text{Mid-focus, } a_d &= -(a_{220} + \tfrac{1}{2}a_{222}) \\ \text{Petzval focus, } a_d &= -(a_{220} - \tfrac{1}{2}a_{222}) \end{aligned} \tag{3.73}$$

The first two cases are obvious from examination of the coefficients of the x or y directions, since one direction of the aberration blur is zero in each case. Figure 3.33 indicates the relative appearance of these two line focus locations. The third focus choice yields the same magnitude coefficient, with opposite signs for the x and y directions indicating that a circular zone of rays in the pupil becomes a circular zone in the image, with the exception that the direction of rotation of the ray pattern in the pupil at the image is opposite to that for the pupil. Geometrically, the appearance of the ray distribution on the image surface is identical to a simple defocus. (As will be seen later, this similarity in appearance to defocus is not true for diffraction images.) This third focus choice will be midway between the two line foci.

Figure 3.34 is a diagram of the location of these focal positions relative to the paraxial reference location. The distance from the reference position to the sagittal focus is seen to be given by a_{220}. The distances between the Petzval, sagittal, mid-focus, and tangential foci are all equal. There is a

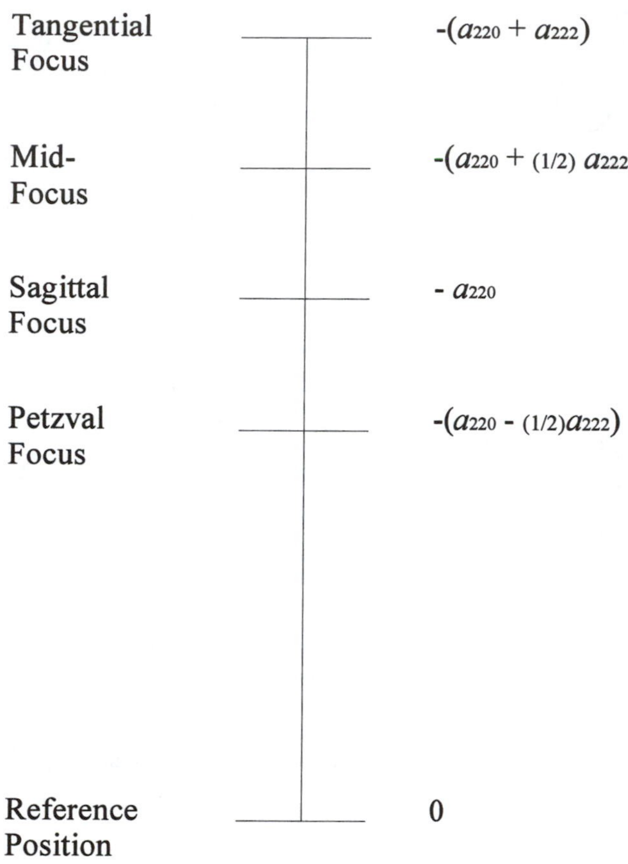

FIGURE 3.34 Relative locations of the critical focal positions associated with astigmatism.

three-to-one ratio of the distances from the Petzval focus to the sagittal and tangential foci. This three-to-one symmetry is characteristic of third-order astigmatism. The Petzval focus is mathematically determined from the location of the astigmatic foci.

The last focus choice, named the Petzval focus (after its discoverer, Josef Petzval), requires more explanation. The image at this focus position is described by

$$\epsilon_x = \frac{-\lambda}{n'u'}\,2(\tfrac{1}{2}a_{222})\eta^2\rho_x$$
$$\epsilon_y = \frac{-\lambda}{n'u'}\,2(\tfrac{3}{2}a_{222})\eta^2\rho_y \qquad (3.74)$$

which has the characteristic that a perfect image is located here if the astigmatism term a_{222} is zero. This position does not necessarily lie at the paraxial reference position, but at a position away from the paraxial position defined

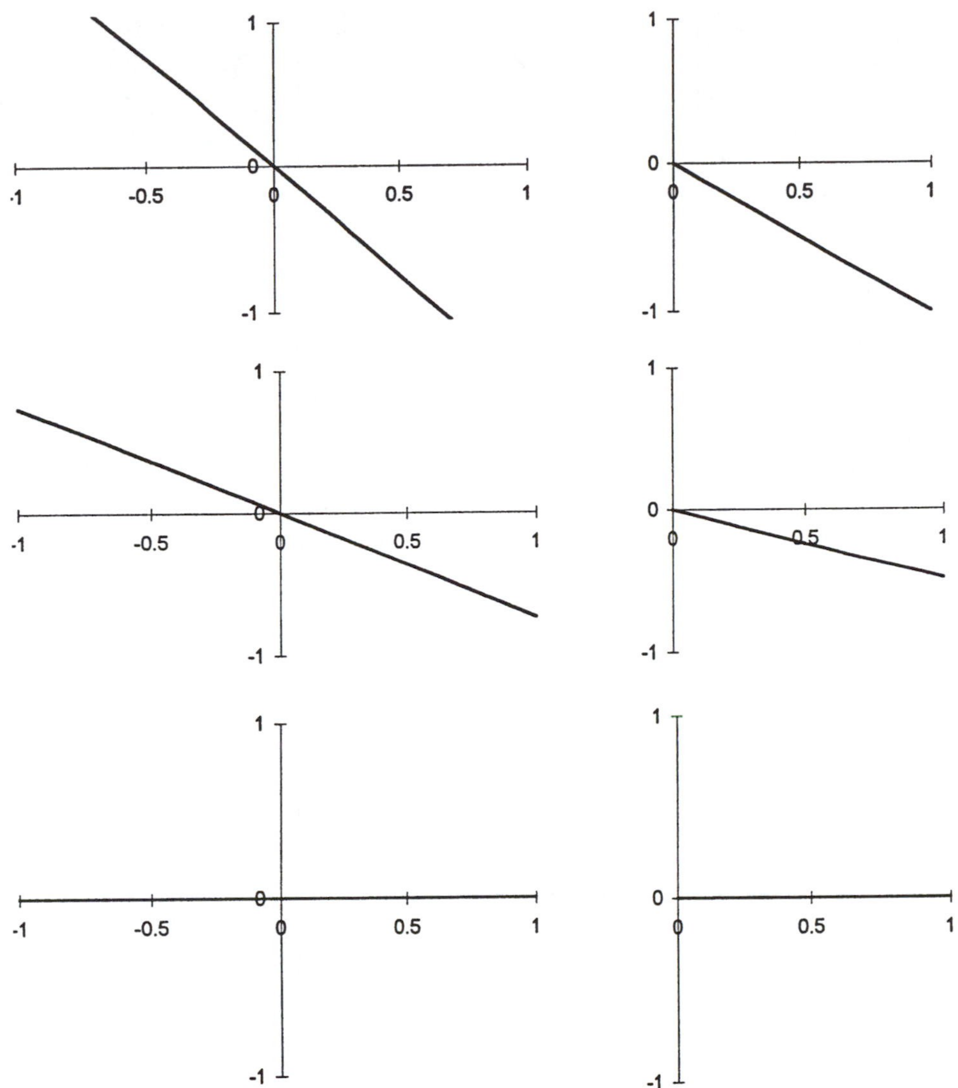

FIGURE 3.35 Ray intercept plot for third-order astigmatism and field curvature.

by the basic power distribution in the lens. An understanding of the location of this surface is essential to the design of flat-field objectives.

The ray intercept plots for astigmatism are shown in Figure 3.35. The focus choice is the paraxial focus surface. The different focus locations for the tangential and sagittal foci can be identified by the different slopes of the ray intercept plots. Figure 3.36 shows the OPD plots for astigmatism and field curvature, again referenced to the paraxial image location. The most familiar summary plot for the locations of the astigmatic fields is the field curvature plot shown in Figure 3.37. Here the second-order behavior of the tangential and sagittal field are clearly indicated. Because astigmatism is an

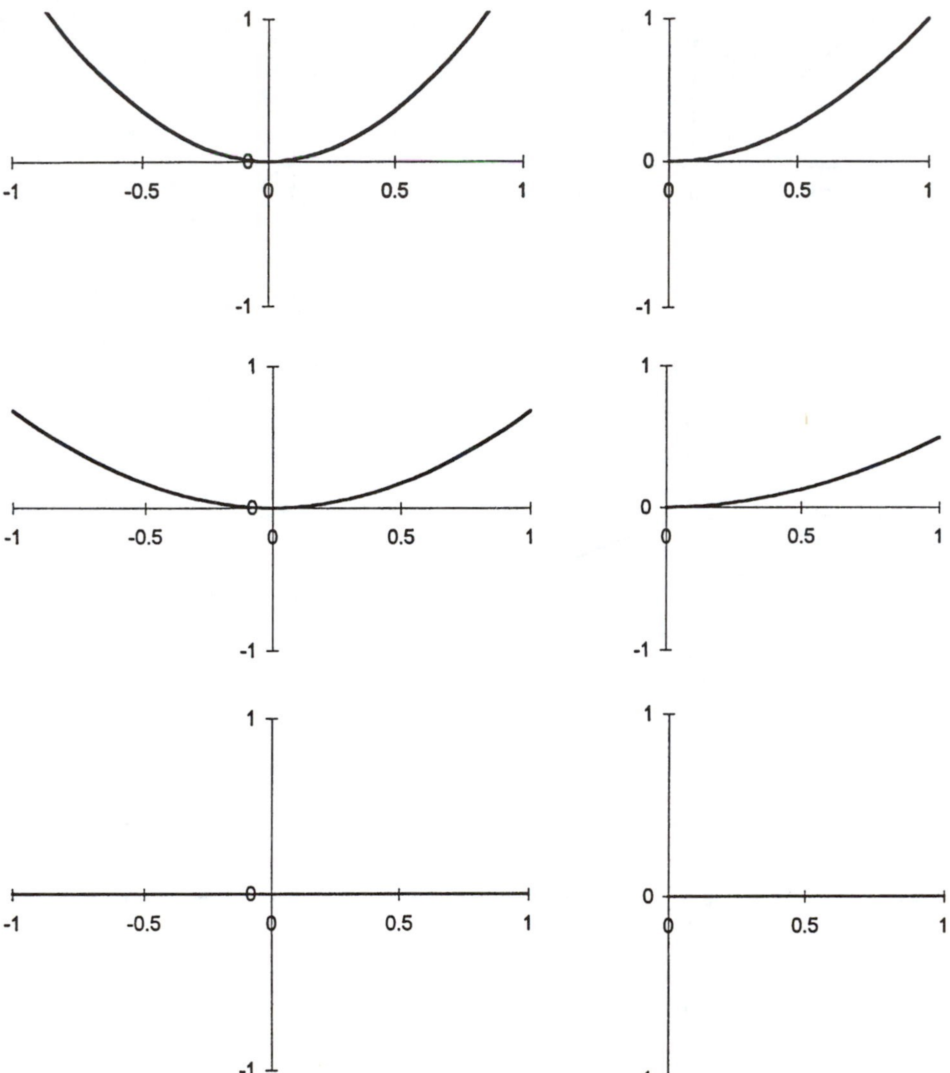

FIGURE 3.36 The OPD plots associated with the astigmatism and field curvature shown in the previous figure.

aberration with even pupil coordinate symmetry, the aberration effect can be reduced or balanced by a focus shift. Examination of the field plots indicates that this cannot be done for all field positions simultaneously.

The final third-order aberration to be considered is distortion. The wavefront and ray aberrations associated with distortion are

$$\begin{aligned} W &= a_{311}\eta^3 \rho \cos\psi \\ \epsilon_x &= 0 \\ \epsilon_y &= \frac{-\lambda}{n'u'} a_{311}\eta^3 \end{aligned} \tag{3.75}$$

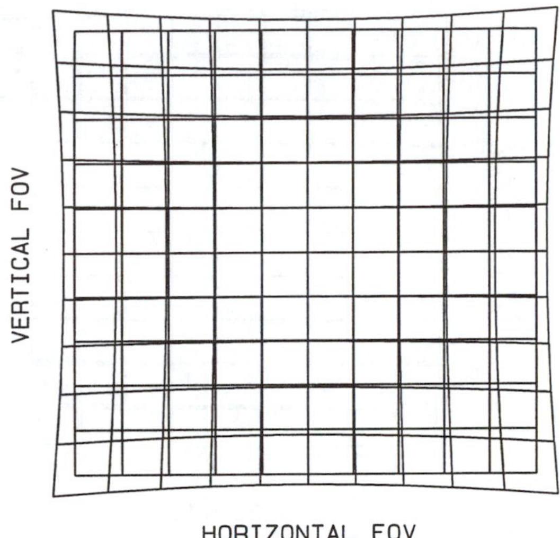

FIGURE 3.37 The appearance of an image of a grid showing positive distortion in comparison to a distortion-free grid image.

There is no displacement in the sagittal direction, and a shift of the entire point image in the tangential direction from the paraxial reference point dependent on the cube of the field. It is usual to refer to this displacement as a percentage distortion defined by

$$\text{Percent distortion} = 100 \frac{-\lambda}{n'u'h'} a_{311}\eta^2 \tag{3.76}$$

where h is the paraxial reference image height at the edge of the field. Note that sometimes the distortion is referred to as a fractional distortion, which does not include the factor of 100.

The meaning of distortion appears to be quite simple. Plots of the ray intercept curves conventionally fail to indicate the presence of distortion, as the reference location for the curves is the chief ray intersection at the reference wavelength. (Some programs contain an option to include distortion in the plots.) Usually the distortion is plotted separately, and must be examined by the designer to ensure that the desired distortion condition exists in the lens. It is quite easy to see constant improvement in image quality in a design while the requirement to control distortion is ignored and the distortion continues to grow as the design proceeds.

One familiar method of plotting distortion is to trace the intersections of a square grid in the object through the lens. In Figure 3.37 the grid traced through a lens with 5% third-order distortion is compared with what the object grid would appear to be if there were no distortion. This figure is for positive or so-called pincushion distortion. Figure 3.38 is a plot for a

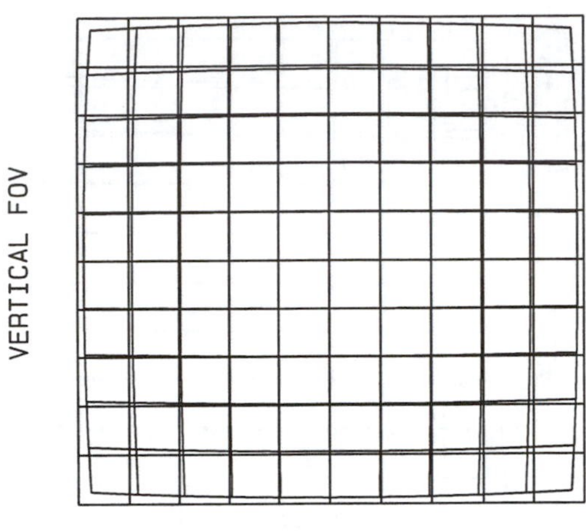

FIGURE 3.38 The appearance of an image of a grid showing negative distortion in comparison to a distortion-free grid image.

similar amount of negative distortion. The comparison of the original object grid makes the presence of distortion quite obvious. The significance of distortion is that it produces a warped version of the object. Examination of the figures shows that the scale relating object and image dimension and shape is not constant, but changes across the field. This means that the size of an object will vary with position in the field. Because distortion is associated with coordinates of the image plane, the shape of an object will not be preserved and will appear to be different depending upon the position in the field.

Rarely will each of these aberrations exist in isolation. Figure 3.39 shows the ray intercept plots and Figure 3.40 the OPD plots for a combination of third-order aberrations in which all of the aberrations are present in one-wavelength units. It is possible to pick out the characteristics of each of the aberrations in this mix from the plots. Figures 3.41 and 3.42 show an attempt to compensate the effect of these aberrations by a focal shift which balances the spherical aberration. The characteristic shapes of the ray intercept curves are still present, but are rotated with respect to the axes by the introduction of the focus shift. The OPD plots have been balanced by the addition of a compensating focus shift.

Obviously, these third-order aberrations do not provide the entire story regarding the aberration content of a lens. It will be seen later that appropriate settings of the third-order aberrations are used to balance the effect of higher-order aberrations, much as the focus adjustment can balance some of the third-order aberrations.

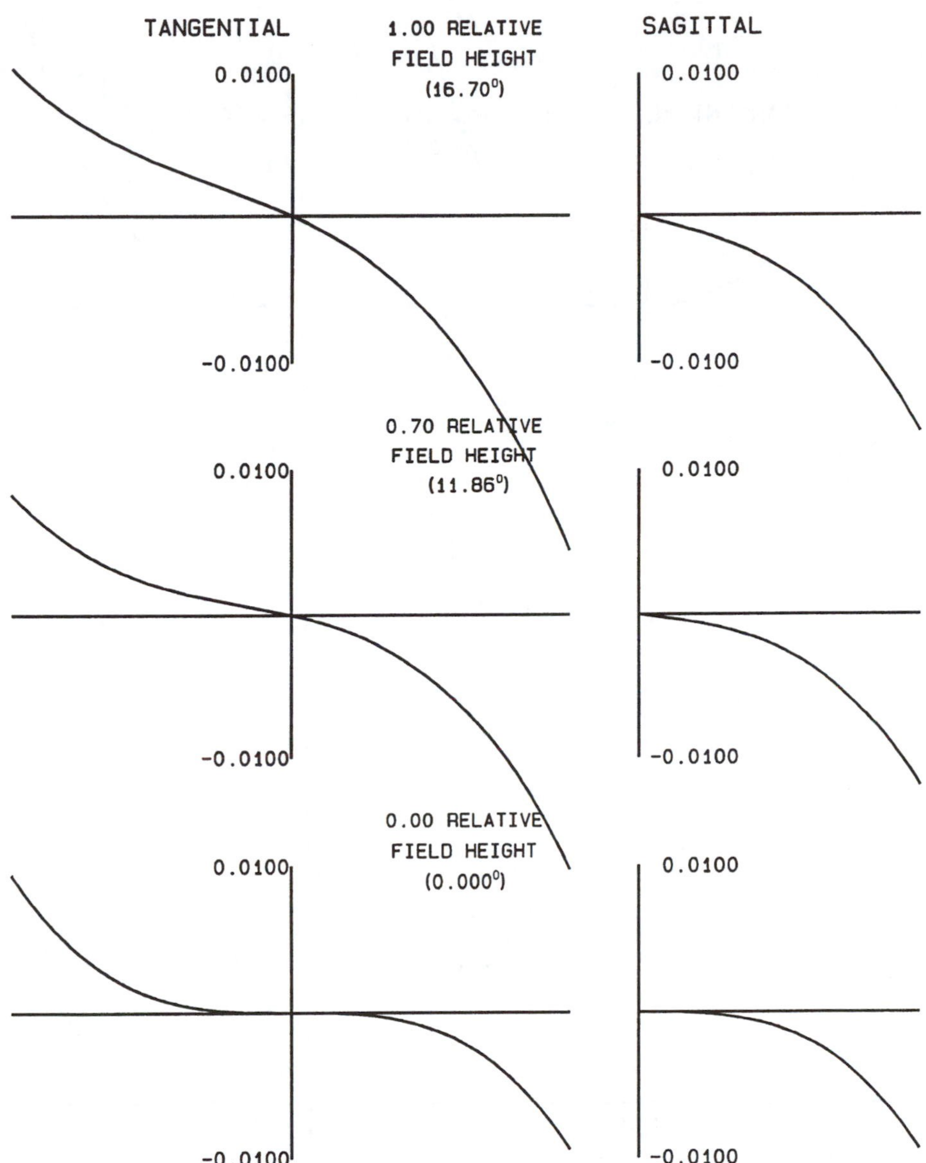

FIGURE 3.39 Ray intercept plot for a lens containing all third-order aberrations (wavelength 546 nm).

3.1.1.2.2 EVALUATION

Coefficients describing these aberrations in a lens system can be determined in several ways. A derivation following the general procedure for defining the spherical aberration can be carried out. The optical path difference change introduced at the surface is computed by evaluating the reference sphere coordinates in intercepts on the surface. The optical path differences are

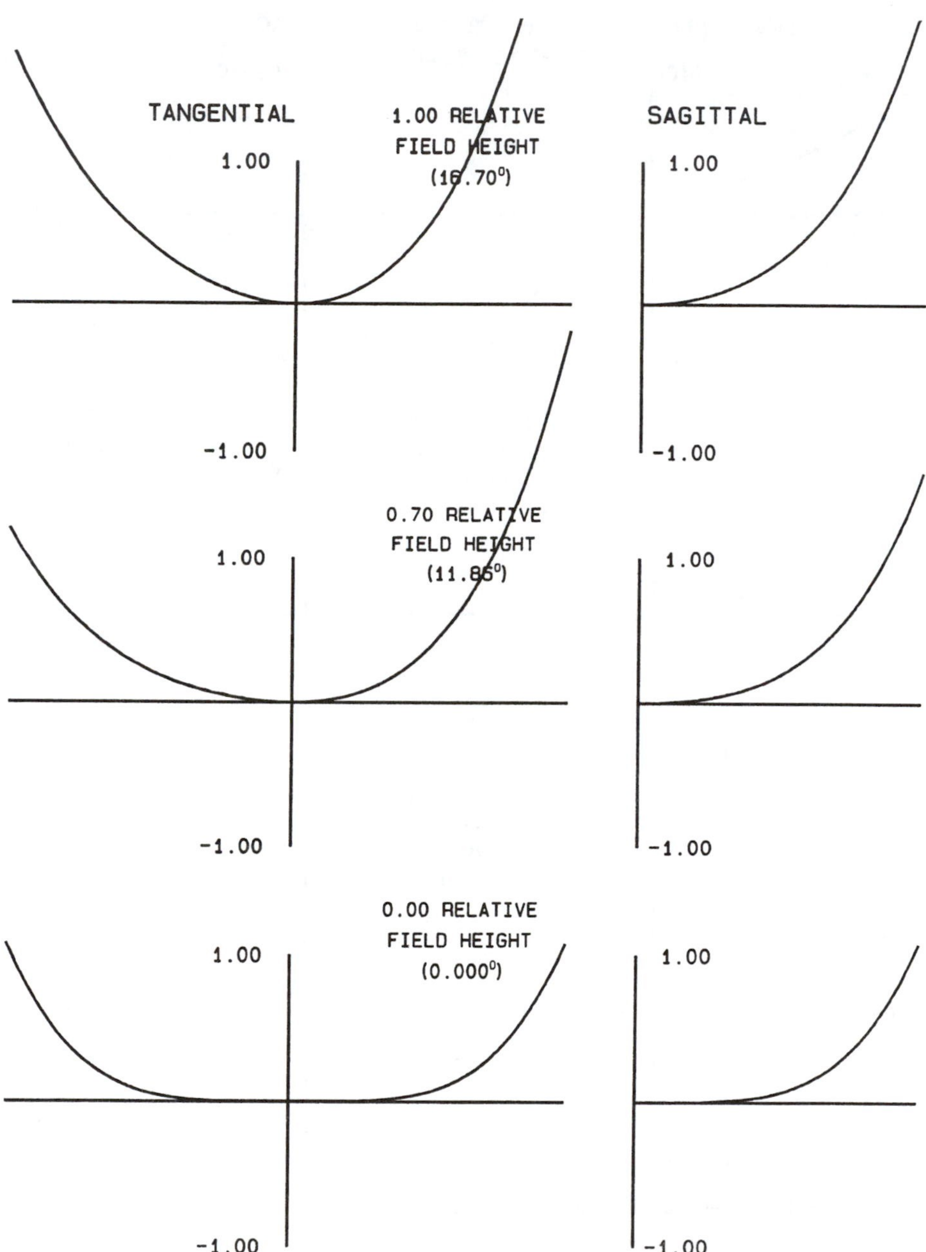

FIGURE 3.40 The OPD plot associated with the previous figure.

obtained by confining the points to lie on a surface, described to the required accuracy. The principal difference for the off-axis aberrations is the inclusion of two parameters on the surface, the marginal paraxial ray height, as in the spherical aberration derivation, and the paraxial chief ray height, which determines which portion of the surface the actual ray bundle passes through.

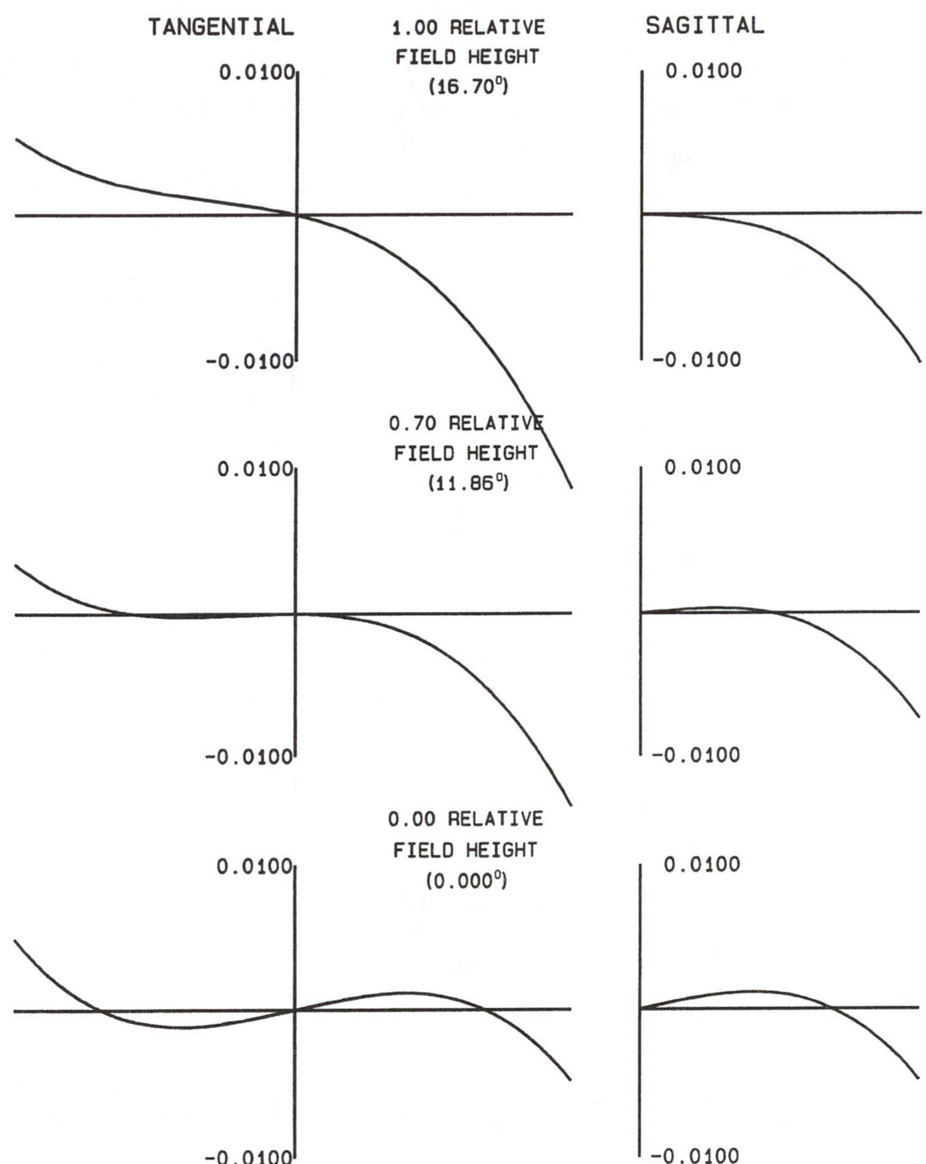

FIGURE 3.41 Ray intercept plot using the same aberration data but adjusted for optimum focal position.

An accessible derivation is included in the book by Welford, which directly provides all of the important third-order aberration coefficients in terms of system parameters and paraxial base ray paths (Welford 1986). Other derivations are found in Hopkins (1950) and in Born and Wolf (1959). Various other ways of getting to the aberrations can be located in Slyusarev (1984) and in Malacara and Malacara (1995). With all of these options for deriva-

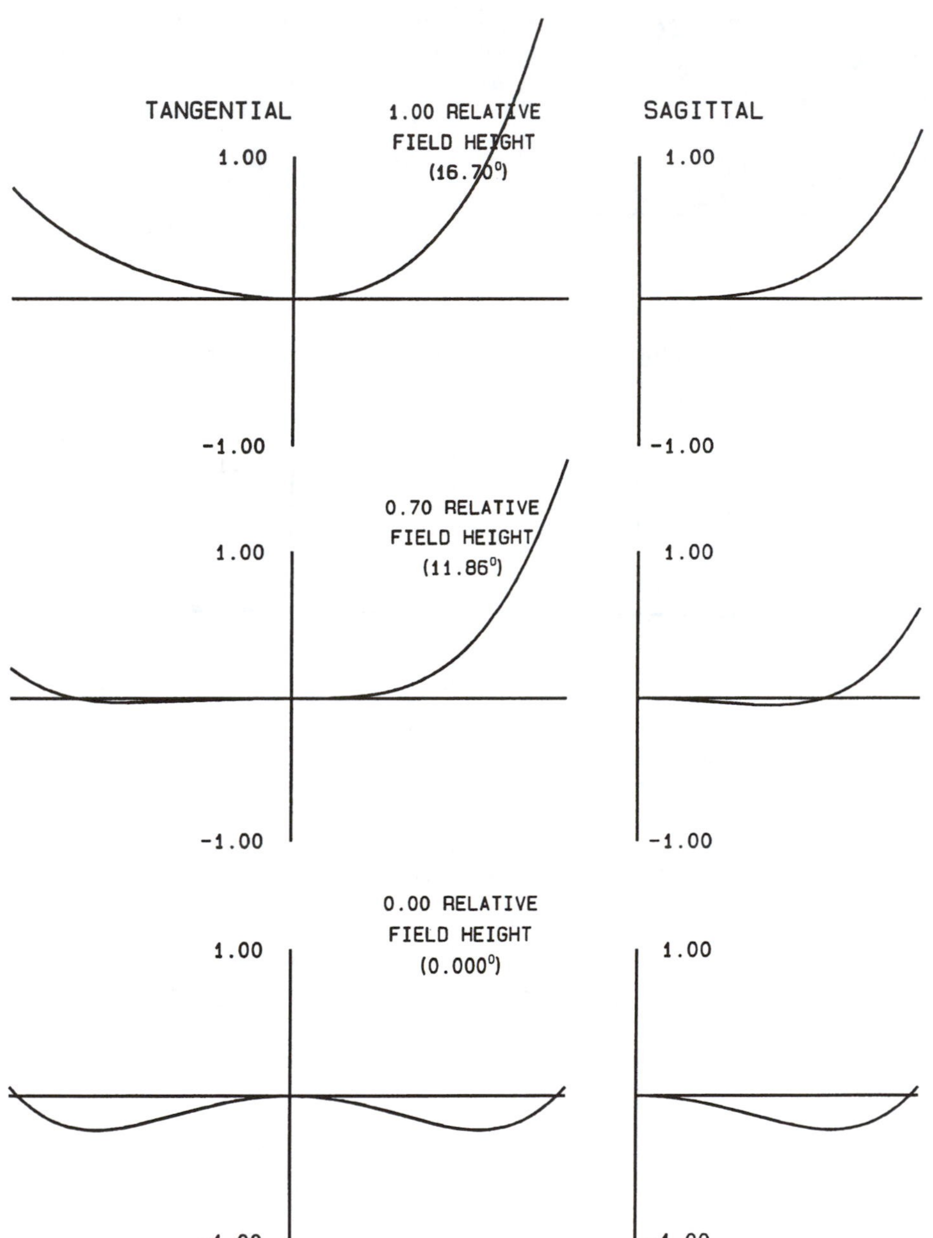

FIGURE 3.42 The OPD plot for the situation shown in the previous figure.

tions available, it is redundant to carry out a detailed derivation here. Each reader will likely find his or her own favorite at some time.

As with the earlier spherical aberration discussion these derivations are usually based upon examination of a drawing of the imaging situation. Important distances are abstracted from the drawing and used in constructing required algebraic functions. The relation between the pupil coordinates and the coordinates on the actual optical surfaces is not always quantified. For obvious reasons, these derivations all presume some reasonable set of conjugates and may need modification to apply generally. An alternate approach is to define the aberrations with respect to an auxiliary axis of the spherical surface that passes through the object point, the image point, and the center of curvature of the surface. Such an axis must exist, because a direct straight and undeviated ray will pass from the object to the image through the center of curvature of the surface. There is an apparent problem with generality, in that the object, and the image also, could be located in the vicinity of the center of curvature, which would appear to make such a derivation invalid. However, the use of a diagram to select suitable dimensions for entry into a derivation is really only diagrammatic, and the real initial and final coordinates will be attached to the chosen paraxial image and object space reference positions in any event. Therefore, any derivation that aids understanding is useful, and may be verified against reality by carrying out tracing of real rays.

Each of the derivations of the third-order aberrations mentioned above usually includes a discussion about the accuracy of the formulae that have been obtained. Because there are approximations involved in representing the real ray path and the real ray coordinates in the pupil, the usual expression is that the accuracy of the equations is "good enough for practical purposes." This means that useful starting points can be discovered and the general imaging properties of the lens determined. This also means that detailed ray tracing will eventually be required for evaluating the actual image.

Another approach to the computation of aberration coefficients is through the work carried out by Buchdahl. In his book *Optical Aberration Coefficients* (Buchdahl 1968), he develops a method for obtaining aberration coefficients to successive orders by a method of iteration. The first-order ray paths are used as a base in determining the third-order aberrations, which are intrinsic to each surface. Fifth-order aberrations are obtained by using the third-order paths as a base. This is an algebraic iterative process, in which there are two components of the aberration produced at each surface, an intrinsic component depending on the first-order ray path and a transferred component which includes the influence of aberrations arising at preceding surfaces. The aberrations now include the effect of pupil aberrations and are expressed in terms of coordinates in the entrance or exit pupils. The approach used provides more generality in obtaining coefficients under all conjugate locations and magnifications. Unfortunately, the complexity of the mathematics involved in the iterative process derivations limits the audience to those who can understand complicated mathematics. A good discussion of the issues is contained in Buchdahl's later book on Hamiltonian optics (Buchdahl 1970).

The work by Buchdahl was of very great importance in introducing the concept of high-order aberration coefficients into the design process in the 1960s. Tracing of rays was a laborious process prior to the introduction of stored-program calculators, and the knowledge gained by evaluating aberration coefficients and the change of coefficients with system parameters was extremely important. As in any derivation of aberration coefficients, the presence of large aberrations makes the actual convergence of these coefficients in a real system frequently not well known. Buchdahl's work encouraged much wider understanding of concepts such as high-order field curvature and the methods of balancing oblique spherical aberration. Alternate approaches to obtaining high-order coefficients have been developed by several workers in recent years, but the existence of a programmable set of Buchdahl coefficients has inhibited wide use of any of these other approaches.

Except for the Buchdahl coefficients, which were converted into accessible forms for programming by Rimmer (1963), none of these high-order aberration approaches has developed into a major force in design methods. The Buchdahl coefficients are available in most design programs today, but have become of less importance in the actual design process. The ability to trace thousands of rays very inexpensively has decreased the need for using high-order coefficients to shorten the computation time. The aberration coefficients can also be very poorly convergent to actual aberrations under many practical conditions. There are no intrinsic indications of when the convergence will be poor. Therefore almost all designers ultimately depend upon ray intercept and OPD evaluation.

Because of the access to several alternate derivations, only a sketch of the derivation of the complete set of third-order aberration coefficients will be presented here. The starting point is a diagram, as in Figure 3.43. The problem is to obtain the change in aberrations at a single surface in terms of the pupil coordinates associated with the passage of a bundle of rays across that surface. The desired description of ray passage is in terms of the paraxial ray intercepts at the surface. The diagram indicates a choice of reference points for the object and image. (Both are conveniently taken on the same side of the surface to make the diagram either more or less confusing, depending upon your point of view.) The diagram indicates one set of reference points, pupil sphere, and sphere located at the spherical surface for object and image reference locations. This path can be read off the diagram as the distance along the appropriate rays in object or image space

$$W = n'(\bar{P}\bar{A}' - QA') - n(\bar{P}\bar{A} - QA) \tag{3.77}$$

From the drawing this can be rewritten as

$$W = n'P'Q - npQ \tag{3.78}$$

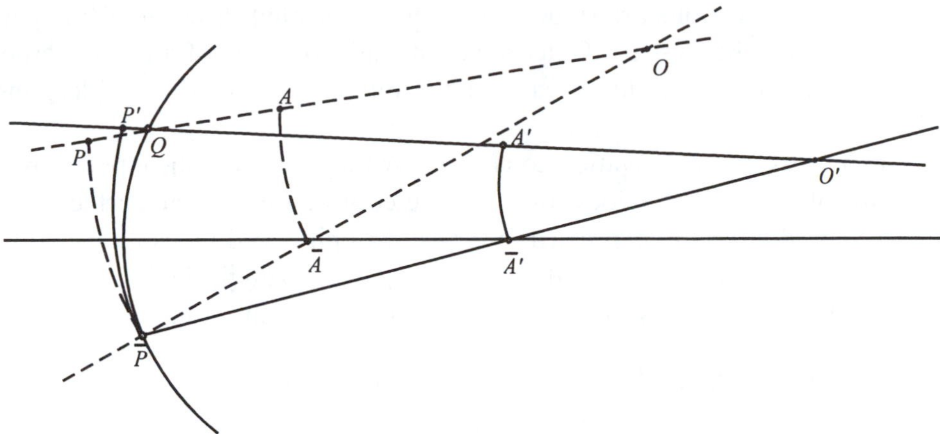

FIGURE 3.43 The important parameters occurring in the discussion of the origin of the third-order aberration coefficients.

Adding and subtracting the distances from the pupil reference spheres to the reference points allows rewriting the path difference introduced at the surface in terms of the reference locations and the intersect on the surface

$$W = n'(\bar{P}O' - QO') - n(\bar{P}O - QO) \tag{3.79}$$

These quantities can be evaluated in terms of the reference point locations and the coordinates on the surface. The coordinates of Q are (x_s, y_s, z_s) on the surface for the intersection point of the lines from the reference locations. The coordinates of $\bar{P}$ are the intersection point of the chief ray on the surface $(\bar{x}, \bar{y}, \bar{z})$. This definition of the aberrations in terms of the surface intersection is of little use in adding up the aberrations through the system. Therefore, these locations will eventually be written in terms of the pupil coordinates and relative field position multiplied by the paraxial values for the marginal and chief rays on the surface. The reference points are given by the (x, y, z) coordinates of the object and image positions for the surface. Generally these will be taken to be the paraxial image reference coordinates.

The aberrations will then be determined in terms of surface coordinates. For some applications this may be useful. In general this is not especially useful in adding the successive OPD errors to obtain a final representation of the aberrations. The resulting OPD values need to be interpreted in terms of pupil coordinates at each surface. The pupil coordinates in turn need to be expressed in a relation between successive surfaces in order to obtain a final correct set of values for the aberration polynomials.

Fortunately, it can be demonstrated, or at least made plausible, that the pupil errors for the third-order aberrations produce contributions greater than third order. Thus it is sufficient to accept the approximation that the third-order aberrations are sufficiently well represented in terms of the paraxial data for the system. Determination and addition of the fifth-order aberrations will require knowledge of the pupil errors to third order, and so on.

The process of the derivation consists of writing out the optical path errors in terms of the reference locations and the coordinates on the surface. The surface coordinates are expressed in terms of pupil coordinates by using the first-order ray trace data describing the ray paths through the lens.

The length $\bar{P}O$–QO is then found using the usual length equation

$$\bar{P}O - QO = [(l-\bar{z})^2 + (h-\bar{y})^2]^{\frac{1}{2}} - [(l-z_s)^2 + (h-y_s)^2 + x_s^2]^{\frac{1}{2}} \tag{3.80}$$

The lower orders are retained by expanding the square root and neglecting orders beyond the fourth. Rearranging and subtracting the terms leads to an expression that includes both the first-order and third-order terms. Accepting the reference locations to be placed at the paraxially predicted image locations eliminates the first-order (second-order in wavefront) terms. The remaining terms are the third-order aberration terms which are now expressed in terms of paraxial data and system constants.

Much algebra then ensues which can cover several pages, and there are many ways of getting to a final useful set of third-order aberration equations. The general discussion here is likely to be sufficient for a practical understanding of the derivation of the aberrations. For details, see the previously mentioned text by Welford (1986).

The result of any of the derivations of the third-order aberrations can be summarized in terms of a set of equations for the third-order aberration coefficients, usually called the Seidel aberration coefficients. One notation that has become more-or-less standard is that introduced by H. H. Hopkins (1950). This notation has been used by many other authors, such as Welford (1986). In a consistent set of forms

$$\begin{aligned}
S_{\mathrm{I}} &= -\sum (ni)^2 y\left(\frac{u'}{n'} - \frac{u}{n}\right)\\
S_{\mathrm{II}} &= -\sum ni\, n\bar{i} y\left(\frac{u'}{n'} - \frac{u}{n}\right)\\
S_{\mathrm{III}} &= -\sum (n\bar{i})^2 y\left(\frac{u'}{n'} - \frac{u}{n}\right)\\
S_{\mathrm{IV}} &= -\sum (nhu)^2 c\left(\frac{1}{n'} - \frac{1}{n}\right)\\
S_{\mathrm{V}} &= -\sum \left[\frac{(n\bar{i})^3}{ni}\, y\left(\frac{u'}{n'} - \frac{u}{n}\right) + \frac{n\bar{i}}{ni}(nhu)^2 c\left(\frac{1}{n'} - \frac{1}{n}\right)\right]
\end{aligned} \tag{3.81}$$

are the complete set of third-order Seidel aberration coefficients when summed over contributions from all of the surfaces of the optical system. The c is the curvature of the current surface, and nhu represents the Lagrange invariant, calculated in the object space and (of course) applicable throughout the entire lens system. The wavefront errors associated with this set is

$$\begin{aligned} W(\rho_x, \rho_y, \eta) = & \tfrac{1}{8} S_{\mathrm{I}} \rho^4 + \tfrac{1}{2} S_{\mathrm{II}} \rho_y \rho^2 \eta + \tfrac{1}{2} S_{\mathrm{III}} \rho_y^2 \eta^2 \\ & + \tfrac{1}{4} (S_{\mathrm{III}} + S_{\mathrm{IV}}) \rho^2 \eta^2 + \tfrac{1}{2} S_{\mathrm{V}} \rho_y \eta^3 \end{aligned} \tag{3.82}$$

which, of course, leads to the ray aberration symmetries discussed above.

There is another common notation that is frequently used in the literature (Born and Wolf 1959), or in U.S. lens design literature (R. Hopkins 1962) in which the coefficients are labeled differently. This is readily translated by using the relationships

$$\begin{aligned} B &= -S_{\mathrm{I}} \\ F &= -S_{\mathrm{II}} \\ C &= -S_{\mathrm{III}} \\ P &= -\frac{S_{\mathrm{IV}}}{(nhu)^2} \\ E &= -S_{\mathrm{V}} \end{aligned} \tag{3.83}$$

The choice of notation is certainly not critical, as long as it is used consistently. There is really no superiority of either notation for the discussion of aberration effects, and the choice of which to use is left to the reader. Either will be used as seems most appropriate for the application in this book.

The aberration contribution equations are quite readable. The first three aberrations appear to be related, and are due to the spherical aberration of the surface. The intrinsic third-order spherical aberration of the surface is determined by the factor involving the change in slope angle over index of refraction before and after the surface. This spherical aberration is mixed into the coma and astigmatism by the choice of the chief ray. Reference to section 2.1.4 indicates another method of looking at the coma. In this case, the conclusions relate to the third-order coma term alone.

The S_{III} coefficient appears as spherical aberration of the chief ray with respect to the auxiliary axis of the surface. The astigmatism appears in the image because the intrinsic spherical aberration introduced by the surface is aperture selected by the choice of the chief ray and the working beam aperture on the surface. The S_{II} term is a mixture of the spherical aberration terms and arises because of the asymmetric cut of the spherical aberration selected by the choice of the chief ray. Changing the selection of the chief ray by shifting the location of the stop will alter the amounts of coma and astigmatism that are introduced at the surface. This interpretation provides a different manner of looking at the astigmatic contribution than described in section 2.1.5. In this case it is the third-order contribution of the astigmatism that is being described.

The Petzval field curvature term, S_{IV}, does not depend upon the path of the rays through the lens, but only on the curvatures and the indices. The invariant serves as a scaling factor to place this into the wavefront aberration.

The distortion, S_V, appears to be quite complicated. The two terms indicate that the distortion is introduced both because of the intrinsic spherical aberration of the surface and the Petzval field curvature of the surface. The distortion contribution is related to all of the aberrations, and thus is very difficult to control at the same time as controlling all of the other aberrations in a lens system.

The formulae as stated in equations (3.73) are quite graphic in their interpretation, but are actually not very well suited for computation. There are cases where division by zero can occur, especially in the distortion coefficient. These formulae can be algebraically massaged into different forms that are more applicable to computation. One method of rewriting these equations leads to a set of relations

$$\begin{aligned}
S &= n\left(\frac{n}{n'} - 1\right)y(i + u') \\
\bar{S} &= n\left(\frac{n}{n'} - 1\right)y(i + u') \\
B &= Si^2 \\
F &= Si\bar{i} \\
C &= S\bar{i}^2 \\
P &= \frac{-(n' - n)}{nn'}\,c \\
E &= \bar{S}i\bar{i} + I\left(\frac{n}{n'} - 1\right)\bar{i}(\bar{u}' + \bar{u})
\end{aligned} \tag{3.84}$$

These equations look different and more complicated, but are actually only rearrangements for computation of the same information as in the earlier equations. This method of writing the equations does not lead to special cases for the computer program to contend with. There are several possible arrangements of these equations that might be used in computations, but all should pass the test of accuracy and no special cases. Examination of the distortion coefficient shows that these forms are immune to errors due to zeros in the denominator, but the previous form of the equations is actually quite a bit more transparent in terms of the origin of the coefficients.

The third-order, or Seidel, aberrations serve as a good model for the lowest order of aberration that is encountered in a lens. In solving design problems, these aberrations can be controlled by the proper selection of parameters and aberrations to be corrected. Bending of a lens component affects the spherical aberration and coma. If the stop is moved, all aberrations are affected. In this sense, the third-order aberrations are considered controllable. There are limits upon the possible aberration corrections that are available in various lens configurations, that will be discussed later.

3.1.1.2.3 ORIGIN

The form of the equations for computing the third-order aberrations obtained in the last section serves as a basis for understanding the origin of the aberrations. The aberrations are computed from the data associated with tracing the paraxial marginal and chief rays through the lens. Some insight toward the origin of aberrations can be obtained from examination of the equations. A better way of obtaining a grasp of the origin of aberrations and the limits upon correction is found by using the equations and real ray trace data together to examine how aberrations behave.

Examination of the symmetries in these equations shows that the first three coefficients, spherical aberration, coma, and astigmatism, have a similar origin. Each of these terms contains a y, which specifies the aperture size and scales the error to the focal surface, and a factor involving the power of the surface through the paraxial marginal ray angles before and following the surface. The progressive change in the terms is in the roles of the angle of incidence of the marginal paraxial ray and the angle of incidence of the paraxial chief ray. The spherical aberration term depends upon the square of the marginal ray incidence angle, the astigmatism upon the square of the chief ray angle, and the coma upon a product of the two. The Petzval term does not depend upon the path of the rays through the lens, but only on the curvatures and the indices. Finally, the distortion has two terms. The first distortion term has an appearance related to the astigmatism, with an added factor of the ratio of the chief ray angle of incidence to the marginal ray angle of incidence. The second distortion term depends upon the Petzval terms and the same factor of the angles of incidence.

A summary of the effects is that the first three aberrations are related to the spherical aberration content of the surface, interpreted in terms of the marginal or chief ray direction. The Petzval term is dependent only upon the geometry of the surface, and the distortion is related to the constraint that the chief ray pass through the location of the sagittal focus location, as obtained from the astigmatism and Petzval terms.

This will become somewhat more evident by examining the construction of off-axis images by real rays at a single surface. First consider the case of astigmatism. Previous discussion of the close skew ray equations showed that the astigmatic effect was related to the angle of incidence of the chief ray. If the angle of incidence is zero, there is no astigmatic component. Figure

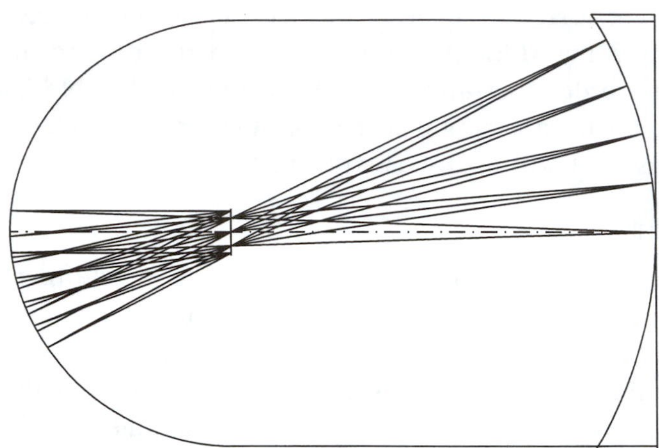

FIGURE 3.44 Single surface system for illustrating the origin of the aberrations. Stop is located at the center of curvature of the surface. The image reference surface is curved as indicated (CODE V).

3.44 shows an arrangement in which the stop is located at the center of curvature of a single surface. The example is a solid Bk7 glass block, with a spherical surface of 100 mm radius of curvature as the first surface. The image is examined within the glass so that we see the effect of a single surface. The second surface is located at the position of the (immersed) paraxial image location. Bundles of rays are traced for several field angles, up to 28°. Because all of the chief rays have zero angle of incidence on the surface there should be no astigmatism. The image for all fields is formed on the curved surface indicated.

Figure 3.45 is a plot of the spot diagrams for the image formed on this spherical surface, and indicates the presence of spherical aberration only. A progressive change in the spot diagrams is noted, due to the change in obliquity of the ray bundles with respect to the stop. The absence of astigmatism indicates that this surface is the Petzval surface for the single surface lens; the curvature is determined by the obvious symmetry of the situation. Plots of the astigmatic fields and distortion are shown in Figures 3.46 and 3.47. In Figure 3.46, the plots are made relative to the curved surface that corresponds to the Petzval surface. In Figure 3.47 the plots are relative to the flat paraxial reference surface. The sagittal and tangential surfaces coincide at the Petzval surface. Reference to the coefficients for astigmatism and coma in the equations shows that these two aberrations are zero when the angle of incidence of the chief ray is zero. The third-order prediction of aberration is supported, and obviously relates to all orders in this specific real ray case.

The stop is then moved to the surface of the lens, as shown in Figure 3.48. The paraxial image surface and the Petzval surface are evidently unchanged.

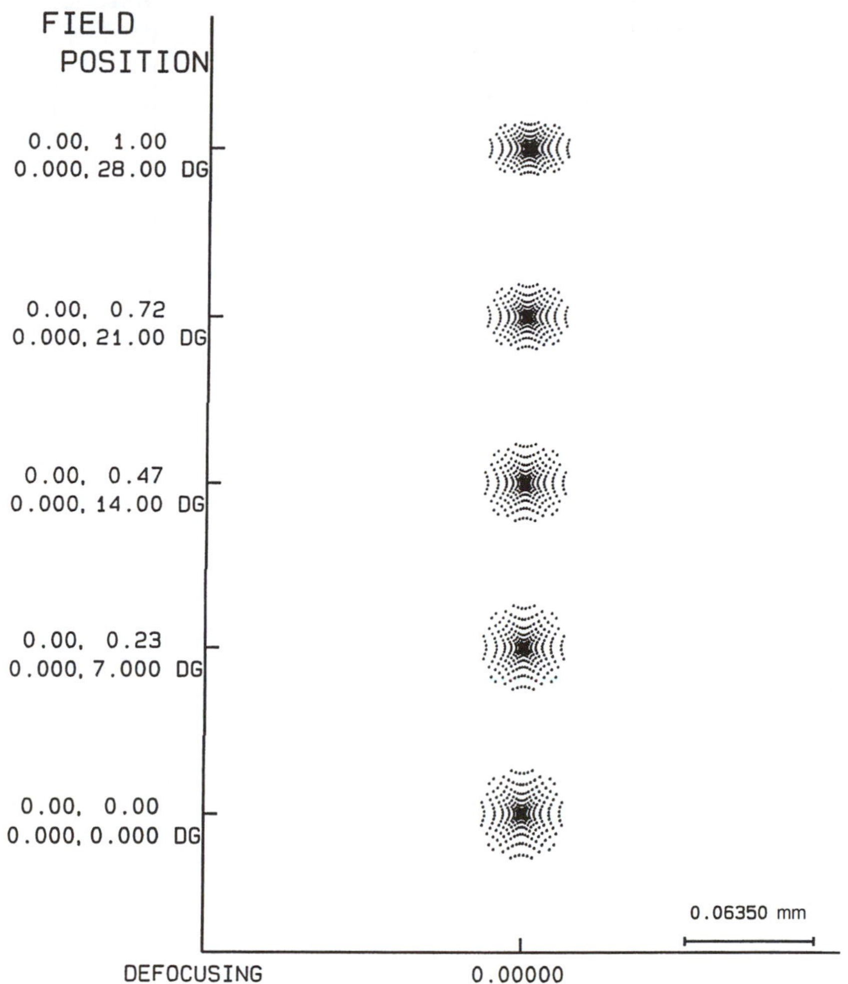

FIGURE 3.45 Spot diagrams for the image with stop at the center of curvature (CODE V).

Examination of the individual ray fans at each field position indicates that the tangential rays cross at a distance further from the Petzval surface. Careful examination indicates that the rays do not appear to cross symmetrically at their new focus location. Figures 3.49 and 3.50 show the same two methods of plotting the astigmatic fields as before. Figure 3.49 shows the sagittal and tangential surfaces with respect to the curved Petzval surface. A three-to-one ratio of distances from the Petzval surface is evident. Figure 3.50 is with respect to a flat image surface, and although it is not conventional to plot the Petzval surface, its location can be estimated from the locations of the tangential and sagittal surfaces. The presence of distortion is now noted in both plots because of the movement of the sagittal focus location from the Petzval location.

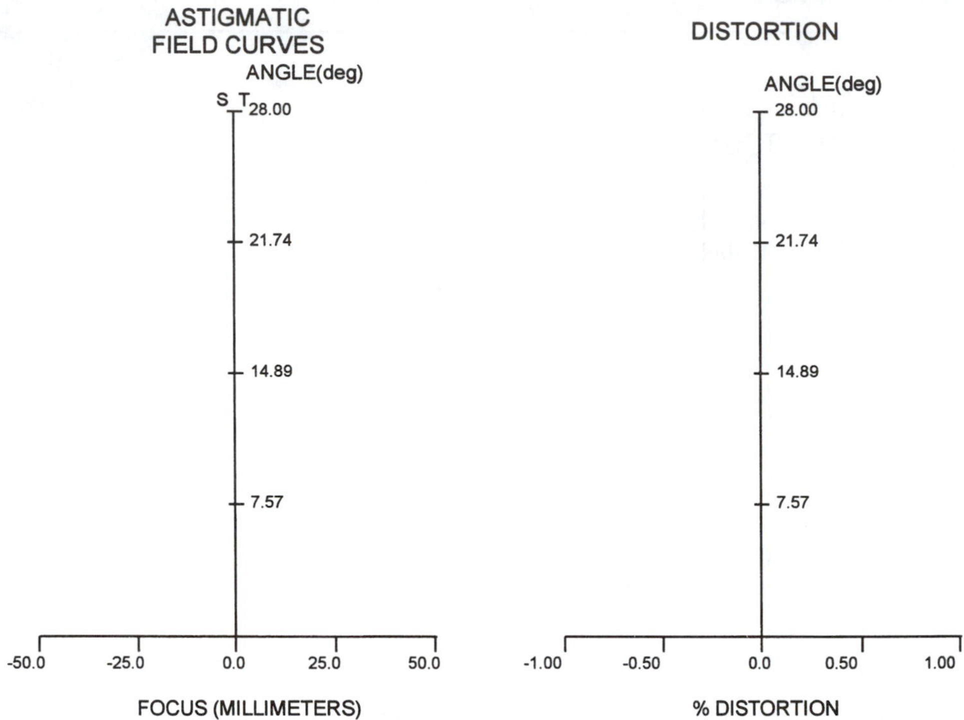

FIGURE 3.46 Plot of the astigmatic fields and distortion for the situation in the previous figure (CODE V).

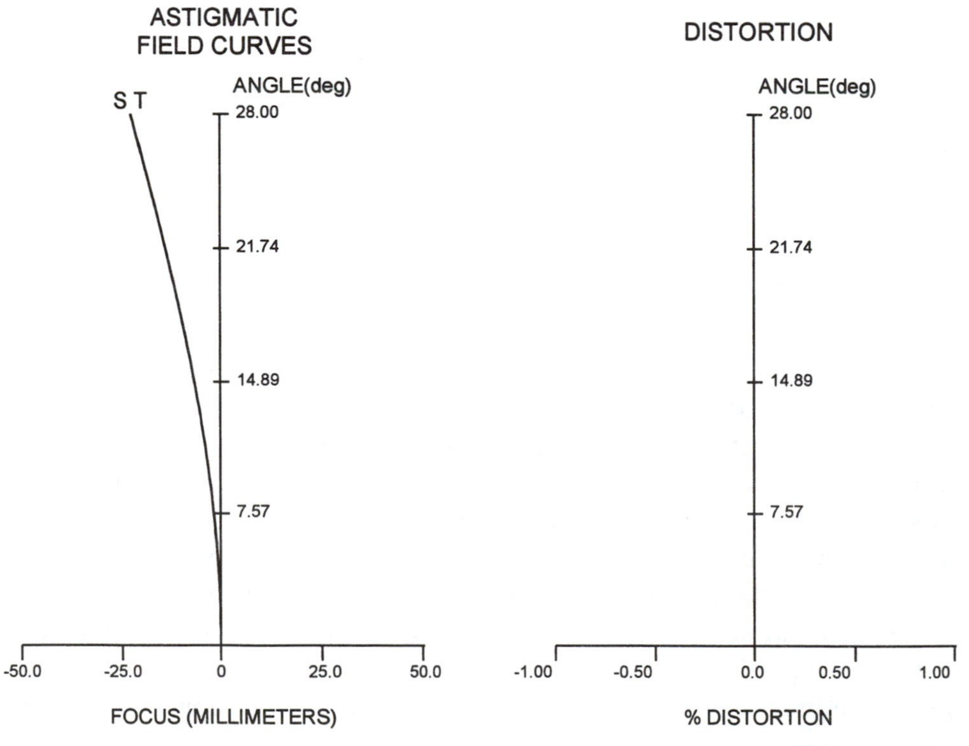

FIGURE 3.47 Plot of the astigmatic fields and distortion for a flat image reference surface. The sagittal and tangential surfaces are shown to coincide with each other in the left diagram (CODE V).

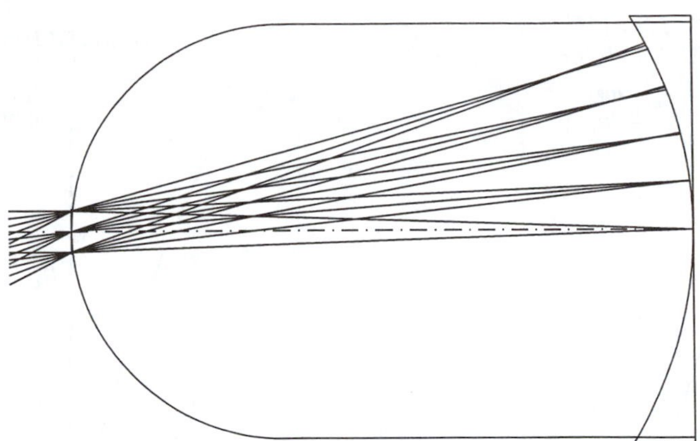

FIGURE 3.48 Single surface system with the stop located at the surface. The tangential focal surface can be seen to lie inside the curved image reference surface (CODE V).

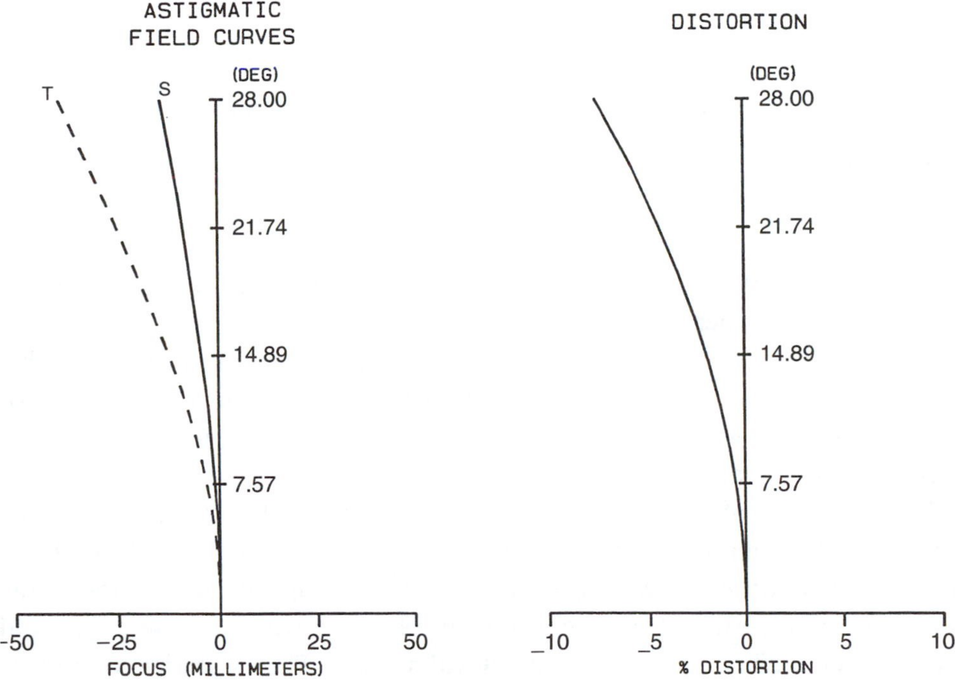

FIGURE 3.49 Plot of the astigmatic fields (——, sagittal; – – –, tangential) and distortion with respect to the curved image reference surface (CODE V).

Figures 3.51 and 3.52 are two sets of spot diagrams obtained on a curved focal surface with Petzval radius but adjusted to 8 and 24 mm inside the original focus location. These correspond to the locations of the sagittal and tangential focus locations for the 21° field. Because spherical aberration and

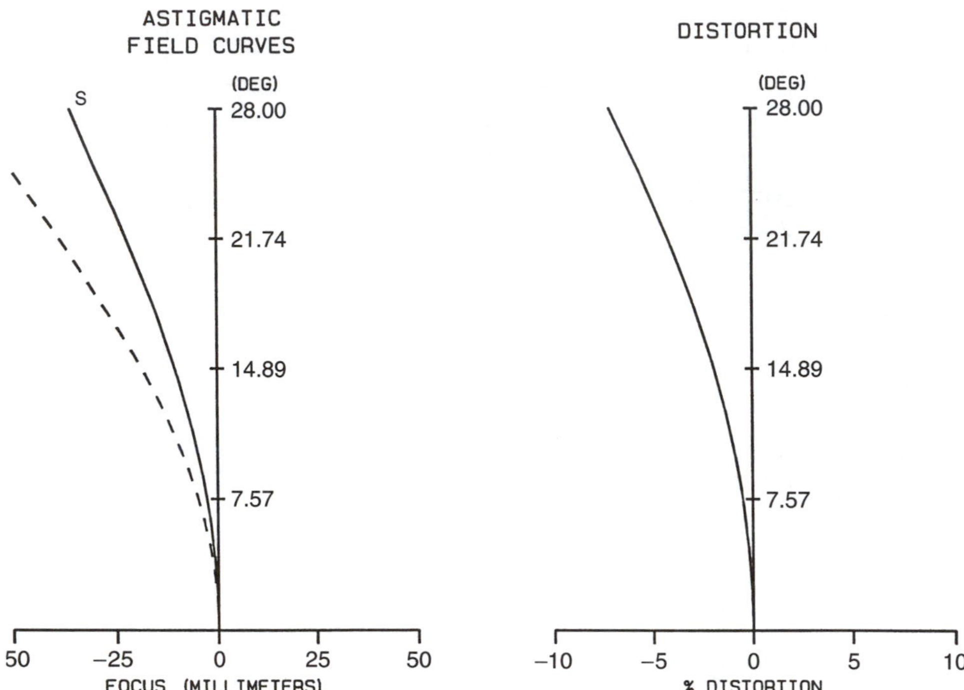

FIGURE 3.50 Plot of the astigmatic fields (——, sagittal; – – –, tangential) and the distortion with respect to a flat reference image surface (CODE V).

coma are present, the images are not pure astigmatic line foci, but are complicated by the other aberrations.

Finally, Figure 3.53 shows a set of rays distributed over the available aperture of the lens, but all originating at the 28° field angle. The effect of moving the stop will select one of these rays as a chief ray, and the adjacent rays as upper and lower rim rays from that object, or field angle, location. The progression of the tangential focus with selection of stop location can be visualized by examining the figure. The selection of the chief ray through the center of curvature of the surface will provide an image at the location of the Petzval surface. Any other selection of the chief ray will cause the tangential focus to be farther away from the Petzval surface. The similarity in appearance of the ray bundle in Figure 3.53 to the spherical aberration ray plots in Figure 3.16 is easily noted. The variation in intersection point on the flat reference surface as the different rays are chosen as chief rays is further evidence of the variation of distortion with stop location and how it is intimately associated with the other aberrations.

The above figures are all for a single surface. The purpose is to provide some link between the equations which describe image formation and the actual passage of rays. Many other simple systems can be developed using a

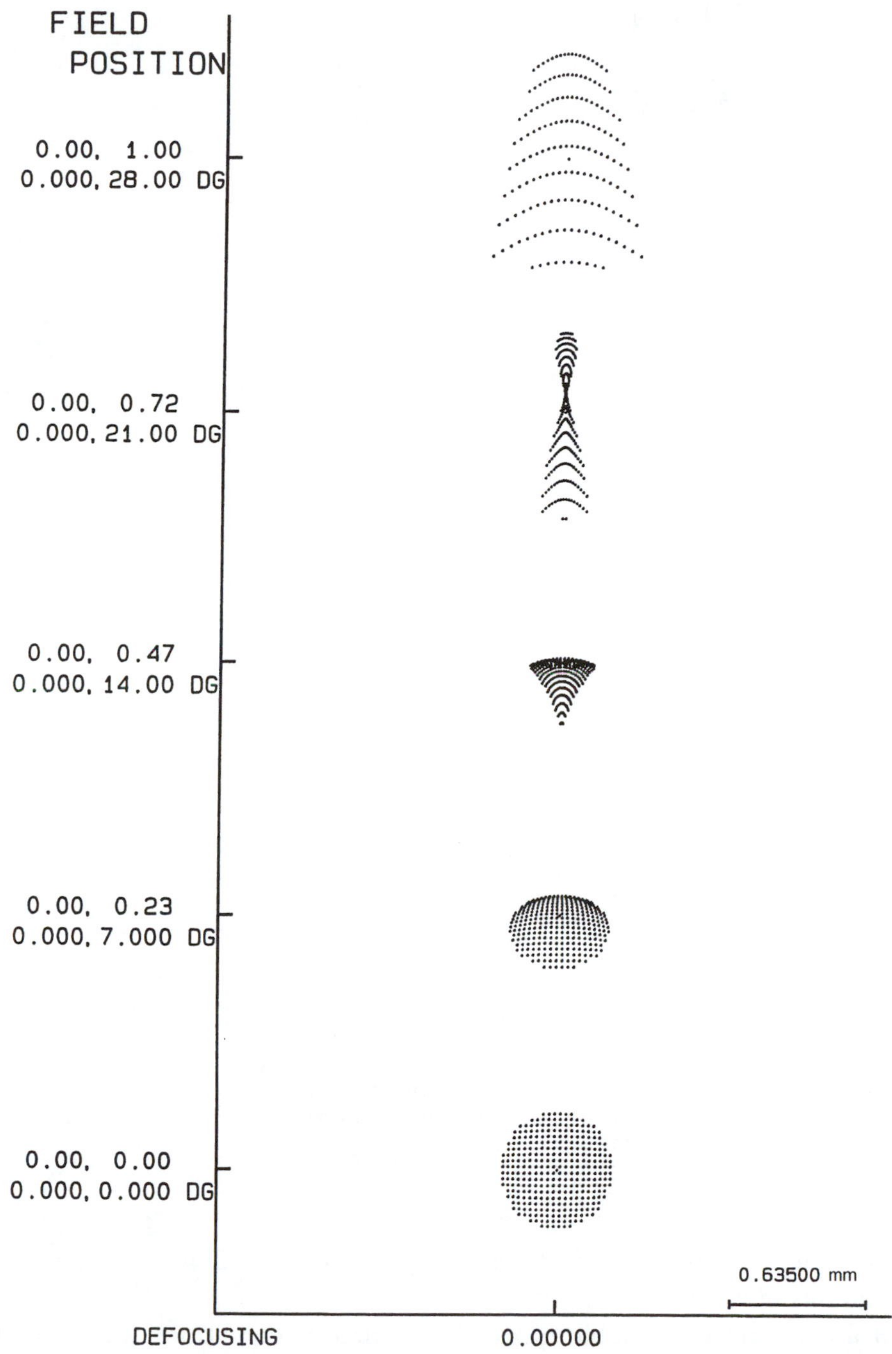

FIGURE 3.51 Spot diagrams for a flat image reference surface shifted 8 mm inside the paraxial focal location. Note the sagittal focal position, compounded with spherical aberration and coma, at the 21° image height (CODE V).

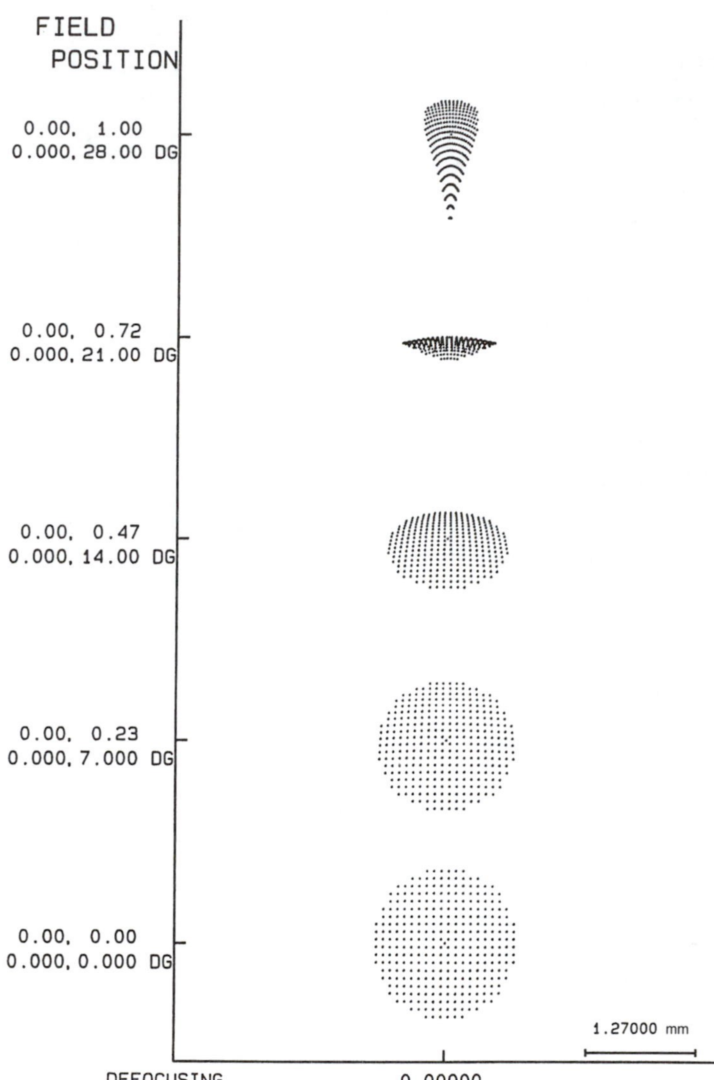

FIGURE 3.52 Spot diagrams for a flat image reference surface shifted 25 mm inside the paraxial focal location. Note the tangential image mixed with spherical aberration and coma located at the 21° image height (CODE V).

lens design program which may be useful in understanding the behavior of aberrations in a lens.

Multiple surfaces are subject to similar symmetry conditions, but the addition of the aberrations is likely to be complex, and higher-order effects may be present as well. The pupil aberrations will change the distribution of rays in the pupil, and will affect the resulting ray aberration on the image surface. Examination of ray diagrams such as these using a lens design program can

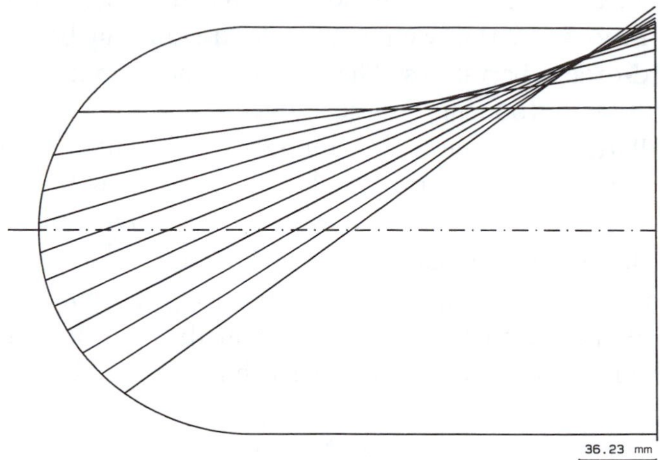

FIGURE 3.53 Single surface system tracing a wide bundle of rays at a fixed field angle. The systematic change in the location of the tangential focus, as well as the asymmetry due to the coma is evident from examination of the rays (CODE V).

help to explain the limits imposed upon aberration correction by various types of lenses.

3.1.1.3 Control of Lens Aberrations

The aberrations produced by a lens are principally determined by the power, glass type, and bending of the element, and the location of the component in the lens system relative to the location of the stop. The thin lens is analytically tractable and has been used as the source of basic design wisdom for many years. Therefore, the aberrations of components within a lens system can best be understood by beginning with an understanding of a thin lens or power element, which has zero thickness. The changes in the basic aberrations that result from converting this ideal lens into a real lens will then be examined.

The reason for spending valuable time considering the thin optical element is the development of insight into the properties of lenses. The major contribution of a lens to the aberrations comes from the power of the element, with a secondary contribution from the thickness of the element. There are exceptions to this general statement, in that the thickness is the major contributor in certain cases. However, the differences in such cases are readily understood once the basic power contributions are determined. A systematic examination of the effect of lens shape and position about the pupil can be carried out which is applicable to most applications. The experience so obtained can be used as a guide in the setting up of a lens system.

The concept of a thin element has been discussed earlier concerning the chromatic aberrations. In this section, the discussion will be extended to include the third-order aberrations. The extent of the change in the aberrations as a result of introducing physical thickness into these thin power components will then be discussed. Although some relatively simple and neat formulae can be obtained, the trends and indications from these formulae are more important than the formulae themselves. Although it is possible to develop some formulae for the high-order aberration content of a series of thin lenses, the major gain in knowledge is obtained from an understanding of the third-order aberration contributions and the stability of high-order aberration contributions once a third-order aberration control has been obtained.

The paraxial power of a lens is described by the power equation

$$\phi = (n-1)(c_1 - c_2) - \frac{t}{n} c_1 c_2 \tag{3.85}$$

which, when t tends to zero, becomes simply

$$\phi = (n-1)(c_1 - c_2) \tag{3.86}$$

The first-order aberrations due to focal shift and magnification error have been discussed previously, and shown to be dependent upon the power distribution in the lens elements. The design of a lens requires that the power distribution be chosen to provide a desirable correction of these aberrations. The shape of the lens element will influence the amounts of the monochromatic aberrations.

The lens can be bent into different shapes by conserving the difference of the curvatures, which maintains a fixed power, ϕ. The shape may be described by either using the first curvature, with the second curvature computed to maintain the power, or by a bending parameter

$$B = \frac{c_1 + c_2}{c_1 - c_2} \tag{3.87}$$

When $B = 0$ the curvatures are equal and opposite, and the lens is equiconvex (or equiconcave, if a negative lens). The values of $B = -1$ or $+1$ describe a plano-convex or convex-plano lens, respectively.

Specification of the first curvature, with the second adjusted to maintain the power is an application of an angle solve. This has more generality than the use of a bending parameter, as the concept may easily be extended to lenses that have thickness, although analytic formulae are not possible.

A conjugate parameter is also useful, as the aberrations will be found to depend upon the conjugates at which the lens is being used. There are several ways to define this, but one common form is

$$C = \frac{u' + u}{u' - u} \tag{3.88}$$

Here the denominator is constant, as it expresses the fixed power of the lens. A value of $C = 0$ implies equal conjugates, or a magnification of -1. When $C = -1$ or $+1$, the image or the object, respectively, is at infinity. The magnification of the lens is

$$m = \frac{u'}{u} = \frac{C + 1}{C - 1} \tag{3.89}$$

The aberration contribution for the lens can be obtained by applying a paraxial ray trace through the lens, and inserting the appropriate algebraic values for the paraxial values into the contribution formulae for the aberration. There are several possible algebraic manipulations that can be carried out, but all lead to similar conclusions. For a lens of zero thickness, or a thin lens, the opportunity to use the same incidence height on both surfaces leads to considerable simplification in the formulae, and a closed form results.

If the stop is located at the lens, a set of easily understood equations are obtained. Two examples follow. Using the bending parameter, a convenient form for the aberrations can be found. There are several forms for this equation (Conrady 1929; Hopkins 1962; Kingslake 1978; Smith 1990) but the one developed by Welford seems to be the most attractive, and that general form will be followed here (Welford 1986). The spherical aberration contribution for the thin lens is a complicated-looking formula depending upon the index of refraction of the lens:

$$S_{\mathrm{I}} = \frac{y^4\phi^3}{4n_0^2}\left[\left(\frac{n}{(n-1)}\right)^2 + \frac{n+2}{n(n-1)^2}\left(B + \frac{2(n^2-1)}{n+2}C\right)^2 - \frac{n}{n+2}C^2\right] \tag{3.90}$$

showing a quadratic dependence of the aberration on the bending and the conjugate parameter.

The coma contribution for the stop located at the lens is

$$S_{\mathrm{II}} = -\frac{y^2\phi^2 I}{2n_0^2}\left[\frac{n+1}{n(n-1)}B + \frac{(2n+1)}{n}C\right] \tag{3.91}$$

where the added I is the invariant applicable to the lens. A linear dependence of the aberration on bending and conjugate is observed.

The astigmatism and Petzval contributions are rather simply given by

$$S_{\mathrm{III}} = \frac{\phi I^2}{n_0^2}$$

$$S_{\mathrm{IV}} = \frac{\phi I^2}{n_0^2 n} \tag{3.92}$$

respectively. The interpretation of the sign of these is rather important. The convention of the positive sign indicates inward-curving astigmatic and Petzval surfaces for a positive lens.

The distortion is simple, as a chief ray incident on the center of a thin lens is undeviated, giving

$$S_{\rm V} = 0 \tag{3.93}$$

or no distortion for the stop-in-contact case.

To complete the set of formulae for the aberrations, the first-order chromatic longitudinal and lateral aberrations for the stop in contact are, using the results in section 3.1.1.1.2:

$$a = \frac{y^2\phi}{V}$$
$$b = 0 \tag{3.94}$$

the interpretation being that the longitudinal chromatic aberration depends upon the power and the aperture; the lateral color will be zero for a thin lens with stop in contact as the chief ray is undeviated when passing through the center of the lens.

It will be useful to relate the following discussion of the aberrations to a specific choice of lens. The example used will be a 100 mm focal length, F/5 lens, with object at infinity. The entrance pupil diameter is 20 mm. The field of view that will be examined is a nominal 15° half field. The optical invariant applicable to this problem is

$$I = n'h'u' = 1.0\tan(15^\circ)(-0.1) = -2.6795 \tag{3.95}$$

where −0.1 is the paraxial numerical aperture or marginal ray slope angle for F/5.

If the stop is moved from the lens, then the aberrations produced by the lens combine to produce a new set of mixtures. These are described by a set of stop shift equations. There are several useful parameters for defining the location of the stop relative to the lens. The stop shift parameter defined earlier is the relative height of the new chief ray in the old pupil or stop. For a single lens, this can be interpreted as the location of the stop relative to the first surface of the lens. (For a thin lens, this will obviously be the location of the stop relative to the lens.)

Figure 3.54 shows the important quantities involved in defining a stop shift parameter, Q. The fundamental definition of the parameter Q is the ratio of the shift of the new chief ray in the scaled old pupil. For any specific surface, moving the stop from a surface the situation looks more complicated, but for the paraxial ray trace, the basic definition above will apply to all surfaces in the optical system.

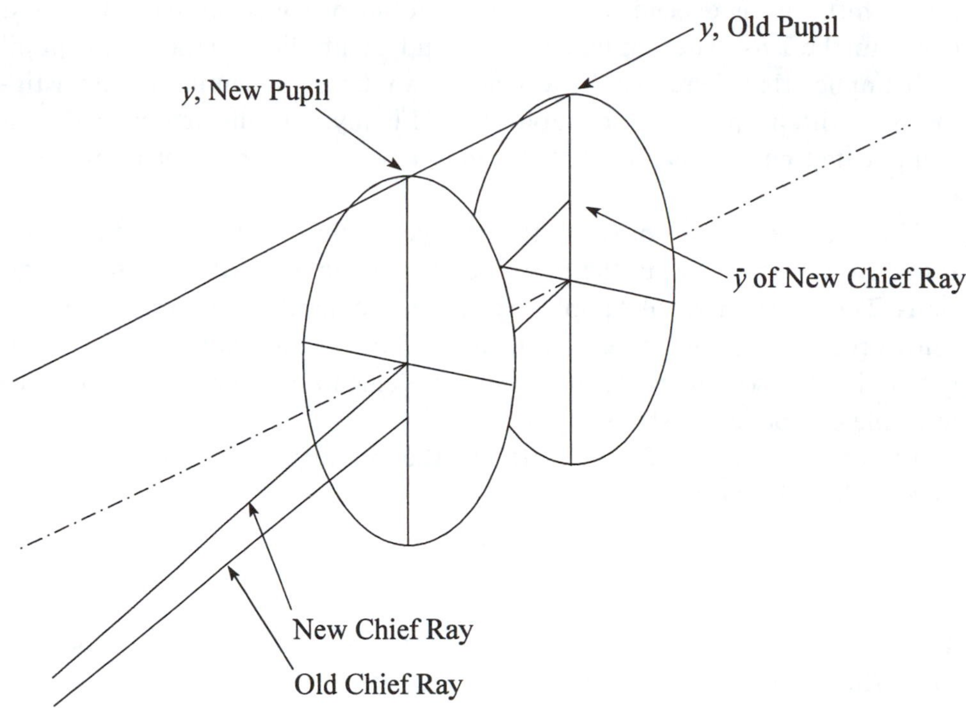

FIGURE 3.54 The concept of stop shift changing the choice of the new chief ray in the old pupil location.

The stop shift effect is best stated as changes in the basic aberration of the component. This amplifies the role of the added contribution due to the pupil shift. The change in each of the

$$\begin{aligned}
\delta S_{\mathrm{I}} &= 0 \\
\delta S_{\mathrm{II}} &= QS_{\mathrm{I}} \\
\delta S_{\mathrm{III}} &= 2QS_{\mathrm{II}} + Q^2 S_{\mathrm{I}} \\
\delta S_{\mathrm{IV}} &= 0 \\
\delta S_{\mathrm{V}} &= Q(3S_{\mathrm{III}} + S_{\mathrm{IV}}) + 3Q^2 S_{\mathrm{II}} + Q^3 S_{\mathrm{I}}
\end{aligned} \tag{3.96}$$

aberration contributions from the lens is then shown in equation 3.96. As this reads, it says that the spherical aberration, S_{I}, and the Petzval, S_{IV}, are independent of pupil shift. The coma, S_{II}, varies linearly with the pupil shift, and is dependent upon the magnitude of the spherical aberration in the lens. Therefore, both positive and negative values of coma can be obtained, and coma can probably be eliminated by an appropriate choice of the pupil location, as long as there is a sufficient amount of spherical aberration in the lens. The astigmatism, S_{III}, has a complicated change with pupil shift, and is dependent upon the amount of spherical aberration and coma in the lens. The combined linear and quadratic variation with pupil shift implies that there may be a local minimum or maximum of the astigmatism with respect to the pupil shift. Finally, the distortion, S_V, is a complicated cubic function that depends upon all of the aberrations in a lens.

This set of equations provides the changes in third-order aberrations that result from a stop shift in the presence of a stated amount of these aberrations. Therefore, these rules apply to any lens, no matter how complex, whose third-order aberration content is defined. The conclusions that will be suggested in this section based upon single elements can be carried over to multiple component systems.

To complete the set of pupil shifts for the thin lens, the chromatic aberrations follow the rules

$$\begin{aligned} \delta a &= 0 \\ \delta b &= Qa \end{aligned} \tag{3.97}$$

These two rules state that the axial chromatic aberration is independent of pupil shift, and the lateral color varies linearly with pupil shift. This should not be a surprise, as the same conclusion was obtained in Chapter 2.

A top-level observation is that if there is no spherical aberration or coma, the astigmatism will be independent of pupil location. The distortion depends upon all of the aberration content, including the amount of Petzval curvature, and can be difficult to eliminate without a known balance of all of the other aberrations. A set of specific equations can be written which combine the thin lens aberration contribution formulae with the stop shift to provide a complex description of the aberrations using power, bending, conjugate factor, and stop location as parameters. In principle, these equations provide a complicated set of cubic equations that could be solved to find a solution for the thin-lens system.

This is generally an unprofitable approach to take. Thin-lens aberration relationships can be used to develop an understanding of the properties of lenses. In most lens designs, there are boundary conditions of space, complexity, and cost that would make an analytical solution very difficult to obtain. Solution of actual lens problems is best carried out using numerical optimization processes. The basic aberration relations do provide links that drive the solution process. The conclusions to be made shortly will apply to the aberration contributions from single elements or a complete optical system, and also apply to the case of thick lenses. No closed-form equations exist for those realizable cases.

These equations are stated for the third-order aberrations only. A larger set of equations can be developed for the fifth- and higher-order aberrations.

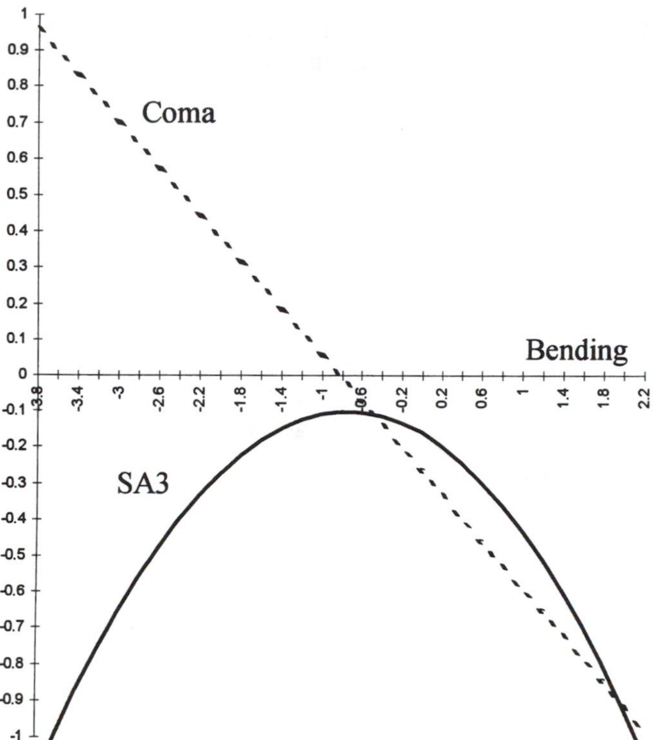

FIGURE 3.55 Coma and spherical aberration as a function of the lens bending for a singlet lens with stop in contact.

The basic principles apply to these higher-order aberrations, but the complexity of the relationships motivates the development of a graphical approach to understanding the change of aberrations with pupil shift rather than an understanding based upon mathematics alone.

The starting point is for the lens with stop in contact. The simplest issue is that of the spherical aberration introduced by this single element. The example in Figure 3.55 is for an F/5, 100 mm focal length lens with a semifield of 15° and an object at infinity. Bk7 glass will be used in the lens. This figure shows a plot of the amount of spherical aberration and coma for various bendings of the lens. The spherical aberration varies continuously with the lens shape, and has a minimum value at a specific shape. Selecting this minimum position leads to the ray intercept plot shown in Figure 3.56. The coma turns out to be close to zero at the shape providing the minimum spherical aberration, and the dominant effect of spherical aberration is evident. Because the spherical aberration ray blur varies with the cube of the aperture, changing the aperture diameter while keeping the lens shape and the power constant will cause a rapid change in the extent of the spherical aberration blur.

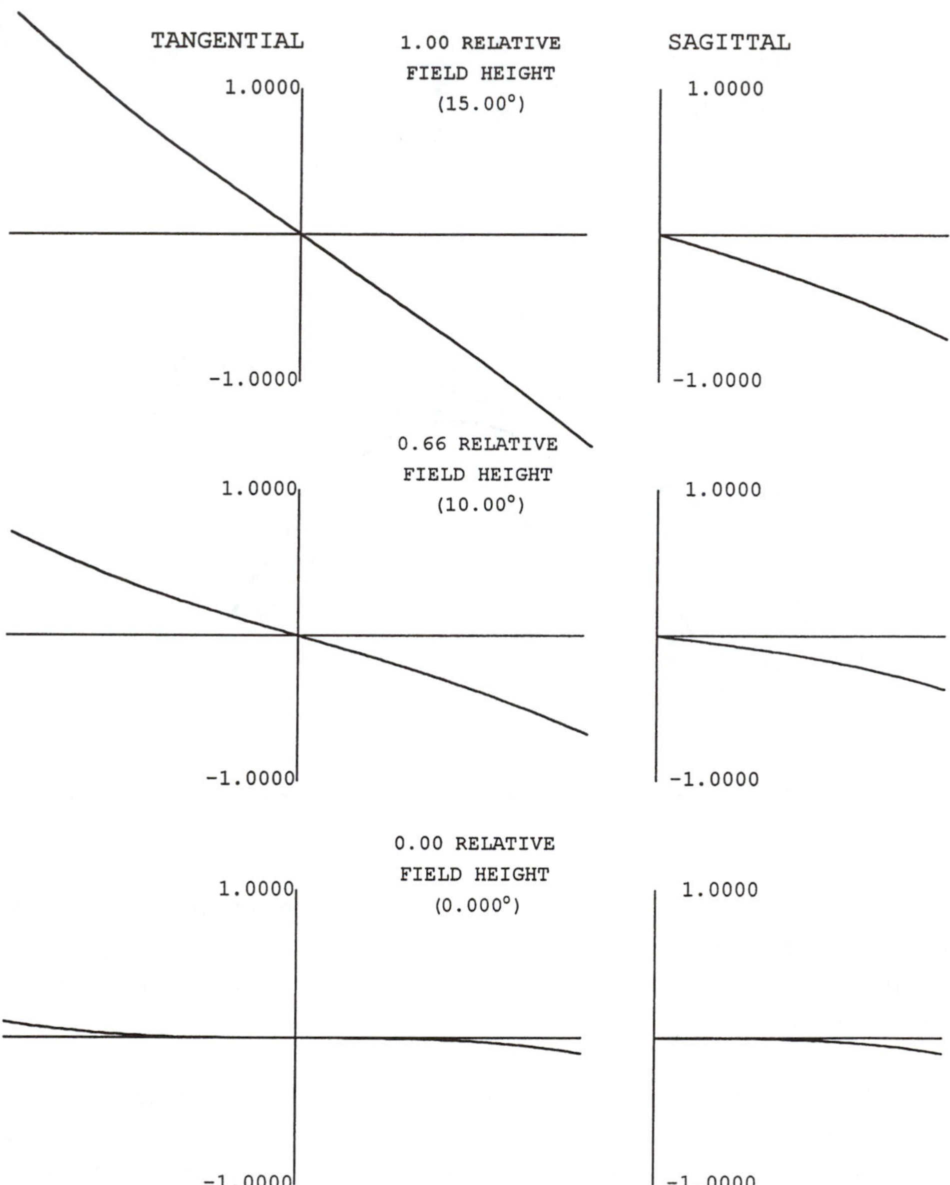

FIGURE 3.56 Ray intercept plots for a single lens with stop in contact, bent for zero coma (wavelength 587.6 nm).

The image errors are dominated by the spherical aberration for the center part of the field, but astigmatism is evident at the outer parts of the field. The spherical aberration is essentially constant over the field, but the focus in the sagittal and tangential direction is shifted. The plot of the astigmatic fields in Figure 3.57 clearly shows the presence of a significant amount of focal shift at the edge of the field.

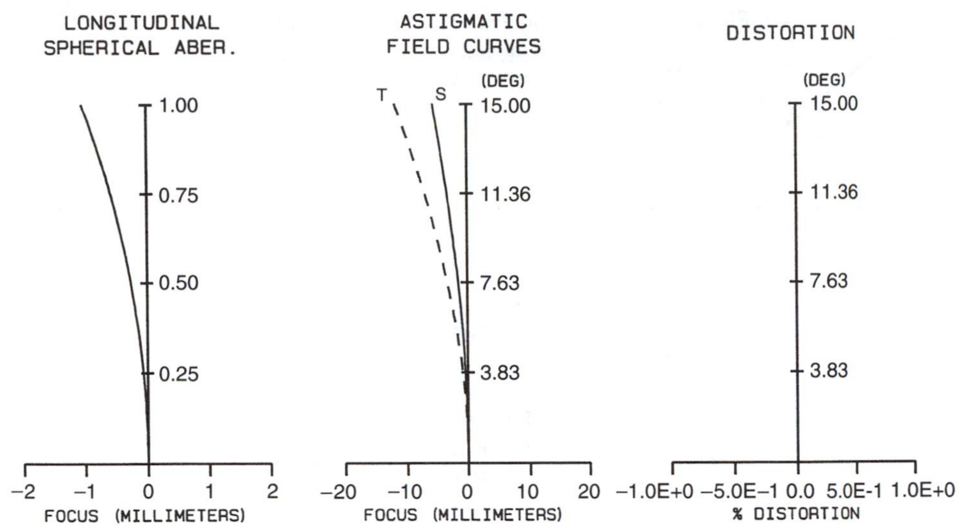

FIGURE 3.57 Longitudinal spherical aberration, astigmatism (——, sagittal; – – –, tangential) and distortion plots for the lens in the previous figure (CODE V).

Adjustment to the best focus on axis leads to the ray intercept curves in Figure 3.58. The extent of the image blur is decreased by about a factor of four on axis, but the field curvature is still evident. Choice of best focus across the field would be more difficult as the axial image would become considerably larger.

The next design option is the shifting of the stop from the lens to change the amount of astigmatism. An interesting relation occurs for the change in astigmatism. Taking the derivative of the change of astigmatism with stop shift, leads to

$$\frac{\partial \delta S_{\mathrm{III}}}{\partial Q} = 2S_{\mathrm{II}} + 2QS_{\mathrm{III}} \tag{3.98}$$

When this is at a maximum, the derivative is equal to zero and the stop shift producing this maximum is $Q = -(S_{\mathrm{II}}/S_{\mathrm{I}})$. The coma at this stop location is

$$S_{\mathrm{II}} + \delta S_{\mathrm{II}} = S_{\mathrm{II}} + Q_k S_{\mathrm{I}} = S_{\mathrm{II}} + \left(\frac{-S_{\mathrm{II}}}{S_{\mathrm{I}}}\right) S_{\mathrm{I}} = 0 \tag{3.99}$$

which says that the coma will be zero when the stop is located at the position causing the maximum amount of field flattening. This stop location is referred to as the "natural stop position" for the lens. The actual value of the astigmatic fields will be found by adding the astigmatism obtained for that stop shift to the amount of the Petzval curvature.

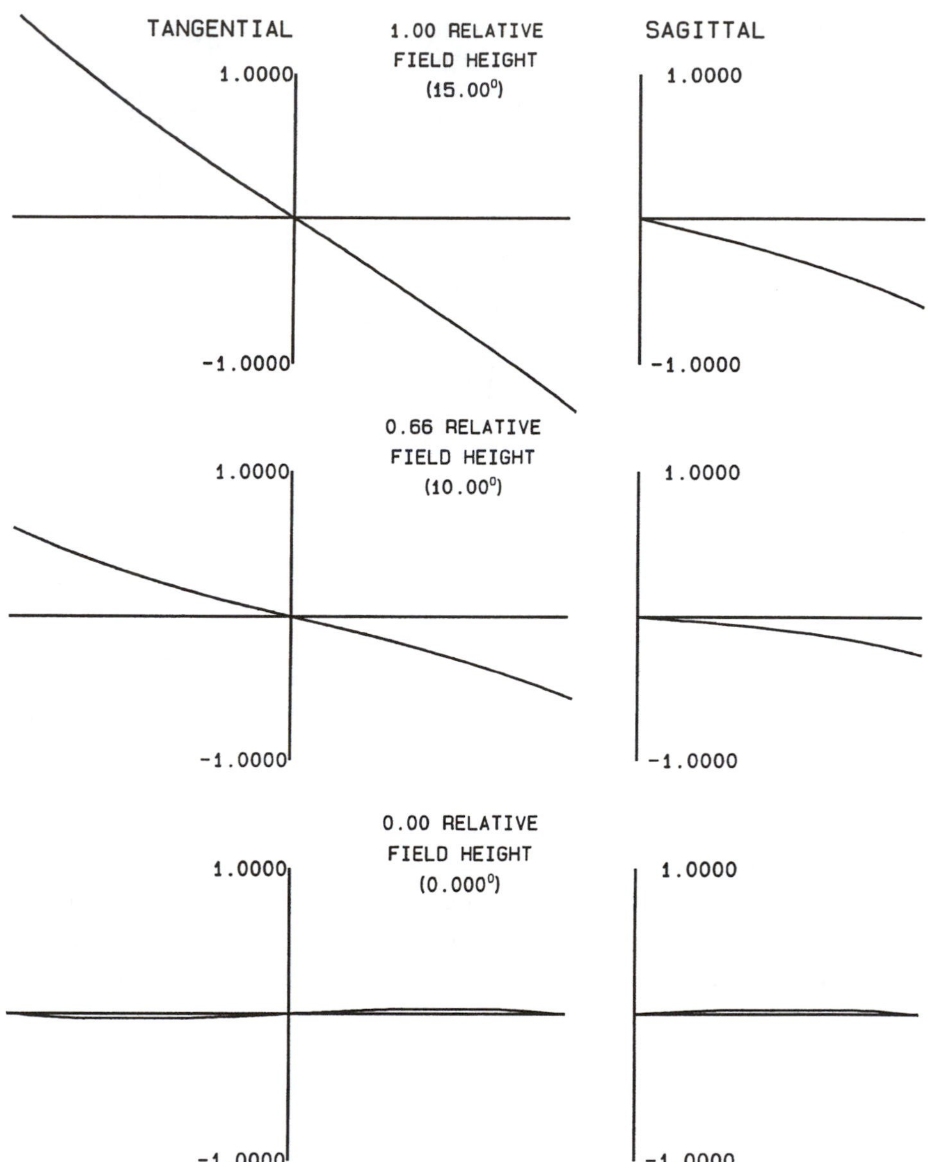

FIGURE 3.58 Ray intercept plots for the lens in the previous figure adjusted for optimum focal position on axis.

The Petzval is determined by the index and power of the lens. The actual amount of the astigmatism is changed by the amount of spherical aberration, and varies quadratically with the amount of stop shift. The shape of the lens is altered by bending to introduce more spherical aberration, and the stop is moved to the natural stop position for each bending until the maximum amount of field flattening is obtained. The ray intercept curves for the lens bent to introduce enough astigmatism to approximately flatten the tangential

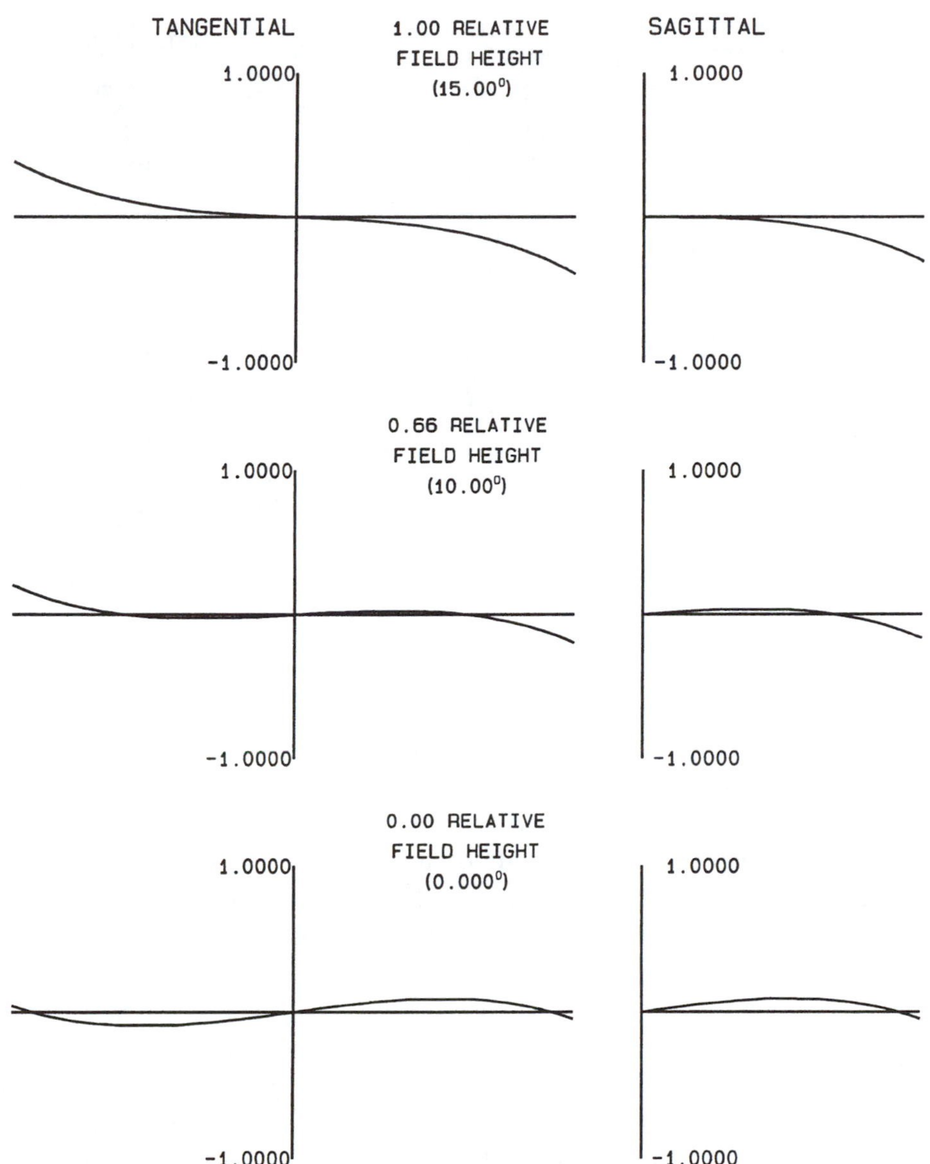

FIGURE 3.59 Ray intercept plots for the singlet lens bent to introduce additional spherical aberration and coma, with the stop shifted to produce zero coma. The reference focus position is shifted to the best focus position at the 0.7 field position.

field are shown in Figure 3.59. Because the variation of astigmatism with stop is quadratic, there will be two locations for the two solutions for the values of Q. Only the solution for the stop in front of the lens is shown here for the thin-lens solution. Examination of the ray intercept curves shows a fairly uniform image across the field. Examination of the field plots in Figure 3.60 indicates that the reduction in the astigmatic field sag is accompanied

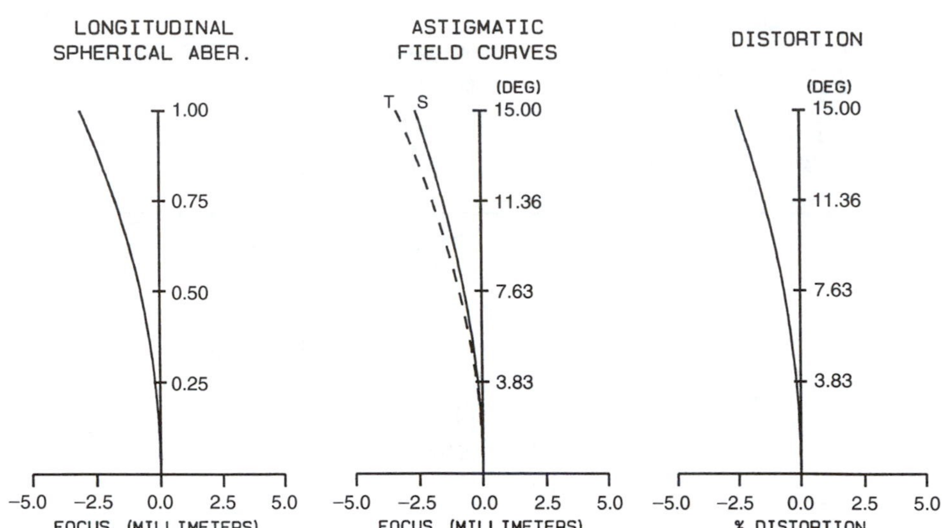

FIGURE 3.60 Longitudinal spherical aberration, astigmatism (——, sagittal; – – –, tangential) and distortion plots for the singlet lens in the previous figure (CODE V).

by an increase in the amount of spherical aberration. The solution now also shows the presence of distortion. The distortion is a consequence of using the spherical aberration to flatten the fields.

It is noted that the tangential field is not quite flat. It turns out that a flat tangential field cannot be obtained for a zero thickness lens. Some finite thickness is required in order to attain a flat tangential field. Therefore, an element which is concave toward the stop will produce some field flattening and will permit a zero coma solution to be found.

The chromatic aberration effects also need to be examined. The amount of longitudinal chromatic aberration is constant for a thin lens no matter what the bending. It will be almost constant for a lens of small finite thickness. As long as the stop is located at the lens, the lateral chromatic aberration will be small, or zero. Shifting the stop to the natural stop position will lead to either positive or negative lateral color, depending upon which solution is selected.

The discussion so far has dealt with a power, or thin, element, and only with third-order aberrations. The effect of thickness and of high-order aberrations is not analytically tractable, and must be evaluated by numerical methods. The number of cases and options introduced is so great that it is difficult to provide a simple summary of these effects. A summary of some possibilities is given in Figures 3.61 and 3.62. The values of the important aberrations are plotted for a single lens of finite thickness. Figure 3.61 is a plot of the sag of the tangential and sagittal fields from a flat image surface.

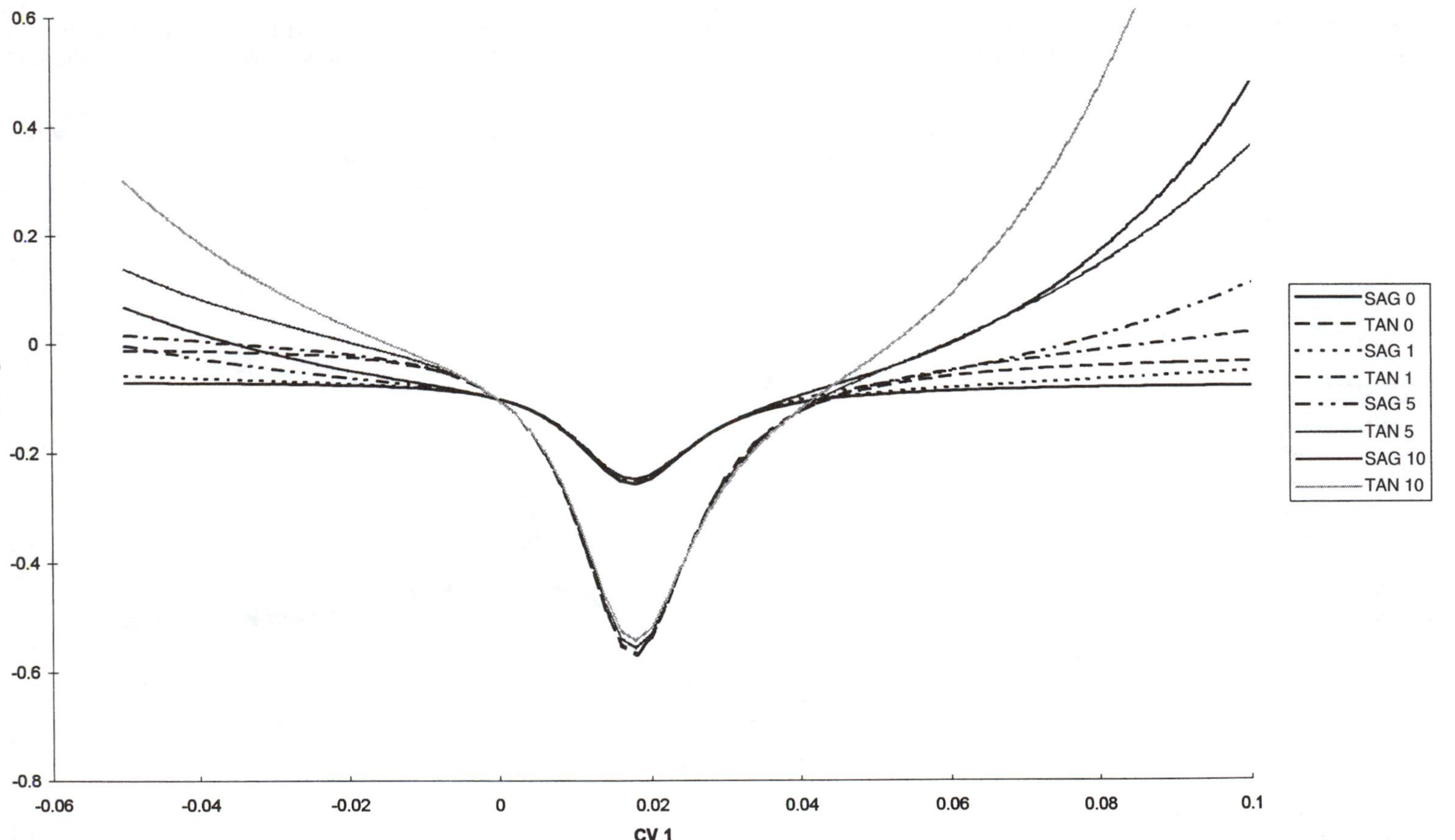

FIGURE 3.61 A plot of the behavior of the location of the astigmatic fields against bending of the lens with the stop always set at the zero coma position. The behavior of the field with increased element thickness is shown. The numbers in the legend indicate the thickness as a percentage of the focal length of the lens.

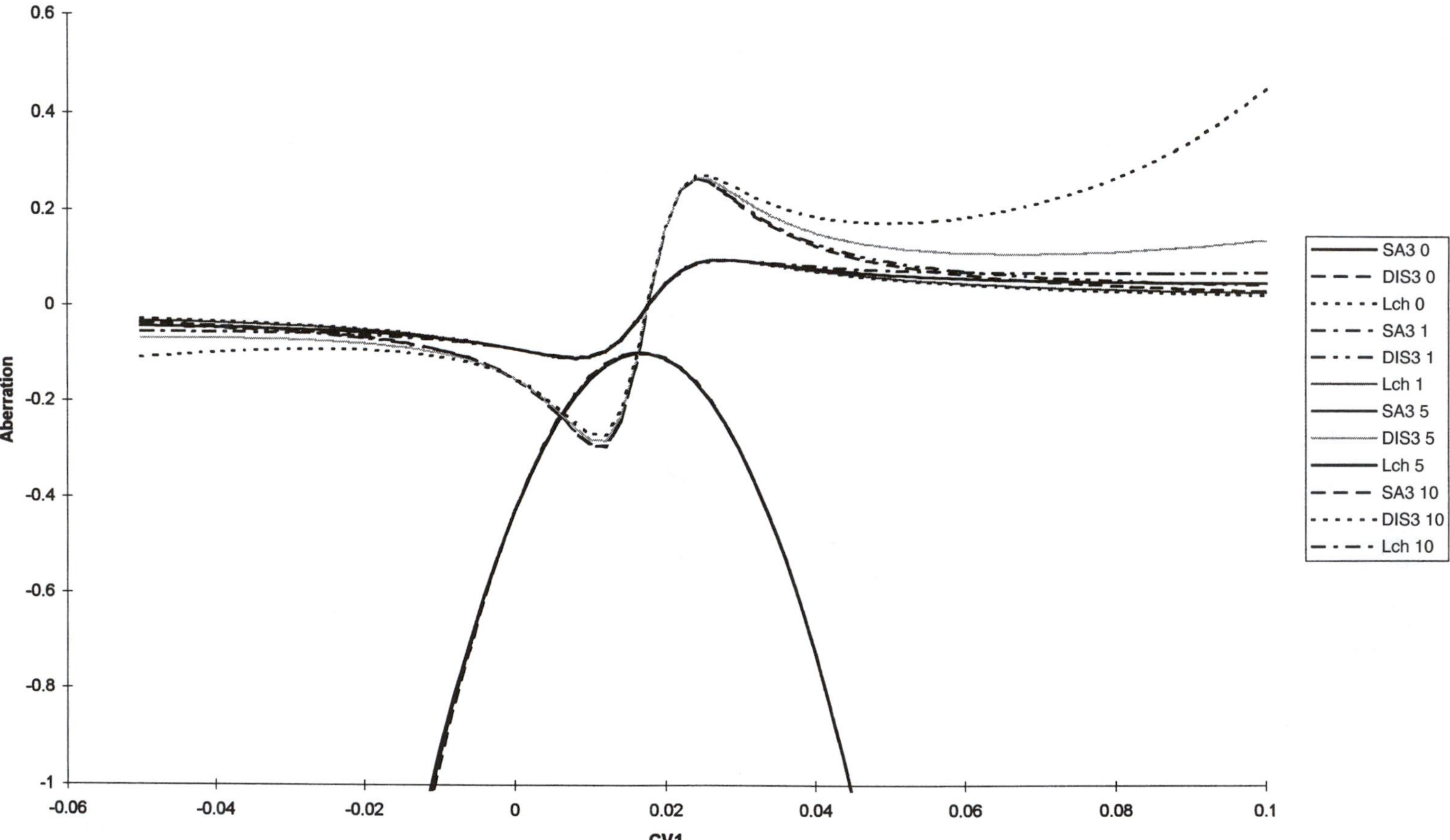

FIGURE 3.62 Plots of the behavior of the spherical aberration (SA3), distortion (DIS) and lateral color (Lch) as a function of the bending of the lens. The stop is always adjusted to be at the zero coma position. As in the previous figure, the number in the legend is the thickness as a percentage of the focal length.

Each of the curves plotted is for a different lens thickness of 0, 1, 5, or 10 mm, all for a focal length of 100 mm. In each case the stop and the lens bending are adjusted to have zero third-order coma. Since the lens has finite thickness the simple bending parameter is less useful than plotting against the curvature of the first surface (CV 1). Comparison of this plot with Figure 3.62, which shows the amount of spherical aberration, distortion, and lateral color for the same bendings and stop locations indicates that for the region near minimum spherical aberration, and zero lateral color and distortion, the astigmatic fields have the maximum amount of inward curvature.

Bending the lens in either direction increases the spherical aberration, and eventually places both the tangential and sagittal fields on the Petzval surface. The amounts of distortion and lateral color increase, and then reach a generally stable value, the sign of which depends upon the choice of the bending. As the bending and the spherical aberration continues to increase, the tangential field approaches flatness for the thin lens, and becomes backward curving for the lenses with increased thickness. The inclusion of a necessary amount of lens thickness greatly affects the astigmatic field correction produced by the lens.

These observations are for a single element. The characteristics of the aberration difference formulae resulting from a stop shift are such that even symmetry exists for longitudinal aberrations such as longitudinal chromatic, spherical aberration, and astigmatism. Odd symmetry exists for lateral color, coma, and distortion. Thus a powerful design approach is to use elements on both sides of a stop. Such quasi-symmetric lenses can provide field flattening with a wide range of possible solutions for controlling coma and distortion. Because the power is split between two elements, and the spherical contribution is proportional to the cube of the power of the individual elements, the spherical aberration of a two element design can be reduced.

3.1.2 HIGH-ORDER ABERRATIONS

Life would be relatively simple if the control and elimination of the third-order aberrations were all that is required. The values to which the third-order aberrations will eventually be set in a lens are determined by the balancing of residual values of the high-order aberrations. The process is analogous to the improvement in image quality that is possible by balancing third-order spherical aberration with a focus shift. In the case to be discussed now, the high-order aberrations will be defined, and the process of selecting low-order balancing aberrations will be described.

Expansion of the wavefront function to the next higher order introduces a number of new aberration coefficients. Some of these are higher-order versions of the third-order aberrations. Other aberrations are new forms of aberrations that are not observed in the third-order aberrations. However, these new aberrations do fall into symmetry classes which can be balanced by low-order aberrations. In general, the understanding of how to control the higher-order aberrations is not as extensive as the understanding of the origin and control of the third-order aberrations. Therefore, the process of design usually requires identification of the magnitudes of the high-order aberrations that are intrinsic to the lens, and adjustment of the lens to balance the effect of the high-order aberrations with the proper amount of the low-order aberrations.

Because the relationship between lens parameters and aberrations is non-linear, this is a dynamic, rather than a static process. Fortunately, the magnitude of the high-order aberrations changes much more slowly with changes in parameters than the low-order aberrations. This leads to local stability, and local optimums in aberration correction. Large changes in the lens parameters will lead to high-order aberration content that may be significantly reduced, resulting in a design with lower residual aberrations.

An example of this was already observed in the balancing of various orders of spherical aberration. A rule defining the optimum balance of the aberrations is selected, and used as the basis for the balancing. The rule selected may be as simple as the compensation of ray or wave aberration at some location within the aperture, may be the mean square residual error, or may be a specific image quality function such as the MTF at a particular frequency. In general, the rules for optimum balancing of aberrations are based upon matching the symmetries of low- and high-order aberrations. An understanding of these symmetries is required.

The next higher order of aberration for a rotationally symmetric lens will be expressed by the sixth-order wavefront symmetry by writing out the allowable terms of the expansion

$$\begin{aligned} W = {} & a_{060}\rho^6 + a_{151}\eta\rho^5\cos\psi \\ & + a_{240}\eta^2\rho^4 + a_{242}\eta^2\rho^4\cos^2\psi + a_{331}\eta^3\rho^3\cos\psi + a_{333}\eta^3\rho^3\cos^3\psi \\ & + a_{420}\eta^4\rho^2 + a_{422}\eta^4\rho^2\cos^2\psi + a_{511}\eta^5\rho\cos + a_{600}\eta^6 \end{aligned} \tag{3.100}$$

This equation was obtained by writing out the allowable terms to order six, and then grouping the terms in order of decreasing field dependence.

The nature of the fifth-order aberrations can best be understood by separating the terms into groups that generally simulate the symmetries of third-order aberrations that have already been described. Table 3.2 summarizes this classification.

The associated equations for the ray displacements can be obtained from differentiation and algebraic manipulation to be

TABLE 3.2 Classification of fifth-order aberrations

Term	Classification
a_{600}	Phase error with field, no ray aberration
a_{060}	Fifth-order spherical aberration
a_{151}	Fifth-order linear coma
a_{420} and a_{422}	Fifth-order field curvature and astigmatism
a_{511}	Fifth-order distortion
a_{240} and a_{242}	Sagittal and tangential oblique spherical aberration
a_{331} and a_{333}	Cubic elliptical coma

$$\begin{aligned}\epsilon_y = & \frac{\lambda}{u'}[\rho^5 \cos\psi(6a_{060}) + h\rho^4((3+2\cos 2\psi)a_{151}) \\ & + h^2\rho^3\cos\psi(4a_{240} + 2a_{242}(1+\cos^2\psi)) \\ & + h^3\rho^2((2a_{331} + \tfrac{3}{2}a_{333}) + (a_{331} + \tfrac{3}{2}a_{333})\cos 2\psi) \\ & + h^5 a_{511}]\end{aligned} \tag{3.101}$$

and

$$\begin{aligned}\epsilon_x = & \frac{\lambda}{u'}[\rho^5 \sin\psi(6a_{060}) + h\rho^4((2\sin 2\psi)a_{151}) \\ & + h^2\rho^3\sin\psi(4a_{240} + 2a_{242}(\cos^2\psi)) \\ & + h^3\rho^2((a_{331})\sin 2\psi) \\ & + h^4\rho\sin\psi(2a_{420})]\end{aligned} \tag{3.102}$$

These complicated-looking equations can be best understood by considering the various terms in sequence. In any real system these terms always appear in some combination, along with lower-order aberrations. The symmetry and effect on image appearance is an important guide to understanding the limits upon image formation. It will be seen that the presence of these aberrations can be deduced from examination of the ray intercept plots.

The first term in a_{060} is the coefficient for sixth-order spherical aberration. This term, and similar higher-order terms, are conventionally balanced by adjustment of the low-order spherical aberration.

The second term, linear in field, with a fourth-order aperture dependence, is fifth-order linear coma. The linear field dependence marks it as being of the same family as the third-order coma. Because this aberration is included in the first-order variation with field, it will be subject to the conditions imposed by the extent to which the sine condition is fulfilled.

The appearance of this aberration is somewhat similar to the previously discussed third-order coma. The symmetry differs a bit, as can be seen from the three-plus-two factors in the symmetry terms for the ray displacement. Because the angular dependence is the same as that of the third-order coma, once a balance of the fifth- and third-order linear comas is established, it will be constant over the field. The resulting aberration is a mixture of two different symmetries, so the resulting image appearance will be a bit more complicated.

Skipping the next term for a moment, consider the term which is cubic in field and quadratic in aperture. This complex term is a cubic coma term, involving the a_{331} and a_{333} aberration coefficients. If the a_{333} coefficient is zero, the remaining a_{331} term has the same x and y aberration symmetry as the third-order coma. The aberration will appear as three-to-one symmetry coma which varies cubically with field. Balancing this aberration against the third-order linear coma will lead to correction at one field point, but altering to plus and minus coma inside and outside the coma cancellation field point.

If any a_{333} coefficient is present, the behavior becomes more complicated. For the case of pure a_{333} coma, sometimes called "line coma," it will be noted that there is no x-direction aberration at all. All of the rays through the aperture will form a line image, with a distribution of intensity along the line. Although this might at first appear to be similar to a sagittal astigmatic focus image, any changes of focus or field will exhibit considerably different behavior. If a mixture of the a_{331} and a_{333} coefficient coma are present, the image has a more familiar comatic asymmetry with respect to the chief ray, but the inclusion angle of the symmetry will depend upon the relative amounts of the two coefficients. The resulting aberration appearance is usually called "elliptical coma."

The presence of this aberration is linked to the degree of symmetry about the stop of the elements of the lens. The relative amount of linear versus cubic coma can be assessed by examining the ray intercept curves across the field. These terms, along with the linear coma terms, constitute very disturbing aberrations because of the systematic lateral blur appearance of points in the image. Because this blur is radially oriented with respect to the center of the field, this systematic change in image appearance is quite noticeable.

The aberration terms that are second order in field and cubic in aperture indicate a variation of spherical aberration with field, referred to as "oblique spherical aberration." Because there are two aberration coefficients involved, this oblique spherical aberration will show many characteristics that differ from third-order aberration. The quadratic field behavior indicates that it can be balanced against third-order astigmatism, but the different aperture symmetry prohibits cancellation by astigmatism. This aberration results from an imbalance of the spherical aberration as a consequence of the obliquity and

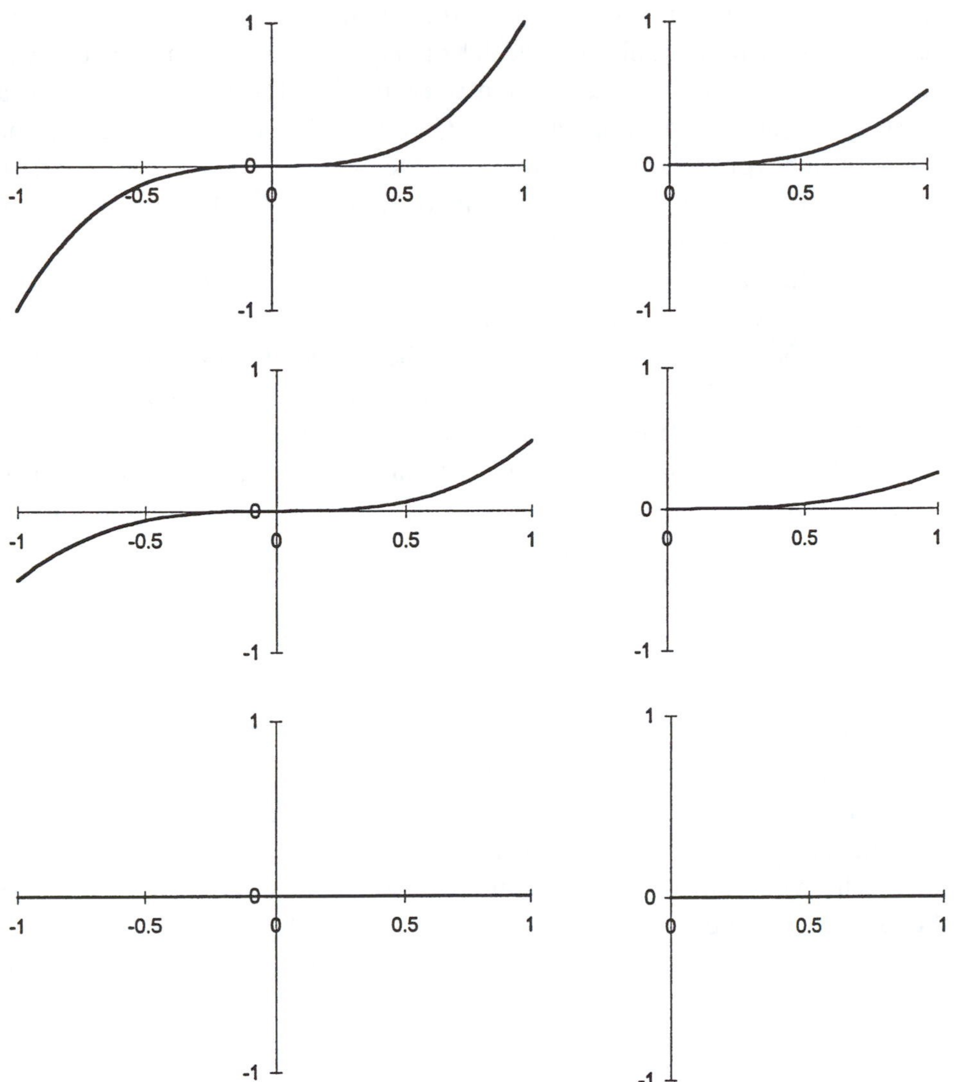

FIGURE 3.63 Ray intercept plots for the case of oblique spherical aberration alone.

anamorphism of the bundles of rays traversing the lens of axis. In most cases, this aberration will constitute the limit upon field angle that can be achieved.

The appearance of this aberration can be quite complicated. If the a_{242} coefficients is zero, the a_{240} term alone is easy to understand. Examination of the x and y aberration intercepts indicates a normal third-order spherical aberration that varies quadratically with field. The aberration is symmetric and has the properties that were associated with third-order spherical in section 3.1.1.2.1.

An example of the ray intercept plots that could be obtained in the presence of oblique spherical aberration are shown in Figure 3.63. The amount of the spherical aberration varies as the square of the field, and will be different in the tangential and sagittal directions. It will be indicated later that the source of this aberration is intrinsic to most types of lenses, and is a consequence of the axial spherical aberration balance being upset by the natural anamorphism of the obliquity of the beams in the off-axis bundles. This systematic change of spherical aberration with field limits the image quality that can be obtained off axis.

Because the aperture symmetry is matched by the third-order spherical aberration, which is constant across the field, the oblique spherical aberration can be balanced by introducing some spherical aberration into the axial image. Figure 3.64 shows this effect. The axial image is worse, but the image at the seven-tenths field zone is balanced and the edge-of-field image improved. The net image quality of the lens across the field is thus much better. This technique is used either explicitly or implicitly in almost all merit functions to find a local optimum for the lens. Improvement beyond this level requires searching for a solution which intrinsically has less oblique spherical aberration.

The presence of the a_{242} coefficient greatly changes the appearance of the image. Many types of image forms are possible with this aberration. The a_{242} coefficient alone produces a four-lobed pattern. Together with the a_{240} coefficient, a range of possible forms appears, with appearance ranging from a four-pointed star to a star with complex loops and, for majority content of a_{240}, an elliptical form of spherical aberration zones. The behavior through focus is likewise quite complicated, and will not be reviewed here. Figure 3.65 shows some of the possible spot diagram forms that can be found with different mixtures of the sagittal and tangential oblique spherical aberration coefficients.

This growth of spherical aberration with field is usually the field-limiting aberration in lenses as it indicates a fundamental instability in the manner in which spherical aberration is balanced across the field. The source of this aberration can conceptually be understood from the following argument. On axis, the spherical aberration is cancelled (or at least controlled) by the balance of contributions across the lens. As the field angle is increased, the shape of the beam traversing the lens becomes elliptical due to the off-axis pupil projection. The balance that produced the spherical aberration is then upset, as the off-axis bundle encompasses a smaller width in the tangential direction, while changing incidence angles and maintaining almost the same dimension in the sagittal dimension. The reduced beam diameter thus encounters less of the built-in natural undercorrection of the elements in the tangential direction, and thus will generally become overcorrected with increasing field angle.

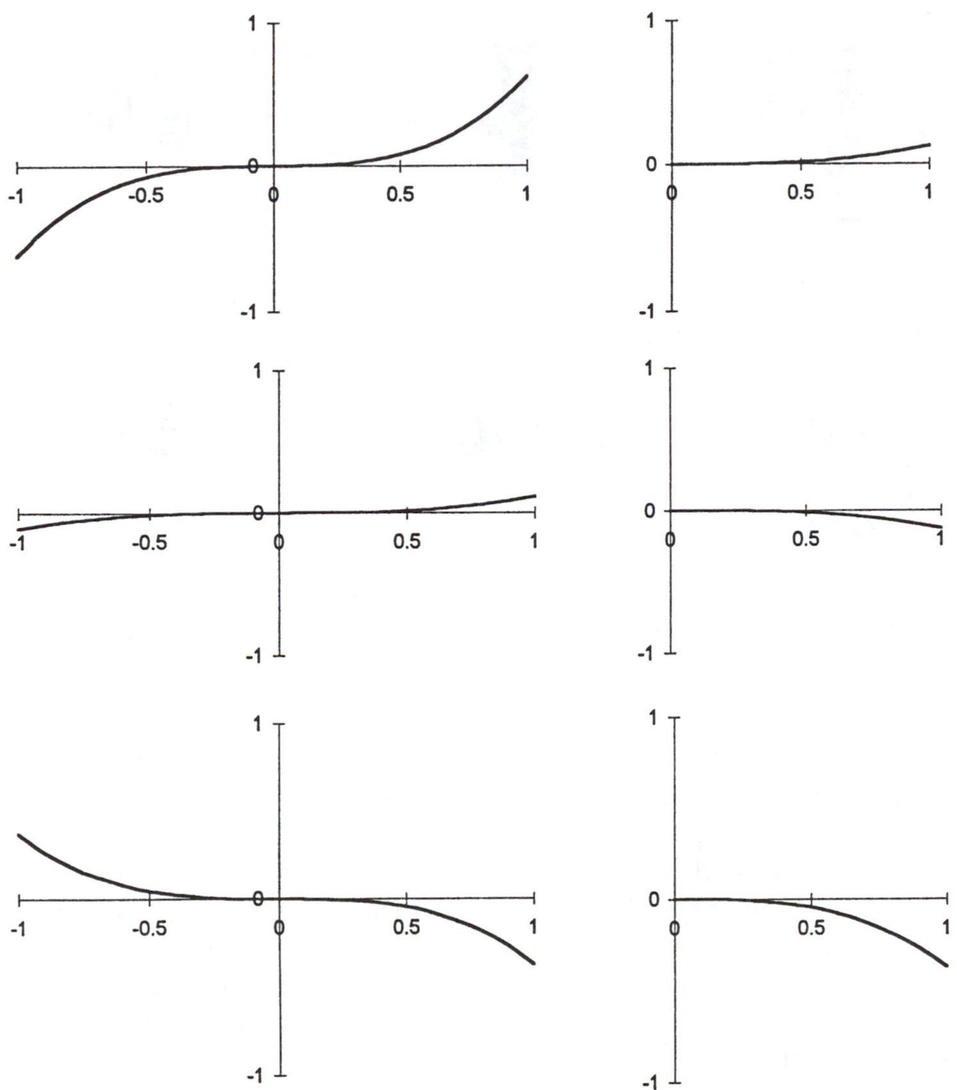

FIGURE 3.64 Ray intercept plots when some third-order spherical aberration is deliberately introduced to offset the growth of oblique spherical aberration across the field.

Fortunately, the natural tendency of the astigmatic fields is to produce inward field curvature, thus moving the best focus location for the overcorrected spherical aberration toward the flat focal reference surface. Because the amount of oblique spherical aberration in the sagittal and tangential directions usually differs, it is desirable to adjust the astigmatic fields so that the best focus for each principal azimuth in the aperture is located at or near the flat focal surface. This may or may not be possible, depending upon the oblique spherical aberration coefficients intrinsic to the lens. The form of sagittal and tangential field surfaces is driven for the most part by the

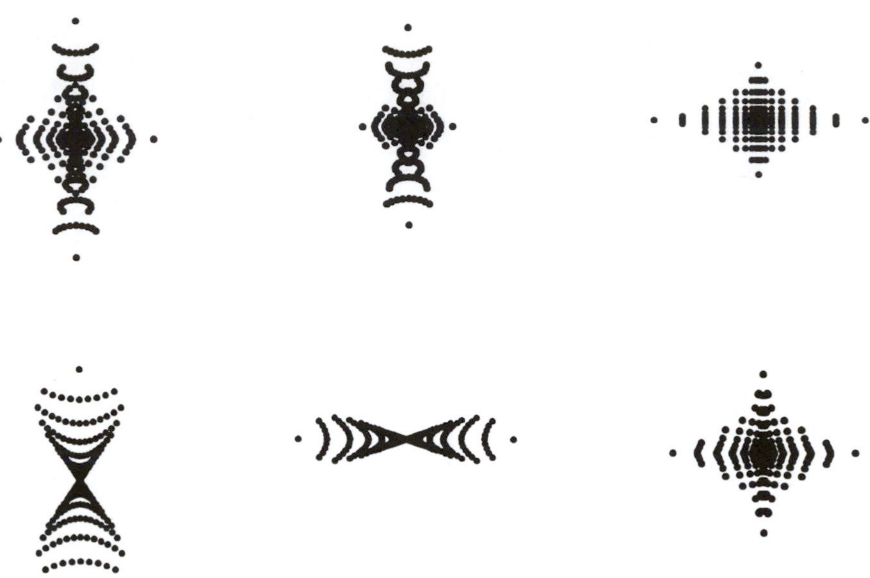

FIGURE 3.65 Some sample spot diagrams that can be expected from various mixtures of the sagittal and tangential oblique spherical aberration terms

amount of oblique spherical aberration that must be balanced, and not by the astigmatism alone. As will be seen in the next paragraph, the actual astigmatic field curves are more complicated than so far discussed.

The terms in fourth power in the field represent high-order astigmatic field curvature. The interpretation of the error is similar to the third-order astigmatism, with the exception that there is a five-to-one ratio between the location of the fifth-order sagittal and tangential curves and the fifth-order Petzval curvature. The ray intercept equations for the combination of third- and fifth-order astigmatism can be rewritten as field focus locations

$$z_{tangential} = \frac{1}{u'}[(3(\mathrm{AST3}) + \mathrm{PTZ3})h^2 + (5(\mathrm{AST5}) + \mathrm{PTZ5})h^4]$$

$$z_{sagittal} = \frac{1}{u'}[(\mathrm{AST3} + \mathrm{PTZ3})h^2 + (\mathrm{AST5} + \mathrm{PTZ5})h^4]$$

Here the values of AST and PTZ are lateral blur values for the aberrations, and the u' appears because of the conversion to axial focal shift. The definition of the third-order Petzval sum was discussed earlier, and represents the location of the zero astigmatism image, and is dependent only upon the curvatures in the lens. A similar interpretation of the fifth-order Petzval sum is possible, in which this surface is a high-order correction to the third-order value. The high-order Petzval curvature is not solely dependent upon the curvatures of the lens, but will vary to some extent, depending upon

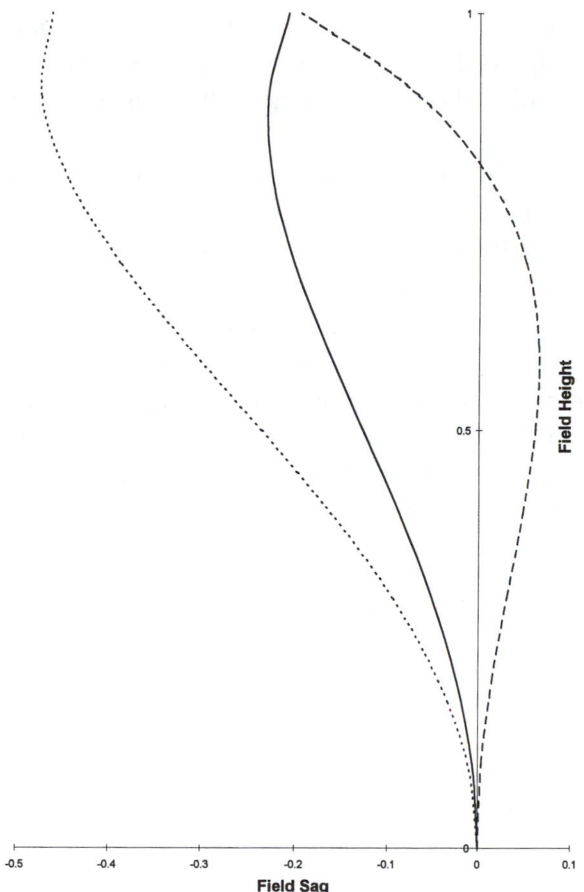

FIGURE 3.66 Plot of the third- and fifth-order sagittal (——), tangential (– – –), and Petzval (· · · · · ·) surfaces that can be expected for a common type of field curvature balance.

the level of the third-order aberrations. As mentioned before, the variation is slow with aberration change, so that in most cases the fifth-order Petzval can be considered as almost a constant determining the astigmatic behavior.

An example of the type of field plots that can be expected is shown in Figure 3.66. A plot of the actual field curvature behavior is established by the positions of the field curves determined from close skew ray traces. This plot will actually include all orders in the field behavior since the astigmatic foci are traced along a real chief ray through the lens system. The presence of a node for astigmatism is observed in many types of lenses. In order for such a node to exist, the third- and fifth-order aberrations have to fall within certain limited ranges. In any complex lens, the field plots as computed from the high-order coefficients will not precisely match the field plots determined by the close skew ray traces. The actual field curvatures will be modified by the low-order aberration content, the pupil mapping changes due to pupil aberrations and the possible content of higher-order field curvature. Thus the analysis of the high-order field curves is primarily useful in furnishing a guide as to the possible aberration balances that can be expected in a complex lens.

In a real lens, the aberrations will occur in many combinations. In addition, all of the aberrations can vary with wavelength. In most cases the lower-order aberrations will have a more rapid variation with color than will the high-order aberrations. Therefore the optimum balancing of aberrations, once achieved at a particular color, will not hold across the spectral regions. Different tactics need to be applied to provide balancing in the presence of intrinsic dependence of the image upon wavelength. The construction of an appropriate merit function should include all of these effects.

The understanding of the aberrations of various orders can be used to explain the imaging characteristics of a lens and provides an avenue for discussing the limits upon the image quality that can be achieved. As an example, consider the double Gauss type lens. The aberration residuals are dominated by high-order spherical aberration and oblique spherical aberration. These aberrations are balanced by the introduction of lower-order spherical aberration and by the proper amount of astigmatism. The high-order astigmatic field curves change rapidly at the edge of the field and prevent complete balancing of the oblique spherical aberration. There is some high-order coma asymmetry visible as well. The choice of the final compromise focal surface is made to balance the effect of the residual aberrations across the field.

The ray plots in Figure 3.67 illustrate this example. The axial image is limited by third- and fifth- (or higher) order spherical aberrations. The focal position is set at about the best compromise focus of the axial image. The seven-tenths field plots indicate a large amount of sagittal oblique spherical aberration. The focus for the tangential direction is located at the same focal position as indicated by the small slope of the ray intercept plot. The focus of the sagittal direction has been set to be close to the best focus, but has a relatively large overcorrected spherical aberration tail. At the edge of the field, the inward-curving tangential focus is a good balance for the oblique spherical aberration, but the effect of the large amount of tangential oblique spherical aberration is limited by the choice of vignetting. The major problem at the edge of the field is the large amount of overcorrected sagittal oblique spherical aberration.

Figure 3.68 is a listing of the Buchdahl high-order aberrations expressed as lateral aberration errors. Examination of this table supports the qualitative analysis in the previous paragraph. The numbers are interesting, but should be taken as a guide, and not as fully descriptive of the full state of correction of the lens. Fixing the individual high-order aberration coefficients to desired

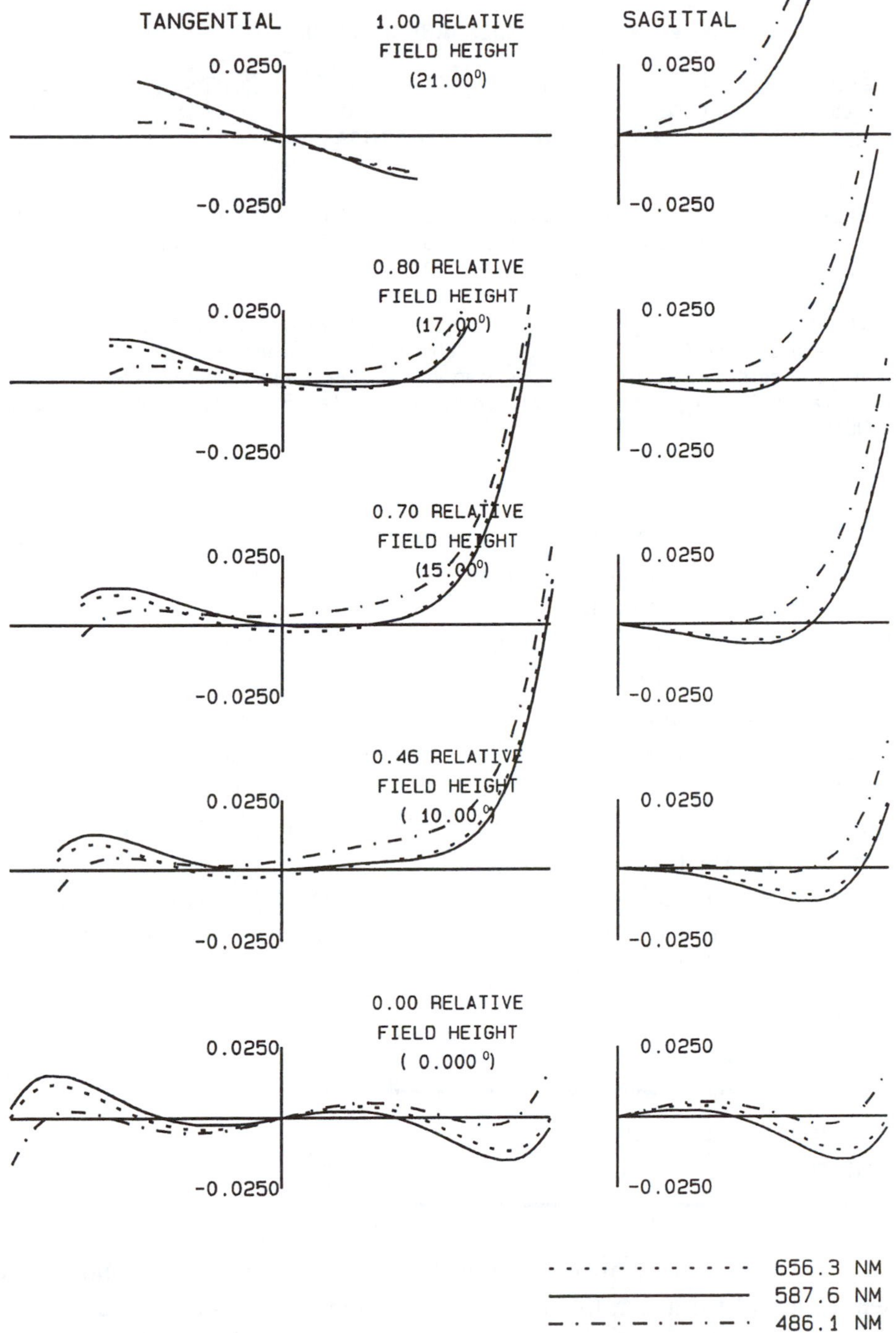

FIGURE 3.67 Ray intercept plot for the double Gauss lens example, showing the effect of low- and high-order aberrations, the important effect of oblique spherical aberration, and the use of vignetting to eliminate some of the most objectionable parts of the aberration (CODE V).

THIRD AND FIFTH ORDER ABERRATIONS

	SA3 SA5 SA7	TCO TCO5 ECOM	TAS TAS5 OTSA	SAG SAG5 OSSA	PTZ PTZ5	DST DST5
SUM	-0.101996	0.062350	-0.062109	-0.116407	-0.143556	-0.254504
	0.059359	0.000034	0.043869	0.113406	0.130791	-0.141049
	0.017500	-0.029266	0.215221	0.154856		

FIGURE 3.68 Program listing of the third- and fifth-order aberration coefficients for the lens in the previous figure. These coefficients were obtained from calculation of the Buchdahl aberration coefficients, rearranged into values describing the transverse aberration contribution of the aberration. The tangential and sagittal oblique spherical aberration terms are labeled OTSA and OSSA (CODE V).

values is not usually possible. The reasons for this involve the intrinsic nature of some of the aberrations, such as oblique spherical aberration, as discussed above. The amounts of the high-order aberrations are also changed by adjustments of the low-order aberrations. Changes in the design will also change the pupil aberrations that determine the detailed mapping of the aberrations on the image surface. While the aberration coefficients are useful as a guide to the correction of the lens, they are not necessarily convergent to the actual ray intercept errors. The ray intercept plots contain all of the high-order aberrations, described in terms of the actual pupil coordinates and on the desired focal surface.

This lens is reasonably well balanced across the field. Any major improvement will require searching globally for a design with a smaller amount of oblique spherical aberration.

3.2 Measurement and Specification

The calculation of aberrations is important in design. The measurement and specification of the aberrations in a completed lens is of importance in determining whether the system has been constructed properly. In addition to the aberrations introduced in the design, additional classes of aberrations can be created as a result of alignment or fabrication errors in the lens. These may need to be combined with the design aberrations to predict the ultimate performance of the lens.

The usual process of verifying the performance of a lens is to measure the aberrations, or functions associated with the aberrations for comparison with similar data computed on the lens. Thus there is a need for understanding the method of specifying and measuring the aberrations.

There are several methods for measuring the aberrations in a lens. Since all imagery is ultimately limited by diffraction and physical optics, the geometrical optics must be sorted out from the observed image data. There are four general approaches used in doing so. The first is the use of an interferometer, to obtain the wavefront error at a specific wavelength and field position. The second is the use of image processing or phase restoration to derive the phase information from the intensity distribution in the image of a point source, or other target. The third is a geometrical measurement in which the dimensions of the point spread function are measured as the image field and focus position is changed in a specific manner. The fourth is the comparison of derived quantities such as MTF or energy concentration curves obtained from measurement and by calculation.

There is, of course, the evaluation of the final measure of quality in the lens, such as resolution or subjective image quality that can be used as a criterion for the degree to which the residual aberrations of the lens meet the user's needs. This is generally a "pass or fail" test rather than a diagnostic test. If the system fails, then other tests are required to obtain information as to how to realign or refabricate the system to meet the goals. (Presuming, of course, that the basic design is acceptable in this regard!)

Interferometry is the principal means used today to sort the geometrical aberrations from the effects of diffraction. The basic data collected is a matrix of wavefront phase information obtained from an interferometer. The interferometric data should be acquired by focusing the detector in the interferometer on the pupil of the lens. In most cases, the interferometer requires a double pass of the wavefront through the lens, so that the wavefront errors are the sum of the two passes. If the aberrations on the lens are small, then the wavefront phase is essentially half of the double-pass value. If the aberrations per pass are large, usually of the order of several wavelengths, the single pass may not be accurately represented by a simple halving of the interferometer phase data.

This data may also be available as a fitted set of Zernike coefficients, plus a statement of the amount of residual error that was not fitted by the order of Zernike coefficients specified. In either case, the usual procedure is to combine this measurement with the wavefront errors predicted by the lens design data, and then compute the interesting image quality functions, such as the point spread functions. The combination can sometimes be quite complicated, if the lens being tested has any significant amount of design aberration. Because the interferometer measures only one wavelength, the evaluation of

TABLE 3.3 Functional forms of Zernike polynomials from the FRINGE set

Number	Polynomial
1	1
2	$2R\cos\theta$
3	$2R\sin\theta$
4	$\sqrt{3}(2R^2 - 1)$
5	$\sqrt{6}R^2\cos 2\theta$
6	$\sqrt{6}R^2\sin 2\theta$
7	$\sqrt{8}(3R^2 - 2)R\cos\theta$
8	$\sqrt{8}(3R^2 - 2)R\sin\theta$
9	$\sqrt{5}(6R^4 - 6R^2 + 1)$
10	$\sqrt{8}R^3\cos 3\theta$
11	$\sqrt{8}R^3\sin 3\theta$
12	$\sqrt{10}(4R^2 - 3)R^2\cos 2\theta$
13	$\sqrt{10}(4R^3 - 3)R^2\sin 2\theta$
14	$\sqrt{12}(10R^4 - 12R^2 + 3)R\cos\theta$
15	$\sqrt{12}(10R^4 - 12R^2 + 3)R\sin\theta$
16	$\sqrt{7}(20R^6 - 30R^4 + 12R^2 - 1)$
17	$\sqrt{10}R^4\cos 4\theta$
18	$\sqrt{10}R^4\sin 4\theta$
19	$\sqrt{12}(5R^2 - 4)R^3\cos 3\theta$
20	$\sqrt{12}(5R^2 - 4)R^3\cos 3\theta$
21	$\sqrt{14}(15R^4 - 20R^2 + 6)R^2\cos 2\theta$
22	$\sqrt{14}(15R^4 - 20R^2 + 6)R^2\sin 2\theta$
23	$4(35R^6 - 60R^4 + 30R^2 - 4)R\cos\theta$
24	$4(35R^6 - 60R^4 + 30R^2 - 4)R\sin\theta$
25	$3(70R^8 - 140R^6 + 90R^4 - 20R^2 + 1)$
26	$\sqrt{12}R^5\cos 5\theta$
27	$\sqrt{12}R^5\sin 5\theta$
28	$\sqrt{14}(6R^2 - 5)R^4\cos 4\theta$
29	$\sqrt{14}(6R^2 - 5)R^4\sin 4\theta$
30	$4(21R^4 - 30R^2 + 10)R^3\cos 3\theta$
31	$4(21R^4 - 30R^2 + 10)R^3\sin 3\theta$
32	$\sqrt{18}(56R^6 - 105R^4 + 60R^2 - 10)R^2\cos 2\theta$
33	$\sqrt{18}(56R^6 - 105R^4 + 60R^2 - 10)R^2\sin 2\theta$
34	$\sqrt{20}(126R^8 - 280R^6 + 210R^4 - 60R^2 + 5)R\cos\theta$
35	$\sqrt{20}(126R^8 - 280R^6 + 210R^4 - 60R^2 + 5)R\sin\theta$
36	$\sqrt{11}(252R^{10} - 630R^8 + 560R^6 - 210R^4 + 30R^2 - 1)$
37	$\sqrt{13}(924R^{12} - 2772R^{10} + 3150R^8 - 1680R^6 + 420R^4 - 42R^2 + 1)$

the effect of chromatic aberration will require more specific information as to the source of the measured error. In an all-reflective system, this is obviously not a problem.

One set of the Zernike polynomials that is commonly used in describing test data on lenses is shown in Table 3.3. This set, named the FRINGE set after the interferometric data analysis program in which it first appeared, is not a complete set through a specific order. It is a low-order complete set supplemented with radial polynomials of higher order to permit better fitting of the type of zonal errors that are commonly encountered as residuals when fabricating large aspheric optical components.

Aberration measurement can also serve as a predictive tool. If the individual components are completely measured, the errors on each component can be added to the lens design data to predict the likely final wavefront error through the lens system. In most cases, the lens design program can be used to predict the final adjustments that will be required to optimize the image quality in the lens.

More traditional methods are available for measuring and specifying aberrations. A visual or electronic measurement of the dimensions of the point spread function can be used to quantify the amount of specific aberrations. Usually, the focus must be changed and several measurements made in order to determine a specific aberration from the observed data.

For example, the spherical aberration may be measured by locating the paraxial focus, and then measuring the diameter of the point spread function. The test engineer must remember that the geometrically perfect image of the point object must be very small relative to the point spread function being measured. Alternately, the position of the focus point of rays from the center of the aperture and some specified zones can be measured, and the spherical aberration quantity obtained from this longitudinal measurement. This may be carried out by reference to known appearance of the point spread function in the presence of aberration, or by masking portions of the aperture to select the effective ray positions that are to be measured.

Somewhat geometrical test methods, such as the knife edge test and Hartmann test, can be used to effectively trace rays by means of their geometrical shadow or averaged position in a measurement system. A modification of the Hartmann procedure suggested by Shack involved the use of a position-sensitive array detector to measure the effective ray displacements in real time. This type of sensor has been used to measure the aberrations, sometimes including atmospheric phase distortions, and feed the data back into active control of the lens system. The availability of inexpensive high-speed computers makes significantly complicated active systems feasible.

REFERENCES

Born and Wolf 1959	Born, M. and Wolf, E. *Principles of Optics*, Pergamon Press, London
Buchdahl 1968	Buchdahl, H. A. *Optical Aberration Coefficients*, Dover Publications, New York
Buchdahl 1970	Buchdahl, H. A. *An Introduction to Hamiltonian Optics*, Cambridge University Press
Conrady 1929	Conrady, A. E. *Applied Optics and Optical Design*, republished 1992 by Dover, New York
Cox 1964	Cox, A. *A System of Optical Design*, Focal Press, London
Herzberger 1958	Herzberger, M. *Modern Geometrical Optics*, Interscience Publishers, New York
Hopkins 1950	Hopkins, H. H. *Wave Theory of Aberrations*, Clarendon Press, Oxford
Hopkins 1962	Hopkins, R. E. *Optical Design, MIL-HDBK-141*, available from Sinclair Optics, Fairport, NY
Kingslake 1978	Kingslake, R. *Lens Design Fundamentals,* Academic Press, New York
Malacara and Malacara 1994	Malacara, D. and Malacara, Z. *Handbook of Lens Design*, Marcel Dekker, New York
Rimmer 1963	Rimmer, M. *Optical Aberration Coefficients*, MS Thesis, University of Rochester
Slyusarev 1984	Slyusarev, G. G. *Aberration and Optical Design Theory*, Adam Hilger, Bristol
Smith 1990	Smith, W. J. *Modern Optical Engineering*, McGraw-Hill, New York
Welford 1986	Welford, W. T. *Aberrations of Optical Systems*, Adam Hilger, Bristol

IMAGE ANALYSIS

4.1 Image Quality Measures

The ultimate measure of acceptable image quality from a lens is customer or user satisfaction with the image that is produced. This measure can be the result of a quantitative evaluation, such as signal-to-noise ratio under certain conditions. In many cases, it may be a subjective evaluation as to the acceptable quality of the image as determined by the viewer. In any case, the goal for the lens designer needs to be expressed in some quantitative values that can be computed by the designer from the lens data. Only in this way can the designer know that the design task is completed.

The goal of lens design is to produce a system which will provide images of acceptable quality for a specified user. Image quality is frequently very subjective, based upon the opinion of a user as to whether the appearance of the image is pleasing or informative. In some applications the image quality can be determined in very objective ways, such as the level of contrast of certain fine details exceeding some specified threshold value. In either cases, there are physical quantities describing the image structure that can be used to evaluate the probable degree of acceptability of an image produced by a lens design.

"Image quality" is a somewhat elusive quantity. The quality of an image may be defined by its technical content or pictorial content. Quantifying the technical content or image structure is easier than attempting to quantify the pictorial content of an image. Specifying acceptable image quality in terms of the contrast of fine details is a common method of quantifying quality, but this usually requires correlation with methods of assessing the acceptability of an image by a viewer. The goal of image analysis is the development of numerical descriptions of image content that can provide some numerical description of image structure, and which can be used to infer the level of image quality.

4.1.1 VISUAL

The acceptability of an image by a user can be a qualitative matter, which involves the interpretation of a physical image by psychometric methods. The image must appear to be pleasing, or acceptable to a viewer, and must contain sufficient detail to convey the required message to the user.

Numerically describing the image requires that the contrast or details of the image structure be described or predicted by a physical model of the lens system. The contour of intensity in the image can be computed and applied to specific imaging situations. In some obvious cases this will be the signal-to-noise ratio, with respect to a specific detector under specified illumination conditions. This ratio should exceed some acceptable threshold in order to provide an acceptable image.

The definition of an acceptable threshold may be based upon subjective criteria involving human observers. The threshold can also be quantitative, as in a document reader or a computer-controlled target tracking system. In an attempt to be general, the threshold may be stated as a function of spatial frequency, rather than for a specific target.

The relation between the quantitative assessment of image structure and the psychophysical perception of image quality is sometimes very difficult, and frequently only approximate. The contrast or other characteristic of the image must be calculated, and matched to some psychometric scale describing the acceptability of the image. There are useful models of image structure that can be used to specify the image quality. In most cases, a simplification is used in which the user of the system states an acceptable level of contrast for a stated spatial frequency. The designer's task is to demonstrate that the lens will produce this specified level of contrast once the lens is fabricated.

The most common attempt to quantify a subjective evaluation of an image is to specify the resolution. This is the highest spatial frequency target of a specific contrast that can be observed in the image. The process of observing a resolved target is intuitively simple, but involves the detection of image modulation in the presence of limiting noise. Other attempts involve the classification of imagery with respect to its intended application. Such rating scales are commonly applied to technical photography for intelligence gathering or mapping.

A statement of aberrations is rarely of interest to a user. The user wants to know how the image structure will meet his or her needs. The Optical Transfer Function (OTF) provides the most generally useful link between the aberration content of a lens and the questions asked by a user.

In this book, the emphasis will be upon computation of the image structure, and the prediction of image quality using some generally accepted criteria. The optical designer will generally be able to apply these general concepts to specific problems. The basis for the computation and prediction is the optical transfer function, which can not only be used to compute to good accuracy the details of light distribution in the image, but can be used directly as a comparative indicator of image quality.

4.1.2 TECHNICAL

The technical measure of image quality is usually very specific. The concentration of light within a specific region on a detector is one example. Another is that the distortion and the boresight of the optics with respect to a specified direction are known to within some tolerance. In some cases the technical need will be for a certain irradiance level and uniformity to be achieved on the image surface.

Another favorite is resolution, which is a statement of the highest spatial frequency in the image that can be detected by an observer or other analysis method. This resolution involves the technical descriptions of the image formed by the lens, the object shape and contrast, and the detector response and noise content.

These quantitative values can usually be interpreted numerically to form a merit function for the design. The subjective visual image quality must also be interpreted in some set of reasonable quantitative values for guidance to the lens designer.

4.1.3 SUMMARY MEASURES

It would be nice if there were some general function describing the image that would unambiguously state whether the optical system would meet a certain need. In particular, some function that states whether the picture produced is "good," "excellent," "acceptable," or "not acceptable." Unfortunately, no such function exists. There are some image descriptors that can provide a relative assessment of the acceptability of the image formed by a lens. If the ultimate sensor is an electronic device, such a target discriminator or tracker, the description of image quality may be very close to flawless. For subjective evaluations, such as pictorial imagery, the descriptor will be quite subjective.

In section 4.4 some generally accepted summary measures will be described. Relating these to the merit function that provides a target for the aberration correction in the lens design can be accomplished, but may be somewhat artistic rather than precise.

4.2 Optical Transfer Function Basics

4.2.1 OTF DEFINITIONS

The Optical Transfer Function has become the function most commonly used for evaluating the image quality in lens design. The OTF provides an assessment of the image structure by stating the contrast and spatial phase of the image of a sinusoidal object for a range of spatial frequencies. Application of Fourier analysis permits the description of any object as a sum of sinusoidal components (Goodman 1968). Therefore the information conveyed by the Optical Transfer Function can be used to construct the detailed form of an image. Because the OTF provides the reconstruction link between the object, lens aberrations, and the image, this function may also be used by itself to provide a criterion for the ability of the lens to form an acceptable image. In communication theory terms, the OTF is a description of a lens as a low-pass-band limited spatial frequency filter of object content.

There is a dual definition of this function. The OTF can be obtained directly as a ratio of the contrast of the image contrast to object contrast. Alternately, the OTF can be obtained as Fourier transform of the point spread function. This latter approach serves as the basis for computing the OTF from lens design data. The source and nature of illumination are also important. The spectral content, coherence properties, and spatial extent of the source, as well as the spectral sensitivity of the detector, will all affect the OTF.

For the majority of lens applications it is sufficient to assume that the illumination is incoherent, usually the case with a broad-wavelength band-pass source of large extent. This implies that there is not a fixed phase or coherence relationship between light from adjacent points in the object. The irradiance on the image surface will sum linearly under the case of fully incoherent imagery. The imagery is then linear in intensity. The OTF is also presumed to apply to a region of the image that is large in comparison to the finest details to be examined in the image. The OTF will be considered as stationary over that region. The intensity is always positive, thus the point spread function and the intensity distribution cannot take on negative values. This positive signal limitation, along with the symmetries imposed by the lens place some theoretical and practical limits upon the range of values that the OTF can attain.

The OTF is used to obtain information about the contrast in the images of specific objects. The basic object used in defining the OTF is a sinusoidal intensity distribution as in

$$I(x) = 1 + m\cos(2\pi\nu + \phi(\nu)) \tag{4.1}$$

Two parameters define a sinusoidal object or image, the modulation contrast and the spatial phase. The contrast of the sinusoid of period p or spatial frequency $\nu = 1/p$ is

$$m = \frac{I_{max} - I_{min}}{I_{max} + I_{min}} \tag{4.2}$$

as shown in Figure 4.1. The spatial phase of the object (or image) is the shift, usually expressed in radians (or sometimes in degrees), of the location of the peak of the sinusoidal signal from the origin. The relation between the spatial phase angle and the actual linear shift is

$$\phi = 2\pi\,\frac{\delta x}{p} = 2\pi\nu\,\delta x \tag{4.3}$$

in which δx is the linear phase shift in lens units, p is the period and ν the spatial frequency of the pattern in lens units.

This first definition of the Optical Transfer Function is demonstrated by using Figure 4.1. The sinusoidal object is imaged through the lens, and the ratio of the contrasts and the shift in phase taken for a specific spatial frequency where the Modulation Transfer Function (MTF)

$$\mathrm{MTF}(\nu_x, \nu_y) = \frac{m_{image}}{m_{object}} \tag{4.4}$$

is the ratio of the modulation contrasts in the object and image of a sinusoidal pattern.

Because the description of the image involves both contrast and phase, the complete OTF is usually expressed as a complex function

$$\mathrm{OTF}(\nu_x, \nu_y) = |F(\nu_x, \nu_y)|e^{i\phi(\nu_x,\nu_y)} = (\mathrm{MTF})e^{i(\mathrm{PTF})} \tag{4.5}$$

where the first factor is the modulus of the function, usually called the Modulation Transfer Function (MTF) and the second factor represents a spatial phase shift ϕ, and is referred to as the Phase Transfer Function (PTF). The phase shift indicates a change in the location of the peak of the pattern relative to fixed set of coordinates. The two dimensions in the frequency values indicate that the image is intrinsically two-dimensional in nature. The two frequency coordinates are, of course, aligned with the x and y coordinates.

Both the contrast and the phase can be functions of spatial frequency and direction on the image surface. As long as the imaging characteristics are constant over the image surface, a single function can specify the entire image surface. If these functional relations change with location on the image sur-

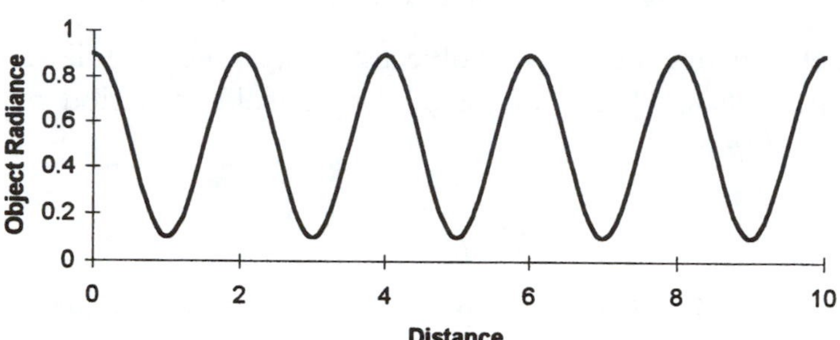

Image Irradiance

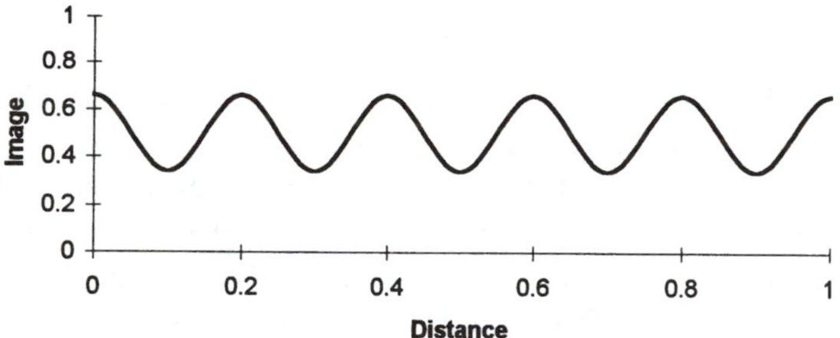

Modulation Transfer Function

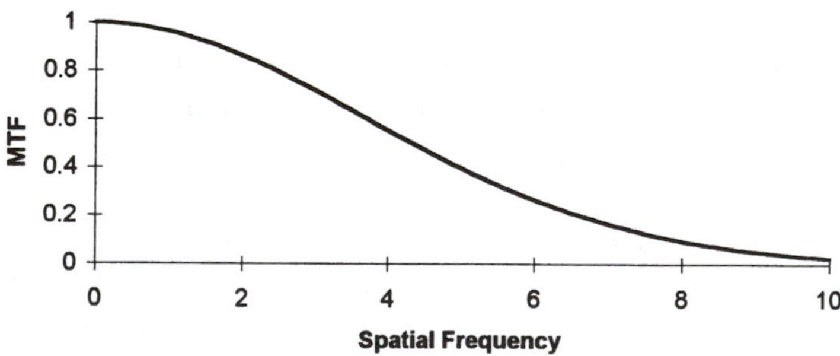

FIGURE 4.1 The Modulation Transfer Function is found by taking the ratio of the contrast ratios in the image and object at each spatial frequency.

face, then a single function defining the MTF will not suffice for the entire image surface.

There are some conditions that apply to the most general usage of the Optical Transfer Function in describing imagery. The two important conditions are stationarity and linearity. These two conditions require that the imaging characteristics of a lens can be represented by a single function for the entire focal surface of interest and over the entire irradiance and detection range of interest.

These conditions are very useful mathematically, but are subject to a reality check when dealing with practical lenses and imaging systems. No real physical system has characteristics that are totally independent of intensity level or of location over an entire image. Some approximations and assumptions need to be made to make the imaging problem tractable. A realistic imaging problem is approached by stating that the imaging process is stable over a region called an isoplanatic patch that is "large with respect to the characteristic size of the spread function, but may be small with respect to the entire field of the lens." Similarly, it is usually assumed that the system will be linear in intensity over a "sufficiently wide" range as to make OTF analysis useful.

In reality, neither of these acceptable regions is well defined for a lens and detector system. Nor are there any acceptable standards for the degree of linearity or stationarity. It is perhaps a bit amusing that the practical limits of applicability for such an important topic have never been examined in detail. In section 2.1.4, the extended sine condition was described. This states the conditions under which the aberrations are first-order stable with image position. This condition needs to be met in order that an isoplanatic patch of any reasonable size will exist in the image. The second-order variation of wavefront, as treated in section 2.1.5, indicates that even if the sine condition is satisfied, there may be higher-order variation of the MTF across the field.

An example of the MTF for a realistic lens is given in Figure 4.2, for the case of the double Gauss lens previously used as an example. The figure is a plot of the MTF versus spatial frequency for three field positions. It can be seen that the lens is not well represented by a single MTF. Another view of this variation with field is indicated in Figure 4.3. This is a plot of the contrast at 20, 40, and 60 lines per millimeter across the field. This plot represents a single direction in the field, so that the area represented by the MTF is greater for regions at the edge of the field. The area influenced by the various MTFs will also be influenced by the actual shape of the detector surface. In this case the normal 35 mm film format would indicate that for field angles beyond about 17°, there will be limitation of image area by the image format. Therefore, the MTF that is poorest at the edge of the lens field actually contributes only to the corner of the format, and is not, on the average, as important in image formation as the MTF at about 15°.

From the preceding discussions of aberrations, it can be seen that some criterion can be developed for defining an isoplanatic patch on the image surface. One possible approach may be to require that the point spread function does not change by more than a specified amount over the patch.

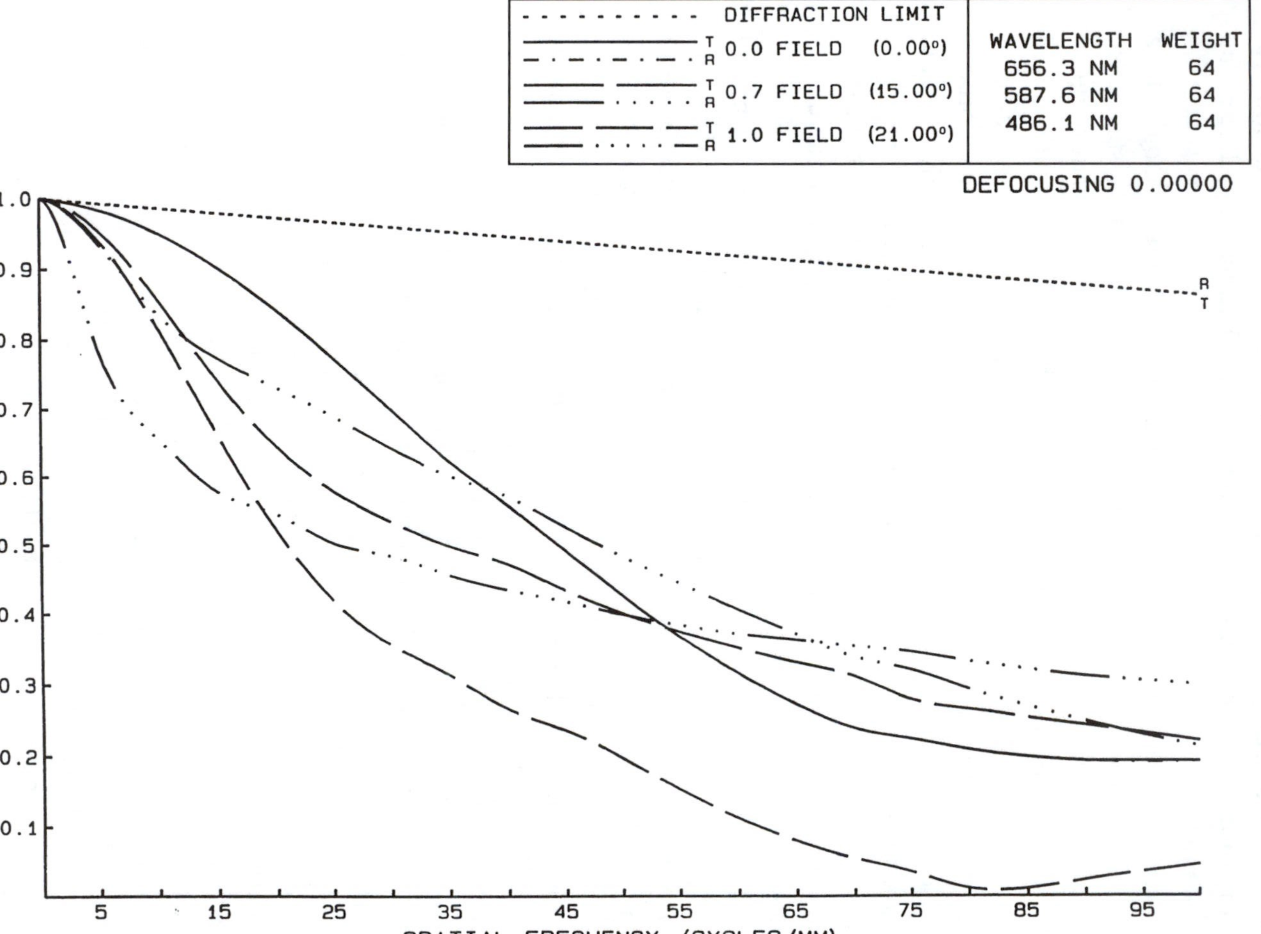

FIGURE 4.2 A plot of the MTFs for a sample lens, for axis, 0.7 and full field (CODE V).

This will translate to some allowable change in the magnitude of the aberration coefficients, or in the total amount of wavefront aberration. The designer needs to be aware of this condition upon the calculation of the MTF in order to ensure that a lens is properly characterized. A visual examination of the data in Figure 4.3 indicates that there are perhaps four regions within which the MTF is reasonably constant. This indicates that about four sets of radial and tangential MTF curves are adequate for describing the behavior of the lens at a selected focal position. The criterion for how constant the MTF needs to be over each of the isoplanatic regions is determined by the intended application for the lens.

When stationarity and linearity apply to an acceptable level, a second very useful definition of the OTF applies. The point spread function, $f(x, y)$ is the intensity distribution on the image surface of a point object having some spectral distribution including the source distribution and the sensitivity of the detector. To make the calculation easier, the total energy or volume under $f(x, y)$ can be set equal to unity. The Fourier transform of the point spread function is then the Optical Transfer Function,

$$\text{OTF} = F(\nu_x, \nu_y) = \int_{-\infty}^{\infty}\int_{-\infty}^{\infty} f(x, y)e^{-i2\pi(\nu_x x+\nu_y y)}\, dx\, dy \tag{4.6}$$

subject to the condition that the total volume under the point spread function is normalized to unity:

$$\int_{-\infty}^{\infty}\int_{-\infty}^{\infty} f(x, y)dx\, dy = 1 \tag{4.7}$$

This relationship is shown in Figure 4.4. This specific normalization is not a limitation, but a convenience, as it causes the value of $F(0, 0)$ to be unity, which makes considerable sense. This definition of the OTF will be consistent with the first definition when this normalization is applied. This normalization also implies that this function is independent of the actual intensity of the object. In these transforms, the limits are formally set at infinity. It is obvious that no practical system will permit integration over an infinite range. The user of these techniques will have to assure the accuracy of the calculation over a limited range. Usually the range of interest will be the size of the isoplanatism patch. Therefore, this computation will have to be repeated for several field positions.

The contrast at each spatial frequency is given by the values of the OTF at the specific frequency. The actual irradiance of the variation of the sinusoidal component at any spatial frequency is the modulation multiplied by the average intensity incident on the image surface.

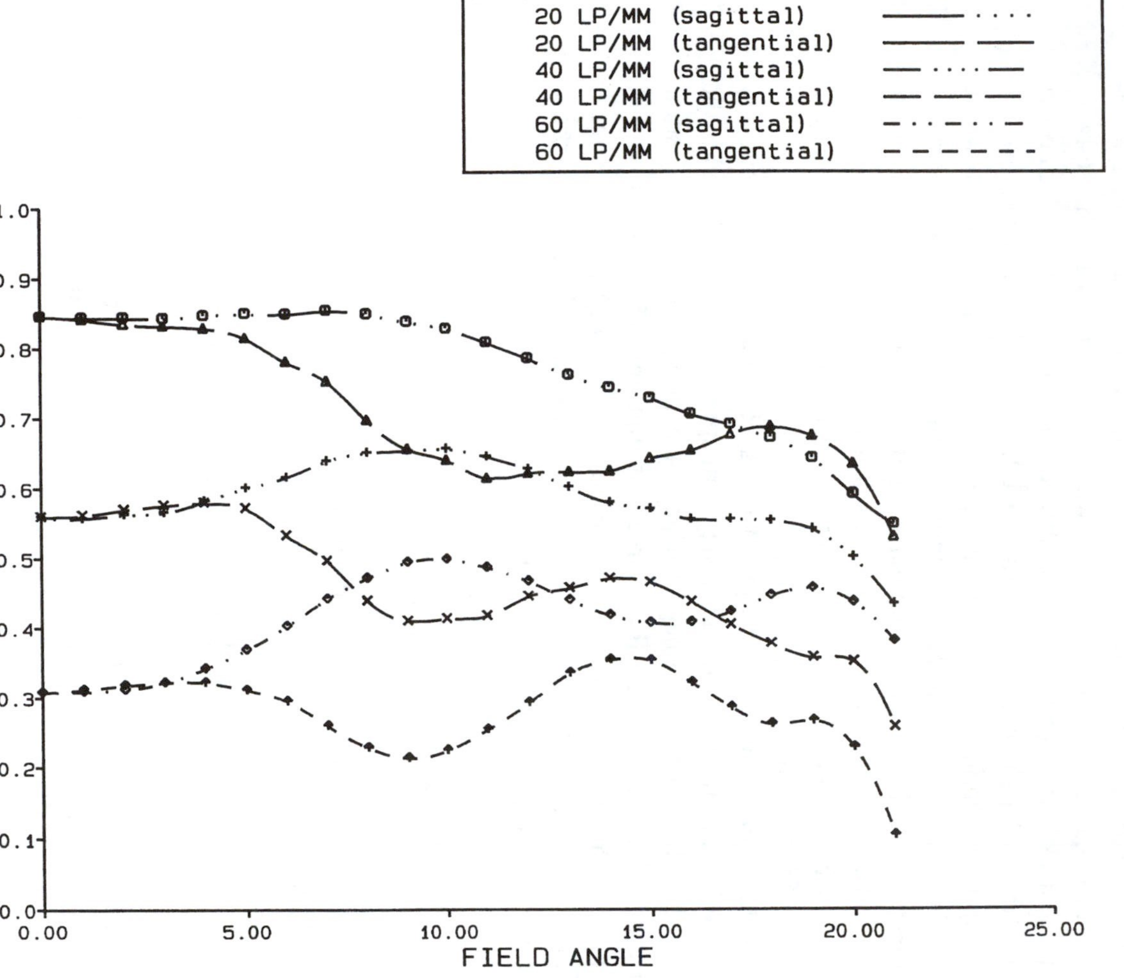

FIGURE 4.3 An evaluation of the variation of the MTF for selected spatial frequencies as a function of the field angle (CODE V).

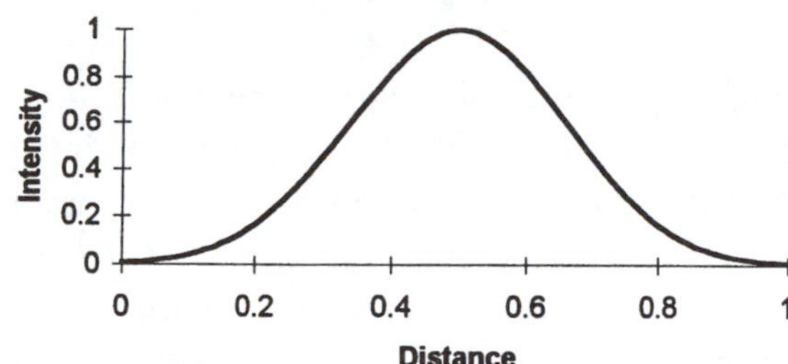

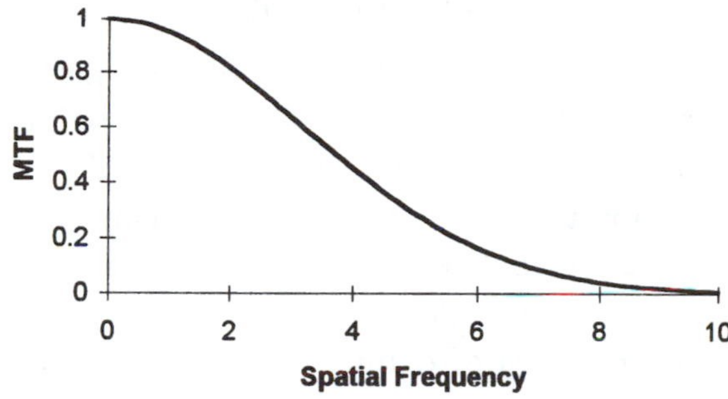

FIGURE 4.4 In the case of a linear stationary system the MTF is found by taking a Fourier transform of the Point Spread Function.

Another option in the calculation is the use of the line spread function, which is the one-dimensional intensity distribution in the image of a fine line object. This can be calculated from the point spread function by selecting a direction for the slit source and integrating along the line. The OTF for spatial frequencies perpendicular to the length of the slit is then the one-dimensional Fourier transform of the line spread function. This will be the OTF for sinusoidal objects oriented at a specified direction on the focal surface.

The importance of the OTF which satisfies the linear and stationary conditions is that knowledge of the OTF permits the prediction of the intensity distribution in the image of any arbitrary object. This application of Fourier optics has been discussed in many books; especially useful are those by Goodman (1968) and Gaskill (1978). In short, this useful property occurs as a consequence of the imaging property of a stationary system in which the intensity distribution in the image is obtained by integrating or summing the intensity distribution the object weighted or smeared by the point spread function:

$$i(\boldsymbol{x}) = \int_{-\infty}^{\infty} o(\boldsymbol{x} - \boldsymbol{u}) f(\boldsymbol{u}) d\boldsymbol{u} \tag{4.8}$$

This integral is called a convolution integral. The object $o(\boldsymbol{x} - \boldsymbol{u})$ is convolved with the point spread function $f(\boldsymbol{u})$. The variable of integration takes all possible values over the range of integration. In any practical system, the use of integration over an infinite space as limits is formal only. The actual range is determined by the physical size of the region of stationarity on the image surface. When this holds, the Fourier transforms of the object and image can be used along with the Fourier transform of the point spread function to yield a simple multiplication relationship of the transformed functions:

$$I(\boldsymbol{\nu}) = F(\boldsymbol{\nu}) O(\boldsymbol{\nu}) \tag{4.9}$$

The functions I and F are Fourier transforms of the image and object intensity distributions and F is the Fourier transform of the point spread function. This is the OTF for the region of interest.

By using OTF techniques any arbitrary object can be modeled and the light distribution in the image determined. The effect of a detector in further reducing the image contrast or adding noise or sampling errors can also be obtained. In general, this level of detail is not used in evaluating the image quality of a lens. Measurements or estimates of the threshold required for satisfactory resolution of detail are obtained, and the evaluation of the image quality only requires comparison of the MTF level obtained with the required threshold.

4.2.2 LIMITING PROPERTIES

The OTF cannot assume arbitrary values. The spread function is formed as the result of a physical process of diffraction, and is subject to the symmetries and magnitude constraints that arise in such processes. Constraints upon the allowable levels and shapes of the OTF are provided by the aperture size and shape, the wavelength, and, of course, the aberrations present in the lens. In Chapter 2, the physical optics and diffraction calculations were reviewed. These principles show that there is a link between the aberration in the exit pupil and the form of the point spread function. This physical relationship leads to limitations upon the possible values that the OTF can take.

The physical limits upon the OTF can be observed somewhat more easily by using the Fourier relationship to convert the transform of the point spread function into a convolution integral over the pupil. The basic approach is to note from Chapter 2 that the amplitude point spread function, $a(x, y)$, can be computed as a Fourier transform of the pupil function. The intensity point

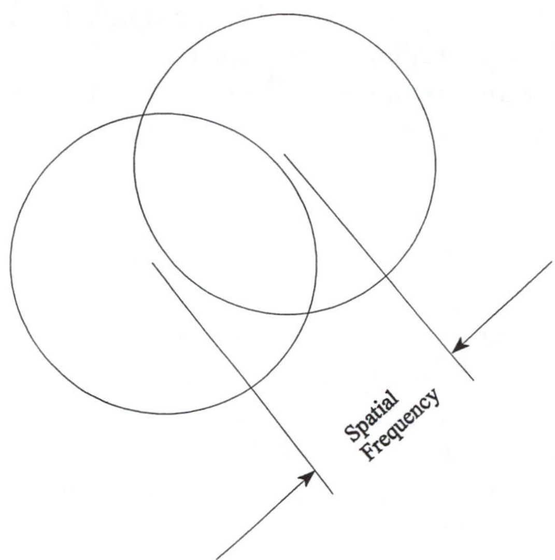

FIGURE 4.5 The concept of aperture convolution to compute the Optical Transfer Function.

spread function, $f(x, y)$, is the complex square of the amplitude point spread function. Symbolically this can be written as

$$\begin{aligned} a(x, y) &= F\{A(u, v)e^{ikW(u,v)}\} \\ f(x, y) &= a(x, y)a^*(x, y) \\ F(\nu_x, \nu_y) &= F\{f(x, y)\} \end{aligned} \tag{4.10}$$

Here the symbol $F\{...\}$ indicates the operation of a Fourier transform. Application of a theorem in Fourier analysis permits writing the Fourier transform of a function that is product of the Fourier transforms of another function directly as the convolution of the original functions. Symbolically again then the OTF can be found by a pupil convolution:

$$F(\nu_x, \nu_y) = A(u, v)e^{ikW(u,v)} \otimes A(u, v)e^{-ikW(u,v)} \tag{4.11}$$

The operation represented by the curious symbol $\otimes$ is a convolution that is written in integral form as

$$\begin{aligned} F(\nu_x, \nu_y) = \int_{-p}^{p}\!\!\int A\left(u - \frac{\nu_x}{2}, v - \frac{\nu_y}{2}\right) e^{ikW\left(u - \frac{\nu_x}{2}, v - \frac{\nu_y}{2}\right)} \\ \times A\left(u + \frac{\nu_x}{2}, v + \frac{\nu_y}{2}\right) e^{-ikW\left(u + \frac{\nu_x}{2}, v + \frac{\nu_y}{2}\right)} du\, dv \end{aligned} \tag{4.12}$$

The symbolic limits of $-p$ to p indicate that the integration is taken over the intersecting areas of the shifted pupils. The area of integration for this function for a circular aperture is shown in Figure 4.5. The amount of shift of the pupils in the integration is proportional to the spatial frequency. The integral is zero outside the intersection of the pupils.

The most important limitation is that optical systems are low-pass spatial filter systems. The maximum frequency present in an OTF is, for a circular aperture, given by

$$\nu_{max} = \frac{1}{\lambda(\mathrm{F/number})} = \frac{2(\mathrm{NA})}{\lambda} \tag{4.13}$$

where NA is the numerical aperture of the lens. The MTF will be zero outside this upper frequency cutoff. If the image is formed in a medium with index other than unity, the appropriate wavelength must be inserted. If the aperture is not circular, then the maximum spatial frequency will not be constant for all directions on the focal surface, but will inversely follow the diameter of the aperture in a specified direction. In many cases it is customary to express the OTF spatial frequency in terms of the ratio of the actual spatial frequency to the maximum, or cutoff, spatial frequency.

Most lenses operate over a wide wavelength band. In this case the limiting spatial frequency will vary with wavelength. The magnitude of the MTF is also limited. For the most common case of a circular aperture, the diffraction-limited MTF in any direction is

$$\begin{aligned} F(\Omega) &= \frac{2}{\pi}[\arccos(\Omega) - \Omega\sqrt{1-\Omega^2}] \\ \Omega &= \frac{\nu}{\nu_{max}} \end{aligned} \tag{4.14}$$

when expressed in a normalized spatial frequency Ω, or ratio of actual spatial frequency to the cutoff frequency. A plot of this important function is shown in Figure 4.6. Note that in some literature, the choice of spatial frequency normalization places the normalized spatial frequency in the range from 0 to 2. This is related to the discussion of coherence; see for example Gaskill (1978). Of course, the choice of normalization is available to the designer to use in any way, depending upon the point to be made. One normalization choice may be to plot versus actual spatial frequency for a stated aperture and wavelength. Other observations are that the slope of this function at the cutoff is zero, and the OTF approaches the axis quadratically for a circular aperture. The slope at the zero frequency end is finite and is determined by the aperture shape.

The height of the Point Spread Function at the origin of coordinates is found from the Fourier transform relation

$$f(0,0) = \iint F(\nu_x, \nu_y) d\nu_x \, d\nu_y \tag{4.15}$$

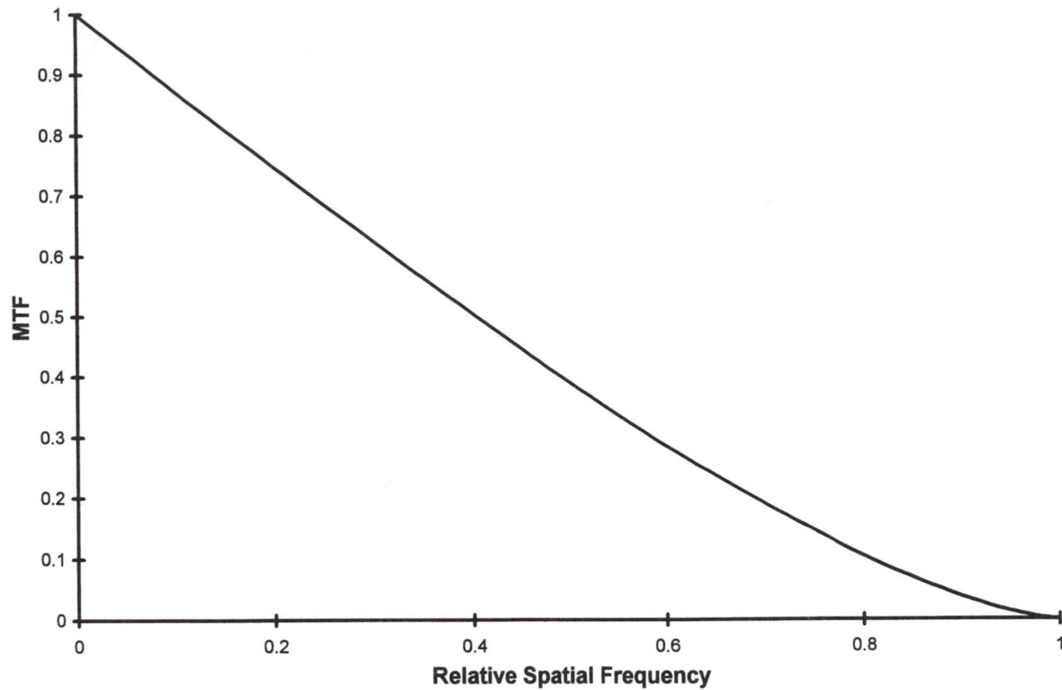

FIGURE 4.6 MTF for a lens with a circular aperture and no aberrations.

so that the peak intensity in the spread function is proportional to the volume under the OTF. The total integrated energy under the point spread function is proportional to the value of the OTF at zero frequency, which has been normalized to unity in the discussion above.

This analytic form applies to an unobscured circular aperture. Simple analytic forms for the MTF are possible only for a restricted set of aperture shapes. If the aperture shape changes, or if obstructions or transmission changes are made in the aperture the MTF will change. The most common aperture blockage is found in reflective systems, in which the shadow of the secondary blocks the central portion of the aperture. Figure 4.7 shows a family of MTF plots for different amounts of aperture blockage. It can be seen that there appears to be a slight gain in contrast at the high-frequency end over the unobscured aperture, but at a reduction in contrast for the mid-spatial frequencies.

Vignetted apertures may have a complex shape that cannot usually be expressed analytically. Numerical computation of the MTF will be required. In extreme cases of vignetting, the shape of the edge of the aperture can be quite complicated and is not sampled well by the matrix used in the numerical computation. In such cases, the mesh must be adjusted to ensure that a good sampling occurs. In most cases, the most efficient approach is to redo the

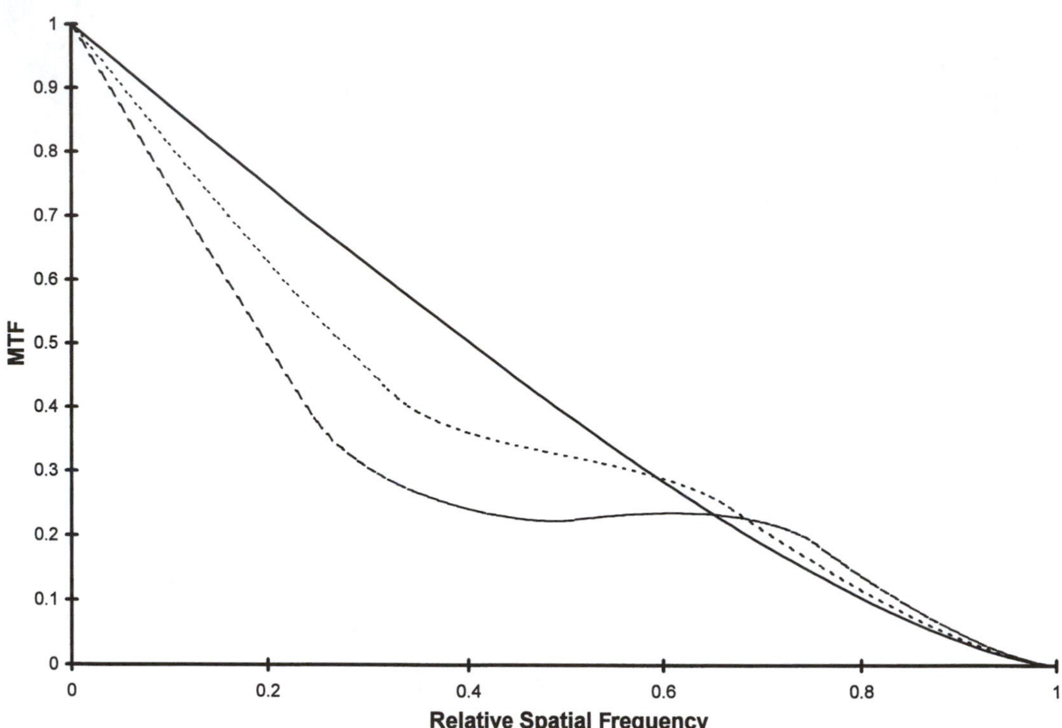

FIGURE 4.7 MTF plots for apertures with circular central obstructions. Ratio of the diameter of the obstruction to the full aperture: ——, unobscured; – – –, 0.5; ······, 0.33.

computation of the MTF with successive finer sampling meshes until a stable set of values for the MTF is obtained.

The obscured-aperture OTF shows an apparent increase in contrast for high spatial frequencies as the obscuration is increased. Although this effect certainly does occur, it is a change in contrast, not in total signal level. The actual signal presented to the detector requires multiplication by the average level of object intensity. Thus it must be remembered that the loss of total energy due to the aperture blockage may reduce any apparent gain in contrast by a loss in actual signal-to-noise ratio at the high spatial frequencies. In cases where obscured apertures must be used, the loss of total signal must be considered in system analysis.

The shape of the diffraction MTF can be altered in a controlled manner by shaping of the transmission function in the aperture. This concept, called apodization, or removal of the diffraction rings, can have some effect upon the contrast at various spatial frequencies, but the effects are usually not dramatically large. An example is shown in Figure 4.8, in which the effect of a variation in edge transmission at the edge of the pupil is demonstrated. This example is for a Gaussian weighting of the aperture and presumes no aberration.

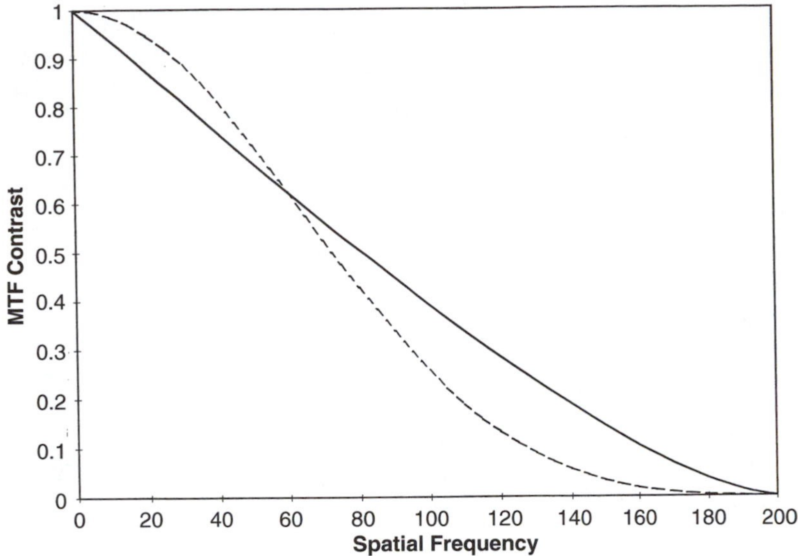

FIGURE 4.8 MTF for a Gaussian-weighted aperture: ——, full aperture; – – –, 1% Gaussian.

In a sense, the vignetting that is used to shape the off-axis aperture to remove damaging amounts of aberration can be considered a sort of "apodization by elimination" of highly aberrated portions of the pupil function. The OTF computation provides a method of quantifying this effect.

Phase errors due to aberrations in the pupil will have a significantly greater effect upon image contrast than will the modification of pupil transmission. The interaction of pupil shape and aberration is important as well. The first cases to be examined will be the effect of aberrations present in a lens with an unobscured, circular, uniformly illuminated pupil.

The presence of aberrations will always reduce the MTF below the diffraction-limited level. A simple way of demonstrating this is to note that aberrations will cause a distribution of energy outside the perfect point spread function on the image surface. The basic rules of Fourier transforms show that an increase in the width of the spread function will lead to a reduction in the width of the Fourier transform of the spread function.

The reduction of contrast due to aberrations can be deduced by examination of the effect of phase modulation on the convolution integral which is used to obtain the OTF. Introduction of aberrations into the pupil will produce oscillations in the value of the argument of the integral, reducing the value of the integrand at each spatial frequency.

A more complicated effect occurs for the most frequent case of lenses that operate over a wide spectral bandwidth. The irradiance at each wavelength is incoherent with the radiation for all other wavelengths. The point spread

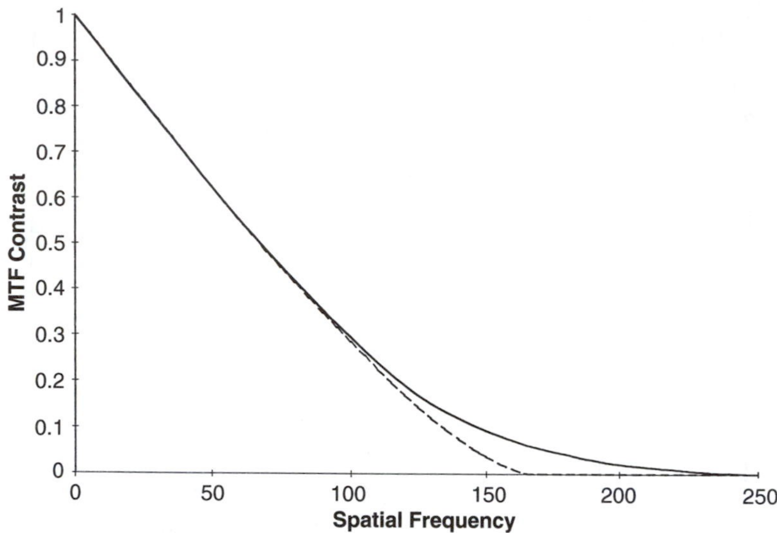

FIGURE 4.9 MTF for lens with no aberrations, but covering a wavelength band of a factor of two from low- to high-wavelength cutoff with uniform spectral weighting: ——, 0.4–0.8 μm; – – –, 0.6 μm.

function for such a lens will be the sum of the point spread functions of all of the individual colors in the point object. The intensity of each of the color components is weighted by the spectral distribution of the source, multiplied by the transmission of the optics and the sensitivity of the detector. The last step, of course, assumes a linear detector with constant spectral characteristics. If the detector has a transfer function that is wavelength dependent, this will have to be included in the computation.

The intensity linearity of the systems permits the integration to be carried out either in the spatial domain, by summing point spread functions, or in the frequency domain, but summing Optical Transfer Functions. Since the upper frequency limit for the transfer function is determined by the wavelength, the computation must be carried out in absolute frequency space. The normalization of the computation must be such that the total spectral weighting is unity. Formally the computation is

$$F(\nu) = \int t(\lambda) F(\nu, \lambda)\, d\lambda \tag{4.16}$$

where the spectral weighting function $t(\lambda)$ is the product of the individual spectral weighting functions for the source, transmission medium, lens, filter, and detector, and is subject to the weighting

$$\int t(\lambda)\, d\lambda = 1 \tag{4.17}$$

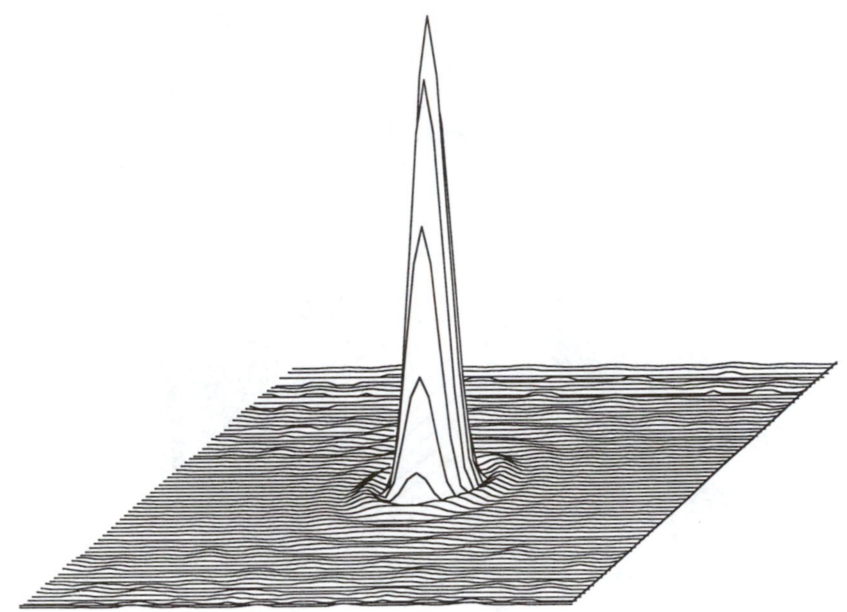

FIGURE 4.10 Point Spread Function Intensity for a perfect lens, computed using a Fourier transform of the pupil function (ZEMAX).

over the applicable spectral range. A simple example is provided in Figure 4.9 for a perfect F/5 lens with a circular aperture and no aberrations having a uniform spectral weighting over three different spectral ranges, compared with a monochromatic OTF.

Any imaging system is, of course, two-dimensional in nature. The discussion so far has treated the MTF plots as a set of one-dimensional cuts into the full two-dimensional function. The methods of computing the OTF using two-dimensional techniques will be discussed in some detail in section 4.3.2. It is important here to discuss the concept, and provide some examples of the full two dimensional MTF.

The concept follows from the one-dimensional discussion held so far. The point spread function of a lens can be computed using diffraction methods. The coordinates are x and y distances on the image surface. The vertical coordinate is intensity. Figure 4.10 is a plot of such a point spread function. Knowledge of this information permits representation of the point spread function as a two dimensional matrix of points representing the intensity of the diffraction image.

The Fourier transform of that point spread function is the full two-dimensional OTF. This OTF is plotted in Figure 4.11. The coordinates are now spatial frequency in the x and y directions across the image surface. The vertical coordinate is the MTF, or contrast of the OTF function. The band-limited nature of the OTF is illustrated by the well-defined cutoff in

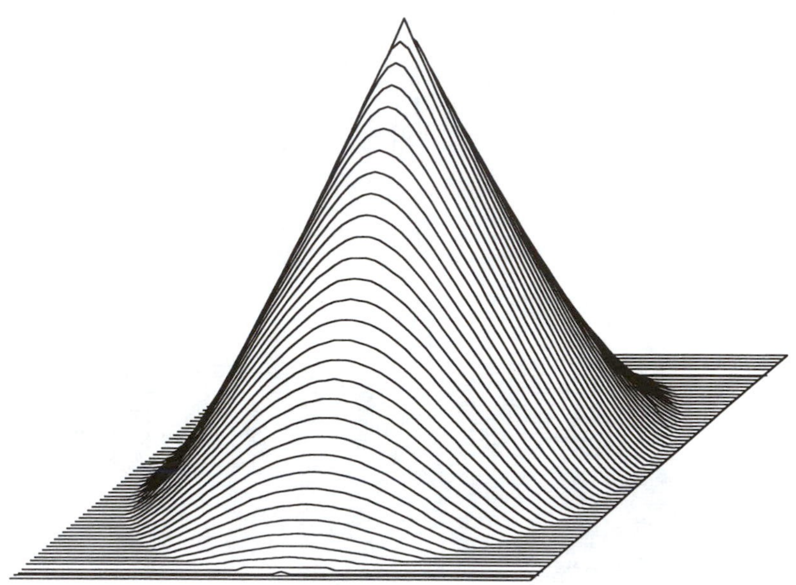

FIGURE 4.11 Surface MTF for a perfect circular aperture. The cutoff frequency is clearly indicated in the plot (ZEMAX).

the spatial frequency coordinates. This cutoff is, of course, circular and independent of direction on the image surface because the pupil is circular in extent. The shape of the surface representing the MTF is seen to be similar to the diffraction-limited MTF plot in Figure 4.6.

This example is for a rotationally symmetric point spread function. The presence of nonrotationally symmetric aberrations will cause the two-dimensional MTF surface to deviate from rotational symmetry. There will be a symmetry through the origin produced by the positive-definite nature of the intensity point spread function. Some examples are shown for coma and astigmatism in section 4.2.3.

One other concept regarding the OTF should be mentioned here. The OTF discussed so far is the result of a true diffraction-imaging process. The calculation is quite complicated (although trivial to carry out on today's high-speed computers). A simpler calculation using geometrical optics can be carried out. In this case, the distribution of rays on the image surface is used to approximate the light distribution that is obtained. The coordinates of the individual ray displacements are Fourier transformed to obtain a Geometrical Optical Transfer Function (GOTF). This type of OTF will approximate the MTF that can be expected in the case of large aberrations. Because of the local caustics that can occur in geometrical images, the geometrical model does not converge smoothly to the diffraction model. Both the high speed of contemporary computers and the development of efficient computation algorithms have led to a general abandonment of the GOTF for describing images.

4.2.3 OTF VERSUS ABERRATION

The effect of aberration on the OTF can be quite complex. The effect of focal position must be included, as well as the variation of the function across the field of view. A systematic investigation can be carried out to provide a basis for interpretation of the effects of aberrations.

The OTF of a perfect optical system is determined by the shape of the aperture and the numerical aperture of the lens system. For initial considerations, it is easiest to presume a circular, unobstructed pupil. When aberration is present, the magnitude of the contrast will be reduced and spatial phase errors may be introduced.

The effect of aberrations on the OTF can be classed into three regions. The first is that of small aberrations, when the RMS aberration level does not exceed 0.07 to 0.1 waves. The second, and most complicated region, is that of aberrations ranging from about 0.07 to 0.25 waves RMS. The large aberration region, when the RMS wavefront aberration exceeds about 0.25 waves, is sometimes called the geometrical region, because geometrical considerations can provide a reasonably good approximation to the anticipated OTF level, at least at low spatial frequencies.

A good starting point is the calculation of the effect of focus error on the OTF. As discussed earlier, focus aberration is a quadratic wavefront error, simply represented by the rotationally symmetric wavefront aberration

$$W(\rho_x, \rho_y) = a_{020}(\rho_x^2, \rho_y^2) = a_{020}\rho^2 \tag{4.18}$$

where the value of a_{020} is equal to the number of wavelengths of defocus wavefront error at the edge of the pupil. This can be related to the physical defocus value by

$$a_{020} = \frac{-\delta_z}{8\lambda(\mathrm{F/number})^2} \tag{4.19}$$

and is usually referred to as the peak wavefront focus error. A useful set of parametric graphs can be developed for several amounts of focus error. Additional insight will be developed by remembering that the relation between the peak wavefront defocus error and the RMS error is

$$W_{RMS} = \frac{a_{020}}{3.5} \tag{4.20}$$

to obtain the RMS wavefront focus error (see section 4.4.2). Conversion to physical values of defocus and spatial frequency can of course be made by knowing the actual wavelength and aperture for the system of interest.

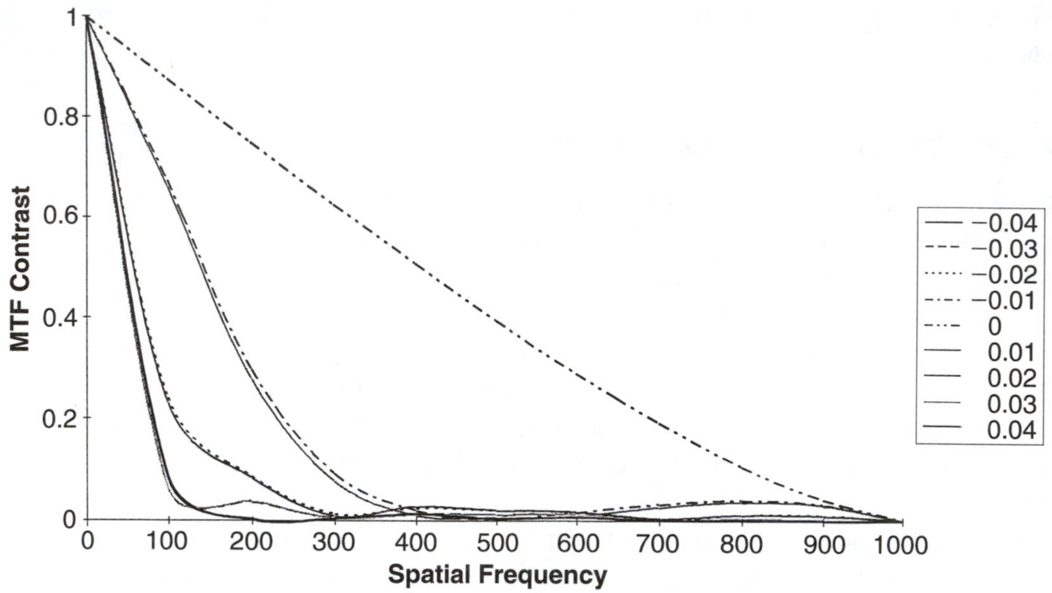

FIGURE 4.12 MTF plots for an F/2 lens at 0, 0.57, 1.14, 1.71, and 2.28 wavelengths of defocus (λ=0.00055).

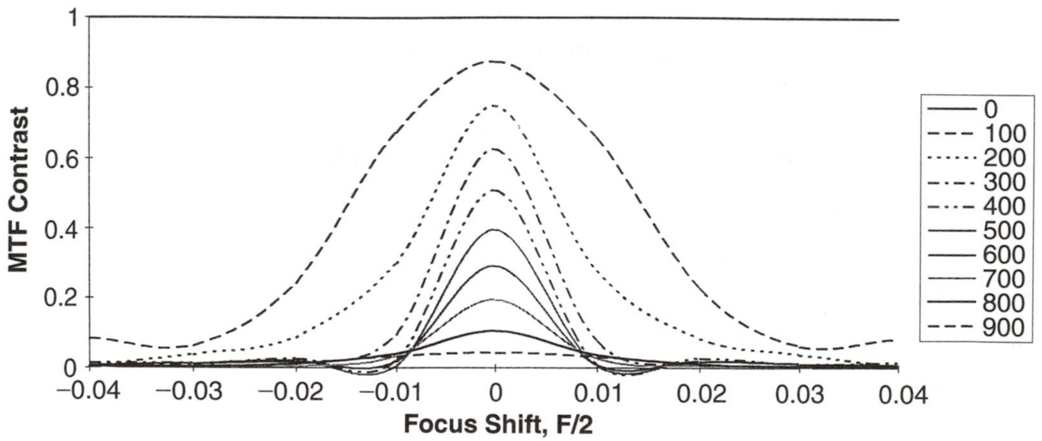

FIGURE 4.13 A through-focus plot of the MTF for several spatial frequencies.

Figure 4.12 is a frequency-normalized plot of the MTF for various amounts of defocus in peak wavelength units. Figure 4.13 is another plot of this same data in which the MTF is plotted for some specific relative spatial frequencies versus the amount of defocus, again in wavelength units.

Some observations can be made. Since the aberration is rotationally symmetric, the only spatial phase shifts allowable are 0 or π. The π phases can be represented as negative values of the MTF. The MTF values for 0.25 waves of focus error, or about 0.07 waves RMS, are about 80% of the values for an

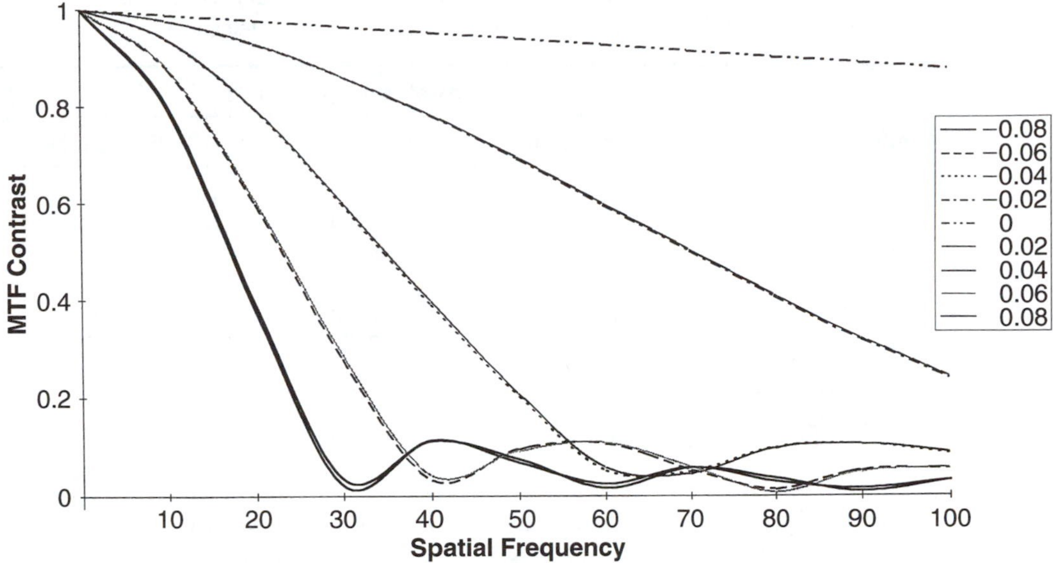

FIGURE 4.14 The geometrical MTF for several focal positions for an F/2 lens.

aberration-free system. The MTF values change approximately quadratically with the amount of aberration at least to this level.

As the peak wavefront aberration increases from 0.25 waves to about 0.5 waves, the MTF drops rapidly, with the mid-spatial frequency contrast approaching zero at about 0.5 waves of focus. Increasing amounts of aberration demonstrate an increasingly more complex behavior. As the aberration approaches one wave of defocus, the shape of the MTF begins to look like the geometrical MTF given by

$$F(N) = \frac{2J_1(\pi Nd)}{\pi Nd} = \frac{2J_1(8\pi\Omega a_{020})}{8\pi\Omega a_{020}} \tag{4.21}$$

where d is the diameter of the geometrical out of focus circular blur. The second half of the equation is a rewrite in terms of the relative spatial frequency and the waves of defocus. The geometrical relation applies well only in the low spatial frequency region for large amounts of defocus. Figure 4.14 shows some comparative plots for the spatial frequency region below $\Omega = 0.1$.

The plotting of OTFs and focus shifts in terms of wavelengths of wavefront error and scaled spatial frequency is quite useful for understanding the basic physics involved in image formation. In any real problem, the engineer will need to know the allowable physical extent of the focus shift. The equations given above provide the link between the theory and the practical values. Some critical values for the axial focus shift for a few typical combi-

TABLE 4.1 Critical values of axial focus shift with wavefront error for different combinations of F/number and wavelength

	Axial focus shift at		
Wavefront error	F/5, 0.5 μm[a]	F/2, 0.5 μm[b]	F/5, 2.5 μm[c]
0.25 peak	0.005	0.002	0.025
0.50 peak	0.010	0.004	0.050
1.0 peak	0.020	0.008	0.100
0.07 RMS	0.005	0.002	0.025
0.15 RMS	0.011	0.004	0.054

[a] $N_{max} = 400$ lines/mm.
[b] $N_{max} = 1000$ lines/mm.
[c] $N_{max} = 80$ lines/mm.

nations of F/number and wavelength are in Table 4.1. Additional values can of course be computed, but the range of applicable values can be estimated from this table. The allowable tolerance upon the location of the focal location can be assessed by selecting the spatial frequency of interest and the required contrast level at that frequency.

It is obvious from this discussion that adjustment of focus has, as expected, a very significant effect upon the MTF. Conversely, in the presence of other aberrations, proper adjustment of the focus position can balance the effect of the aberration. This will be demonstrated shortly for spherical aberration. In any lens, there is usually only one focus position allowable for the entire field of the lens. Therefore, the balancing of focus position against the field-dependent aberrations, such as astigmatism and field curvature is very important.

A lateral shift of the reference location will also influence the OTF. However, the introduction of a shift introduces a linear spatial phase shift and does not change the modulus of the OTF, but will change the spatial phase portion of the OTF. The linear phase shift that is introduced by a lateral shift of the coordinates is given by

$$\phi = 2\pi N \delta_y \tag{4.22}$$

where the lateral shift, δ_y, and the spatial frequency, N, are in corresponding units, such as millimeters and lines per millimeter. The 2π converts the phase shift to radians. Use of 360 as a multiplying factor, instead of 2π, would provide the value in degrees. The interpretation of the linear phase-shift term is that it provides the shift of the origin in units of the spatial wavelength at any particular frequency. The slope of the linear phase-shift component is proportional to the distance of the centroid of the point spread function from the origin of coordinates.

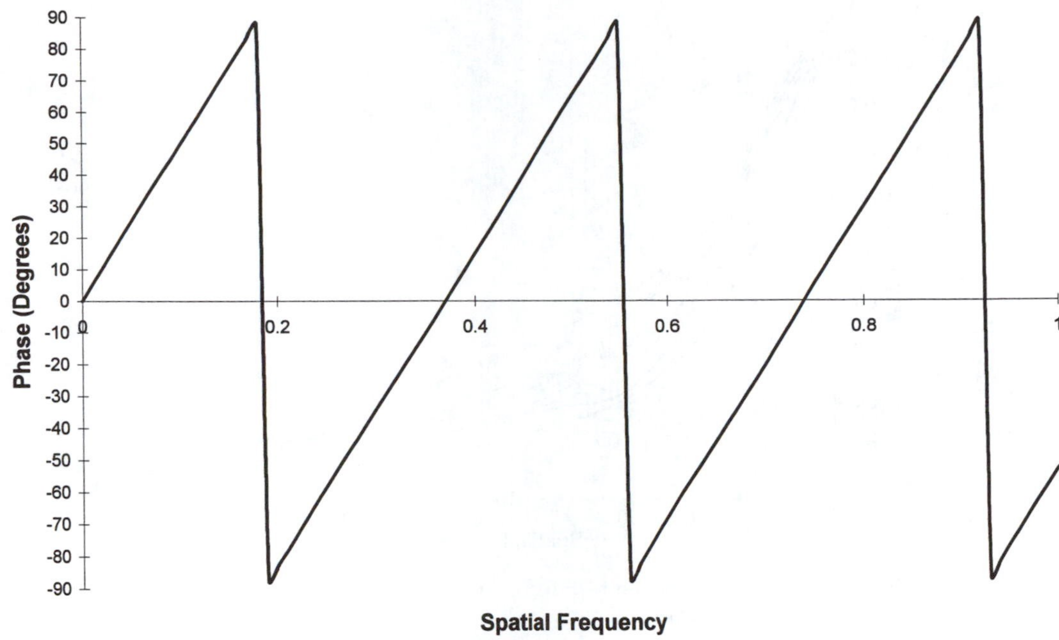

FIGURE 4.15 A plot of the Spatial Phase Transfer Function for a perfect lens with a lateral shift of the reference position. The jumps associated with the principal value of the arc-tangent function are obvious.

Although the linear phase shift can be large, the usual convention is to plot the phase error in the principal value of the angle for the range $\pm 90°$. This leads to some peculiar-looking plots, such as in Figure 4.15. When lateral aberrations, such as lateral color or coma are present, nonlinear phase-shift components can be present, providing more complex interpretation to the error.

The focus shift or lateral shift can be eliminated by a shift of the image reference coordinates. If aberrations are present, the image degradation can only be balanced or minimized by introduction of focus change or lateral shift of the origin.

Before discussing the complex effects that can occur in the presence of high-order aberrations, it is instructive to evaluate the effect of chromatic aberrations on the OTF. First-order chromatic aberration is the variation of the focus position or lateral position of the point spread function as a function of wavelength. The complete effect of these aberrations on the MTF can be obtained by integrating over the spectral range, including the weighting of the source and detector. In most cases, all of the aberrations will be dependent upon wavelength, but a simple model provides an estimate of the reduction in image contrast that will occur in the presence of chromatic errors.

Figure 4.16 illustrates the effect of first-order longitudinal chromatic aberration. The integration over the wavelength range reduces the MTF because

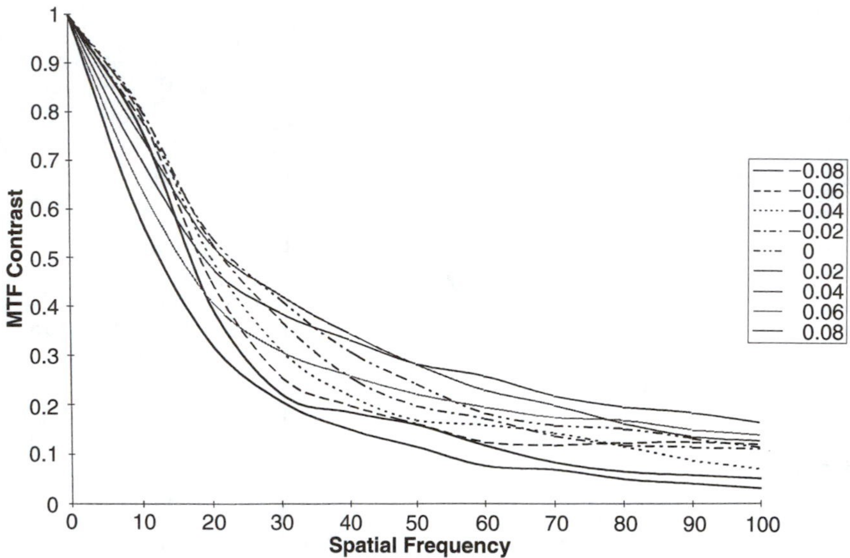

FIGURE 4.16 An example of the MTF for a lens with longitudinal chromatic aberration. The function is shown for several focal positions.

of the addition of out-of-focus components. The effective depth of focus is apparently increased because of the spread of focal positions for the various colors, and a longer focal region appears to be obtained, as shown in Figure 4.17. This will be true for a "color blind" detector. A color-sensitive detector will narrow the allowable focal range. In addition, the blur will depend upon the color of the object. Therefore the interpretation of the MTF values for this lens requires an understanding of the imaging process that created it. Other samples of chromatically balanced aberrations will be found in the detailed design examples in Chapter 7.

When spherical aberration is present a more complicated situation occurs. The introduction of a focus shift cannot cancel the aberration, but can be used to optimize the focal position. The evaluation of a lens which contains spherical aberration should include an evaluation of the MTF at the optimum focus position. The net wavefront error will be

$$W = a_{040}\rho^4 + a_d\rho^2 \tag{4.23}$$

as previously discussed. This is the type of aberration that can be expected in a single lens, or as a residual aberration in a more complex lens system. Examination of the MTF values for several levels of spherical aberration is instructive in understanding some of the limitations in image formation.

The example used here is based upon the use of an aberration module which contains only third-order spherical aberration, and is available in the CODE V program. The MTFs are for an F/2 lens, at 0.500 μm wavelength,

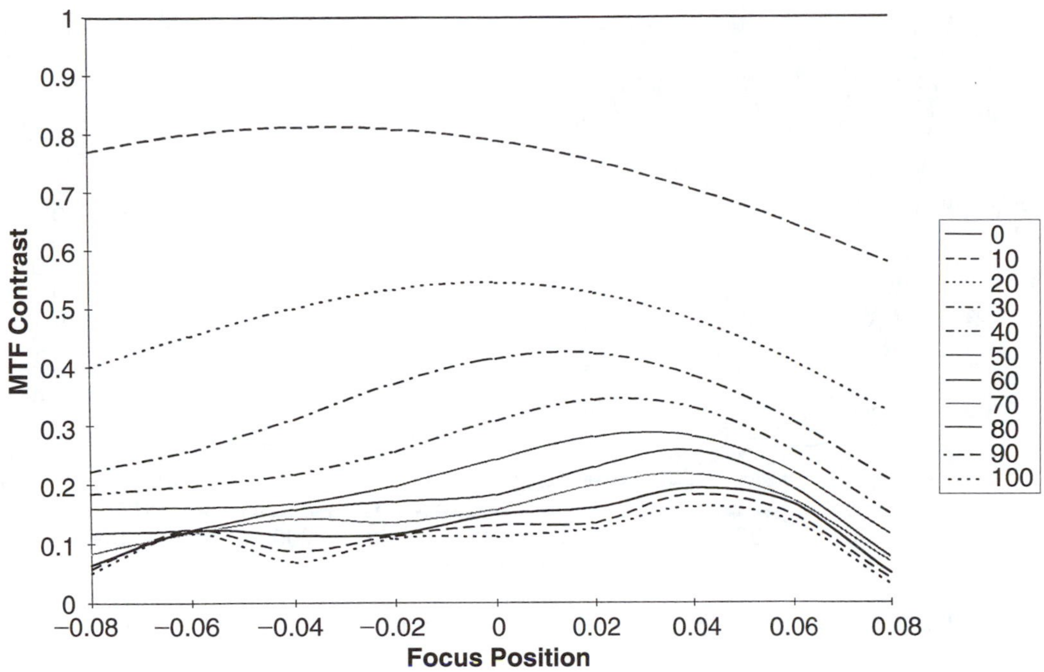

FIGURE 4.17 A through-focus plot for the example with longitudinal chromatic aberration, displaying the effective increase in focal depth but with a reduced contrast of the MTF.

and having a focal length of 100 mm. Since the aberration is defined in terms of wavelengths, the focal length is irrelevant. Scaling of these considerations to any other wavelength or F/number can be accomplished by scaling the spatial frequency and the dimensions appropriately. For this case, the cutoff spatial frequency is at 1000 lines per mm, and a focal shift of one wavelength corresponds to an axial displacement of 0.008 mm. (As an example, scaling to F/5 at 0.6 µm wavelength would lead to a cutoff frequency of 333 lines per mm and a shift of 0.024 mm corresponding to one wavelength of focus.)

Reference to Figures 4.12 and 4.13 indicates the effect of focus in the absence of any aberration. The symmetry of the MTF on both sides of focus is obvious. The peak contrast at all spatial frequencies is located at the same focal position, the paraxial focus location. The focus region over which there is only a small reduction in the contrast is quite small, with the peak contrast at each spatial frequency being reduced to about 50% of its peak value at a focus shift of 0.005 mm for 500 lines per mm, and 0.015 mm for 100 lines per mm.

The presence of one wavelength of spherical aberration leads to Figure 4.18 and to the focus plot in Figure 4.19. The MTF is reduced at all focal positions, and is no longer symmetrical around any focus position. The best

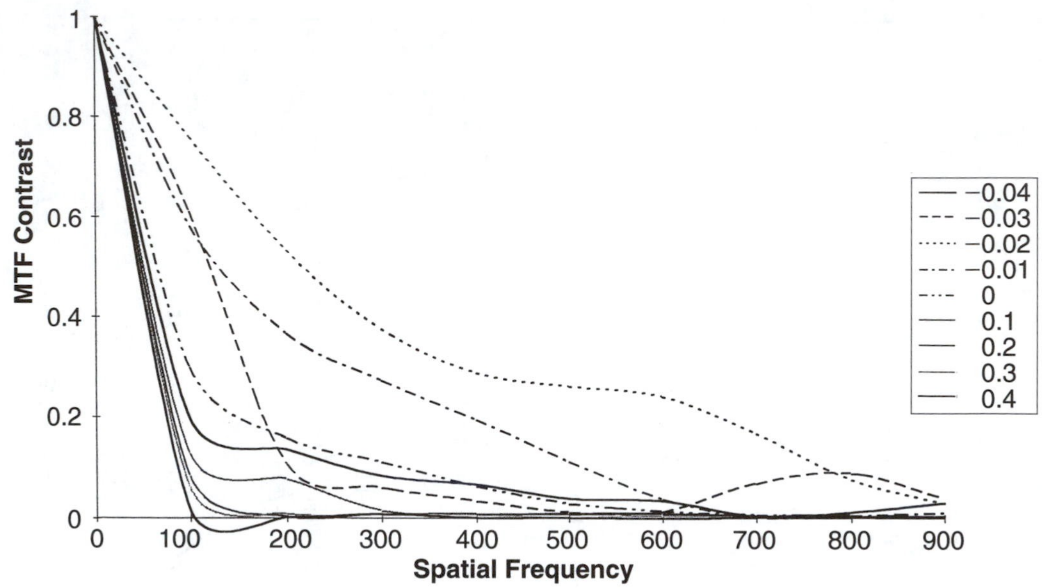

FIGURE 4.18 MTF for a lens with one wavelength of third-order spherical aberration.

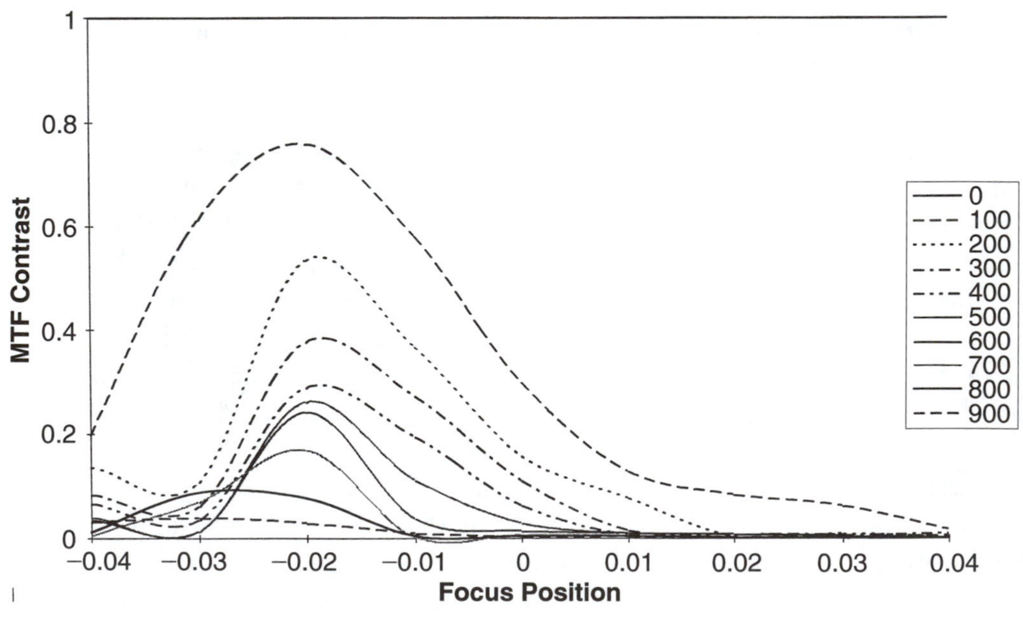

FIGURE 4.19 A through-focus plot for the case of one wavelength of third-order spherical aberration.

focus position is located at a focus shift of −0.018 mm toward the lens from the paraxial focal location. The amount of reduction of the MTF would likely be noticeable for any imaging situation, dependent, of course, upon the nature of the detector.

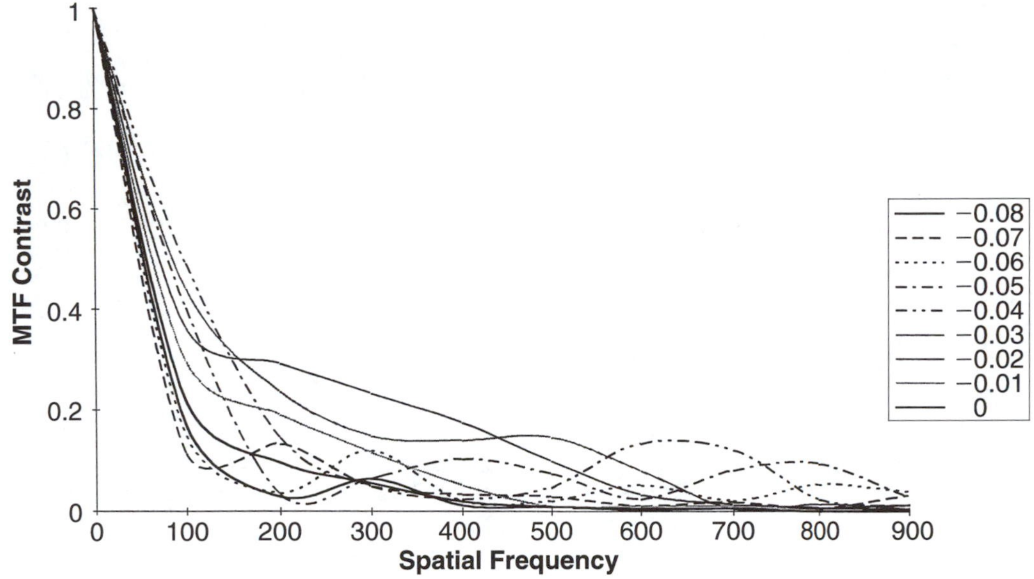

FIGURE 4.20 MTF for two waves of third-order spherical aberration.

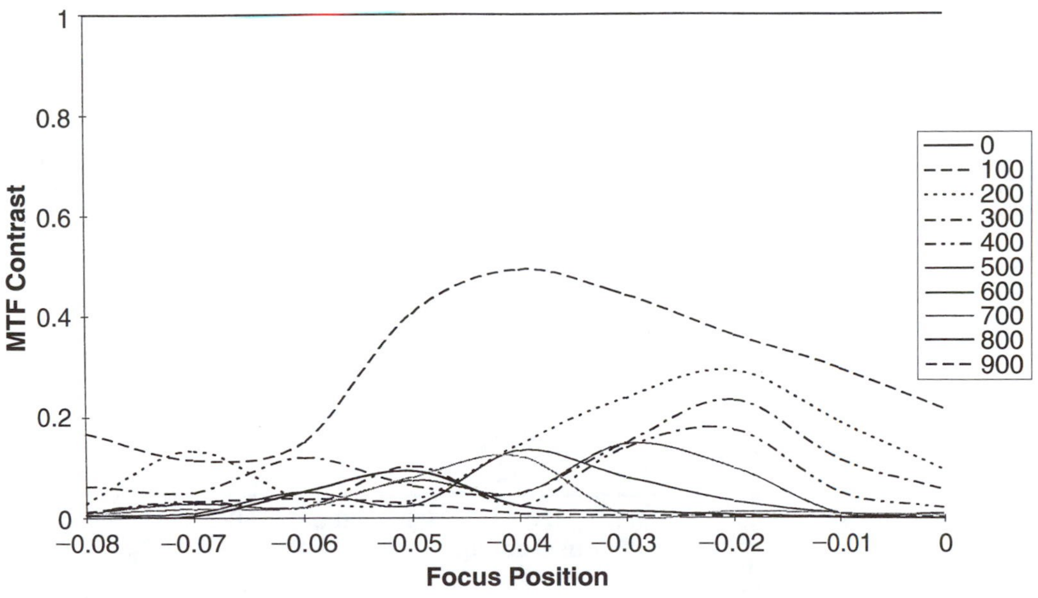

FIGURE 4.21 Through-focus MTF for two waves of third-order spherical aberration.

Doubling the spherical aberration to two wavelengths leads to the plots in Figures 4.20 and 4.21. A further reduction in the MTF contrast occurs. The best focal location moves farther from the paraxial focus location. Increasing asymmetry indicates that the best focus location is quite different for each

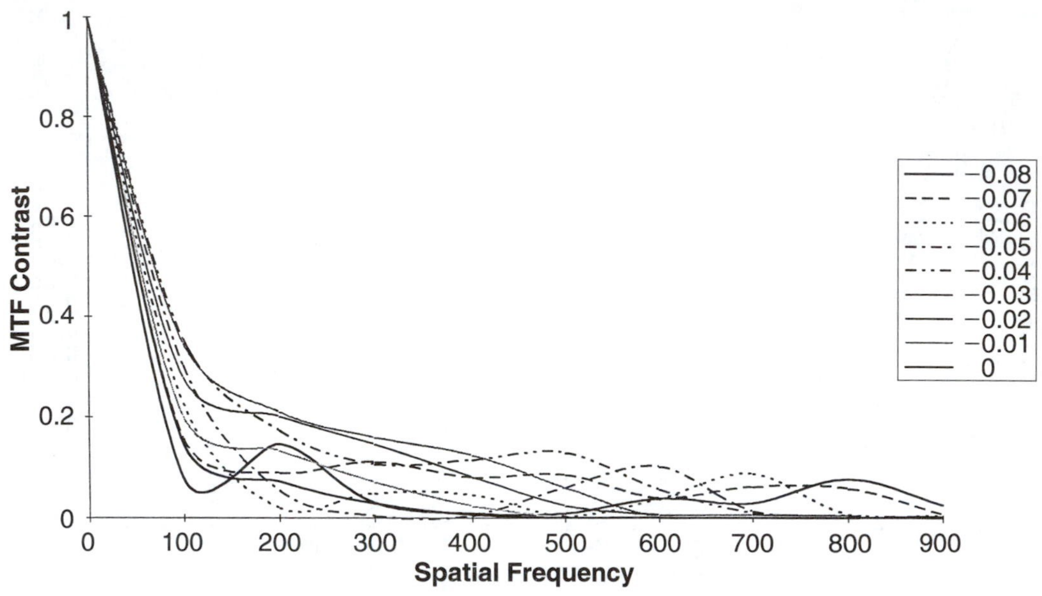

FIGURE 4.22 MTF for three waves of third-order spherical aberration.

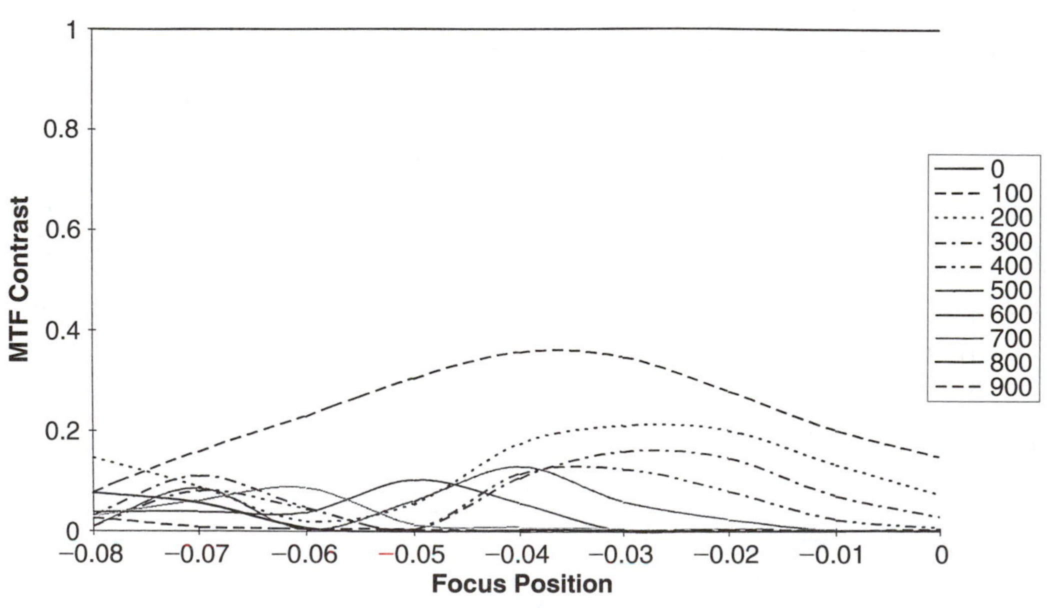

FIGURE 4.23 Through-focus MTF for three waves of third-order spherical aberration.

spatial frequency. The definition of best focus location now requires some knowledge of the intended object and detector. The effect of three wavelengths of spherical aberration shown in Figures 4.22 and 4.23 shows an increased trend in the reduction of contrast.

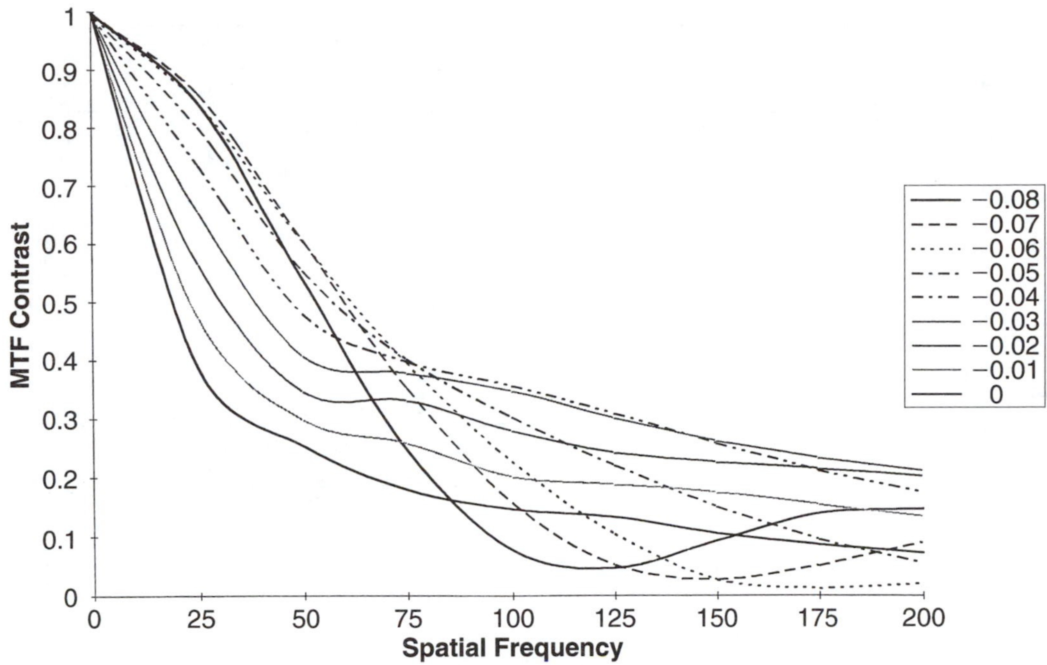

FIGURE 4.24 MTF for three waves of third-order spherical aberration, reduced spatial frequency range.

Since the focal range plotted in the above examples has been held constant there are some general trends that are noticeable. The contrast is reduced, with good contrast available only at lower spatial frequencies, below about 20% of the cutoff frequency for the last case. The focal depth appears to have been increased as a result of introducing spherical aberration. This is evident, but is a consequence of the necessity of accepting reduced contrast over all spatial frequencies. The presence of significant levels of aberration complicates the image analysis problems because the nature of image formation is more dependent upon the details of the aberration residual.

Large amounts of spherical aberration are clearly applicable to imaging situations in which the usable spatial frequency region is limited. Figures 4.24 and 4.25 are plots of the MTF for the spatial frequency region up to 200 lines per mm. This shows the details at lower spatial frequencies, and indicates two regions of best focus, one for good contrast at low spatial frequencies at about −0.06 mm, and another for lower contrast at low spatial frequencies but increased contrast at higher spatial frequencies at about −0.025 mm. This pattern of behavior is characteristic of the presence of spherical aberration, and is important in optimizing the design including the effect of a specific detector. This is even more evident in Figures 4.26 and 4.27 for the case of four waves of spherical aberration. Note that here the spatial frequency range has again been reduced.

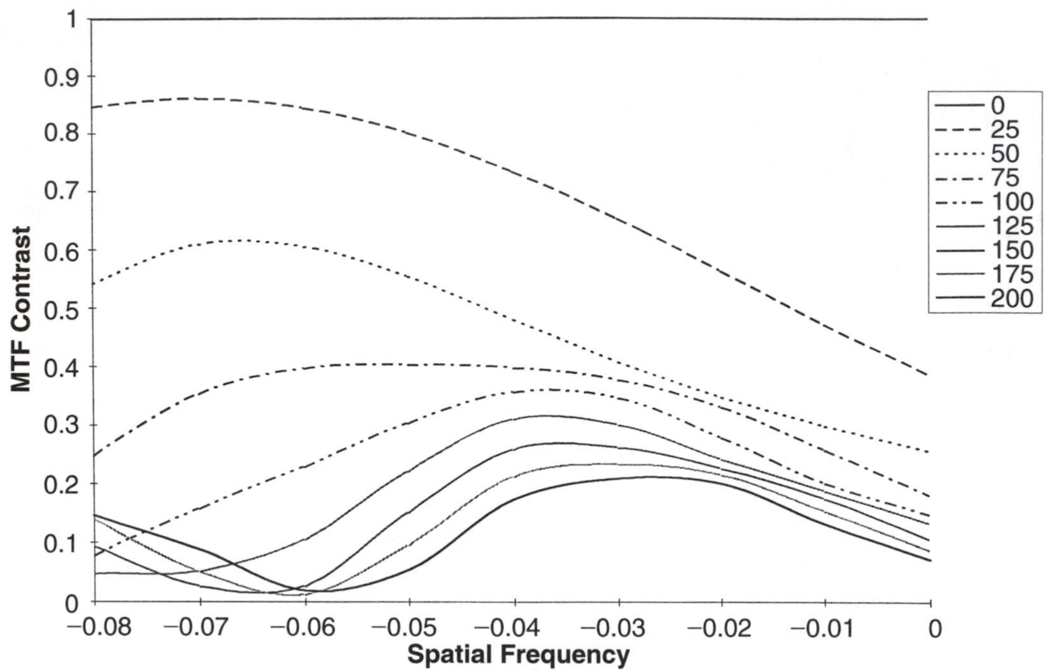

FIGURE 4.25 Through-focus plot for three waves of third-order spherical aberration. The lower spatial frequency shows more complex detail about the change of MTF with focal position.

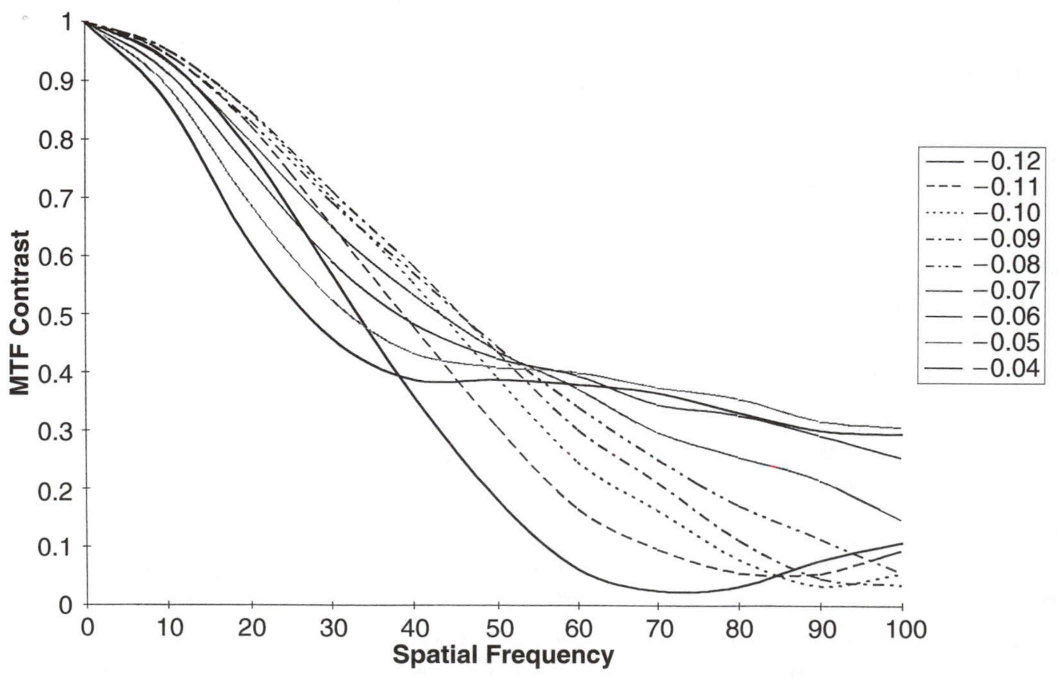

FIGURE 4.26 MTF for four waves of third-order spherical aberration.

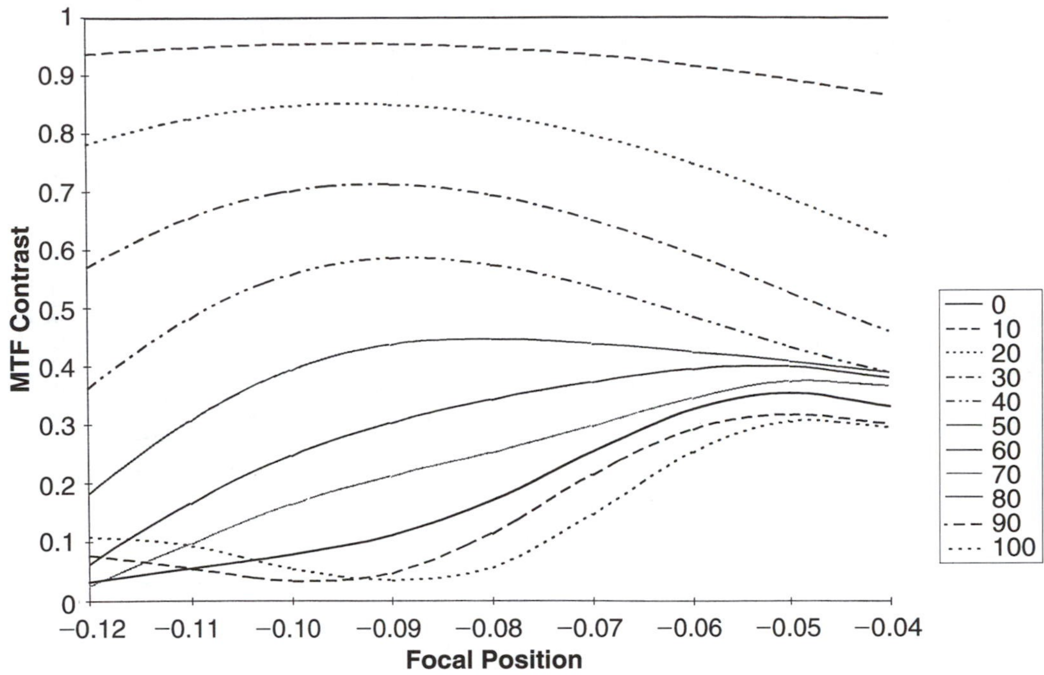

FIGURE 4.27 Through-focus MTF for four waves of third-order spherical aberration.

The final examples are for four waves of spherical aberration at a reduced spatial frequency range and lengthened focal range. In Figures 4.28 and 4.29, the characteristic effect of spherical aberration is evident. The presence of the spherical aberration produces asymmetric behavior with focal position. The selection of best focus location will be very definitely dependent upon detector characteristics and object detail. The best focus choice depends upon the spatial frequencies selected as important. In any analysis problem, the weighting of the response by the expected scene content and the detector properties is necessary.

This asymmetry has led to the concept of designing either for "high contrast" or for "high resolution." Examination of the MTF curves shows that one focal position will provide the highest contrast at low spatial frequencies, but at the cost of reduced contrast at the high spatial frequencies. Selection of another focal position will reverse this relation, and is the location of the high-resolution focus. It is very important to notice that the ability to tailor this MTF behavior is not an unrestricted opportunity, but is a consequence of the spherical aberration that is present in the lens.

In section 3.1.1.2.1, it was demonstrated that the effect of high-order spherical aberration can be reduced by the introduction of an appropriate amount of low-order spherical aberration. The usual choice for the balance is

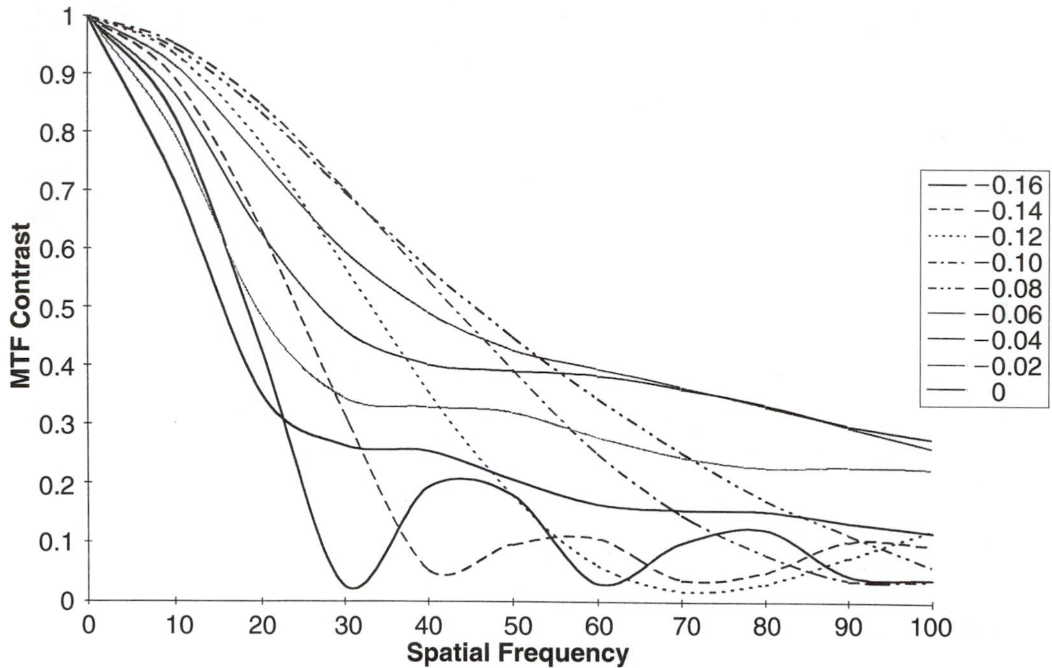

FIGURE 4.28 MTF for four waves of third-order spherical aberration. The focal range has been lengthened.

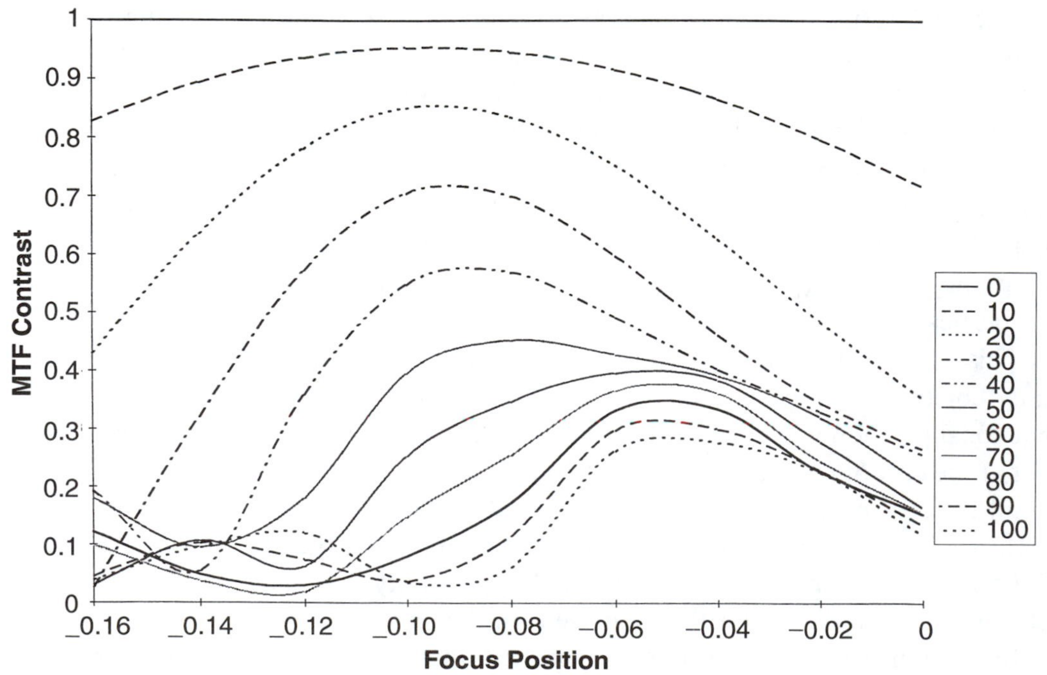

FIGURE 4.29 Through focus MTF for four waves of third-order spherical aberration with an extended focal range.

edge-corrected spherical aberration. The MTF behavior for this type of aberration is very important to understand, as it provides an estimate of the best correction that can be expected in a lens with an intrinsic content of residual high-order aberration. In the third-order spherical aberration examples above, the spherical aberration has been balanced by defocus. The best focus is found approximately at the position where the RMS residual wavefront error is at a minimum. From the approach used in section 4.4.2 it is found that the ratio between the RMS wavefront error and the peak wavefront error for third-order spherical is 13.4; it can be seen that the plots for one wavelength of third-order spherical correspond to about 0.07 waves RMS. For two, three and four waves, the RMS error is about 0.14, 0.21, and 0.28 waves. Inspection of the MTF plots indicates that for the case when the RMS wavefront error is about 0.07, the imaging is still almost perfect at the best focus. When the 0.14 waves RMS case is examined, there apparently begins to be a significant deviation from perfection. For larger errors, the behavior of the aberration and MTF become more complicated, and the RMS wavefront error does not suffice as a single figure of merit.

The next example is for fifth-order spherical aberration, balanced by third-order spherical aberration according to what will lead to the minimum RMS wavefront error. This balance is such that the ray intercept for a real ray at full aperture matches the paraxial intercept. This is, of course, the edge-balanced spherical aberration referred to in section 3.1.1.2.1. In this case it is convenient to refer to the amount of residual spherical aberration in terms of the effect on residual RMS wavefront aberration at best focus. The examples to be discussed are 0.07, 0.14, 0.22, and 0.29 waves RMS. These correspond to 3.7, 7.4, 11.6, and 15.3 waves of balanced fifth-order spherical aberration.

Figure 4.30 is a set of MTF plots for various focal positions for the 0.07 waves RMS error. Figure 4.31 shows the through-focus behavior. Comparison with the one-wavelength case of third-order spherical shows a similar behavior. A major difference here is that the location of the best focus position lies considerably closer to the paraxial focus, and the symmetry of the MTF curves with focus is somewhat better. Continuing on with more aberration, the 0.14 waves RMS case in Figure 4.32 shows that perfect MTFs cannot be achieved at any focal position. This is supported by the plots in Figure 4.33. For larger amounts of aberration, the frequency range is reduced in order to have a meaningful plot. Figures 4.34 through 4.37 show that the behavior with focus shows less asymmetry than for the case of large amounts of third-order spherical aberration.

Spherical aberration is a very important residual aberration. In a real lens, spherical cannot be considered alone. The best focus position and limit on image quality must include the effects of other aberrations. Before looking at

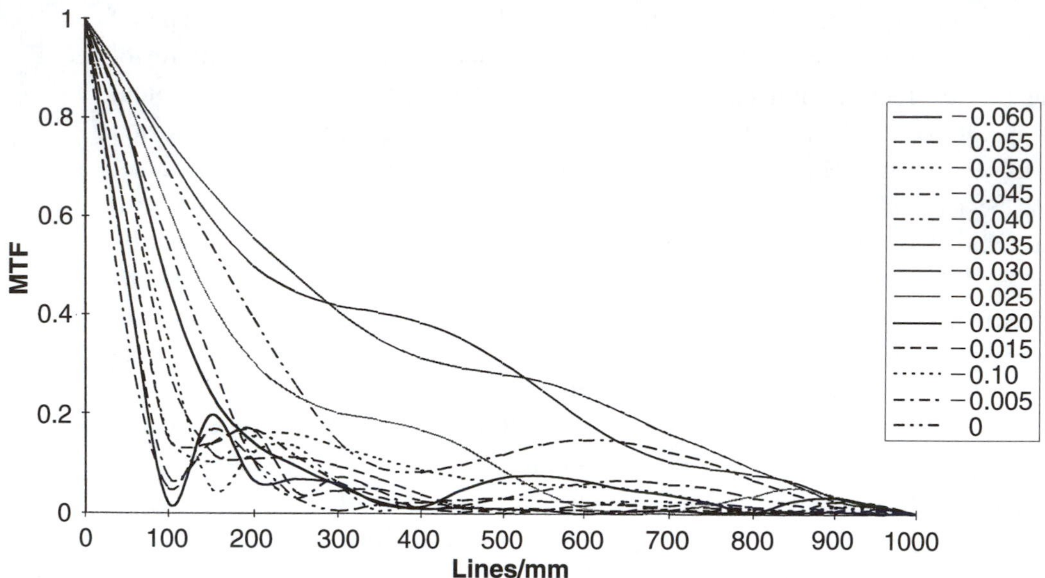

FIGURE 4.30 MTF for edge balanced third- and fifth-order spherical aberration having an RMS wavefront error of 0.07 at best focus.

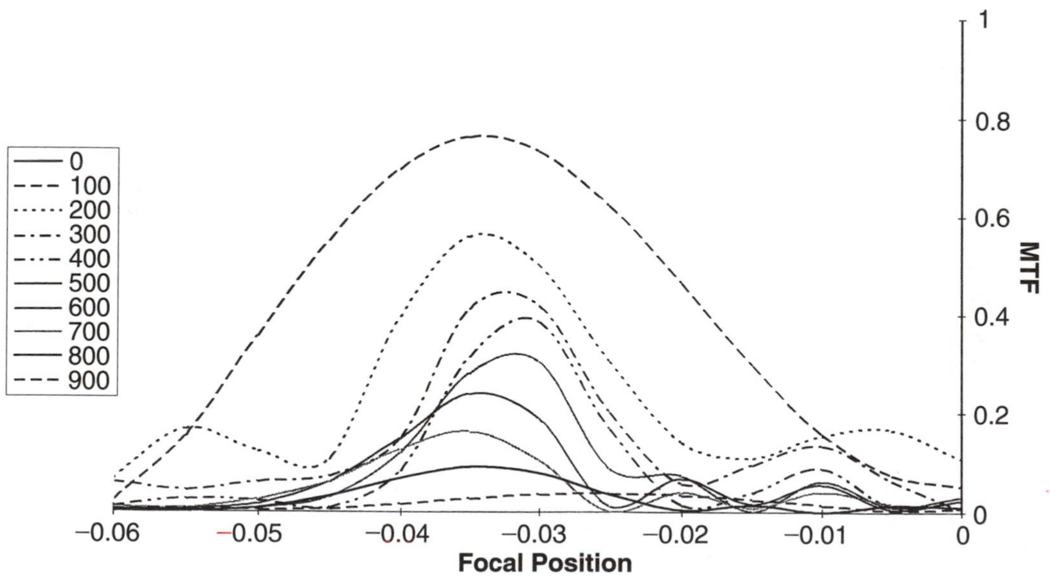

FIGURE 4.31 Through-focus MTF for the 0.07 waves RMS case of balanced spherical aberration.

the integrated effect of aberrations it is useful to consider each of the primary aberrations alone, in order to gain an understanding of the effect upon image quality and image symmetry that will occur.

The first off-axis aberration is coma. The origin and symmetry of the aberration have been discussed in Chapter 3. The effect on the MTF can

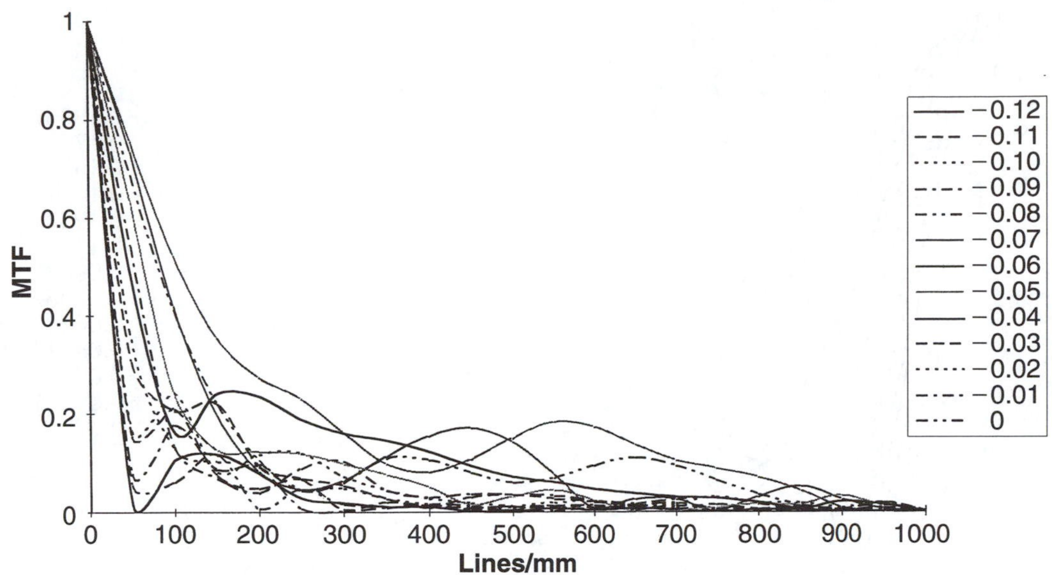

FIGURE 4.32 MTF for 0.14 waves RMS of balanced spherical aberration.

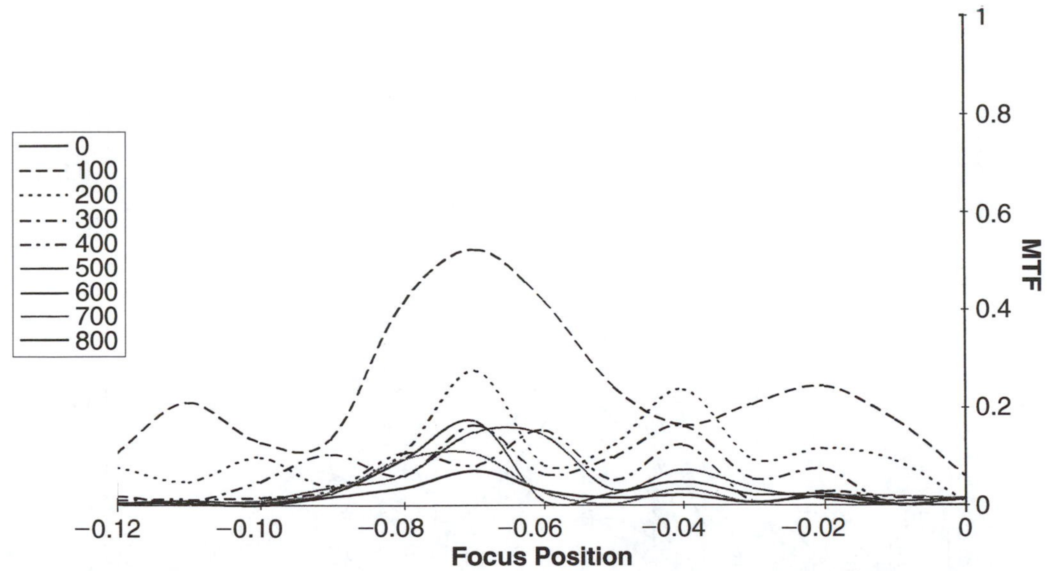

FIGURE 4.33 Through-focus MTF for 0.14 waves RMS of balanced spherical aberration.

be quite complicated. The MTF for differing amounts of coma is indicated in Figure 4.38. The effect is seen to be asymmetric with respect to azimuth on the image surface, since different values of the MTF are obtained in the radial and tangential directions.

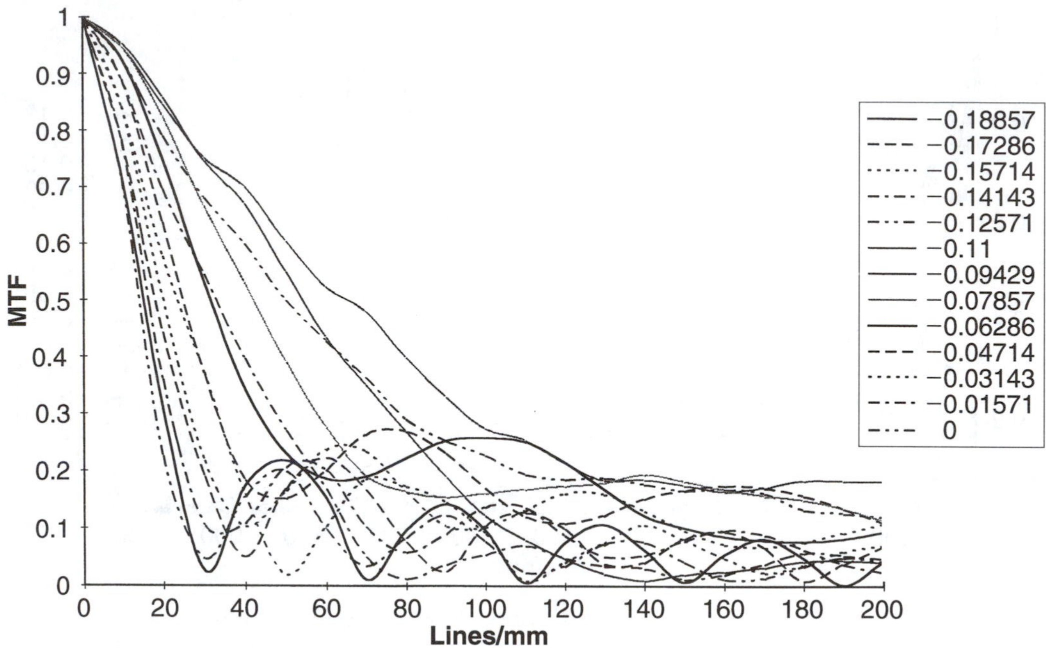

FIGURE 4.34 MTF for 0.22 waves RMS of balanced spherical aberration.

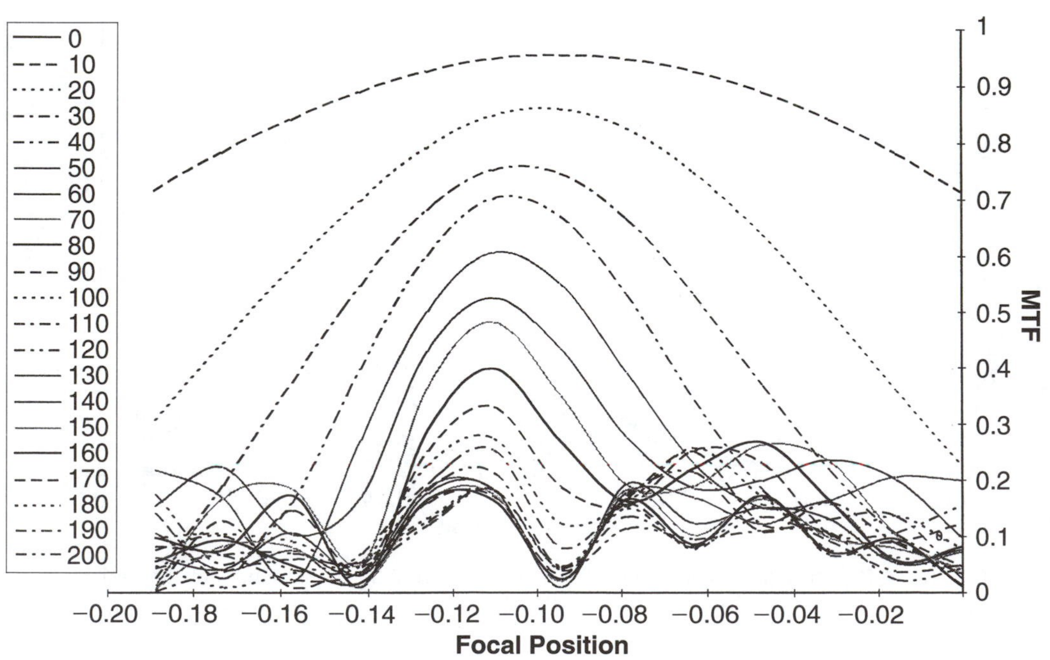

FIGURE 4.35 Through-focus MTF for 0.22 waves RMS of balanced spherical aberration.

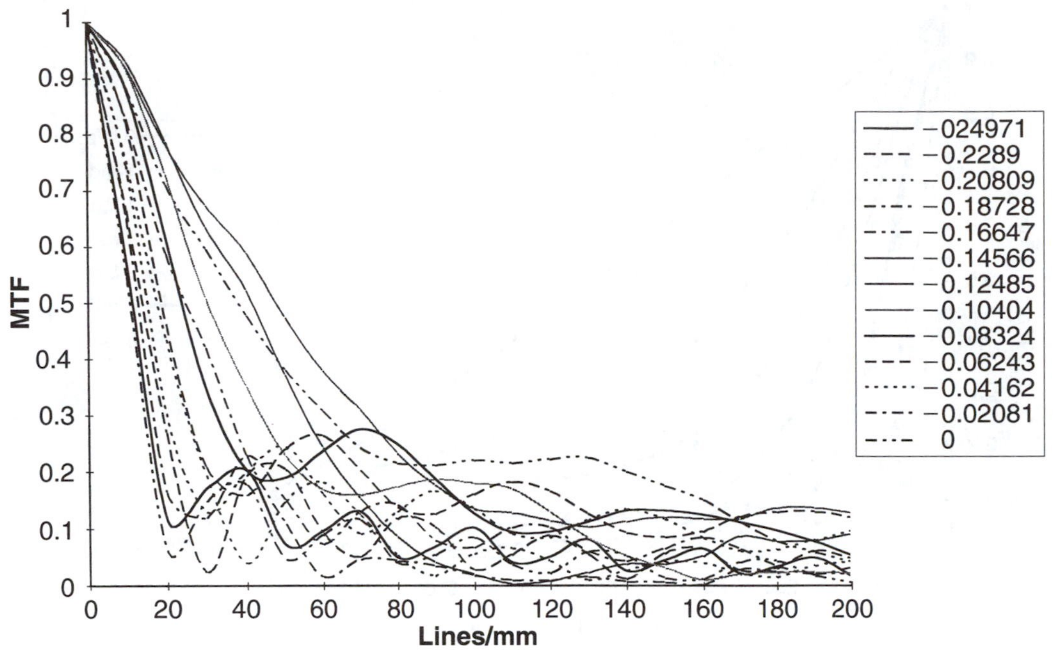

FIGURE 4.36 MTF for 0.28 waves of balanced spherical aberration.

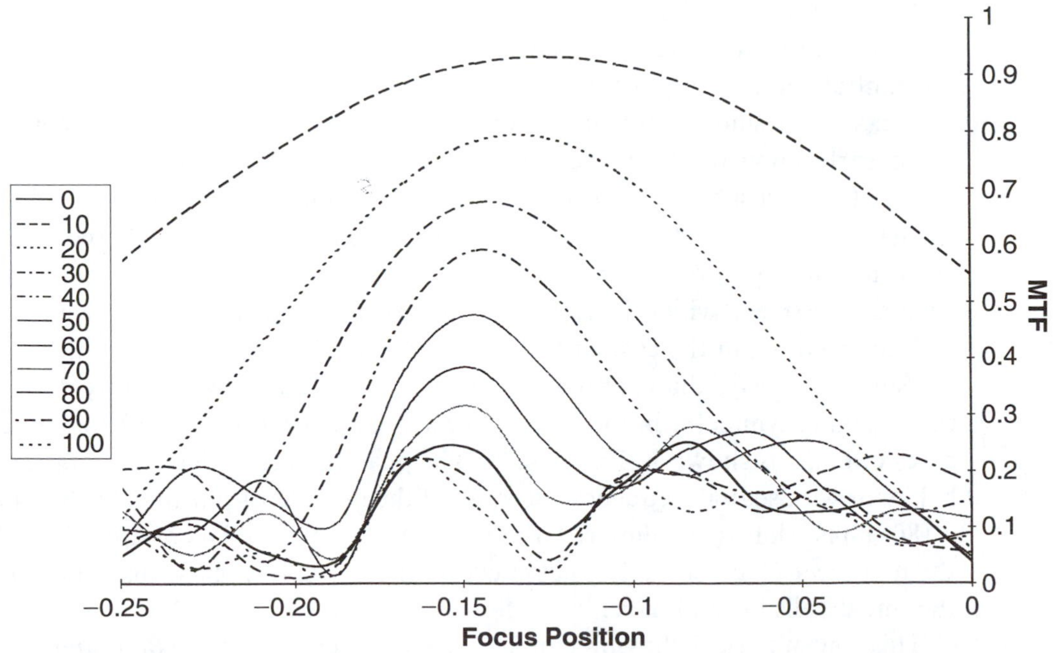

FIGURE 4.37 Through-focus MTF for 0.28 waves RMS of balanced spherical aberration. Note the extended range of focal position.

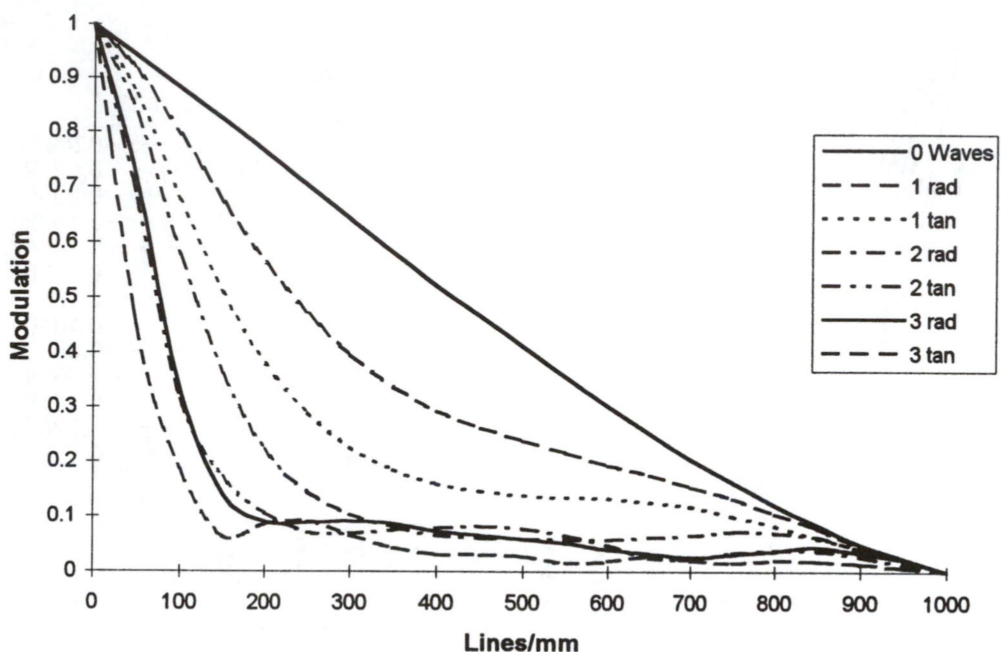

FIGURE 4.38 Radial and tangential MTF for zero, one, two, and three waves of third-order coma.

The asymmetric behavior of coma also produces a spatial phase shift. The effect upon the image of a single sinusoidal component is not significant, as the contrast of the component is given by the modulus of the MTF. However, the peak of the sinusoid is shifted from the origin by an amount that depends nonlinearly upon the frequency. The effect upon any complex object is to change the intensity distribution in the image. Coma is especially disturbing in that the shape of the image of any point on an object is altered in a consistent and progressive manner with respect to the axis of the lens. An asymmetric smear which increases linearly with the distance from the axis will be produced in the case of third-order coma. Because the orientation of the smear is always directed toward the axis, an observer will readily detect the changing symmetry in the image. The spatial phase function for the same cases of coma is shown in Figure 4.39. The phase is seen to have a nonlinear behavior with spatial frequency. Because of the method of plotting, each time a 180° phase shift is encountered, there is a phase jump, and a continuation of the nonlinear phase shift. The net effect on the image is a phase distortion of the fine details in the image.

This asymmetric behavior can be better understood by examining the MTF solid. Figure 4.40 is a solid plot of the MTF surface for two waves of coma. The symmetry of the function is defined by the 60° asymptotic angle of the coma spread function. Examination of the geometrical spot diagram as

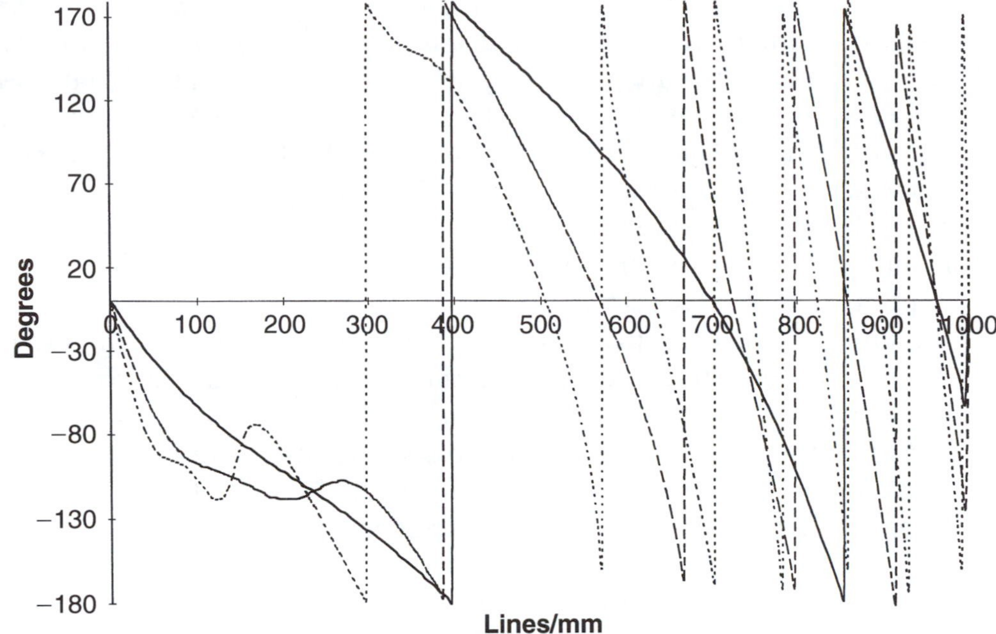

FIGURE 4.39 Plot of the spatial phase transfer function for the cases of coma presented in the previous figure: ——, one wave; – – –, two waves; ······, three waves.

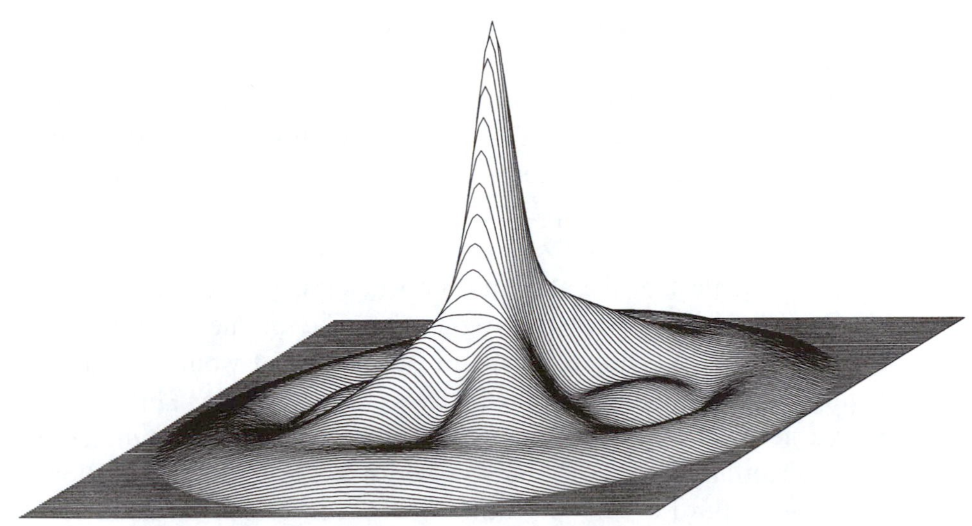

FIGURE 4.40 Surface MTF for the case of two waves of coma. The symmetry typical of the 60° asymptotic behavior of the coma is clearly seen (ZEMAX).

well as the point spread function indicates that there are two lines in the image that form narrow peaks separated by a 60° angle. These lines or peaks serve to produce high-frequency ridges in the MTF for spatial frequencies perpendicular to these peaks. Thus there is a 120° separation between the regions indicating high-frequency content in the MTF. The normal method of measurement and specification of the MTF, or of the resolution, in a lens generally uses the sagittal, tangential, and 45° orientations of the spatial frequency targets. Therefore, this property of a comatic image is often missed in practical image specification.

Shifting the focus in the presence of coma will not improve the image quality. The peak response will be located at the nominal paraxial focal location.

Astigmatism is the next aberration to consider. As shown in the sections on aberrations, astigmatism is a splitting of the focus for the sagittal and tangential foci. The symmetry of the aberration provides a shift in the focus dependent upon the azimuth of the spatial frequency being examined. In the case of rotationally symmetric systems, the orientation of the sagittal and tangential directions is established by the axis of the lens. The nominal best focus for astigmatism is at a focus halfway between the two focal lines. A demonstration of the MTF for various amounts of astigmatism can be obtained by using a lens design module to insert three waves of astigmatism into a lens. Selection of appropriate field heights and setting of the image surface curvature to follow the best focus surface provides a set of MTF plots, as in Figure 4.41.

Figure 4.42 is a through-focus plot for a spatial frequency at about 0.1 of the cutoff spatial frequency. This set of plots shows a successive shift of the peaks in the radial and tangential directions on each side of the mid-focus. This is a somewhat nontypical example of the effect of astigmatism alone. These plots are illustrative of the effect of astigmatism, and can be used to estimate the effect of the aberration on images. Figure 4.43 is a plot of the MTF surface for one wave of astigmatism. The symmetric, but not rotationally symmetric behavior is a type of behavior that would be seen in a lens having astigmatism alone. If the focus is changed, this MTF solid changes to have a near-perfect MTF appearance in either the sagittal direction or the radial direction, with a very narrow width in the opposite direction.

The final third-order aberration to be considered is distortion. As an aberration, this seems quite simple. The chief ray for the system does not intersect the image surface at the paraxially predicted image location. An otherwise perfect point spread function is located at a position that differs from the paraxial location by an amount proportional to the cube of the distance from the axis. Because this shift is nonlinear with field, the shape of any object will vary with the distance from the axis. The MTF effect of

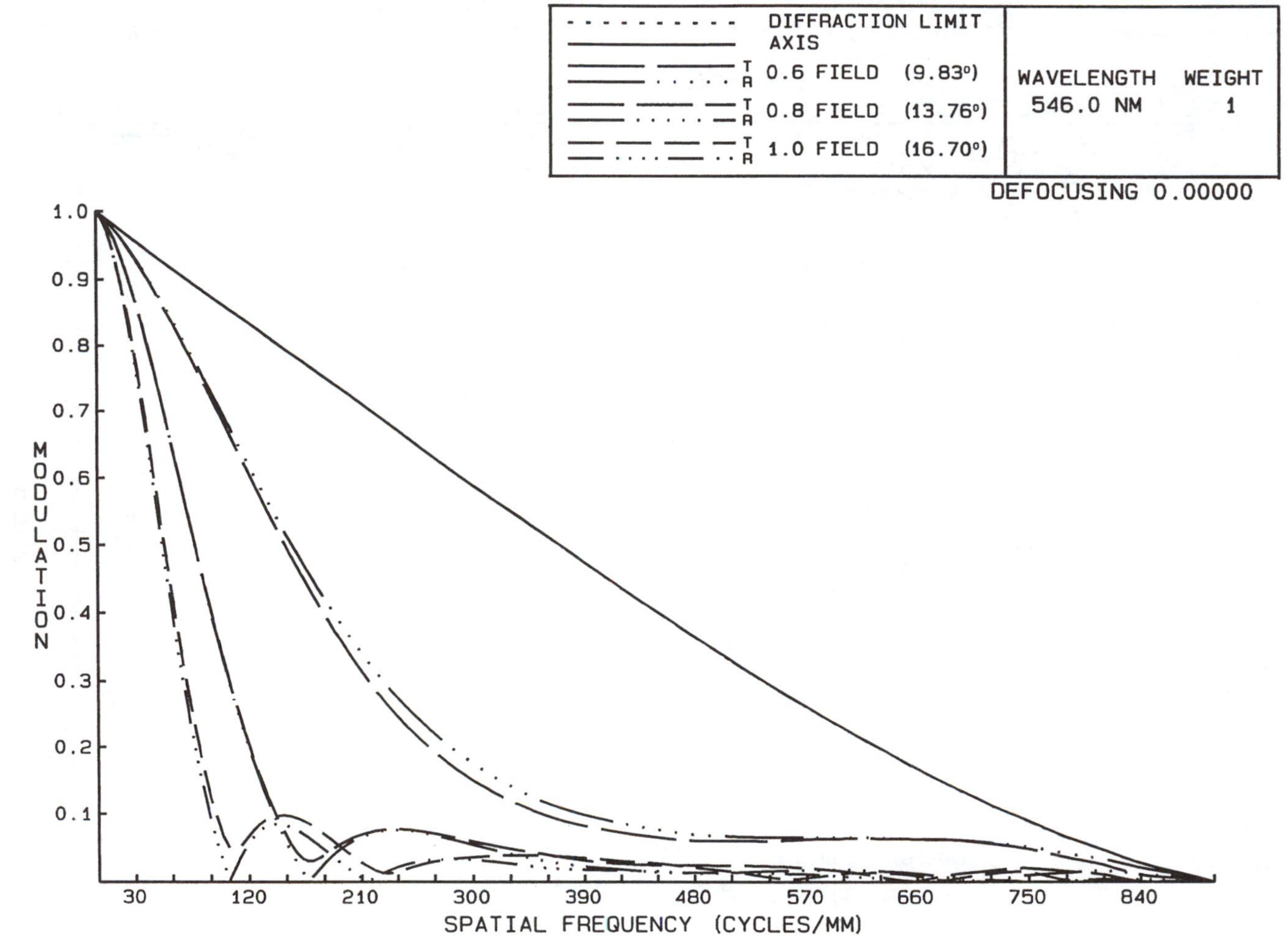

FIGURE 4.41 MTF for zero, one, two, and three waves of astigmatism, with the focus set at the mid-focus position for all cases (CODE V).

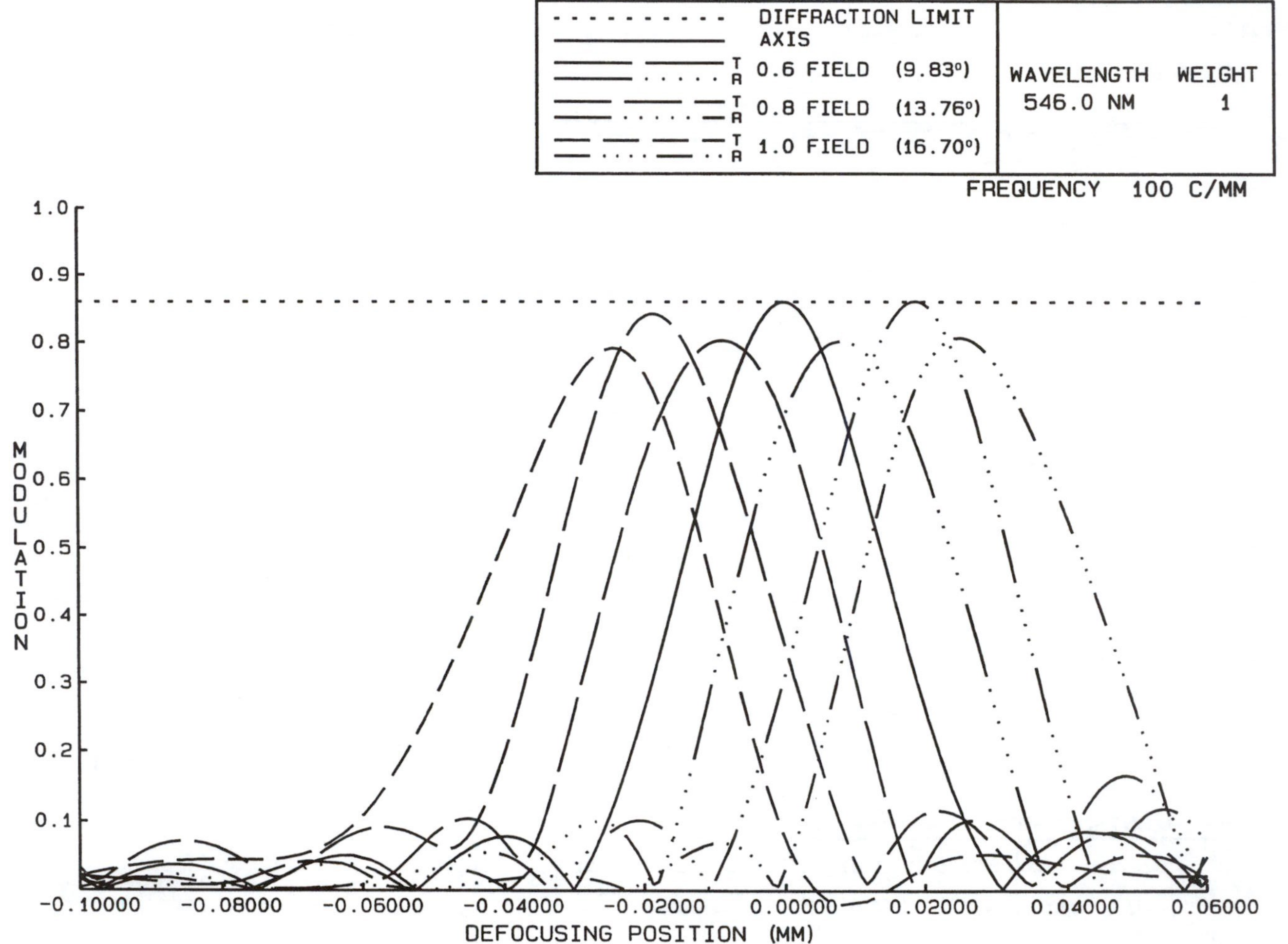

FIGURE 4.42 Through-focus MTF for the cases of astigmatism in the previous figure. The radial and tangential peaks of contrast are seen to split equally on each side of the mid-focus location.

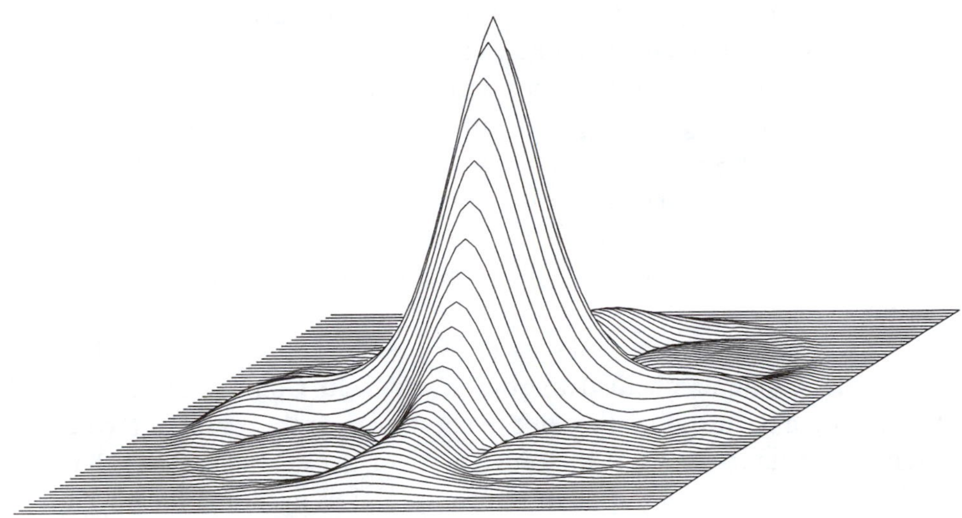

0.5500 TO 0.5500 MICRONS AT 0.0000, 0.0000 DEG.
SIDE IS 376.38 INVERSE MM, PEAK IS 1.0.

FIGURE 4.43 Surface MTF for one wave of astigmatism, with the focus set at the mid-focus position. The fourfold symmetry imposed by diffraction from the saddle-shaped wavefront function is clearly seen. The MTF is seen to be lower in the 45° azimuths than in either the radial or tangential directions (ZEMAX).

distortion is the introduction of a linear with frequency spatial phase shift, the amount of which varies with the distance off axis.

The linear spatial phase shift was earlier shown to have no effect upon the image appearance. However, distortion does change the image appearance because of the anamorphism introduced. The local magnification will not be the same in the radial and the tangential directions. A small square object will appear to be a rectangle or even a trapezoid, depending on the orientation of the object relative to the radial direction from the lens axis. This can be quite disturbing in high-precision imaging situations, such as lenses for producing microcircuits.

On a larger scale, distortion will make the shape of an object change dependent upon its location on the image plane. Photogrammetric applications will require special image mapping in order to match the images of similar objects. In photolithographic applications, microcircuit elements printed with distortion will fail to match circuit elements produced with different lenses. Adjustment of the magnification can reduce the effect of distortion, but it cannot eliminate it.

In pictorial imaging, distortion will cause images of large straight objects to appear curved. The amount of curvature depends upon the location in the field. In certain cases, such as extreme-wide-angle lenses, distortion is an intrinsic part of imaging, as it is not possible to image extreme wide angles, such as 180°, onto a flat focal surface without distortion. In such cases, distortion as usually defined can be as large as 20 or 30%.

Each of the above aberrations has been considered individually. However, in most practical imaging situations, aberrations appear in combinations. A lens covering a reasonable field of view must have the aberrations balanced so as to permit one choice of best focal plane to cover the entire field. This is accomplished by adjusting the balancing aberrations against the residual high-order aberrations so that the optimum balance across the field is obtained.

The astigmatism and Petzval curvature must be adjusted so as to balance the growth of oblique spherical aberration with field. Compensating spherical aberration can be introduced to shift the best focus in the center of the field to match the optimum focus at the outer parts of the field. The MTF balance obtained will be the result of all of the compromises in setting the aberration balance.

4.3 OTF Computation

The computation of the OTF for a lens can be carried out following the basic integral definitions for the function. The difference is that in the case of a real lens, the data is represented as a set of sampled wavefront and image plane information. In both the pupil and the image, the region of integration is finite in extent. The computation is thus an application of finite Fourier analysis. The data is represented as a two-dimensional matrix of information describing the pupil wavefront errors or image plane intensity information. The evaluation of the functions must be accomplished over a finite wavelength interval for most lenses.

The application of the OTF to imaging systems requires that the conditions of stationarity and linearity be met, or at least reasonably closely met. The first step is defining pupils that meet the isoplanatic condition over some portion of the image surface. The wavefront error is then computed at sampled points in the pupil, and used in calculating the pupil function, which is inserted into the diffraction integral. The task is made easy by recognizing that the scale relation between the pupils will cause a regular grid of rays in the entrance pupil to be a scaled regular grid of rays in the exit pupil.

Pupil definition appears quite straightforward, except that in a real optical system, the apertures defining the pupil may be distributed throughout the lens, and may be different at differing field angles. The paraxial pupil definitions may be quite inadequate for representing the actual path of rays through the lens. Pupil aberrations may be sufficiently large that the actual pupils are significantly distant from the paraxially defined pupils.

The OTF can be correct only if an adequate sampling of the aberrations in the pupil is carried out. The OTF for an aberration-free lens has only two parameters, the aperture size and the wavelength. When aberration is introduced, considerably more information is required to describe the wavefront errors. In general, the sampling must be chosen so that the wavefront error does not change by more than 0.25 waves between sampling points in the pupil. In practice the sampling required can be less, subject to a sanity check on the complete computations.

4.3.1 PUPIL DEFINITION

The paraxial entrance and exit pupils for a lens system are easily defined as the locations of the image of the aperture stop of the lens in object and image space. The axial location of the pupil is located by the crossing of the chief ray across the axis in object or image space. The sizes of the pupils are defined by the height of the marginal paraxial ray at the location of the entrance or exit pupil. The ratio of these values defines the magnification of the paraxial pupils.

The paraxial pupil concept is useful for understanding the ideal or first-order properties of the lens. In any real finite pupil the location and dimensions of the pupil must be determined by tracing a sufficient number of rays that the periphery of the pupil can be determined. The actual chief ray is then determined by the center of the coordinates describing the actual pupil. In some cases, especially catadioptric lenses, the chief ray may not actually get through the system, and the selection of coordinates is made about the inferred location of the center of the pupil.

The effective pupils for calculation of image formation are the spherical reference spheres located in the pupils and centered upon the chosen, usually paraxial, reference locations in the object and the image. In the case of isoplanatic imagery, the coordinates in these spherical pupils are related to each other by the magnification of the pupils. The intrinsic anamorphism that relates to the pupils is established by the angle of the chief ray in object and image space.

The pupil definition is somewhat more complex in the case of real rays as aberrations and vignetting change the portion of the aperture that is accessible for image formation and may change the location of the pupil along the axis. A lens which is locally isoplanatic will show a direct mapping between a grid of rays traced into the entrance pupil and the corresponding grid of rays located by the same rays passing through the exit pupil. This greatly eases the problem of obtaining the OPD values at a uniform set of points in the exit pupil to permit computation of the OTF and Point Spread Function.

The pupil is defined by the clear apertures of the lens system. Rays are traced from an object point and the upper and lower tangential rays, as well as the maximum sagittal ray passing through the lens are located. The chief ray is the specific ray that passes through the entrance pupil midway between the upper and lower edge rays. The approximate shape of the pupil can generally be thought of as elliptical in shape, with the eccentricity of the ellipse determined both by the relative angles of the chief rays in object and image space, and the amount of vignetting. In a real system, the actual shape of the vignetted pupil might not be well approximated by an ellipse. It also is quite possible for either, or both, of the pupils to be located at infinity.

The aberration content of the lens is described by the pupil function, which is obtained by tracing a grid of rays in the entrance pupil, and computing the optical path error associated with each ray passing through the exit pupil. This optical path difference for each ray relative to the chief ray is calculated on the exit pupil reference sphere.

The pupil function is complex,

$$P(x_p, y_p) = A(x_p, y_p)\, e^{i2\pi W(x_p, y_p)} \tag{4.24}$$

with the phase being the OPD error and the amplitude being the transmission factor for the pupil. The wavefront error in the pupil is the phase of the pupil function measured in wavelengths. For areas outside the vignetted clear apertures, the amplitude is zero, as it is for regions of the pupil that are obscured.

The wavefront error function can be obtained from a number of sources. The most common method in optical design is tracing a two-dimensional set of rays with coordinates equally spaced in the pupil. The Optical Path Difference is computed for each ray, providing a two-dimensional matrix of values of the aberration function. If the lens satisfies the isoplanatism condition, then selection of an equally spaced set of coordinates in the entrance pupil will result in an equally spaced set of coordinates in the exit pupil. This matrix of data can then be used to compute the image plane intensity through the use of a Fourier transform.

Alternate methods of obtaining the pupil aberration function involve the use of polynomials to fit a set of Optical Path Differences obtained from ray trace data. Several polynomial forms can be used, the most common being the Seidel wavefront terms or some set of the Zernike polynomials. The wavefront error can then be computed at any desired set of data grids in the pupil. Other sources of wavefront aberration data are experimental measurements, usually involving an interferometer, which can result in either a polynomial representation or a set of measured optical path errors. Most lens design programs permit importation of this type of test data into the program as a special surface error. Computation of the MTF then can include a hybrid of design data and interferometric data. In many cases, this can be used as the basis of a process to vary the design parameters to improve the alignment of the components in a lens to balance the measured fabrication error found to be in the components of the lens.

4.3.2 FFT METHODS

The fast Fourier transform (FFT) method of computation is actually a straightforward application of Fourier analysis (Shannon 1970). The computation begins with calculating the pupil function at a set of sample points. This function is represented as a two-dimensional matrix of complex numbers. For the computation of an incoherent imaging system, the pupil matrix is embedded in a zero matrix with twice the diameter of the pupil. Because any computation is an application of the Fourier method, the accuracy and sampling considerations discussed here will apply to all methods of computation.

The actual computation that is carried out by a computer program will permit the replacement of the continuous Fourier transform by a discrete Fourier transform. This is most familiarly identified as a Fourier series representation of the physical problem. The diffraction integral to be evaluated can be written as

$$a(\epsilon_x, \epsilon_y) = \iint_S A(u, v) \exp[-2\pi i\phi(u, v; \epsilon_x, \epsilon_y)]\, du\, dv \tag{4.25}$$

where (ϵ_x, ϵ_y) are coordinates in the image surface and (u, v) are exit pupil coordinates. It is required that the integration element $du\, dv$ be a metric of the entering wavefront in order that a uniform mapping of the wavefront energy be obtained. In the case where the extended sine condition applies, this condition is automatically met and an accurate rectilinear mapping of the entrance pupil ray coordinates into the exit pupil occurs.

The phase error in the argument of the complex exponential can be rewritten as

$$\phi(u, v; \epsilon_x, \epsilon_y) = W(u, v) - W'(u, v; \epsilon_x, \epsilon_y) \tag{4.26}$$

where $W(u, v)$ is the pupil function phase error discussed earlier computed at a set of points in the pupil. The function W' is the added phase encountered when the lateral position of the image plane coordinates (ϵ_x, ϵ_y) is varied. From Figure 2.39, the value of the function W' is

$$W' = A - B = \frac{(A^2 - B^2)}{2R} = \frac{(A^2 - B^2)\cos\theta}{2L} \qquad (4.27)$$

This leads after some evaluation of the coordinates as shown in section 2.2.2 to

$$a(\epsilon_x, \epsilon_y) = \iint A(u, v) \exp[-2\pi i W(u, v; H)] \exp[2\pi i(u\epsilon_x + v\epsilon_y)/R]\, du\, dv \qquad (4.28)$$

where the (u, v) coordinates form a natural set of coordinates for the computation, and the second complex exponential forms a Fourier kernel for the computation.

The geometry of the transformation is described in section 2.2.2 on computation of diffraction images. The choice of the coordinates is between pupil planes that are parallel. The amplitude computation is in the specified image surface. Should the amplitude on a tilted or curved image surface be required, then the orientation of the pupil and image planes should be tilted to correspond with the local image surface.

Additional insights can be gained by writing the (u, v) coordinates into a form that is useful in computation. The axial-paraxial pupil semidiameter is defined by the quantity u_m. The region occupied on axis in (u, v) space can be described as a circle in the normalized region (u', v'):

$$\left.\begin{aligned} u &= u' u_m \\ v &= v' u_m \end{aligned}\right\} \text{ wherein } u'^2 + v'^2 \leq 1 \qquad (4.29)$$

Inserting this into the argument for (u, v) and dividing by λ to obtain a phase value leads to a complicated-looking string of equations:

$$\begin{aligned} \frac{2\pi}{\lambda R}[u\epsilon_x + v\epsilon_y] &= \frac{2\pi}{\lambda R}[u' u_m \epsilon_x + v' u_m \epsilon_y \cos\theta] \\ &= \frac{2\pi}{\lambda}[u'\epsilon_x(\lambda \mathrm{NA}) + v'\epsilon_y(\lambda \mathrm{NA}\cos\theta)] \end{aligned} \qquad (4.30)$$

Relative to the axial numerical aperture $\mathrm{NA}_0 = u_m/L$ the final part of the equation becomes

$$\frac{2\pi}{\lambda}[(\lambda \mathrm{NA}_0 \cos\theta)u'\epsilon_x + (\lambda \mathrm{NA}_0 \cos^2\theta)v'\epsilon_y] \qquad (4.31)$$

For convenience, and perhaps clarity, the image plane coordinates can be written as

$$\begin{aligned} x' &= \epsilon_x(\lambda \mathrm{NA}_0 \cos\theta) \\ y' &= \epsilon_y(\lambda \mathrm{NA}_0 \cos^2\theta) \end{aligned} \qquad (4.32)$$

The formula so obtained is then written as

$$a(x',y') = \int\int A(u',v') \exp\left[\frac{-2\pi i}{\lambda} W(u',v')\right] \exp\left[\frac{-2\pi i}{\lambda}(u'x'+v'y')\right] du'\,dv' \tag{4.33}$$

which is a straightforward Fourier integral transform. The geometry of the coordinate in the pupil and on the image surface is contained in the factor relating the pupil coordinates and the image coordinates to the variables defined. This transform is amenable to computation in many ways, with the actual coordinates being unwrapped after the oblique pupil geometry is taken into account.

The evaluation of this function on a computer requires the application of sampled data. The pupil function is sampled at a two-dimensional set of points, selected to be uniformly spaced in the pupil. For reasons that will become clear shortly, the choice of the sample values must be made in the following manner. Assume a matrix of $N \times N$ complex values. The pupil area will be sampled within a central $N/2 \times N/2$ set of values. Usually this is a circular region within a matrix of twice the size as the sample points contained within the pupil. The size of the increment in the pupil region is then

$$\Delta u = \frac{u_m}{(N/4)} = \frac{4u_m}{N} \tag{4.34}$$

Since the integral form of the Fourier transform is being replaced by a discrete Fourier transform, this increment can be considered as being located in an effective frequency parameter space. The upper limit upon the distance in this effective frequency space is given by $2u_m$. In order to compute the image amplitude distribution, the integral is replaced by a sum. Consider increments in $\Delta\epsilon$ and Δu such that

$$\begin{aligned} x' &= j\Delta\epsilon \quad & y' &= k\Delta\epsilon \\ u' &= l\Delta u \quad & v' &= m\Delta u \end{aligned} \tag{4.35}$$

Then the discrete Fourier transform sum applies:

$$a(j,k) = \sum_{N/2}^{N/2}\sum_{N/2}^{N/2} A(l,m) \exp[-2\pi i W(l,m)] \exp[(2\pi i)(jl+km)(\Delta\epsilon\Delta u/\lambda R)] \tag{4.36}$$

In order that the system of equations satisfy a Fourier series, the increments in the effective frequency must be chosen to make the lowest frequency periodic in the total spatial range to be covered. Thus

$$2\pi\left(\frac{\Delta u}{\lambda R}\right)2\epsilon_m = 2\pi \tag{4.37}$$

where ϵ_m is the image plane radius covered by the computed diffraction pattern. Then, since the width of the pupil is $(N/2)\Delta u$ and the distance from the pupil to the image surface is R,

$$\epsilon_m = \frac{\lambda R}{2\Delta u} = \frac{N\lambda(\mathrm{F/number})}{4} \tag{4.38}$$

so that the area extent on the image surface is determined by the number of samples taken in the computation of the transform. The aperture dimension is fixed, so that using more points provides a better sampling of the pupil function. The sampling distance in the image surface is given by

$$\Delta x = \frac{\epsilon_m}{N} = \lambda(\mathrm{F/number}) \tag{4.39}$$

so that the sampling distance in the image surface is fixed by the pupil diameter. The greater the number of points in the pupil, the larger will be the applicable region on the image surface. This makes some useful sense, as large pupil aberrations that require closer sampling will usually extend the spread function over a wider area.

Some physical sense can be made of this effect by calculating the effect of a phase tilt between two adjacent sampling points on the direction of the geometrical ray from that portion of the pupil. Using the relation between the lateral intercept of the ray and the slope of the wavefront error leads to the conclusion that if the OPD between two adjacent sampling points in the pupil is $\lambda/2$, or a half wavelength, the ray will intersect the image surface at ϵ_m, the edge of the area covered by the transform. The conclusion then is that to be safe, and ensure that the geometrical energy is contained within the transform region on the image surface, the OPD between adjacent sample points should be less than a half wave. Using a quarter wave as the criterion would seem to be a safe practice.

The amplitude image is represented by a two-dimensional matrix of real and imaginary parts, which retain the phase relations between the portions of the point spread function. The intensity point spread function is found by applying a complex square

$$f(j,k) = a(j,k)a^*(j,k) = R^2 + I^2 \tag{4.40}$$

at each point in the matrix. The result is a matrix of points which describe the point spread function. Up to now, all steps in the computation are linear, preserve phase, and are reversible. This operation is a nonlinear operation, which causes the phase to be eliminated in favor of an intensity representation of the point spread function.

The two-dimensional OTF can be obtained by applying a Fourier transform to the spread function data. Using the sampled data, the transform is

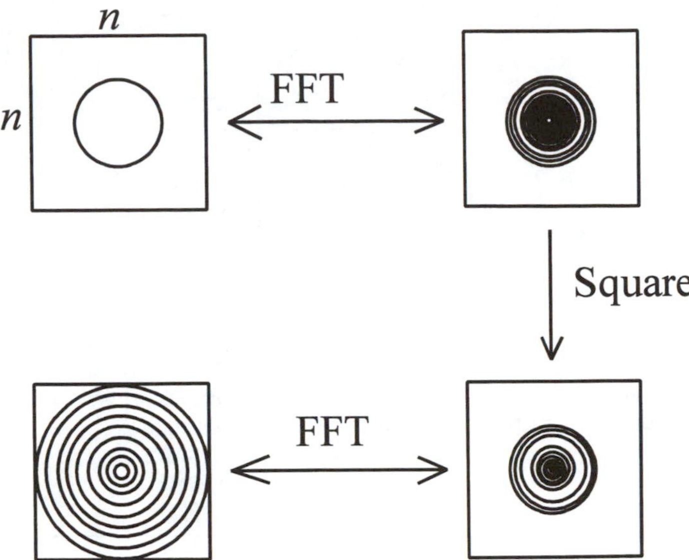

FIGURE 4.44 The process of using a two-dimensional fast Fourier transform technique for computing the Point Spread Function and the Optical Transfer Function. The square block represents a matrix of data which is half filled, as shown by the included circle, with the complex pupil function for the lens.

$$F(l, m) = \sum \sum f(j, k) e^{-2\pi i(lj+mk)} \Delta\epsilon \Delta\nu \tag{4.41}$$

The values of $\Delta\epsilon$ and $\Delta\nu$ are also related by the requirement for a Fourier series transform that $\Delta\epsilon\Delta\nu = 1$. In order to find the maximum spatial frequency included within the transform matrix, set $2\pi(\lambda(\mathrm{F/number})/2)(2\nu_m) = 2\pi$ (here the total frequency width for the matrix is twice the maximum spatial frequency). This results in

$$\nu_m = \frac{1}{\lambda(\mathrm{F/number})} \tag{4.42}$$

for the maximum spatial frequency. This is the same as obtained from the convolution analysis earlier. The spatial frequency region now occupies the entire width of the $N \times N$ matrix, the expansion beyond the pupil being due to the nonlinear operation of calculating the intensity by squaring the amplitude point spread function.

The relation between the mathematical representation of the diffraction process and the physical reality of the process is seen by examining Figure 4.44. The wavefront pupil function is shown immersed in the central portion of the matrix. The point spread function occupies the entire extent of the matrix. The final Fourier transform leads to an OTF which fills the entire matrix. The computation can be efficiently carried out using the fast Fourier transform algorithms that are widely available in the literature. Usually, the

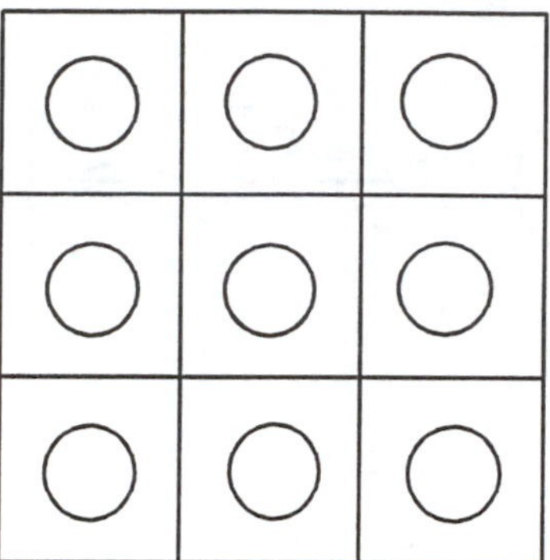

FIGURE 4.45 The array of points show that the sampled aperture data represent a discrete Fourier series representation of the actual diffraction situation.

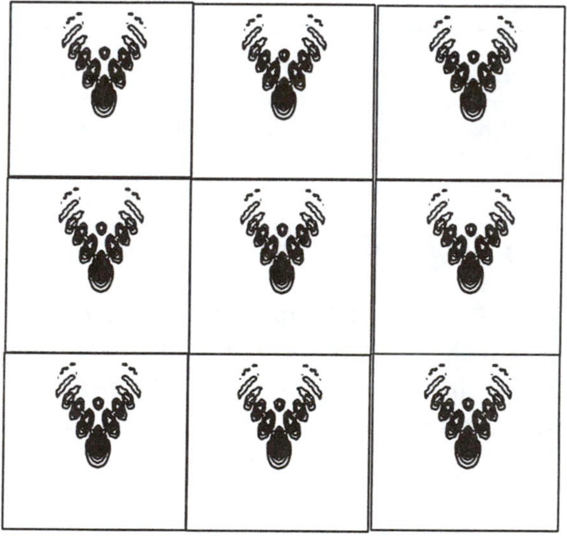

FIGURE 4.46 The intrinsically repeated discrete computation that represents the physical diffraction situation. An example of the nonaliased overlap of the computed function can be seen in Figure 2.48.

value of N is constrained to be a power of 2, with 32×32 or 64×64 being adequate for most optical problems.

Another important concept is the consequence of the discrete sampling of all of the functions involved. The actual computation being carried out is actually a Fourier series representation of the actual physical situation. The function is effectively represented by a pupil that is replicated by the period of the lowest spatial frequency determined by the choice of sampling. In this description, the effective computation uses a repeated pupil function as

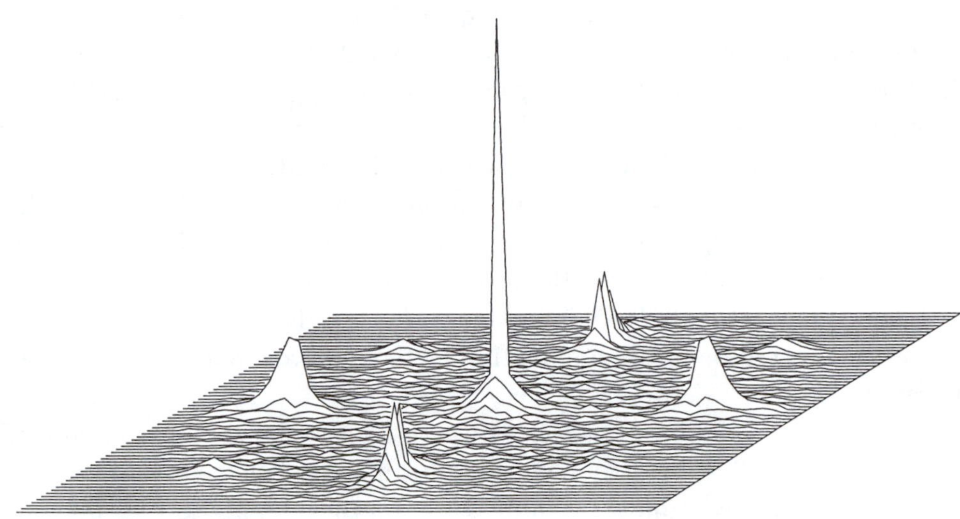

0.5500 TO 0.5500 MICRONS AT 0.0000 DEG.
SIDE IS 1440.14 INVERSE MM, PEAK IS 1.0.

FIGURE 4.47 A surface MTF computation for an image with spherical aberration that is sufficiently spread out that aliasing occurs in the computation. The four peaks out near the cutoff frequency are aliased information due to inadequately fine sampling of the pupil function. In this case the sampling was deliberately left inadequate to illustrate the point. In actual computations, the designer should be sure that this situation does not arise (ZEMAX).

shown in Figure 4.45. The point spread function is likewise represented by one period of a repeated matrix, as shown in Figure 4.46. As long as all of the energy diffracted by the aperture remains within one periodic cell of the calculation, then the discrete computation is correct. If the aberration spread becomes sufficiently large that some reasonable fraction of the point spread function overlaps adjacent cells, then aliasing will occur. The nonlinear operation of computing the intensity in the point spread function causes a loss of the coherent phase information, and the subsequent computation of the OTF will indicate folded or aliased components. An example of this is shown in Figure 4.47. The apparent new peaks at high frequencies are due to aliasing. This effect will occur with all types of computation of the OTF, and must be watched for by the designer using any program for computing the OTF.

4.3.3 CONVOLUTION METHODS

The pupil convolution approach discussed in section 4.2.1 offers an alternate method of directly computing the OTF. This approach is most convenient and efficient when only some spatial frequencies are required, or when only a restricted number of azimuths are to be computed. As pointed out in section 4.2.1, this convolution method is a consequence of application of basic Fourier optics methods.

Historically, use of the convolution method preceded application of the FFT method. There have been many approaches to the evaluation of the convolution integral, primarily driven by the state of the art of computation methods that were available at the time (Hopkins 1956; Barakat 1962; Minnick and Rancourt 1968; Hopkins 1983). Several of the options available for numerical integration in the convolution method are covered in an article by Barakat (1980).

Since the convolution approach is a consequence of the application of Fourier theory, all of the comments about sampling and accuracy of representation of the wavefront error in the FFT approach apply to this mode of computation. Unfortunately, the sampling errors that may occur in the computation of a convolved OTF along one azimuth are not as easily identified by visual inspection of the OTF plots. Therefore any wise user should be careful to examine alternate computations with different aperture sampling to ensure that the computed results are meaningful. A good procedure is to methodically increase the density of sample points describing the pupil function until a stable result is obtained for the computation. This is not guaranteed to converge, however, as different computer programs may use different approaches to defining the mesh of points in the pupil or of interpolating between data points, and there can be constant errors as a result of the numerical method used in the integration. Errors of this type are frequently encountered when there is a large amount of residual aberration in the wavefront.

4.4 Summary Measures

The trend in optical design is to use the MTF as the mechanism for specifying the required image quality. For historical reasons, as well as some sensible technical reasons, this is not always the appropriate approach. The description of image quality is often carried out by using some other summary measures. Some of these, such as Strehl ratio and energy concentration are quantitative. Some, such as resolution, are semiquantitative. Others are related to a subjective evaluation of image quality.

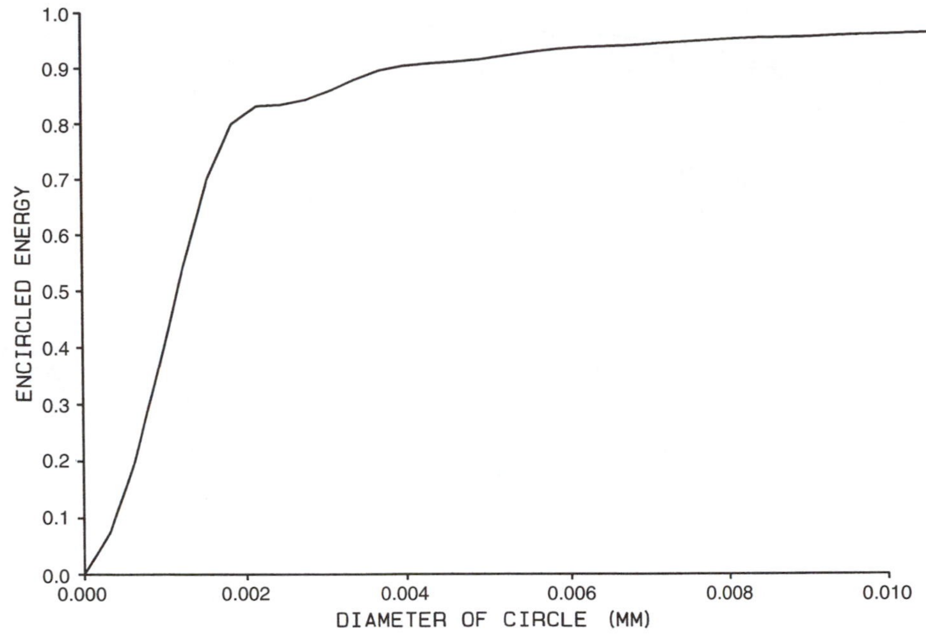

FIGURE 4.48 Encircled energy for a perfect lens (CODE V).

4.4.1 ENERGY CONCENTRATION

The fraction of energy within a specified diameter region on the image surface is a measure of the energy concentration. Computation of this is carried out by choosing a center of coordinates and then carrying out the integration over the point spread function. The choice of the center of coordinates can be chosen in several ways. For symmetric spread functions, there is an obvious center. For point spread functions that are asymmetric the center can be chosen to maximize the energy within each energy area, or in one specified area:

$$E(\rho) = \int_{-\pi}^{\pi} \int_{0}^{\rho} f(r, \theta) r \, dr \, d\theta \tag{4.43}$$

The presentation of the data is as in Figure 4.48. This data plot is for a perfect, unaberrated circular pupil which shows that about 84% of the energy is contained within the Airy radius of 1.22λ(F/number). Figure 4.49 is a plot of the energy concentration for several focal positions, showing the spread of energy concentration with increasing focus aberration. The advantage of using the energy concentration for describing an image is that a single number can be obtained, which states the diameter of a circular area on the image

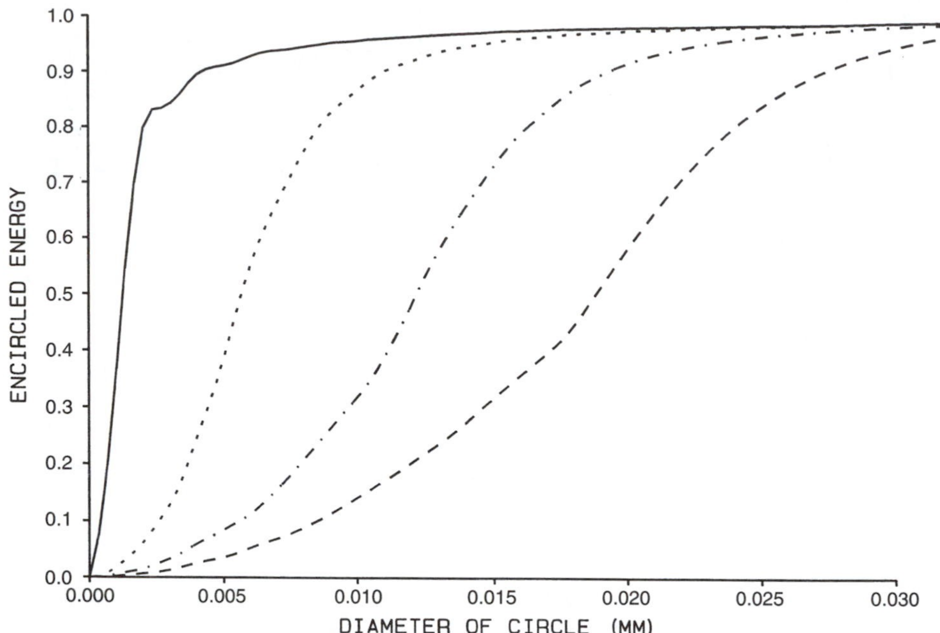

FIGURE 4.49 Encircled energy for zero (——), one (· · · · · ·), two (· – · – ·), and three (– – –) waves of defocusing (CODE V).

surface containing a specified fraction of the energy passing through the pupil. Alternately the amount of energy within a specific area provides a measure of the efficiency of the lens in placing light from a point object on a specified detector dimension. The details of the intensity distribution in the point spread function are ignored in this specification. Thus images with coma, astigmatism, or other general aberrations are described by a single number. For some users this is a convenience. In some cases, this is an unwarranted simplification.

4.4.2 STREHL RATIO, RMS WAVEFRONT ABERRATION

A simple theoretical description of the degree of perfection in an image is the Strehl ratio. This is the ratio of the intensity at the center of the diffraction image to the intensity that would be present if no aberration were present. This has long been a measure for perfection in telescope and microscope systems. Although the evaluation of the quality of an image by an observer does not correlate very well with this measure, it is a good indicator of the degree of perfection for lens systems that are very close to diffraction limited. Images in which the aberration is large are not well represented by the Strehl

TABLE 4.2 Ratios of RMS to peak wavefront aberrations

Aberration	Peak to RMS	RMS to Peak
Defocus	3.5	0.286
Third-order spherical	13.4	0.075
Fifth-order spherical, balanced with third order	57.1	0.017
Third-order coma	8.6	0.116
Third-order astigmatism	5.0	0.200
Smooth random errors typical of fabrication errors	≈5	0.200

Note: all ratios (except defocus) include adjustment for optimum focus and lateral shift.

ratio because the intensity at the center of the point spread function is then dependent upon the exact nature of the aberration.

A very useful relationship between the Strehl ratio and the RMS wavefront error is the Marechal relation (Marechal 1947):

$$i_p(0,0) \cong 1 - \left(\frac{2\pi}{\lambda}\right)^2 (W_{RMS})^2 \tag{4.44}$$

The RMS wavefront error is computed by

$$\text{Define } \langle W^n \rangle = \frac{\int (W(u,v))^n du\, dv}{\int du\, dv}$$

$$\text{then } W_{RMS} = \sqrt{\langle W^2 \rangle - \langle W \rangle^2} \tag{4.45}$$

where the function $W(u, v)$ is the wavefront error function described earlier. This definition of the RMS error is normalized to the area of the pupil, and adjusts the mean value of the residual wavefront error to zero. This relationship is a good fit to reality as long as the value of W_{RMS} is less than about 0.1. A particularly important value is when $W_{RMS} = 0.07$. For this value, the Strehl ratio $i_p = 0.8$. This is conventionally accepted as defining the limit of error in a "perfect" or diffraction limited system. This ratio is also referred to as a measure of "the power in a bucket" (Holmes and Avizonis 1976) or a measure stated as a multiple of the diffraction-limited dimension. The Strehl ratio and its associated RMS wavefront error have come to be accepted for describing the behavior of high-quality laser-beam control and tracking systems. Table 4.2 lists some ratios of RMS to peak wavefront aberrations.

A plot of the Strehl ratio as a function of the amount of defocus wavefront error is shown in Figure 4.50. The actual computed value, the Marechal

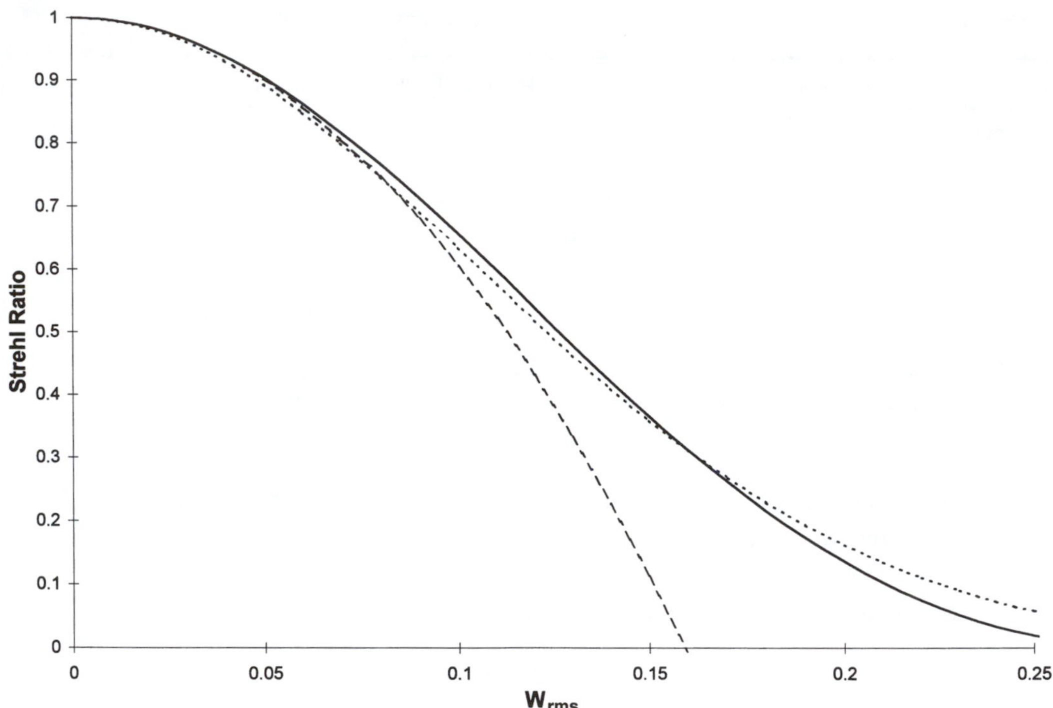

FIGURE 4.50 The Strehl ratio as a function of the RMS wavefront error: —, computation; – – –, Marechal relation; ······, approximation.

relation and an empirical simplified relationship are shown on this plot. The useful empirical approximation is

$$i_p(0,0) = \exp -\left(\frac{W_{RMS}}{0.148}\right)^2 \cong \exp(-46W_{RMS}^2) \tag{4.46}$$

The OTF can be related to the RMS wavefront error as well. In general, the effect on the MTF for given values of the RMS wavefront error is approximately the same up to about 0.1 or 0.2 waves RMS. The RMS wavefront error would appear to have significance as a single number criterion for the state of correction of the optical system.

It is possible to exploit this dependence on RMS wavefront error by developing a model for the approximate MTF for given amounts of error. There are several approaches to carrying this out based upon use of an expansion of the MTF (Scott 1964; Hufnagel and Stanley 1964). Another approach (Shannon 1994) uses an empirical assumption that the approximate MTF can be represented by a product of the aberration-free, diffraction-limited MTF and an aberration transfer factor (ATF):

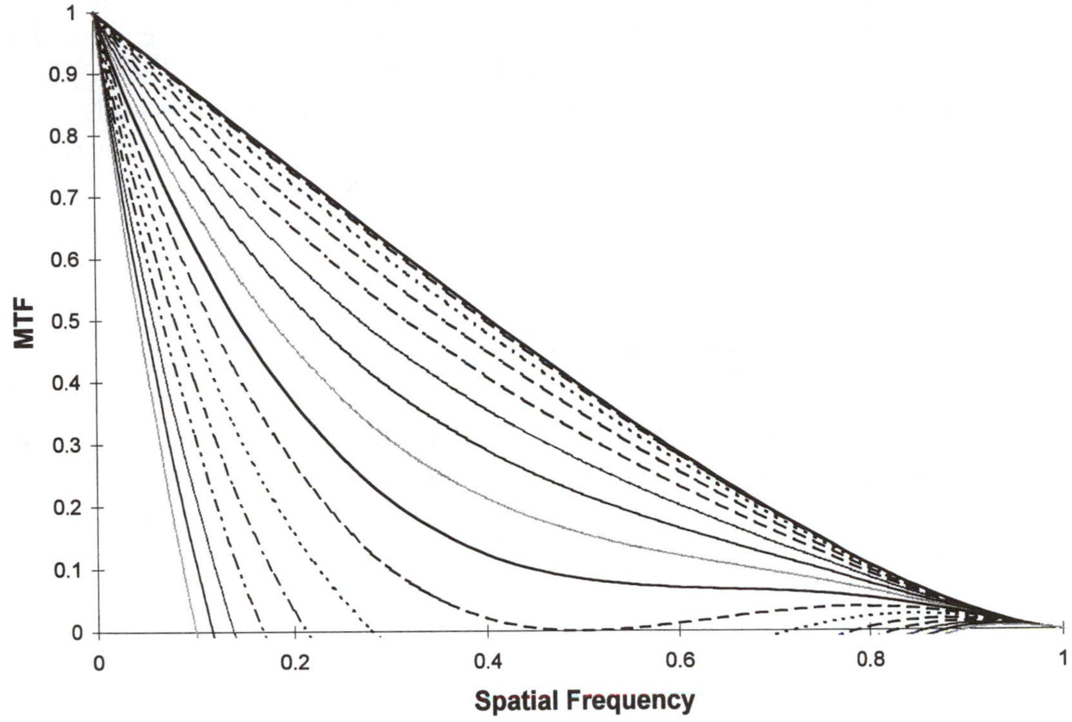

FIGURE 4.51 Approximations of MTF for 0 to 0.3 waves RMS in steps of 0.02.

$$\begin{aligned}
\mathrm{DTF}(\Omega) &= \frac{2}{\pi}\left[\arccos(\Omega) - \Omega\sqrt{1-\Omega^2}\right] \\
\Omega &= \frac{\nu}{\nu_{max}} = \nu\lambda(\mathrm{F/number}) \\
\mathrm{ATF}(\Omega) &= \left[1 - \left(\frac{W_{RMS}}{0.18}\right)^2 (1 - 4(\Omega - 0.5)^2)\right] \\
\mathrm{MTF}(\Omega) &= \mathrm{DTF}(\Omega)\mathrm{ATF}(\Omega) \qquad (4.47)
\end{aligned}$$

Figure 4.51 is a plot of the MTF, the product of these two functions. This is, of course, an empirical model, and values for the MTF that are negative are ignored.

This approximate model is very useful for estimating the amount of residual RMS wavefront error that can be allowed in a design. First, the wavelength characterizing the system and the F/number to be used are selected. The use of the empirical approximation is accomplished by deciding upon the requirements for MTF and spatial frequency. These can easily be interpreted to provide an estimate of the allowable residual RMS wavefront error for the system.

The RMS wavefront aberration has considerable use in the tolerancing of optical systems. The empirical MTF discussion indicates that the effect of a change in the aberration can be predicted. If the errors are presumed to be uncorrelated with the aberrations, then the RMS wavefront errors will sum in a mean-square sense. A level of allowable degradation of the image quality is established, and the errors permitted in the lens are budgeted against this residual. Details of this procedure are discussed in Chapter 6.

4.4.3 RESOLUTION AND SUBJECTIVE OPTICAL QUALITY

Often the quality of an optical system is expressed in terms of resolution. A user might specify that "the optical system must resolve 45 lines per millimeter." The meaning of this is that a user can distinguish the presence of detail in the image separated by 1/45 millimeters. As it stands, this is only partially useful to the designer, as the conditions under which this occurs need to be stated. The contrast of the object, the description of the target to be used for testing the resolution, the characteristics of the detector and the illumination need to be stated.

Resolution is a threshold detection process. A periodic object of stated shape and contrast is usually chosen. The aerial image formed by the lens can be described by the process laid out in Figure 4.52 (Goodman 1968; Gaskill 1978). The object is Fourier transformed to obtain the object spectrum. This is multiplied by the lens MTF, resulting in an image spectrum. The spectrum multiplication step must be a complex one, in which any spatial phase transfer function is included. (Other disturbing imaging functions such as scattering, atmospheric turbulence and image motion can be incorporated as additional frequency spectrum factors.) The inverse Fourier transform of the image spectrum yields the image intensity distribution.

An example of this is shown in Figure 4.53. The appearance of a three-bar high-contrast object is shown for several different imaging conditions, in which the cutoff of a perfect lens is set at successively lower spatial frequencies. The plot of the object spectrum is shown in Figure 4.52, along with plots of the several different perfect transfer functions that are used in the computation. The imaging process for this object is seen to be a bit complicated, because the object contains a wide range of spatial frequencies.

Comparison of this image intensity distribution with an appropriate level of the detector noise provides information as to the probability of detection of the signal. Depending upon the needs for the system, the resolution requirement may be to either identify the object or to detect its presence and location. Usually the detection criterion will be proportional to the

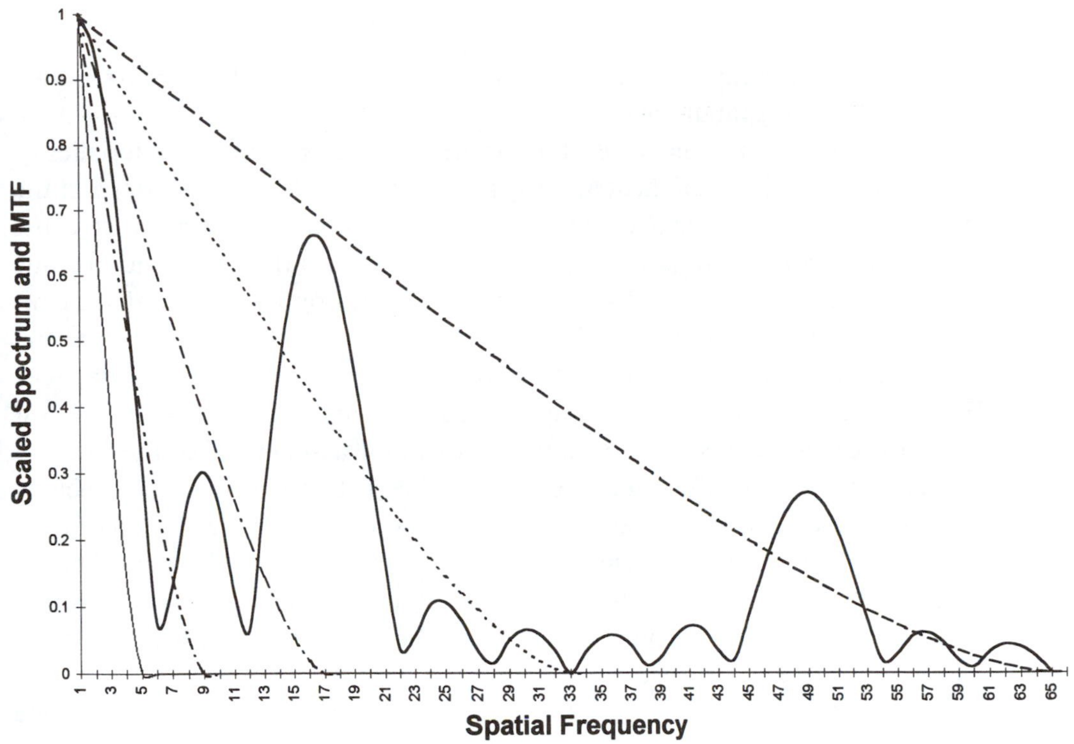

FIGURE 4.52 Plot of the spatial frequency spectrum of a three-bar chart. Overlaid are the MTFs for five different aperture values.

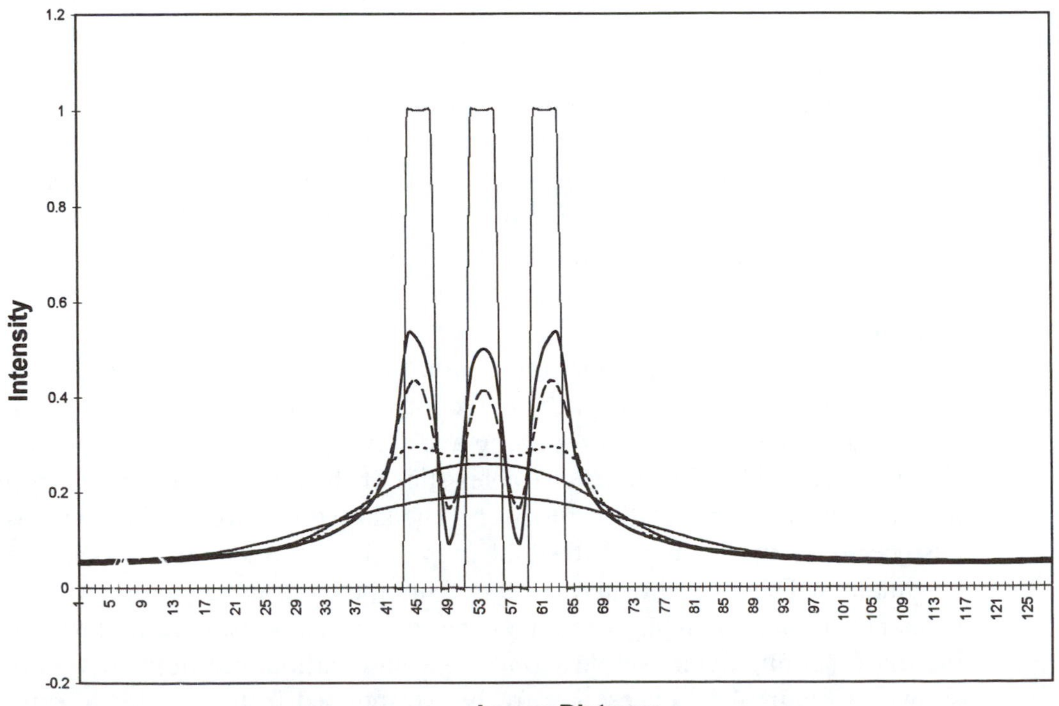

FIGURE 4.53 A plot of the original object intensity distribution for the three-bar chart object. Added are plots of the intensity distribution in the image for five different MTF corresponding to several aperture diameters and resulting cutoff frequencies. This shows the reduction in image contrast and change in image shape as a consequence of the low-pass characteristics of the lens MTF.

signal-to-noise ratio obtained by averaging the signal and the noise power over a specified spatial region.

If an electronic scanner is used to examine the image, the characteristics of the scanner can be quantified, and the rules for detection used to interpret the usefulness of a given image contrast. When a human observer is used, the resolution problem is quite complicated, and requires the application of one of many statistical models summarizing the operation of the visual system.

Frequently a "standard" USAF three-bar target is chosen for visual inspection. The period of the target should be related to the desired distance on the image by the magnification. The contrast and illumination of the test target needs to be specified. Once this is done, the actual OTF of the lens can be used to construct the image intensity profile. If this exceeds some specified threshold, the target will be resolved. The threshold involves the contrast sensitivity and the noise characteristics of the detector. A performance threshold curve for a detector can be developed by determining the actual image contrast that will match the noise threshold and be detected by an observer. If the targets used in determining the threshold curve are sinusoidal, then the intersection of the threshold and the MTF curve will determine the resolution. If the contrast of the object is less than unity, the MTF curve can be multiplied by this contrast producing an effective image contrast curve, and the intersection of this curve with the threshold will determine the resolution.

Usually this system analysis will be carried out by the user. The designer can then be given a specific requirement upon the MTF at a specified spatial frequency. Based upon logic similar to that in the preceding paragraph, the designer can assume that the lens will meet the user's needs if the MTF requirements are met. Usually, the user will expect that the MTF goals will be met after fabrication. The designer must then provide sufficient margin in the design that the image contrast will be met in the presence of expected tolerance errors in the lens.

When the detector is photographic film, the nonlinear recording characteristics need to be included in the detection threshold evaluation function. Presuming some "standard" processing and viewing conditions for the material, the threshold and transfer characteristics of the film can be lumped into a single threshold modulation function. The simplest approach for the lens designer is to interpret the threshold modulation as required aerial image modulation. This will, of course, be based upon the object being a regulation three-bar resolution target (although other objects could be used in developing this function). Some sample aerial image modulation threshold curves are shown in Figure 4.54. (These are freely extrapolated from samples of published data.) The interpretation is that the product of the object spectrum and the lens MTF must exceed the threshold at the desired spatial frequency in

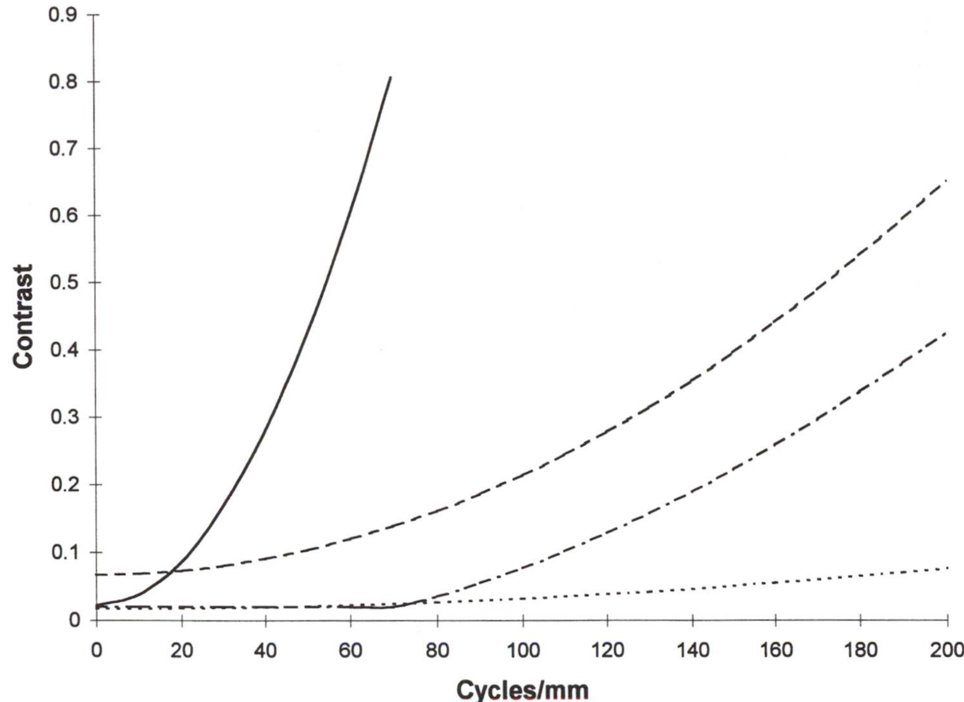

FIGURE 4.54 Aerial image modulation (AIM) threshold curves for several different photographic films: ——, high-speed color; – – –, high-speed B&W; ······, very slow B&W, ·—·—·, low-speed B&W.

order that the target be resolved. The contrast of the object is included by multiplying the lens MTF curve by the appropriate contrast ratio of the object. Conversely, the threshold modulation may be divided by the contrast ratio to obtain a family of threshold curves, as indicated in Figure 4.55. Reading of the intersection of the lens MTF curves with the threshold curves can then provide a family of information about the capabilities of the lens to provide resolution of various contrast targets.

A more general definition of image quality is the subjective quality factor (SQF) (Granger and Cupery 1972). This factor is based upon the total space available for image modulation above the noise threshold of the detector. Many factors can be included, such as the MTF of the eye and the visual system when examining an image at a particular scale. The general concept is to compute the integral of the distance between the lens MTF and the threshold over the spatial-frequency region of interest. The image scale may be included by using the product of the lens MTF and the anticipated MTF of the eye and visual system. This value is then a relative measure of the acceptability of the image to an observer.

Subjective quality factors such as this are demonstrated to be effective through statistical correlation of the quantity with observer judgments of

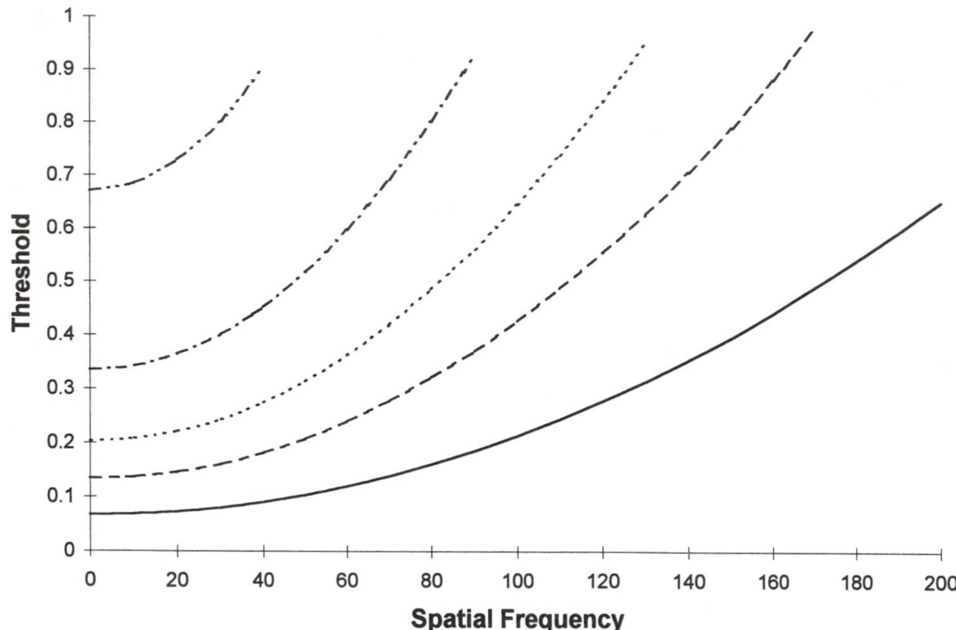

FIGURE 4.55 Aerial image modulation threshold curves for one selected film but with different object contrast values: ——, 1; – – –, 0.5; ······, 0.33; · — · — ·, 0.2; —· · —, 0.1. The MTF for the lens is plotted on these curves. The intersections of the MTF plot with these thresholds provide an estimate of the variation of resolution with object contrast.

image quality. For a lens designer, the main use is to provide a measure of the sensitivity of the user's perceptions to the design parameters of the lens. In almost all cases, the requirements for a design will be expressed in terms of a required MTF at a specified spatial frequency, although, of course, other summary measures may be preferred by the customer for the lens.

This chapter provides some basic ideas about the interpretation of the image quality from a lens. In many cases, the designer will be given a goal for the MTF or the energy concentration (usually after application of manufacturing tolerances). For special systems, such as infrared tracking systems or industrial automated inspection systems, the detection threshold requirements will vary. In almost all cases the capable designer will make his or her job considerably easier by extending these techniques to understand the customer's problem in designer terms.

4.5 Sampled Data Systems

Sampled data systems using two-dimensional CCD detector arrays are the heart of most modern imaging systems. The information capacity and image quality from such sampling array detectors continues to grow each year, but has yet to match the full quality and detail of photographic materials. Of course, many cameras using photographic film will continue to be a primary image recording medium for consumer memory recording and in graphic arts. The basic methods of describing the requirements for array detectors and photographic materials are quite similar. Threshold curves for resolution or image acceptability can be developed in a manner similar to that described in section 4.4. In most cases, this will be sufficient for the lens designer in setting the goals for the imaging system being designed.

The major difference between these arrays and a continuous detector, such as photographic film or the human eye, is that the data is recorded as a set of spatially discrete pixels. The image, when displayed, also appears as a set of pixels which are examined as a discrete set by another optical system, usually the human eye. The image thus contains spurious information, which is related to the pixel dimensions. Many books and papers have been written about the display of sampled information; the comments in this chapter are intended to be sufficient for the lens designer to comprehend the issues, and include this information in the design process.

The first concept is that of aliasing, which is the introduction of spurious information in the recorder image due to the pixel sampling. This aliasing can be reduced or eliminated by replaying the image through a suitable low-pass filter to remove the aliased frequencies. Suitable shaping of the point spread function in the display can also mitigate the visual effect of aliasing. Another approach is to modify the point spread function with a spatial "antialiasing" filter which more or less provides a duplicated representation of the basic spread function with the periodicity of the sampling detector array. This spatial filtering is accomplished via a birefringent stack which is usually located just following the lens in image space. The details of this device are described elsewhere (e.g. Greivenkamp 1992). The major requirement upon the lens designer is to include space for this filter by including a surrogate block of glass of appropriate thickness and index of refraction into the lens design. This type of antialiasing filter is most commonly used in color CCD video cameras, and can be seen in one of the zoom lens examples in Chapter 7.

The MTF associated with a discrete array detector is determined by the individual pixel shape and dimensions. The lens MTF is multiplied by the detector MTF to obtain the net MTF for the system. Image reconstruction is possible in the same manner as described in section 4.4. In most cases, the image reconstruction process is the responsibility of the customer, and the lens designer is requested to meet certain MTF requirements, or may be given a set of threshold modulation curves and instructions to provide certain intersections with these curves.

In many cases, the customer will not be worried about the details of the MTF but will be interested in the percentage of the incoming energy from a target element that is collected by an individual detector element. This is most common when the optical system is a tracking system in which image position and signal to noise ratio is the important information. The encircled energy described in section 4.4.1 can be used. Alternately, some computer programs permit the designer to define a particular size of detector box for calculation of the energy concentration.

An example of the usual specification of this sort will be that "70% of the energy entering the aperture from a point object will be contained within a 20 μm diameter." The designer will have to evaluate the energy concentration from the design, and then include a tolerance for errors in the system, such as focus shift. This is an easy image quality requirement for the designer, and usually quite unambiguous. The designer can simply accept such an easy statement, or can evaluate the problem to determine whether this is a sufficiently good description of the requirements.

The complete model of the image collected by a detector array and displayed to an observer is possible. More specific information, such as the details of contrast and signal-to-noise ratio in an image to be used in a specific application such as targeting or product inspection can be carried out. The principal information required from the lens design is the MTF of the lens for the spectral conditions, field angle, and focus position that are applicable. It is important for the designer to determine whether limited information, such as the MTF in the usual two-directions is sufficient, or whether a full two-dimensional image model is required.

REFERENCES

Barakat 1962 Barakat, R. Computation of the transfer function of an optical system from the design data for rotationally symmetric aberrations. Part I. Theory, *J. Opt. Soc. Am.* 52, 985

Barakat 1980 Barakat, R. The calculation of integrals encountered in diffraction theory, Chapter 2 in *The Computer in Optical Research*, ed. Frieden, Springer-Verlag, Berlin

Gaskill 1978 Gaskill, J. D. *Linear Systems, Fourier Transforms and Optics*, Wiley, New York

Goodman 1968 Goodman, J. W. *Introduction to Fourier Optics*, McGraw-Hill, New York

Granger and Cupery 1972 Granger, E. M. and Cupery, K. N. An optical merit function (SQF) which correlates with subjective image judgments, *Phot. Sci. Eng.* 16, 221

Greivenkamp 1992 Greivenkamp, J. E. Color dependent optical prefilter for the suppression of aliasing artifacts, *Appl. Opt.* 29(5), 676

Holmes and Avizonis 1976	Holmes, D. A. and Avizonis, P. V. Approximate optical system model, *Appl. Opt.* 15, 1075
Hopkins 1956	Hopkins, H. H. The frequency response of optical systems, *Proc. Phys. Soc.* B, 69, 562
Hopkins 1983	Hopkins, H. H. Canonical and real space coordinates used in the theory of image formation, Chapter 8 in *Applied Optics and Optical Engineering,* ed. Shannon and Wyant, Academic Press, New York
Hufnagel and Stanley 1964	Hufnagel, R. E. and Stanley, N. R. Modulation transfer function associated with image transmission through turbulent media, *J. Opt. Soc. Am.* 54(1), 52
Marcehal 1947	Marechal, A. *Rev. d'Optique* 26, 257
Marechal and Francon 1960	Marechal, A. and Francon, M. Diffraction Structure des Images, *Rev. d'Optique*, Paris
Minnick and Rancourt 1968	Minnick, W. A. and Rancourt, J. D. Transfer function calculation techniques for real optical systems, *Proc. SPIE* 13, 87
Scott 1964	Scott, M. The present state of the art with regard to detection and recognition, *Appl. Opt.* 3, 13
Shannon 1970	Shannon, R. R. Some recent advances in the specification and assessment of optical images, in *Optical Instruments and Techniques 1969*, ed. Dickson, Oriel Press, Newcastle
Shannon 1994	Shannon, R. R. Optical specifications, Chapter 35 in *Handbook of Optics*, McGraw-Hill, New York

5 DESIGN OPTIMIZATION

It is not essential that a designer understand all of the details of the process by which a design program carries out the optimization. Successful designers will, however, understand the principles and options that are available. Much time and effort has been expended by program developers to make the process as bulletproof and transparent to the user as possible. The past few years have seen an incredible improvement in the ability to control the modification of lenses by a program, and to explore new regions for solutions.

A basic comprehension of the important issues and procedures used is needed by any successful designer. This chapter provides enough insight to permit the designer to make the decisions necessary, but does not provide enough information to write design optimization programs. For detailed information the reader is referred to papers by Levenberg (1944), Wynne (1959), Rosen and Eldert (1954), Spencer (1963), and the summary by Kidger (1993). Discussion of newer techniques for optimization are found in papers by Kuper, Harris, and Hilbert (1994), Forbes and Jones (1993), and, of course, the various program manuals.

Optimization consists of adjusting the parameters of a lens to meet as closely as possible the requirements placed on the design. Current design programs have achieved a high degree of sophistication, and can rapidly search the design space for the closest approach to the design goals. The process of optimization requires the selection of a starting point and a set of variables. The goal for the design must be interpreted as a merit function in terms of allowable aberrations, usually involving both aberration coefficients and rays. The merit function may also contain lens parameters, such as weight or size, that must be held within stated targets. The merit function is evaluated at each stage of the design to monitor the progress of the design. The magnitude of the merit function is basically the square of the distance from the current aberration correction to the desired aberration function in whatever space the designer needs to solve the problem. Reducing the magnitude of the merit function should indicate that the design is closer to the desired solution.

Since the aberrations are generally related in a quite nonlinear manner to the variables, finding the direction to move toward a solution is not simple. The merit function is also usually a simplified abstraction of all of the requirements on the lens. Therefore there may exist several solutions with similar merit function values that are quite different in actual construction parameters or in image quality. Conversely, similar image quality may be obtained from systems with quite different merit function values.

Any lens must exist within some boundaries of space, weight, material choice, or other properties. In all practical design problems, cost ($, £, ¥, or otherwise) is also either a design goal or a target. In many cases the lens must operate within a specified environment, or fit to a particular mechanical interface. Each of these properties can be added to the design goals or boundary conditions. The sensitivity of the design to fabrication errors can also be used in determining the direction of the optimization process.

The optimization process seeks to locate an acceptable configuration within prescribed boundaries of the lens parameters that meets the image quality goals. In all cases, the variables are changed, derivatives of the merit function constituents are computed, and the resulting equations solved to determine a set of variables that should reduce the value of the merit function. The nonlinearity of the relations between variables and merit function constituents indicates that the distance of each successive step in variable space should be the smallest possible to move toward improvement in the design. Therefore a process called damped least squares is used to encourage the computer program to find the closest useful solution.

This process works quite well, but has the problem that a better solution may exist at another location in variable space. Therefore methods of exploring the design space by methodical perturbation of the parameters into different design regions have been developed. These global optimization processes take several forms, but the goal is to produce a library of possible solutions that can be searched to locate the best design. The use of global optimization is subject to the use of an appropriate merit function to ensure that each solution that is located is a likely candidate to meet the image quality requirements.

The general process of computation begins with definition of the variables, along with boundary specifications. The merit function is constructed from sets of aberrations and ray displacements, along with system requirements. Each variable is changed by a small amount and the change in each of the aberrations in the merit function computed.

These changes are used to construct a two-dimensional derivative matrix each element of which is the ratio of the change in the aberration to the change in the variable. This matrix is used, usually in a set of normal equations to define a least mean-square change in the variables to meet the aberration targets. The matrix is solved or inverted to obtain a set of suggested derivatives defining a direction of change for each of the variables.

The intrinsic nonlinearity of the design problem then suggests that the solution be projected along the indicated direction in variable space, until a new minimum of the aberrations along that direction is located. A value for the new merit function is evaluated.

The process begins again, with a new calculation of the derivatives at the new solution location. The process is repeated again, and iteration is stopped when a solution within the aberration residual tolerances is reached. If no solution is found, the iterations proceed until the solution stagnates, and the merit function does not change, or when the designer chosen maximum number of iterations has been carried out.

The designer observes the general course of the solution. Using judgments as to the rate of improvement and other indicators about the state of the design, the designer will make an alteration in the variables or the merit function, and return to iterating.

The designer also has to be aware of the possibility that the request made for the design is not physically achievable with the variables that are being used. The optimization procedure itself does not tell the designer this. It is necessary to examine the convergence and behavior of the process to ascertain when stagnation is due to physical limitations and not just a poor choice of variables or aberrations.

5.1 Variable Definition

The starting point for the design is usually a layout containing the anticipated number of surfaces, spacings, and glass choices. These are obtained from the initial considerations of the first-order properties of the lens. In some cases the starting point may be an existing design that is to be modified to meet a new set of requirements.

5.1.1 CHOICE OF VARIABLES

Any parameter describing the lens could be used as a variable. Usually only a subset of the available variables is used in order to maintain some control over the properties and configuration of the lens. Some variables may be used as specific paraxial (or real ray) solves to control certain first-order properties of the system. These become indirect variables and indicate that the system derivatives that are calculated are subject to certain side conditions. Since these solved variables do not need to be explicitly computed, proper use of solves can speed up the design process. In multiconfiguration problems, such as zoom lenses, there are global variables which must maintain the same values in all configurations, while some variables may be different in the various configurations. All variables are naturally subject to some boundary limits imposed by physical or economic reality.

The most important variables are the curvatures of the surfaces. Usually the designer will elect to use the "radius of curvature," which sounds redundant, but is easier to visualize, as it is the physical quantity that will be measured in building the lens. The curvature determines the first-order power, and consequently the path of each bundle of rays through the lens. The initial values selected for the curvatures should be reasonable, in order that light makes it through the lens without total internal reflection. The allowable values for the curvatures should be bounded to remain within an acceptable region. Usually this is such that the ratio of the diameter of the surface to the radius of curvature is much less than unity.

There are physical limits that define the range that may exist for the curvature. For example, the radius of curvature of a spherical surface cannot be less than the semidiameter of the surface. Because this quantity determines the power of the surface, requirement of an optical power to lie within some range on a certain glass will result in limiting the available values for the curvature.

When setting up the lens, it is frequently appropriate to use a paraxial ray angle to define the surface curvature. Such angle solves are very important in design in that the focal length or final magnification of a lens can be established by use of a paraxial marginal ray solve on the last surface. A methodical way of bending a lens element is set by using a marginal ray angle solve on the second surface of the element. Variation of the first surface will modify the lens shape while keeping the lens power constant. Additional solves can further constrain the meaning of the available variables.

Other variables are available on each surface that can be used to solve particular problems. The surface may be aspheric in many ways. The simplest is to use a rotationally symmetric aspheric, which will maintain the paraxial power of the surface. Changing the surface shape will change the aberrations that are introduced while maintaining the system power. As described earlier, there are many possible forms for the aspheric. The most frequently used is the polynomial aspheric. The logic is that the aberrations come in orders described by a polynomial. Therefore alteration of the surface shape by a given power term will principally change the aberration content of that power. For spherical aberration, this is reasonably true. The addition of higher-order terms permits correction of higher orders of the intrinsic spherical aberration produced at the surface. The effect on other aberrations is not quite as clear because the effect depends upon the passage of the chief ray through the surface.

Addition of aspheric terms is effective only if appropriate aberration terms are used in the merit function. Therefore specification of variables should only be done with some thought in mind as to the purpose of the added surface variable.

Other potential variables may be used in modifying the contribution of surfaces to aberrations. Diffraction gratings may be added to the surface to modify the transmitted or reflected wavefront. There are many flavors of diffractive or holographic optical structures that may be added to a surface. Each of these has physical limitations as to what it may accomplish, and is subject to the need to ensure that the component serves a useful purpose in the design.

The next type of variable is the separation between optical surfaces. This can be the thickness of the element, or the airspace between the elements. In some cases, the thickness can be zero, indicating the intention to contact or cement surfaces. In general, if left unbounded, lenses will usually expand to fill all of the available space during a design. Therefore the thickness variables should always be bounded. The aberrations will usually change slowly with changes in axial thickness, so thickness may or may not be a useful variable. The axial thickness of glass elements should be such that the elements are buildable, usually in the range of one-tenth to one-quarter of the diameter. However, the edge thickness of an element determines the feasible central thickness. Most design programs include a feature to constrain the element thickness so that a nonzero edge thickness is maintained. For negative lenses, the opposite is true, the central thickness must be maintained reasonably small to avoid extreme edge thicknesses.

The separation of elements is usually an effective variable. As seen before, the aberrations introduced by an element change markedly with separation from the stop. Usually it is necessary to introduce some boundary conditions to encourage the separations to stay within some reasonable range. When elements come close together, the allowable axial separation must be such that the edge separation is no less than some specified value. Therefore the boundaries that are allowable for separations and thickness are intimately related to the curvatures bounding those separations.

In most cases, the goal of the design is a symmetric lens that produces as symmetric as possible image quality across the field. A construction parameter that can sometimes be profitably used as a variable is the tilt or decenter of the surface. Small offsets will be used later in determining the sensitivity of the lens for use in establishing assembly tolerances. Use of a component with a large tilt or decenter introduces an offset in the aberrations, that can be matched with another offset of another element. This can be used to produce a good image from an unobscured reflecting system. Sometimes the deliberate use of tilts and decenters can be used to achieve particular distortion or magnification effects.

The optical properties of the materials used in a lens can obviously be variables. Usually the optical glasses used will be established at the beginning for cost, delivery or environmental reasons. For special apochromatic applications, the selection of the glass may be dictated by basic physical parameters. In many cases, the glass choice may be open, and the computer optimization process may be permitted to search for the best choice. The glass, or other refractive material, will have three basic variables, the index, dispersion, and partial dispersion. As seen from the discussion of optical glasses in section 2.3.1, only the first two are actually free and useful variables. The physical properties of optical glasses require that these glass variables be changed only within some allowable boundaries on the map of glass properties.

The finite number of available glasses leads to some special conditions on making glass a variable. Although the glass optical properties may be allowed to vary continuously within a specified boundary, the result will be a glass that is unavailable commercially. These materials must be replaced by the closest available material and an optimization run completed with the specified material in order to have a physically viable lens. Optimization programs include all of these possibilities as optional ways of handling the material variable.

The analogy to aspherics in refractive materials is gradient index. A glass can be represented by specific constrained sets of variables of index with position. A general gradient may be used, or selection made from a specific family of gradients.

A similar comment about fitting to available supplies can be made for surface curvatures in some applications. It may be desirable to fit the surface to a set of available test glasses. The optimization process is similar, in that the closest test glass radii are chosen, and then the remaining variables, usually only the thicknesses and separations, are used to carry out final optimization. If the result is not satisfactory, a systematic setting of one or more curvatures to a continuous variable will be necessary. Another case in which the curvature may be bounded is if a surface is near-plano. Cost indicates that substitution of a plane surface would be desirable, and the designer should always investigate this possibility.

There may be some well-hidden boundaries upon the curvature of surfaces. The control of ghost images or of narcissus (reflection of the detector back onto itself) in infrared systems places some subtle limits upon the allowable distribution of curvatures. Maintenance of the distribution of curvatures to permit multiple blocking of lenses during fabrication may also be required.

There is another concept to be considered in selecting variables for use in a design. It is possible to set up a lens in which two (or more) variables have almost the same effect upon the aberrations. The matrix describing the relation between the variables and the aberrations becomes "ill conditioned." This is familiar when solving simple linear equations in which the solution is indeterminate because two variables are closely correlated. These variables can fight for domination of the solution, and either very slow or oscillatory progress toward the goal is observed. The designer should be conscious of this possibility, and should observe the trend of variables occasionally during the optimization process.

Finally, the location of the focus surface can be a variable. The discussion of aberrations in Chapter 3 references the aberrations to the paraxial location. In most cases, it is shown that the best focus is located at some offset from the paraxial focus. Thus the location of focus can be a variable. If this is used, the merit function must contain some constraints, perhaps not explicit, that will avoid continuous wandering of the focal surface.

5.1.2 REPRESENTATION OF VARIABLES

The variables can be represented as their corresponding physical quantities. In some cases the optical effect may be a useful parameter to use. The angle solve converts a variable into a constrained variable that can take on values determined by a specific mathematical condition. Thus the surface curvature will change, but only in response to other changes in the optical system. A similar effect can be accomplished by introducing the desired values for the paraxial quantities into the merit function, and using the variable directly. The result may be the same, but the course of progress to a solution may be quite different.

5.2 Boundary Determination

5.2.1 VOLUME LIMITS

The most common boundaries on a variable are those in which the specific variable must remain within some stated limits. For example, the thickness might be required to be greater than 1 mm and less than 30 mm. Sometimes this is absolute, as when a mechanical structure must be fitted into a lens.

More often, the designer would like to have the variable boundary maintained, but would like the actual value of the variable to wander around to see if there is a better solution in the neighborhood. In that case the boundary might be described as a target for the design and added to the merit function as a desirable aberration correction.

In general, all variables should be bounded in some manner. The designer should be willing to relax the boundaries to explore adjacent regions of solution. Some programs provide for "elastic" boundary conditions which permit temporary violation of the limits followed by correction back into the allowed region.

5.2.2 PRACTICAL LIMITS

The setting of boundary conditions is further complicated by the existence of practical limits on variables. Curvatures may not exceed reasonable values that can be fabricated, usually a ratio of radius of curvature to diameter of greater than four or five. In some cases, this ratio will be established by the need to block multiple elements at one time during fabrication.

Thickness of elements, minimum length of airspaces and the weight of elements are usually limited for practical reasons. The depth and slope of aspheric surfaces are always subject to very practical limits by fabrication and testing. The choice of materials is always limited by availability and the expected environment.

5.3 Merit Function Definition

The merit function is composed of a set of defects or aberrations that are to be corrected or minimized to certain values. The choice of the aberrations in the merit function is extremely important in defining the goals and course of design of the lens. Because it is generally physically impossible to simultaneously set all aberrations in a lens to zero, the choice of the aberrations and the selection of the targets for these aberrations is a critical part of the design process. There are cases where a specific set of aberrations can be corrected to zero. The residual aberrations will then be a measure of the degree of correction of the lens, and must be included in a rational merit function if it is to be a measure of the state of correction of the lens.

"Aberration" is a generic term to describe the possible list of defects that may be used in describing the required lens systems performance. The constituents of the aberrations may be drawn from many sources. The process of correction may involve absolutely correcting the aberration to a specified target. This is a specific constraint on the design. In the most usual case, it may require the minimization of the distance of an aberration from some target value, with some tolerance and weight of importance assigned to the minimization. In this book the term "aberration correction" is used generically to cover both choices of constraint or minimization. Minimization of an aberration with a very large relative weight is similar to a constraint, and the judgment as to whether to minimize or constrain is sometimes not clear.

The selection of aberrations and weights in a merit function can be carried out in detail by the designer, or it can be relegated to a default selection provided by the program provider. Both approaches are justified depending upon the requirements. Usually the most successful design will be a hybrid of both approaches to setting up a merit function. The default merit functions available in modern design programs are a result of considerable experience and experimentation over several decades. The basis for selecting the aberrations is an understanding of the nature of lenses and aberrations. The exact choice of rays and balancing aberrations is largely an empirical process. In most cases the default merit function alone will not be adequate to optimize a lens. In some cases, especially of well-defined lens types, the best merit function will turn out to be very close to the default merit function. These merit functions leave room for altering the extent and number of rays in the aperture, and setting of the relative weights and positions of the field angles. Knowledge of each design program's "canned" merit function is available in the program manual. In most cases, the designer will have to supplement this description with some experimentation.

In the following discussion, a logical progression of the construction of a general merit function will be discussed. The purpose will be to explain the nature of construction and modification of such a function. There are several ways to approach the problem. In this case the approach will be to build upon increasing complexity in describing the image forming process.

The simplest aberrations to be corrected are the paraxial requirements on the lens. The focal length, or the reciprocal of this, the power, is usually required. This can be inserted as an aberration. Alternately, the magnification may be important and needs to be the defining aberration for the power of the lens, presuming that the overall conjugate distance for the optical system is fixed as well. Another approach is to fix the object at whatever distance the program likes as optical infinity (or a prescribed object distance) and define the field angle. The paraxial image height could be used as an aberration that will fix the resulting focal length. The focal length could also be fixed by using a marginal ray angle solve on the last surface, which would fix the power (and the magnification) without actually explicitly incorporating the focal length into the merit function. Another approach is to carry out the design and then scale the lens to the desired focal length.

Each of these choices can be used to produce the desired result, but each will produce a different path in the design process. The choice depends upon the specific design problem. One warning, however; choosing two of these options simultaneously can lead to severe problems in finding a solution. If the two requests are competing, the matrix to be solved during optimization can become ill conditioned. This means that two variables have almost the same effect upon the aberrations, and the solution process can find a large number of essentially equivalent solutions, and will range over large variable changes searching for the best solution. In the case of simple linear equations, this means that the solution is undefined. In the nonlinear optical design case, it leads to a singular matrix and eventually leads to a stall or a divergence in the optimization process. The exact result will depend upon the choices made in programming the optimization algorithm.

The problems referred to in the previous paragraph often occur in as simple a case as setting the focal length target. Accidental construction of redundant or close-to-redundant aberrations of a more general nature can easily occur. Some of these may not be readily evident to the designer constructing the merit function.

Once the basic paraxial properties are defined, the designer has to select the set of actual image aberrations to be minimized or corrected. Basic to almost all lenses is chromatic correction. The discussion of first-order chromatic correction in section 3.1.1.1.2 demonstrates the requirements for glass selection. The aberrations to be corrected are the axial and lateral first-order chromatic aberrations. In most programs these are calculated as aberration coefficients. These may be added to the merit function simply by referring to them by some shorthand notation as PAC, FCHY, AXCL, or some other designation chosen by the programmer.

The correction of these quantities depends upon the provision of appropriate variables in the lens. Examination of the discussion in section 3.1.1.1.2 shows that not all combinations of variables will permit these aberrations to be corrected to zero. Furthermore, appropriate optical materials have to be specified in order that the correction be obtained. Using the glass as a variable can be useful, but it is evident that there are many glass choices that will provide similar corrections for just the first-order chromatic aberration. Several possible solutions may exist, and other factors will define the allowable glass regions.

The first-order aberrations are complicated by the presence of secondary residual chromatic aberration. Simple addition of a requirement to provide simultaneous correction at other than the two specified wavelengths will usually lead to problems. Section 3.1.1.1.2 shows that only certain glass combinations can be used to obtain secondary color correction. An attempt to use both specifications in a condition that cannot physically be met will lead to a stagnation (at best) in the solution process. In some cases this can lead to complete divergence of the solution process as the program attempts to find a solution that does not exist.

The Petzval field curvature is a third-order aberration, but usually is added to the first-order setup options for a lens. As shown in section 3.1.1.2.1, the Petzval curvature is fundamental to flat-field correction, and is determined by the curvatures in the lens. Since the power distribution among the elements is also determined by the same parameters, the inclusion of the Petzval in the merit function provides an immediate restriction to the possible range of variables. It is tempting to simply set the Petzval curvature at zero, and hope for later correction to zero of the astigmatism. This temptation should be resisted because of the relations between astigmatism, coma, spherical aberration, and stop position referred to in section 3.1.1.2.3. This will restrict the possible range of variables that are required for correcting the other aberrations. Additionally, the high-order aberration conditions discussed in section 3.1.2 show that some balancing against the high-order aberration will be required.

Choice of an appropriate target for the amount of Petzval curvature is a complex decision. The best generic statement that can be made is that the Petzval curvature should be chosen small enough that it is possible to balance the aberrations across the field sufficiently well to achieve the required image quality, but no smaller. In addition, the discussion of section 3.1.1.2.3 shows that there is a natural field curvature to a refracting surface, so that a basic negative or inwards Petzval surface is usually preferred. The usual choice is to start with setting the Petzval radius at three to five times the focal length, with a minus sign for the inward curvature. This can be directly specified in some programs, in others some mathematical gymnastics is required in order to define this Petzval criterion in terms of image blur at the edge of the field. On the other hand, the image blur description is useful to use in comparison with the intended residual image errors. Another choice is whether to correct the Petzval to a fixed value or to let it minimize to some value during design. In the latter case, the weight assigned to the Petzval will define the narrowness of the likely range of the design.

Once the first-order chromatic options and the Petzval have been defined the next step in design is to add the other aberrations to be corrected (or minimized). The third-order aberrations are next to be considered. These can easily be added to the merit function in most programs simply by using the names SA3, CMA, TAS, DIS3, or other choices as given in the program documentation. It is easy to set these as targets to zero. In most cases a solution is readily obtained. Examination of the solution usually indicates that it is not a good one in terms of total aberration content. Ray trace evaluation shows that there are residual high-order aberrations, as well as chromatic variation of the third-order aberrations that need to be balanced.

Development of a logical merit function can then proceed to construction of sums of the third- and fifth-order aberration coefficients to produce a balanced aberration residual. The fixed targets for the third-order aberrations are replaced by a combined aberration in which the sum of the third- and fifth-order aberrations is targeted at zero. The intrinsic aberration content of the lens is included in this improved merit function and the design then should be the best balance of the aberrations, at least of those included in the calculation, or at least to the extent that the fifth-order aberrations represent the actual aberration content of the lens.

If a solution is found, the image quality is examined in order to see if the design goals are met. Usually, there will be larger aberration residuals than are desirable. These may be due to a poor choice of balancing of aberrations, or due to the failure of the aberration coefficients to adequately represent the aberrations. For this reason, the aberration coefficients are generally not used in making up the final merit function used in design. A set of rays is selected which sample the aperture and field at key locations. Aberrations which describe the symmetry behavior of the lens are established by combining the ray intercepts (or the optical path differences) for the rays in an appropriate manner. These aberrations are weighted according to the importance of the aberration and the importance of each field point in the design.

The selection of the aberrations follows the understanding of the aberrations shown in Chapter 3. The axial image aberration balance can be determined by tracing a single ray at the edge of the aperture and correcting the ray intercept error to be zero at the paraxial image surface. Tracing a ray at about 0.7 of the aperture yields a measure of the residual aberration after edge ray balancing of the spherical aberration has been attained. Tracing two rays at the extreme wavelengths of the spectral region at 0.7 of the aperture height provides information about all orders of axial chromatic aberration. Correction of the difference in the ray intercept height of these latter two rays at the paraxial focal location to zero provides a balance for the spherochromatism. The actual selection of the ray heights can be used to adjust the aberration balance to meet an optimum balance for the aperture and spectral range. The best focus for the lens on axis will not necessarily lie at the paraxial location, but will be at a location that minimizes the aberration blur according to some desired image quality criterion. As has been seen in Chapter 3, the choice of the optimum aberration balance and focal position will change with the magnitude of the residual aberration.

It is instructive to examine the options involved in setting the aberrations at the zero field point. This will serve as a model for the approach that can be used in the more complicated off-axis case. There are many ways of specifying the optimum balance, just as there are many ways of describing the image quality. The usual goal is to provide a contrast value in the MTF at some spatial frequency that exceeds some threshold target. Since computation of the MTF at each step in the iteration is not economical, some model for setting the aberrations is usually chosen. Some simple models are the minimum RMS geometrical spot size or the minimum RMS wavefront error. Others are weighted combinations of the aberrations that better predict the MTF at specific spatial frequencies. All of these summary aberration statements assume that the best focus position has been located as part of using this criterion.

The designer input to setting up such criteria is the selection of the extent of sampling of the aperture and the weights to be applied to the aperture height (and, of course to the fields, to be discussed later). The optimization iterations are carried out. The image quality is then examined. If acceptable, it may be time to stop the design process. If not, the sampling and weights are adjusted in an attempt to improve the desired image quality. This is frequently an empirical process, in which the designer tries a few new values and examines the imagery. In most cases a rapid convergence occurs. Since the actual spectral weighting, aperture settings, and spatial frequency requirements are different for each problem, the designer needs to use experience in refining the design at this stage.

At this point some of the ART (Adjustment of Residual Terms!) in designing appears. It is the designer's responsibility to match the requests in the merit function to the needs of the customer. This has to be done in a realistic manner, including consideration of the physical limits of the imaging process.

The problem of defining an appropriate merit function becomes more complicated for off-axis images. The same principles apply as for the axial image, but the situation is complicated by probable asymmetry of the image. The best approach to use is to model the selection of rays and aberrations on the knowledge of the likely symmetry. Selecting rays to correct coma, astigmatism, and field curvature as shown in section 3.1.1.2.1 permits addition of these aberrations and includes all orders of error. The specification of ray heights in the pupil will change the emphasis of the balancing to use aberrations over different portions of the pupil. The setting of the sampling area for rays in the aperture is naturally established by the allowable vignetting. In using default merit functions for most programs, the specification of allowable vignetting automatically inserts this sampling, ensuring that the aberrations are evaluated over the functional portion of the aperture.

The most effective merit function for the beginning of the design is the one that uses the fewest rays to define the aberrations. Since ray tracing consumes most of the time in carrying out an iteration, this permits the most rapid exploration of design space. The minimum number of rays that need to be traced at each off-axis field point then would seem to be four, a chief ray, upper and lower meridional rays, and a sagittal ray. These rays can be used to construct the changes of aberrations with field. Addition of three more rays at about 0.7 of the aperture in the meridional and tangential direction would permit aberration constructions similar to those on axis.

The number of field points chosen to be used in the merit function will also influence the speed of computation and iteration. General discussion of aberration theory suggests that three field points are needed for any reasonable description of the imagery across the field. For very small field angles, and reasonable numerical apertures, only two field points, the center and edge of the field, may suffice. In general the selection of the center of the field, about 0.7 of the field and the edge of the field will provide a reasonable coverage of the aberrations for the merit function. The logic in this choice is that the usual limiting aberration for a lens is oblique spherical aberration, which varies as the second power of the field. The usual requirement is that the imagery be balanced to be as good as possible at the 0.7 field height, with some degradation allowable for the extreme field. Obviously this choice is generic, and does not take into account the intended use of a specific lens. Examination of the image quality over other portions of the field is usually desirable. Changes in the choice of the field locations for the merit function can be made after such an examination.

Most programs allow a specification of the weight attached to the aberration at each field point. In some cases the weight can be applied to the total x and y dimensions of the image or to the sagittal and tangential directions separately. This permits the designer to specify the balance of the aberrations and emphasize the correction over a region of the field. In a visual system, where the user will be able to move the pointing of the optical system to acquire an object in the center of the field, a very low weight can be applied to the field edge. In a photographic system where a picture may be enlarged from any portion of the negative, it is important to make the imagery as uniform as possible across the field. The adjustment of the field weights will also change the aberration balance that results. This selection of weights is another example where the judgment of the designer must be utilized.

The procedure followed in design is to carry out a series of optimizations with this minimum ray set, usually chosen as the default merit function. Examination of the ray plots and image quality for several weights upon the merit function components will yield an estimate of significant residual errors in the lens. The designer then adds to the basic merit function specific rays which address special aberration correction problems. Experimentation with the weights upon these added aberrations is necessary to ensure that the design will converge to a satisfactory solution. Examples of added ray aberration may be the upper or lower aperture rays which show specific aberration problems at the edge of the field. Frequently it is necessary to add some more field points, which are placed near the edge of the field in order to gain control of high-order field aberrations.

The design is carried out until no improvements in the merit function can be found. In order to ensure that the best possible solution has been located, the designer should then carry out a methodical search of design space by varying the starting parameters and letting the iterations take place. If the solution keeps returning to the same location, there is at least a general local minimum for the merit function. Another possibility is to employ the global search approaches available in most lens design programs. The basis for these will be described in section 5.4.2. It is important to note that the choice of components of the merit function drives the solution. A poorly constructed merit function will not ensure that a better image quality solution will be found during any global search activity.

The final design requires minor adjustments to the lens to meet the best possible image quality requirements within the constraints of the lens. In some cases it may be appropriate to add values proportional to the curvatures on specific surfaces in order to coax the design program to move the region of solution to one in which the curvatures are reduced, and the lens is expected to be more tolerable to fabricate. The goal may be to eliminate some elements by forcing the curvatures of some of the elements to become zero. This frequently works, but the designer needs to adjust the merit function to new expectations from the system with fewer elements. The designer may also have access to a library of test plates that can be used to reduce cost in manufacturing, forcing the lens to fit as many surfaces as possible to existing test equipment.

The preceding discussion is, of necessity, somewhat general. Each problem is a bit different. The logic that needs to be followed is evident.

5.4 Optimization Procedures

The iterative cycle used by lens design programs for optimization is divided into four parts. The first step is the computation of derivative increments linking the changes in the lens parameters to each of the aberrations. The second is the construction of a derivative matrix which can be solved or inverted to provide a vector in the variable pointing toward the desired solution, with a low probability of providing a divergent process. The matrix is then solved for a new solution vector. The last step is the determination of the best possible solution based upon the merit function that can be squeezed out without recomputation of the derivative matrix.

It was previously noted the relations between variables and aberrations are very nonlinear. All of the procedures used presume that there is some range of linearity over which the derivatives computed are linear and valid. Several techniques have been used to extend the range of this linear modeling by including some second-derivative information. Most procedures attempt to restrain the amount of change in each step so that the presumptions of linearity are somewhat maintained. However, such restraint leads to the possibility of the system being stalled or captured within a local minimum rather than a perhaps better global minimum. Recently, several methods have been developed to permit a global search to be carried out. There have been some demonstrations of success in this regard.

There are other issues of numerical accuracy, truncation of computations, and time saving that must also be taken into account in developing a successful optimization procedure.

5.4.1 LOCAL

5.4.1.1 Damped Least Squares

The earliest successful solution procedure is the damped-least-squares process. Most programs use this procedure or some minor variant of it as the core of the optimization process.

The value of the merit function is defined by the root sum square of the weighted aberration values

$$\psi = \sqrt{\sum \omega_j^2 (\alpha_j - \alpha_{jt})^2} \tag{5.1}$$

in which ω is the weight applied to the jth aberration, here called $\alpha_j - \alpha_{jt}$, the difference of the aberration from the target value. The state of correction of the system is then stated to be equal to the merit function value ψ. The goal is to make this function zero. The larger the value, the farther the lens is from a desired solution. The larger each weight, the more importance that is placed upon the specific aberration. The designer (usually along with the computer program) selects the sets of appropriate aberrations and weights. These are changed from time to time as the design proceeds in order to force the solution to proceed in a desirable manner.

These aberrations are perhaps better called "defects" because the components of the aberrations can be any quantity. It is typical to mix ray aberrations, optical path errors, paraxial and aberration coefficients, and construction parameters in a useful merit function. The importance of each of the defects is defined by the selection of each of the weights upon the aberration.

The merit function sum as shown above serves as a summary measure of the status of the design. The individual components of the merit function are used in establishing the equations that are used in carrying out the iterative optimization steps. The fundamental equations to be solved at each step in the process are

$$(\alpha_i - \alpha_{it}) + \left(\frac{\partial \alpha_i}{\partial x_j}\right) x_j = 0 \tag{5.2}$$

where the x_j are to be found. If the system were linear, then one step would yield a solution. In matrix form this can be written symbolically as

$$\boldsymbol{\alpha} + \boldsymbol{A}\boldsymbol{x} = 0 \tag{5.3}$$

where the bold type indicates that $\boldsymbol{\alpha}$ and $\boldsymbol{x}$ are vectors and $\boldsymbol{A}$ is a matrix relating the aberrations and variables. The merit function will be minimized if the changes $\boldsymbol{x}'$ are the solution of

$$\boldsymbol{A}^{\mathrm{T}}\boldsymbol{A}\boldsymbol{x} = -\boldsymbol{A}^{\mathrm{T}}\boldsymbol{\alpha} \tag{5.4}$$

This obtains the best mean-square solution of the equations, presuming that the derivatives in the matrix are actually linear. In practice, the relationships change with the length of the step in variable space, so that an improved solution is not guaranteed by this process.

It was suggested by Levenberg (1944) and Wynne (1959) that the equation should be damped to inhibit the region of variable change. This is accomplished by minimizing the altered merit function

$$\psi' = (\boldsymbol{\alpha} + \boldsymbol{A}\boldsymbol{x})^2 + p^2\boldsymbol{x}^2 \tag{5.5}$$

where p is a damping constant. The larger the value of p relative to the size of the merit function the smaller will be the change of variables in the direction of the solution. The solution of this new equation is

$$(\boldsymbol{A}^{\mathrm{T}}\boldsymbol{A} + p^2\boldsymbol{I})\boldsymbol{x} = -\boldsymbol{A}^{\mathrm{T}}\boldsymbol{\alpha} \tag{5.6}$$

where $\boldsymbol{I}$ is the diagonal unit matrix, indicating that the damping factor adjusts the diagonal terms of the equation to be solved. The process of optimization using this damped-least-squares technique is to use some value for the damping factor that permits progress toward the supposedly linear solution, but keeps the variables in the region of the current solution. Trial values can be used which minimize the size of the subsequent merit function. In this manner the quasi-linear predicted solution direction is used until it provides no more gain in the actual solution. The differentials in the $\boldsymbol{A}$ matrix are recomputed and the process repeated. This apparently simple process is quite effective.

The actual application in a lens design program is considerably more complicated. It is often necessary to solve some aberrations to precise values, such as maintaining a focal length, for example. Other aberrations cannot be solved exactly but need to be minimized in the least-squares sense. To accomplish this the procedure of Lagrange multipliers can be used. In this case, if the following equations are to be exactly solved:

$$\boldsymbol{Bx} = \gamma \tag{5.7}$$

Here $\boldsymbol{B}$ is a matrix containing a set of derivatives to be solved for the quantities in γ. The pair of matrix equations to be solved jointly are now

$$\begin{aligned}(\boldsymbol{A}^{\mathrm{T}}\boldsymbol{A} + p^2\boldsymbol{I})\boldsymbol{x} + \lambda\boldsymbol{B}^{\mathrm{T}} &= -\boldsymbol{A}^{\mathrm{T}}\alpha \\ \boldsymbol{Bx} &= \gamma\end{aligned} \tag{5.8}$$

where λ is a set of scalar multipliers that constrain the solution to the required exact values. These multipliers never need to be solved for explicitly, as only the new values of $\boldsymbol{x}$ are required. These exact solutions can be troublesome if used wrongly, as they may not actually be achievable, and will drag the equations, and the design, into a divergent region. An alternate, and frequently more desirable, approach is to minimize the conditions required exactly, but with an extremely high weight on those specific aberrations. The course of the design will be different under these two approaches because of the basic nonlinearity of the system.

The actual implementation of these solution procedures in any program requires some additional steps. The derivatives relating the aberrations and the variables need to be computed. Most frequently this is accomplished by using finite differences. The step length used in computing the derivative is important because of the probability of local nonlinearities. It would be most useful to use a derivative step that is comparable with the required variable change. Unfortunately, this change is not available prior to the computation. The usual process is to make a variable step that produces approximately the same amount of aberration change for each of the variables. In the case of curvatures, this would be accomplished by calculating the curvature step to provide a few wavelengths of optical path change at the edge of the clear aperture. Other derivative variable increments can be computed on a similar basis.

One result of that approach is to make all of the derivatives more or less the same size, or at least within a couple of orders of magnitude of each other. This will provide the most accuracy in equation solving. In some programs derivatives are computed for more than one step length, and the information so gained used in evaluating the region over which the lens will be more or less linear in nature. There are also hidden constraints in the variables. Although each variable is computed independently, the lens may have some paraxial solves on some surface or spacing so that the actual derivatives obtained are taken with respect to certain side conditions. This is actually a very potent advantage of finite-difference derivative taking, as the image plane can easily be constrained to be placed a specified distance from the paraxial image location, for example.

Derivatives can be computed using one of several formulae. While this may appear to be more rapid in computation, the probable cross-coupling between some variables is quite complicated to include in such computations. With present high-speed computers, the programming of finite-difference derivatives usually is superior to other apparently clever computation schemes.

Incorporation of second-derivative information into the computations is accomplished in some programs (Dilworth 1978). The multiple step lengths described above can lead to the development of a set of pseudo-second derivatives that may be used to alter specific damping parameters to introduce a sort of response to the second derivative changes. Actually, searching for an optimum damping parameter is usually the most efficient approach.

5.4.1.2 Orthonormalization

When a large number of variables are used, it is possible that some pairs of variables may be quite closely coupled in their aberration derivative behavior. When this is the case, the equations can become ill conditioned and very large changes in these redundant variables are required to make any substantial changes in the aberration correction. The lens can thus be forced well out of the region of any local linearity. Grey (1963) implemented an approach to avoid this by constructing new orthogonal sets of variables out of linear combinations of the basic lens variables, which provide a new set of equations in the orthogonal space. These new variables are orthogonal and independent over some range. The resulting equations are well conditioned and readily yield new solutions. The orthogonal basis set may be used several times to find new solutions until the improvement in the merit function is negligible. Then a new basis set can be constructed and the process repeated. This approach is basically a different organization of the damped-least-mean-squares process.

5.4.1.3 Steepest Descent Methods

Another approach to solving the equations is to seek a solution vector that will provide the maximum change in the merit function. This is somewhat similar to the orthonormalization concept except that there is no attempt to find an orthogonal combination of variables. Instead the direction of change is calculated for a large number of combinations of the variables. The vector providing the steepest drop in the merit function value is chosen, and followed until no reduction in the merit function is noted. The search begins again for a new steepest descent direction and the procedure is followed again.

The procedure is not very efficient, as nonlinearities require that the step length taken at each cycle be very small. The number of steps to make a change is very large, and the procedure is very slow. Besides, the approach can become entirely bogged down by a local minimum of the merit function.

5.4.1.4 Adaptive Methods

Adaptive methods are basically the constant changing of the targets and the weights during the solution process. The process of solution is similar to the least-squares method. The trick used to maintain the solution vector within a local linear region is to adjust the targets for the aberration correction to be only a fraction of the distance from the current state of the system to the desired state. It is reasonable that moving part-way to the goal will cause fewer nonlinear perturbations in the solution. A series of steps are made, moving the aberration targets toward the desired regions as each local solution is obtained.

The mechanics of this approach were described by Glatzel and Wilson (1968). The approach has been used successfully by a number of designers. As in the maximum gradient method, the discovery of new, perhaps better, minima is not likely to occur. Use of the procedure will require systematic changes of the starting point to explore the design space. Most design programs which allow the development of specialized macros to run the program can be used to apply this technique.

There have been a number of other approaches to solution that have been tried over the past few decades. Close variants of damped least squares have been the survivor of much experimentation and trial of new approaches.

5.4.2 GLOBAL SEARCHES

The problem of becoming stuck in a local minimum of the merit function, and missing a nearby (or distant) much lower minimum has always been a concern in lens design. In the past few years, several approaches have been tried to avoid this problem. Some have been reasonably successful. The basic approach in globally searching for a solution is to take a giant step in some direction, and then search locally for another solution, which may or may not be a better one. The resulting lens is saved and compared with the original design. In most cases, only a better solution, in terms of merit function value, is retained, and worse solutions are discarded. Traditionally, the skill of the designer has been the driver in determining the direction of search for a solution. New additions to lens design programs have been made to allow the program to make many of the decisions in searching an extended design space.

There are several approaches to global searching. Probably the most inefficient is to take a random step in the variables, and see if a better solution is obtained, and then randomly search again. In some designs this can be instructive, as clever use of this technique by a knowledgeable designer can lead to identification of approaches to better solutions. Also not useful in most cases is a patterned scan of the design space. In any reasonable lens, with six or more variables, the number of trial solutions is enormous. In recent years, two developments have occurred which appear to have basically solved the global minimum problem in optical design. These approaches to global optimization are best used after the design has been tamed by the use of basic damped-least-squares optimization. The impact of the global optimization process on lens design is just beginning to be felt. Obviously, the quality of the answer is dependent upon the skill of the designer in stating the requirements of the design in an appropriate merit function, as well as selecting a reasonable starting point.

An important recent development by Kuper et al. (1994) uses a very potent search technique in which information about the existing solutions is used to predict profitable directions for the next step in the search. This systematic approach to searching the design space has been shown to lead to significant improvements in many cases of otherwise well-designed lenses. It may not produce any plausible new designs from a starting point in some cases, dependent upon the choice of the merit function being used.

Simulated annealing is another systematic approach of moving through design space. The general approach here is to perturb the starting solution, and only accept solutions with improved merit functions (Forbes and Jones 1993). By discarding any intermediate solutions which increase the merit function, the solution will progress eventually to the lowest point. Frequently quite successful, the process is necessarily lengthy, and requires that the starting solution be near another minimum in order to find it in a reasonable period of time. As in any global optimization process, the designer must be very careful about construction of the merit function in order to obtain a successful outcome. This approach has also been used successfully in fields other than optical design.

The successful designer will use these global-search approaches as an aid in exploring the possible design space. But, as will be seen in some of the problems examined in Chapter 7, the designer must use creativity in recognizing trends in the search. By this mechanism the designer can add or eliminate elements as required to reach a goal. The designer can also select designs with simpler configurations or loosened tolerances that will eventually result in a superior outcome. In almost all cases, simply allowing the global-search approach to run loose without guidance will not ensure the best possible solution. However, the probability of finding a good solution has been greatly enhanced through the availability of these techniques.

REFERENCES

Dilworth 1978 Dilworth, D. C. Pseudo-second-derivative matrix and its application to automatic lens design, *Appl. Opt.* 17, 3372

Forbes and Jones 1993 Forbes, G. and Jones, A. Towards global optimization and adaptive simulated annealing, *Proc. SPIE* 1354, 144

Glatzel and Wilson 1968 Glatzel, E. and Wilson, R. Adaptive automatic correction in optical design, *Appl. Opt.* 7, 265

Grey 1963 Grey, D. S. Aberration theories for semiautomatic lens design by electronic computers. II. A specific computer program, *J. Opt. Soc. Am.* 53, 677

Kidger 1993 Kidger, M. T. Use of the Levenberg-Marquardt optimization method in lens design, *Opt. Engng.* 32, 1731

Kuper, Harris, and Hilbert 1994 Kuper, T. G., Harris, T. I., and Hilbert, R. S. Practical lens design using a global method, *International Optical Design Conf. 22*, 46, Optical Society of America, Washington, DC

Levenberg 1944 Levenberg, K. A method for the solution of certain non-linear problems in least squares, *Quart. J. Appl. Math.* 2, 164

Rosen and Eldert 1954 Rosen, S. and Eldert, C. Least-squares method for optical correction, *J. Opt. Soc. Am.* 44, 250

Spencer 1963 Spencer, G. H. A flexible automatic lens correction procedure, *Appl. Opt.* 2, 1257

Wynne 1959 Wynne, C. G. Lens design by electronic digital computer, *Proc. Phys. Soc.* 74, 316

6 TOLERANCE ANALYSIS

No design task is complete until the tolerances have been evaluated. The lens design alone, while interesting to the designer, needs to be fabricated within some range of realistic tolerances in order to be of interest to the user. A full tolerancing of a lens may become a more difficult task than the original design of the lens. The specified tolerances must be sufficient to ensure that the image quality goals are likely to be met. The tolerances required for fabrication are the major drivers in determining the cost of actually building and assembling a lens.

Before proceeding to carry out tolerancing the designer must decide upon the allowable degradation in the image. Despite some claims to the contrary, no system will ever be built absolutely perfectly with no deviations from the specified parameters. Therefore the imagery produced by a real system will differ from that of a perfectly fabricated system. During the design stage the designer should have considered this problem and designed into the lens sufficient margin that some errors in the lens parameters can be allowed. The balance between design margin and allowable tolerance loss is frequently an important economic issue. It is also important to remember that some margin usually must be assigned to operational considerations, such as setting the focal position, or to environmental effects such as temperature changes.

There are three steps to actually assigning tolerances to the system parameters. First, the relations between errors in the parameters of the lens and defects in the image are obtained. This information is then used to construct a table of the allowable changes in each lens parameter that will keep each of the defined image errors within an acceptable range of values. Then the allowable tolerances must be distributed among the parameters with some regard to the difficulty in fabricating and assembling the lens. The final allowable error ranges of the parameters must be communicated in a useful and verifiable form to those responsible for fabricating the lens.

Tolerancing is inherently statistical; the statement of the tolerances is a bet that if the lens is fabricated within a specified "three-sigma" error range, the result will be 98% (or more) likely to be an acceptable system. There is the added complication that there may be offsetting or compensating errors produced by groups of parameters. For example, leaving focus setting as a compensator greatly loosens the tolerances within the lens. This effect can also be used to loosen the tolerance specified on an individual parameter by using judicious adjustments to overcome cumulative error buildup during assembly. In lenses there is an additional complication that a simple error in a mechanical dimension, such as a radius of a surface, will have an effect on the image quality that varies with the field angle. Therefore decisions must be made about the relative importance of image errors at various points across the field. Setting the tolerances wisely and economically requires knowledge of the ability of the fabrication shop and an understanding of the production processes to be used.

6.1 Tolerance Definition

Defining the problem is always the first step. The designer must decide what are the significant image quality functions that need to be considered, and how much variability from nominal will be allowed. This will directly influence the size of the tolerances eventually placed upon the variables.

6.1.1 IMAGE ERRORS

The image errors used in obtaining tolerances are frequently based upon the merit function used in designing the lens. The simplest option is to require that the merit function not change by more than a given percentage due to likely errors in the lens parameters. Other factors describing the image may be limited by specific numerical considerations, such as distortion or boresight pointing error. Image-quality-specific characteristics that can be used in establishing tolerances include the MTF, energy concentration, field curvature, and depth of focus. In many cases, these detailed image characteristics are summarized in terms of the allowable RMS wavefront error for the lens. This provides a one-dimensional parameter to serve as a tolerance driver.

The fundamental requirement upon the lens is that it produce a certain level of image quality. This subject was discussed in detail in Chapter 4. There it was indicated that specifying image quality can sometimes be quite difficult. Ensuring that the qualitative impression of an image is acceptable or pleasing to a user can be quite arbitrary. For this reason the designer establishing tolerances is usually faced with maintaining some physical description of the image within an acceptable level.

The most frequent requirement is that the MTF be maintained above some specified value at one or more spatial frequencies. Some qualifiers on this statement are usually needed. For example, the requirement might be that the MTF be greater than 0.35 at 30 lines per millimeter across 90% of the field. This is satisfying, except that the interpretation of 90% needs to be stated. Is it area or linear dimension? The allowable focal adjustment over which this tolerance must be maintained is also important. Some permissible depth of focus is needed to establish the method of setting the focus. It may be that the actual requirement is that some focal location can be found in which the imagery meets the requirement across 90% of the area of the format. In some applications (probably most applications!) the optical system can be pointed at the object, so that the specification might better include the possibility of meeting the requirement in the central half of the area of the field with some specified allowable loss in the outer half of the field. These details in specification setting are important because of the intrinsic field dependent nature of the aberrations that need to be corrected. The design battle can be lost, or at least made very difficult, at this point if an unrealistic specification is accepted.

Setting specifications for a lens system which is supposed to be close to diffraction limited is actually quite simple. As noted in Chapter 4, any aberrations reduce the OTF below the aperture-determined diffraction limit. The specification is then set by stating the allowed degradation of the system. The RMS wavefront error is a one-dimensional guide to this degradation, as long as the RMS error does not exceed about 0.2 wavelengths. The model described in section 4.2.2 is useful here.

Lenses that are required to have lower levels of image quality are more difficult to tolerance. The possible combination of aberrations introduced by changes in the lens parameters can provide a large number of possible aberration conditions. It is likely that some errors will actually improve the image in some portions of the field. Such complications become difficult in that the usual need is to maintain the image quality above some specified level but within an allowable range of focal plane adjustment. Usually only one choice of focus position and tilt adjustment of the system is possible.

6.1.2 FABRICATION LIMITS

The usual assumption is that tolerance errors produce only small changes in the lens, and this produces small errors in the aberrations. A change table is generated which links the change in every parameter to a change in each component of the merit function. The process of tolerancing is the inverse

TABLE 6.1

Item	Tolerance	Expressed as	Measured as
Radius	±Length	Fringes	Fringes
Center thickness	±Length	Length	Length
Edge thickness	Length	Length	Result of edging diameter
Axial spacing	±Length	Length	Mount-determined separation
Index of refraction	$\pm n_d$		Melt data
Dispersion	$\pm V_d$		Melt data
Tilt	±Arc seconds	Angle	Runout
Decenter	±Length	±Length	Runout
Irregularity	Fringes	Surface fringes	Fringes
Asphericity	±Coefficient	Surface fringes	Fringes
Wedge	±Angle	Angle	Fringes
Flatness	±Curvature	Fringes	Fringes

of design, in that the change table is used to determine the maximum change of each of the parameters can be made that will not change the merit function beyond a specified value. This value of the change of the parameter is then the sensitivity of the parameter.

Tolerances can be established by assuming that each parameter acts individually, and that the likely total error of the system is the result of the statistical addition of each independent error. This is not strictly true for most lens systems, as compensators for the aberration can usually be found. These compensating effects, when properly accounted for, can greatly loosen the level of required tolerances.

A simple example is the tolerance upon the back focal length of a lens. If the lens is to be placed into a mechanical mount such that a fixed distance must be maintained between the lens and the image, a very tight tolerance on the back focal length exists, as the image position must be maintained within the depth of focus for the lens. A good mechanical design will permit some adjustment of the final focal surface, so that the best focus can be accessed after assembly of the lens. Then the tolerance upon the back focal length is greatly loosened, and may possibly be as great as the allowable range of focus in the mechanical design. In that case, other requirements, such as the change in aberrations with error in the lens will become significant in determining the allowable back focal length.

Table 6.1 lists some of the lens parameters that commonly are toleranced, and the units in which they are commonly expressed, or measured. Allowable parameter errors must be measurable, and expressed in terms that the optical fabricator or engineer can understand and relate to system needs. In general the errors in curvature or flatness of surfaces can be expressed in fringes of departure from a nominal value. Radius of curvature is also expressed in ± radius units. The approximate relation to the surface sag error is that

$$\frac{\delta r}{r} = \frac{\delta z}{z}$$

$$\delta z \approx \frac{1}{8}\left(\frac{D}{r}\right)^2 \delta r \tag{6.1}$$

so that for cases of nearly flat surfaces, the size of δr can be very large for a small value of δz. Therefore one must be cautious in thinking that a "standard" radius tolerance of, say, ±0.001 inches, should be applied to all radii. It is better to consider the radius tolerance as a fraction or percentage of the radius.

Thicknesses and separations are usually referred to their central values. Even for a rotationally symmetric component, the measurement of thickness is somewhat of an indirect measurement, as the center of the element has to be located. The edge thickness and the central thickness are obviously related by subtracting (or adding) the sag of the surfaces at the edge of the element. In a lens mount, the determining values are the separations of the locating rings for the elements, which, of course, includes the actual thicknesses and radii of the lens in determining the resulting airspaces. The thicknesses are determined by the actual errors on the curvature as well as the spacers. Usually, the tolerance that must be held on a curvature is sufficiently small that the sag error produced by the curvature error is negligible when compared with the allowable thickness or separation tolerance.

Tilt and decenter are complicated by the fact that the wedge in an element is the result of a combined tilt and decenter of the surfaces of the element. If an element has spherical surfaces, the wedge can be removed by properly aligning the element by edging or centering the element. A common approach is to precision-edge the element and then locate it in a cylindric barrel. The tilt, decenter, or wedge of a symmetric element is usually obtained by measurement of the runout of the element. If an element is placed on a ring or a three-point support and rotated, the runout measured on the second surface is a measure of the tilt and decenter of the second surface. The element can be centered by sliding along the first surface until the runout is eliminated to within a stated tolerance. Some programs provide runout tolerances as a part of the normal tolerance output. Others leave the data in tilt and decenter form.

The above considerations apply to lenses with spherical surfaces. Aspherics differ in that a single defined center of symmetry for the surface exists. Usually this is best specified as an additional tolerance on the component. Cylindric surfaces provide another problem in that the axis about which the surface is rotated must be defined in space, both in location and orientation.

TABLE 6.2

Quality level	Wavefront residual (no. waves)	Thickness (mm)	Radius (%)	Index	V-number (%)	Homogeneity	Decenter (mm)	Tilt (arc sec.)	Sphericity (no. fringes)	Irregularity (no. fringes)
Commercial	0.25 RMS 2 peak	0.1	1.0	0.001	1.0	0.0001	0.1	60	2	1
Precision	0.1 RMS 0.5 peak	0.01	0.1	0.0001	0.1	0.00001	0.01	10	1	0.25
High precision	<0.07 RMS <0.25 peak	0.001	0.01	0.00001	0.01	0.000002	0.001	1	0.25	<0.1

Tolerances on the optical materials place boundaries upon the acceptable values for optical and physical properties. Usually this is done by reference to the supplier's specifications. The optical parameters that need consideration in tolerancing are the index, dispersion, and homogeneity. Other parameters that need to be stated as tolerances are the birefringence, as it affects the internal stress content of the material, and the various physical properties. Usually these are a consequence of the glass manufacturing process, and must be accepted rather than specified.

Table 6.2 shows a set of reasonable tolerances that can be maintained on various parameters in a fabrication process. The table presumes three levels of quality, based upon the usual needs for optical systems. These are estimates that seem reasonable, based upon experience. The actual experience with any specific shop may differ somewhat. This table indicates that the tightness of the tolerances that may be expected depends upon the fabrication process to be used. The tolerances that can be maintained in production lots will likely differ for a specific system and will be different from those applicable to prototype fabrication. These are starting points for the tolerancing process. The greatest mistake that a designer can make is to simply accept these numbers as defaults. Within a lens, surfaces have different effects upon the image, and the tolerances can vary widely. The actual distribution of tolerances will depend at least partially upon the designer's estimate of the difficulty of fabricating the surface.

Another issue is selecting reasonable specifications for the optical characteristics of a lens. Table 6.3 (Shannon 1994), lists some of the important optical properties for a lens and some likely values that can be expected in normal cases. These can be held more closely, but will usually require special care in fabrication and assembly to do so.

Unlike the image quality, these parameters can be specified and measured as specific numerical quantities. The required values for these quantities must be included in the tolerancing activity.

6.2 Tolerancing Procedures

No matter what the tolerancing problem may encompass, the procedures are somewhat similar. The effect of changes in the parameters on the output, usually an image, needs to be obtained. Allowable changes in the parameters must be allotted according to some rule. The probability of best case/worst case scenarios must be considered. The coupling and compensation of errors

TABLE 6.3

Parameter	Accuracy target	Importance	How verified
Focal length	1 to 2%	Determines focal position and image size	Lens bench
F/number	$< \pm 5\%$	Determines irradiance at image plane	Geometrical measurement
Field angle	$< \pm 2\%$	Determines extent of image	Lens bench
Magnification	$< \pm 2\%$	Determines overall conjugate distances	Trial setup of lens
Back focus	$\pm 5\%$	Image location	Lens bench
Wavelength range	As needed Set by detector and source	Describes spectral range covered by lens	Image measurement
Transmission	Usually specified as $> 0.98^n$ for n surfaces	Total energy through lens	Imaging test, radiometric test of lens
Vignetting	Usually by requiring transmission to drop by less than 20% or so at the edge of the field	Uniformity of irradiance in the image	Imaging test, radiometric test of lens

needs to be considered. Finally the whole proposition has to be evaluated and justified against the capabilities of the organization(s) that will produce the system. In some cases, the recommendations regarding tolerances have to be translated into a language that will be understood by the organization fabricating the system.

6.2.1 GENERAL

Each tolerancing process begins with a set of assumptions. There is little point in requesting a shop to produce items that are more tightly toleranced than the system is capable of delivering. Therefore a starting table of default tolerances is proposed, usually as the initial data for a tolerance analysis program. These values are obtained from experience with fabrication of optics, and are rules to be kept in mind when carrying out any design.

Some set of data similar to this will be the starting point of the procedure. A change table is computed by the program which determines the amount of change in the image due to these tolerances. Specification of the image change may be made against a merit function that was used in the original design, or against some other appropriate function, such as the OTF at a specified spatial frequency. The result is a table of values of change in the image, say MTF, for each parameter. Each of the parameters can be scaled to an amount that just matches the allowable amount of MTF degradation. This is a basic set of individual tolerances.

An important possibility in tolerancing is a compensator which can offset some of the effect of a parameter error. The most frequently used compensator is the choice of focal position, and possibly of focal plane tilt to adjust for asymmetric errors. This can be included by evaluating the change table with respect to an adjustment of the focal surface. The basic set of tolerances then can include a most probable required focal adjustment as a result of each of the individual tolerances being applied. This is quite realistic as almost all optical systems permit focus adjustment after assembly.

The table of individual tolerances would be satisfactory except for the obvious fact that there are many possible error sources in a lens. It is necessary to presume that each parameter will have an error whose magnitude is distributed over the range of possible errors. If all of the errors were independent, and it is assumed that the errors are distributed around the expected mean values with a normal probability distribution, then the central limit theorem indicates that the probable error in the entire system will be the root sum square of the individual errors.

If each tolerance is considered as a one-sigma value for each distribution, then the most probable result of assembling the lens within these rules would be that there is a one-sigma probability that the final assembly will meet the allowable degradation of the system.

6.2.2 COMPUTER ANALYSIS

Computer programs mechanize the approach to tolerancing. Each program has some tolerancing model built into the program. The approach used is to develop and distribute tolerances based upon allowable changes in some function describing the imagery by the lens. In some cases, special merit functions are developed to set the tolerances. Usually there is an easily used default approach. In most practical cases, the defaults will have to be adjusted to obtain a reasonable set of tolerances. In all cases, sample cases should be run to ensure that the designer understands the meaning of the tolerances, and that they are reasonable.

One of the major examples is the approach used in the CODE V program. The tolerancing option permits determination of tolerances by three methods of image error specification. First is ray-trace-based tolerancing, which is based upon RMS wavefront error or MTF degradation. Second is tolerancing of the distortion or boresight errors in a lens. Third is tolerancing based upon changes in the first- and third-order aberrations. In other programs, the ability to obtain tolerances based upon changes in the merit function is added, and is similar to the last approach in the CODE V program.

The ray-based tolerancing calculates a table of the amount of each tolerance that will cause a change of 0.01 in the RMS wavefront error at the worst field and zoom position. In order to speed up the computation, the calculation is done by presuming small derivative increments and computing the wavefront error using a calculated derivative of the wavefront error. This provides a derivative that can be used to relate changes in each parameter to the net RMS wavefront error. If the designer has not entered specific instructions regarding items and groups to be toleranced the program sets up a comprehensive default set of parameters and uses focus position as a compensator for the image errors. This provides the basic set of data for tolerancing. It is a rather large table relating a change in each parameter to the change in the RMS wavefront error at each field and zoom position. The tolerances suggested by this procedure are usually quite tight.

The program then adjusts each parameter to the fabrication limit described in a default table of errors. Using the derivative model permits rapid computation of the likely amount of RMS error at each point in the field if the default tolerances are applied. Assumption of independence in the parameters leads to a statement of the one-, two-, and three-sigma image errors that may occur.

The ray-based tolerance program has some other options regarding the image errors. Because the derivative of the wavefront error is computed for a specific ray set, the effect upon the MTF can be estimated from using the same data. The designer input then includes the critical spatial frequency at which tolerancing is to take place. The program can apply similar rules to obtain the likely MTF image error resulting from applying default fabrication tolerances.

Much more input can be used by the designer to explore the tolerance problem. The designer can add compensators, presuming that specified steps may be chosen during assembly to allow optimization of the lens during assembly. The default tolerance ranges on each parameter can also be adjusted, and the weight on field and zoom position changed to suit the imaging needs for the lens. Systematic changes of this sort can be used to lead to the loosest possible set of tolerances.

The distortion and boresight characteristics of the lens are a special consideration. The application may require that the image show distortion below some limit. These errors are a change in direction of the chief ray with changes in each of the components in the lens. This will generally lead to a set of tolerances that differs from the allowable changes in image quality. A symmetric error in image quality may be acceptable. In some cases, the image mapping and image quality requirements may both be specified, and the resultant tolerance set will include both types of errors. The boresight error has two consequences. One is that the center of the image field is offset from the center of symmetry of the image. The other is that the image is essentially unchanged, but the pointing direction of the lens is different from the nominal direction. Either may be the case, and the designer must be careful to ensure that the specification and tolerances are assigned to the proper definition.

The first- and third-order aberration tolerancing methods would appear to be considerably more primitive than the ray-based tolerancing discussed above. This is certainly the case, and generally this approach to defining the tolerances is not heavily used. There is a justification for examining a system in this regard. The amounts of the low-order aberrations in a lens system are established by the balancing of high-order aberrations. In general, these low-order aberrations change more rapidly with changes of parameters than do the high-order aberrations. The basis of this tolerancing is that the changing of these aberrations upsets the balance established in the design. This method of examining tolerances is not as direct as the ray-based approach, but the computation of the change table relating parameters to errors is considerably more rapid. Expressing the change of the third-order aberrations in terms of RMS wavefront errors can provide considerable insight into the importance of various parameters.

This understanding of tolerances can provide some new views upon how to evaluate a lens during fabrication. For example, the astigmatism balance is established by the spherical aberration of the elements located away from the stop. The amount of the astigmatism is shown in Chapter 3 to depend upon the stop distance. Therefore changes in the overall length of the lens will lead to changes net in field curvature. For photographic type lenses, shortening the overall length will usually cause an inward-curving net field surface. Other basic conditions that are related to the third-order aberrations are the basic distortion of a well-corrected lens. It was shown in Chapter 3 that the intrinsic distortion from adding thickness to a parallel plate is positive. Thus the distortion balance is easily upset by a change in total glass thickness. These items can be important in debugging a lens assembly.

Similar comments apply to the tolerancing based upon changes in the merit function. This is also a very rapid method of tolerancing because the computer program is optimized to provide a very rapid computation of this function. The designer has to first determine the average amount of image error due to a change in the merit function. This can usually be accomplished by making some trial changes and observing the actual change in the MTF values. The tolerancing change table can then be scaled by the amount of allowable merit function change.

It is important to note that this is an average change. The weights imposed during design are applicable so that there is an existing weighting of the importance of various portions of the field in the definition of the tolerances. Addition of appropriate focal position solves will provide the compensating parameter.

6.2.3 STATISTICAL COMBINING OF ERRORS

One of the most difficult parts of tolerancing is taking advantage of the statistical aspect of error distribution. It is easy to adjust each of the individual tolerances such that the image is always acceptable at every point in the field. This is a meaningful tolerance if there is only one error in the system. However, if there are two errors, they may add to produce an unacceptable image error. Or, perhaps, they may cancel, leading to no error at all. Or there may be some partial compensation or addition. In principle, every possible combination of errors could be investigated to model the effect of tolerance errors. This is possible only in the simplest cases. The number of possible combinations that can exist is extremely large. For a triplet that has about 20 toleranceable parameters, the number of two-at-a-time combinations is only about 400, but there are about two million combinations of four parameters. The problem becomes difficult very rapidly.

The most frequently used approach is to presume that all of the errors are independent and follow a normal probability distribution. If this were true, then the overall distribution is another normal distribution whose variance is the sum of the variances of the individual tolerances. The one-sigma width is the square root of the variance. If the stated tolerances are taken as the one-sigma values for the individual distributions, then the probability of meeting the goal can readily be estimated. In a practical example, if there are four tolerance errors, each of which is described by a one-sigma value that just meets the acceptable value of, say, change in the RMS wavefront error, then the result of all four parameters acting together would likely be twice that of an individual error. If there are twenty parameters, then the combination probability would be described by a sigma value 4.5 (square root of twenty) times the individual error probability.

The assignment of combination tolerances can be considered as the inverse problem. If the calculation of the individual tolerances provides a list of errors, each of which is individually at the limit of acceptability, then the tolerance to be assigned which should provide a one-sigma probability of meeting the image quality goals would be obtained by dividing the size of each tolerance by the square root of the number of parameters.

This sounds good, but the large number of possible parameters in a lens and the likelihood that some combinations of errors will actually reduce the net wavefront error in some portions of the field makes this a somewhat severe criterion. This approach can provide sets of very tight tolerances that may well make a lens unable to be built, or at least impractical. For example, presume that separation errors of ± 0.02 mm in a triplet are found to be acceptable for their individual effect upon the image quality. Using the rule above would mean that the thickness allowance would drop to ± 0.004 mm, a considerably harder tolerance for a production shop to hold. What initially appeared to be an almost commercial assembly task would be forced into a more expensive tight-precision task by the tyranny of tolerance distribution.

In some cases the tolerances can be loosened considerably by leaving a critical component for final adjustment. One example of this is the assembly of a long-focal-length, zero-distortion astrometric lens (Vukobratovich et al. 1994). It was found by experimentation on the computer that the decenter and tilt tolerances on four of the five elements could be loosened by almost an order of magnitude if the last element were allowed to be adjustable to compensate the residual asymmetric errors from the lens assembly. The amount of adjustment in three directions that was required could be predicted from the measured amount of aberration. Computer experimentation also indicated that if the elements were within tolerance and the image quality met the desired levels, the distortion would be within acceptable limits. This was used as the lens assembly process since there was no method to verify the distortion to the desired accuracy prior to installation on the telescope.

6.2.4 EXAMPLES OF TOLERANCING

Some examples of tolerancing a lens will now be described. The selected examples are chosen to demonstrate some of the important points that have been made in this chapter. In most real tolerancing problems, the number of variables and items to be toleranced will be much greater than shown here. The basic principles are the same but the amount of computer and personal effort required to carry out the tolerancing is usually much greater.

Doublet no.9 for Tolerance Example

	RDY	THI	GLA
OBJ:	INFINITY	INFINITY	
STO:	55.89991	8.000000	BK7_SCHOTT
2:	-44.45122	0.000000	
3:	-44.55636	3.500000	SF2_SCHOTT
4:	-172.34341	93.748098	
IMG:	INFINITY	-0.074248	

SPECIFICATION DATA

EPD	28.60000		
DIM	MM		
WL	656.30	587.60	486.10

INFINITE CONJUGATES

EFL	100.0000
BFL	93.7481
FFL	-98.7422
FNO	3.4965

FIGURE 6.1 Design data for the doublet used in the tolerance study.

6.2.4.1 Airspaced Doublet

An airspaced doublet is selected as a simple example for a small field. In this lens there are only seventeen parameters to be toleranced. The requirement is that the MTF on axis be better than 80% of the design value at half of the cutoff spatial frequency. The consideration of only one field greatly simplifies the tolerancing requirements and permits a simplified view into some of the tradeoff involved in tolerancing.

The lens chosen as an example is a reasonably good doublet given by the data in Figure 6.1. We will be concerned in this example only with the best axial MTF that is obtained after adjustment of the focus position as a compensator. Figure 6.2 is a plot of the design value for the MTF of the lens. The goal will be to maintain the MTF of the fabricated lens as better than 50% of the nominal value of the MTF at 100 lines/mm. For the design the MTF at 100 lines/mm is calculated as 0.414, and the resulting acceptable MTF after degradation is taken to be 0.2, or a larger value. This is arbitrary, and a user would probably like to have a higher value, usually about 80% of the design value. The requirements are normally set either by a decree of the customer, or by agreement between the designer and the user.

A tolerancing run on CODE V generates a list of thirty-five possible fabrication and assembly errors in this very simple lens! The list includes:

The figure or irregularity error on four surfaces
The radius of curvature on four surfaces
Thickness of two elements and one airspace

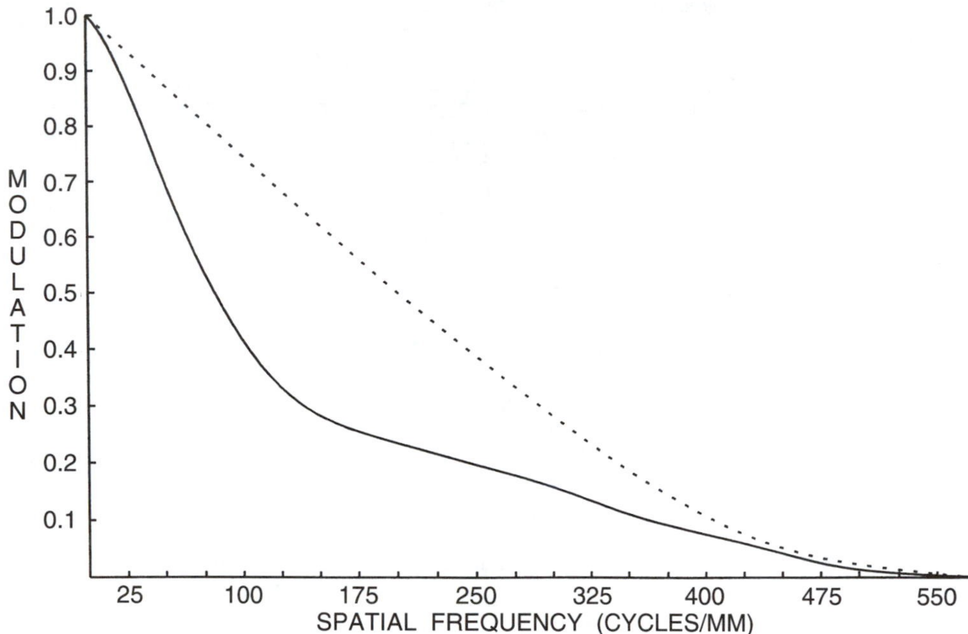

FIGURE 6.2 MTF for the doublet as designed. ——, axis; ······, diffraction limit.

Four possible errors in index and *V*-number

Eight errors of cylindrical deformation in two azimuths each on four surfaces

Twelve errors of wedge, tilt, and lateral shift in two directions on two elements

It may seem that there are more errors than are likely to occur, but the table is inclusive. It is likely that the designer will determine that some of these errors are more important than others, or may choose to ignore some possibilities as being unlikely. This is an important step, as the distribution of errors among a large number of sources can lead to quite tight tolerances on some items.

The program is asked to carry out an inverse tolerancing approach, wherein the changes in each parameter that will produce some level of error are determined. The first conclusion is that the lens cannot be fabricated to the desired quality, at least not considering the default limits upon the tolerances. This usually happens because defaults in any computation scheme are designed to cover a wide range of possibilities. The designer must exercise some judgment in order to relieve the tolerances.

To understand some of the problems, consider the following simplifications: The elements have been manufactured to nominal specifications, and

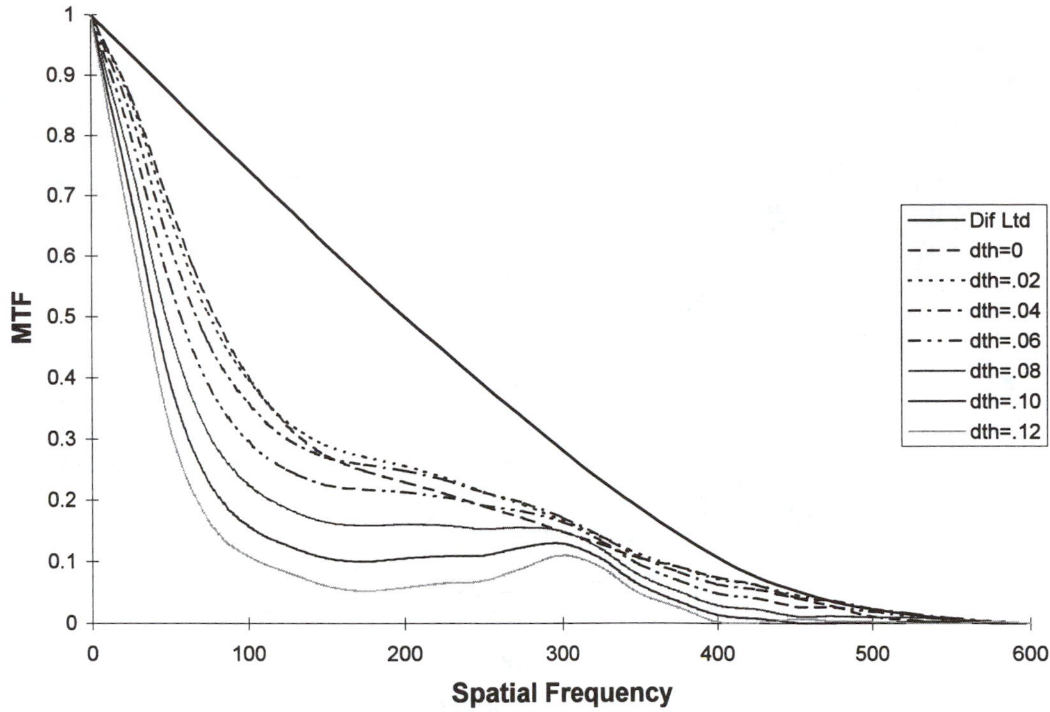

FIGURE 6.3 MTF for various values of the spacing between the elements.

the assembly can have only two errors that need to be investigated. Although this is a simple case, the process of thinking about the tolerances will be instructive.

The effect of errors in decenter of the first element and errors in the air-space between the elements can be investigated in detail. Figures 6.3 and 6.4 are MTF values for the lens with stated errors in each of these two parameters, with the focal surface adjusted to optimize the MTF. A somewhat similar behavior is noted in the two cases, except that the decenter introduces an asymmetric error which shows as different values for the sagittal and tangential direction in the MTF. Examination of these figures (or, of course the tabulated values) at 100 lines/mm would yield acceptable errors in each of the parameters.

Figures 6.5 and 6.6 are plots of the variation of the MTF at 100 lines/mm for each of the parameters being studied. It can be seen that there is a smooth change of MTF with parameter error. Also plotted on these graphs is the polychromatic RMS wavefront error for each parameter. Figure 6.7 is a scatter diagram resulting from the computer experiment. Each point is an MTF value plotted against the corresponding RMS wavefront error value for this lens. The solitary point at the origin is placed there only to note the

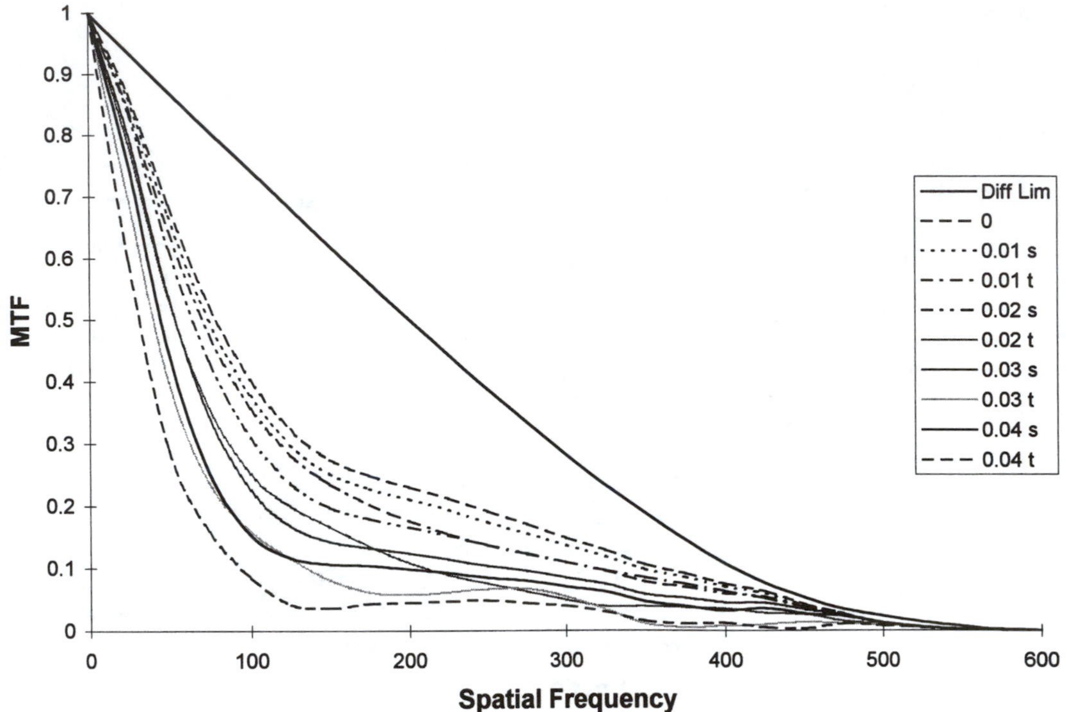

FIGURE 6.4 MTF for decenter of the two elements.

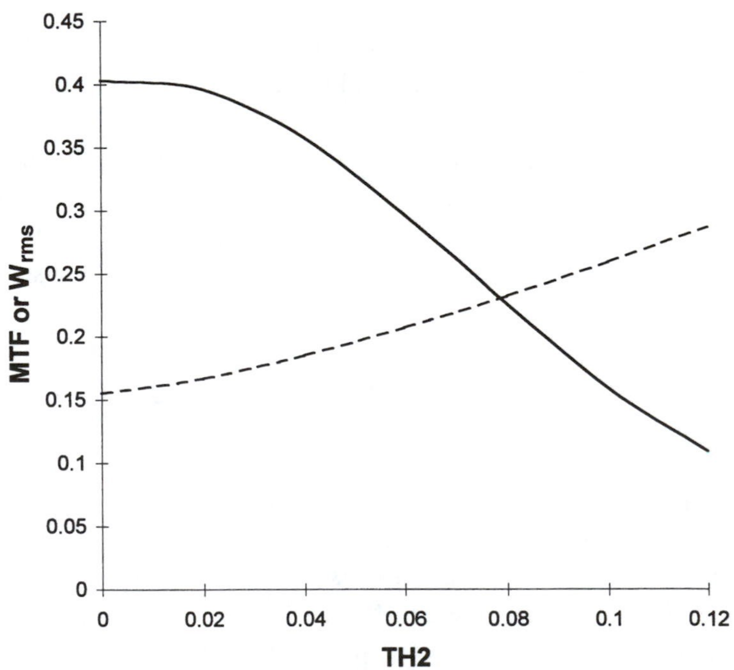

FIGURE 6.5 MTF at 100 lines/mm (——) and amount of RMS wavefront error (– – –) as a function of the separation error.

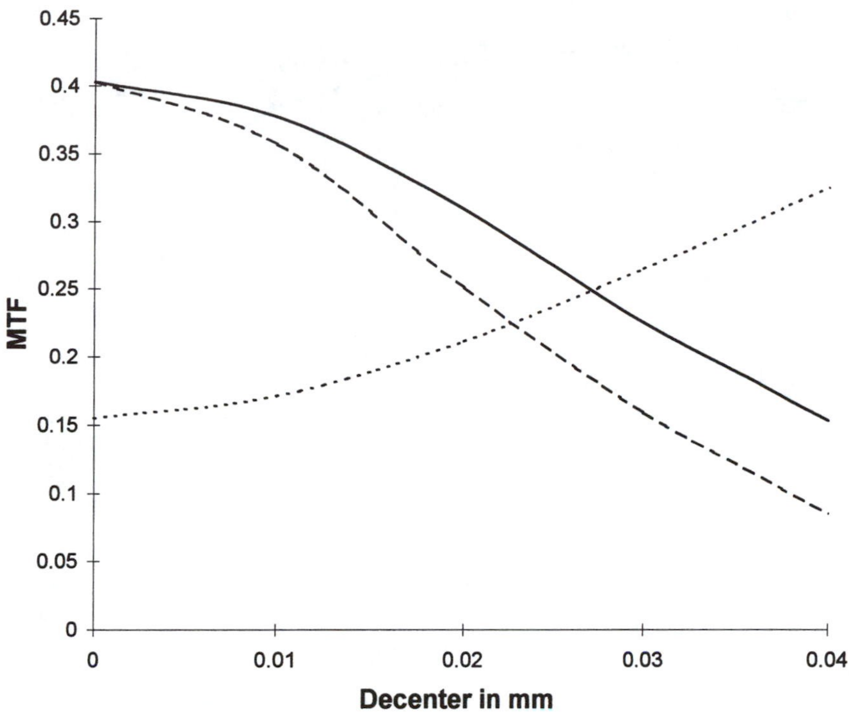

FIGURE 6.6 MTF at 100 lines/mm in sagittal (——) and tangential (– – –) directions as well as RMS wavefront error (······) as a function of the element decenter.

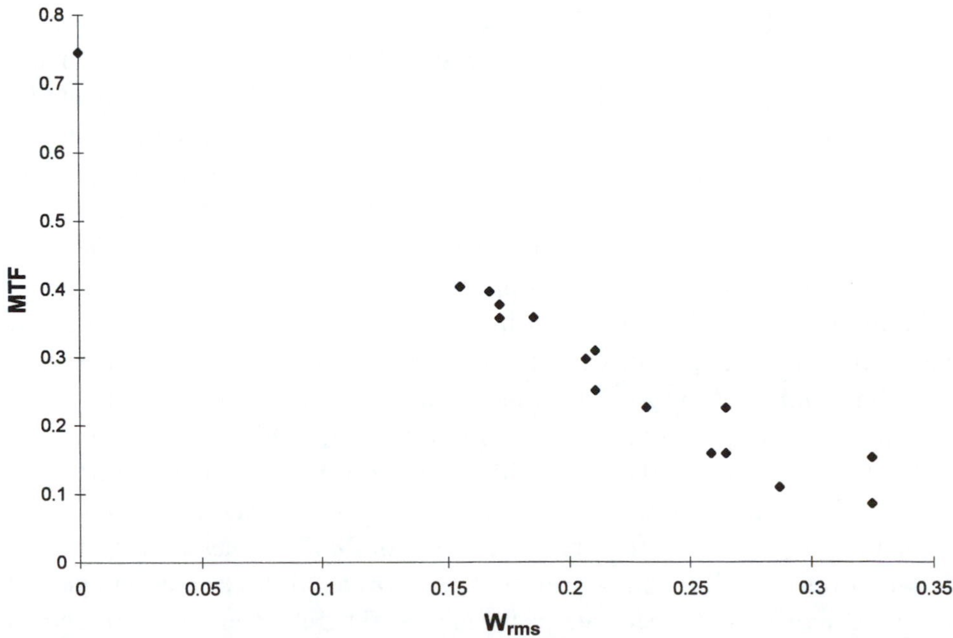

FIGURE 6.7 MTF at 100 lines/mm versus RMS wavefront error for a number of combinations of separation and decenter error.

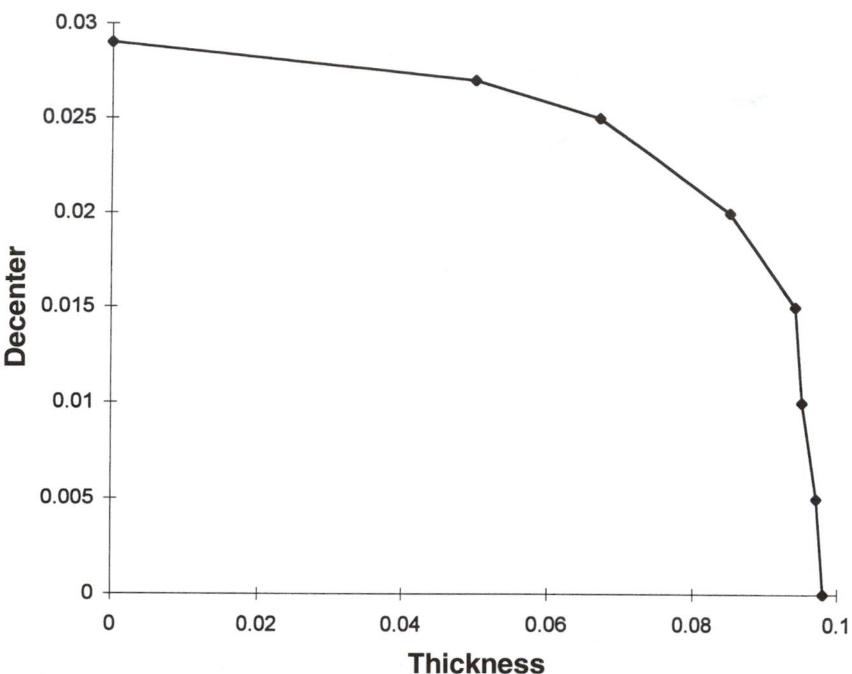

FIGURE 6.8 Allowable joint range of separation error and decentering. Any values within the contour are acceptable within the tolerance budget.

diffraction limit that can be obtained at 100 lines/mm. The highest MTF attainable by the lens under this condition is shown for about 0.15 waves RMS, which corresponds to the design residual aberrations. Tolerance errors in these components reduce the MTF as shown. From this diagram, it can be seen that a limit of about 0.25 RMS wavefront error would be acceptable. This is a change of an additional 0.1 waves of RMS wavefront error. The word "about" applies here, as the RMS wavefront error serves only as an index of the likely error. Any combination of the thickness and decenter errors that meet this index would appear to be acceptable.

Because these errors are different in nature it is appropriate to assume that the effect of individual error will add in a mean-square-sum manner. It is easy enough to evaluate how good an approximation this is in this case. Figure 6.8 is a plot showing eight different combinations of the thickness and decenter error that all add up to the same RMS wavefront error. Actual calculations indicate that the resulting lens shows 0.279 ± 0.009 RMS wavefront error when these eight cases are applied. The set of MTF values that occur are shown in Figure 6.9. It can be seen that *on average* all these cases will yield a lens with *substantially* the same performance. Therefore either approach to tolerancing can be useful. Note that the RMS wavefront error being used here is for an average over a wavelength interval.

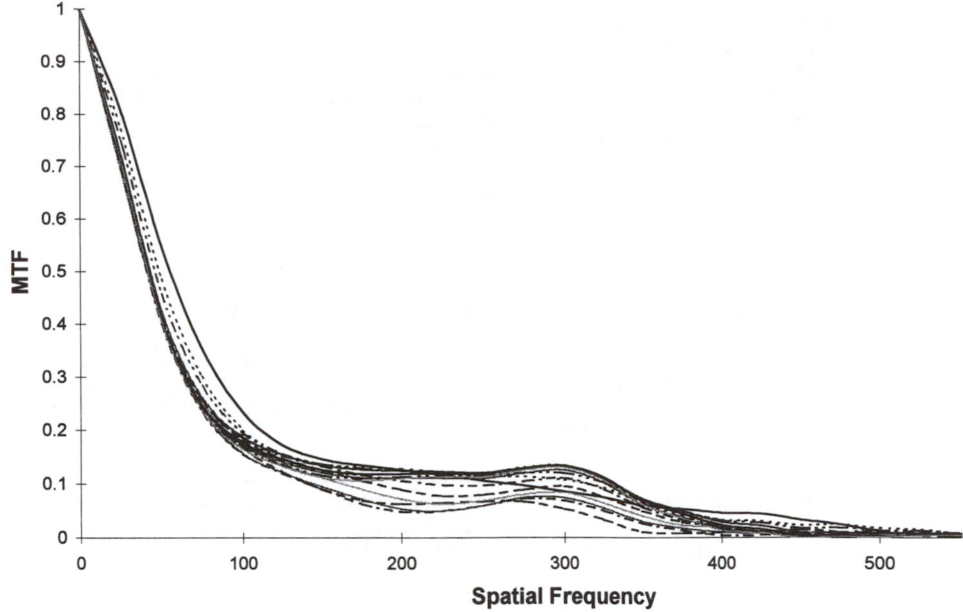

FIGURE 6.9 Family of MTFs resulting from a number of trials of the joint spacing and decenter errors.

The case just described was for only two errors, but similar reasoning applies to any other errors. The problem becomes complicated very quickly, so that the handling of all of the important errors is best done with the aid of a good computer program. In order to complete the task with this lens, only the most significant errors will be considered. These include the radius tolerances, surface figure thickness and spacing and the wedge, tilt, and decenter of the two elements.

When this tolerancing is carried out using the normal limits on the errors, the conclusion is that the image quality requirement of a change of 0.1 RMS wavefront error will be met only 84% of the time. That is, about 16% of the lenses made presuming a normal tolerance error distribution will show errors greater than desired. Examination of the tolerancing program output reveals that some of the errors are at their minimum limits as determined by the setup of the tolerances. Changing the allowable decentering minimum error from ±0.02 mm to ±0.01 mm, and reducing the lower limit on the surface irregularity tolerance changes the distribution so that the image quality goal will be met 99% of the time.

This last judgment call by the designer is extremely important. The recognition is that in order to meet the image quality goals consistently, the mounting and centering of the elements must be carried out with extra care. Figure 6.10 gives a list of the final data for the tolerances on the lens

The probable change in RMS assumes a uniform distribution of manufacturing errors over the range for all parameters except tilt and decenter which have a truncated Gaussian distribution in X and Y

CUMULATIVE PROBABILITY	CHANGE IN RMS
50.0 PCT.	0.040
84.1 PCT.	0.065
97.7 PCT.	0.087 *
99.9 PCT.	0.107

* If it is assumed that the errors can only take on the extreme values of the tolerances, the 97.7 percent probable change in RMS is 0.186

CENTERED TOLERANCES

SUR	RADIUS	RADIUS TOL	FRINGES POW/IRR	THICKNESS	THICKNESS TOL	GLASS	INDEX TOL	V-NO (%)
1	55.89991	0.2000	4.0/ 1.00	8.00000	0.55000	BK7	0.00100	0.40
2	-44.45122	0.0200	2.0/ 0.50	0.00000	0.01000			
3	-44.55636	0.0200	2.0/ 0.50	3.50000	0.85000	SF2	0.00150	0.40
4	-172.34341	1.3000	2.0/ 0.50	93.74810				
5			-0.07425					

Radius, radius tolerance, thickness and thickness tolerance are given in mm.
Fringes of power and irregularity are at 546.1 nm. over the clear aperture
Irregularity is defined as fringes of cylinder power in test plate fit

TOLERANCE LIMITS

	MINIMUM	MAXIMUM	
* RADIUS	0.0200		mm.
* SAG	0.0020	0.0500	mm.
** POWER	1.0	12.0	Fringes
IRREGULARITY	0.10	3.00	Fringes
THICKNESS	0.01000	5.00000	mm.
INDEX	0.00010	0.00200	
V-NUMBER(%)	0.20	0.80	%

DECENTERED TOLERANCES

ELEMENT NO.	FRONT RADIUS	BACK RADIUS	ELEMENT WEDGE TIR	ELEMENT WEDGE ARC MIN	ELEMENT TILT TIR	ELEMENT TILT ARC MIN	EL.DEC TIR	EL.DEC mm.
1	55.89991	-44.45122	0.0200	2.5	0.0084	1.0	0.0114	0.0100
2	-44.55636	-172.34341	0.0040	0.5	0.0082	1.0	0.0047	0.0100

Radii are given in units of mm.

For wedge and tilt, TIR is a single indicator measurement taken at the smaller of the two clear apertures. For decenter and roll, TIR is a measurement of the induced wedge and is the maximum difference in readings between two indicators, one for each surface, with both surfaces measured at their respective clear apertures. The direction of measurement is parallel to the original optical axis of the element before the perturbation is applied.
TIR is measured in mm.

Decenter is measured perpendicular to the optical axis in mm.

TOLERANCE LIMITS

	MINIMUM	MAXIMUM
TIR	0.0020	0.1000
TILT	0.0003	0.0050
DECENTER	0.0100	0.2500

FIGURE 6.10 Output of a tolerancing evaluation (CODE V).

that could be presented to the fabrication organization. At the top is a summary of the expectations if the tolerance values are held. Below are tolerance values and a listing of the limits negotiated with the program by the designer for each of the tolerance types.

6.2.4.2 Complex Lens

A six-element Gauss type objective is a typical example of a complex lens to be toleranced. The particular lens that will be examined is the final double Gauss design from section 7.1.3. The fabrication specifications for this lens are shown in Figure 6.11. The tolerancing for this example will be carried out on CODE V. The procedure is first to run a tolerancing with the default values for the allowable errors in the lens. The result of this run will then be used as a guide to alter the allowable tolerances to obtain a set of tolerances which lead to the likelihood of an acceptable level of image quality.

The choice is to tolerance for the level of MTF to be expected after the tolerable errors are applied. The designer needs to supply a set of goals for the tolerancing. In this case, five fields were used in order to ensure that the image quality is fairly represented across the field of view of the lens. The program was told to weight the extreme edge of the field less than the center of the field, a reasonable requirement.

When a tolerance run was called for, the program identified ninety-five possible tolerances and compensators. The involved radius, thickness, and index of refraction error, as well as tilt and decenter tolerances upon individual surfaces and upon elements. The only compensator used was the focal position. This first computer run yielded many pages of detailed information providing the probable amount of change of MTF at each field with a change in the parameter being toleranced. While it is tempting to fatten up this book with the detailed listings, each problem is sufficiently different that little would be gained by doing so. These detailed listings are useful to the designer in that they provide a direct relation between the size of the tolerance and the effect upon, in this case, the MTF. Locating those tolerance errors that have the major effect upon the image quality can be carried out using this information. Since this lens covers a wide field, this detailed table also provides the designer with useful information regarding which errors primarily influence the edge of the field, and which mainly influence the center of the field. The coefficients represented by the detailed data set are used as the basic information in distributing tolerances, much as the derivative matrix is used in automatic correction.

A summary of the result is the output shown in Figure 6.12. The values of the assigned tolerances are indicated, along with a summary of the probable effect on the image. The degradation is stated in terms of a two-sigma value.

```
                              FABRICATION DATA

                                    RRS

15-Oct-95
Modified Double Gauss Lens

ELEMENT   RADIUS OF CURVATURE                      APERTURE DIAMETER
NUMBER    FRONT          BACK        THICKNESS      FRONT      BACK        GLASS
-----------------------------------------------------------------------------------

OBJECT           INF                 INFINITY
  1      30.0684 CX    76.4989 CC      4.0094      29.0679    29.0679    LAF2 Schott
                                       0.5000
  2      19.4556 CX    66.1682 CC      7.2068      24.4571    24.4571    LAF2 Schott
  3      66.1682 CX    12.6067 CC      1.3030      24.4571    20.0000    SF3 Schott
                                       7.0000
                                   APERTURE STOP        14.5182
                                       7.0000
  4     -16.1338 CC    60.7320 CC      1.3030      20.0000    22.0000    SF64A Schott
  5      60.7320 CX   -22.4764 CX      5.2422      22.0000    22.0000    LAF2 Schott
                                       2.2990
  6      84.6316 CX   -49.7526 CX      4.8000      28.0000    28.0000    LAF2 Schott
           IMAGE DISTANCE =           26.0621
 IMAGE           INF                                    38.3864

-----------------------------------------------------------------------------------

 NOTES - Positive radius indicates the center of curvature is to the right
         Negative radius indicates the center of curvature is to the left
       - Dimensions are given in millimeters

       - Thickness is axial distance to next surface

       - Image diameter shown above is a paraxial value,
         it is not a ray traced value
```

FIGURE 6.11 Specification data for a sample photographic type lens (CODE V).

SUR	RADIUS	RADIUS TOL	FRINGES POW/IRR	THICKNESS	THICKNESS TOL	GLASS	INDEX TOL	V-NO (%)	INHOMO-GENEITY
1	30.06838	0.0200	4.0/ 2.00	4.00937	0.02000	LAF2	0.00100	0.20	
2	76.49891	0.1000	8.0/ 3.00	0.50000	0.02000				
3	19.45557	0.0200	2.0/ 1.00	7.20684	0.02000	LAF2	0.00030	0.20	
4	66.16820	1.0000	12.0/ 3.00	1.30304	0.02000	SF3	0.00040	0.20	
5	12.60665	0.0200	2.0/ 1.00	7.00000	0.02000				
6				7.00000	0.02000				
7	-16.13375	0.0200	2.0/ 1.00	1.30304	0.02000	SF64A	0.00040	0.20	
8	60.73196	0.4000	12.0/ 3.00	5.24225	0.02000	LAF2	0.00040	0.20	
9	-22.47641	0.0200	2.0/ 1.00	2.29903	0.04000				
10	84.63157	0.2000	12.0/ 3.00	4.80000	0.08000	LAF2	0.00100	0.20	
11	-49.75262	0.1000	12.0/ 3.00	26.17918					
12				-0.11705					

ELEMENT NO.	FRONT RADIUS	BACK RADIUS	ELEMENT WEDGE TIR	ELEMENT WEDGE ARC MIN	ELEMENT TILT TIR	ELEMENT TILT ARC MIN	EL. DEC/ROLL(R) TIR	EL. DEC/ROLL(R) mm.
1	30.06838	76.49891	0.0040	0.5	0.0087	1.0	0.0117	0.0200
2	19.45557	66.16820	0.0020	0.3			0.0205	0.0200 (R)
2- 3	19.45557	12.60665			0.0060	1.0	0.0066	0.0200
3	66.16820	12.60665	0.0020	0.3				
4	-16.13375	60.73196	0.0020	0.3			0.0319	0.0200 (R)
4- 5	-16.13375	-22.47641			0.0060	1.0	0.0052	0.0200
5	60.73196	-22.47641	0.0020	0.3				
6	84.63157	-49.75262	0.0080	1.0	0.0084	1.0	0.0179	0.0200

POLYCHROMATIC MODULATION TRANSFER FUNCTION

RELATIVE FIELD	FREQ L/MM	AZIM DEG	WEIGHT	DESIGN	DESIGN + TOL *	COMPENSATOR RANGE (+/-) * DLZ S12
0.00, 0.00	30.00	TAN	1.00	0.612	0.389	0.253952
0.00, 0.46	30.00	TAN	1.00	0.538	0.354	0.253952
0.00, 0.70	30.00	TAN	1.00	0.467	0.323	0.253952
0.00, 0.80	30.00	TAN	1.00	0.460	0.240	0.253952
0.00, 1.00	30.00	TAN	1.00	0.176	0.061	0.253952

FIGURE 6.12 First output of tolerances for the lens in the previous figure using nominal allowable individual tolerances for the components (CODE V).

That is, if the individual tolerances are maintained with the stated ranges, there is a 97% likelihood that the lens will meet the stated degradation in image quality. The values were judged to indicate too large a drop in the MTF to be acceptable.

The next step was to examine the actual tolerances assigned to each parameter of the lens. Those errors having the greatest effect upon the MTF degradation were tightened experimentally to require that these critical tolerances would be held as tight as possible. The particularly bad actors for this lens turned out to be the tilt and decenter of the inner menisci and the thickness of these elements. The allowable tolerances on these items were tightened, and the entire tolerance run carried out again. This time the output shown in Figure 6.13 was obtained. The expected degradation in the MTF is reduced somewhat. The reallocation of tolerances on the surfaces by the program indicates a slightly surprising result that the radius and irregularity tolerances on some of the elements could be loosened as a result of tightening the tilt and decenter values. This seems reasonable. The actual values that are suggested may not seem quite so reasonable when compared with the normal values obtained in fabrication and listed in Table 6.2.

The result of shifting the burden from one set of tolerances to another is similar to the conclusions about tolerance budgeting on the doublet in the previous section. With the large number of allowable tolerances, the problem rapidly becomes intractable to manual considerations, and the designer begins to rely upon the algorithms in the computer program. Examination of the new tolerances indicates that much of the tolerance burden has been shifted to the mechanical alignment of the lens elements in the cell. This cycle of adjusting the tolerances is repeated again, until an acceptable distribution of tolerances and system degradation is obtained.

The model used in deriving the tolerances uses an approximation to the actual MTF. The model for distributing the tolerances is a statistical model presuming certain likely distributions of the errors that may occur. Therefore, it is important that the designer try a few sample runs using errors of the type suggested by the output of the program in order to ensure that the model does indeed provide a reasonable result. Some design programs provide a method for carrying out a number of "Monte-Carlo" variations of a lens in order to provide a table of likely system errors.

6.2.4.3 Cassegrain Telescope

This lens has a narrow field of view and close to diffraction-limited imagery. The system has only two reflecting surfaces, both containing aspherics. The requirement is that the wavefront error be no greater than 0.1 waves RMS.

SUR	RADIUS	RADIUS TOL	FRINGES POW/IRR	THICKNESS	THICKNESS TOL	GLASS	INDEX TOL	V-NO (%)	INHOMO-GENEITY
1	30.06838	0.0050	4.0/ 2.00	4.00937	0.01000	LAF2	0.00100	0.20	
2	76.49891	0.1200	8.0/ 3.00	0.50000	0.01000				
3	19.45557	0.0100	2.0/ 1.00	7.20684	0.01000	LAF2	0.00030	0.20	
4	66.16820	0.1700	12.0/ 3.00	1.30304	0.01000	SF3	0.00040	0.20	
5	12.60665	0.0100	2.0/ 1.00	7.00000	0.02000				
6				7.00000	0.02000				
7	-16.13375	0.0100	2.0/ 1.00	1.30304	0.00500	SF64A	0.00040	0.20	
8	60.73196	0.1300	12.0/ 3.00	5.24225	0.00500	LAF2	0.00040	0.20	
9	-22.47641	0.0050	2.0/ 1.00	2.29903	0.02500				
10	84.63157	0.1600	12.0/ 3.00	4.80000	0.02500	LAF2	0.00100	0.20	
11	-49.75262	0.0540	12.0/ 3.00	26.17918					
12				-0.11705					

ELEMENT NO.	FRONT RADIUS	BACK RADIUS	ELEMENT WEDGE TIR	ELEMENT WEDGE ARC MIN	ELEMENT TILT TIR	ELEMENT TILT ARC MIN	EL. DEC/ROLL(R) TIR	EL. DEC/ROLL(R) inches
1	30.06838	76.49891	0.0050	0.6	0.0087	1.0	0.0023	0.0040
2	19.45557	66.16820	0.0015	0.2			0.0051	0.0050 (R)
2- 3	19.45557	12.60665			0.0060	1.0	0.0033	0.0100
3	66.16820	12.60665	0.0010	0.2				
4	-16.13375	60.73196	0.0025	0.4			0.0060	0.0037 (R)
4- 5	-16.13375	-22.47641			0.0060	1.0	0.0013	0.0050
5	60.73196	-22.47641	0.0030	0.5				
6	84.63157	-49.75262	0.0050	0.6	0.0084	1.0	0.0089	0.0100

RELATIVE FIELD	FREQ L/MM	AZIM DEG	WEIGHT	DESIGN	DESIGN + TOL *	COMPENSATOR RANGE (+/-) * DLZ S12
0.00, 0.00	30.00	TAN	1.00	0.612	0.495	0.163513
0.00, 0.46	30.00	TAN	1.00	0.538	0.473	0.163513
0.00, 0.70	30.00	TAN	1.00	0.467	0.407	0.163513
0.00, 0.80	30.00	TAN	1.00	0.460	0.369	0.163513
0.00, 1.00	30.00	TAN	1.00	0.176	0.123	0.163513

FIGURE 6.13 Revised output of the tolerances for the lens after rearranging the error budget (CODE V).

The alignment of the components and the surface errors on the components need to be specified, and the effect of fabrication errors included.

This problem is a bit different from the previous problems, as each telescope is an individually produced and assembled system. The optimum adjustment of the surfaces can be made to produce the best image, as long as the location of the image is not rigorously fixed. It is even possible to incorporate corrective errors in one component, such as the secondary, to offset residual errors in the primary. The aspheric portion of each surface can be toleranced separately from the base spherical power of the surface, but the centering requirements are now more complex. Sensitivity analysis of the alignment of the components can be used to indicate to mechanical engineers what may be an optimum structure for holding the optics.

A place to start is to identify the possible sources of error, and develop a tolerance budget for the wavefront error contributions of each of the assigned errors. These errors can be "static," such as would be obtained from fabrication or initial assembly, or "dynamic," as might occur during use of the telescope. The primary mirror location will be taken as the reference location, and it is assumed that the image surface will be moved and tilted to optimize the image during the fabrication and alignment stages. Errors due to vibration or displacements during use generally will be more severe due to the inability to compensate with the image position.

The following can be identified:

Power of primary
Power of secondary
Aspheric correction of primary
Aspheric correction of secondary
Irregularity error on primary
Irregularity error on secondary
Decenter of secondary
Tilt of secondary
Spacing between primary and secondary

A spreadsheet of possible errors can be used to distribute the allowable wavefront error under the assumption that all of the individual errors sum together in a mean square manner. The relation between the allowable errors and the allowable errors in the systems are determined from a tolerance computation. The finite-difference relation between changes in each component and the effect upon the wavefront error is calculated.

The compensators used here are very important. The fabrication process will be to make the primary and secondary separately, and then assemble them with a spacing such that the best image is obtained. Once the compo-

TABLE 6.4 Demonstration of error budget in Ritchey Chretian Cassegrain where an equal contribution from each error source is assumed

Error source	Relative error	RMS waves	Value and units
Power of primary	1	0.033	33.33 spherical fringes
Power of secondary	1	0.033	333.33 spherical fringes
Aspheric correction of primary	1	0.033	0.47 surface fringes peak
Aspheric correction of secondary	1	0.033	0.47 surface fringes peak
Irregularity error on primary	1	0.033	0.03 surface fringes RMS
Irregularity error on secondary	1	0.033	0.03 surface fringes RMS
Decenter of secondary	1	0.033	0.08 mm
Tilt of secondary	1	0.033	13.20 arc seconds
Spacing between primary and secondary			0.31 mm
Total RMS wavefront error	3	0.1	

Compensators allowed are focus position and axial separation during assembly.
Data from an OSLO tolerance computation.

nents have been assembled into the telescope, then the tolerance on the separation of the components must be maintained within a sufficiently close tolerance that a good image is maintained. In each case, it is decided that the location of the best image focus will always be located. In this example, the final focal length of the system is not held to be constant, but a result of the fabrication operations. This permits much of the radius or power tolerance on the mirrors to be used to obtain the best image quality.

Table 6.4 is based upon an assumption that all of the errors introduce an equal amount into the image. This leads to very loose tolerance upon the spherical power of the mirrors, with extremely tight tolerances upon the aspheric shape of the mirrors.

Table 6.5 is a result of the designer applying a distribution of errors on the components. The spherical power requirements on the mirrors are tightened, and the resulting budget applied to loosening the aspheric accuracy of the primary and secondary.

This last table represents only one possible choice of the distribution. Different choices of the distribution can be used to provide tolerance relief in certain errors, or to respond to known errors in particular errors. The assignment of the operational allowances of tilt, decenter, and spacing to the mechanical assembly are particularly important as these numbers drive the approach needed for the mechanical mounting of the system.

TABLE 6.5 Demonstration of error budget in Ritchey Chretian Cassegrain where distribution of errors is selected by the designer

Error source	Relative error	RMS waves	Value and units
Power of primary	0.2	0.005	5.39 spherical fringes
Power of secondary	0.05	0.001	13.46 spherical fringes
Aspheric correction of primary	0.5	0.013	0.19 surface fringes peak
Aspheric correction of secondary	0.5	0.013	0.19 surface fringes peak
Irregularity error on primary	2.5	0.067	0.07 surface fringes RMS
Irregularity error on secondary	2	0.054	0.05 surface fringes RMS
Decenter of secondary	1	0.027	0.06 mm
Tilt of secondary	1	0.027	10.66 arc seconds
Spacing between primary and secondary	1	0.027	0.25 mm
Total RMS wavefront error	3.71382552	0.1	

Compensators allowed are focus position and axial separation during assembly.
Data from an OSLO tolerance computation.

Other possible options include finishing the aspheric figuring of the secondary to compensate an increased amount of error on the primary. This will greatly affect these two components, but will not change the operational tolerances. One option that can be useful in choosing options for mechanical design is the existence of a "neutral point" about which the combined tilt a decenter of the secondary will have a minimum effect upon the image quality. This point exists since the tilt and decenter both generate axial coma as the principal image degradation from the mechanical errors. Therefore a small tilt can be offset by a small decenter. Examination of the coma produced shows that this neutral point exists at 664 mm behind the secondary. A mechanical design that constrains the motion of the secondary to this form would permit increased mechanical errors within the allowable image error specifications.

Consultation with the fabrication and test personnel will permit refining the tolerance budget so that the requirements are maintained within the capabilities of the organization producing the system. For any lens, especially one of this type, the mechanical and fabrication engineers should be a part of the tolerancing investigations.

6.3 Tolerance Specification

The design data is usually provided in the form of a listing of the system parameters and associated tolerances. A drawing is prepared for each element and the lens assembly that carries these tolerance specifications to the fabrication facility. Each industrial organization, and each fabrication shop will likely have its own set of standards regarding communication of this information in the form of drawings.

Recently, there has been a move toward an international standard for optical drawings. This new standard, called ISO 10110, *Optics and Optical Instruments – Preparation of drawings for optical elements and systems*, is rather long and involved; it is presently in the process of final revision and is proceeding through international approval. A good summary of the basic concepts is found in Kimmel and Parks (1995). It is possible that there will be sufficient economic pressure to force use of this standard in the near future.

The first task of the designer is to provide the data for input to the drawings. This will take the form of a table of allowable tolerances. This would then be transferred to a drawing by an individual. In most cases in the future, the transfer will be done automatically by the lens design computer program. Such direct transfer has the great advantage of an error-free transfer. However, the designer is responsible for every number that is supplied, and the direct transfer may be so easy that the tolerances do not get a good review for thoroughness or sanity before appearing on a drawing. The human intervention of a designer reviewing the tolerances is extremely important. As a practice, this is best done in cooperation with a mechanical or fabrication engineer attached to the project, who may actually have to live with the designer's tolerances.

Figure 6.14 is a sample drawing produced using the standard, taken from Kimmel and Parks (1995). It is obvious that much information is being conveyed in the shorthand language behind the standard.

6.4 Environmental Analysis

There is another aspect of fitting a lens into the environment in which it is to live. The designer may keep the room in which the design is done to a comfortable 72°F (22°C). The lens will probably not spend its life in such

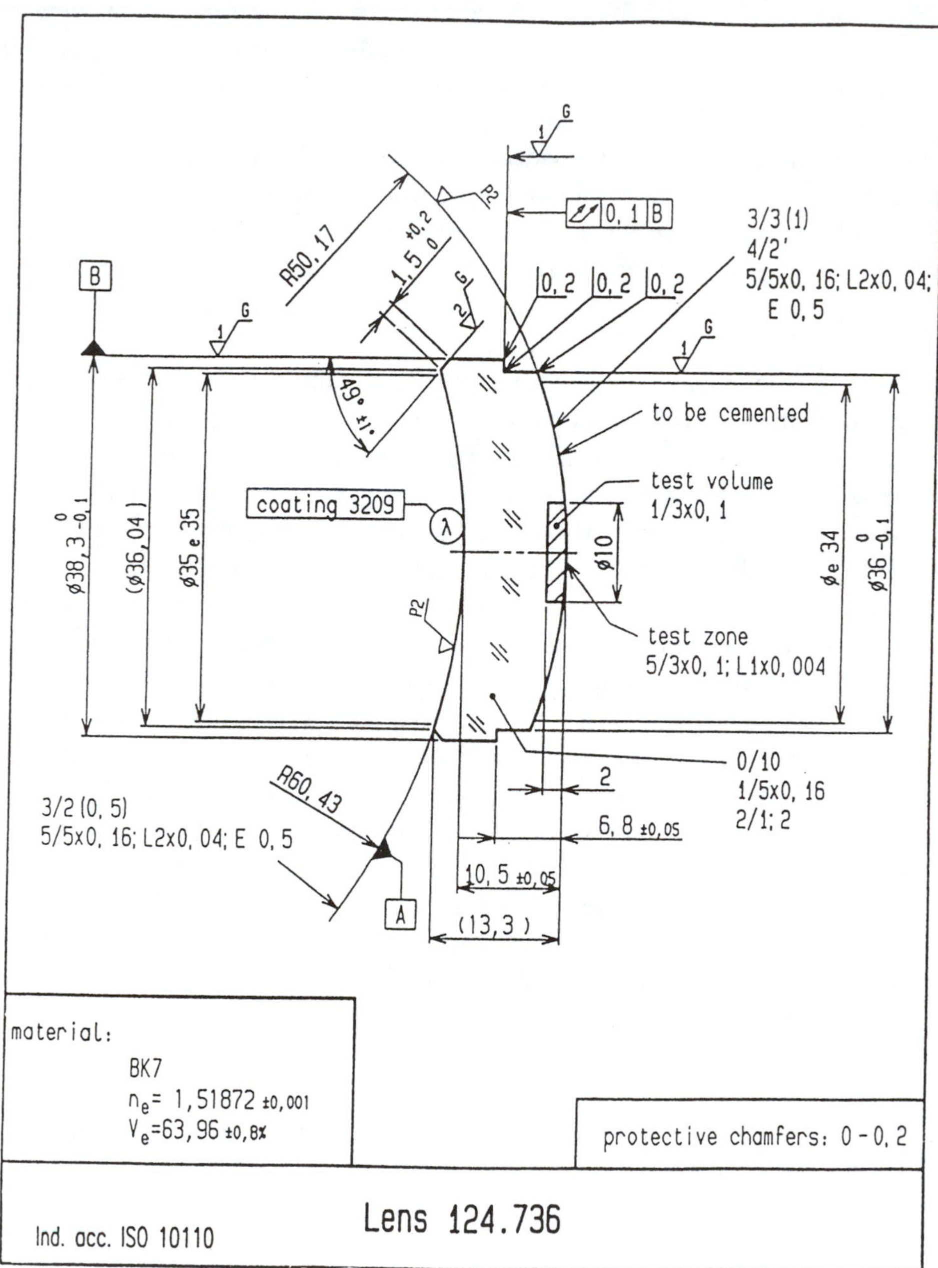

FIGURE 6.14 Sample ISO 10110 lens drawing.

a pleasant environment. The temperature may change over wide ranges, and the lens may be subjected to mechanical shocks and other unpleasant activities. Lens design programs provide the ability to determine the effect of temperature changes on the performance of a lens. The shock and vibration environment is not quite as simple, and will require the cooperation of the mechanical engineers to provide information as to the expected displacements that will occur in the lens.

The thermal model for a lens uses the expansion coefficient and index of refraction coefficient with temperature to compute the new lens shape and inserts these changes into the data stored in the program. The separation of the elements is obtained by applying the temperature change to the lens separation using an expansion coefficient representative of the mounting material. All of the usual optical quantities can then be computed using this new data.

Usually the most profound change is that of focal location with temperature. Secondarily, the image quality may change with temperature. In most cases, the mode used by the lens design program is a simple "elastic model" which does not account for gradients through or across the lens element, or for deflections introduced by the squeezing or loss of defined location of the lens element by relative contraction or expansion of the mounting material. With these limitations, a quite accurate concept of the effect of ambient environment can be calculated.

REFERENCES

Kimmel and Parks 1995 — Kimmel, R. K. and Parks, R. E. *ISO 10110 Optics and Optical Instruments – A User's Guide*, Optical Society of America, Washington, DC

Shannon 1994 — Shannon, R. R. Tolerancing techniques, Chapter 36 in *Handbook of Optics*, McGraw-Hill, New York

Vukobratovich et al. 1994 — Vukobratovich, D., Valente, T., Shannon, R., Hooker, R., and Sumner, R. Design and construction of an astrometric astrograph, *Proc. SPIE* 1752, 245

7 DESIGN EXAMPLES

We will now deal with the application of the concepts discussed in the previous chapters by applying these methods to the design of a number of lenses. These examples have been selected to provide a view of the techniques used in design at several greater levels of complexity. The majority of this chapter will deal with the detailed design of somewhat traditional, or basic, design problems, as they actually provide the basis for most of the optics done in this world. Some of the newer opportunities for design will be discussed in the latter portion of the chapter.

The goals set for the designs are somewhat arbitrary, but realistic, versions of those encountered in real life. Expanding beyond these goals will serve as a good learning experience for any individual who wishes to increase his or her knowledge of the methods of lens design.

Several different design programs will be used in these examples. A difficulty with this is that it may lead to some confusion as to the meaning of various program-dependent features that are necessary in working with a computer. The great advantage is that the essential nature of the optical problems will emerge. The design programs used, CODE V, OSLO, and ZEMAX, happen to be those used by the author for many years in teaching a course in lens design at the Optical Sciences Center.

In general, the progress of lens design methods has followed the requirements for lenses. Any lens must meet the image quality, irradiance, and field coverage requirements that the user needs. These will naturally vary for different applications. An excellent review of the many applications and requirements for lenses can be found in the book by Ray (1988), and the reader is referred to that book for application discussions. In this chapter on design examples, the requirements will be stated for each system, without delving too deeply into the origin of the specifications. The books by Smith (1992) and by Laikin (1995), as well as a portion of the book by Cox (1964) provide many examples of lenses. The various available lens design programs provide libraries of patent and other designs drawn from the literature that can be used both as samples and starting points for future designs.

There are no fixed rules regarding the approach to the use of lens design programs. Each lens design problem can be approached in a number of ways. In the examples discussed here, methods are suggested, and in some cases some alternate approaches demonstrated. A good design program will have literally hundreds of possible options and thousands of combinations of options for controlling the operation of the program. The basic principles of design can be conveyed without getting lost in this sea of possibilities, as will be shown in these examples. Certainly users of the programs will develop their own approaches and set of "best ways" to do design.

Program operation problems can be controlled by referring to the specific program manuals for the program that is being used. Since most programs are updated several times a year, the content of these manuals is a moving target, and will likely be different in detail by the time the reader approaches this book. Those who make a career of lens design will therefore find that their reading enjoyment of program operation manuals is varied and unending. New approaches to the use of menus and graphical features has made interacting with the programs somewhat less painful, but the need for the designer to understand the basic concepts involved in design will always remain.

7.1 Basic Design Forms

The vast majority of lens designs are based upon the use of spherical or aspheric surfaces to meet specified imaging targets. In some cases the fabrication of these lenses may use decidedly new fabrication techniques, such as direct glass or plastic molding, or mounting techniques that do not depend upon the usual centering and cementing techniques. The design specifications usually presume a symmetric field and a circular aperture.

Sometimes these design forms are called "classical" or "traditional" forms. This is a result of the understanding of the principles driving the design having developed over many years. In all cases, the specific forms of lenses achieved today are considerably different from those that were in use only a few years ago, and generally provide greatly superior image quality. In the sections that follow, the path travels from a detailed investigation of the simplest basic forms to approaches required for extremely complex lenses.

7.1.1 DOUBLETS

The achromatic doublet represents the simplest type of corrected objective. The use of two materials with differing dispersive properties permits control of chromatic aberrations. The shape or bending of the individual elements permits control of the spherical aberration and coma. The astigmatism and Petzval curvature are intrinsically fixed by the power of the lens. Because the stop is located at the elements, there are not enough parameters available to permit correction over a wide field of view. The design of a doublet provides a simple basis for a thorough investigation of the basic issues involved in balancing of aberrations. The design concentrates on balancing of the axial aberrations, that is, the spherical and chromatic aberrations. The narrow field limit imposed by the field curvature and astigmatism permits correction of the lateral color and coma to be carried out as correction of the first- and third-order aberration alone. In this section, a typical doublet will be designed, and the options and limitations imposed by the design configuration examined.

The goal will be to design a 100 mm focal length, F/2.8 doublet for the nominal "visual" spectral range. The range is between the *F* and *C* wavelengths, and will use the *d* wavelength as the reference wavelength. The nominal field that can actually be covered is small, and the effect of the astigmatism will be evaluated after balancing the aberrations in the central portion of the field. The object will be located at infinity.

The starting point is the thin-lens power distribution obtained in Chapter 3 for the first-order design of doublets. A pair of glasses that have a wide difference in *V*-numbers is a reasonable starting point. Examination of glass catalogs suggest that two easy-to-work, inexpensive, chemically stable glasses from the Schott Catalog are Bk7 (517642) and F2 (620364). The difference in *V*-number is 27.8, which results in 2.3 times excess power being required for the Bk7 element.

The solution for the powers is found by applying equation 3.40 and leads to

$$\begin{aligned} \phi_a &= 2.30827\phi = 0.0230827\,\text{mm}^{-1} \\ \phi_b &= -1.30827\phi = -0.0130827\,\text{mm}^{-1} \end{aligned} \tag{7.1}$$

for the two elements. The curvatures of these elements are

$$\begin{aligned} c_a &= 0.044664\,\text{mm}^{-1} \\ c_b &= -0.021099\,\text{mm}^{-1} \end{aligned} \tag{7.2}$$

This thin-lens power solution can be inserted directly into a lens design program, with zero-thickness elements. An initial decision upon the shape of the lenses needs to be made. Because automatic correction will be used, there is no need to be meticulous about this, as the computer program will rapidly seek a solution. The only requirement is that the starting point be a reasonable one. (There is also no need to use very many significant figures in the lens setup, but, as they are there, why not?) A useful arbitrary starting point is to make the crown element equiconvex, with curvatures of $0.022332\,\mathrm{mm}^{-1}$ on the first surface, and of $-0.022332\,\mathrm{mm}^{-1}$ on the second surface. This corresponds to radii of 44.778 mm on each surface. The first attempt at design will presume that the elements are cemented together, thus the curvature of the first surface of the second lens will have to match the curvature of the back surface of the first lens. This would lead to curvatures of $-0.022332\,\mathrm{mm}^{-1}$ and $-0.001238\,\mathrm{mm}^{-1}$, giving a radius of −807.754 mm on the rear surface.

Placing these into a lens design program also requires establishing the aperture size of the bundle entering the lens. The object distance is at infinity so that the entrance beam and entrance pupil diameter will simply be the focal length divided by the F/number, or 35.7142 mm. The location of the stop also needs to be specified. For this narrow-angle design there is no reason not to take the first surface as the location of the stop.

Optical infinity is defined by a zero slope angle of the entering marginal ray, and is a well-defined concept. In order to feed the computer's desire for finite numbers, and still retain the possibility of very distant objects, almost all programs treat infinity as a finite but very large number. An object distance of 10^{10} mm or greater is usually taken as equivalent to infinity. In most cases, this is quite adequate.

The other entry data required is a semifield angle. Nominally this can be taken as 5°, as long as the aberration correction to be carried out is specified only in the region of the axis. Programs usually require a statement of a specific object height, usually taken as negative to make the image height come out positive in the image space. Some programs permit insertion of a field angle and compute the appropriate height, also inserting a default value for the object distance.

The wavelengths to be used are also inserted as starting data for the lens. The wavelengths chosen usually cover the range of sensitivity of the detector to be used, and one of these must be identified as the reference wavelength. The choice for this doublet is the d wavelength as the reference wavelength and the F and C wavelengths as the ends of the spectral range of interest. This selection can later be refined as the actual spectral response of the detector is introduced. Initially, we will presume that all wavelengths have equal weight.

The paraxial heights and angles are computed by the program for the selected reference wavelength. The location and size of the entrance and exit pupils is established. The clear apertures required on the surfaces are computed using the paraxial ray data, although real rays will eventually be specified for setting the apertures.

The insertion of glass types into a program is usually accomplished by stating the name of the glasses. The program searches a stored glass catalog, obtains the dispersion coefficients, and calculates the needed indices of refraction. In some cases the designer will have to insert the known indices of the materials directly. There are also approaches where a glass material model is used for some computations.

The order of data insertion into the computer program usually starts with basic lens data, such as entrance pupil diameter, field angle, wavelengths, and object distance. The data for each surface is inserted in sequence: radius, separation, glass type. Added information, such as clear apertures, aspheric surface type, and so on may be required, but is not crucial for this problem at the early stages. The designer must also select some surface to serve as the location of the aperture stop. For this doublet we will, as noted, take the first surface as the stop location.

The last surface inserted will be the image surface. In modern design programs the image surface is treated as a normal surface with the exception that there is no thickness of medium following the surface, and the program knows that all image data should be computed on that surface. The image surface should initially be placed at the location of the paraxial image surface. In some programs, an offset or adjustment for the specified reference focal location will be attached to the image surface. In other programs, a dummy surface can be used to identify the paraxial focal location, and the defocus adjustment to the desired focus location introduced as the last space prior to the image surface.

Each program will have some set of default quantities that will substitute for entries that are not explicitly stated. Usually, for wavelength, these are the F, d, and C wavelengths. In most cases, defaults of an infinity object, a nominal field angle, and nominal entrance pupil diameter and surface location are built in. Note that not all programs have graceful default numbers, however. For each surface, plano or zero curvature, zero thickness, and air as the medium following the surface are the usual three defaults. (There is a possible minor exception. Some programs may explicitly use the first surface as the entrance pupil surface, making surface 2 the first lens surface. Obviously this is program specific and attention must be paid to the program manual in order to discover these points and keep all the surface data straight.)

If the system parameter numbers are correctly inserted into a lens design program, the output verifies the paraxial data for the lens. It is also useful to verify that the desired color correction has been achieved. A sample calculation is shown in Figure 7.1 for the paraxial data, as well as the chromatic and

```
    Book Doublet no. 1
    Paraxial values and 3rd order transverse aberrations -      587.6 nm
          EFL            = 100.000000         IMAGE F/NO    =  2.800022
          IMAGE DISTANCE = 100.000000         IMAGE HEIGHT =  1.000000

SURF   Y(MARG)    U(MARG)    NI(MARG)   Y(CHIEF)    U(CHIEF)   NI(CHIEF)
       SPH ABER   TAN COMA   TAN ASTIG  SAG ASTIG    PETZVAL   DISTORTION
       AX COLOR   LAT COLOR  PETZ CURV
                   (U is after refraction)
 EP  17.857000    0.000000              0.000000    0.010000


STO  17.857000  -0.135872    0.398783   0.000000    0.006593    0.010000
     -0.712268  -0.053583   -0.002023  -0.001127   -0.000679   -0.000028
     -0.211831  -0.005312   -0.007609

  2  17.857000  -0.101800   -0.810963   0.000000    0.006173    0.010000
      0.879286  -0.032527    0.000485   0.000217    0.000084   -0.000003
      0.423067  -0.005217    0.000938

  3  17.857000  -0.178570   -0.200585   0.000000    0.010000    0.010000
     -0.232819   0.034821   -0.001778  -0.000621   -0.000042    0.000031
     -0.211191   0.010529   -0.000472

IMG   0.000000  -0.178570               1.000000    0.010000

SUM  -0.065801  -0.051290   -0.003316  -0.001531   -0.000638    0.000000
      0.000044   0.000000   -0.007142
```

FIGURE 7.1 Paraxial ray trace showing chromatic and third-order aberration data for thin doublet (CODE V).

```
    Book Doublet no. 1
         POSITION 1
                  587.6 NM                                  656.3 NM
                  486.1 NM
 COORDINATES   RAY ABERRATIONS 587.6 NM          RAY ABERRATIONS  656.3 NM
               RAY ABERRATIONS 486.1 NM
   X      Y      DELTA X    DELTA Y      OPD       DELTA X    DELTA Y       OPD
                 DELTA X    DELTA Y      OPD
FOCUS = (   0.000000)
 1.00   0.00     0.205060   0.000000   -7.535-ET    0.184870   0.000000   -6.192
                 0.289967   0.000000  -17.208
 0.80   0.00     0.039790   0.000000   -0.747-ET    0.034253   0.000000   -0.806
                 0.080135   0.000000   -4.283
 0.60   0.00     0.001058   0.000000    0.204-ET    0.001622   0.000000   -0.068
                 0.018952   0.000000   -1.020
 0.40   0.00    -0.002371   0.000000    0.092-ET   -0.000132   0.000000   -0.076
                 0.004647   0.000000   -0.277
```

FIGURE 7.2 Ray intercept data for initial doublet setup.

third-order aberrations. The designer should examine this listing to ensure that the lens is optically reasonable. Another computation, shown in Figure 7.2, provides data on tracing of ray fans. These will usually be quite large at the beginning of the design.

The next step will be to change the lens variables to correct the third-order spherical aberration and the chromatic aberration. In all programs this requires the definition of variables and aberration targets for the merit function.

There are several approaches to selecting variables. At first we will select only the three available curvatures of the doublet. These will be used to correct the spherical aberration, the chromatic aberration, and the power, or focal length. The focal length can be stated explicitly as an aberration target, or may be included as a specific solve within the lens data.

The surface or separation solves are important aids in controlling the behavior of a lens in a design program. These permit the designer to specify desired paraxial optical properties on a parameter, and have the program set the actual surface parameter to the desired value. An important solve that is usually required is one which sets the paraxial image location as the final surface, or the image surface. The program permits the inclusion of a paraxial thickness solve on a surface by stating some declared code which will differ between programs. Usually something like **3 py 0.0** or **thi s3 hmy 0.0** is used on the surface in place of a statement of the thickness. In more recent programs, the checking of a box on a spreadsheet format containing the design data accomplishes the same effect. The surface instructions tell the program to set the thickness following surface 3 to be such that the paraxial y height would be zero on the next surface. In our example, surface 3 is the last surface, making surface 4 the image surface. If a solve preceding the image surface is not used, this thickness will have to be used as a variable in the optimization process to keep the paraxial marginal ray height at zero on the image surface. (At this point it is obvious that there are many ways to accomplish similar things in modern lens design programs. Each designer will find his or her own favorite approach to handling these problems.)

Another solve that may be used is to recognize that the paraxial marginal ray angle leaving the last surface can be used to set the power of the lens. Since the object is at infinity for this problem, the power or focal length can be determined by using a curvature solve on the last surface, so that the focal length is always maintained at the desired value as the first two curvatures are varied. A code such as **3 puy −0.17857** or **cuy s3 umy −0.17857** sets the paraxial marginal ray angle to −0.17857 following the lens. This will mean that an F/2.8 cone is obtained, which will correspond to 100 mm focal length for the proper entrance pupil height. The specified actual last surface of the lens itself is then an adjuster that maintains the focal length and F/number for the lens. This curvature will be removed from the available set of variables, and the aberration target for the power of the lens is no longer required because the focal length is automatically included in the lens setup. (Whether this curvature solve or explicit addition of the focal length to the merit function is used is a matter of choice for the individual designer.)

The variables are usually specified by checking boxes in a data spreadsheet or inserting some codes on each surface such as **v 1 cv 1 −0.07 .08 10** or **var s1** or **ccy s1 0**, or some other program specific notation. These instructions may include statements about the allowable boundary constraints or other limits on use of the variables. Most programs include methods for specifying global variables, such as all curvatures, for example.

The aberration targets need to be specified. Most programs include a default set of aberration targets that have been developed by the authors of the program to meet most general imaging requirements. These "canned" merit functions have been tested over many years and usually provide a convenient option for correcting the lens. In some cases these defaults are either inadequate for the purpose, or may amount to overkill for the needs of the specific designs. For the first step on this doublet we will be asking only to correct both the first-order color and the third-order spherical aberration to zero. These targets can be specified individually in the appropriate manner by stating **ax = 0.0; sa = 0.0** or perhaps **o1 spher sa3-0.0;o2 color fchy-0.0**, or any of a number of different program-dependent input patterns. In some programs it is possible to select the aberration targets from a set of menus. The literal requirement to start with is to choose the first-order longitudinal chromatic aberration to be zero, as well as the third-order spherical aberration at the reference wavelength to be zero. (As previously discussed in this book, we will be searching for a balance of low- and high-order aberrations. For the purpose of learning how to approach a design, we will methodically build up to the canned merit function.)

Once all of the data is inserted, the program is given the instruction to go into automatic solution or iteration mode. The variables are varied, and successive steps in the solution process lead, hopefully, to a system meeting the specified aberration targets. For the starting doublet not much is asked, so the answer comes easily. The aberration theory discussed in Chapter 3 indicates that a solution should be possible. The aberration targets are easily accessible and are corrected to zero under the condition that the focal length is either maintained by the curvature solve for the specified paraxial marginal angle leaving the last surface of the lens, or that the lens focal length is included in the merit function.

The lens is still mathematically a thin lens with zero central thickness. A view of the lens, as in Figure 7.3, is actually a bit amusing. The aberrations seem controllable. Although the specified aberrations are zero, the actual aberration in the image is not zero, as there are high-order aberrations and variation of spherical aberration with wavelength to contend with.

The state of correction can be assessed by examining a ray intercept plot, as in Figure 7.4. Here it can be seen that there is a large amount of residual spherical aberration. Since the third-order aberration is zero, the residual

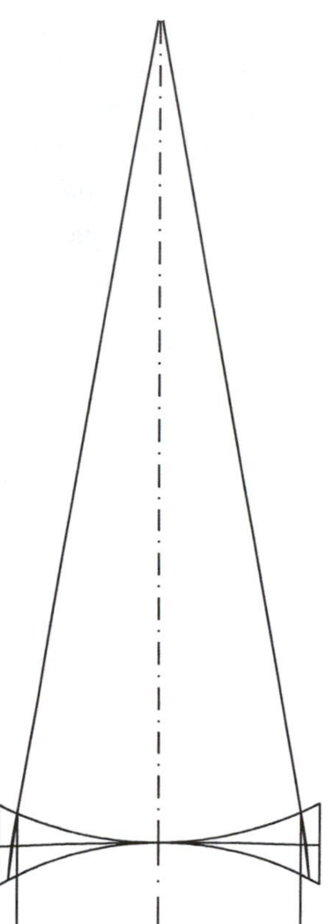

FIGURE 7.3 Layout and ray trace for thin doublet (CODE V).

must be high order (fifth and higher), which needs to be balanced by introducing the appropriate amount of third-order aberration. The spherical aberration remaining in the extreme wavelengths needs to be balanced by setting the best focus in the presence of spherical aberration to a common focus for each wavelength.

Before continuing with refining the aberration correction, it is appropriate to introduce some reasonable thicknesses for the lens, as shown in Figure 7.5. The aberration plots, as in Figure 7.6, are not significantly changed by the introduction of thickness, confirming the validity of beginning with the thin-lens power solution.

There are several approaches to the next step in aberration correction. High-order spherical aberration coefficients and chromatic and spherical aberration coefficients calculated at the extreme wavelengths could be computed and used to construct new aberrations. An easier way is to trace three rays, one at the edge of the aperture at wavelength *d* and two at the 0.707 height in the aperture at wavelengths *C* and *F*. Construction of ray aberrations to make the edge ray have zero aberration in the paraxial focal plane

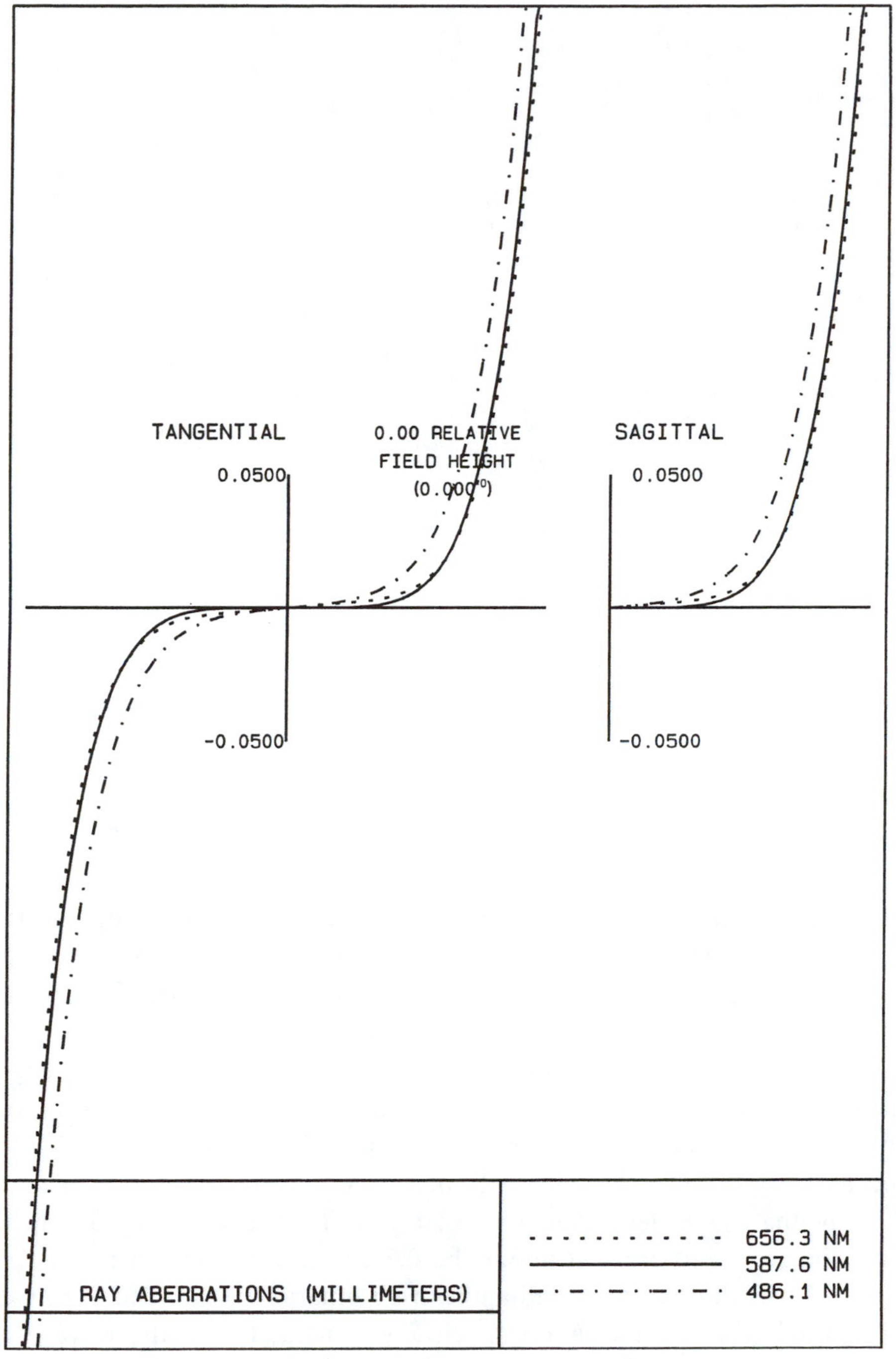

FIGURE 7.4 Axial ray intercept plot (aberrations in mm) for starting doublet (CODE V).

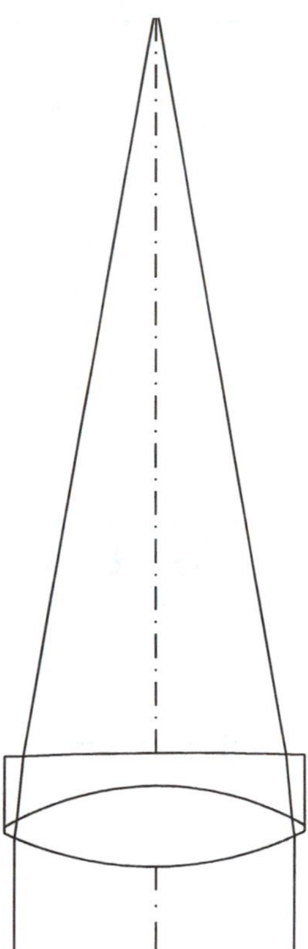

FIGURE 7.5 Layout for doublet with thicknesses inserted (CODE V).

and make the aberration in the *C* and *F* rays cancel at the 0.707 aperture height will accomplish inclusion of these balancing aberrations. These are included as specific aberration constraints or targets in the set of aberration targets. The same variables are used.

The ray intercept plots are now as in Figure 7.7. The reference wavelength, *d*, shows the high-order overcorrection balanced by low-order undercorrection. The ray intercept curves for the *C* (red) and the *F* (blue) wavelengths indicate a variation in the spherical aberration and focus with wavelength, including the low-order spherical aberration. The balance that is chosen causes the curves to cross at about the 0.7 aperture height zone. In effect, this moves the best focus, including both chromatic defocus and chromatic variation of spherical aberration, to closely coincide across the wavelength region. The presence of secondary chromatic aberration prevents the combined focus at all wavelengths from falling at the paraxial focus location.

The zonal focus for the wavelengths at the edge of the spectral region does not coincide with the zonal aberration of the central reference wavelength. This is due to the residual secondary color, resulting from the limitations

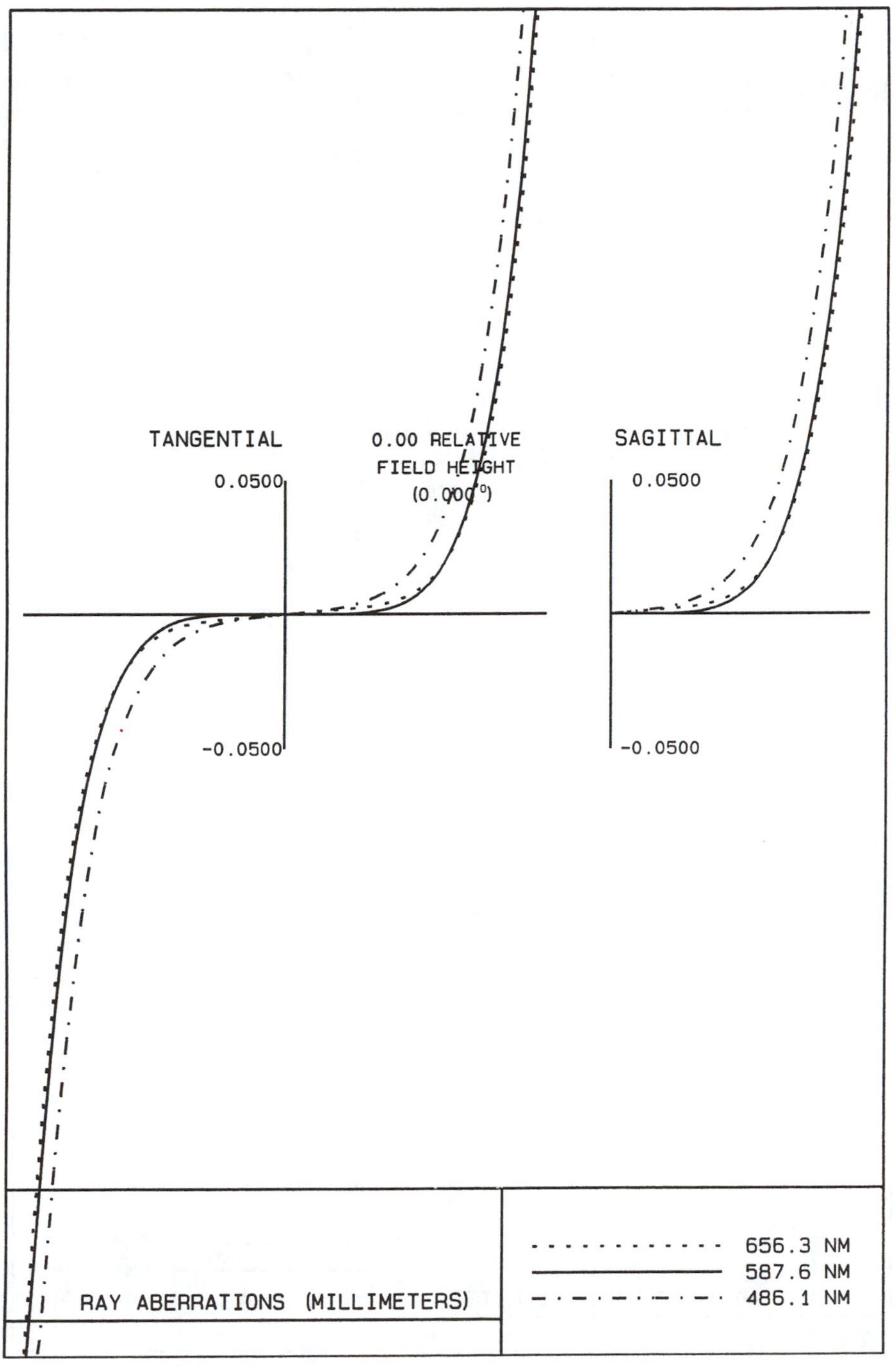

FIGURE 7.6 Ray intercept plot for thick doublet (CODE V).

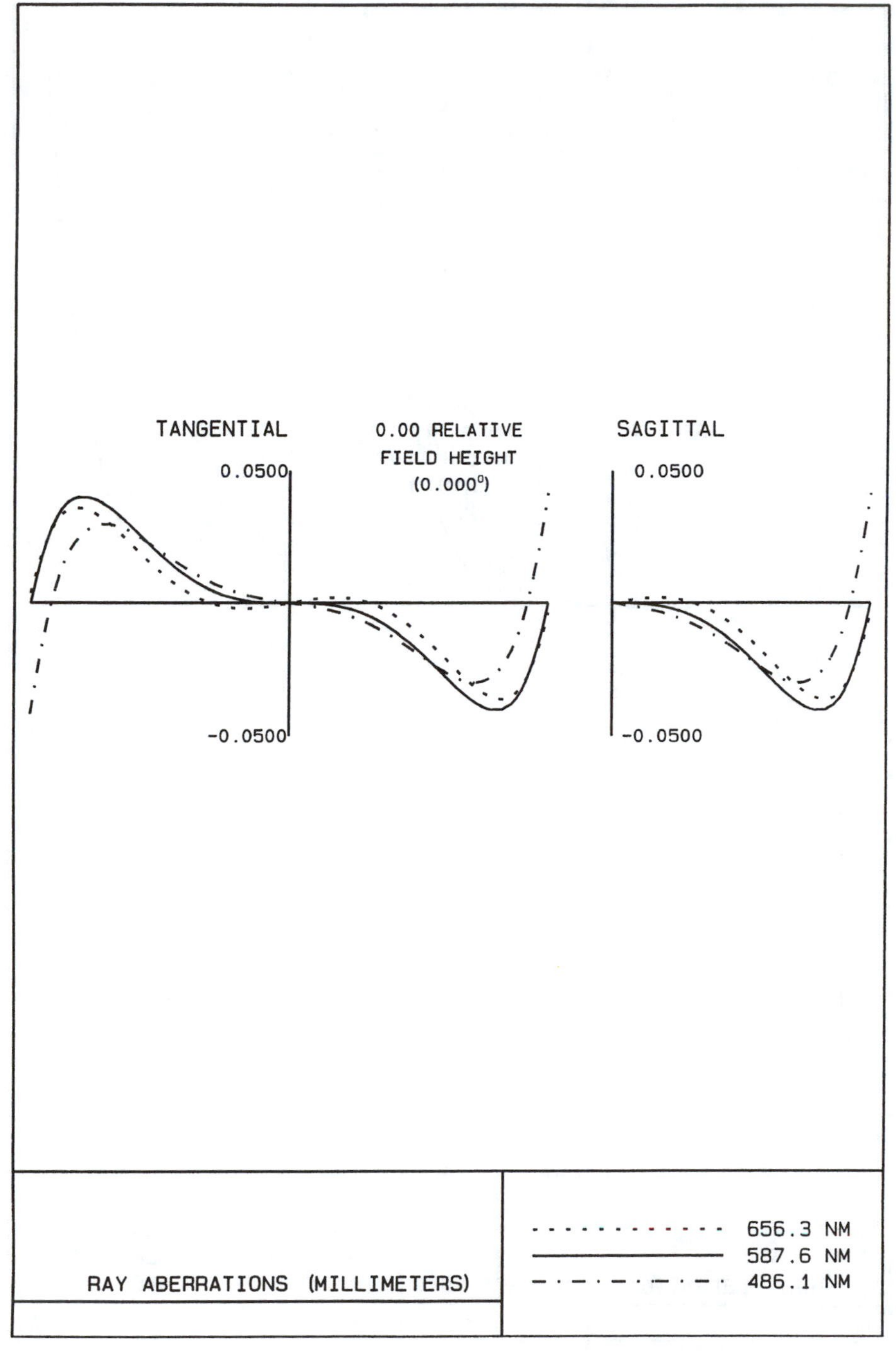

FIGURE 7.7 Ray intercept plot for doublet after correction (CODE V).

imposed by using ordinary glasses for the doublet design. At F/2.8, the spherochromatism and zonal spherical aberration is significantly larger than the effects due to the secondary color.

The high-order aberrations have now been balanced by the introduction of appropriate amounts of primary chromatic aberration, which indicates an improved light concentration in the image. At this point, the variables are exhausted, and no more improvement appears likely. The design has reached a minimum of the merit function. For this approach, the chosen aberrations have been corrected to zero. The measure of the aberration remaining in the lens is the residual aberration left over after the correction, or balancing, has taken place.

In order to evaluate the state of correction it is necessary to include some measure of the residual aberration. This can be done by adding to the merit function some errors describing the residual aberrations. A simple characterizing function is the size of the zonal spherical aberration remaining, given by the amount of aberration of the ray at wavelength *d* at the 0.7 image height. This is proportional to the amount of high-order aberration that is being balanced by the introduction of low-order aberration. Some other useful measures include the amount of high-order aberration, or the amount of third-order aberration that needs to be introduced to balance the high-order errors.

Other more general measures include the RMS wavefront error or the RMS or second moment of the ray distribution left in the image. More complex image quality measures include the MTF at some critical spatial frequencies. This needs to be evaluated to determine whether the image quality obtained from this design meets the customer's requirements.

This residual aberration measurement, such as the zonal aberration, can be added to the merit function, and the iteration continued to see if a better region of solution exists. No new variables exist which permit correction of this residual aberration to a specific target. An attempt to do this will not be successful, because all of the available variables are committed to obtaining a stated aberration balance. Attempts to carry out impossible correction may lead to a complete stagnation or divergence of the design. The lens now appears as in Figure 7.8.

The minimum for the merit function that was located is usually only the closest minimum to the starting point for the design. A search can be carried out to locate the lowest possible, or global, minimum. In order for the remaining size of the merit function to be as significant, it must relate to the effect of the residual aberration on image quality.

The search for alternate minima can be accomplished by the designer by changing the parameters of the lens and starting the optimization process again. (An alternate approach is to use the global search options in the design

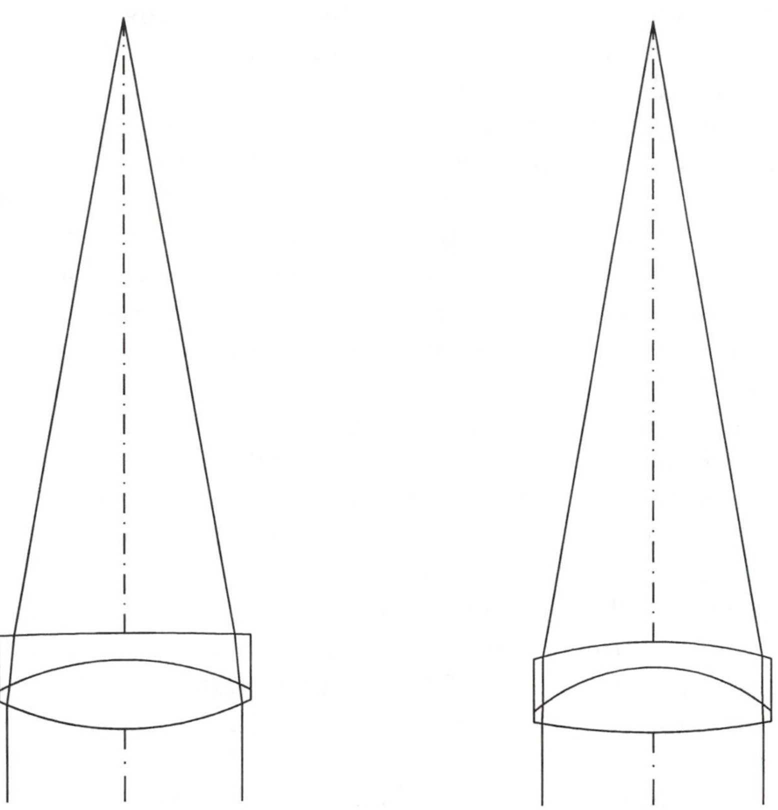

FIGURE 7.8 Layout of corrected doublet (CODE V).

FIGURE 7.9 Layout of alternate solution for doublet (CODE V).

program.) An example of a designer-controlled search for the doublet finds another local minimum. The lens was bent into an extreme form, with the first surface being plano, and the other two surfaces following to maintain the color and the power. The optimization led to another solution shown in Figure 7.9. The aberration was corrected according to the rules for the two selected rays, and is shown in Figure 7.10. Clearly the amount of residual zonal aberration is greater for this solution, so that the designer would likely select the first solution found. Another attempt at a different start with the lens bent extremely to the right leads to the preferred solution directly, indicating that this is likely the best solution within the stated region.

The best solution using this merit function, found after several iterations and attempts at different starting points, is shown in Figure 7.11. In each case the ray intercept plot and lens data are very similar, indicating that this is the best that can be expected for the given lens parameters. The associated ray intercept plots are shown in Figure 7.12. The balance at the edge of the aperture follows the prescribed best choice of the spherical aberration bal-

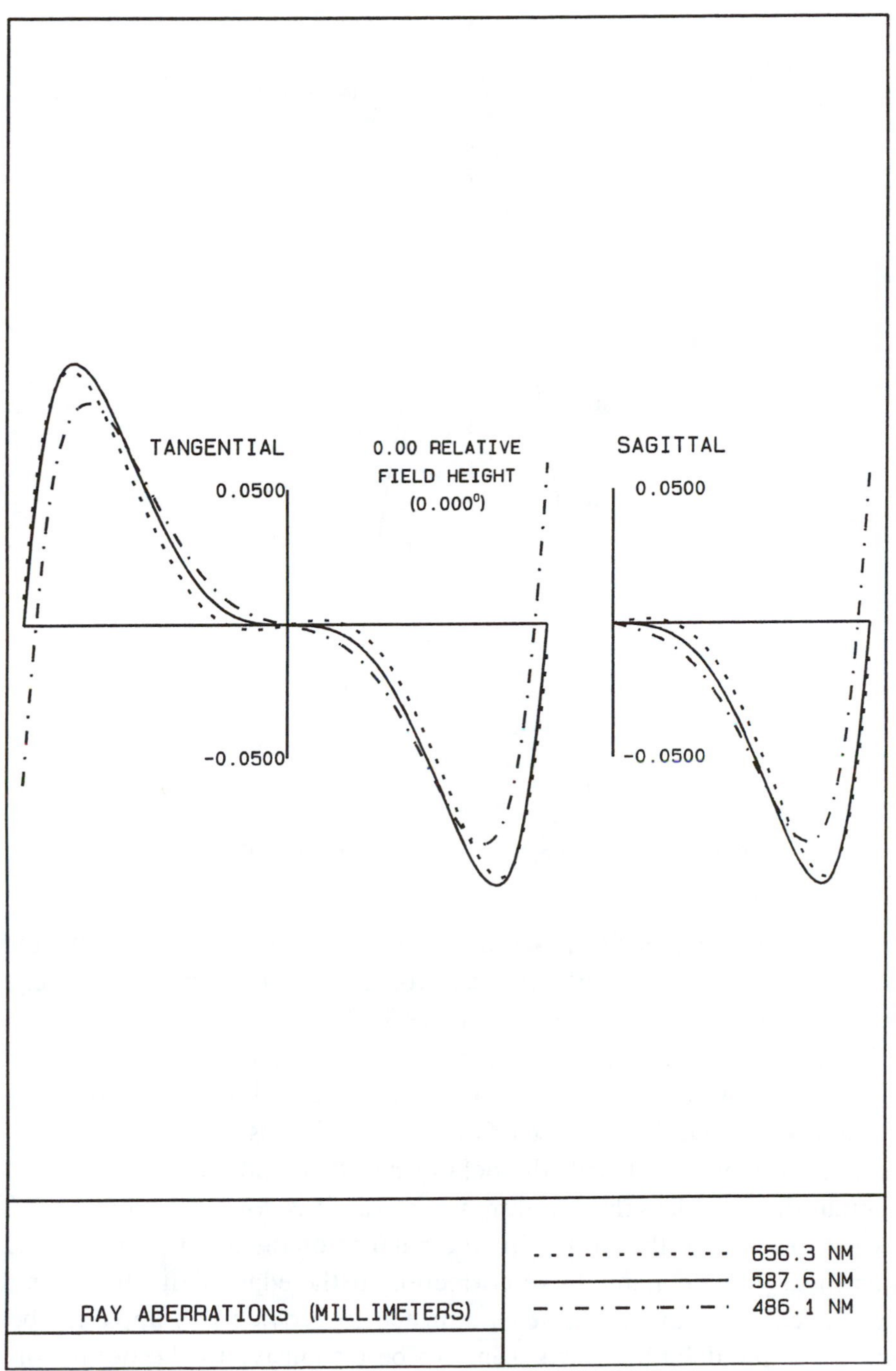

FIGURE 7.10 Ray intercept plot of alternate solution (CODE V).

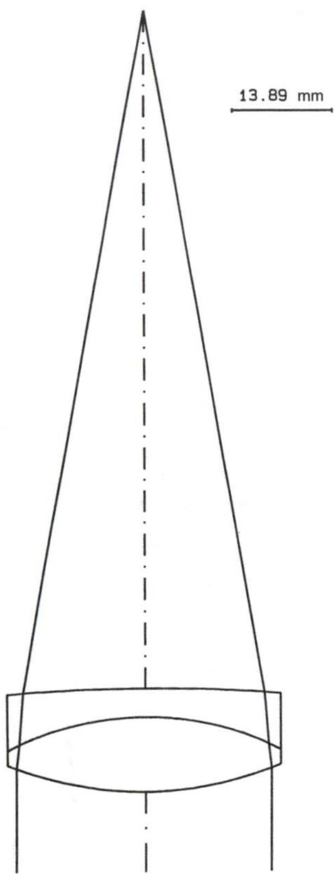

FIGURE 7.11 Layout of optimized doublet (CODE V).

ance. The best focus position will not lie exactly at the paraxial focus but at some small distance inside the paraxial focus. The ray intercept plot for this best focus adjustment is shown in Figure 7.13.

The question is whether another balance of the aberrations can achieve a better solution. Here the option is to choose the default merit function for the program being used (in this case CODE V). The lens is again optimized using the same variables, and with the default merit function and a constraint to maintain the focal length at 100 mm. The result is seen in Figure 7.14. The lens is substantially the same, but the balance of the aberrations is slightly adjusted to provide a small overcorrection at the edge of the aperture. The merit function so far used can be adjusted to provide this apparent better balance, or the default, which seems to be responsive to the desires of the designer, can be used. This default choice of the merit function will be used for further investigations into the design type. (At this point the novice designer may ask, why not use this default merit function at the beginning? This can, of course, be done, but as explained elsewhere in this book, the designer needs to establish control of the design process and understand what

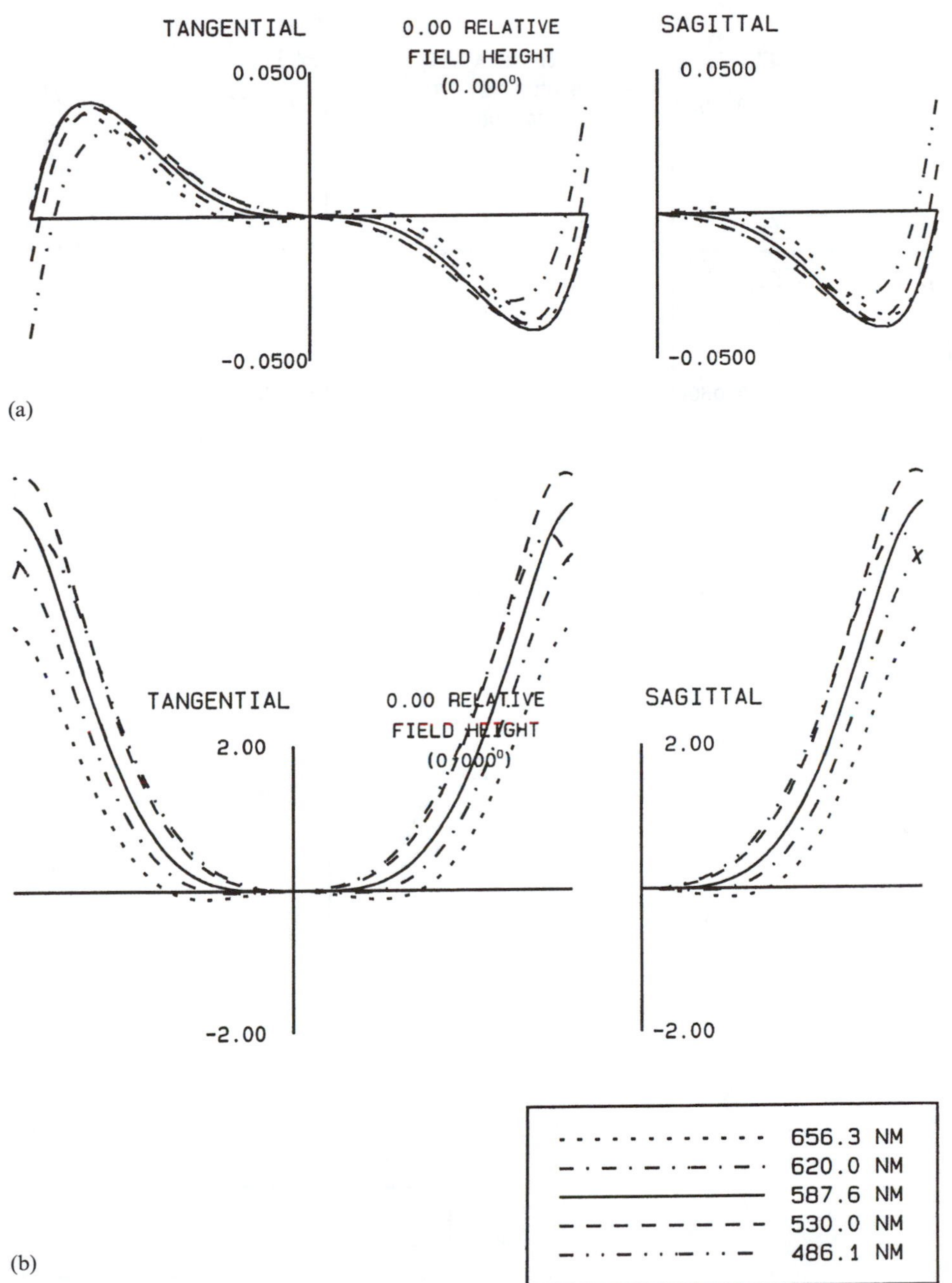

FIGURE 7.12 Optimized doublet. (a) Ray intercept plot (mm); (b) OPD plot (waves) (CODE V).

the merit function is doing. It should be obvious that many approaches will yield similar solutions.)

Since a minimum of the merit function has been obtained, it is now appropriate to evaluate the image quality in detail. A quantitative evaluation of the

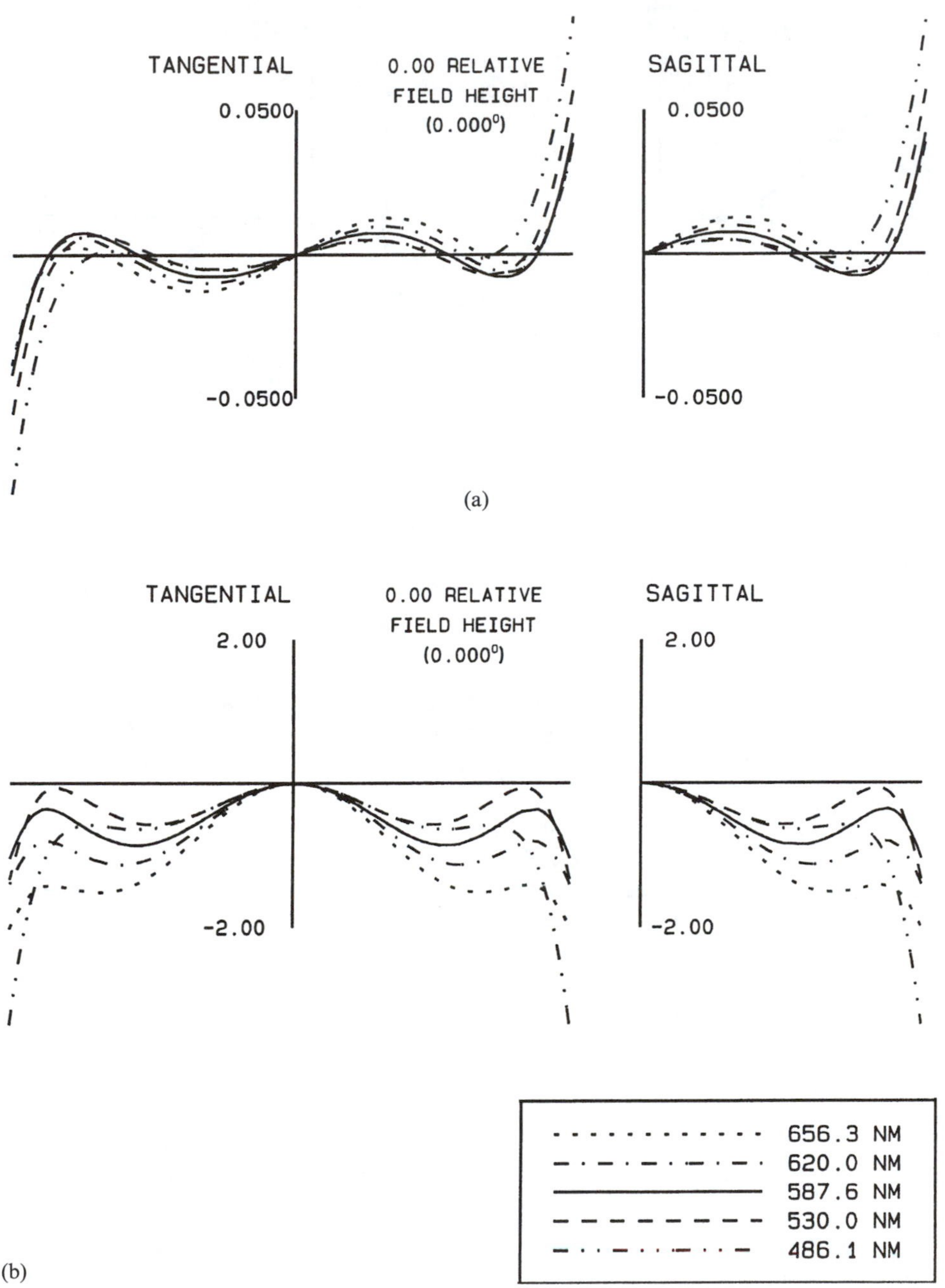

FIGURE 7.13 Optimum focus. (a) Ray intercept plot (mm); (b) OPD plot (waves) (CODE V).

MTF as a function of focal position is shown in Figure 7.15, for a spatial frequency of 30 lines/mm. The best focus will change slightly with the choice of spatial frequency, but an average best choice will be at about −0.15 mm. The MTF versus spatial frequency plot is given at this focus in Figure 7.16.

The design has been optimized for F/2.8, which corresponds to a reasonably large numerical aperture. Once control over the design has been achieved by the choice of the variables and an appropriate merit function, the design could easily be adjusted for a number of different relative apertures. The same merit function has been applied to F/3.0, F/3.5, F/4.0, F/4.5, and F/5.0, with the results shown in Figures 7.17–7.21. The lens improves significantly between F/3.5 and F/4.0, and approaches a diffraction-limited value for smaller numerical apertures. The difference between the results shows the prominence of spherical aberration and spherochromatism as image quality limiters at large relative apertures, and the importance of secondary color for small relative apertures. There are several ways of expressing these results, but a study of this sort provides some perspective on the extent to which the lens can be improved.

Other optical materials may be chosen. Once the design is under control, it is easy to change the materials and compare the results. There have been many studies of glass choice in achromatic doublets in the past, but a good approach when a design program is available is to permit glass variation. As discussed in Chapter 5, the variation of glass types requires a definition of a glass model and boundaries upon the range of allowable materials. The designer must ensure that a real lens results by replacing the glass model with the closest real glasses for the final design. In order to search the possible choices, a few iterations in CODE V were carried out with the contact doublet. After several tries, a suggested improved material choice was SK5 and BaSF54. The ray intercept curves for this lens are shown in Figure 7.22. The MTF at best focus is shown in Figure 7.22. For completeness, both of the designs are listed in Figure 7.23, in the form used by the computer program.

The result shows that some improvement in the MTF has occurred by the better choice of materials. Comparison of the MTF in Figure 7.22 with the original MTF curves in Figure 7.16 indicates some improvement. Whether the gain is worth the use of the new glasses is a decision that must be made jointly by the designer and the customer for the lens.

So far, the design has considered only the cemented doublet, with the restriction that the front curvature of the rear element must match the curvature of the rear of the front element. The three variables are all used in setting the aberrations so far discussed. It would be useful to ensure that the lens covered some finite field with stationary, or isoplanatic, imagery. For this the coma must be corrected. The coma for the contact design is small, but making it zero is desirable, and possible.

The technique used is to add another variable by breaking the cemented contact between the lenses. There are now four curvatures available, and four conditions to be satisfied. The additional design freedom may also prove useful in locating a region of solution in which the residual zonal aberration

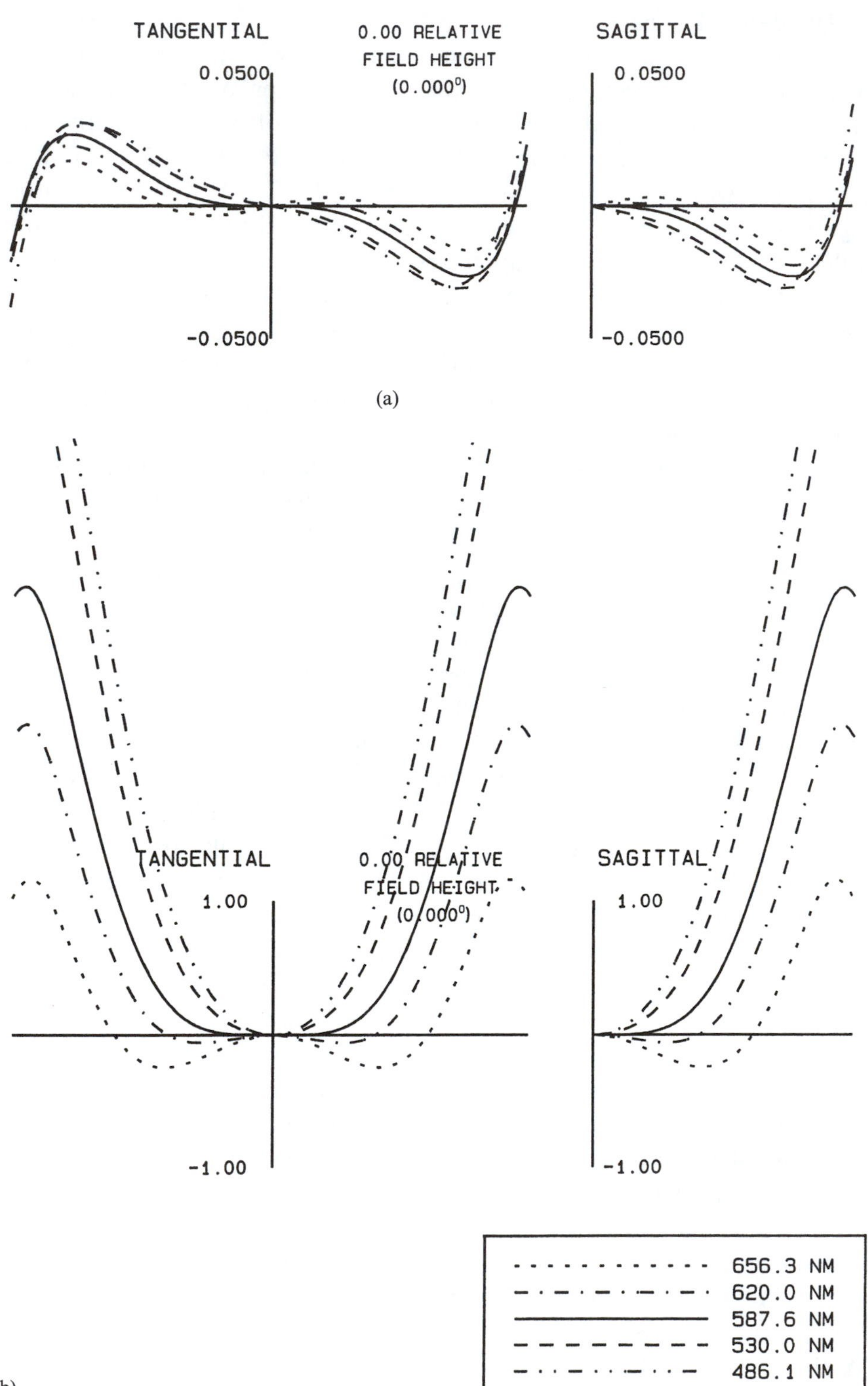

TANGENTIAL
0.00 RELATIVE
FIELD HEIGHT
(0.000°)
SAGITTAL
0.0500
-0.0500
0.0500
-0.0500
(a)
TANGENTIAL
0.00 RELATIVE
FIELD HEIGHT
(0.000°)
SAGITTAL
1.00
-1.00
1.00
-1.00
656.3 NM
620.0 NM
587.6 NM
530.0 NM
486.1 NM
(b)

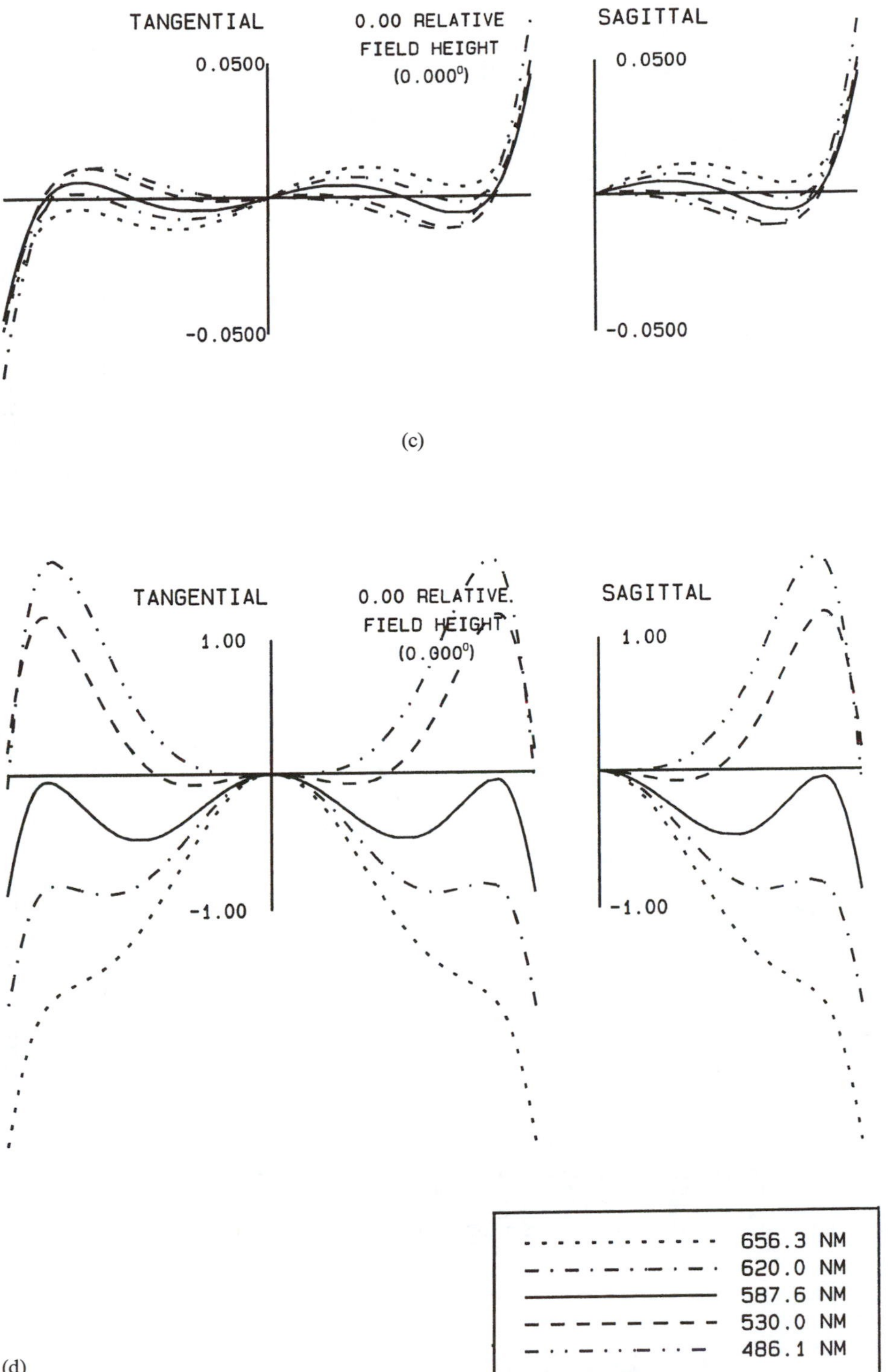

FIGURE 7.14 Doublet after default optimization. (a) Ray intercept plot at paraxial focus (mm); (b) OPD plot at paraxial focus; (c) ray intercept plot at best focus (mm); (d) OPD plot at best focus (waves) (CODE V).

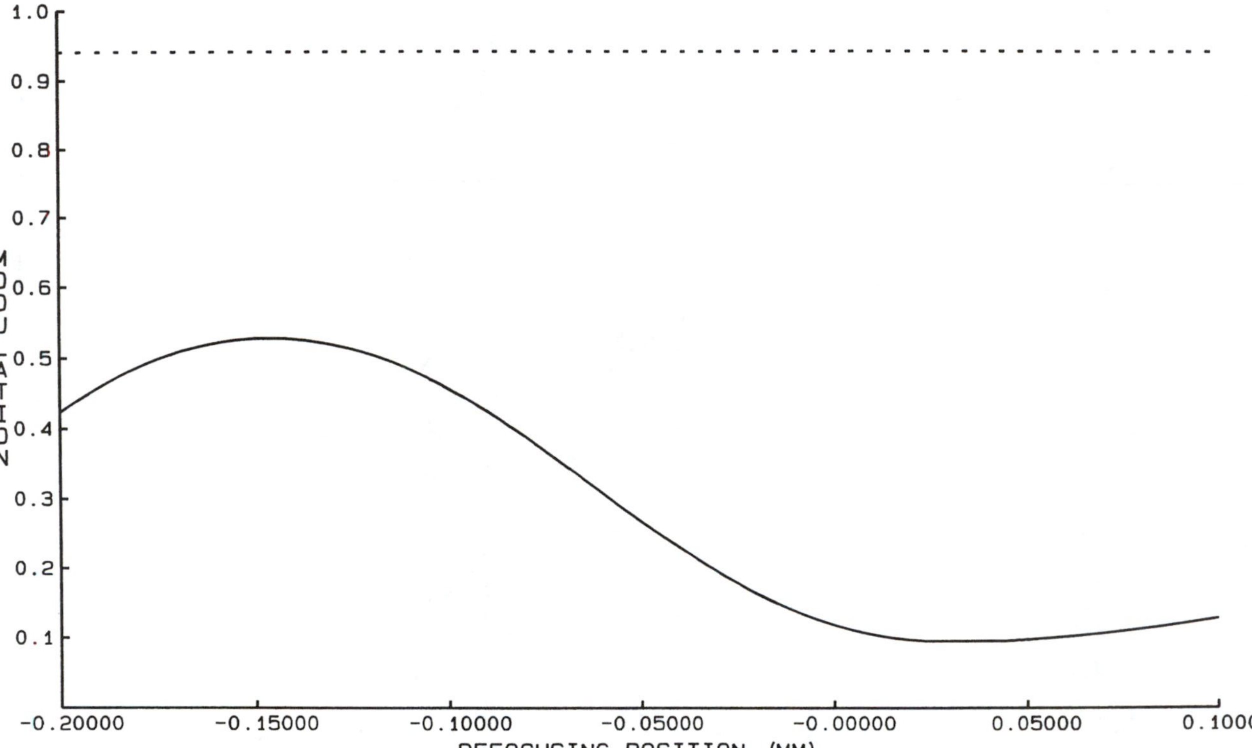

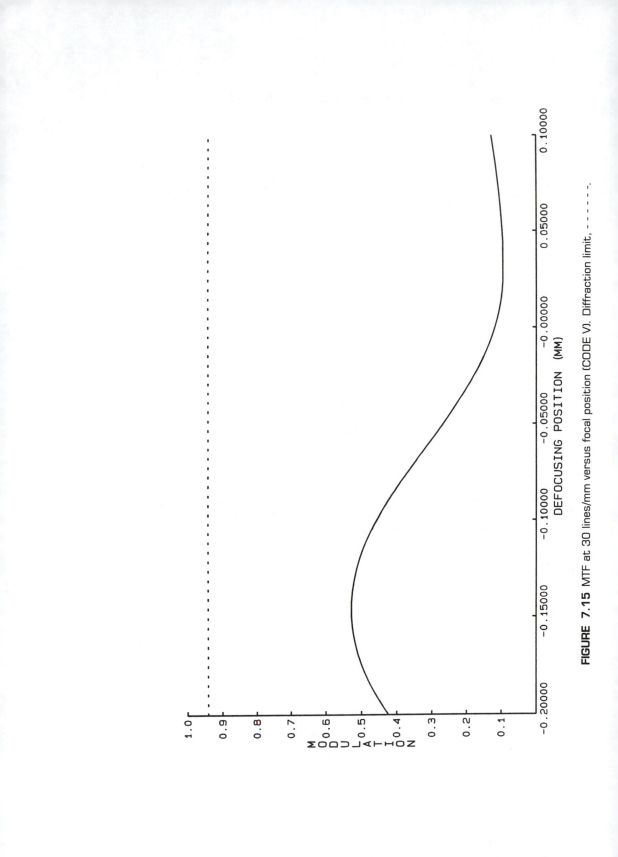

FIGURE 7.15 MTF at 30 lines/mm versus focal position (CODE V). Diffraction limit, - - - - - -.

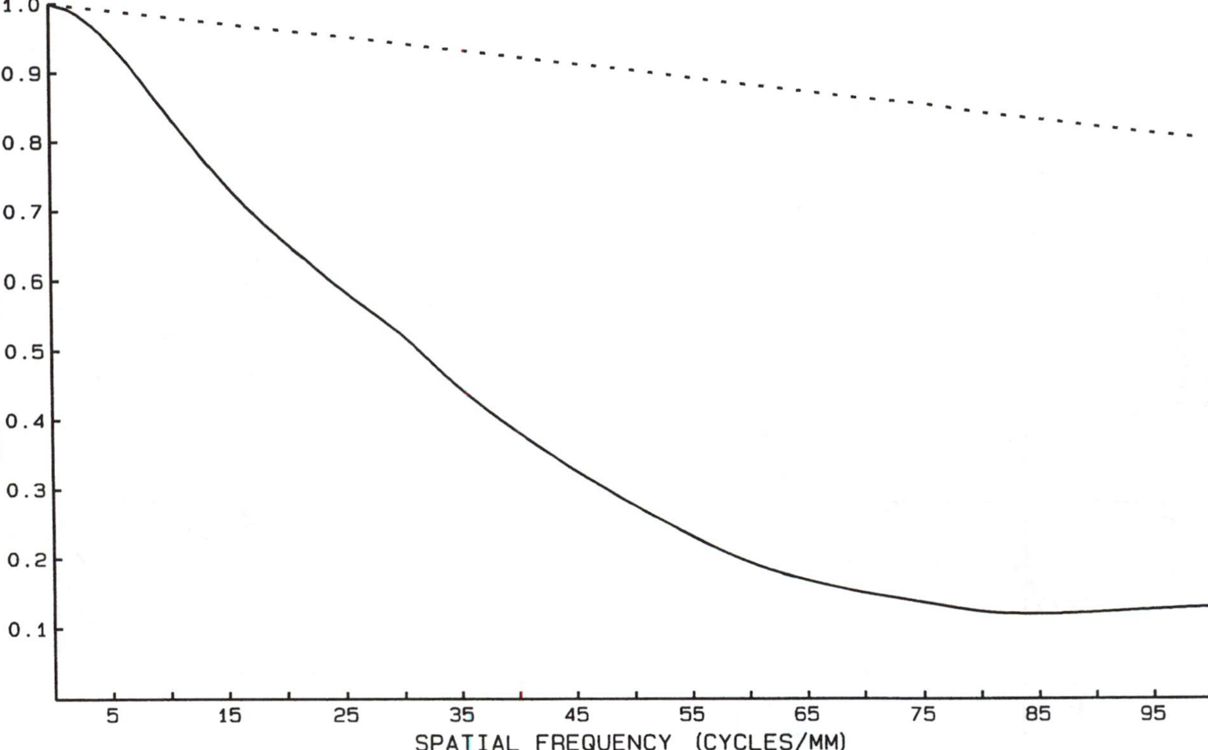

FIGURE 7.16 MTF at best focal position (CODE V). Diffraction limit, - - - - - -.

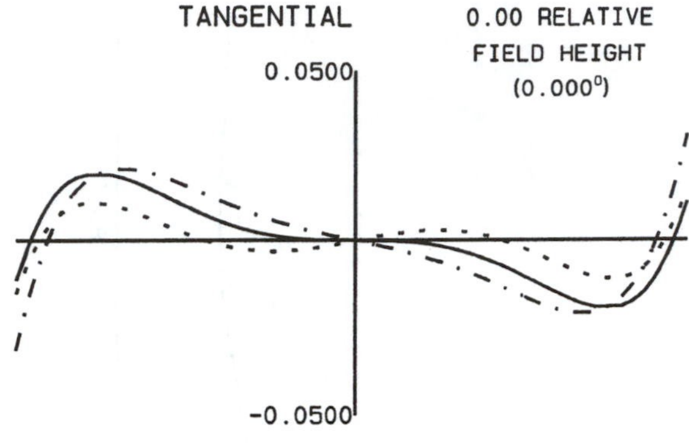
TANGENTIAL
0.0500
-0.0500
0.00 RELATIVE
FIELD HEIGHT
(0.000°)
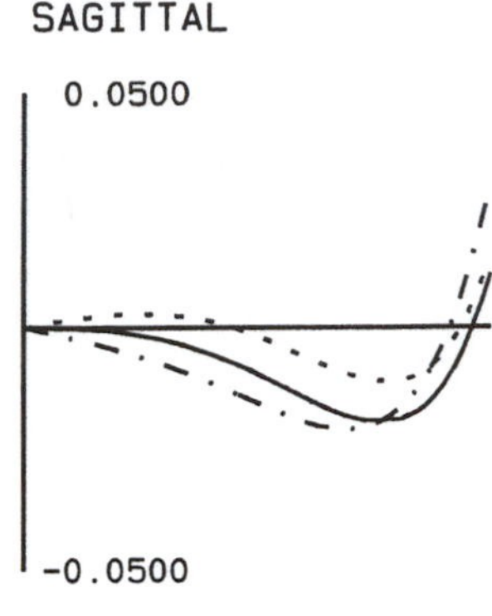
SAGITTAL
0.0500
-0.0500

(a)

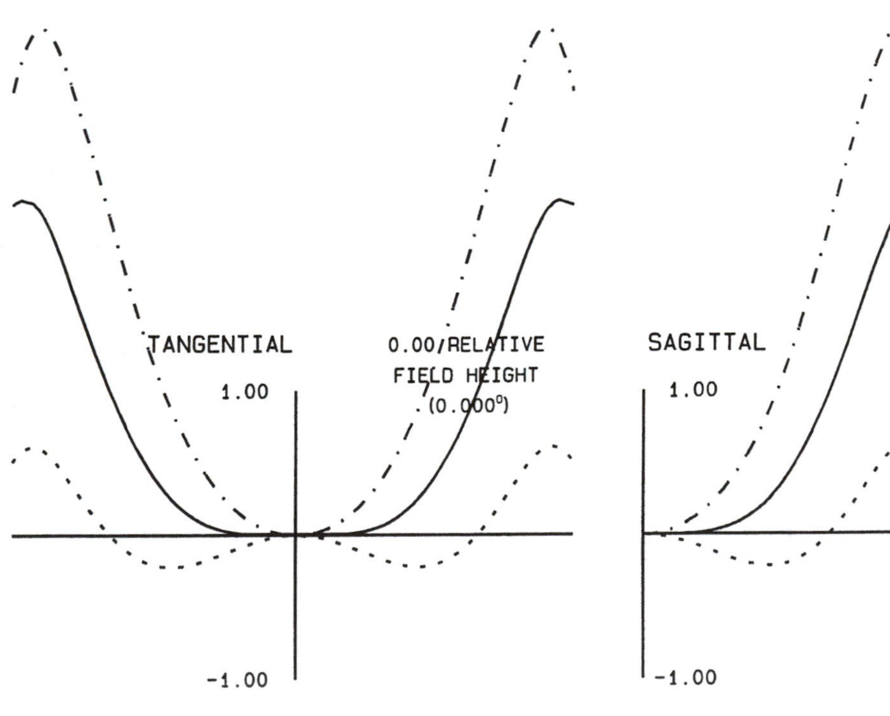
TANGENTIAL
0.00 RELATIVE
FIELD HEIGHT
(0.000°)
SAGITTAL
1.00
-1.00
1.00
-1.00

(b)

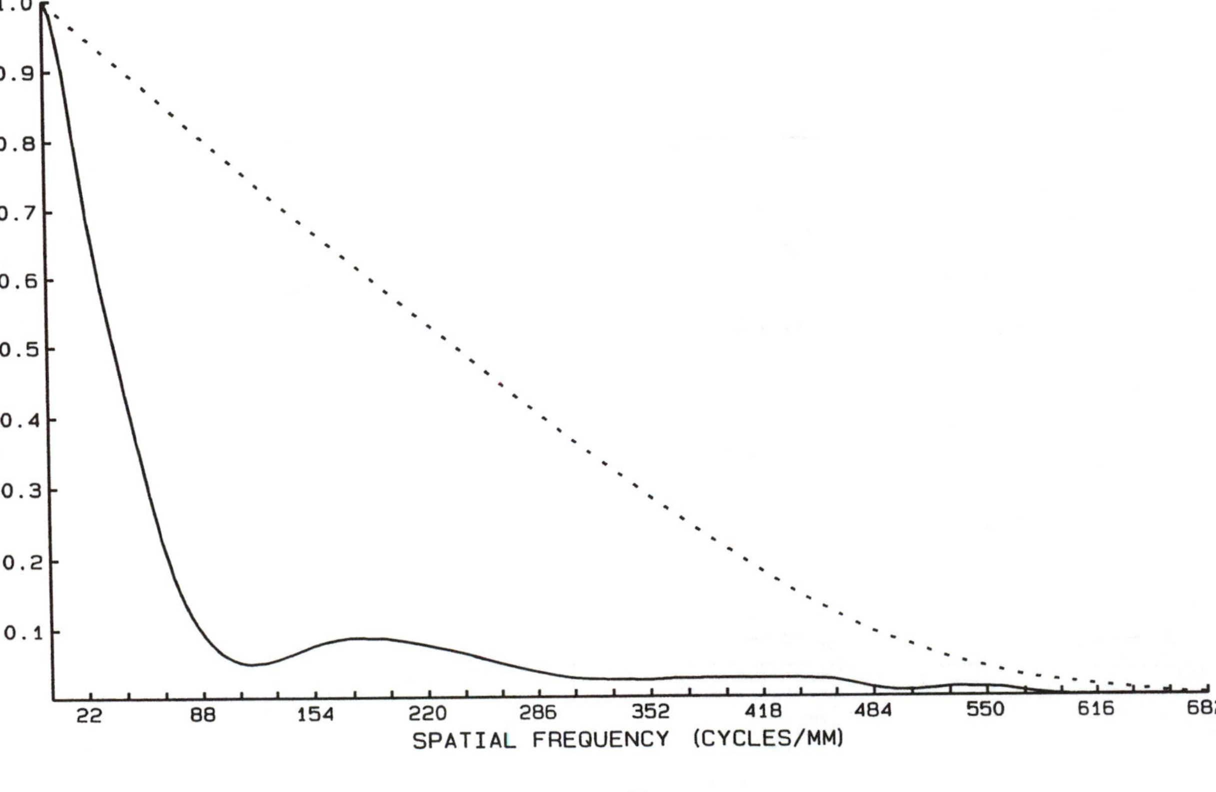

(c)

FIGURE 7.17 Doublet at F/3.0. (a) Ray intercept plot (mm); (b) OPD plot (waves); (c) MTF (CODE V). Diffraction limit, - - - - - -.

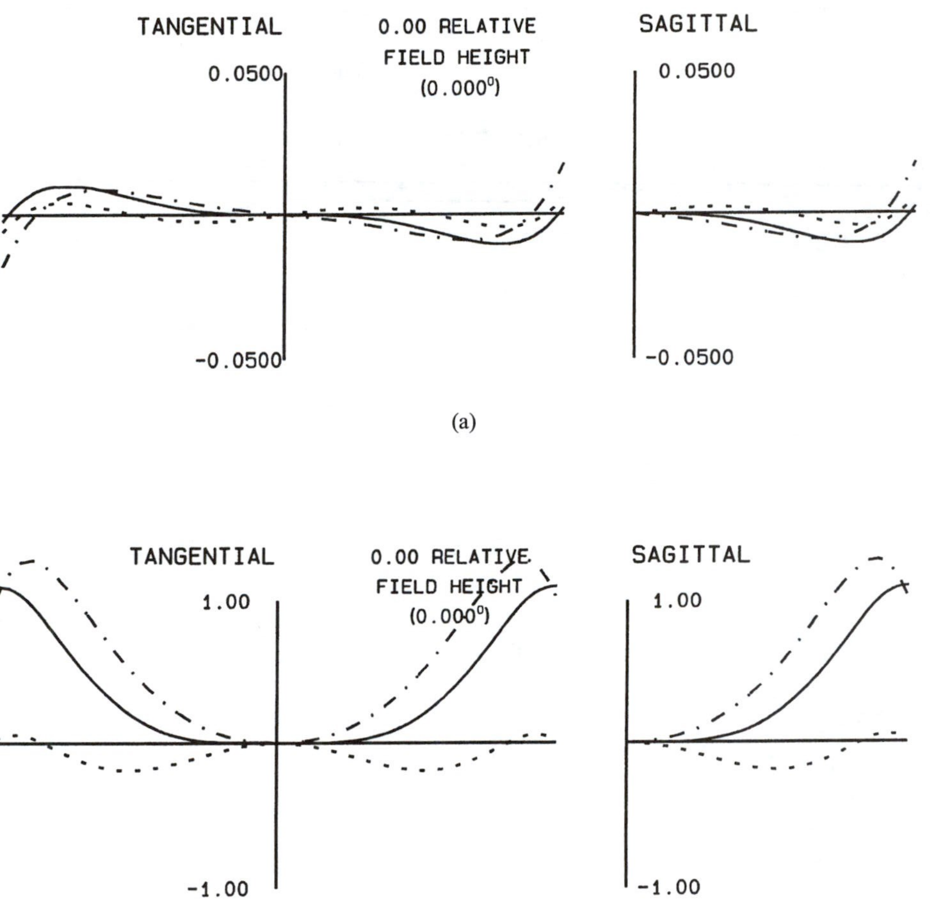
TANGENTIAL
0.00 RELATIVE
FIELD HEIGHT
(0.000°)
SAGITTAL
0.0500
-0.0500
0.0500
-0.0500
TANGENTIAL
0.00 RELATIVE
FIELD HEIGHT
(0.000°)
SAGITTAL
1.00
-1.00
1.00
-1.00

(a)

(b)

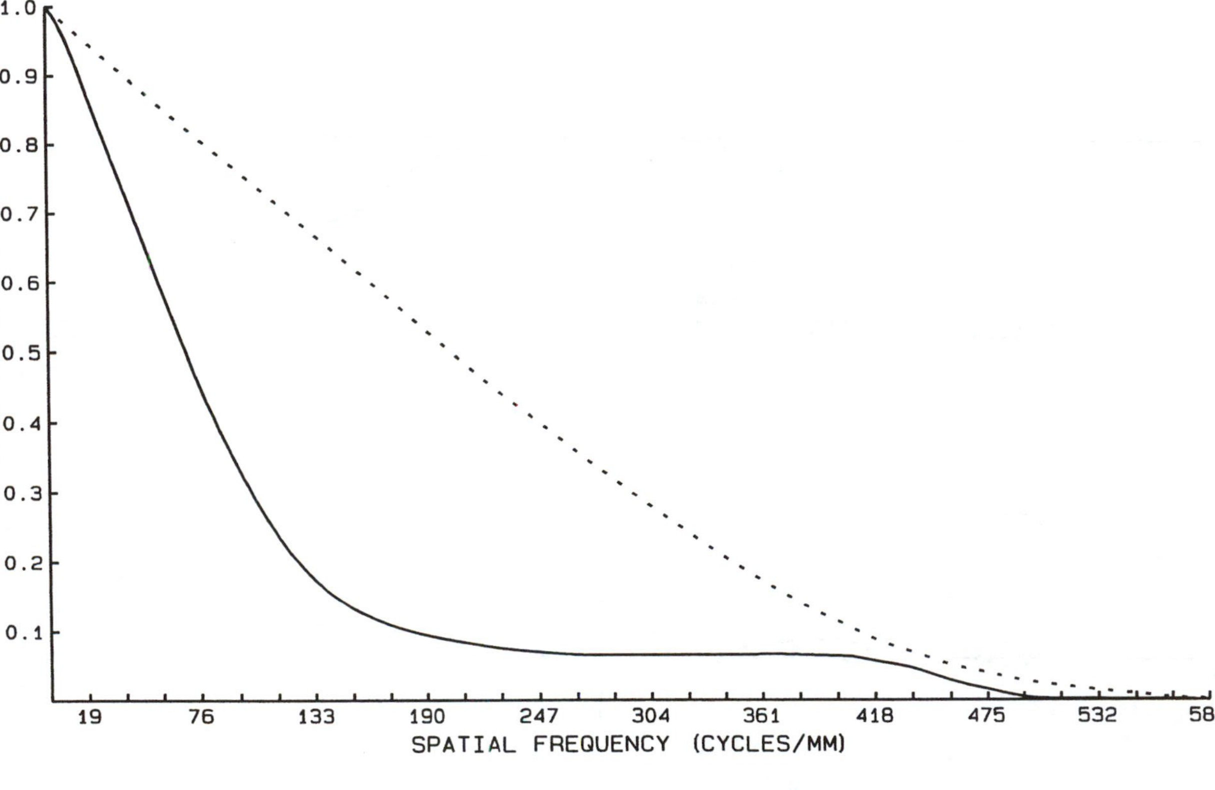

(c)

FIGURE 7.18 Doublet at F/3.5. (a) Ray intercept plot (mm); (b) OPD plot (waves); (c) MTF (CODE V). Diffraction limit, - - - - - -.

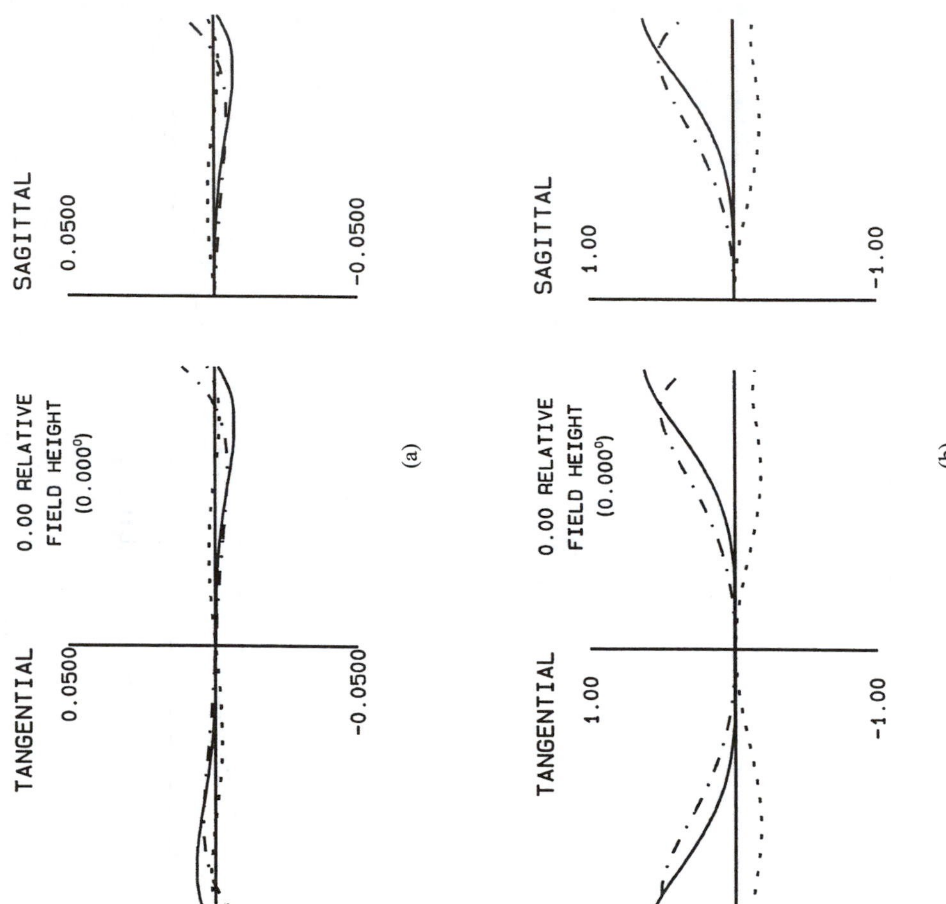

TANGENTIAL
0.0500
-0.0500
0.00 RELATIVE
FIELD HEIGHT
(0.000°)
SAGITTAL
0.0500
-0.0500
TANGENTIAL
1.00
-1.00
0.00 RELATIVE
FIELD HEIGHT
(0.000°)
SAGITTAL
1.00
-1.00

(a)

(b)

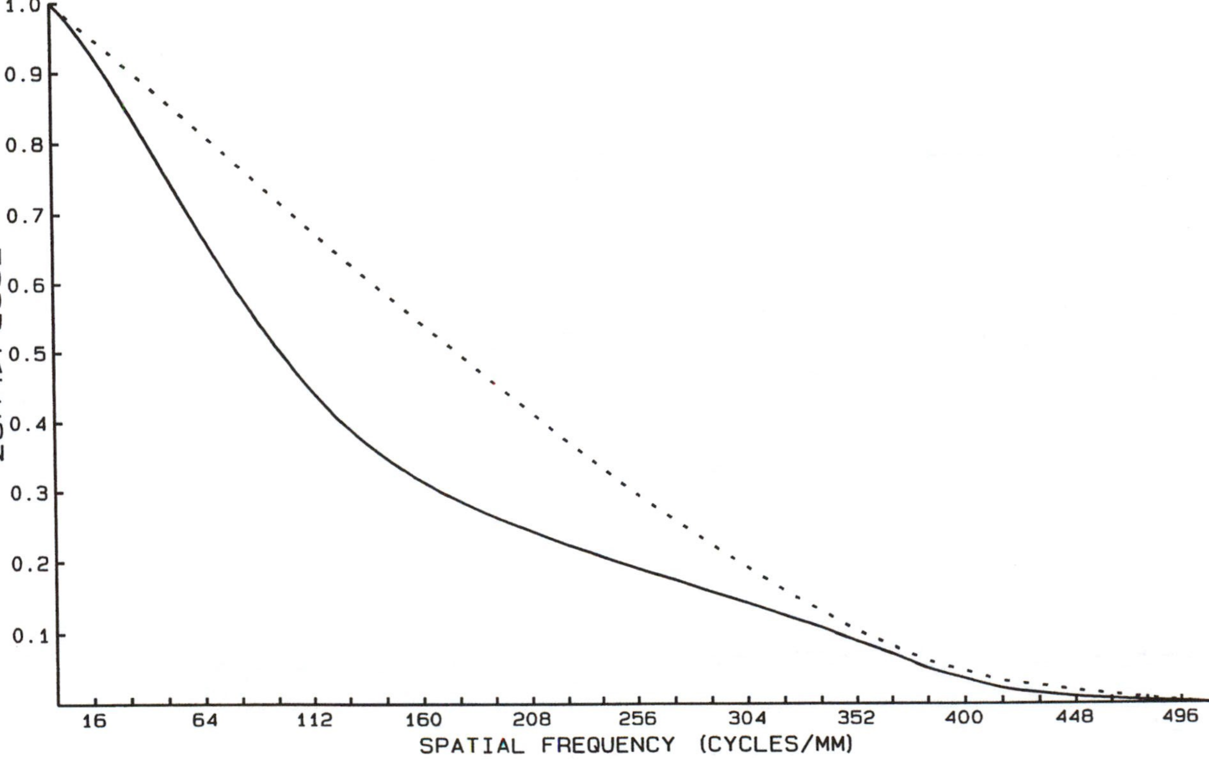

(c)

FIGURE 7.19 Doublet at F/4.0. (a) Ray intercept plot (mm); (b) OPD plot (waves); (c) MTF (CODE V). Diffraction limit, - - - - - -.

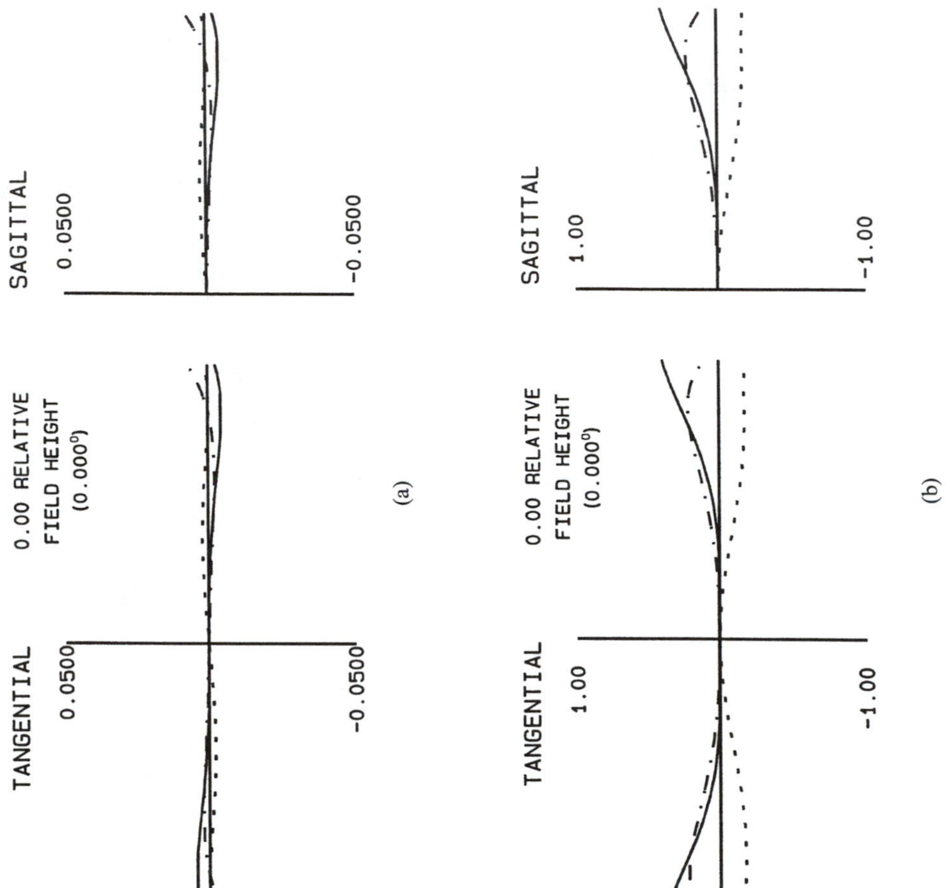
TANGENTIAL
0.00 RELATIVE
FIELD HEIGHT
(0.000^0)
SAGITTAL
0.0500
-0.0500
0.0500
-0.0500
TANGENTIAL
0.00 RELATIVE
FIELD HEIGHT
(0.000^0)
SAGITTAL
1.00
-1.00
1.00
-1.00

(a)

(b)

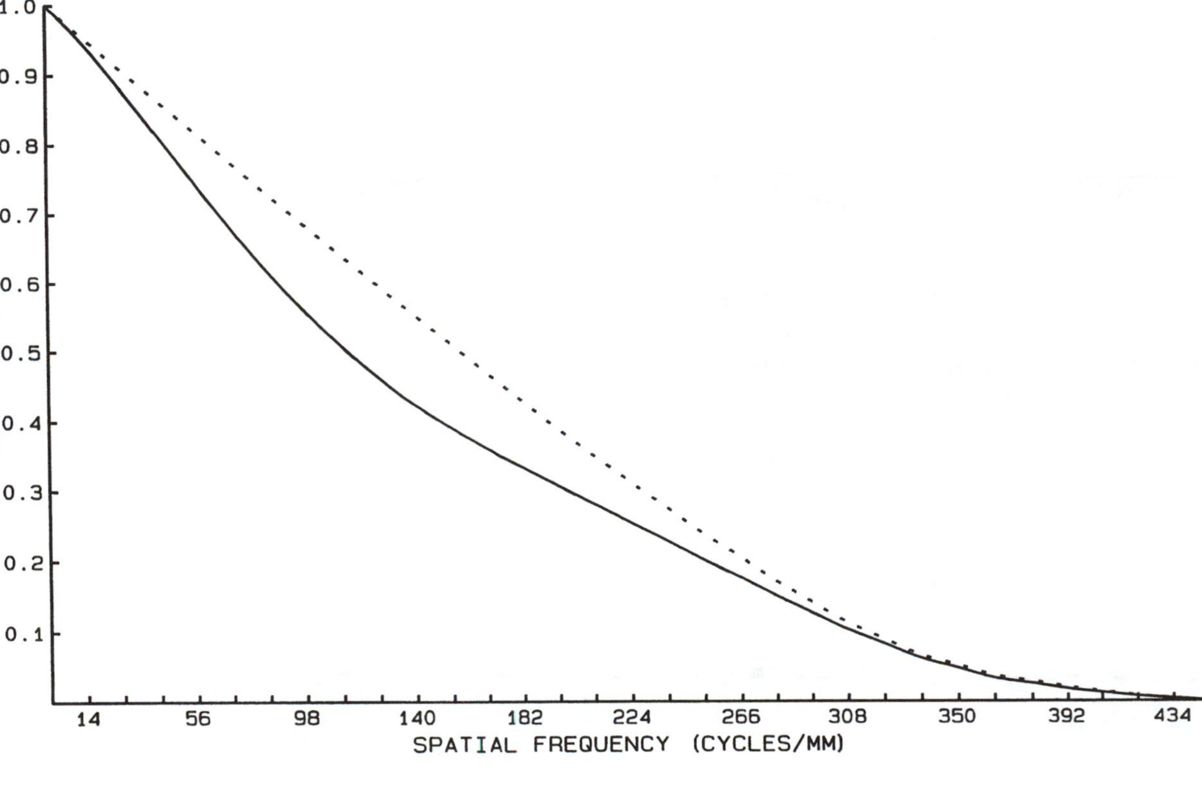

(c)

FIGURE 7.20 Doublet at F/4.5. (a) Ray intercept plot (mm); (b) OPD plot (waves); (c) MTF (CODE V). Diffraction limit, - - - - - -.

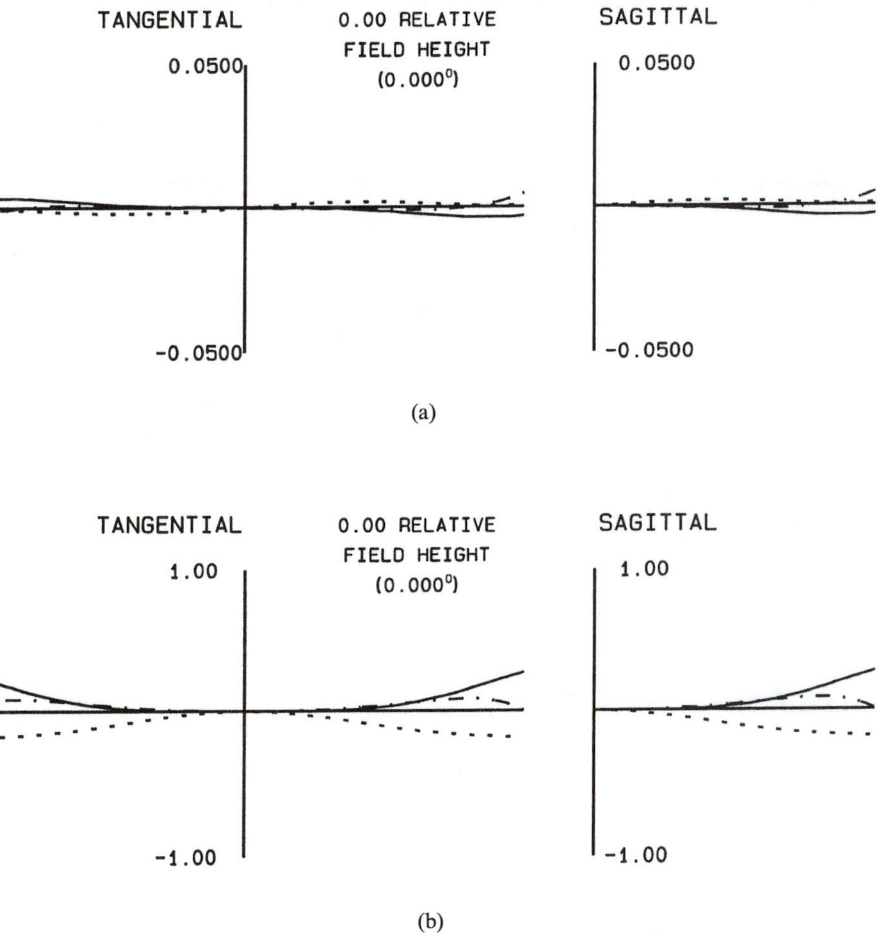

TANGENTIAL
0.00 RELATIVE
FIELD HEIGHT
(0.000°)
SAGITTAL
0.0500
-0.0500
0.0500
-0.0500
(a)
TANGENTIAL
0.00 RELATIVE
FIELD HEIGHT
(0.000°)
SAGITTAL
1.00
-1.00
1.00
-1.00
(b)

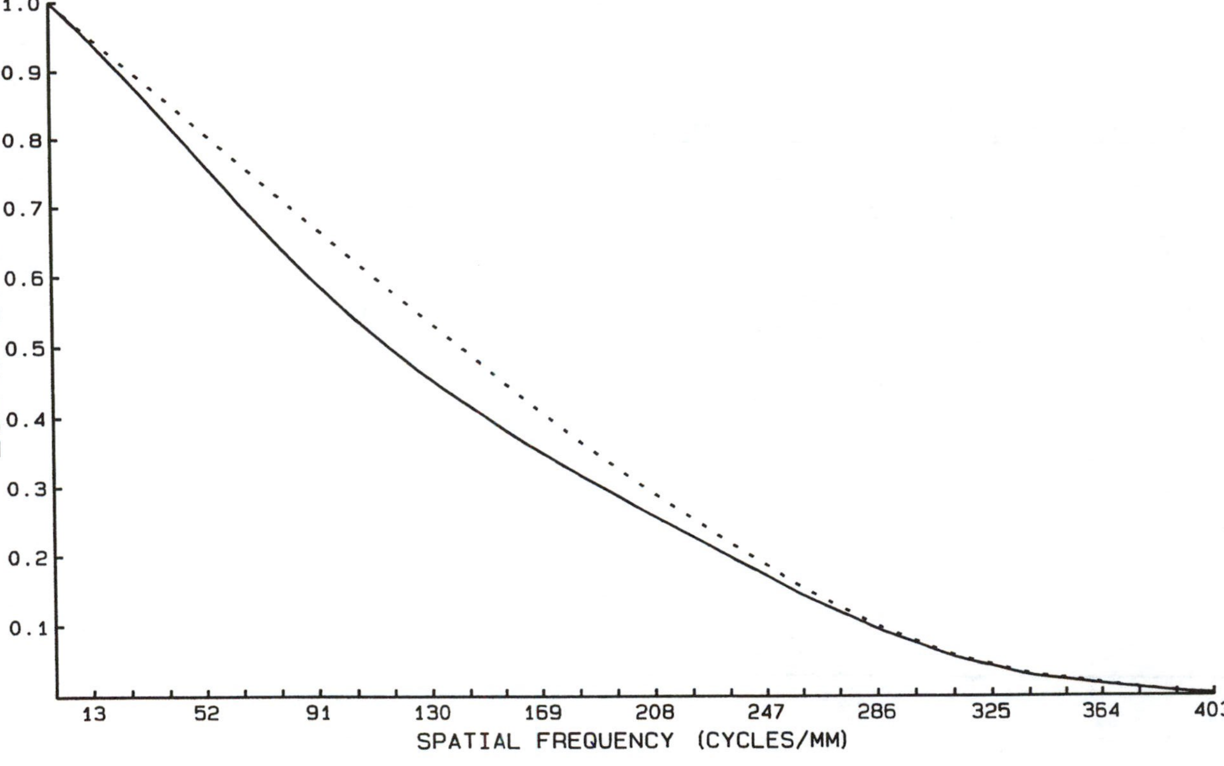

(c)

FIGURE 7.21 Doublet at F/5.0. (a) Ray intercept plot; (b) OPD plot: ······, 656.3 nm; ——, 587.6 nm; ·—·—·, 486.1 nm; (c) MTF (CODE V). Diffraction limit, - - - - - -.

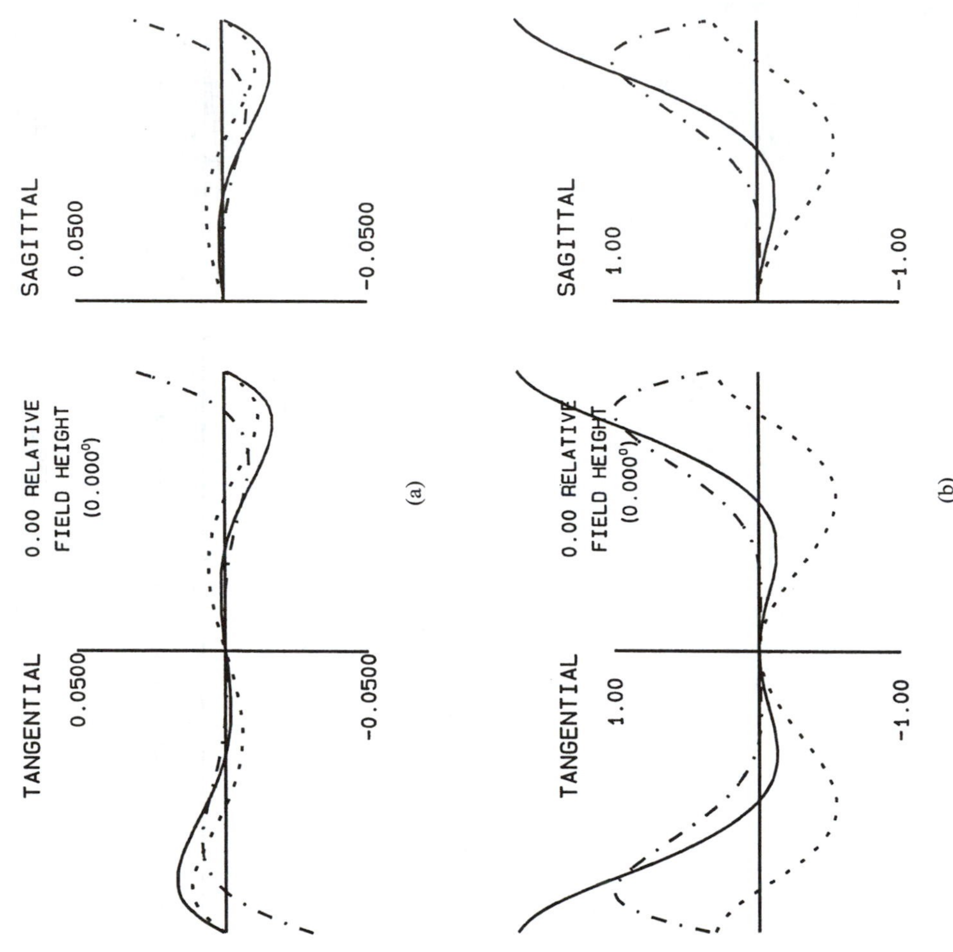
TANGENTIAL
0.00 RELATIVE
FIELD HEIGHT
(0.000°)
SAGITTAL
0.0500
-0.0500
(a)
TANGENTIAL
0.00 RELATIVE
FIELD HEIGHT
(0.000°)
SAGITTAL
1.00
-1.00
(b)

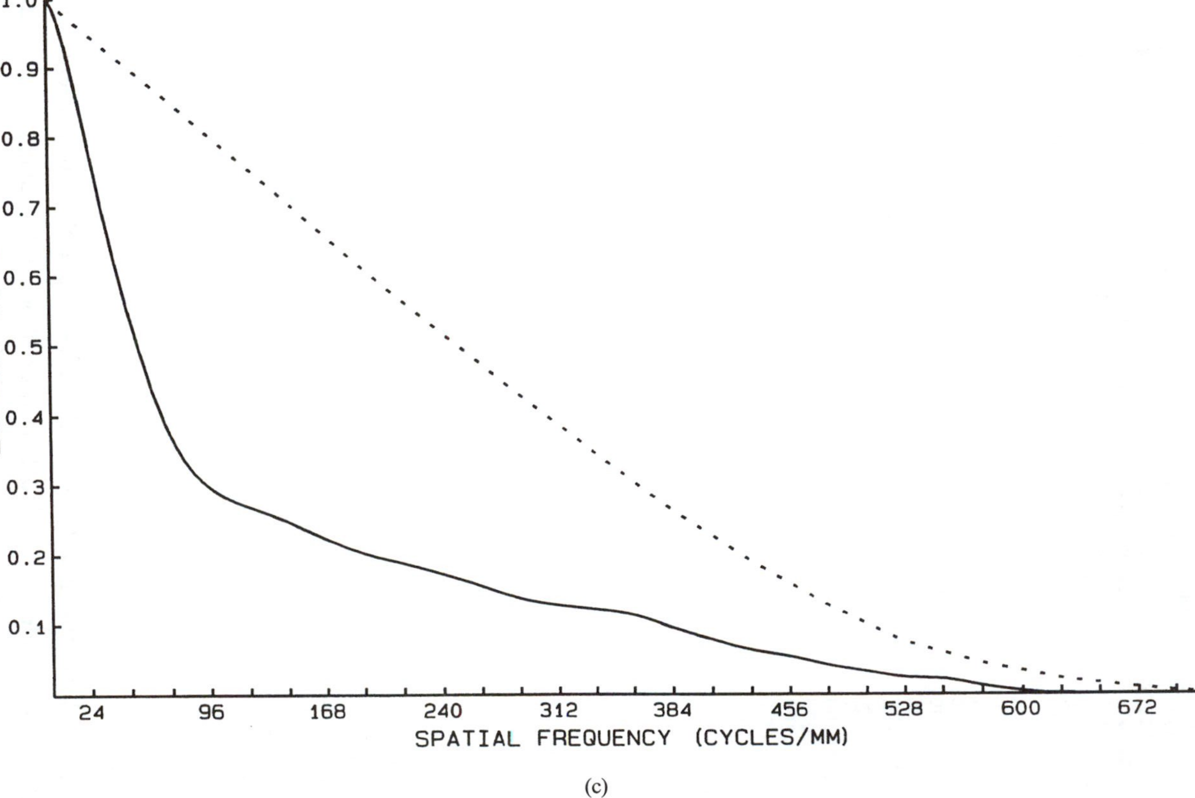

(c)

FIGURE 7.22 Redesign with new glass choice at F/2.8. (a) Ray intercept plot; (b) OPD plot: ······, 656.3 nm; ——, 587.6 nm; ·—·—·, 486.1 nm; (c) MTF (CODE V). Diffraction limit, - - - - - -.

```
CODE V> lis
          Chapter 7 Doublet Analysis
                       RDY              THI       RMD        GLA              CCY     THC    GLC
    OBJ:          INFINITY         INFINITY                                   100     100
    STO:          52.54197        10.000000            BK7_SCHOTT               0     100
      2:         -43.32587         4.000000            F2_SCHOTT                0     100
      3:        -237.45929        91.633676                                     0     PIM
  > IMG:          INFINITY        -0.158459                                   100     100

  SPECIFICATION DATA
     EPD         35.71420
     NFO                7
     FFO         -0.20000
     IFO          0.05000
     DIM               MM
     WL            656.30     620.00     587.60     530.00     486.10
     REF                3
     WTW                1          1          1          1          1
     XAN          0.00000
     YAN          0.00000
     VUY          0.00000
     VLY          0.00000

  REFRACTIVE INDICES
     GLASS CODE                  656.30       620.00       587.60       530.00
     BK7_SCHOTT                1.514321     1.515539     1.516798     1.519584
     F2_SCHOTT                 1.615030     1.617461     1.620037     1.625931

  )LVES
     PIM

  INFINITE CONJUGATES
     EFL          99.9964
     BFL          91.6337
     FFL         -99.1911
     FNO           2.7999
     IMG DIS      91.4752
     OAL          14.0000
     PARAXIAL IMAGE
      HT           0.0000
     ANG           0.0000
     ENTRANCE PUPIL
      DIA         35.7142
      THI          0.0000
     EXIT PUPIL
      DIA         36.0042
      THI         -9.1746
CODE V> out t aaacdb
```

(a)

is reduced. With the addition of the airspace between the elements, the separation of the elements also becomes a possible variable. This must be used with some discretion, as the lateral chromatic aberration will no longer be zero if the elements are separated. Thus a close spacing is all that is acceptable. In many cases the radii of the surfaces surrounding the airspace can be adjusted so that an edge contact just outside the clear aperture is used to establish the separation. This produces a mechanically desirable solution.

Once this variable is added, setting the coma to zero is simple. It is usually sufficient to simply require that the third-order coma be zero. Alternately a constraint can be added using the upper and lower rays for y direction at the edge of the field to force symmetry of the image about the chief ray. The axial

```
CODE V> lis
      SK5 - BaSF54
                 RDY              THI        RMD        GLA              CCY     THC     GLC
   OBJ:        INFINITY         INFINITY                                 100     100
   STO:        61.61542        10.000000            SK5_SCHOTT             0     100
     2:       -51.14868         4.000000            BASF54_SCHOTT          0     100
     3:      -216.51512        92.403361                                   0     PIM
   IMG:        INFINITY        -0.075173                                 100     100

 SPECIFICATION DATA
    EPD          35.71420
    DIM                MM
    WL             656.30     587.60     486.10
    REF                 2
    WTW                 1          1          1
    INI               ORA
    XAN           0.00000
    YAN           0.00000
    VUY           0.00000
    VLY           0.00000

 REFRACTIVE INDICES
     GLASS CODE                        656.30        587.60       486.10
    SK5_SCHOTT                       1.586192      1.589128     1.595811
    BASF54_SCHOTT                    1.729609      1.736265     1.752520

 SOLVES
    PIM

 INFINITE CONJUGATES
    EFL          100.0000
    BFL           92.4034
    FFL          -98.8727
    FNO            2.8000
    IMG DIS       92.3282
    OAL           14.0000
    PARAXIAL IMAGE
     HT            0.0000
    ANG            0.0000
    ENTRANCE PUPIL
     DIA          35.7142
     THI           0.0000
    EXIT PUPIL
     DIA          36.1214
     THI          -8.7367
CODE V> prt
      Printing file SK5.LIS(1)
```

(b)

FIGURE 7.23 Listing of design data for doublets. (a) Bk7–F2 design; (b) SK5–BaSF54 design (CODE V).

image changes somewhat because there are now four available refracting surfaces. The axial image ray intercept plots, shown in Figure 7.24, are clearly improved with a significant reduction in the amount of high-order aberration that needs to be balanced.

In this design a choice has been made that the airspace distance shall be as small as possible. The design that has been obtained is an "open contact" doublet, in which the edge thickness of the airspace is greater than the center thickness. An entire family of solutions exists for different constraints placed upon the airspace.

The high-order spherical aberration content of the lens changes markedly now. Small changes in the separation of the elements will produce many

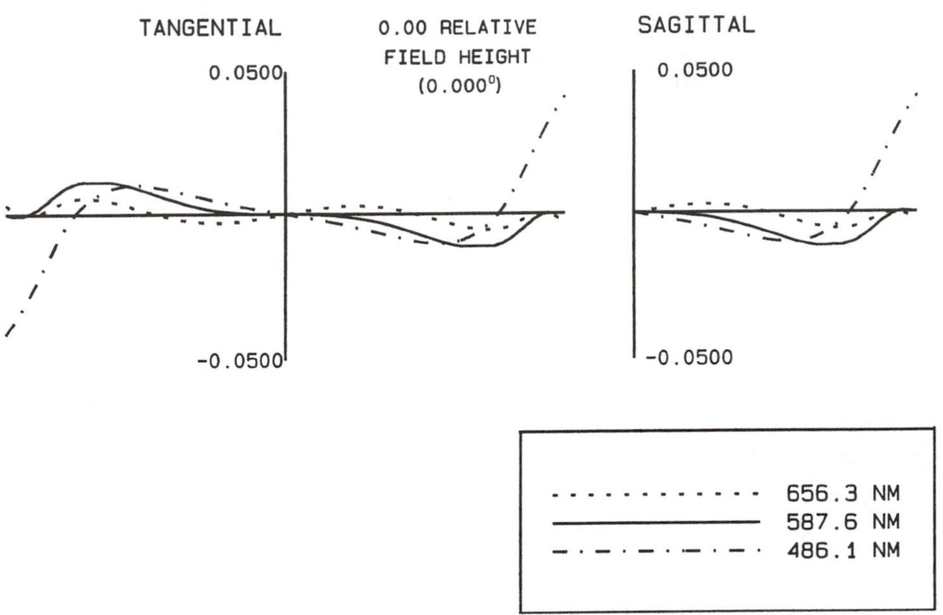

FIGURE 7.24 Ray trace plot for broken-contact aplanatic doublet (CODE V).

different solutions with differing high-order aberration content. The explanation for this is that the angles of incidence for ray at the edge of the aperture are very high on surfaces 2 and 4. This indicates that the direction of the refracted ray is quite nonlinear with respect to the entering angles of the ray. Because the ray is dropping rapidly in the space between the elements, there will also be a considerable change in incidence height for an edge ray at each side of the airspace. In this case, the addition of a broken contact to permit correction of the coma adds quite a wide region of possibilities to the design.

The simple doublet solution becomes dependent upon the thickness of the airspace and may have a number of almost equivalent solutions. The behavior of the high-order aberrations can be quite different for these airspaced lenses. The art in lens design consists of using the options available to obtain the best possible solution. The designer uses the capability of locating an acceptable solution to find the combination of parameters that best solves the problem.

Because the balance of aberrations appears to be very good, it is appropriate now to examine the image quality that can be expected from this lens. When the MTF is calculated at the best focal position, Figure 7.25 is obtained. This shows that although the MTF is improved with this new

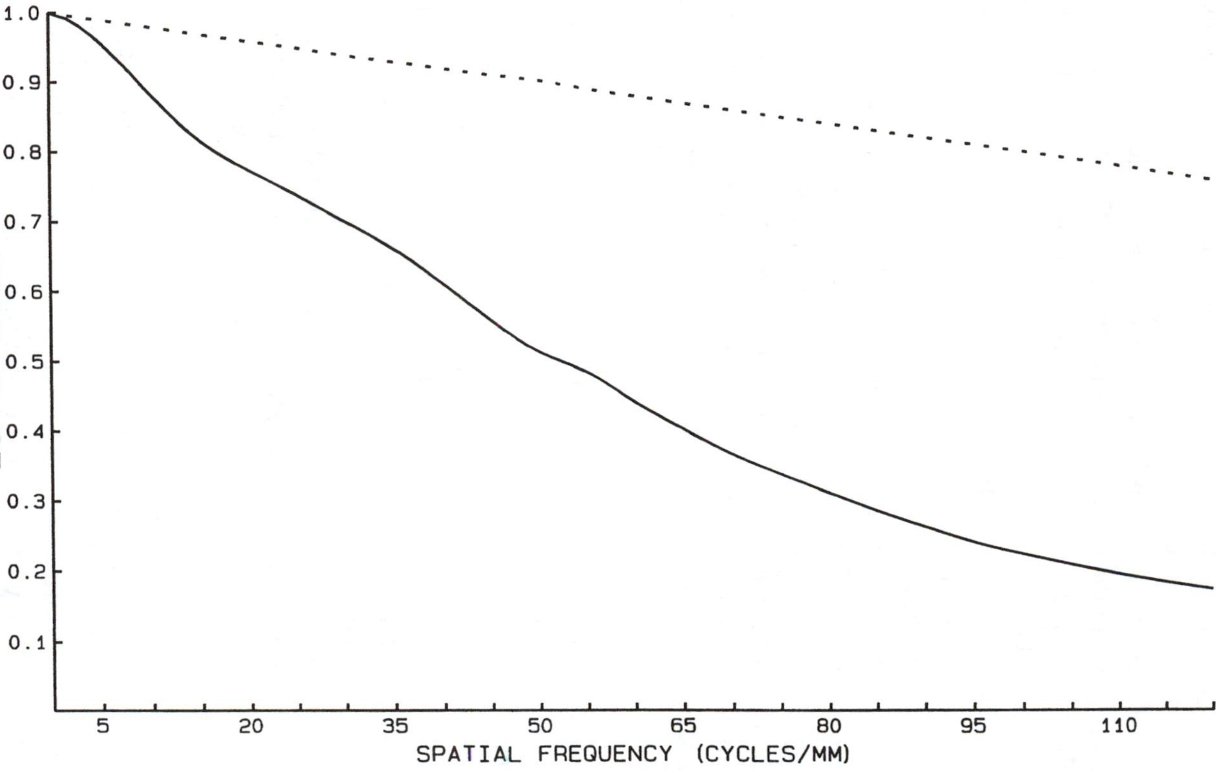

FIGURE 7.25 MTF for aplanatic doublet (CODE V). Diffraction limit, - - - - - -.

balancing of aberrations, the lens is still quite far from being diffraction limited over the wavelength region.

The discussion so far has been directed toward understanding the balancing of aberrations in the central portion of the field. The close-spaced or contacted doublet will contain an intrinsic astigmatism and field curvature which cannot be corrected because the lens has corrected spherical aberration, and the stop is located at the lens. From section 3.1.1.2 the field will be dominated by a residual Petzval curvature and astigmatism. The ray intercept plots across a 5° half field are shown in Figure 7.26. These curves show that in the broken-contact doublet the off-axis ray plots are quite straight, indicating that coma is not an issue. Field focus plots for the lenses examined in this section are given in Figure 7.27. From these it is seen that for a field of 5°, the field curvature becomes overwhelming, with the good imagery limited to a degree or two. Examination of the longitudinal spherical aberration (note the plot scale) and comparison with the astigmatic field curves indicates that beyond about 2°, the astigmatic aberration is greater than the spherical aberration.

The effect upon the image can be seen from plots of the MTF versus field for these lenses. The plots in Figure 7.28 are obtained by selecting the best focus position on axis, and then computing the change in sagittal and tangential MTF over a 5° field. The broken-contact, coma-corrected doublet shows an almost constant MTF over the first degree of field, confirming that the lens is locally isoplanatic in that region. Beyond 2°, this improvement is rendered unimportant as the field curvature becomes dominant.

As a summary of the work done in this section, Figure 7.29 is a comparative plot of the MTF versus field for the contacted doublet. The presence of some additional spherical aberration zone and some small amount of coma is evident in the more rapid decrease of the MTF at all plotted spatial frequencies as the field is increased.

The question obviously arises as to how these lenses could be improved beyond the image quality obtained. Both the image quality calculations and aberration theory from Chapter 3 indicate that nothing can be done about the field curvature. The coma correction shows that local isoplanatism is possible. The only improvements that can be made are in the spherical aberration residual and the change of aberration and focal position with wavelength. The spherical aberration can be reduced by splitting the crown element into two approximately equally powered components. These can be bent so that the amount of high-order aberration is reduced. This will permit some improvement for large numerical apertures, until the chromatic aberration residual limit is reached. A similar improvement can be made with the use of an aspheric on the crown element. For some problems, a gradient index crown element could be used, but this is limited by the size of the available materials.

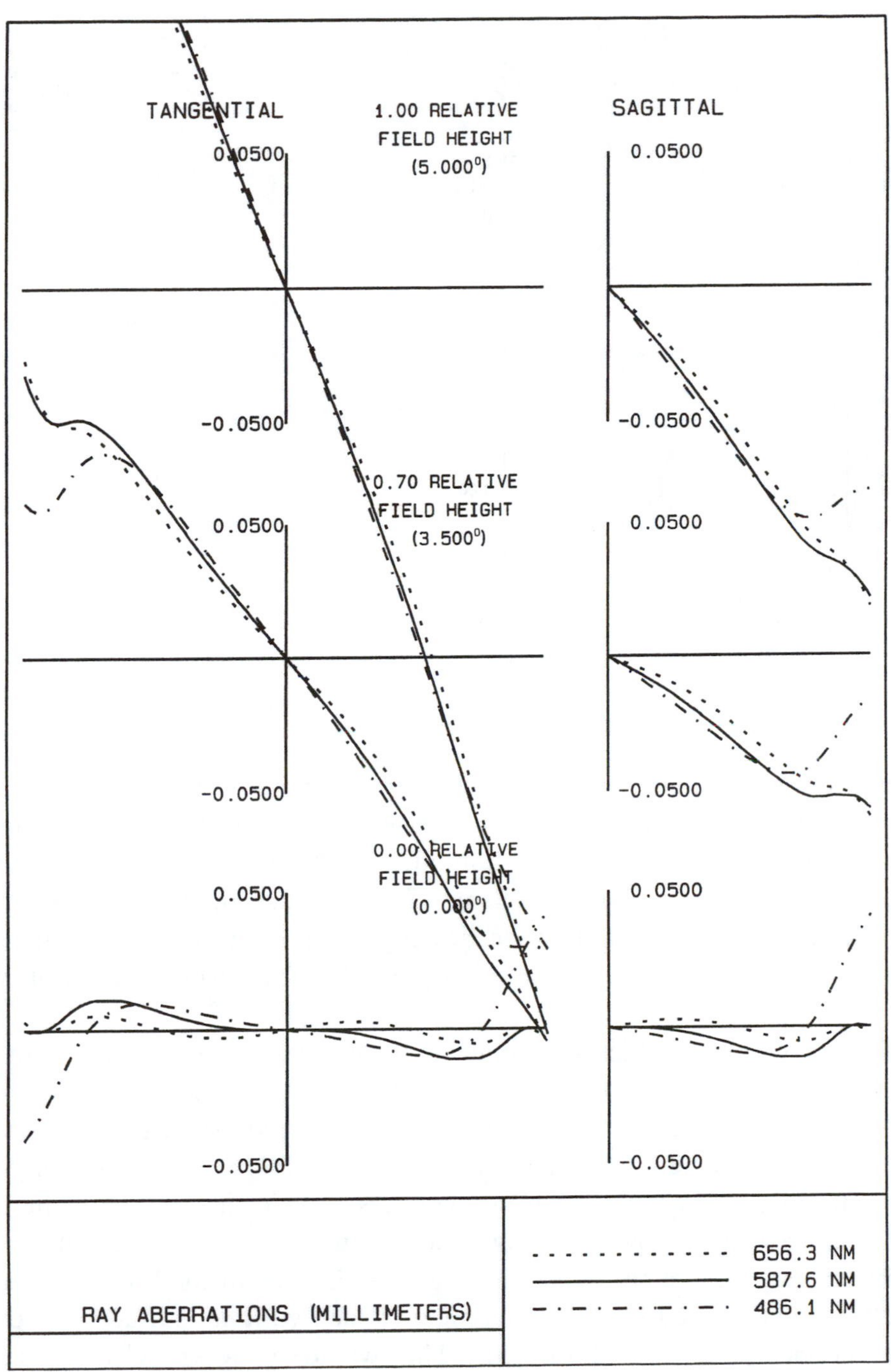

FIGURE 7.26 Ray intercept plots for 5° half field.

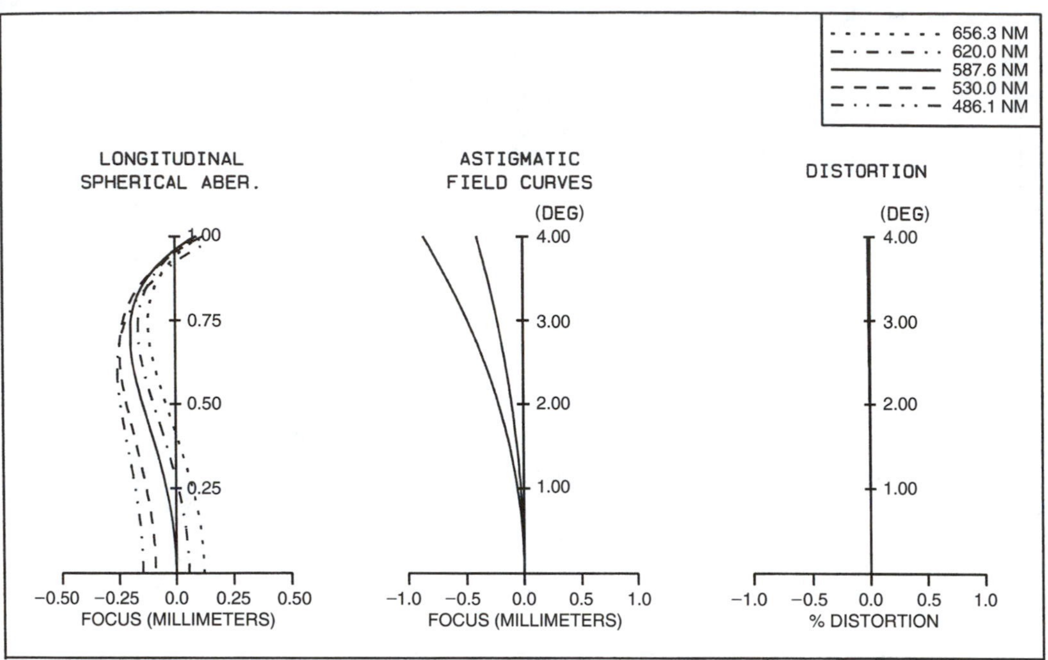

FIGURE 7.27 Longitudinal spherical aberration and field plots for airspaced doublet (CODE V).

Improvement of the chromatic aberration correction is possible only by searching for an apochromatic solution, using special glass types. The gain is not automatic, however, as a solution must be found in which the variation of spherical aberration with wavelength is also reduced. In most cases this apochromatic design will best be accomplished by using three different glasses, although some improvement is possible with clever selection of two glasses.

The achromatic doublet is the simplest corrected lens system. Much can be learned by examining various balances of the aberrations and their effect upon the MTF. Useful design exercises would include such topics as wide spectral ranges, correction at the extremes of the visible range and simplifying lenses for manufacture by reducing tolerances or by using only two different radii in the lens. A specific design exercise would be to obtain the best possible design under the constraint that the airspace in the broken contact doublet is determined by an edge contact of the two inner surfaces just outside the clear aperture of the lens. This would make the elements self-registering and eliminate the fabrication of a precision spacer for assembly.

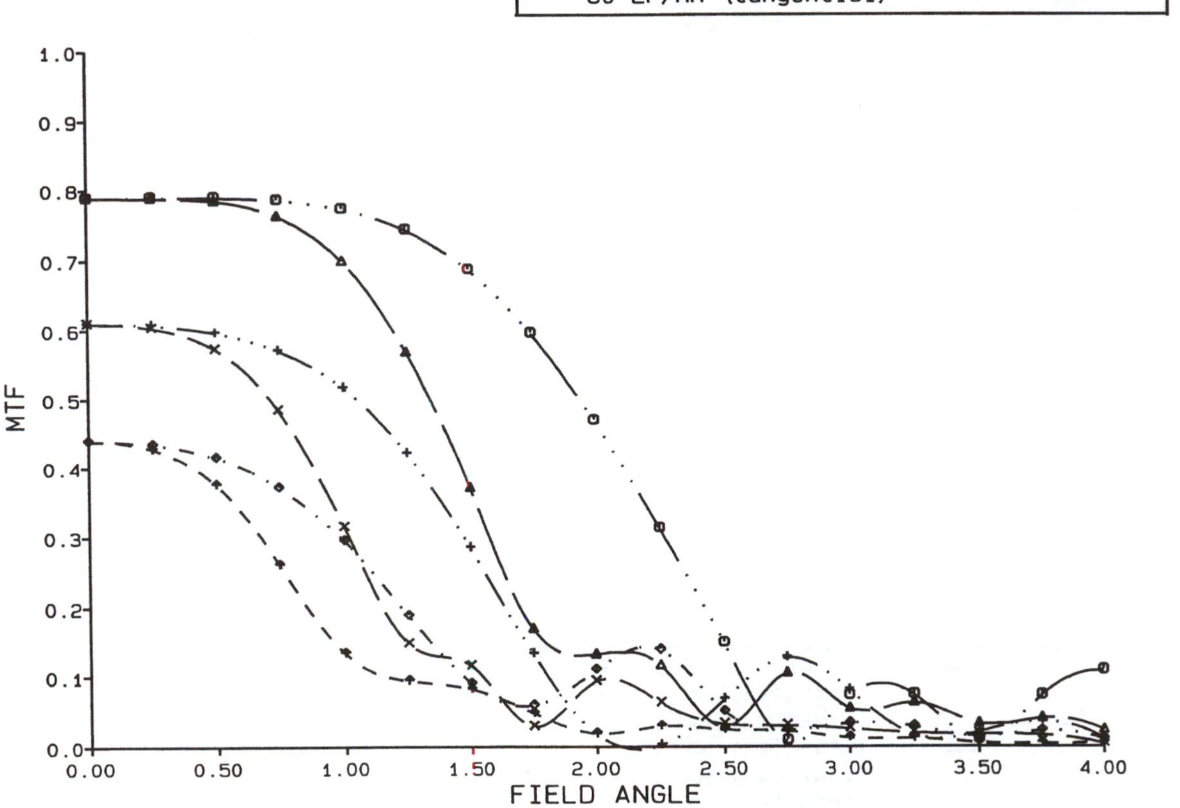

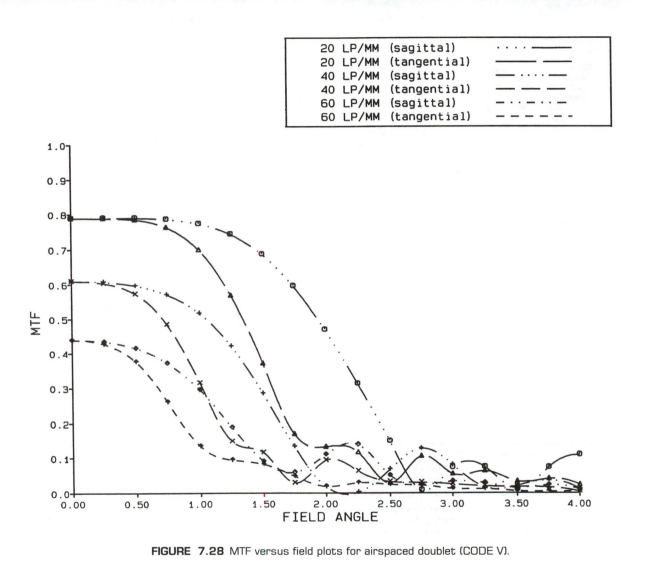

FIGURE 7.28 MTF versus field plots for airspaced doublet (CODE V).

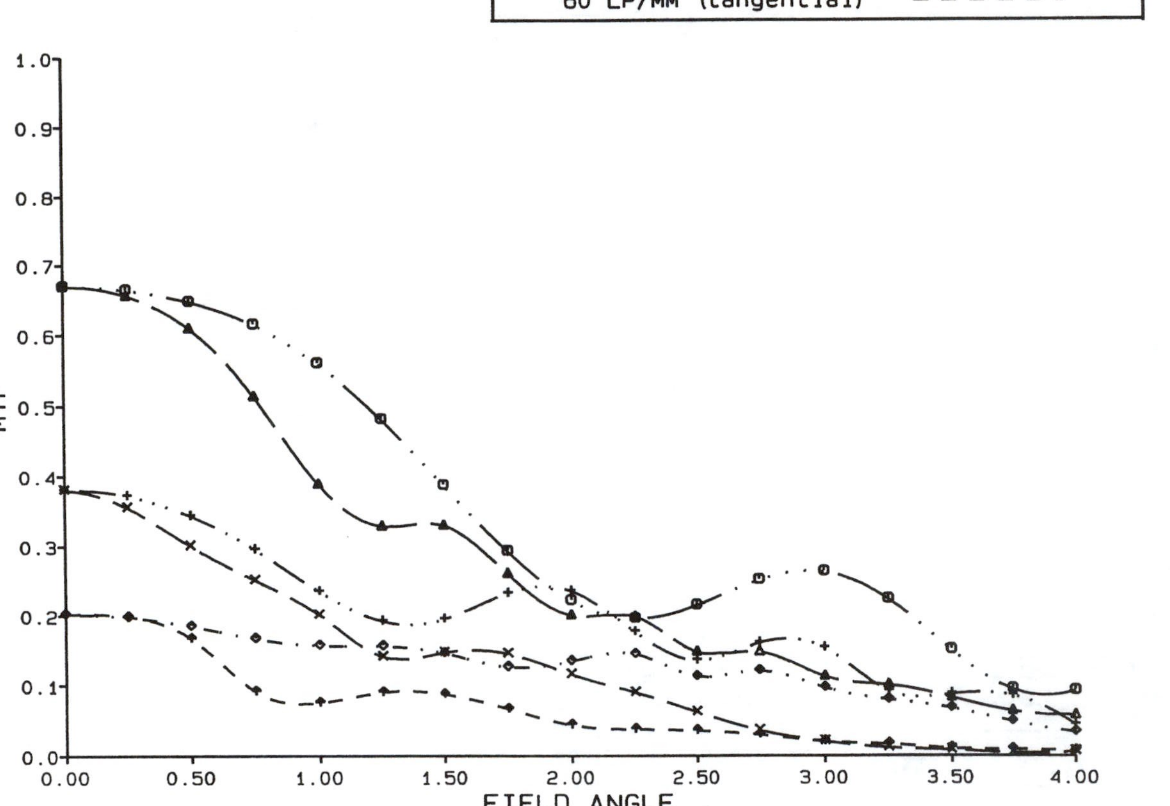

FIGURE 7.29 MTF versus field plots for contacted doublet (CODE V).

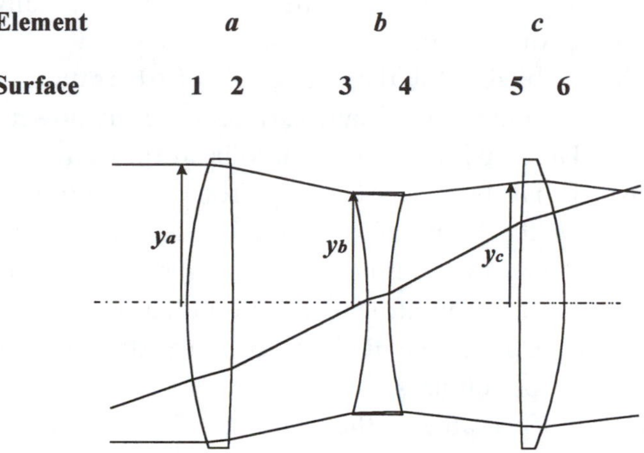

FIGURE 7.30 General layout of triplet type of lens.

7.1.2 TRIPLETS

The next level of complexity for a lens is the triplet form. This type of lens is the simplest lens system form in which all of the aberrations can be controlled over a wide field. The simplicity of the design form does place limits upon the correction of aberrations that can be achieved, because the control of the low-order aberrations is used to balance the effect of rather large high-order aberrations. The triplet form permits beginning from a first-order solution and proceeding all of the way through a design. A significant amount can be learned by working through the details of the design of a triplet.

The MIL handbook *Optical Design* (Hopkins 1962) contains a lengthy discussion of the design of a triplet. Other references are in the book *Lens Design Fundamentals* (Kingslake 1978) and there are several discussions in the *Journal of the Optical Society of America.* The approach to a triplet used here is only one way of explaining the operation of the lens; alternate discussions can be found in the literature. Examples of some triplet lenses may be found in the compilations by Cox (1964), Smith (1992), and Laikin (1995).

An example of the triplet form is shown in Figure 7.30. The general form consists of low-dispersion crown glass elements arranged at some spacing outside a low-index, high-dispersion flint, negative element, with the stop located at this central element. The arrangement of the central negative element and separated outer elements permits control of the Petzval curvature and the astigmatic fields. The starting point is to find a thin-lens power solution for the properties that are independent of the shape of the individual elements. The shapes of the elements control the monochromatic aberrations and the choice of glass and the element thicknesses eventually establish the high-order aberration content of the lens.

The variables available for the correction of aberrations fall into three classes. The first and most significant variables are the powers and spacings of the elements. The stop is initially taken to be at the central element, but for practical reasons may be moved slightly from the central position. These variables establish the basic color correction and field curvature for the lens. The choice of glasses at this point is also an important variable. Usually, at the start, the central flint element is surrounded by a pair of crown glass elements of the same glass type. Refinement of the design can lead to the use of three different types of glass.

The second set of variables is the bendings of the elements. The shape of the three elements permits control of the three principal aberrations, spherical aberration, coma, and astigmatism. These remaining variables, element thickness and modifications of the glass choice, can be used to correct the distortion and search for a solution with a minimum amount of high-order aberration content.

The paraxial properties of this thin-lens form are described using three elements, named a, b, and c, with the stop located at the central element, b. The separations of the elements are t_a and t_b, the back focus will be t_c. The overall focal length of the lens is set to be $f = 1/\phi$. Four equations apply for the focal length and bending-independent aberrations that need to be controlled:

$$\begin{aligned} y_a\phi &= y_a\phi_a + y_b\phi_b + y_c\phi_c \\ a &= \frac{y_a^2\phi_a}{V_a} + \frac{y_b^2\phi_b}{V_b} + \frac{y_c^2\phi_c}{V_c} \\ b &= \frac{y_a\bar{y}_a\phi_a}{V_a} + 0 + \frac{y_c\bar{y}_c\phi_c}{V_c} \\ P &= \frac{-\phi_a}{n_a} + \frac{-\phi_b}{n_b} + \frac{-\phi_c}{n_c} \end{aligned} \tag{7.3}$$

The equations can be solved for the powers of the three lenses, given a starting point for the separations of the lens elements. A direct solution is not easily accessible because there are many interrelations between the paraxial heights, the powers, and the separations of the lens. Therefore the practical solution of these equations is not obvious, and requires much substitution and algebra to become workable. In general (Hopkins 1962), the process of solution involves making reasonable assumptions about the starting point, and then examining the solutions for a class of similar system parameters.

The first equation defines the total power. The second and third are the axial and lateral color, respectively. The fourth equation sets the value of the Petzval curvature. Experience suggests that as a start this should be chosen so that the Petzval radius is about -3 to -4 times the focal length of the lens, rather than correcting the Petzval curvature to zero. An attempt to use too small a Petzval curvature will lead to very high powers in the individual elements, and subsequent failure in the lens design. The usual starting point taken is to select a moderately high index crown for both of the outer elements, and a reasonable flint, usually lower index, for the central lens. Examination of the equations shows that this will maximize the contribution of the central negative element to the field flattening. A good starting choice for glasses is SK16 for the outer elements and SF2 for the flint center element. In the absence of any other indications, it is appropriate to select equal separations on each side of the central element. The total length of the lens will depend upon the glass selection, but a starting point would be about a quarter to a third of the focal length for the sum of the two separations. A back focus of greater than two-thirds of the focal length is workable, and is usually desirable for a photographic objective.

The separations do not directly enter into these equations. They are implicit in the choice of the paraxial y heights on the lens. The entering y height, y_a, will be determined by the required aperture for the lens. The third y height, y_c, will be set by the back focal length requirement, and will usually be about two-thirds of the entering y height. The center lens y height will be determined by the separation and the powers of the elements. For solving the equations, this can be arbitrarily selected, and used to define the requirements upon the powers. About six-tenths of the entering height is a good starting point.

With this combination, the central flint element is of course negative, balancing the chromatic contributions of the outer elements. The lateral color is set to zero, which is achievable because of the balance of aberration contributions from elements located on each side of the stop. Setting of the chief ray heights on the outer elements is not necessary, as the lateral color will be set to zero, and the initial assumption of the equal airspaces will cause these values to be equal and opposite. The axial color can be set equal to zero, but it is frequently best to start with a small residual chromatic undercorrection, which will be required to offset the spherochromatism, which will eventually be identified in the lens. Solving these equations sets the required power distributions for the lens. Addition of sufficient thickness to the elements to provide manufacturable elements is obviously desirable, once the lens has been transferred into a lens design program.

A useful approach is to solve the equations that are most easily accessible as a linear set of equations, those for the power, axial color, and Petzval,

Triplet Starting Point - Thin Lens Solution							
					Glass		
Parameters		kfac	1		na	va	
ya	1	***ta***	0.258131		1.62	60	
yb	0.620216	tb	0.258131		nb	vb	
yc	0.7				1.62	36	
					nc	vc	
Powers derived			angles		1.62	60	
pa	1.471284		-1.47128				
pb	-2.87056		0.309082				
pc	1.870118		-1				
Aberrations			Targets		Diff	wt	
p	1		1				
a	-0.00912		0		-0.00912	10	
b	0.002703		0		0.002703	10	**F.M.T.**
P	-0.29064		-0.28571		-0.00493	1	0.009075
Lens Diagram							
0	1						
0.258131	0.620216						
0.516261	0.7						
1.216261	0						

FIGURE 7.31 Calculation of thin-lens starting point for triplet (Excel).

using a linear equation-solving scheme. Values are chosen for the y heights on the three lenses, and three powers are found. The separations now must be determined after the powers are obtained. The lateral color contribution can then be computed. The fourth condition, that of zero lateral color, can be obtained by changing the y height parameters and recomputing the separations and lateral color. If an iterative equation-solving method is used, the solution procedure is repeated until a set of optical parameters yielding zero lateral color is obtained. In addition, other criteria can be added to the solution, such as a requirement on the ratio of the two airspaces, or that the excess powers of the components be minimized. Selecting different glasses, or using different crown-type glasses can also be considered. There is a rather large solution space that appears to be accessible for any reasonable choice of glasses. Since this is a thin-lens starting solution, it is usually not advisable to expend too much energy on this part of the process.

One useful technique in setting up lenses for design is to use a commercial spreadsheet for working out and solving the equations relating the components. While a designer might reasonably go directly to a lens design program, the use of a general spreadsheet permits the logical inclusion of constraints and relationships that may not easily be available in a design program. Figure 7.31 is such a sheet for finding initial solutions for the above equations, using the EXCEL spreadsheet program. Since the paraxial process is linear, it is not important that choice is made for focal length, field, and aperture diameter. In this example, unity is used for each, and the actual values to be used in a lens design program input can be scaled from these values.

The arrangement of parameters and numbers on the sheet is entirely the choice of the designer. At the top, the desired starting parameters are listed. In this case, the variable parameters will be the first airspace, the marginal height on the second lens, and the ratio of the two airspaces. Below these values the powers for the elements are computed using the paraxial approaches discussed in Chapter 2. The critical aberration formulae are listed in the next lower set of data. At the bottom, the spreadsheet programmer has included a plot of the marginal ray heights through the lens, as a visual guide to what is happening.

The designer could make manual changes and search through the range of possibilities to try to locate a solution. In this example, the designer has stated some required values for the aberrations, zero for chromatic aberrations and a Petzval curvature of $-1/3.5$ times the power of the lens. The difference between these targets and the computed values are listed and combined as a sum of squares to provide a figure of merit (FMT) of the solution. The built-in solver for the EXCEL program can then be used to vary the starting parameters in an attempt to reduce the figure of merit to zero. An exact solution is not needed because the next step will be to introduce the powers found into a design program and search for a thin-lens solution. The data listed in the figure is a reasonable starting point. In principle, the thin-lens equations for spherical aberration, coma, and astigmatism could be included here. However, this is not a useful way for a designer with access to a lens design program to spend time.

The spreadsheet approach can be a powerful design analysis tool. The designer can introduce new glasses and search for possible solutions, always keeping control of the problem. The main point is that the design of a lens of the triplet class can begin from initial first-order considerations.

The most frequently used approach in contemporary lens design is to build the required constraints upon the solution into the input for a lens design program, and let the program charge on toward possible solutions. Such an approach also permits easy introduction of realistic thicknesses into the elements. The form of these equations does, however, provide some significant insight into the interrelation of variables and the form of the lens, so that a separate solution of these equations to find a starting point is very instructive. Once a possible solution is obtained from the spreadsheet approach, this can be inserted into the design program, and design may proceed.

An example of the next step is shown in Figure 7.32. The data from the spreadsheet has been entered into the OSLO program, and scaled for focal length, diameter, and field angle. The lens looks peculiar because a thin-lens starting point is entered. Examination of the ray pattern on the lens drawing shows that the rays actually bounce between surfaces in the outer elements, indicating that these are really positive lenses. As a start, equal curvatures have been selected for the elements. In order to illustrate how thin-lens solutions are actually quite proper in a lens design program, the next few steps will continue with the zero-thickness elements.

The aberration discussion in Chapter 3 indicates that the monochromatic aberrations are controlled by bending of the elements. The spherical aberration, coma, and astigmatism are controllable by solving for the bending of the three elements. The coma is evidently correctable to zero by the balance of the coma from the outer elements against the intrinsic coma of the center lens. The astigmatic correction can be understood by applying the understanding of the meniscus lens separated from the stop. The presence of spherical aberration in the outer elements produces backward-curving astigmatic contributions necessary to control the field curvature aberrations. The center negative element also produces overcorrected spherical aberration that will cancel the undercorrected spherical aberration from the outer elements. Selection of the bendings of all three elements permits simultaneous elimination of coma.

The bendings can be solved such that the tangential field is flat. Because the spherical aberration of the center element varies quadratically with its shape, there are two solutions for the lens, in which the two possible bendings of the central element are used. These are referred to as the right- and left-hand solutions for the lens. For simply third-order aberrations, both solutions are equivalent. In general, one solution will be preferable to the other because of smaller residuals of high-order aberrations.

Figure 7.32 shows a third-order solution of the type requested. The tangential field has been set flat. The ray intercepts indicate the presence of higher-order spherical, coma, and astigmatism. The scale on the ray plots is 2 mm, indicating a large amount of residual high-order error that needs to be balanced. Sufficient thickness has been introduced to make real lenses in the next step, indicated in Figure 7.33. The scale of the high-order aberration residuals is still quite large.

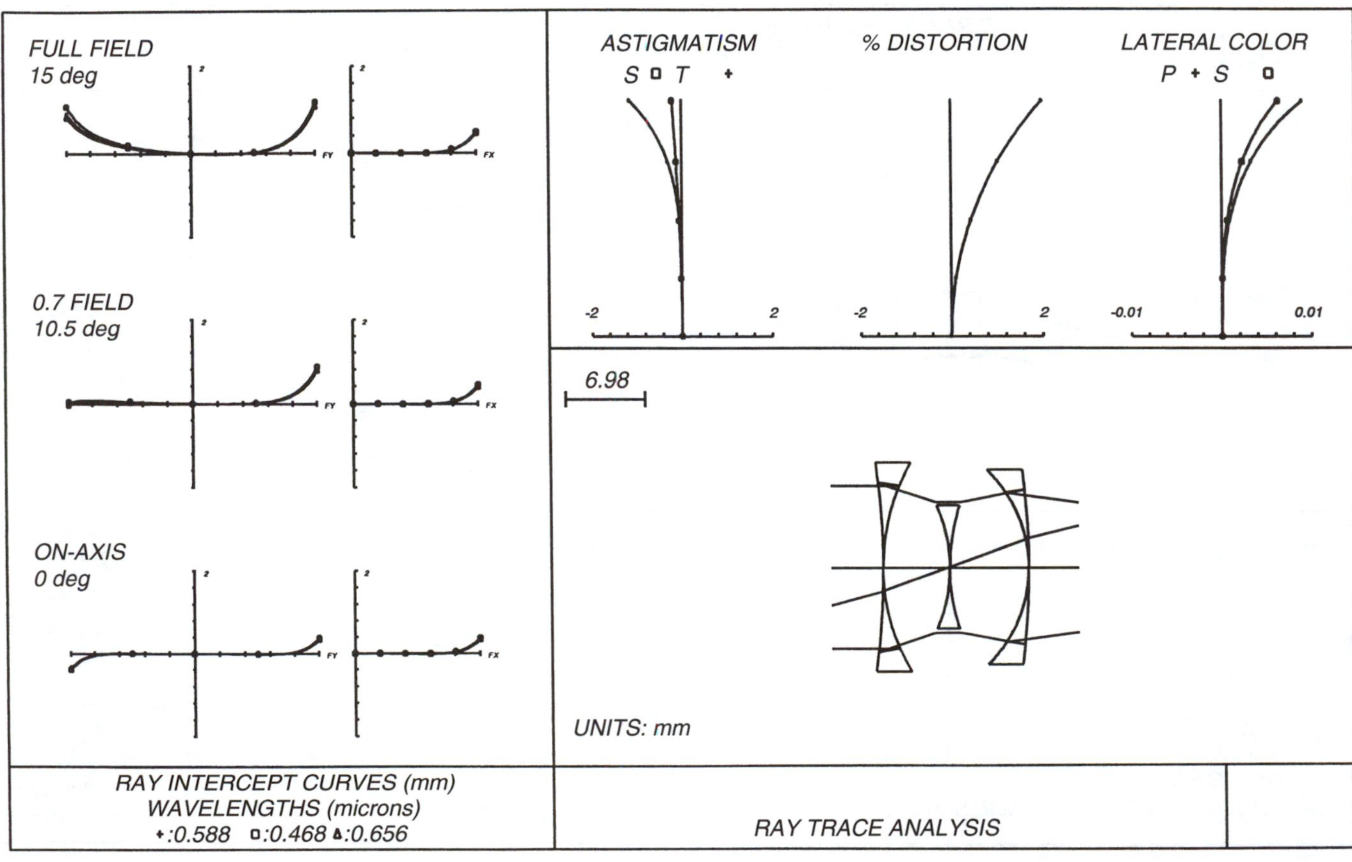

FIGURE 7.32 Third-order solution for thin-lens triplet (OSLO).

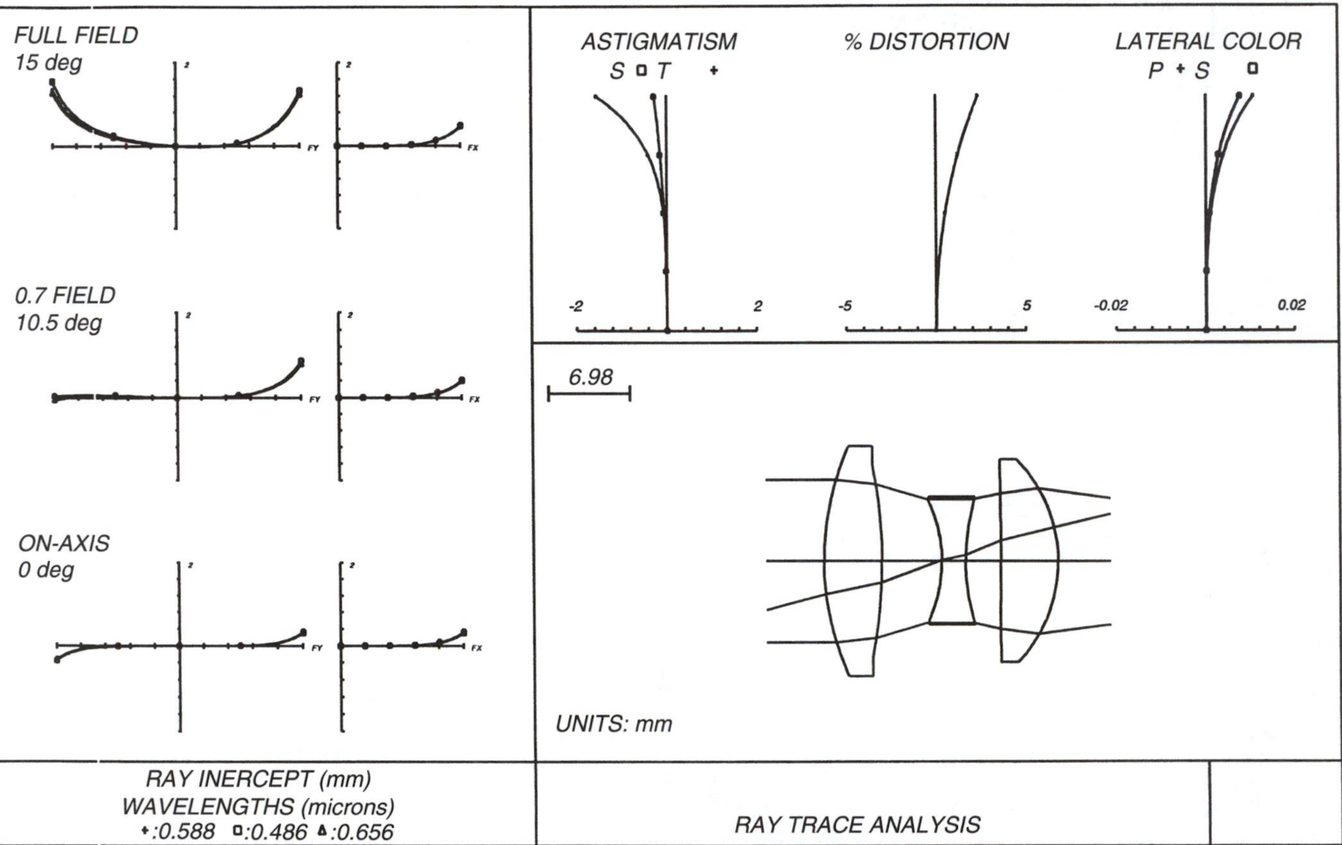

FIGURE 7.33 Third-order solution with thicknesses (OSLO).

The argument about bending given here is easily understood in terms of the thin-lens solution. As element thickness is added, the understanding becomes blurred by the complex change of aberrations with change of shape of a thick element. The general principles do apply, and can be used to explain the likely shapes of the elements that are eventually obtained.

The distortion is not easily controlled because the variables have all been used on the other aberrations. Variation of the ratio of the airspaces from equality, as well as appropriate variation of the element thicknesses will provide a range of solutions that can be used to obtain distortion control while maintaining correction of the other aberrations.

Once control of the third-order aberrations is obtained, these aberrations must be set to balance the high-order aberrations. The axial image can be corrected such that the third-order spherical offsets the overcorrected fifth- and higher-order spherical. Since the lens needs to be optimized over a wide field of view, there is also a need to adjust the actual balance of aberrations on axis to shift the best focus inward from the paraxial focus, in order to offset the inwards curvature of field. Thus, the final choice of axial aberration balance cannot be made until the field aberrations are controlled.

The selection of either the right- or left-hand solution should be made for the lens with the lowest cubic coma content. The astigmatic fields will be set initially for a flat tangential field. The actual field surface balance will be determined by the high-order (mostly fifth) astigmatism. The most desirable solution is one in which the astigmatism is cancelled at some point in the field, usually at 0.7 to 0.8 of the full field. This will occur when the chief ray for that field angle experiences a zero net sum of difference in focus between the sagittal and tangential foci.

Obviously a good lens design program works on all of the aberrations with all of the variables at the same time. The logical discussion here is intended to provide insight for the designer to make decisions about what step to take next in the lens design. Figures 7.34 and 7.35 illustrate the effect of carrying out a balancing for the high-order aberrations. Because of the simplicity of the lens and the large amount of aberration to be balanced, vignetting was allowed. The element sizes have been selected to provide about 50% or so illumination at the edge of the field. At this point, the default merit function for the program is chosen, with weights and numbers of rays adjusted to suit the design requirements. In order to keep the design within the original triplet form, only the curvatures are allowed to vary. The ray trace plots in Figure 7.35 indicate significant improvement, with the scale of the ray trace plots being dropped by a factor of ten. An idea of the level of image quality can be obtained from examination of the MTF plots presented in Figure 7.36. The imagery at the edge of the field is restricted to less than 20 cycles/mm.

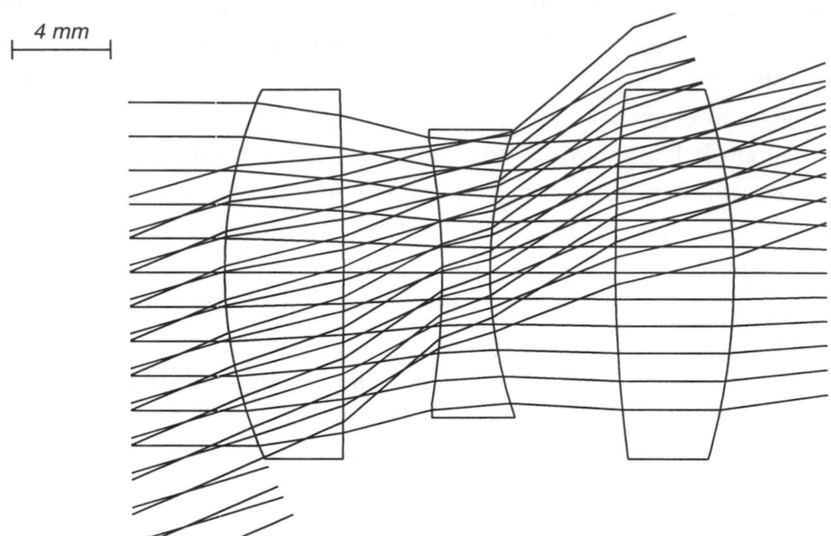

FIGURE 7.34 Triplet with lens diameters sized for vignetting (OSLO).

The design program was then used to improve the design significantly. The glasses were added as variables, and the central airspaces were allowed to vary, but were restricted in the allowable range of variation. The vignetting was maintained at the same level. The default merit function was experimented with by varying the weights on the fields. After several iterations, a new lens was obtained which is shown in Figure 7.37. The lens is slightly shorter, with a different symmetry. The ray intercept plots shown in Figure 7.38 are scaled another factor of four smaller, and now show more details of the image balance. The distortion is slightly larger, now approaching 2% at the edge of the field. Figure 7.39 shows the MTF curves, which are significantly improved over the previous design. Some of the limitations upon the MTF behavior can be seen from the plots in Figure 7.40, in which the MTF at 25 cycles/mm is plotted against focal position. The split between the sagittal and tangential best focus position for the field angles is clearly indicated, and shows that the best focus across the field is necessarily a compromise.

Figure 7.41 shows another view of the aberration content of the lens. The spot diagrams are plotted for several focal positions for three field locations. In addition the energy concentration at three field positions is plotted. The appearance of the off-axis images which indicate the presence of a mixture of aberrations is somewhat typical of the images that can be expected from this type of lens. Figure 7.42 provides a record of the lens that has been examined in the last few plots.

Understanding the design of the triplet objective requires some understanding of the sources of aberrations from single elements. The total

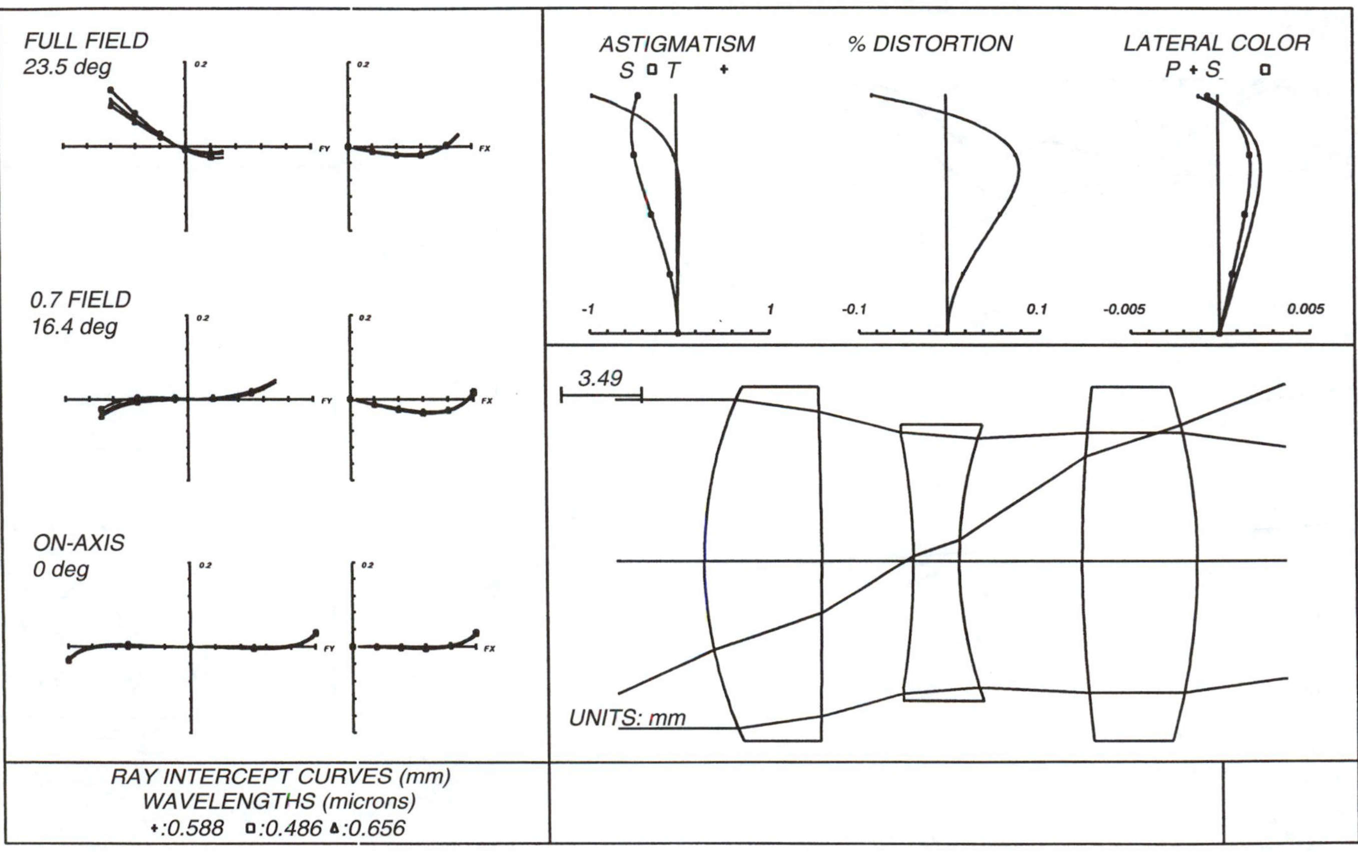

FIGURE 7.35 Optimized design with vignetting (OSLO).

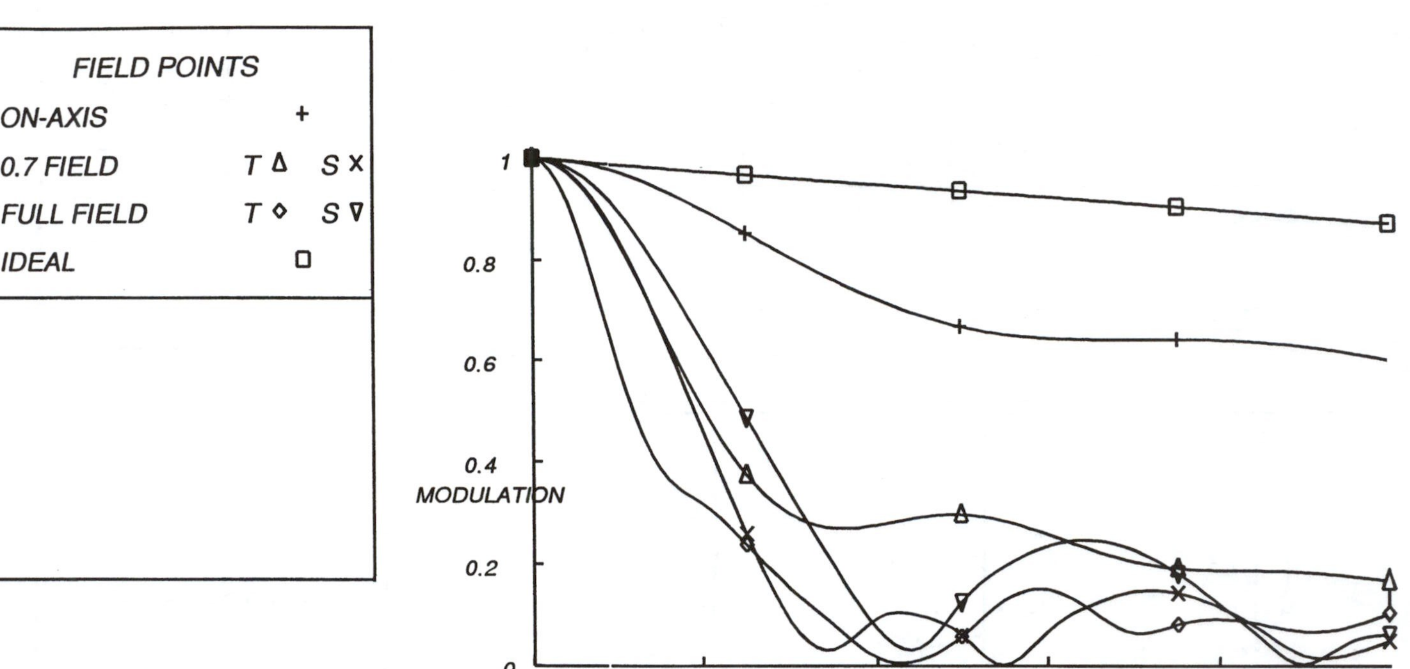

FIGURE 7.36 MTF for the design in the previous figure (OSLO).

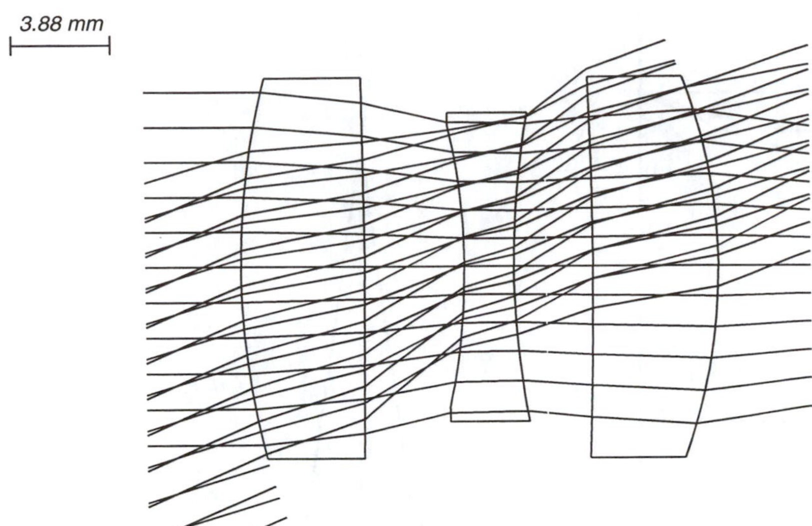

FIGURE 7.37 Triplet after permitting glass variation (OSLO).

power of the lens is obtained by summing y times the power of each lens. In general, the correction axial color of the lens is obtained by balancing the color by a negative high-dispersion lens and positive low-dispersion lenses. The symmetric disposition of the lenses about the stop leads to correction for the lateral color. The large amount of undercorrected spherical aberration from the positive elements is compensated by the overcorrected aberration contributed by the central negative lens. The Petzval curvature is controlled by the choice of the sum of the powers of the lenses.

The astigmatic field correction is obtained by permitting the outer elements to assume bendings that introduce excess spherical aberration. Since the element is located at a distance from the stop, the large spherical aberration generates positive, or backward-curving, astigmatic field components, as well as coma. Since the sign of the coma depends upon the direction from the element to the stop, a balance for corrected coma along with a net field-flattening astigmatism results. The spherical aberration is corrected by choosing an appropriate shape for the central negative element. Since there are two bendings of the central element that produce the requisite amount of positive spherical aberration, there are two regions of solution.

The choice of the amount of third-order aberration to be left in the lens depends upon the amount of higher-order aberration that is present. The balance of the third-order spherical can initially be taken as sufficient to edge-correct the high-order spherical aberration. The chromatic aberration balance is such as to balance the spherochromatism.

The coma is usually set to be zero, as is the lateral chromatic aberration. A better choice is to use these aberrations to balance high-order coma or

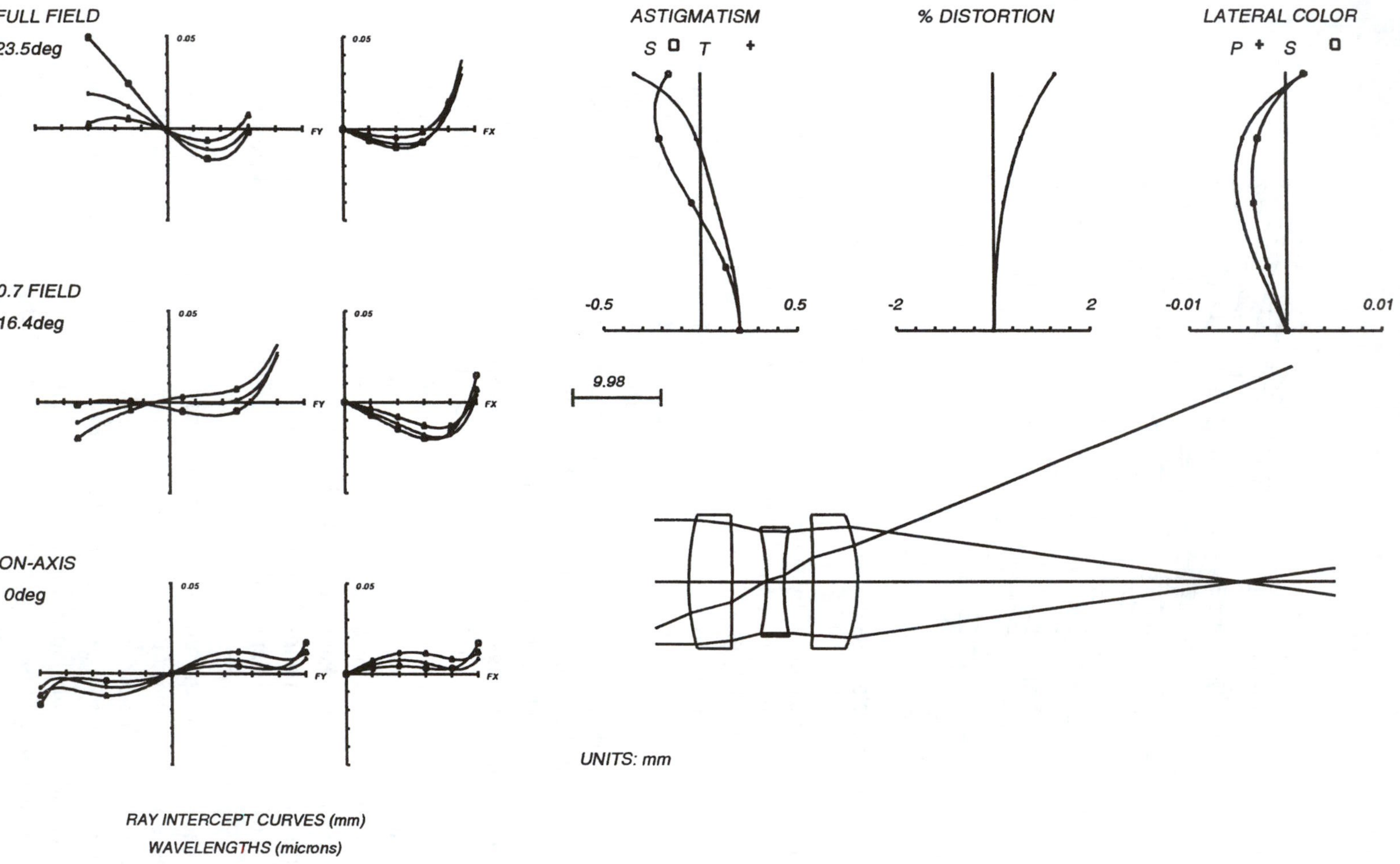

FIGURE 7.38 Triplet after glass variation and optimization (OSLO).

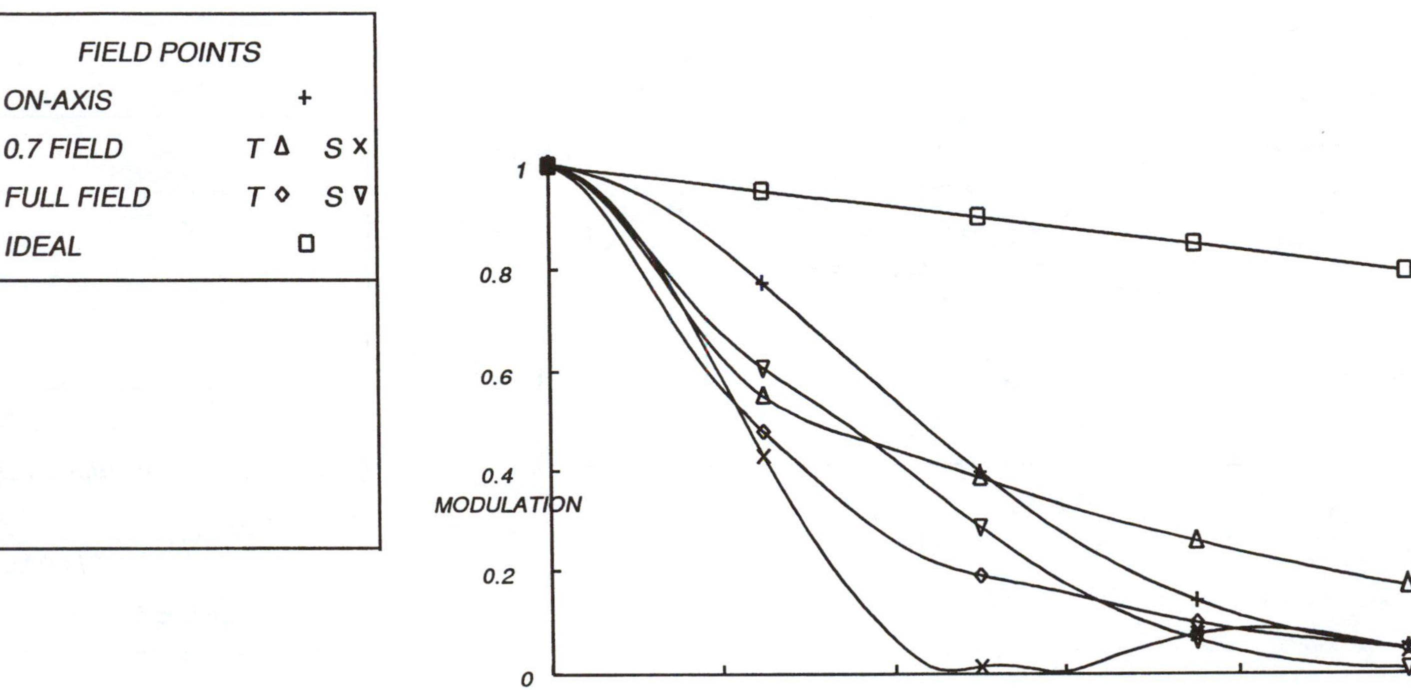

FIGURE 7.39 MTF for optimized triplet (OSLO).

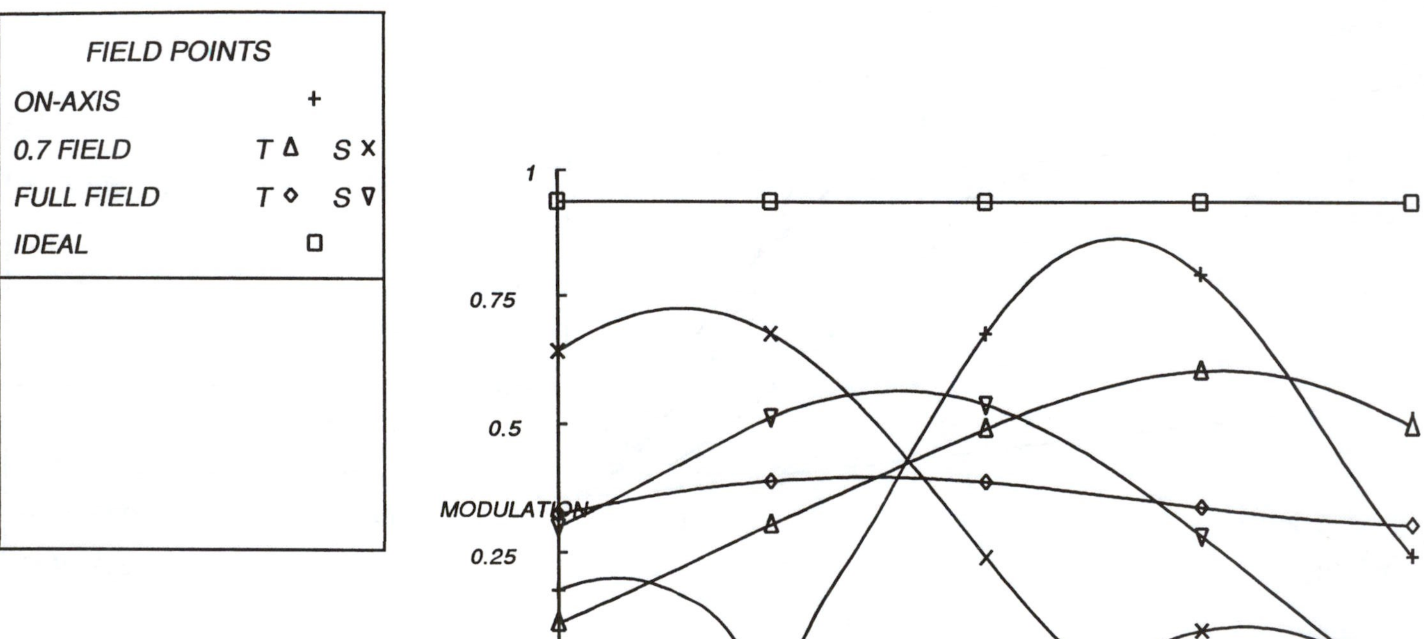

FIGURE 7.40 MTF at 25 lines/mm through focus (OSLO).

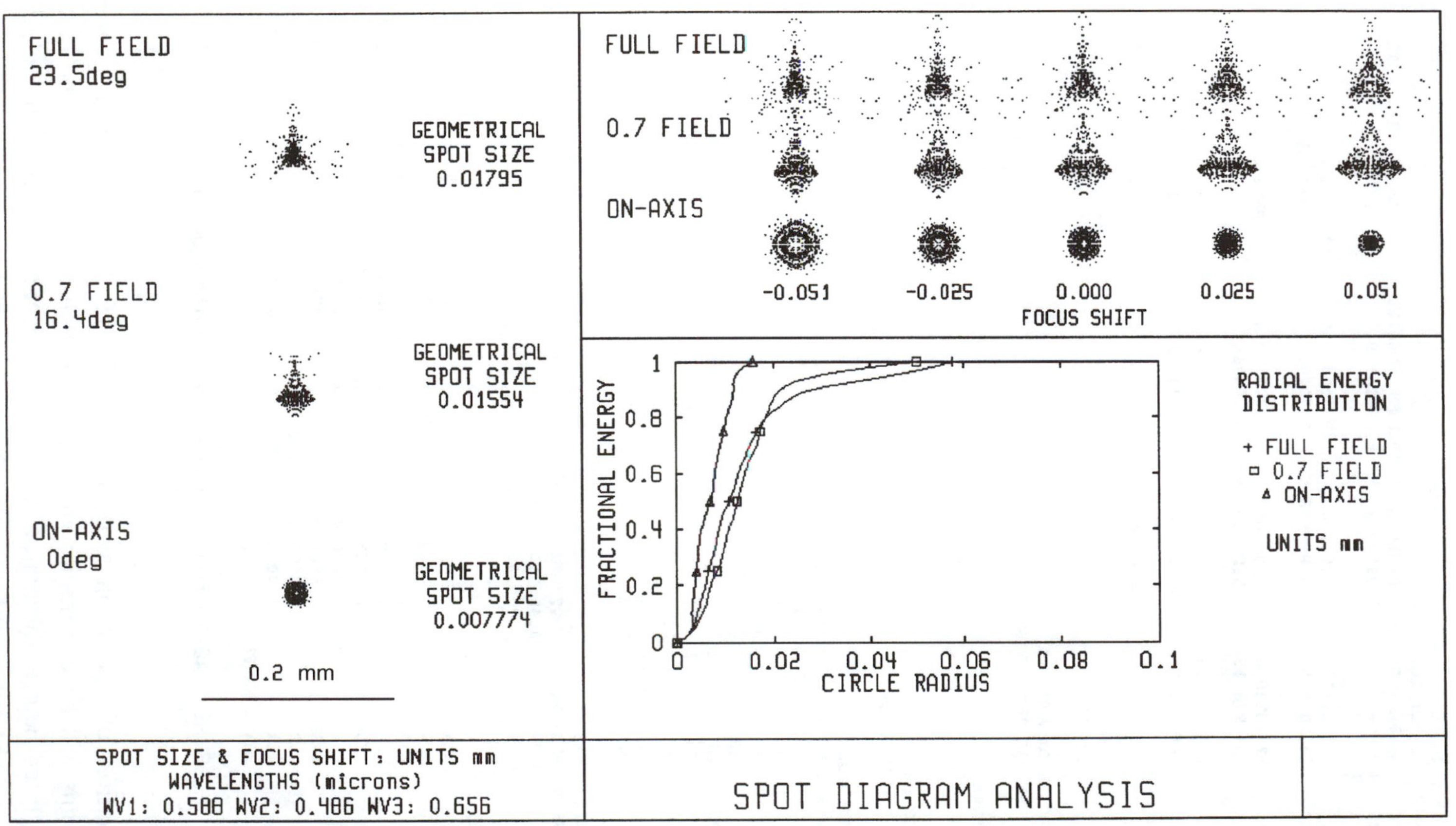

FIGURE 7.41 Spot diagrams and encircled energy for optimized triplet (OSLO).

```
*LENS DATA
Chap 7.1.2 Triplet
 SRF         RADIUS           THICKNESS    APERTURE RADIUS       GLASS  SPE    NOTE
  0            --           1.0000e+20      4.3481e+19            AIR

  1       30.249192 V        5.000000        7.700000 K        LASFN31 C
  2     -343.048758 V        3.999988 V      7.700000             AIR

  3      -26.523065 V        2.000000        5.696179 A           SFL6 C
  4       34.878053 V        3.203507 V      6.243064 S           AIR

  5     -191.890387 V        5.000000        7.700000          LASFN31 C
  6      -21.995266 V       43.456080 S      7.700000 K           AIR

  7            --           -0.200000       21.905619 S           AIR

  8            --              --           21.851206 S

*SPECIAL DATA TYPES
 NO SPECIAL DATA TYPES

*SOLVES
  6      PY          --

*PICKUPS
 NO PICKUP DATA

*APERTURES
 SRF    TYPE APERTURE RADIUS
  0     SPC    4.3481e+19
  1     SPC       7.700000    CHK
  2     SPC       7.700000
  3     CMP       5.696179
  4     CMP       6.243064
  5     SPC       7.700000
  6     SPC       7.700000    CHK
  7     CMP      21.905619
  8     CMP      21.851206

*WAVELENGTHS
CURRENT  WV1/WW1      WV2/WW2      WV3/WW3
   1     0.587560     0.486130     0.656270
         1.000000     1.000000     1.000000

*REFRACTIVE INDICES
 SRF     GLASS             RN1          RN2          RN3          VNBR          TCE
  0      AIR             1.000000     1.000000     1.000000        --            --
  1      LASFN31         1.880669     1.895765     1.874290    41.009121     68.000000
  2      AIR             1.000000     1.000000     1.000000        --        236.000000
  3      SFL6            1.805182     1.827801     1.796093    25.393449     90.000000
  4      AIR             1.000000     1.000000     1.000000        --        236.000000
  5      LASFN31         1.880669     1.895765     1.874290    41.009121     68.000000
  6      AIR             1.000000     1.000000     1.000000        --        236.000000
  7      AIR             1.000000     1.000000     1.000000        --        236.000000
  8      IMAGE SURFACE
```

FIGURE 7.42 Program listing of lens data for the optimized triplet (OSLO).

variation of distortion with wavelength. However, the most successful designs will have a minimum amount of high-order coma to be balanced. As stated before, the major criterion for selecting which of the solutions is likely to be successful is to use the design which has the lowest content of elliptical coma which varies with the cube of the field.

The setting of the balance condition for astigmatism is considerably more complicated. The first thought would be that the deliberate introduction of Petzval and astigmatism would be used to balance the fifth-order Petzval and astigmatism. A good design will require that the setting of the third-order field aberrations be used to balance the oblique spherical aberration as well. Since the oblique spherical aberration varies as the square of the field, this can be used to set the best focus for the spherical aberration across the field to lie on a specified best focus plane.

Because of the intrinsic nature of imagery to squeeze the aperture anamorphically as the field increases, the oblique spherical aberration will tend to provide overcorrection with increasing field angle. As an image improvement to optimize the image quality across the field, the axial spherical aberration can be altered to add some additional undercorrection. This will add to the oblique spherical aberration, producing a bias that will offset some fraction of the oblique spherical aberration. This deliberate introduction of undercorrected spherical aberration will reduce the image quality on axis and force the best focus to lie inside of the paraxial focal plane. Both of these attributes are actually beneficial in making the image somewhat more uniform across the field. When using the default merit function within a design program, this effect is controlled by judicious setting of the weights upon the aberrations at various field heights. Addition of some intermediate fields to the merit function can also provide some opportunity to selectively control the balance of aberrations across the field.

The distortion in a triplet ends up being whatever the design provides. The variables in the lens have all been used up in providing the balance for the aberrations so far discussed. However, variation of the thickness and spacing of the elements will lead to a number of designs, which all have the aberration balance discussed so far, but with different distortion residuals. The art of design here is to use this knowledge to locate a design which has the best choice of balancing aberrations, along with a satisfactory level of distortion.

A well-designed merit function should lead to location of the optimum design including all of the choices for balancing discussed so far. The number of aberrations that can be controlled roughly matches the number of variables. The principal variables are the curvatures and separations of the elements. Secondary variables are the thicknesses of the elements and the choice of glass for the lenses. A count shows six curvatures, two airspace separations, three lens thicknesses, and three glass types. Important design goals that the customer requires are the back focus and the overall length of the lens. The precise location of the stop is also a possible variable.

Optimization of the image quality with so few free variables leads to some rather large aberrations at the edge of the aperture and the outer part of the field. An important control is to remove those aberrations by blocking the heavily aberrated parts of the aperture through vignetting. The designer will usually select some allowable level of vignetting across the field for when the lens is used at full aperture. A change in the aperture will change the effect of the vignetting. Since the removal of aberrated aperture by vignetting is actually accomplished by the clear apertures of the lenses, not every arbitrary vignetting profile is available. The possible vignetting profile is determined by the length of the lens and the field angles. The designer must ensure that the actual vignetting profile is matched by the selected aperture diameters.

The optimum balancing of the aberrations in the lens will depend upon the choices for allowable amounts of aberration at each aperture setting. The usual decision by the designer is to set the aberration balancing so that optimum balance actually occurs for the lens stopped down some specified amount. The imagery will not be optimum for the largest aperture, but the lens will most often be used at a lower aperture, providing the best imagery on the average to the customer. This could be the optimum solution, and the designer must be aware of the opportunity to use such design options as vignetting and balance of aberrations to locate the solution most suitable for the customer.

As a final comment in this section, Figure 7.43 shows five different triplets that have figures of merit that are different. The lenses are all slightly different, but are obviously members of the same family. Figure 7.44 is an overlay of all of the MTF values for these lenses. Clearly all are functionally the same. The message is that there are usually many possible solutions that meet a given figure of merit. The selection of any single design may not be the only possible solution that can be attained. For lenses covering a large field of view, the selection of the "best" solution may not be unique. The solution space is actually quite broad, even though the calculated figure of merit may indicate that there are significant differences between the lenses. There is an art in deciding when and where to stop a lens design.

7.1.3 GAUSS TYPES

The double Gauss type of lens is the most common lens system used for photographic or video applications. This type of lens is capable of covering fields of about 20° to 25° half field at apertures up to F/2, or even larger, and focal lengths from 35 to 100 mm, with image quality suitable for most applications. This type of lens is extremely versatile and may contain from five to eight elements, depending upon the required aperture and field. The lens type derives from the use of symmetric thick elements surrounding a central stop, positive outer elements providing control of the astigmatic fields and a "Gauss airspace" between the outer positive elements and the thick meniscus

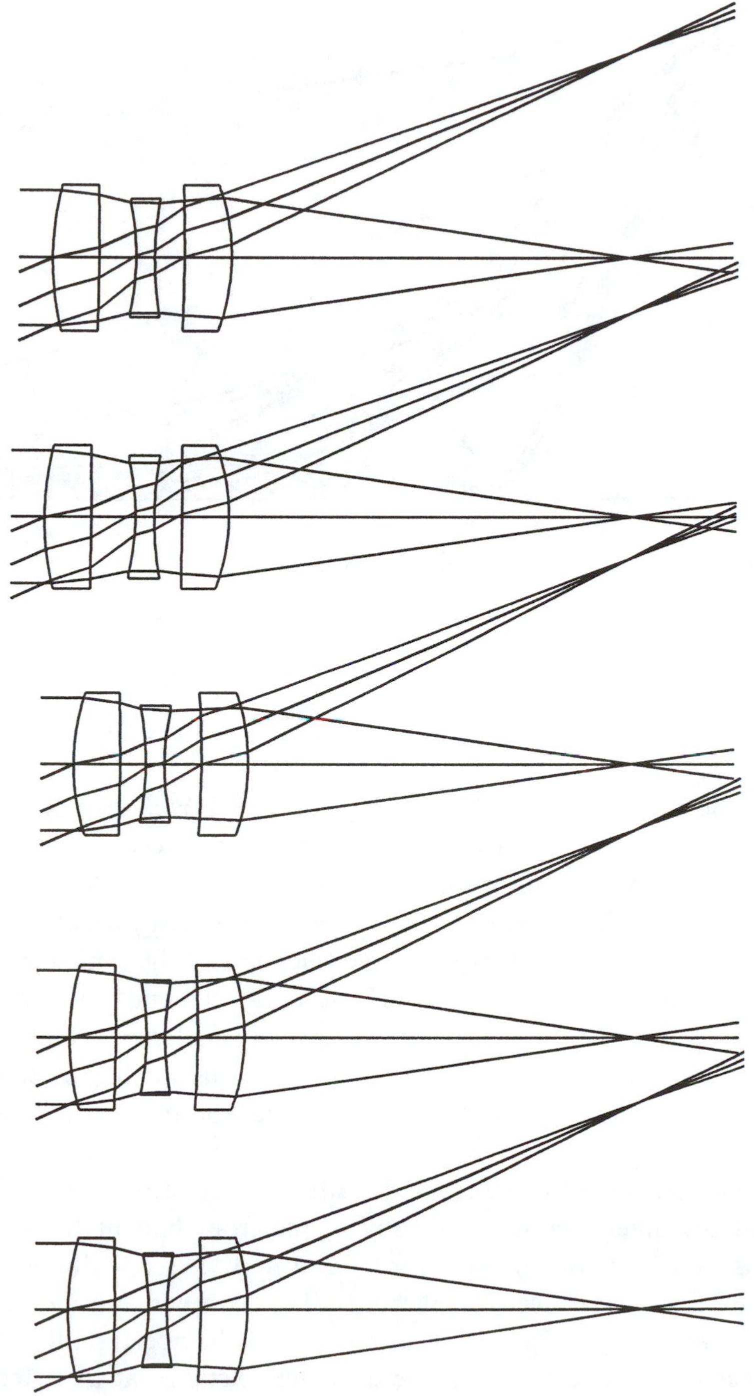

FIGURE 7.43 Layout view of five triplet lenses with independent optimization.

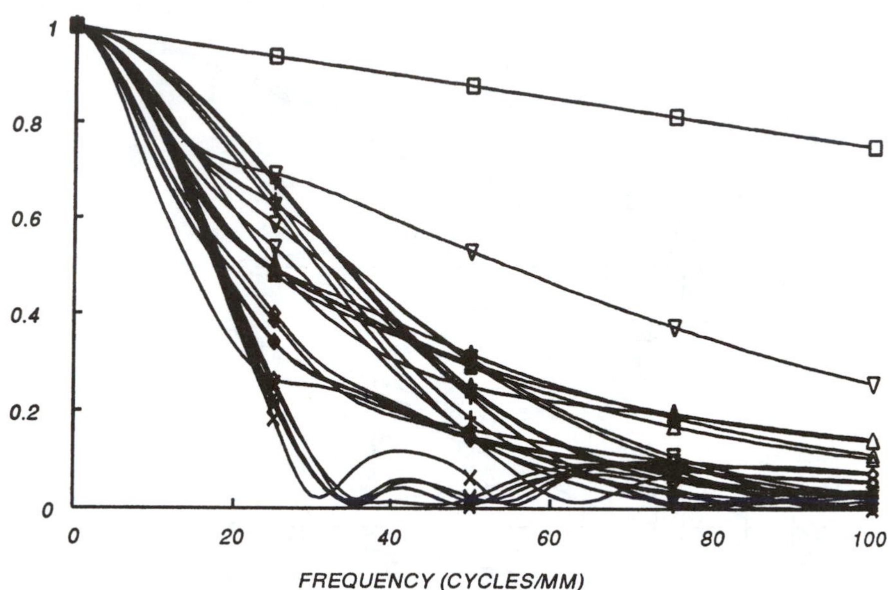

FIGURE 7.44 Summary plot of MTFs for the five lenses in the previous figure (OSLO).

elements. Specific lenses for various purposes can deviate quite a bit from this generic form.

The double Gauss objective historically developed from the double meniscus achromat, in which achromatic doublets were placed about a central stop, with the elements bent toward the stop to flatten the field, as described in Chapter 3. Addition of two outer elements permitted increased aperture for a given field correction. Increasing the thickness of the symmetric pair of inner lenses introduced convergence of the imaging bundle within the elements, providing an introduction of power with a smaller contribution to the inward-curving Petzval surface. The thick menisci surrounding a central stop became a standard approach to increasing the aperture of photographic imaging lenses.

An additional benefit came from the airspace between the outer pair of lenses and the inner menisci, especially in the front half of the lens. This airspace is called a Gauss airspace, first used in a telescope doublet having this type of airspace between the elements. The path between the successive elements increases from the center to the edge of the aperture, thus spherochromatically aberrated rays traverse a greater distance at the edge of the aperture, with the blue rays having a significantly lower height of incidence upon the following surface. The effect of the next surface is to continue to converge the bundle, but with slightly less power for the blue than the red. This tends to compensate some of the variation of spherical aberration with wavelength. The result is that a better balance of the rays can be established across the aperture and spectral region.

The generally symmetric nature of the design indicates that there will be less variation of angle of incidence on the lens surfaces with field angle, which permits better control of the astigmatism. So far all this looks quite advantageous, but there is an intrinsic limit to the imagery that can be obtained from this type of lens caused by oblique spherical aberration. As the bundle of rays moves from axis toward the maximum field angle, there will be an anamorphism introduced in the beam passing through the lens. The tangential direction of the bundle of rays passing through the lens will be compressed relative to the sagittal beam dimension. Therefore the balance of spherical aberration correction established between the inner and outer lens elements will be changed, with the central portion seeing less negative spherical correction in the tangential direction than in the sagittal direction. The result will be the presence of overcorrected oblique spherical aberration which is intrinsic to the lens form, and, of course, varying as the square of the field.

This aberration is balanced by permitting the introduction of the appropriate amount of third-order astigmatism as a balancing aberration. The performance of the lens type is dominated by this residual oblique spherical aberration. Increased aperture is obtained by adding elements on the outer part of the lens. The seven-element double Gauss is a result of splitting the power of the front element, improving the spherical aberration correction. The eight-element Gauss adds another lens in the rear to permit better control of the field correction.

The general nature of the lens can also be explained in analogy to the basic triplet type of lens. If the surfaces are grouped as in Figure 7.45, the outer set of surfaces form a positive pair of elements surrounding an inner negative element, which is effectively immersed in the higher index of the inner elements. The central airspace forms the equivalent of a strong negative, or field flattening, component analogous to the central element of the triplet. This mode of forming the field flattening permits the use of primarily collective or converging surfaces in most of the lens. Except for the inner surfaces of the central menisci, the angles of incidence are moderate, leading to reasonable amounts of high-order spherical aberration to be balanced. The ability to obtain aberration control at relatively larger apertures is a consequence of this effect.

Other general descriptions of the behavior of the Gauss-type lens follow different lines of reasoning. The symmetry of the elements about a central stop may be identified as most important. In other cases the existence of thick meniscus elements is considered as fundamental to the Gauss-type lens.

An article by Mandler provides an excellent detailed summary of the design of double Gauss objectives for photographic and video imaging

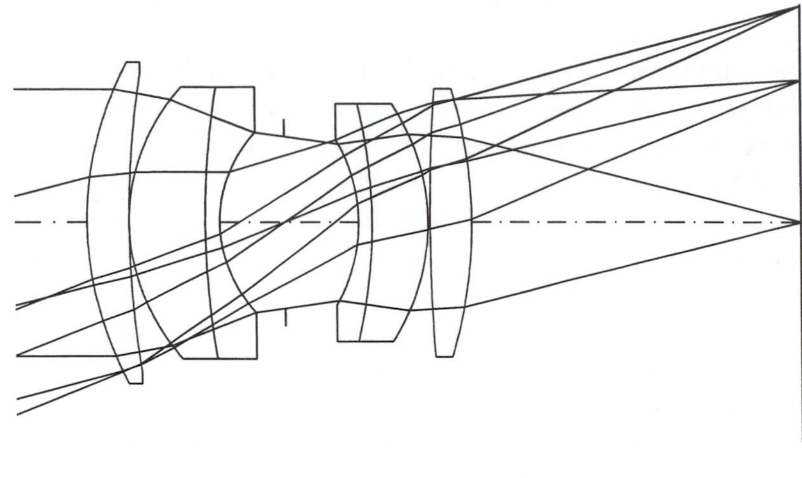

FIGURE 7.45 Layout of a sample double Gauss lens (CODE V).

(Mandler 1980). A summary of the history of the Gauss type objective can be found in an article by Wöltche (1980).

The triplet analogy and the constriction of the beam aperture through the lens contributing to the field flattening have been described in the literature. The constriction is used by Glatzel (1980) as a design principle used in high-resolution, modest field of view, lenses for microlithography. In his article, Glatzel suggests that better imagery with less limitation by spherical aberration is obtained by designing multiple element lenses with more than one constricted region.

Specifically, the Gauss airspace is the open airspace at the front of the lens. This airspace is a technique that can be used to reduce the amount of chromatic variation of spherical aberration. Since the airspace is open, with the air path at the edge of the aperture greater than the axial airspace, there will be an opportunity for the chromatically varying rays from the edge of the aperture to intersect the following element at a dispersion of heights caused by the chromatic aberration of the first element. Since the refracted angle is dependent upon the height of incidence of the ray on the surface, there will be a different amount of angular refraction applied to rays of different wavelength. Clever choice of glasses, curvatures, and separations can use this approach to minimize the spherochromatism.

The field properties of the double Gauss lens are similar to those of the triplet, but generally lead to a lower level of high-order field curvature. The outer elements surround a central airspace which has surfaces concave to the central stop. The meniscus elements surround this region, and provide a glass spacing between the central "air lens" and the outer elements. Since the index of refraction of the central "air lens" is less than that of the surround, the central airspace serves as a negative lens. The correction of the aberrations in this type of lens is similar to that of the triplet, the advantage being that the generally concave-to-stop nature of all of the surfaces leads to a gentler introduction of balancing aberrations and thus less higher-order residual to be balanced at each field angle. The target for Petzval curvature can be reduced over that of the triplet, and Petzval radii of up to five times the focal length are common.

The oblique spherical aberration is an important limiting aberration for this type of lens. The fundamental source of this aberration can be seen from the anamorphic change in the pupil shape with field angle. On axis, a balance for spherical aberration is achieved. Off axis, the bundles of rays traverse the lens obliquely, with the sagittal pupil having essentially the same width as the axial bundle, but with the tangential direction of the pupil undergoing a compression or cosine effect projection on the surfaces of the lens. This change in effective beam diameter is especially noticeable on the inner surfaces of the inner menisci of the lens. Because the aperture is reduced in the tangential direction, the amount of undercorrected spherical aberration produced is reduced. Thus the spherical aberration tends to become overcorrected as the field is increased.

The oblique spherical aberration is more clearly a field limiter in this type of lens than in the triplet because the apertures that are usually used in a Gauss-type lens are larger than in a triplet. Therefore, the cubic aperture dependence of the oblique spherical becomes dominant. In most cases, vignetting is used to remove the extreme tangential aperture portions and reduce the effect of the oblique spherical aberration. Careful choice of vignetting can be used to gain a reasonable balance of the tangential and sagittal aberrations. Since light is removed by vignetting, the uniformity of irradiance over the field suffers, and must be maintained within the allowable dynamic range of the detector. Because most Gauss-type lenses are used at reduced aperture for the majority of applications, the effect of vignetting is felt only at the full aperture and is frequently an allowable tradeoff in the design.

If the lens stop is reduced, say from F/2 to F/4, the total aberration is, of course, reduced. Proper implementation of vignetting will also leave a reduced vignetting factor at the reduced aperture. For most photographic purposes, the dynamic range of the film permits a "half stop" tolerance in exposure, or a reduction of square root of two in amount of irradiance across the field. Thus a usual vignetting loss of 25% to 30% at the edge of the format is generally acceptable.

The setup for a double Gauss lens can be obtained in a manner similar to a triplet, but the more usual approach is to start from an existing design, selected

```
CODE V> lis
      Mandler SPIE V237 1980
                 RDY              THI       RMD        GLA            CCY     THC     GLC
   OBJ:        INFINITY         INFINITY                              100     100
     1:        33.61854         4.009367           LAF23_SCHOTT       100     100
     2:        95.85394         0.000000                              100     100
     3:        19.97667         7.206837           LAFN2_SCHOTT       100     100
     4:        86.04102         1.303044           SF1_SCHOTT         100     100
     5:        13.57171         5.933863                              100     100
   STO:        INFINITY         6.926182                              100     100
     7:       -16.13770         1.303044           SF13_SCHOTT        100     100
     8:       -49.85648         5.242247           LAFN2_SCHOTT       100     100
     9:       -22.03147         0.200468                              100     100
    10:       186.15491         4.009367           LAF21_SCHOTT       100     100
    11:       -45.62660        31.262060                              100     PIM
   IMG:        INFINITY         0.000000                              100     100

 SPECIFICATION DATA
    EPD       25.99999
    DIM             MM
    WL          656.30      587.60      486.10
    REF              2
    WTW             64          64          64
    XAN        0.00000     0.00000      0.00000
    YAN        0.00000    15.00000     21.00000
    VUX       -0.00102     0.03327      0.10581
    VLX       -0.00102     0.03327      0.10581
    VUY       -0.00102     0.17985      0.56121
    VLY       -0.00102     0.30555      0.51082

APERTURE DATA/EDGE DEFINITIONS
    CA
    CIR S1                 14.533956
    CIR S2                 14.533956
    CIR S3                 12.228570
    CIR S4                 12.228570
    CIR S5                  8.820608
    CIR S6                  8.419671
    CIR S7                  8.068851
    CIR S8                 10.624823
    CIR S9                 10.624823
    CIR S10                12.028101
    CIR S11                12.028101
```

FIGURE 7.46 Program listing of design data for the sample double Gauss lens (CODE V).

by the number of elements and the closeness of the starting parameters from the desired final lens parameters. There are many possible starting points in the literature, in patents and stored in the libraries of most lens design programs. In this book, we will choose an arbitrary starting point and explore the nature of the imagery to be expected in these lenses. We will also examine the options of aperture versus field angle tradeoffs and the effect of vignetting.

The understanding of how a double Gauss lens operates can be obtained by examining a representative example from the literature. The example is taken from the article by Mandler in the 1980 International Lens Design Conference (Mandler 1980). This lens is depicted in Figure 7.45. It is an F/2, 52 mm focal length lens covering 21° half field, suitable for a standard "double frame" 35 mm camera format. The lens is quite typical of the design form with the thick meniscus elements and the outer converging elements surrounding a central stop location. Figure 7.46 shows the construction data for this lens.

The nature of the image quality can be examined by looking at the ray intercept curves in Figure 7.47. There is a fairly large residual spherical aberration zone on axis, presumably deliberately introduced to balance the oblique spherical aberration present in the design. The extent of the ray intercept curves off axis is quite shortened because of vignetting of about 20% at 15° and about 50% at the field edge. The balance as shown at the 15° field position is to play the field curvature against the spherical aberration to locate a best focus position at the field position. The effect of the sagittal oblique spherical aberration is quite noticeable. The image quality is permitted to degrade in favor of balancing at the lower fields by the time the extreme edge of the field is reached. This field position is located in the corners of the format and covers a small portion of the image.

If the vignetting is overridden, and rays are traced across the full aperture of the lens across the field, the ray intercept plot shown in Figure 7.48 results. The dramatic growth in oblique spherical aberration is clearly seen. The intrinsic nature of this aberration in the lens type indicates that the only effective method of eliminating this aberration is to eliminate the rays by an appropriate sizing of the lens apertures. The inclusion of appropriate vignetting, and a shift to the best compromise focal position across the field, lead to the ray intercepts shown in Figure 7.49. Figure 7.50 shows the spot diagrams that are associated with the image in this lens. The image form is typical of that expected from a double Gauss. Note the dimensions of the spot diagrams, indicating that the core of the image is contained within approximately a 0.025 mm diameter across most of the field. The field plots for the lens are presented in Figure 7.51. The balancing of the astigmatic fields is evident, and can be seen to be comparable to the focus shift experienced by the spherical aberration. This is another indication that the overall design of the lens requires an appropriate balance between the spherical aberration and the field-dependent aberrations.

In order to quantify the effect of this aberration on the image, the MTFs are calculated across the field. These are presented in Figure 7.52. For most of the field, the contrast at 30 lines/mm is of the order of 0.5, quite acceptable for most purposes. The plot of MTF versus focal position is shown in Figure 7.53, for the contrast at 25 lines/mm through focus. The compromise peak of the MTF response is seen to be at about the −0.1 focus position. The astigmatism and field curvature place the best focus at various fields at slightly different positions. The working depth of focus can be found by assuming some acceptable value of contrast, say an MTF of 0.25. The allowable focus shift to maintain this level over the entire field is about 0.075 mm. If the requirement is over "most" of the field, the depth increases to about 0.15 mm. This is actually quite reasonable for a camera.

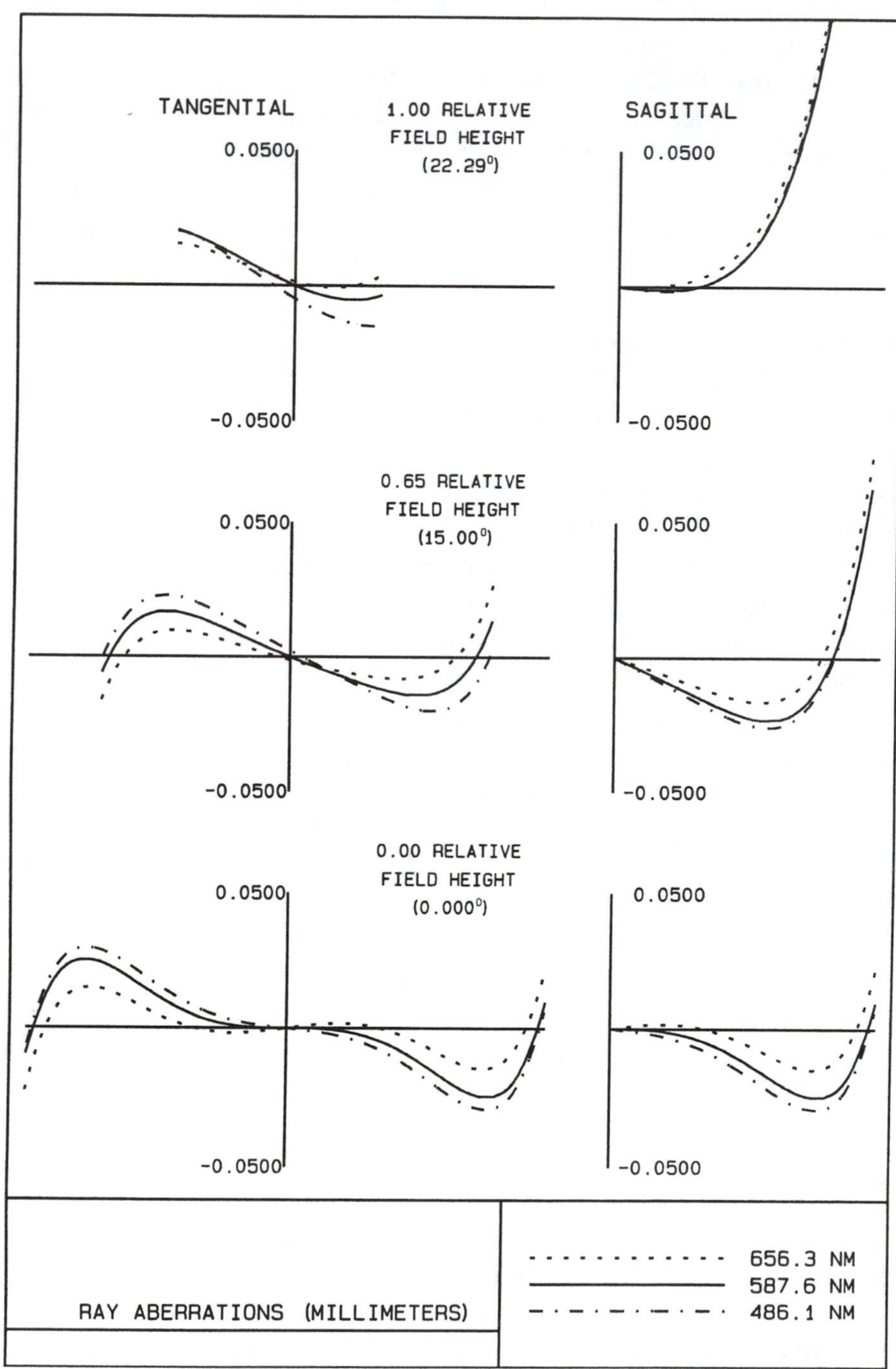

FIGURE 7.47 Ray intercept plot for sample double Gauss lens (CODE V).

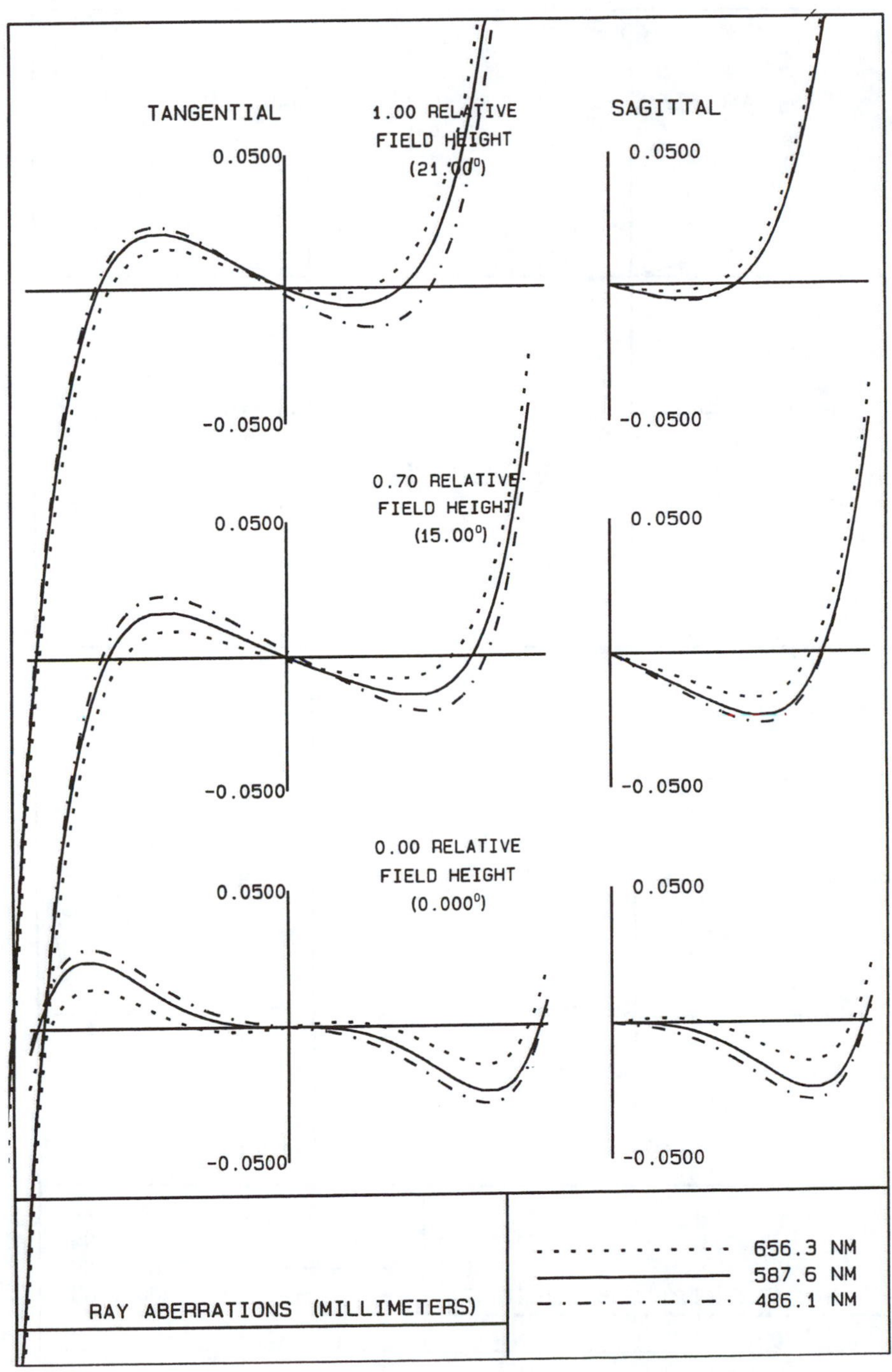

FIGURE 7.48 Ray intercept plot with vignetting removed (CODE V).

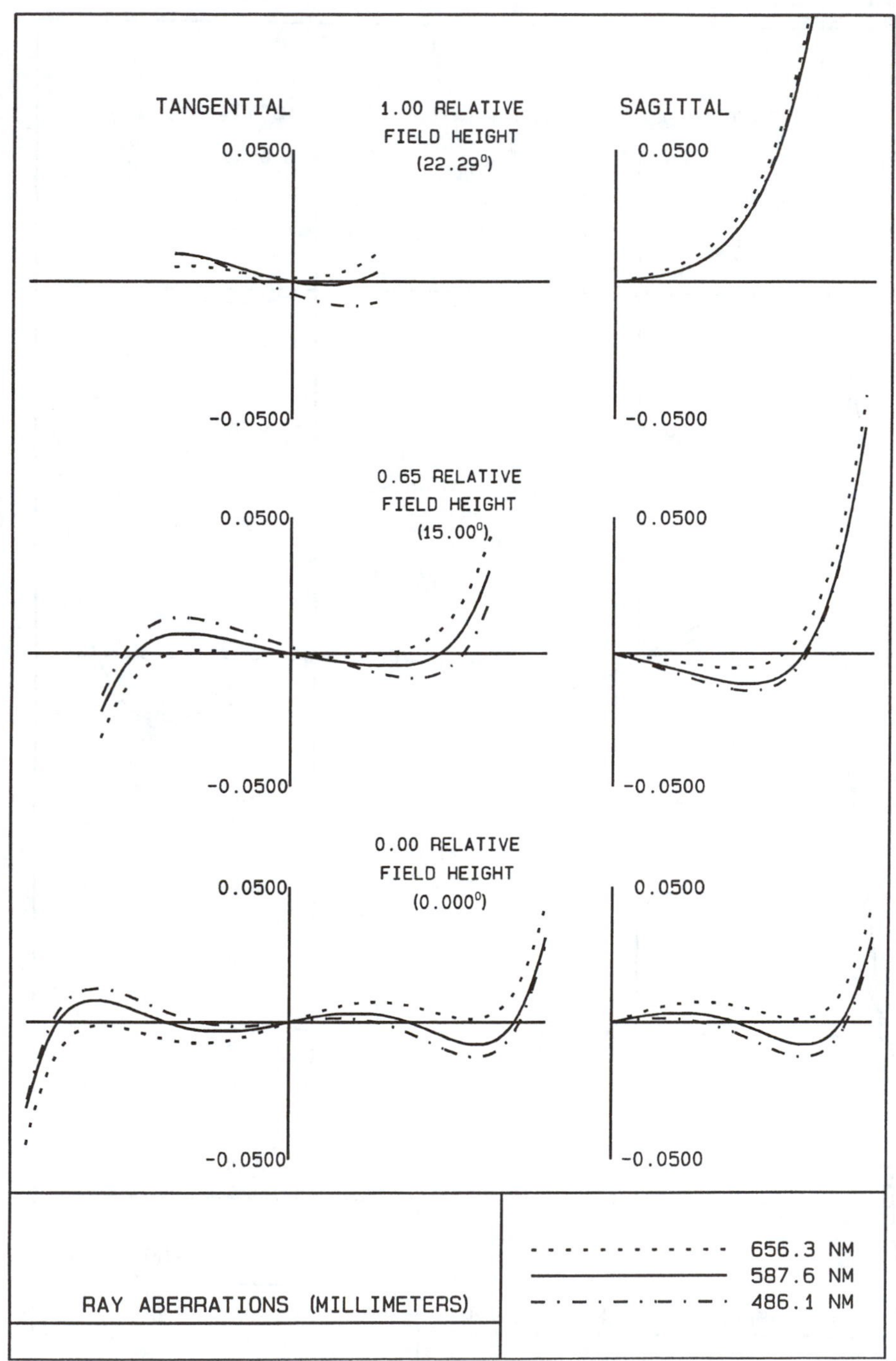

FIGURE 7.49 Sample double Gauss adjusted to best focus (CODE V).

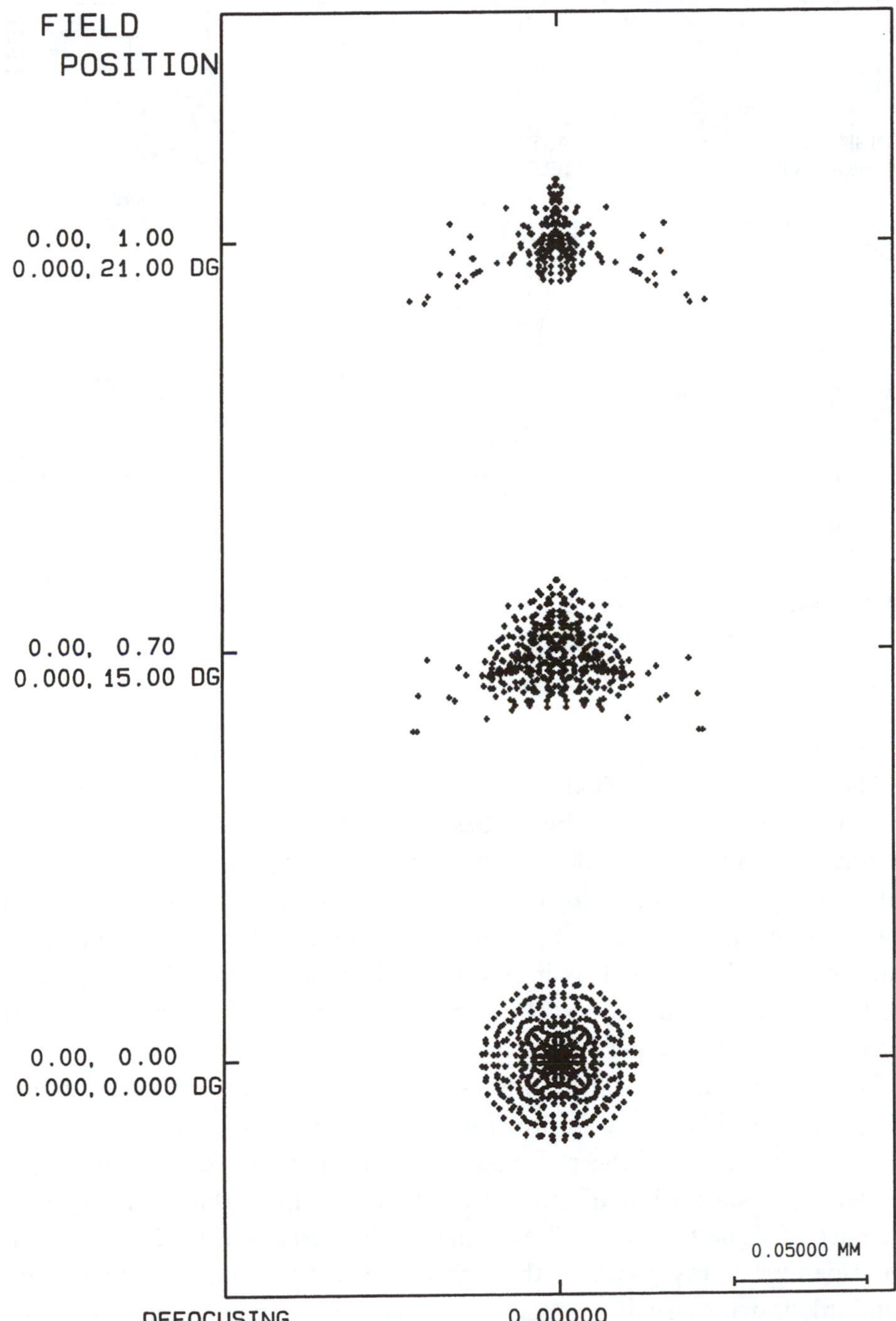

FIGURE 7.50 Spot diagrams for sample lens at best focus (CODE V).

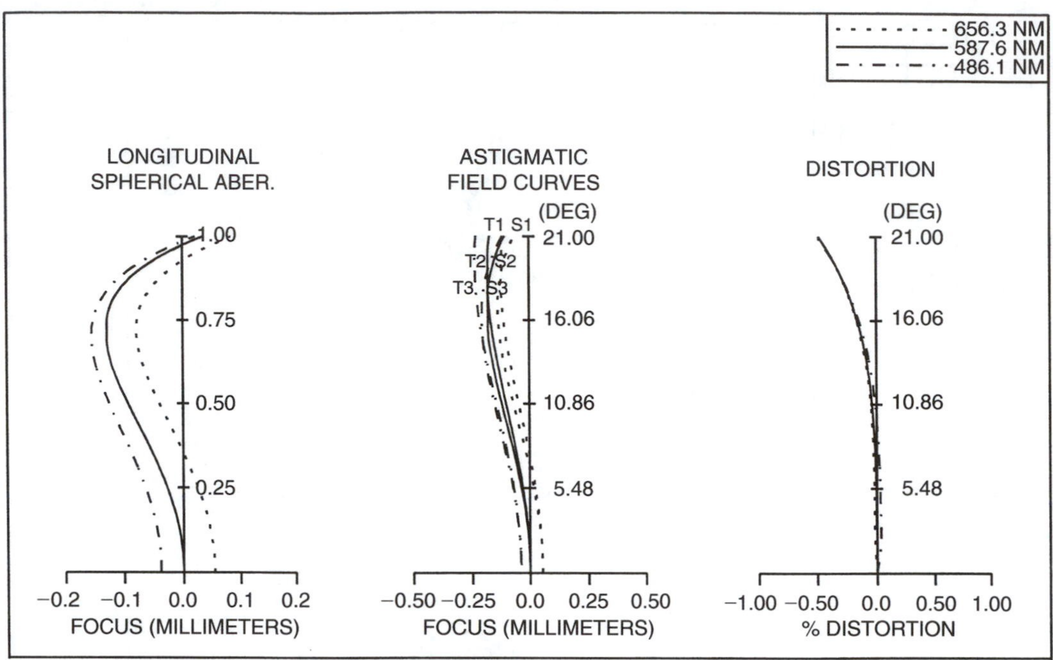

FIGURE 7.51 Longitudinal spherical aberration and field plots for sample lens (CODE V).

The MTF across the field at the best focus position is plotted in Figure 7.54. The response is seen to be the best near the axis, but is actually relatively uniform across the field. These plots provide a very good insight toward the appearance of the image in an actual situation. It is obvious that a full examination of the image formation in the image space can lead to very many plots such as this. It is important to increase the number of points in the field for a computation such as this, in order to be certain that the plot accurately reflects the performance of the lens.

Figure 7.55 is a plot of the ray bundles through the lens for a large number of field points. The visual impression from this plot is important to understand. Examination of the ray passage shows that the edge of the pupil is defined by a succession of surfaces, depending upon the field angle. The presence of vignetting, as well as some pupil aberration, leads to a complicated pattern of ray passages through the lens. Examination of some nonstandard drawings such as this is very useful to the designer in making detailed decisions about required clear apertures, and in providing sufficient clearance of the mounting of lenses and other associated hardware by the mechanical designers detailing the lens assembly. It is obvious from drawings such as this that the allowable vignetting in a lens is not arbitrary, but is limited to a restricted set of possibilities.

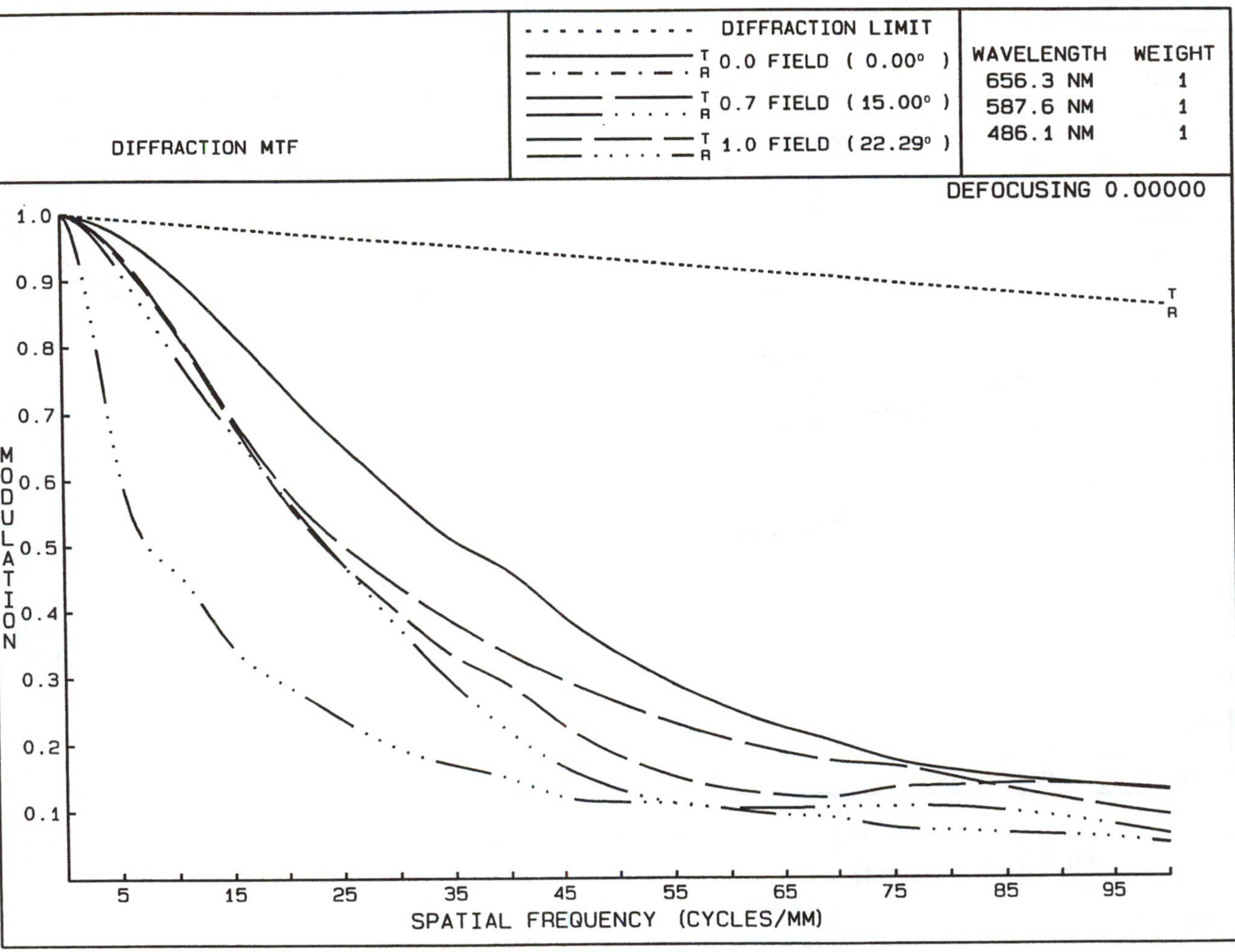

FIGURE 7.52 MTF plots for sample lens (CODE V).

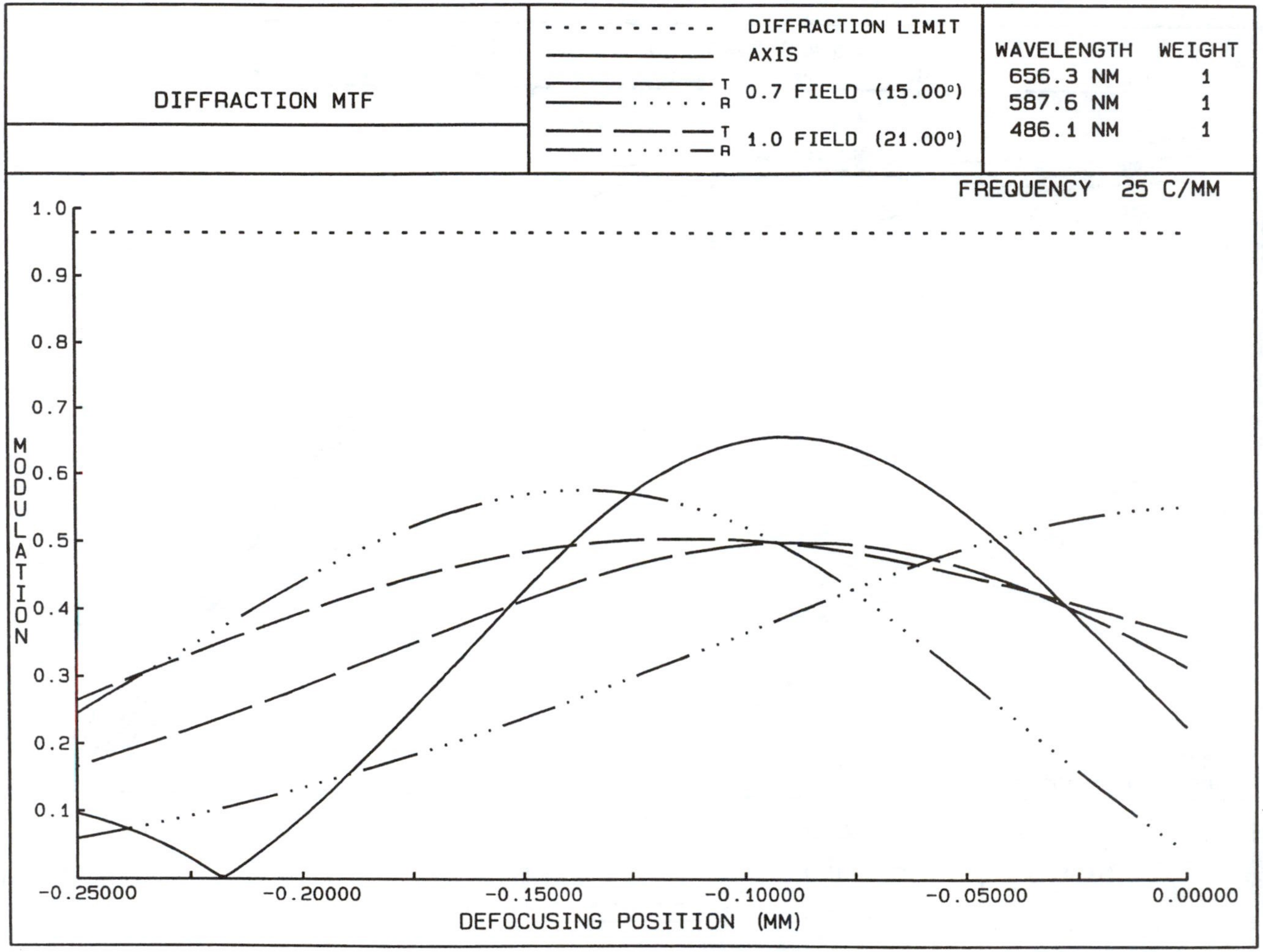

FIGURE 7.53 MTF through focus for sample lens (CODE V).

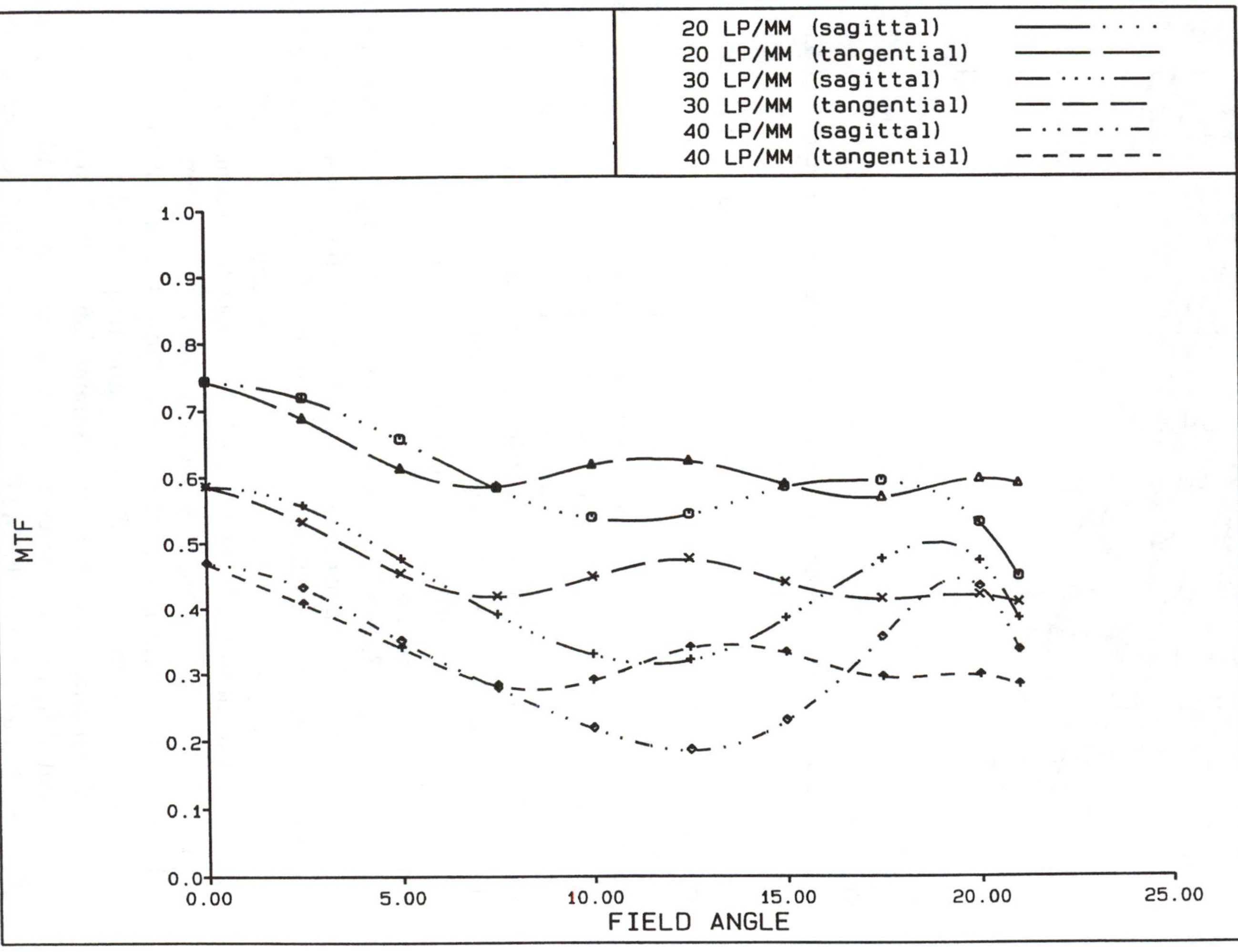

FIGURE 7.54 MTF across field for sample lens (CODE V).

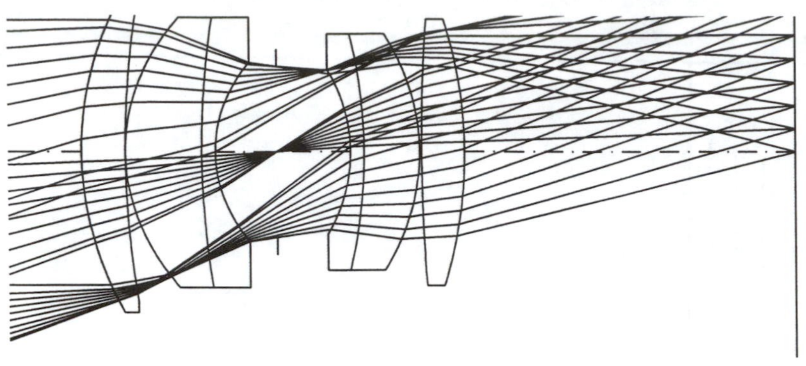

FIGURE 7.55 Layout of Gauss lens example with multiple field ray traces (CODE V).

A comparison with the triplet designs in the previous section can be obtained by stopping this lens to F/3.5. The MTF values for F/3.5 are indicated in Figure 7.56. Although not shown here, the designer can determine that the vignetting is reduced because only a central portion of the aperture is being used for imagery. The MTF values are all better than the triplet, as can be seen by comparison with Figure 7.44, which contains similar information regarding the triplet. The illumination uniformity is also better for the Gauss-type lens. It is evident that six elements should be better than three for image quality, but the design has to be properly carried out to ensure this.

Figures 7.57, 7.58, and 7.59 show another aspect of the behavior of this lens. The MTF is plotted in each case with the diffraction cutoff frequency as the upper limit. The plots for F/2, F/3.5, and F/5.6 indicate that the lens progressively improves when stopped down, and approaches a diffraction-limited lens when stopped down enough. For aperture settings slower than about F/7 or so, the lens becomes, and remains, diffraction-limited across the field. These plots were all taken for a lens adjusted to an optimum focus at each chosen F/number. The diffraction cutoff frequency decreases with the stopped-down lens, so that careful examination of the contrast at some chosen spatial frequency is necessary to completely understand the imaging situation. Examination of the MTF contrast at, say, 100 lines/mm shows a slow increase in contrast with aperture reduction until about F/6.3. The diffraction limit will then dominate, and the contrast will reduce again. Thus the lens has an optimum operating aperture. This can be assessed qualitatively, by inspection, from the information given here. A designer would have no trouble in carrying out the computations to support and accurately identify the behavior at any desired spatial frequency. Such infor-

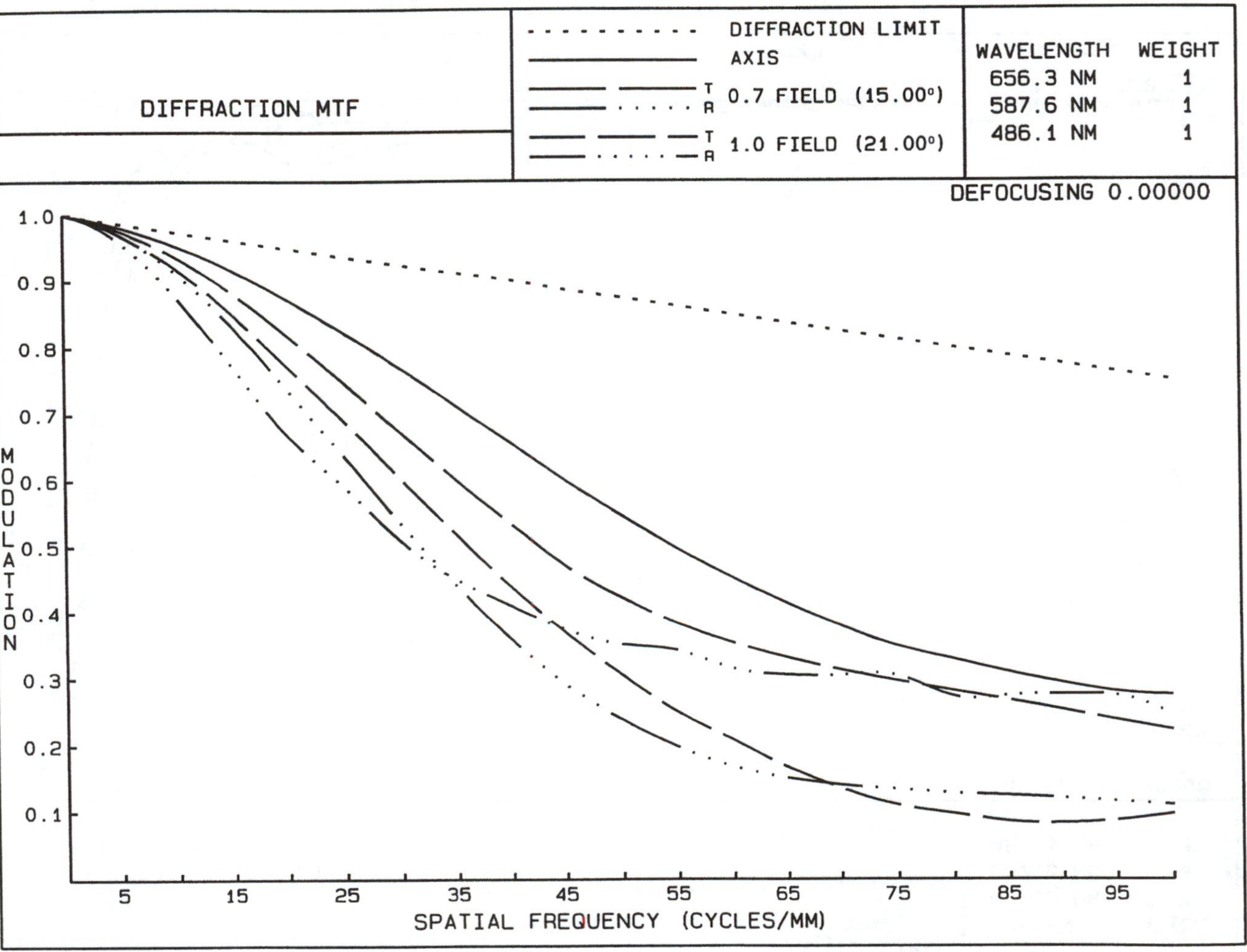

FIGURE 7.56 MTF plots for the Gauss lens stopped down to F/3.5 (CODE V).

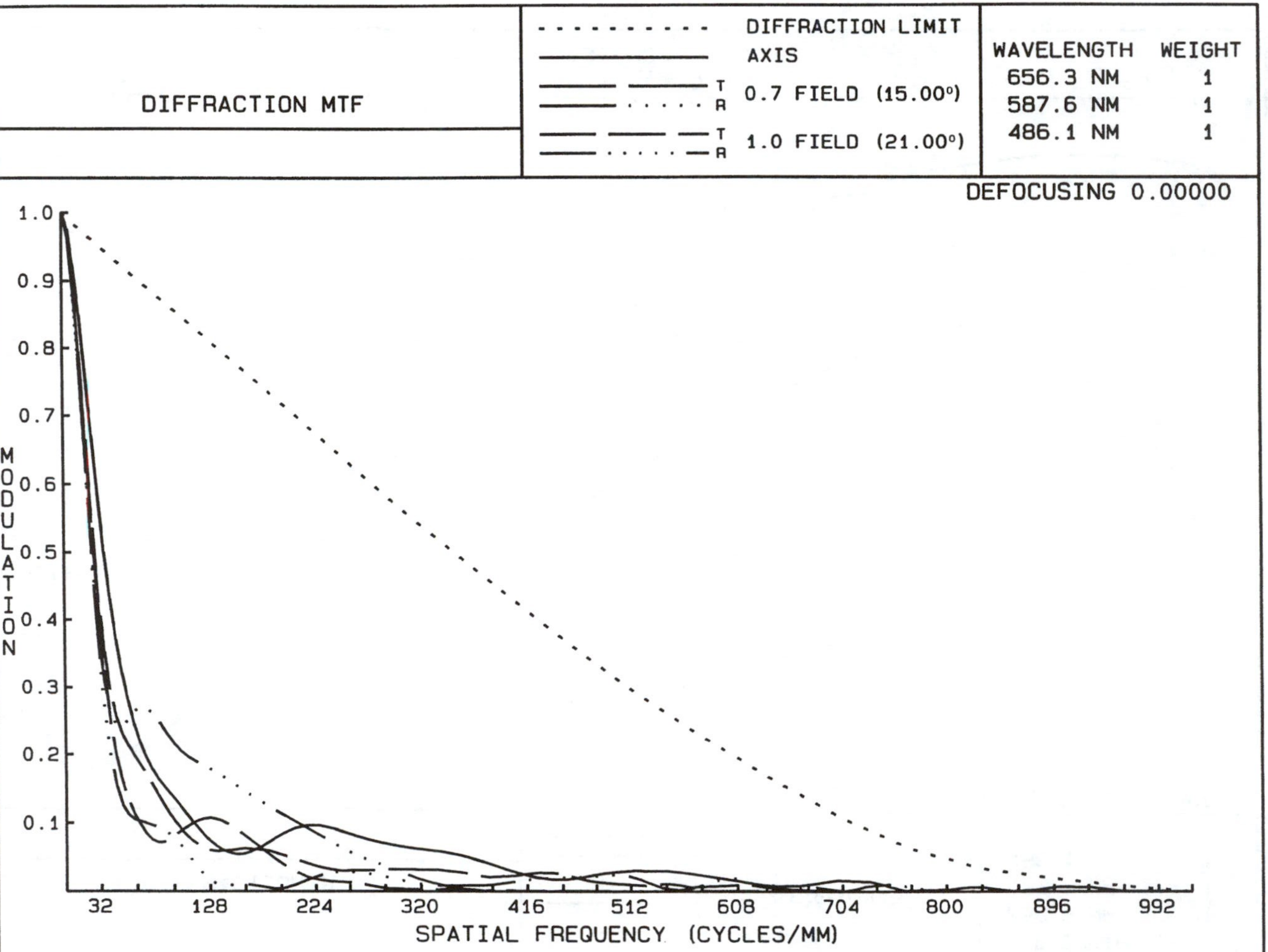

FIGURE 7.57 MTF for full spatial frequency range at F/2 (CODE V).

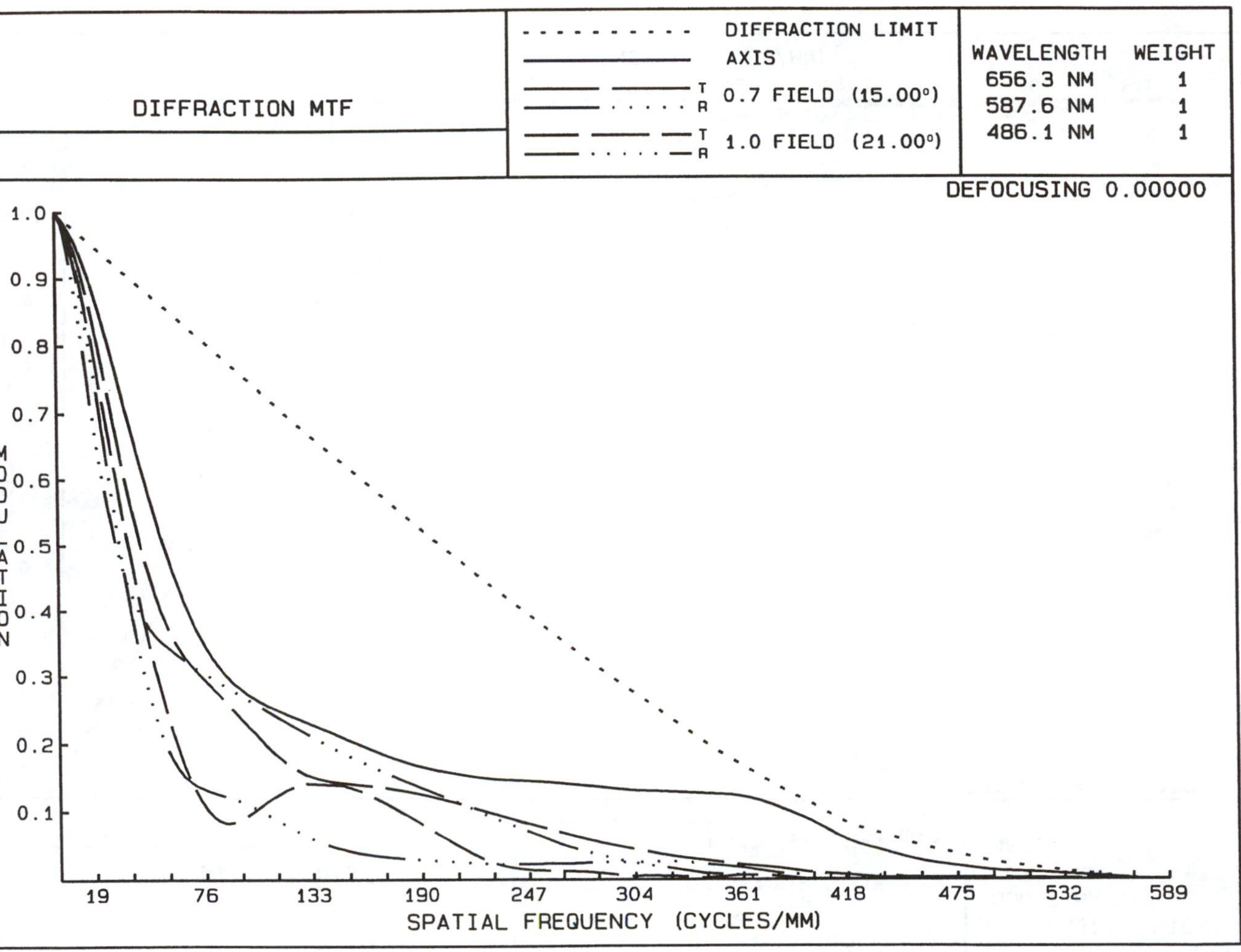

FIGURE 7.58 MTF for F/3.5 (CODE V).

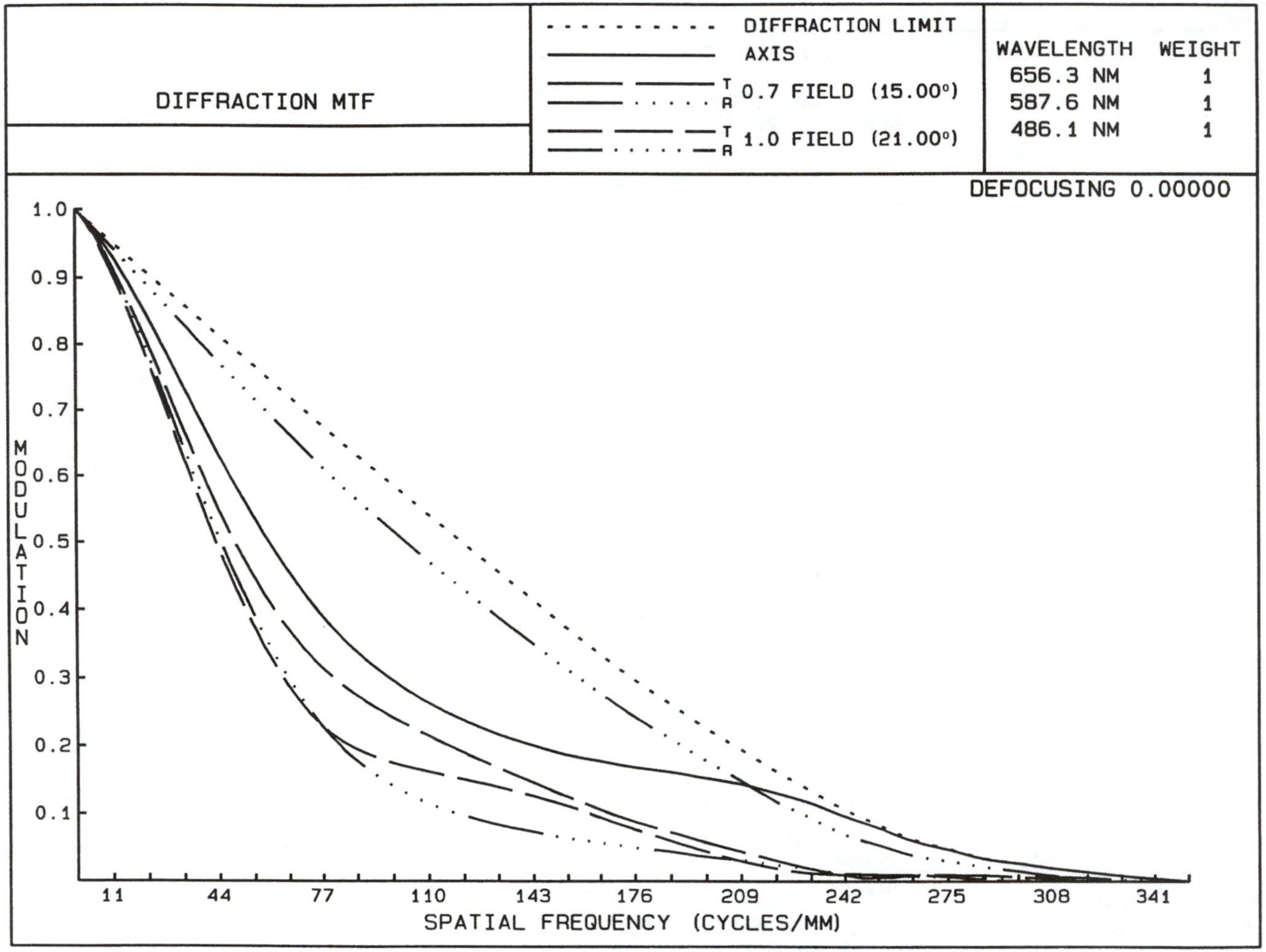

DIFFRACTION MTF

DIFFRACTION LIMIT
AXIS
T/R 0.7 FIELD (15.00°)
T/R 1.0 FIELD (21.00°)

WAVELENGTH	WEIGHT
656.3 NM	1
587.6 NM	1
486.1 NM	1

DEFOCUSING 0.00000

MODULATION: 1.0, 0.9, 0.8, 0.7, 0.6, 0.5, 0.4, 0.3, 0.2, 0.1

11 44 77 110 143 176 209 242 275 308 341

SPATIAL FREQUENCY (CYCLES/MM)

FIGURE 7.59 MTF for F/5.6 (CODE V).

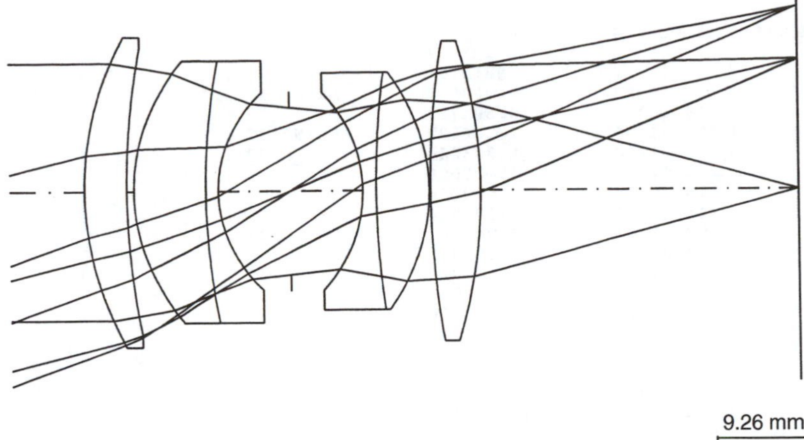

FIGURE 7.60 Layout of modified double Gauss lens (CODE V).

mation can permit the designer to tune the behavior of the lens to be optimum at the aperture most likely to be used in a given application.

The question now can be asked as to what could be done using a lens design program to alter this lens to meet a specific need. Some clues are obvious. Using the lens at a reduced aperture will permit some increase in the field. The oblique spherical aberration intrinsic to the lens shows that this is not an easily won battle, as the reduced-aperture lens will have to be vignetted to about the same relative amount across the field in order to beat the oblique spherical aberration. Increase in the aperture to faster than F/2 is also an option, but will require rebalancing to reduce the axis zonal spherical aberration in order to recapture similar image quality. This will necessarily upset the balance against the oblique spherical aberration, and the allowable field will be reduced. These tradeoffs are intrinsic to the double Gauss lens type. A designer can modify and redesign to meet customer needs within this field-aperture envelope.

Moving outside the envelope requires some fundamental changes in the design. The general uniformity of image quality across the field is a great advantage of the double Gauss lens, so that modifications without a drastic change in lens type are desirable. This can be explored in a design process which eventually led to the modified double Gauss illustrated in Figure 7.60. This is an F/2, 50 mm focal length lens with similar field and vignetting requirements as the previous example. This design was reached by a methodical investigation of design space in which the glasses used in the lens were varied in an attempt to find a better choice. In order to compare the lenses, the back focus requirements also have to be kept similar, as a reduced back focus is an important tool in improving imagery. Not surprisingly, this lens is slightly larger than the previous lens.

```
CODE V> lis
        Modified Double Gauss Lens
                   RDY            THI      RMD      GLA              CCY   THC   GLC
   OBJ:          INFINITY       INFINITY                             100   100
    1:           34.15085       4.009367          LAFN8_SCHOTT         0   100
    2:           78.48747       0.681721                               0     0
    3:           20.54186       7.206837          LAFN2_SCHOTT         0   100
    4:           66.81986       1.303044          SF3_SCHOTT           0   100
    5:           13.84286       7.000000                               0   100
   STO:          INFINITY       7.000000                             100   100
    7:          -15.08790       1.303044          SF64A_SCHOTT         0   100
    8:           90.01661       5.242247          LAF2_SCHOTT          0   100
    9:          -19.98205       0.104825                               0     0
   10:           95.01186       4.800000          LAF2_SCHOTT          0   100
   11:          -52.53383      31.742999                               0   PIM
   IMG:          INFINITY      -0.141130                             100     0

 SPECIFICATION DATA
    EPD        25.99999
    DIM              MM
    WL           656.30     587.60     486.10
    REF               2
    WTW               1          1          1
    XAN         0.00000    0.00000     0.00000
    YAN         0.00000   15.00000    21.00000
    VUX         0.00604    0.04584     0.14055
    VLX         0.00604    0.04584     0.14055
    VUY         0.00604    0.17574     0.40182
    VLY         0.00604    0.35786     0.59240

 APERTURE DATA/EDGE DEFINITIONS
    CA
    CIR S1                      14.533956
    CIR S2                      14.533956
    CIR S3                      12.228570
    CIR S4                      12.228570
    CIR S5                      10.000000
    CIR S6                       8.419671
    CIR S7                      10.000000
    CIR S8                      11.000000
    CIR S9                      11.000000
    CIR S10                     14.000000
    CIR S11                     14.000000

 INFINITE CONJUGATES
    EFL           50.0000
    BFL           31.7430
    FFL          -18.9847
    FNO            1.9231
    IMG DIS       31.6019
    OAL           38.6511
    PARAXIAL IMAGE
     HT           19.1932
    ANG           21.0000
    ENTRANCE PUPIL
     DIA          26.0000
     THI          25.4629
    EXIT PUPIL
     DIA          29.2480
     THI         -24.5031
CODE V> prt
        Printing file MODLEN.LIS(1)
```

FIGURE 7.61 Listing of design data for modified lens (CODE V).

The major change is in the glass types. The listing lens parameters in Figure 7.61 shows that the glasses are generally slightly higher index materials, and are mostly in the right half of the conventional index versus V-number chart. The design process required use of an MTF merit function in the optimization procedure in CODE V. The design tactics involved selecting critical spatial frequencies which were moved up from 15 to 30 lines/mm as the design improved. The final adjustment of the lens was carried out by refining the MTF contrast targets in the optimization. The field weights were adjusted to make the MTF uniform across the field.

The ray intercept plots in Figure 7.62 indicate that the basic oblique spherical aberration is still present. There is a small reduction in the oblique spherical permitting a smaller spherical aberration zone at the center of the field. The field focus plots in Figure 7.63 show a reduction in the astigmatic compensation across the field. The spot diagrams in Figure 7.64 have a familiar shape, but do seem to indicate a slightly better concentration of light in the core of the images.

The comparison with the previous lens can be seen from the MTF plots in Figures 7.65–7.67. These figures indicate an improved MTF contrast at all spatial frequencies. The MTF versus field plots indicate uniformity and better contrast across the field. The through-focus plots indicate a slightly reduced focal depth, however. The reason for this better behavior is probably due to a slightly higher average index of refraction of the glass used, and to a careful choice in setting the merit function components during the design. This is an example of the art of lens design adapting to the needs of the problem.

Nevertheless, this is still a double Gauss, and while the modified design may appear better, it is not necessarily significantly superior when all aspects of the design and tolerancing are considered. The main lesson is that it is a member of the family of double Gausses, and it illustrates the benefits, as well as the limitations, of the design type. It is likely that this design class will continue to be very frequently used because of the general uniformity of image quality across the field, and stability with change in aperture, that can be expected.

As another example, we will examine the design of a finite conjugate lens to meet the requirements posed in Chapter 2 for a viewer. Recall that the requirement was to provide a lens which formed a ten times magnified image on a screen with a total length of 1 m from slide to screen. The first-order layout did not consider the required image quality so that it will be the responsibility of the designer to meet the evident need.

The required lens has a focal length of 82.645 mm, an F/number of 3.17, and works at a field angle of 6.2°. This should be a very simple problem to solve with a lens of the double Gauss type. The difference is that the lens will be operating from a short to a long conjugate, at a fixed magnification. In order to shortcut the starting point, a double Gauss is selected from the CODE V lens library (the design discussed above could also have been used as a starting point). As a first step, this lens is scaled to the desired final focal length. The lens setup uses a description of the entrance pupil

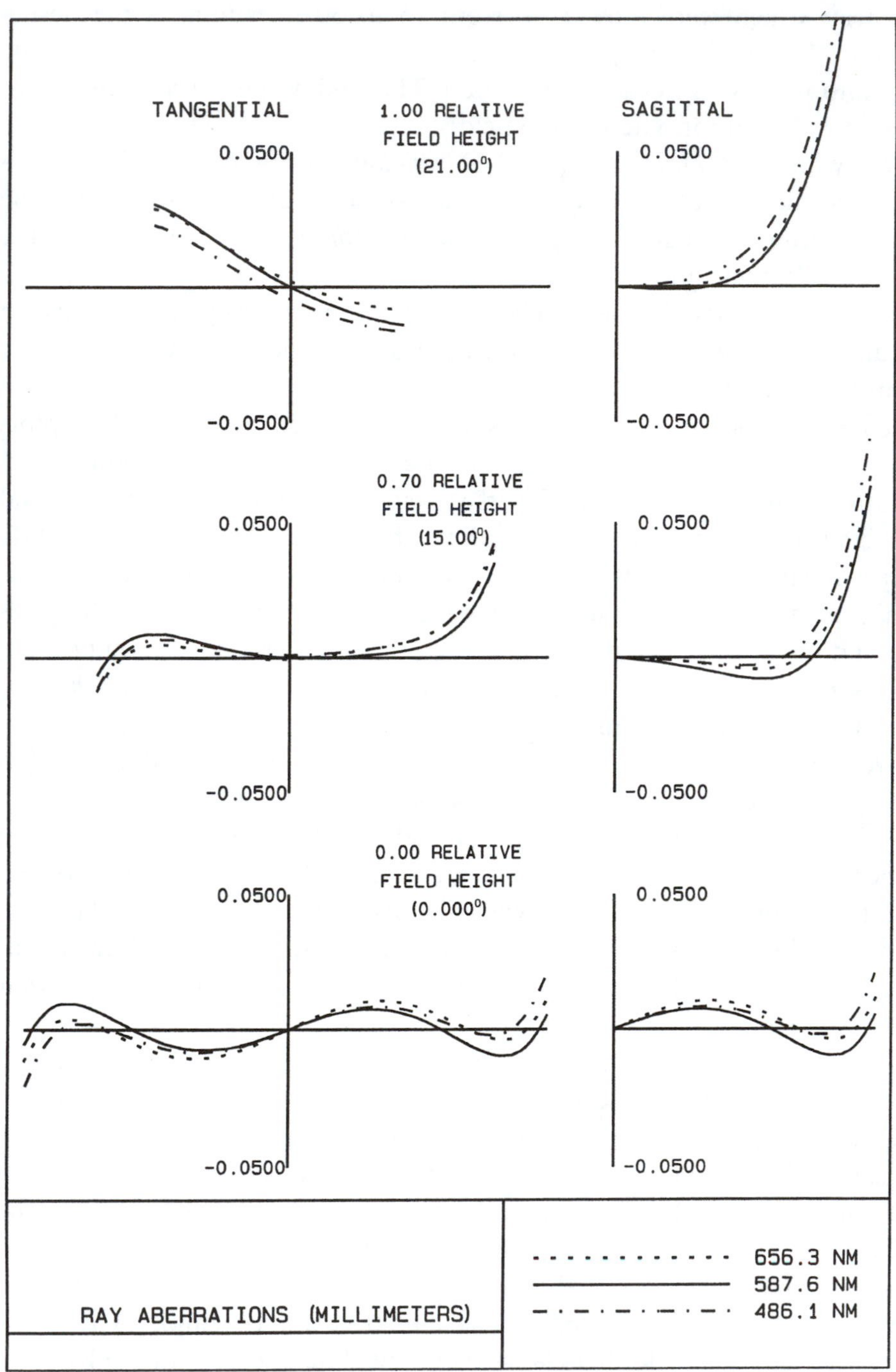

FIGURE 7.62 Ray trace plot for modified lens (CODE V).

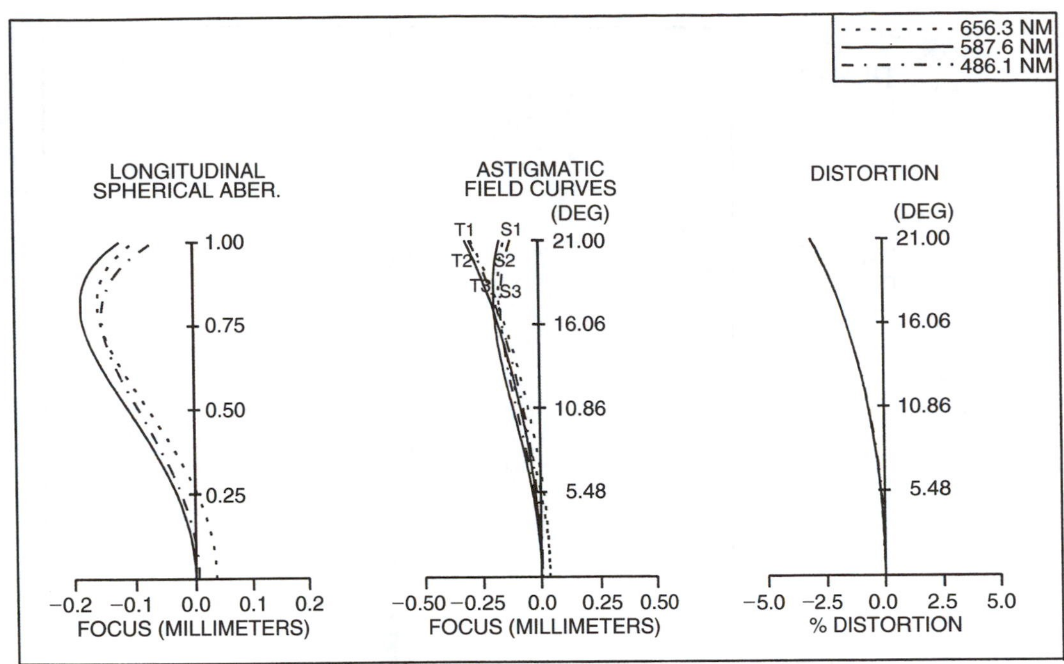

FIGURE 7.63 Field curves for modified lens (CODE V).

diameter, 28 mm, and a specification of the object height at the corner of the slide, 21.6 mm. The conjugates are set to obtain the desired magnification, without worrying about the overall length at this step, because the location of the principal points in the lens is likely not what we want in the end anyway. Setting the conjugates can be handled many ways, using the program to display the image height, for example. CODE V has a setup option for the magnification that adjusts the object distance to meet the requirement at each stage in the design, so this can be used. In other programs, a different mode of establishing the conjugates may be required.

Once the initial setup is obtained, the merit function is set up using the default rays at three field heights, 21.6 mm, 17.0 mm, and axis. A constraint that the overall length (or total track length) be corrected to 1000 mm is added. All of the lens curvatures, thicknesses, and glasses are tossed into the problem as variables in order to see where the design would like to go. Because a six-element double Gauss is likely to be overkill for this problem, a solution is likely.

The optimization proceeds quite rapidly because there are about forty-eight variables available. An examination of the ray plots and MTFs indicates that a reasonable solution region has been reached. The glasses are replaced with the closest Schott principal glasses. It is noted that the two

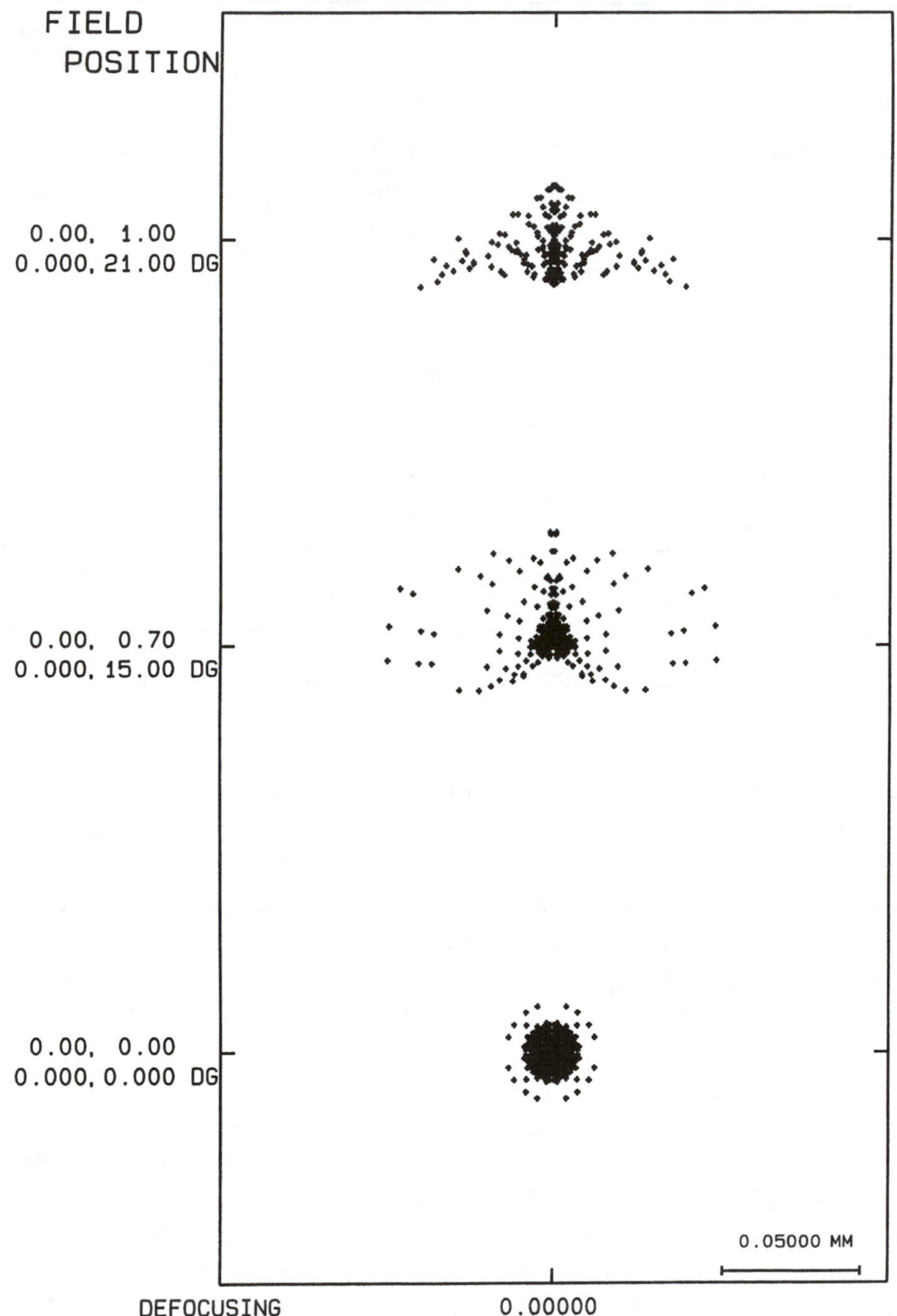

FIGURE 7.64 Spot diagrams for modified lens (CODE V).

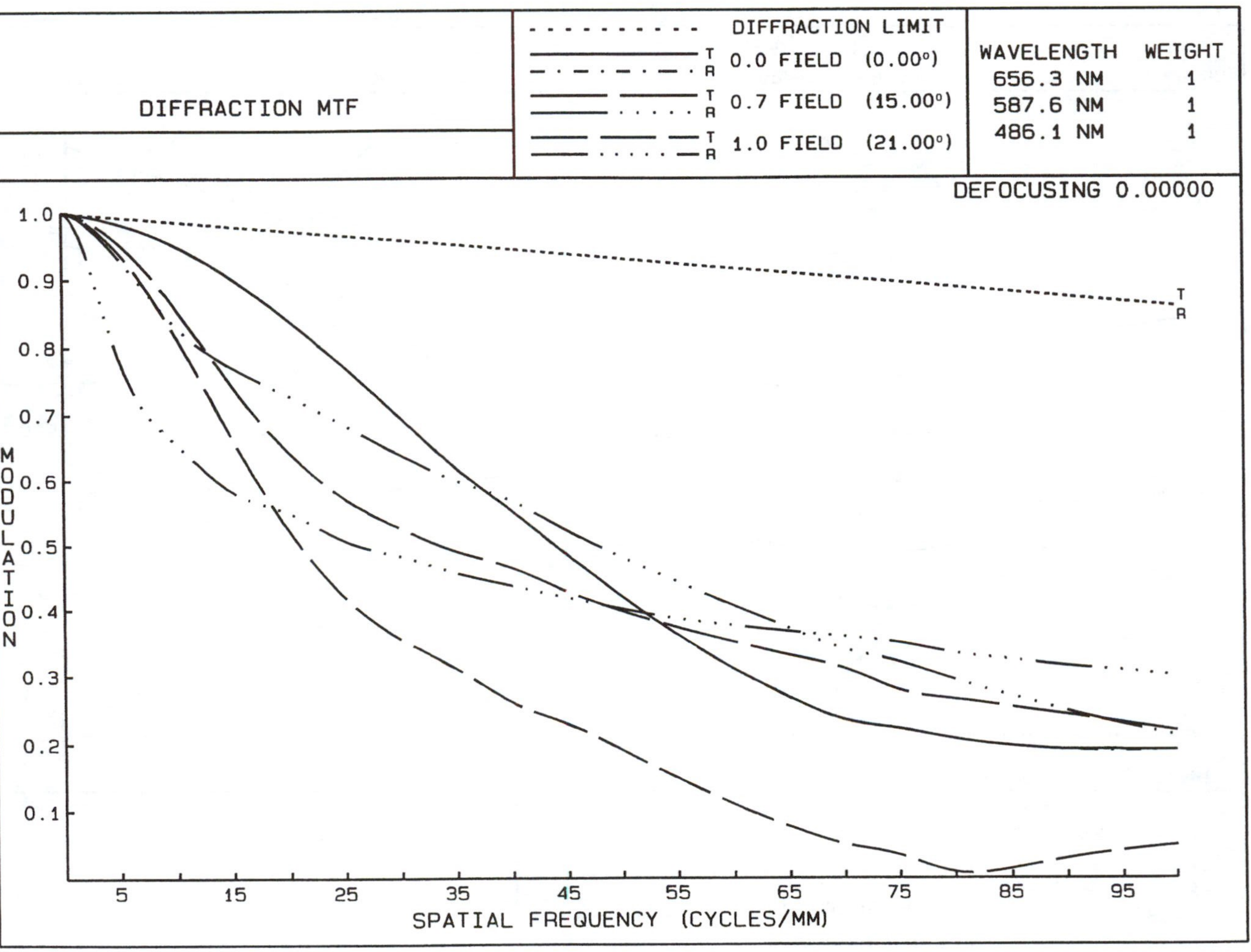

FIGURE 7.65 MTF plots for modified lens (CODE V).

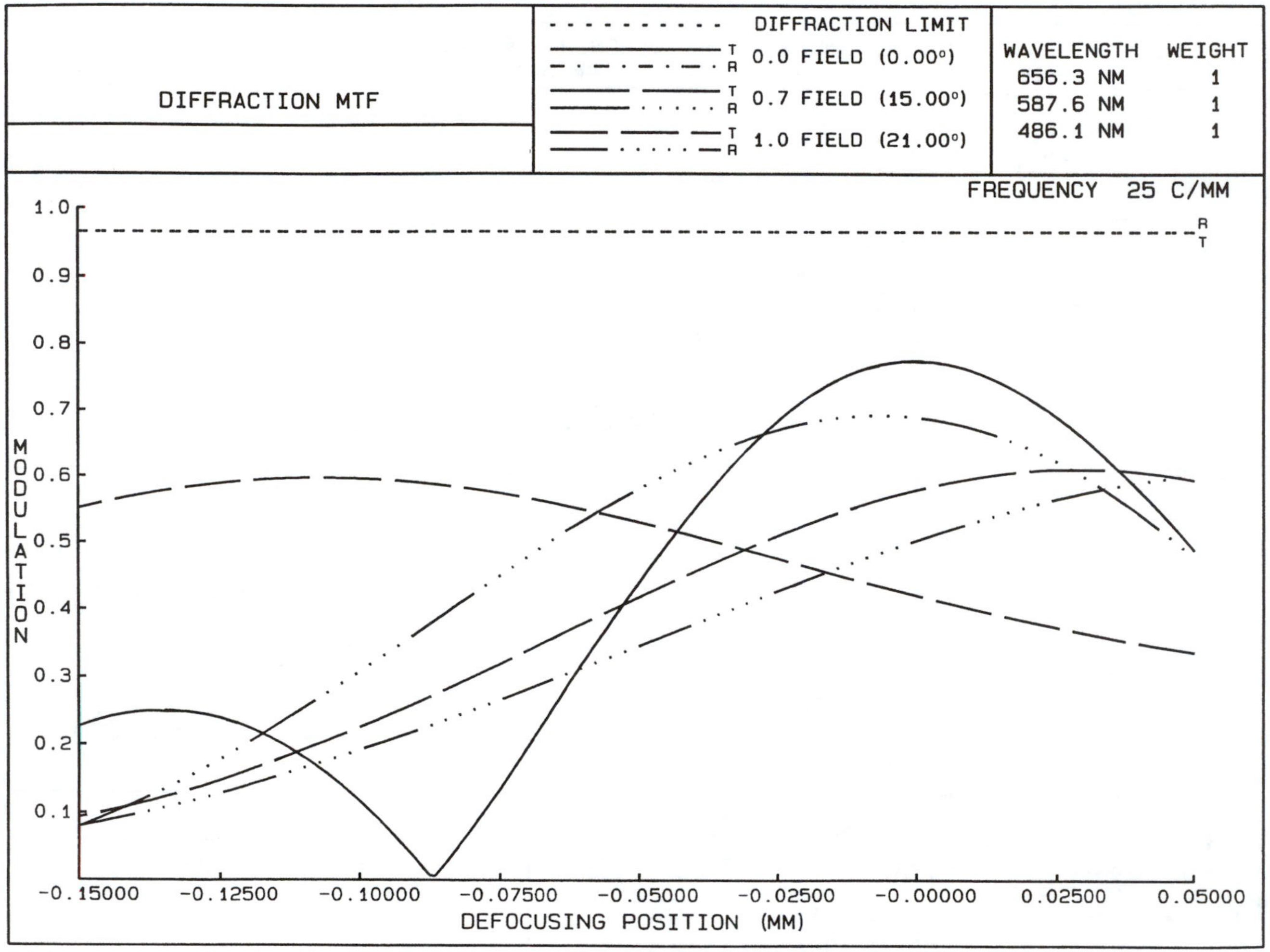

FIGURE 7.66 MTF versus focal position for modified lens (CODE V).

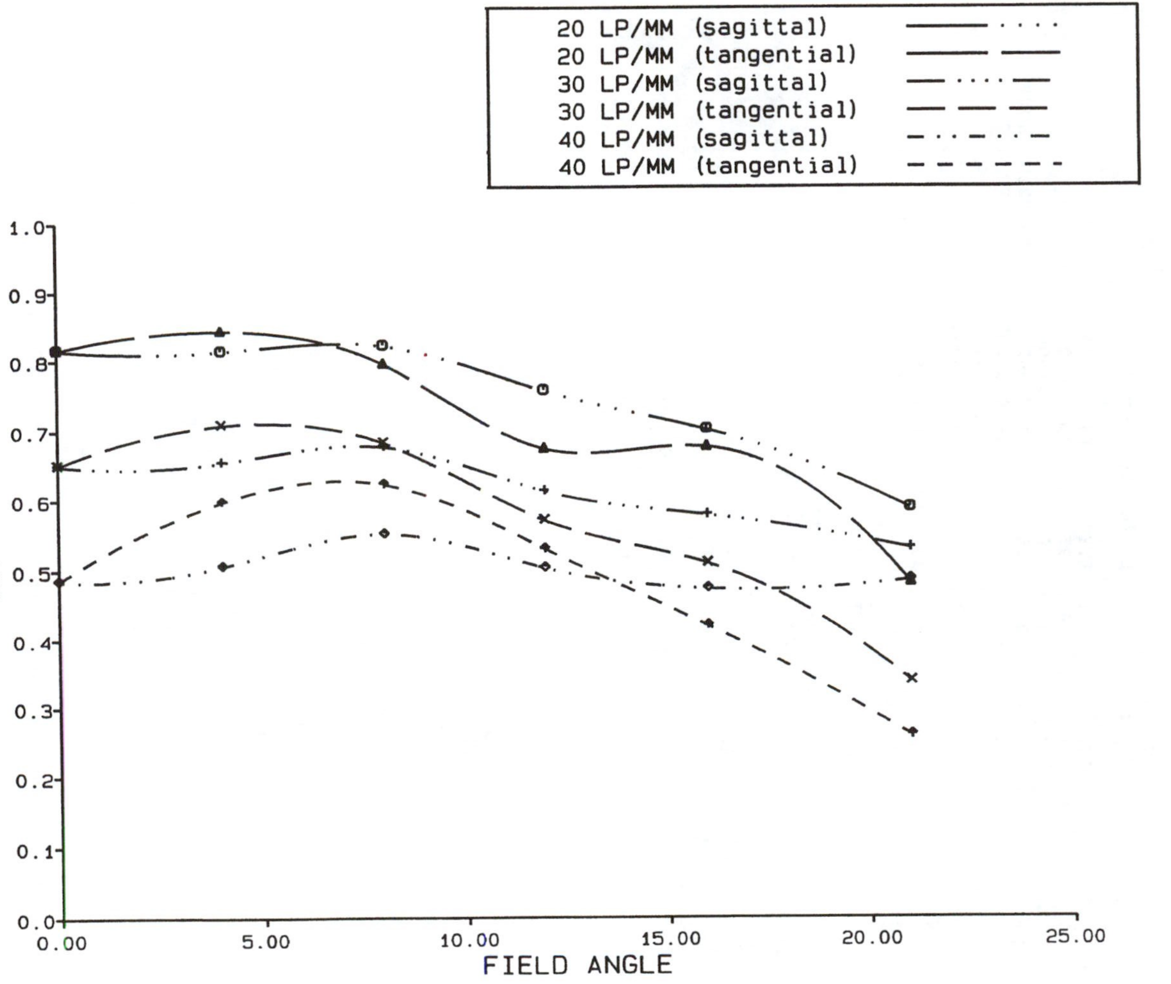

FIGURE 7.67 MTF versus field for modified lens (CODE V).

```
CODE V> lis
      Finite Conjugate for Chapter 7
                  RDY                THI        RMD         GLA
   OBJ:         INFINITY          81.268634
     1:        126.25295          10.000000              SK15_SCHOTT
     2:       -135.02956          27.707868
     3:         27.38416           4.779526              LAKN22_SCHOTT
     4:        -43.54932          10.000000              SF2_SCHOTT
     5:         27.32860           6.383414
   STO:         INFINITY          14.825290
     7:        -17.16764           8.275862              BAF50_SCHOTT
     8:        -33.97835           0.113432
     9:        -69.06520           2.529030              LAFN7_SCHOTT
    10:        -33.68701         834.123613
   IMG:         INFINITY           0.000000

 SPECIFICATION DATA
    EPD          26.00000
    DIM                MM
    WL             656.30       587.60      486.10
    REF                 2
    WTW                 1            1           1
    INI
    XOB           0.00000        0.00000          0.00000
    YOB           0.00000      -15.12000        -21.63330
    VUY           0.00000        0.00000          0.25000
    VLY           0.00000        0.00000          0.25000

 INFINITE CONJUGATES
    EFL           82.6453
    BFL            7.6710
    FFL          -73.0041
    FNO            3.1787
 AT USED CONJUGATES
    RED           10.0000
    FNO           73.4212
    OBJ DIS       81.2686
    TT          1000.0067
    IMG DIS      834.1236
    OAL           84.6144
    PARAXIAL IMAGE
     HT          216.3330
     THI         834.1236
    ANG            6.2066
    ENTRANCE PUPIL
     DIA          26.0000
     THI         109.1833
    EXIT PUPIL
     DIA          11.7943
     THI         -29.8192
```

FIGURE 7.68 Program listing of lens data for finite conjugate lens (CODE V).

glasses in the rear meniscus element are very close, and could be replaced by the same glass, reducing to a five-element lens. This lens is then optimized again using the same merit function. The image quality is sufficiently good that the vignetting is reduced to only 25% at the edge of the field in the final optimization.

The data for this lens is presented in Figure 7.68. Figure 7.69 is a drawing of the lens. The lens layout indicates that the lens is quite long, almost a focal length long. The elements, none of which looks difficult to fabricate, have some large airspaces leading to a fairly long lens. The ray intercept plots in

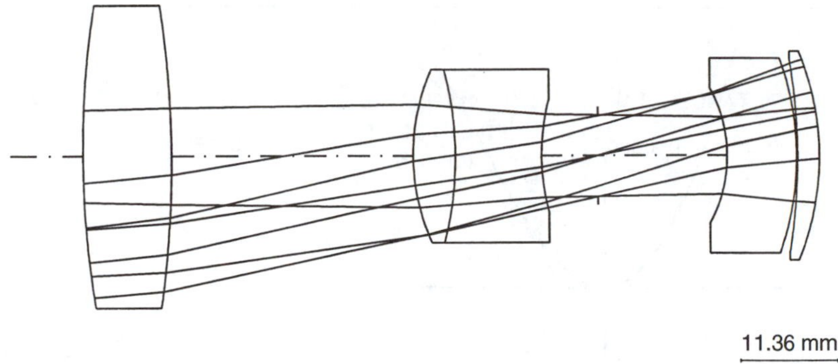

FIGURE 7.69 Layout of first design for finite conjugate lens (CODE V).

Figure 7.70 seem to indicate a large amount of aberration. A plot of OPD values in Figure 7.71 shows that the aberrations are not really large, being less than a quarter wavelength over most of the aperture and field. This apparent discrepancy of ray plot dimension and OPD results from the magnification. This lens is operating at F/35 at the long conjugate end. The image is being formed on a screen for visual examination. Thus resolution of about 10 lines/mm would be adequate for good image quality. This is comparable to the approximately 0.1 mm image diameter which is shown in the ray plots.

The acceptable image quality is confirmed by the MTF plots of Figure 7.72. For most of the field, the lens performs essentially as diffraction limited, with some loss at the corners of the field. Since the eye performance is limited to about 10 lines/mm, the lens looks usable. Plots of the field curves in Figure 7.73 show about 2.5% distortion, which is probably acceptable, depending, of course, upon the intended application for the viewer.

The lens design is not yet complete. It seems to be a bit long for the application, although no requirement was stated. Rather arbitrarily, a constraint to reduce the overall length from 82 mm to 70 mm was added, and the design redone. The lens converged easily, but the merit function value went from 707 to 808, which appears to be quite a change.

The data for the shortened lens are listed in Figure 7.74 and the layout presented in Figure 7.75. The appearance is similar, but the overall length of the lens is now 70 mm. The ray and OPD plots in Figures 7.76–7.78 indicate that the lenses are similar, but with a slight increase in the size of the ray pattern at the center of the field, which could be consistent with a merit function increase. The MTF plots in Figure 7.79 indicate the same difference with the previous lens, but show a bit more uniformity across the field. Therefore, this lens is actually slightly better suited for the application, even though the calculated merit function is lower. (This again shows that

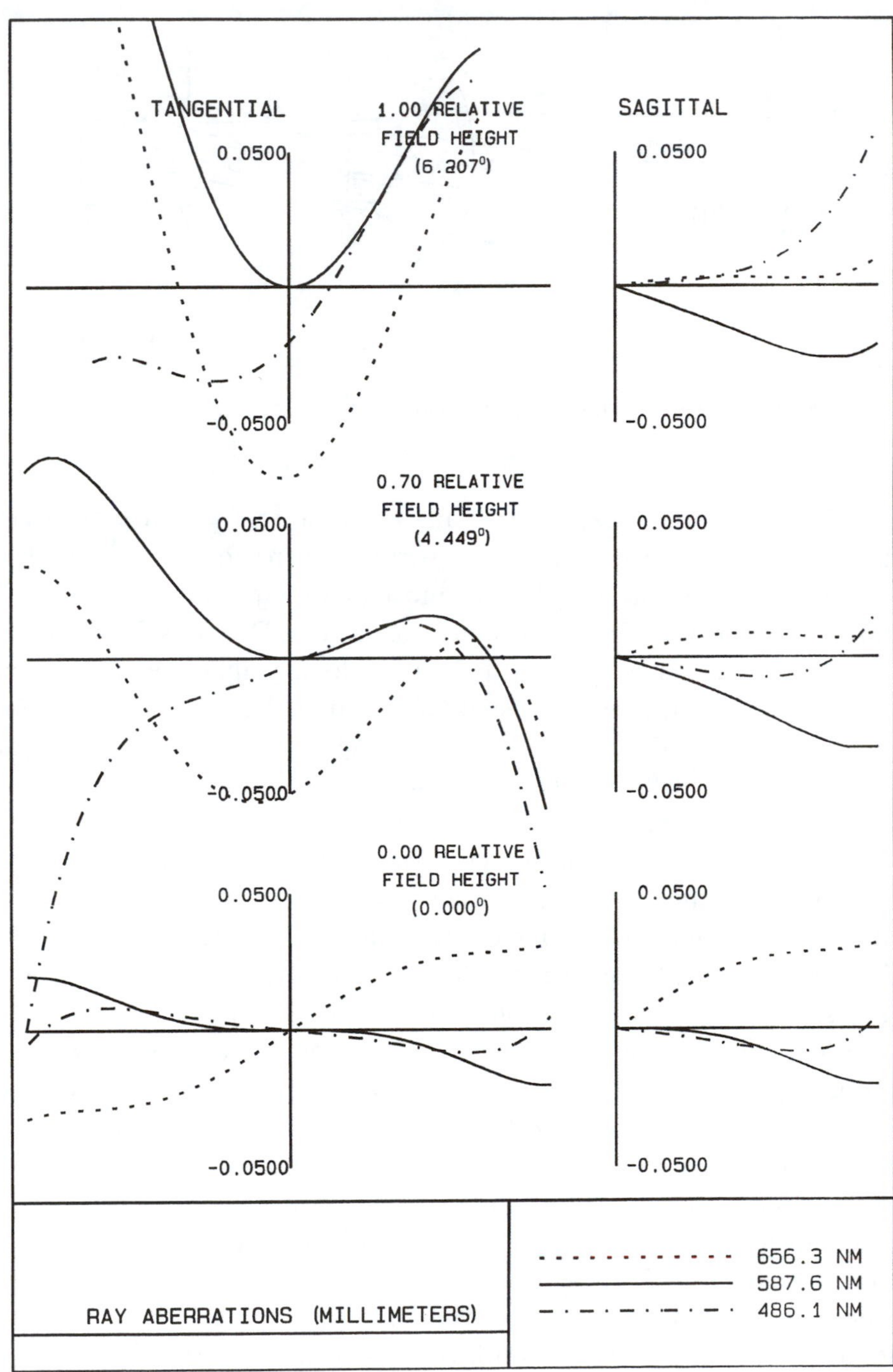

FIGURE 7.70 Ray intercept plots for finite conjugate lens (CODE V).

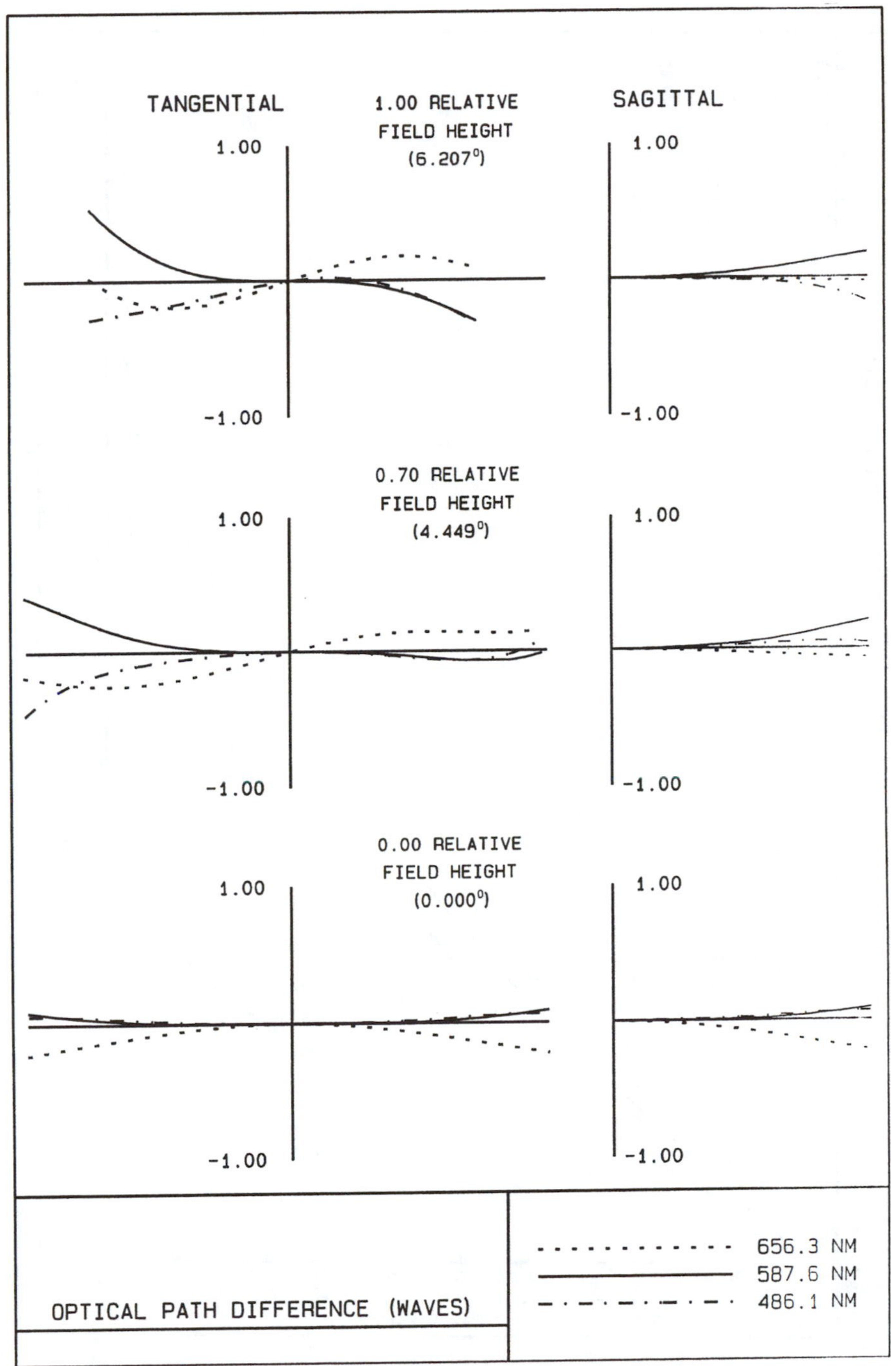

FIGURE 7.71 OPD plots for finite conjugate lens (CODE V).

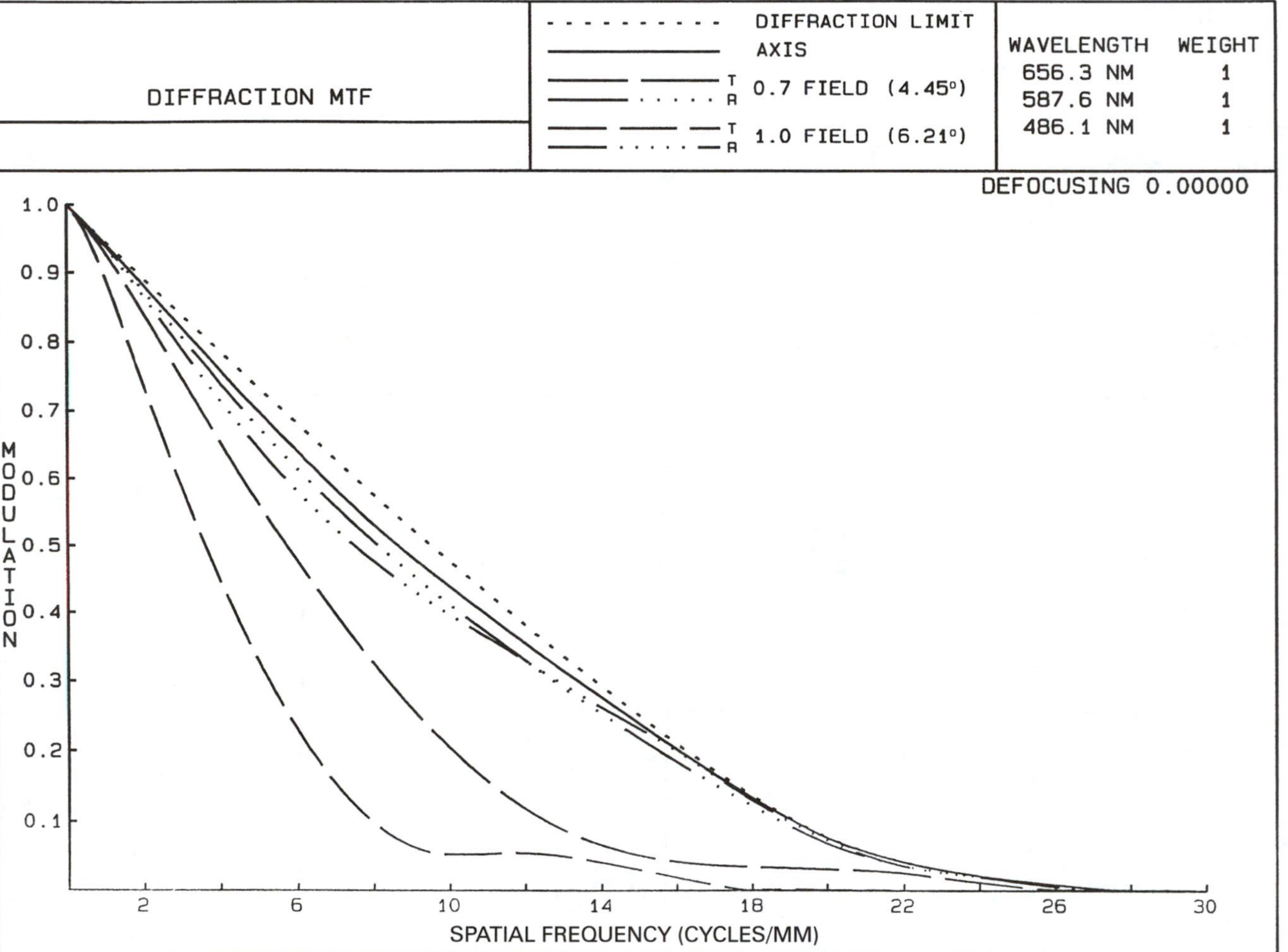

FIGURE 7.72 MTF plots for finite conjugate lens (CODE V).

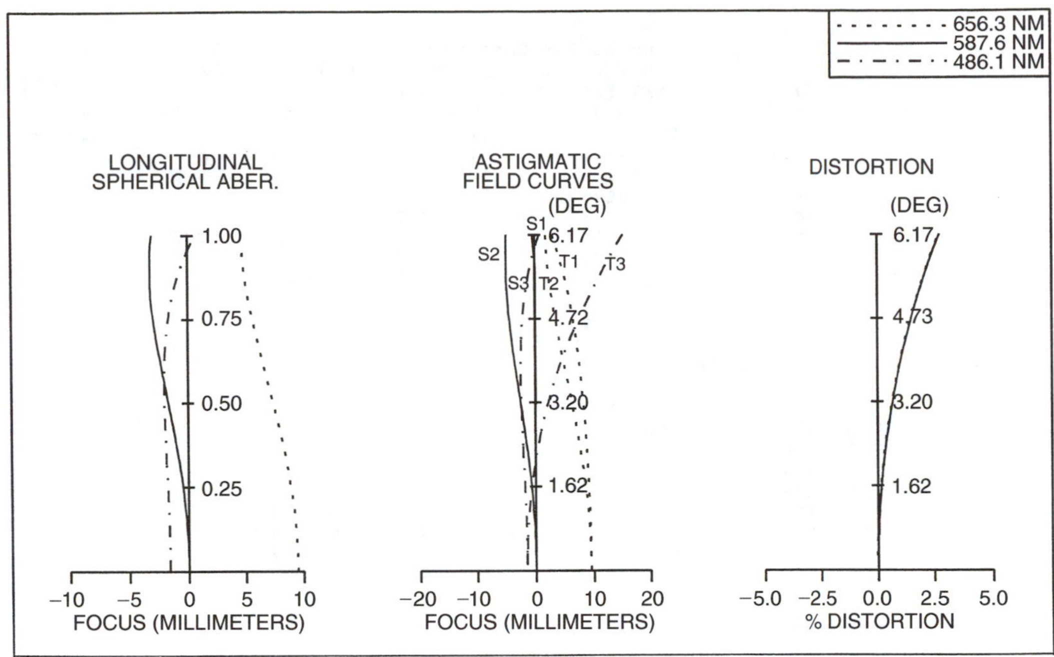

FIGURE 7.73 Field plots for finite conjugate lens (CODE V).

a merit function calculation must be considered along with the calculated image quality to produce a true value judgment on a lens.)

In order to determine whether a significant improvement is possible at this point, this last lens was subjected to a global search using the same merit function, with an arbitrary increase to 75 mm overall length. The results of a five-hour global search on a 486/66 computer led to over a dozen possible designs, a few of which indicated smaller values for the merit function. The best lens indicates an improvement, but not actually large, as the lens is very close to the diffraction limit at most field angles. The listing for this best lens is shown in Figure 7.80 and the layout in Figure 7.81. The ray plots in Figures 7.82–7.84 are similar to the previous designs. The MTF plots in Figure 7.85 show that the lens is substantially similar to the previously discussed lens. Although there does not seem to be a great change in the lens with all of the design investigation, an important result is that the lens is likely at an optimum for the problem. The existence of a number of designs with similar characteristics also permits adding requests to the merit function to allow some loosening of tolerances, perhaps.

Is this lens design complete? Probably not. This design appears to be buildable, but there would be some gain from improving the aberrations at the edge of the field to make the MTF more uniform. For the application of a viewer with a screen, the maximum use will probably be at the center anyway,

```
CODE V> lis
      Finite Conjugate for Chapter 7
                      RDY              THI       RMD        GLA
   OBJ:            INFINITY         87.832369
     1:            80.35322         10.000000          SK15_SCHOTT
     2:          -189.96775         11.685032
     3:            30.58341          4.779526          LAKN22_SCHOTT
     4:           -56.10122         10.000000          SF2_SCHOTT
     5:            26.82574          8.222478
   STO:            INFINITY         15.232174
     7:           -16.87616          8.275862          BAF50_SCHOTT
     8:           -22.69447          0.250000
     9:          -149.89788          2.529030          LAFN7_SCHOTT
    10:           -88.96150        841.194176
   IMG:            INFINITY          0.000000

  SPECIFICATION DATA
     EPD        26.00000
     DIM              MM
     WL           656.30      587.60      486.10
     REF               2
     WTW               1           1           1
     INI
     XOB         0.00000        0.00000         0.00000
     YOB         0.00000      -15.12000       -21.63330
     VUY         0.00000        0.00000         0.25000
     VLY         0.00000        0.00000         0.25000

  SOLVES
     RED      10.000000
     PIM

  INFINITE CONJUGATES
     EFL         82.6447
     BFL         14.7476
     FFL        -79.5679
     FNO          3.1786
  AT USED CONJUGATES
     RED         10.0000
     FNO         61.8876
     OBJ DIS     87.8324
     TT        1000.0006
     IMG DIS    841.1942
     OAL         70.9741
     PARAXIAL IMAGE
      HT        216.3330
      THI       841.1942
     ANG          7.5091
     ENTRANCE PUPIL
      DIA        26.0000
      THI        72.5493
     EXIT PUPIL
      DIA        14.1257
      THI       -30.1529
```

FIGURE 7.74 Program listing for shortened length lens (CODE V).

so that a loss in image quality in the center of the field would be undesirable. There are probable advantages to reducing element thicknesses and searching for better glasses, although the design investigation indicates that the gain is likely to be slight. For this, the designer should consult with the shop that will be doing the fabrication, because the experience so far is that the lens is not sensitive to specific glasses. The distortion residual could also be questioned by the customer, and could probably be reduced. It is not likely that another

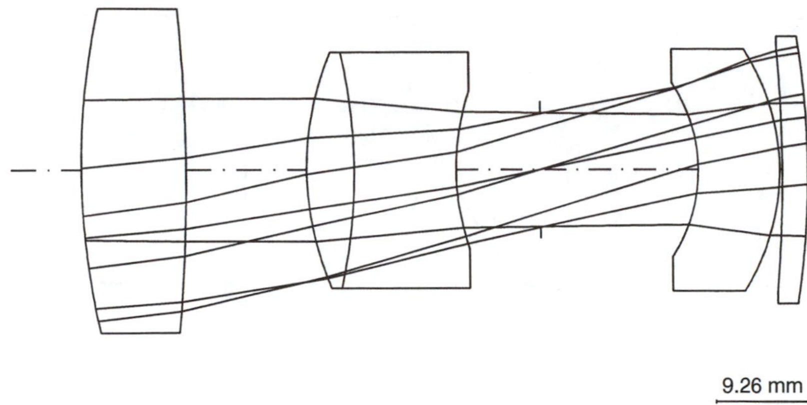

FIGURE 7.75 Layout of shortened lens (CODE V).

element could be removed from the design. It would be interesting to attempt to do so, however.

There is a hidden constraint that likely has an effect upon the image quality and distortion. As part of the problem, fixing the focal length and magnification and total track length of 1000 mm required that the principal point separation in the lens be zero. This will influence the nature of the power distribution in the lens. Possible improvement would be obtained by keeping the magnification and the total track fixed, but letting the focal length float as part of the design. This is an option available in finite conjugate systems that can sometimes be used to advantage. There are problems such as finite conjugate scan lenses where the nodal point location is critical, so placement of the cardinal points may be an issue. (When this happens remember that the distortion makes the nodal point location variable with image height! The constraints can often become quite complicated.)

The resulting lens also does not look like a "double Gauss," at least not in the classical sense. It is quite common that a lens design program instructed to operate with some specific constraints will produce a design that does not look like the intended form. In most cases, this will be quite acceptable. The breadth of possibilities that can lead to simplified or easier to fabricate lenses is quite extensive, and provides an opportunity for a designer to exercise some ingenuity in design.

The possible implications of these added constraints once again is a warning that designs taken from patents and other sources may have evolved to forms that are not evident due to some unstated constraints placed on or by the designer. Judgments about the most desirable matching of the image quality to the application will also alter the direction of a design. This is another example of the art and technology of optical design being closely related.

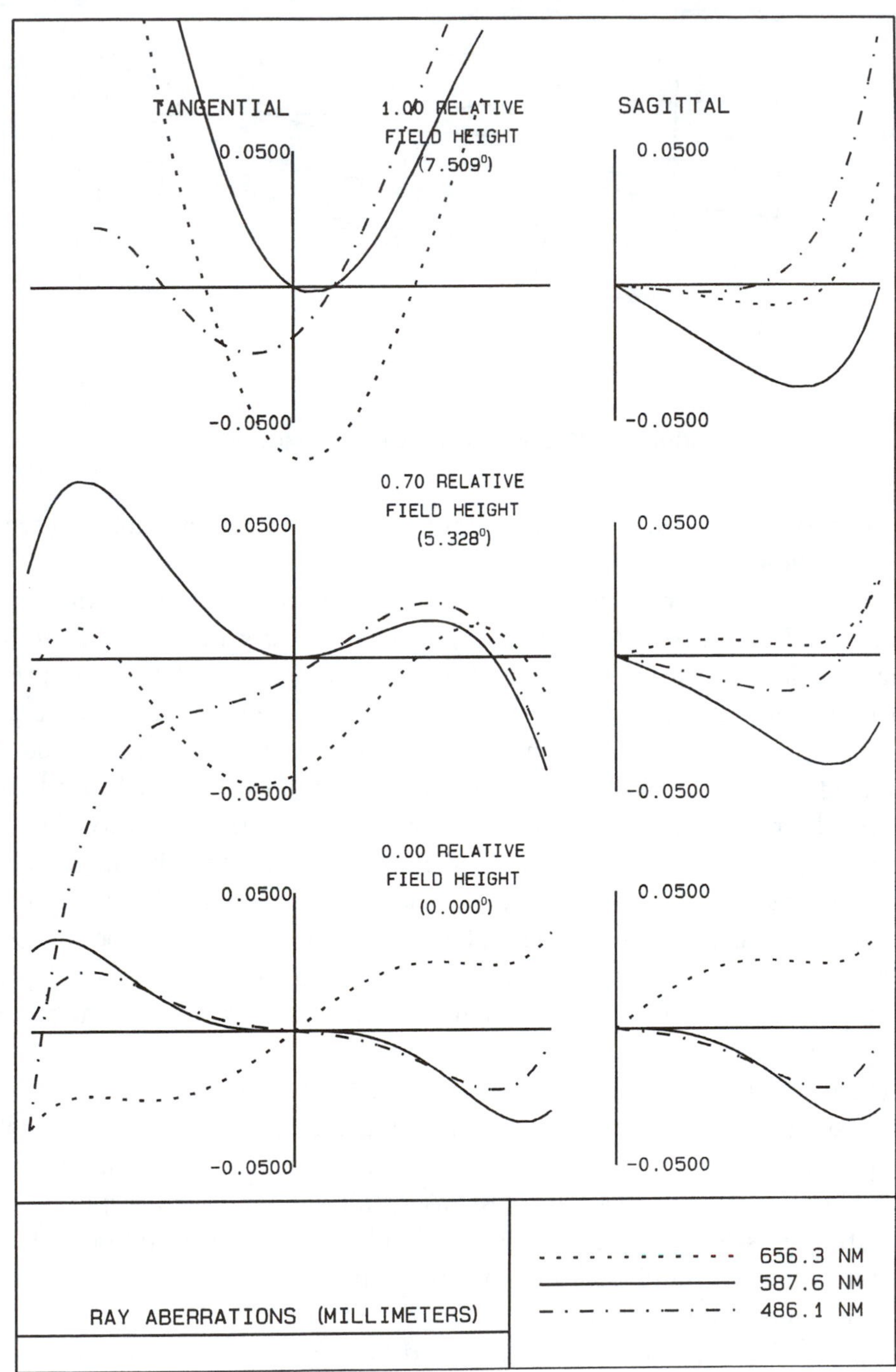

FIGURE 7.76 Ray intercept plot for shortened lens (CODE V).

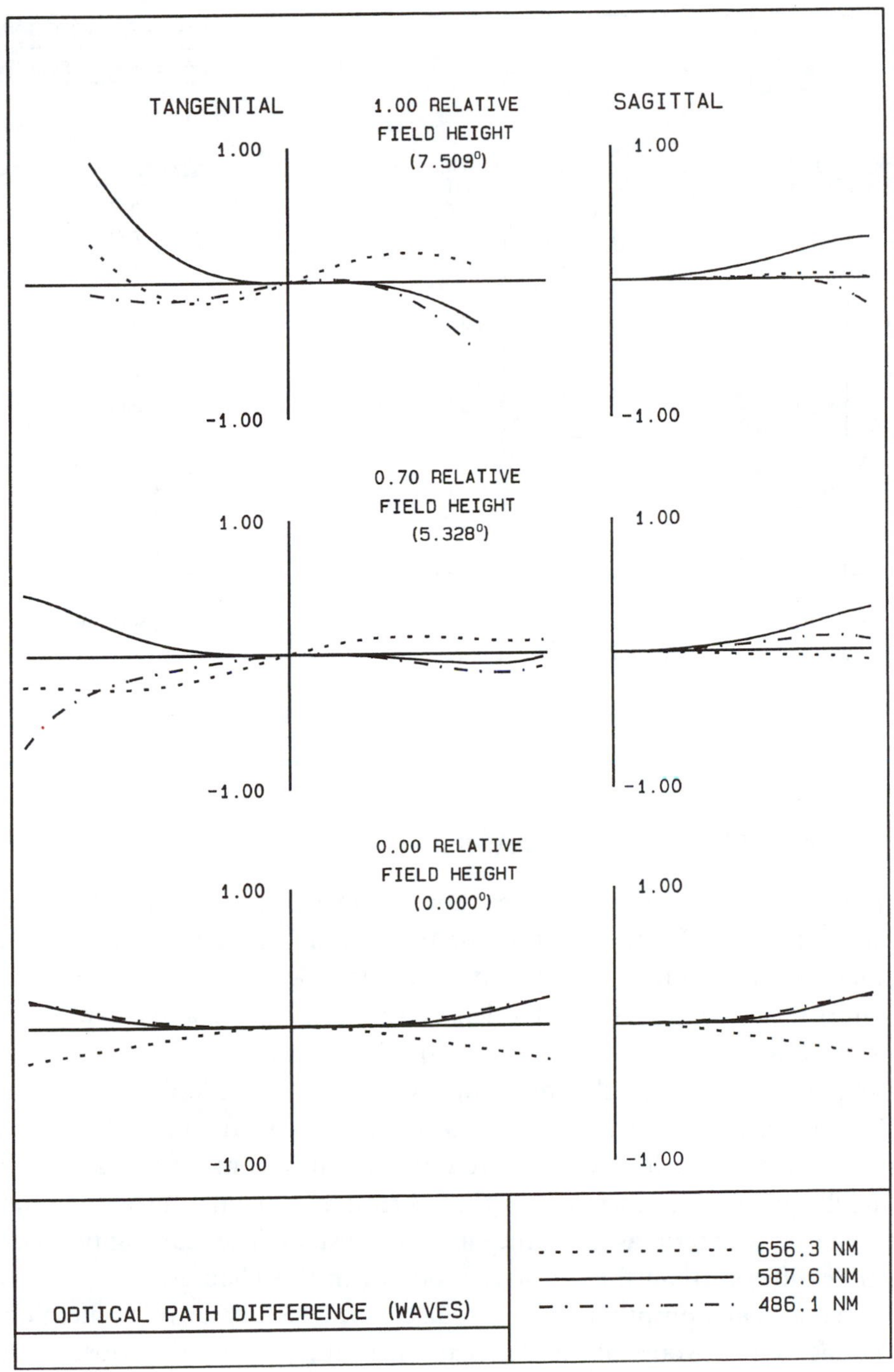

FIGURE 7.77 OPD plots for shortened lens (CODE V).

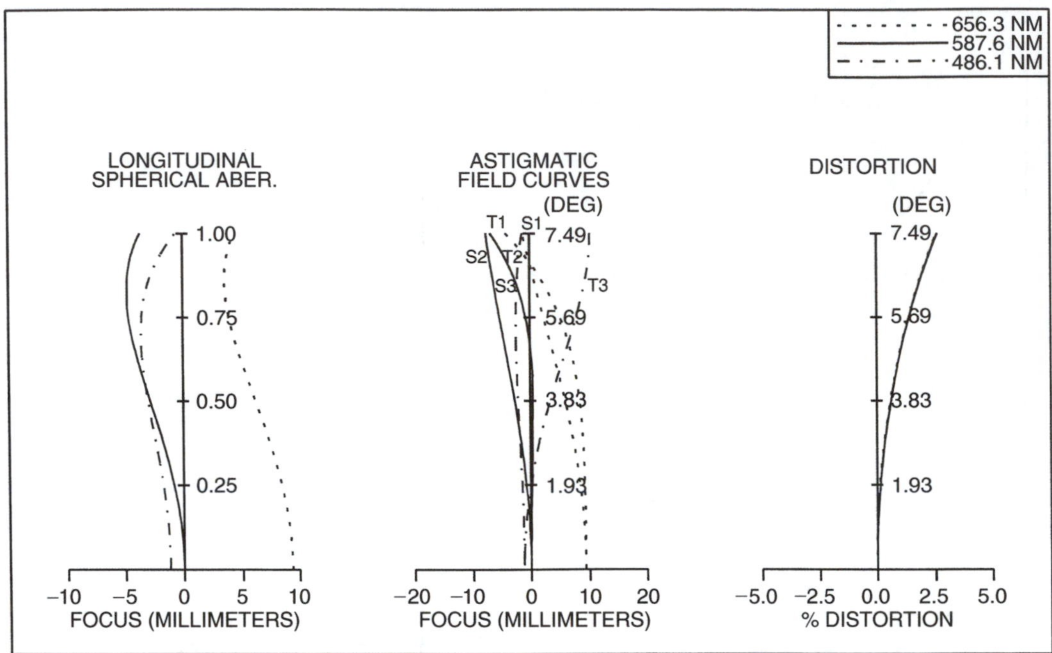

FIGURE 7.78 Field plots for shortened lens (CODE V).

7.1.4 MISCELLANEOUS IMAGING OPTICS

Miscellaneous imaging optics covers a wide range of possibilities. Telephoto, inverse telephoto, Petzval portrait or projection lenses, wide-angle, reprographic, and special types of telecentric and nontelecentric metrication lenses fall into this category. Complex lenses for microlithography can also be considered as miscellaneous, but very special, types of lenses. There are so many possibilities that only an example or two is worth looking at.

The design of these various types of lenses follows the procedures so far demonstrated. In each case there are different constraints and goals applied to the design. The optimization of the aberration correction usually needs to meet the same criteria as any other imaging system. The starting points will often be different than those discussed so far in this chapter.

This is also an appropriate place to discuss the history and classification of lens design types. Many of the classical forms of lenses are now obsolete and certainly not worth detailed discussion. In some cases there are modern derivatives that are useful as starting points. Modern lens design programs have led to an amazing variety of design forms, some of which defy classification into traditional forms. The reader interested in this topic is referred to the book by Ray (1988) which includes a thorough survey of the normal

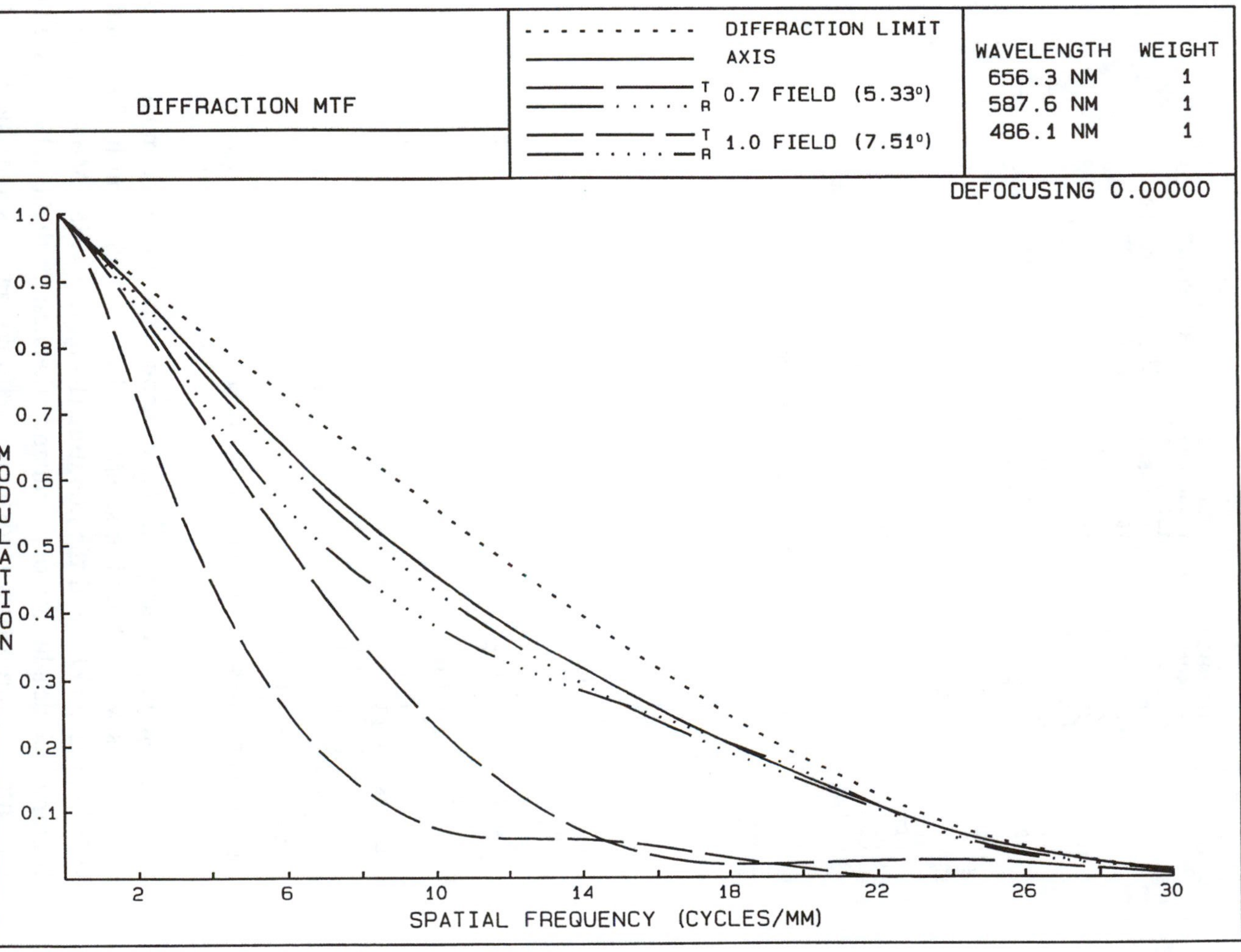

FIGURE 7.79 MTF plots for shortened lens (CODE V).

```
     Finite Conjugate for Chapter 7
                      RDY                  THI        RMD         GLA
> OBJ:             INFINITY             85.324863
    1:             99.40488             10.000000           SK15_SCHOTT
    2:           -175.06765             15.357508
    3:             28.87023              4.779526           LAKN22_SCHOTT
    4:            -55.80568             10.000000           SF2_SCHOTT
    5:             27.43738              8.083035
  STO:             INFINITY             15.657543
    7:            -16.38669              8.275862           BAF50_SCHOTT
    8:            -22.56605              0.250000
    9:           -290.49268              2.529030           LAFN7_SCHOTT
   10:           -115.93293            839.704245
  IMG:             INFINITY              0.000000

SPECIFICATION DATA
   EPD        26.00000
   DIM              MM
   WL           656.30     587.60      486.10
   REF               2
   WTW               1          1           1
   INI
   XOB         0.00000      0.00000      0.00000
   YOB         0.00000    -15.12000    -21.63330
   VUY         0.00000      0.00000      0.25000
   VLY         0.00000      0.00000      0.25000

INFINITE CONJUGATES
   EFL          82.6420
   BFL          13.2845
   FFL         -77.0607
   FNO           3.1785
AT USED CONJUGATES
   RED          10.0000
   FNO          63.1490
   OBJ DIS      85.3249
   TT          999.9616
   IMG DIS     839.7042
   OAL          74.9325
   PARAXIAL IMAGE
    HT         216.3330
    THI        839.7042
   ANG           7.3497
   ENTRANCE PUPIL
    DIA         26.0000
    THI         78.3471
   EXIT PUPIL
    DIA         13.8262
    THI        -30.6624
```

FIGURE 7.80 Program listing of data for lens after global search optimization (CODE V).

types of photographic lenses that are encountered. Other summaries of the classifications will be found in several books (Cox 1964; Smith 1992; Laikin 1995; Malacara 1995) and the literature (Kingslake 1946). Most lens design programs include a library of patent and other designs that can be browsed to find lenses that are worth studying or that may serve as starting points for other designs.

Figure 7.86 is a very brief family tree of the major lens types. The singlet, doublet, and triplet types have been discussed in some detail. There are several branches of design that have evolved from the basic types. The symmetric lenses are built up by placing components having one, two, or three elements more or less symmetrically about a central stop. The double Gauss

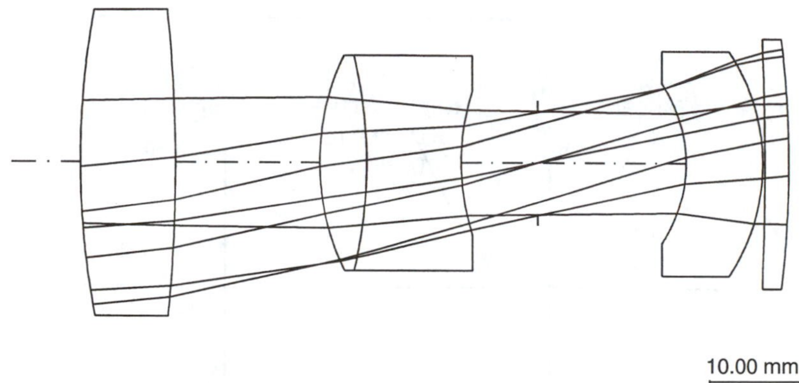

FIGURE 7.81 Layout of design of previous figure (CODE V).

is a sort of hybrid of the symmetric lens and the triplet form, as discussed in the last section. It is important to realize that in almost every case, the imaging requirements are nonsymmetric, usually with infinity as one conjugate. Therefore fully symmetric lenses are not usually the best solution to an imaging situation.

Continued addition of elements more or less symmetrically about the stop minimizes the rapid growth of angle of incidence with field and leads to several forms of very-wide-angle lenses. The nonsymmetric designs fall into some specific categories. The use of two separated positive components leads to a Petzval class of lenses. The distribution of power permits large numerical apertures to be obtained, but the sum of two positive powers leads to large field curvature, which limits the useful field for such lenses.

Combinations of lenses with one negative and one positive power provides some useful optical properties. These lenses are telephoto and reverse telephoto, or retrofocus, lenses. These lenses are required where specific mechanical constraints upon overall length and back focus are specified. The intrinsic large asymmetry of these designs usually leaves a large residual of distortion, which makes these lens type inappropriate for some applications. In an extreme case, the inverse telephoto can be designed to use this distortion to advantage and leads to extremely wide angle lenses called "fisheye" lenses.

The basic methods used in the design of alternate types of imaging lenses follows the approaches outlined for the lens types described earlier. The starting point and boundary conditions are usually selected to maintain the lens within the desired classification. In each case, the designer needs to understand the limitations and advantages of the configuration that is chosen. The general merit functions that have been developed for optical problems will usually require some additional components in order to meet the specialized needs of the user of these systems.

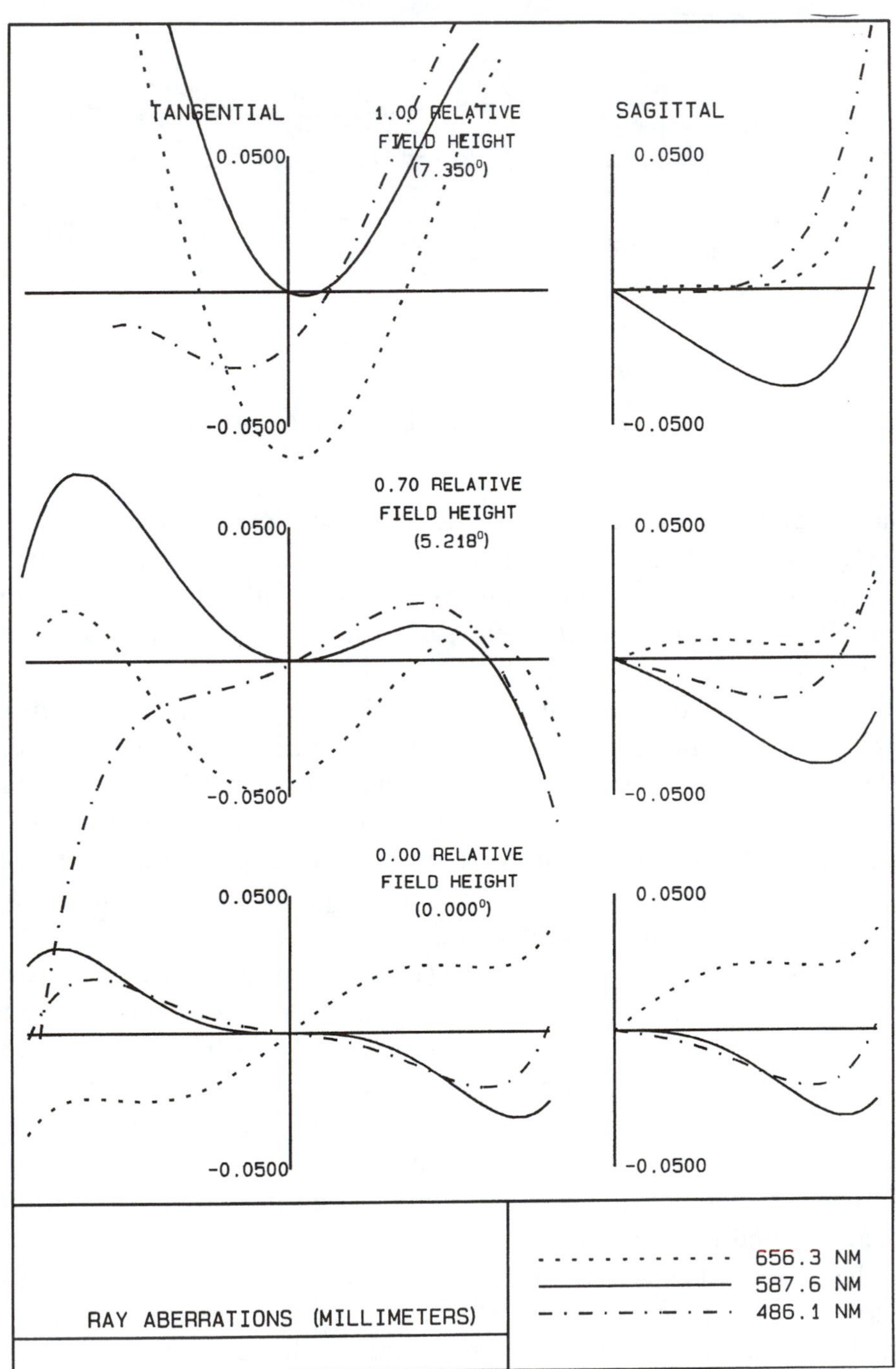

FIGURE 7.82 Ray trace plot for design of Figure 7.80 (CODE V).

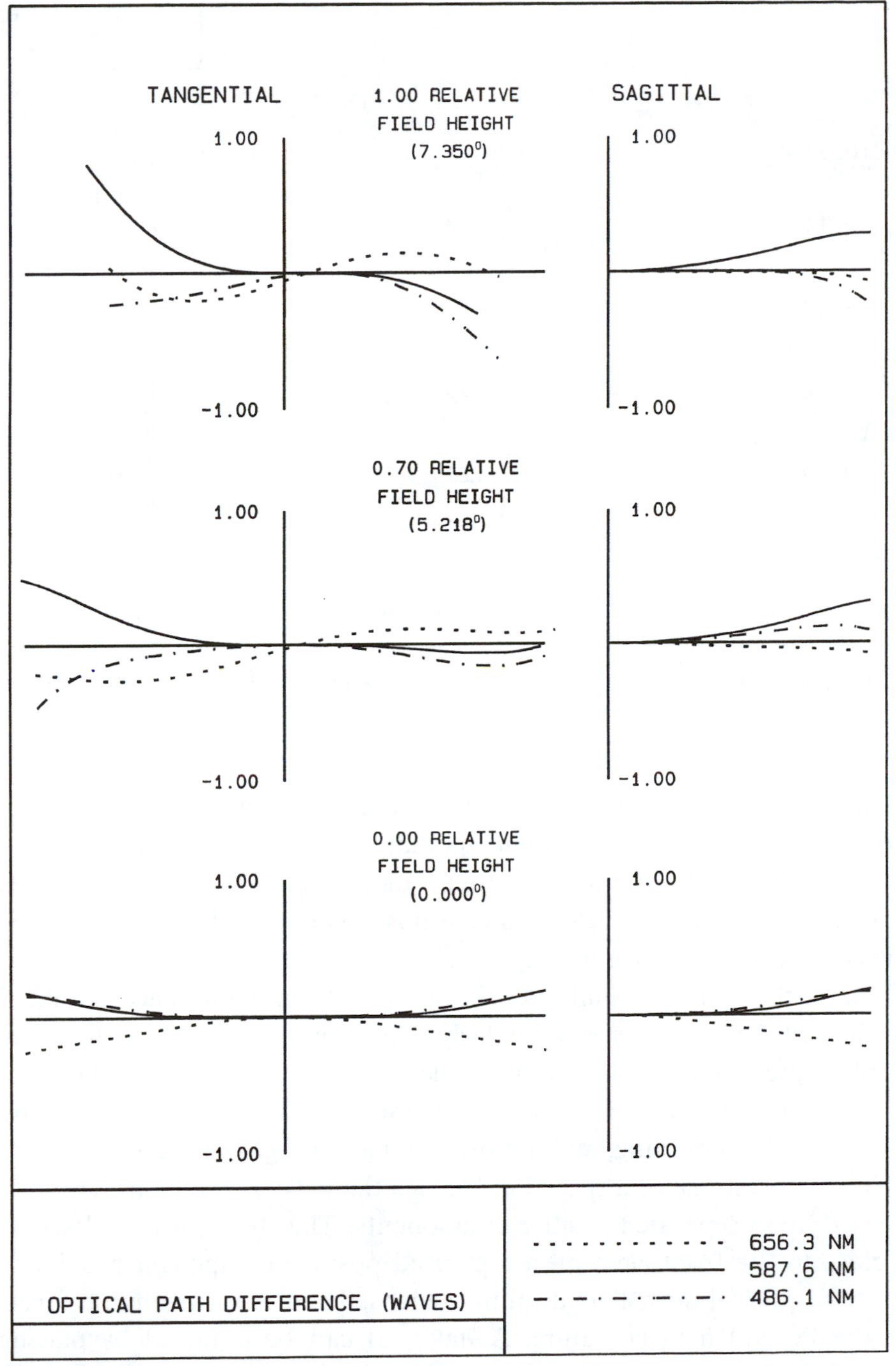

FIGURE 7.83 OPD plot for design of Figure 7.80: ······, 656.3 nm; ——, 587.6 nm; ·—·—·, 486.1 nm (CODE V).

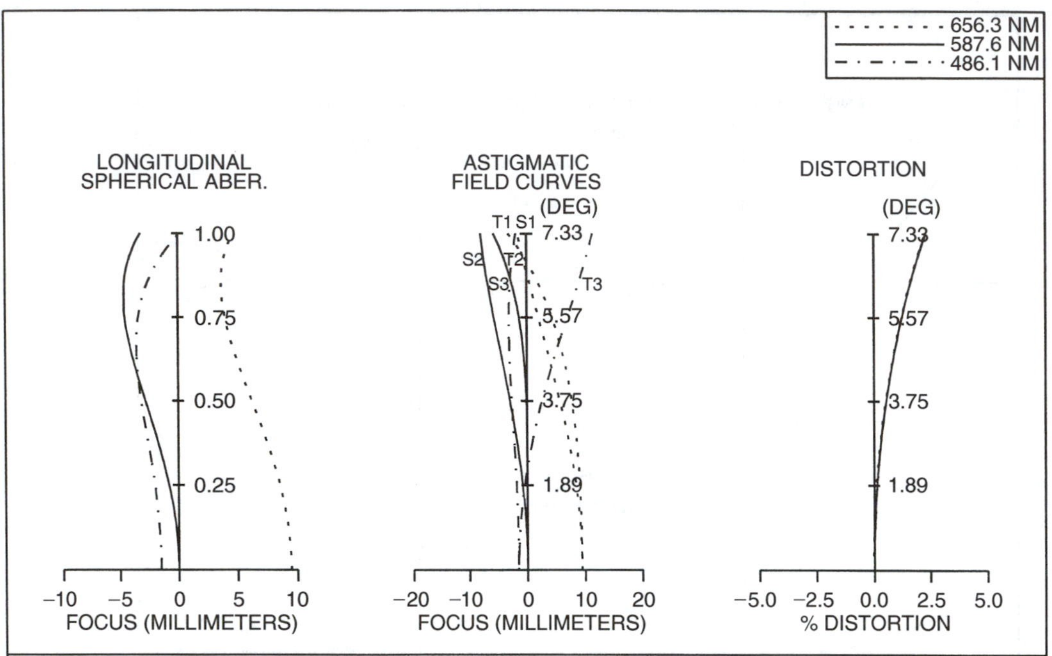

FIGURE 7.84 Field plots for design of Figure 7.80: ······, 656.3 nm; ——, 587.6 nm; ·–·–·, 486.1 nm (CODE V).

The types of lenses falling into this general nonclassification land here because of some specific requirements that are made. While the general criteria for lens performance need to be met, balancing these requirements against special needs of field or image position can lead to compromises in lens complexity or image quality.

One of the common limitations in optical design is the space available in which to fit the lens. This is particularly important for long focal length lenses for photographic cameras. It is desirable to force a lens of, say 400 mm focal length, to fit into a length of perhaps 200 mm or less. This will lead to a lower weight for the mounting, and a more compact design for the system.

The first example of a specialized lens is the Petzval lens. This design type is based upon separated positive components. The stop is usually located at or close to the front element, a separated positive component provides the ability to control the astigmatism in the lens. There is no intrinsic correction for the Petzval field curvature. A flat field can be obtained by placing a negative element as a field flattener very close to the image surface. This lens type is capable of large numerical apertures, F/2 being easily achieved. The lack of correction of the Petzval curvature limits the field to a few degrees, perhaps 10° or 15° half field, unless field-flattening elements are added near the focal surface.

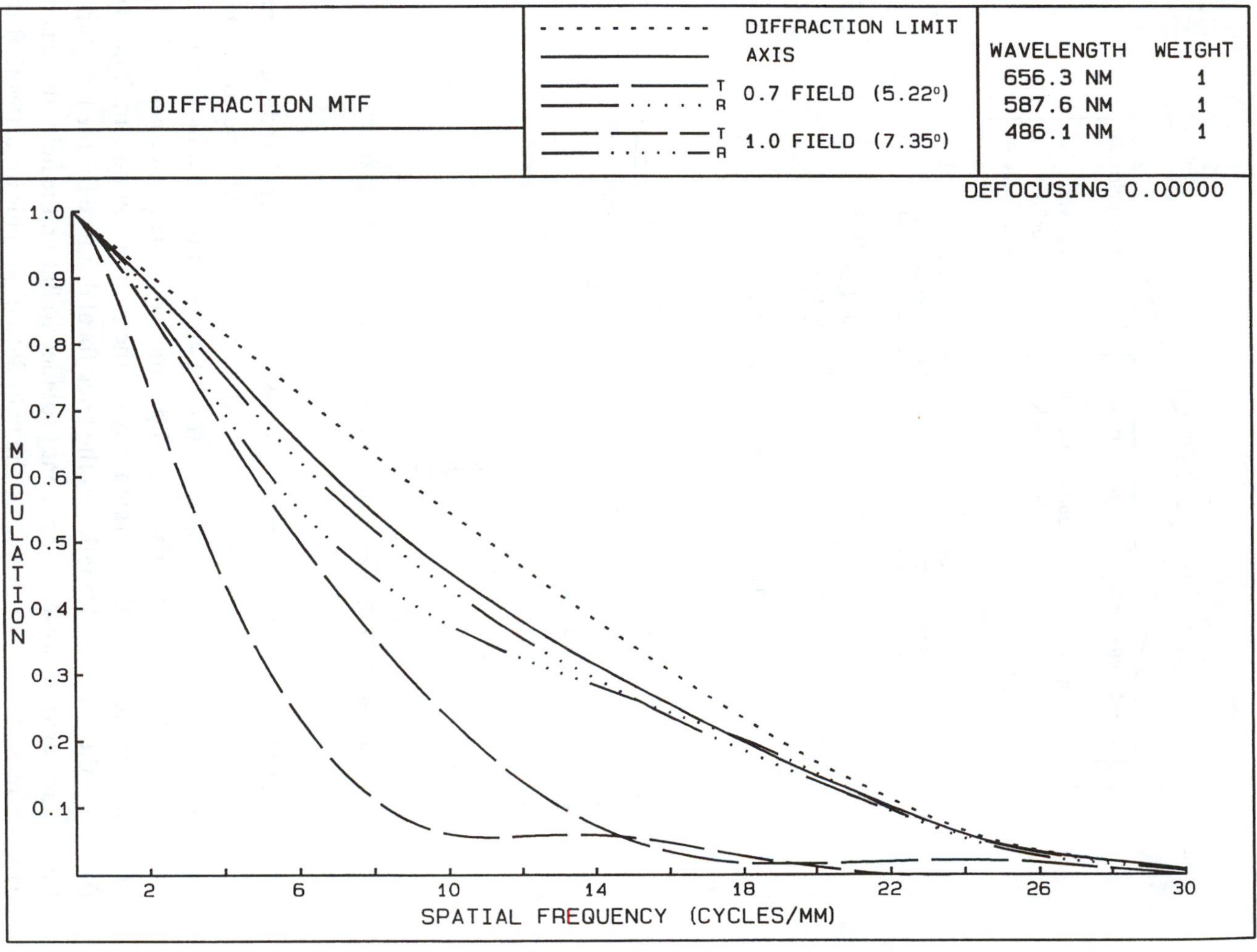

FIGURE 7.85 MTF plots for design of Figure 7.80 (CODE V).

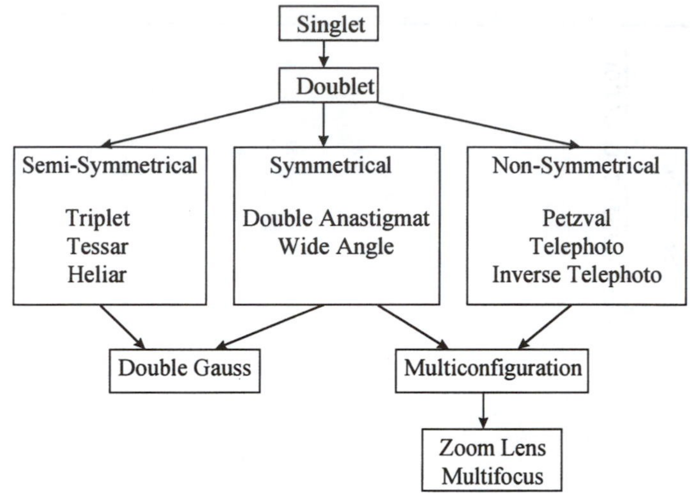

FIGURE 7.86 Family tree of lens types.

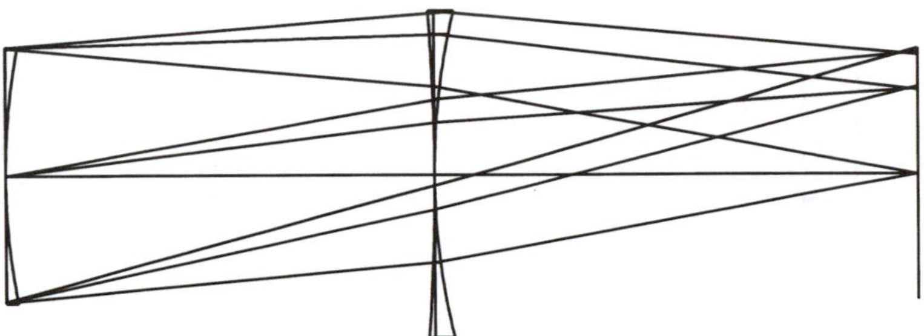

FIGURE 7.87 Thin lens layout of Petzval starting point (ZEMAX).

Consideration of the aberration discussions in Chapter 3 indicates that chromatic correction will require that both components will have to be individually color corrected. The basic form is two separated doublets, sharing the power of the total lens. The critical parameter in the layout is the choice of the relative height of the marginal ray on the second component. This will determine the back focal length and the back focal distance for the lens. Since there is no negative power component provided to flatten the field, there is no excess power required to overcome the field flattening. This indicates that the amount of aberration that will have to be corrected is less than for a lens with internal field-flattening components.

The ZEMAX program was chosen for this example. The starting point was an initial power and separation described by the basic paraxial description of the lens. The process of design is shown in the figures. The starting lens is a thin-lens system shown in Figure 7.87. The input to the program was

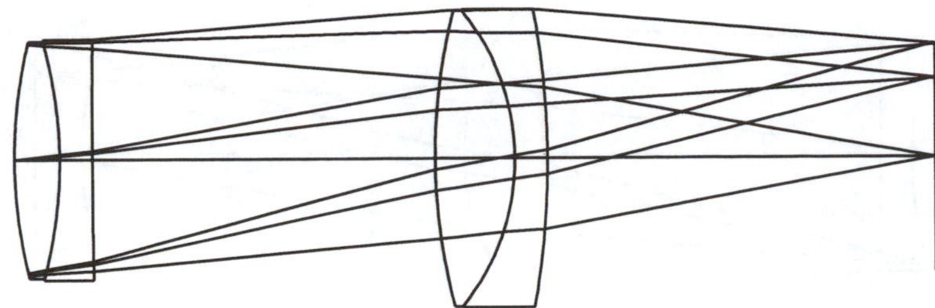

FIGURE 7.88 Petzval lens with realistic thicknesses (ZEMAX).

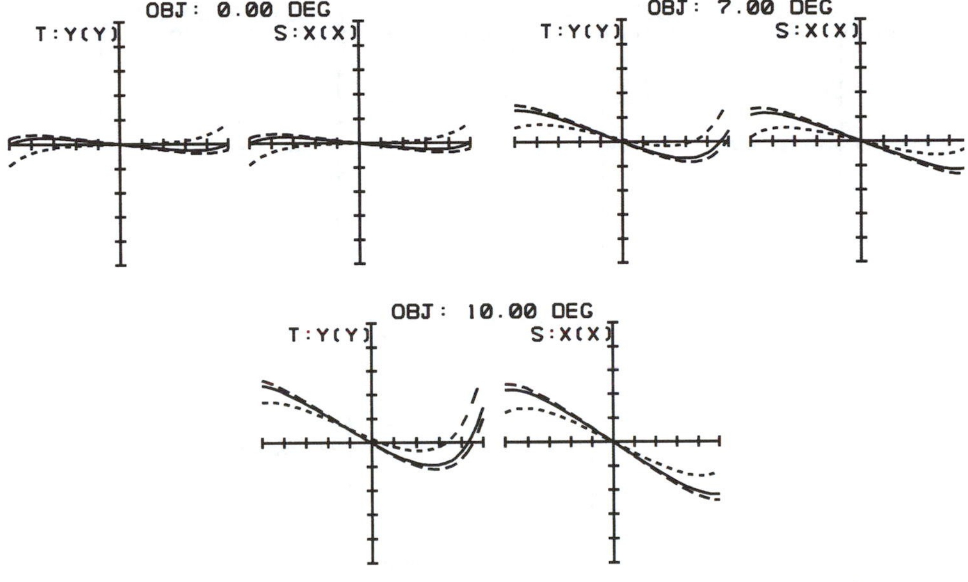

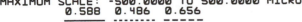

FIGURE 7.89 Ray intercept plots for Petzval lens (ZEMAX).

to set the lens powers by placing half of the power in the first lens, using an angle solve, and then setting the separation to make the marginal ray height 0.7 of the entrance pupil height. The stop was placed at the front lens. The intrinsic inward field curvature can actually be seen in the rays traced in this design.

Thickness was added to make a reasonable set of elements, and the chromatic and third-order aberrations were corrected, as shown in Figures 7.88 and 7.89. So far, only the third-order aberrations have been considered. The intrinsic inward-curving field can easily be seen in the ray trace, and it

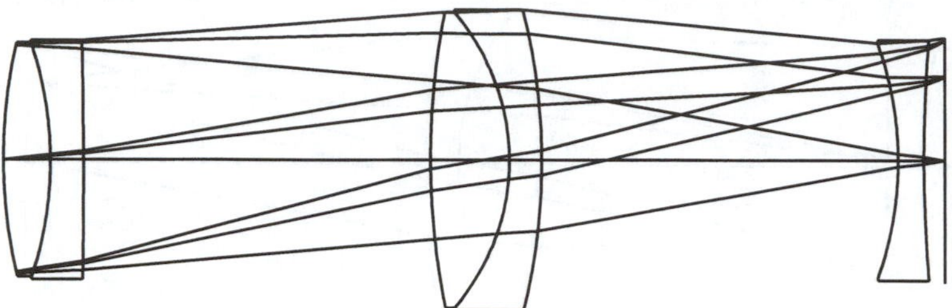

FIGURE 7.90 Petzval lens with field flattener (ZEMAX).

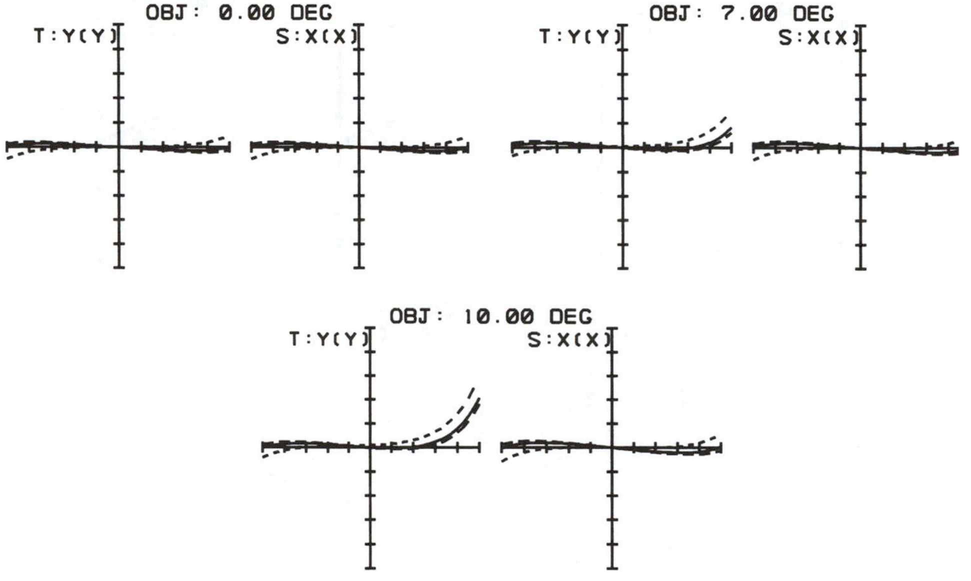

FIGURE 7.91 Ray intercept plots with field flattener (ZEMAX).

dominates the aberrations across the field. The chromatic correction has eliminated both longitudinal and lateral color. The next big step was to add a negative lens near the focal surface to compensate the inward-curving field. This is shown in Figure 7.90, with a ray trace plot for the lens in Figure 7.91. The field is indeed flattened, as shown by the lack of focal error in the sagittal and tangential directions. The lateral color and the third-order distortion were also corrected at this stage. The major problem is an overcorrected high-order coma which shows up principally in the upper portion of the pupil. The angle solves and height solve are still retained. The field flattener was inserted with a height solve to place it in front of the image surface with an axial height of 1.0 mm on the first surface. This technique is suggested as a way of keeping the field flattener from disappearing through the image during the design stage. The actual focal length has moved slightly from 100.0 mm to about 103 mm as a result of adding the field flattener. The system could be scaled to maintain the focal length exactly, if desired.

The default merit function was then added, along with a constraint on the distortion, to continue the correction using ray errors. The lens looks quite similar to the one in Figure 7.92, because the paraxial solve constraints have been maintained. The ray pattern shows a reasonable aberration balance. The MTF plots for this lens are shown in Figure 7.93, and Figure 7.94 shows the specification data. This 100 mm F/2 lens is beginning to look reasonably good. More variables are obviously desirable to improve the lens. There are several approaches that could be taken. The paraxial solves that constrain the layout could be lifted, and a search made for the optimum values. Additional variables could be obtained by separating the contact for the two doublets. Both of these actions were carried out, with a constraint placed on the edge separation of the broken contacts to keep the elements close together. The paraxial constraint solves were removed from the lenses and the thicknesses, and the focal length target was added to the merit function with a very high weight (100) to ensure that the condition is met.

The default merit function was applied, with some improvement in the lens. It was noted that the high-order coma causing the large aberration at the upper edge of the pupil at the maximum field was the major problem. This was resolved by limiting the size of the second element to vignette the extreme edge rays, with a vignetting factor of about 15% at the edge of the field. This has a very beneficial effect upon the aberrations, and led to the configuration shown in Figure 7.95. The stop was also allowed to move around in the lens, and settled just behind the front element as shown. The ray intercept plots also looked considerably better, as shown in Figure 7.96. All of the changes have permitted the lens to relax into a solution which was not far from the starting point that had been arbitrarily, but logically, chosen. The MTF plots for this lens are shown in Figure 7.97. These are about twice as good as the solution for the contacted doublets, and now represent quite a good 100 mm focal length lens. No tolerancing was carried out on this lens, but inspection of the configuration in Figure 7.95 shows relatively gentle curvatures and modest angles of incidence, suggesting that the lens should have somewhat reasonable tolerances. A listing of the final configuration is included in Figure 7.98.

The glasses used in this example are normal Bk7 and F2. The original constraints on the lens have evidently led to a reasonable region of solution. The next step in the design would be to try to improve the lens by examining

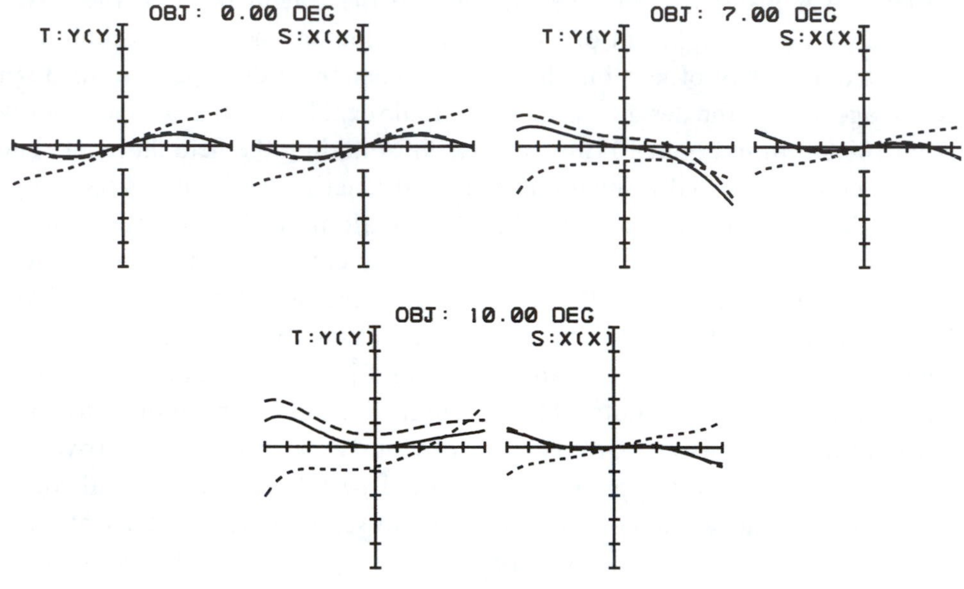

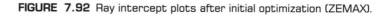

FIGURE 7.92 Ray intercept plots after initial optimization (ZEMAX).

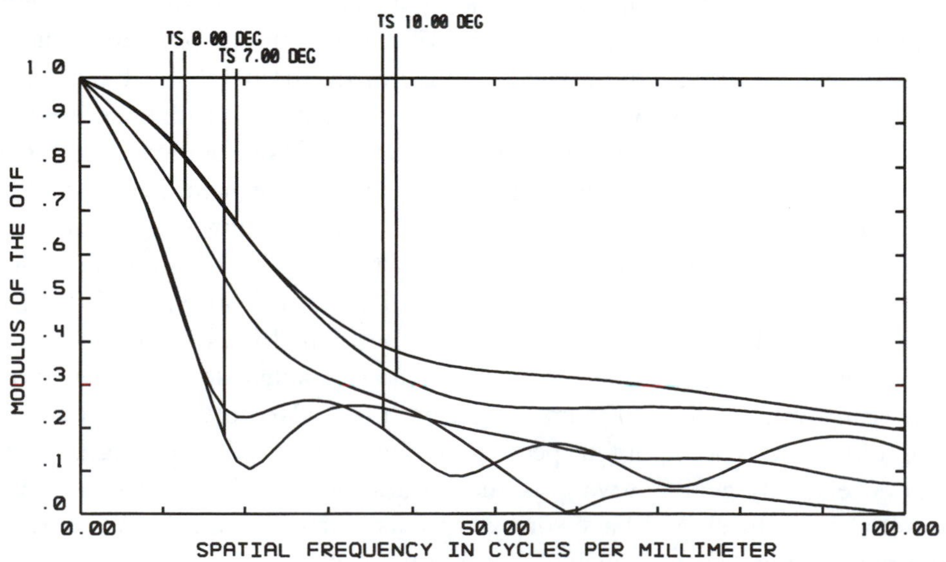

FIGURE 7.93 MTF plots of Petzval (ZEMAX).

System/Prescription Data

Title: Petzval - Field Flattened

GENERAL LENS DATA:

Surfaces : 9
Stop : 1
System Aperture :Entrance Pupil Diameter
Eff. Focal Len. : 108.676 (in air)
Total Track : 143.612
Image Space F/# : 3.04294
Parax. Ima. Hgt.: 19.1626
Entr. Pup. Dia. : 35.7143
Maximum Field : 10.0
Lens Units : Millimeters

Wavelengths : 3
Units: Microns

#	Value	Weight
1	0.587600	1.000000
2	0.486100	1.000000
3	0.656300	1.000000

SURFACE DATA SUMMARY:

Surf	Radius	Thickness	Glass	Diameter
OBJ	Infinity	Infinity		
STO	85.0627	7.0000	BK7	36.4093
2	-65.6827	5.0000	F2	36.5931
3	-1311.1278	53.6137		37.3953
4	88.2113	12.0000	BK7	46.2960
5	-72.8809	5.0000	F2	45.8634
6	-200.5418	53.6133		45.6040
7	-34.5337	4.0000	BK7	35.5686
8	-315.8738	3.3845		37.8528
IMA	Infinity	0		39.2653

FIGURE 7.94 Program listing of data on Petzval lens (ZEMAX).

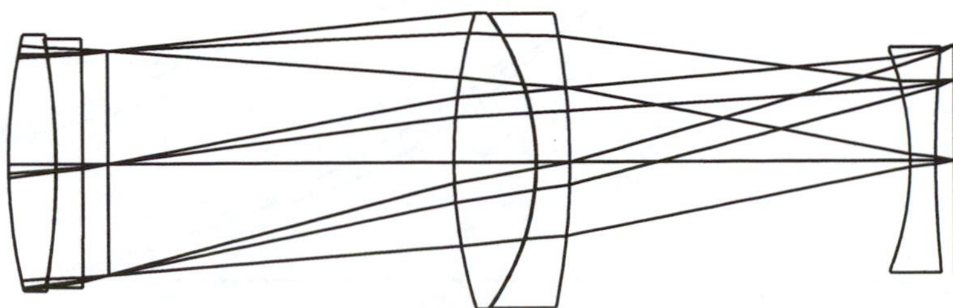

FIGURE 7.95 Layout of Petzval after final optimization (ZEMAX).

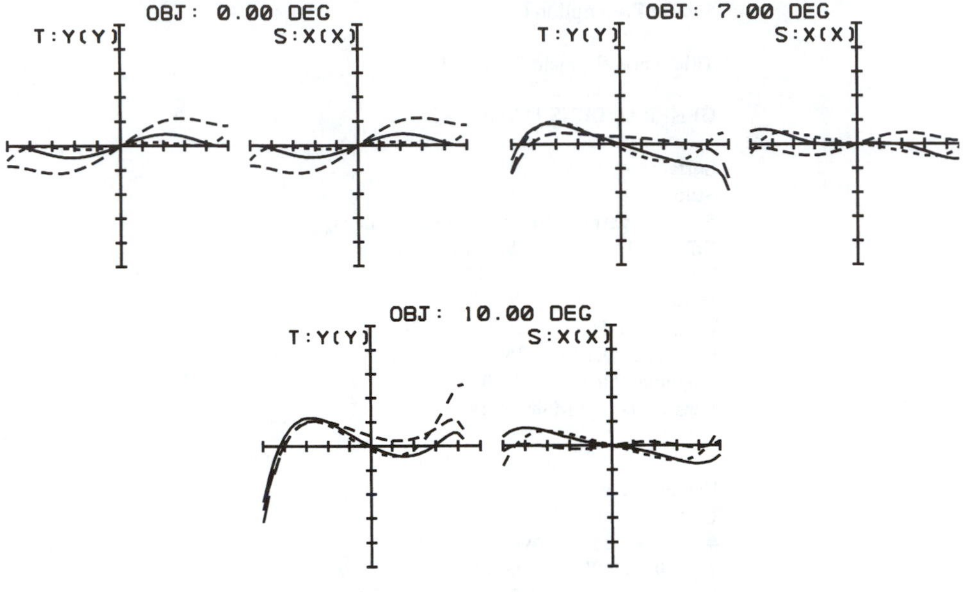

FIGURE 7.96 Ray intercept plot of lens after final optimization (ZEMAX).

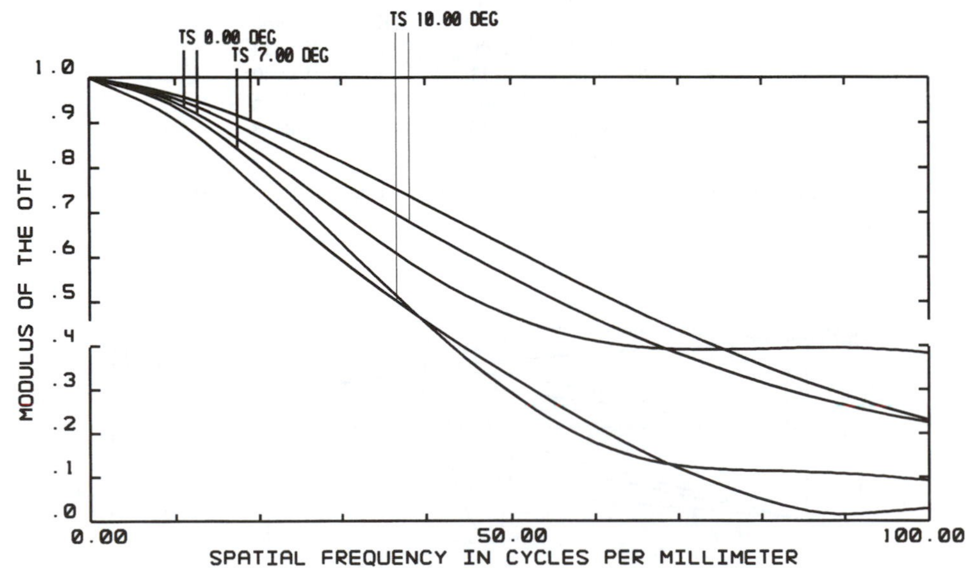

FIGURE 7.97 MTF plots for final optimized Petzval (ZEMAX).

System/Prescription Data

Title: Petzval - Stop moved inside

GENERAL LENS DATA:

Surfaces : 12
Stop : 5
System Aperture :Entrance Pupil Diameter
Eff. Focal Len. : 100 (in air)
Total Track : 143.272
Image Space F/# : 2.8
Parax. Ima. Hgt.: 17.6327
Entr. Pup. Dia. : 35.7143
Maximum Field : 10
Lens Units : Millimeters

Wavelengths : 3
Units: Microns

#	Value	Weight
1	0.587600	1.000000
2	0.486100	1.000000
3	0.656300	1.000000

SURFACE DATA SUMMARY:

Surf	Radius	Thickness	Glass	Diameter
OBJ	Infinity	Infinity		0
1	88.3814	7.0000	BK7	39.0561
2	-124.5867	0.1008173		37.7501
3	-103.5086	4.0000	F2	37.9215
4	-1463.3569	3.791605		36.1703
STO	Infinity	52.21736		34.1269
6	80.8624	12.0000	BK7	45.0412
7	-39.8055	0.09996285		44.9133
8	-39.9235	5.0000	F2	44.6831
9	-115.4620	52.19458		44.8380
10	-45.7793	4.0000	BK7	33.4011
11	182.2222	2.867972		34.3785
IMA	Infinity	0		35.0566

FIGURE 7.98 Program listing of final Petzval design (ZEMAX).

alternate glasses, differing configurations with an additional element in the front or rear group, and experiments with alterations in the weighting and ray choice in the merit function. The back focus distance was chosen to be small to permit the most effective use of a single element as the field flattener. An increase in back focus could be obtained and the field could probably be increased somewhat, but the field flattener would probably have to become a compound lens element to control both the distortion and the lateral color. This example shows how rapidly a lens can converge to a solution from a simple paraxially defined starting point which is given directly to the program.

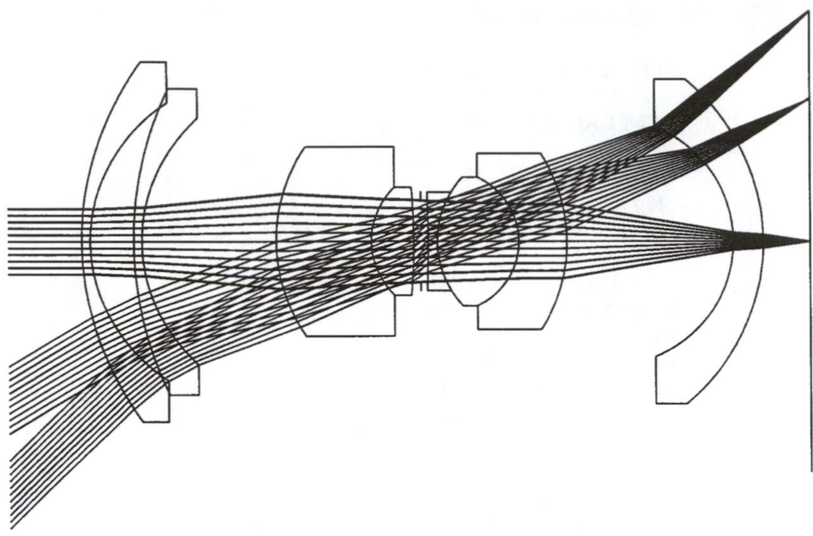

FIGURE 7.99 Example of symmetric wide-angle lens (CODE V).

A second example of special purpose lenses is a symmetric wide-angle lens. The purpose of these lenses is to form an image of a very wide field with low distortion. The central power form of the lens type is a more-or-less symmetric imaging portion similar in nature to a double Gauss lens. This lens is surrounded by one or more pairs of negative outer elements that image the pupil of the inner lens with reduced magnification, thus increasing the field angle. Since the lens is nominally symmetric, the distortion will be low because the chief ray passes almost symmetrically through the lens, with low angles of incidence on all surfaces. The imaging is, of course, not symmetric, as the conjugates are infinity on the object side and close to the final element on the image side.

One example of such a lens is shown in Figure 7.99. It is a "Angulon" type of lens taken from a patent by L. Bertele. The lens covers a 45° half field at F/3.5. In this example it has been scaled to a 21.5 mm focal length fitting a standard 35 mm camera frame. Only moderate vignetting is used here in order to maintain as constant as possible the illumination across the image surface. All of the positive convergent power of the lens is in the center five elements. The outer lenses provide both a flattening Petzval contribution and compress the field angle in the central lens. This has the favorable aspect of limiting the growth of the oblique spherical aberration which is intrinsic to image formation, and adds some favorable aberration to the entrance pupil.

The ray intercept plots for this lens are presented in Figure 7.100. The correction is seen to be balanced across the field. The vignetting reduces the

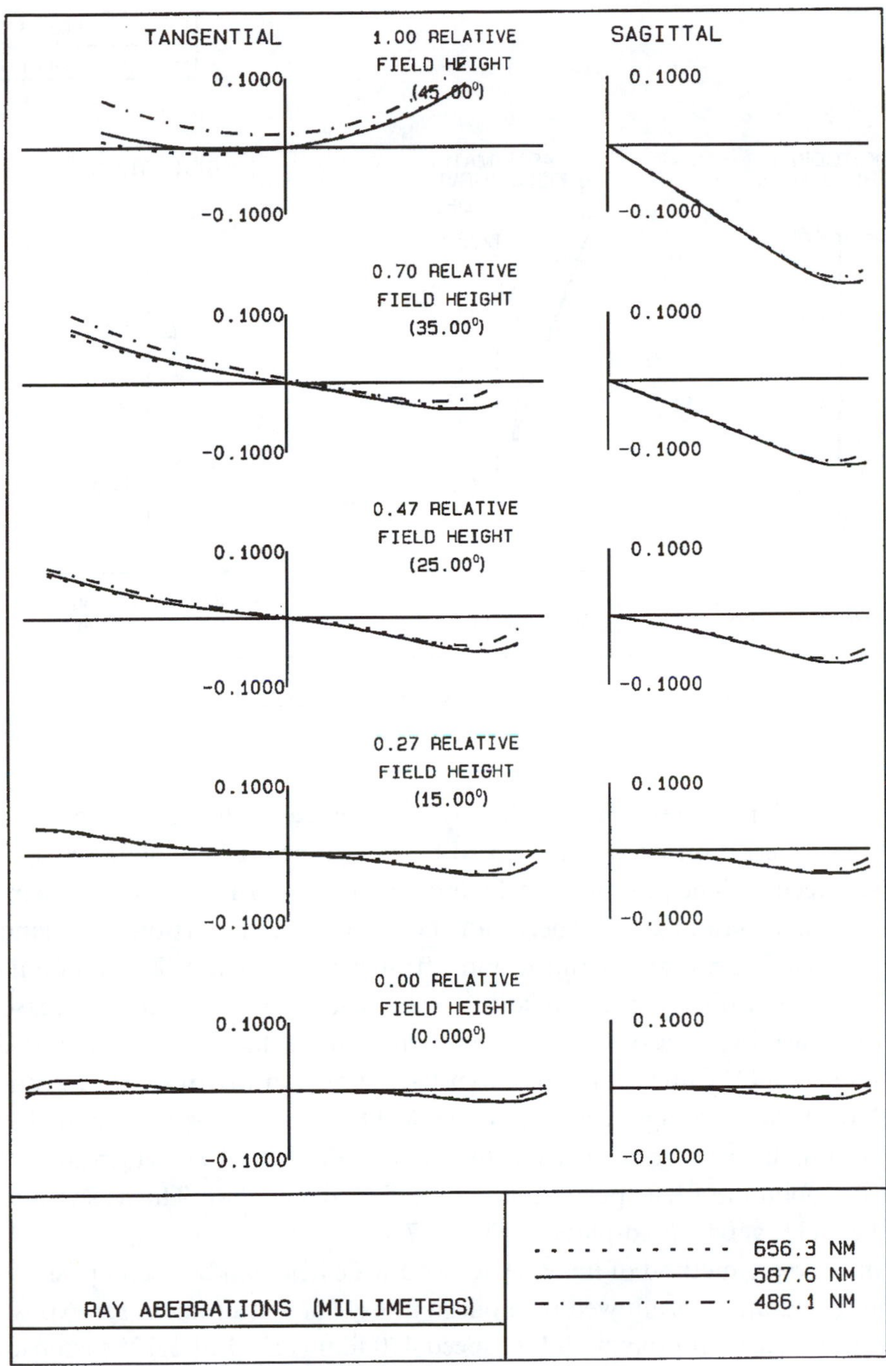

FIGURE 7.100 Ray intercept plot of wide angle lens (CODE V).

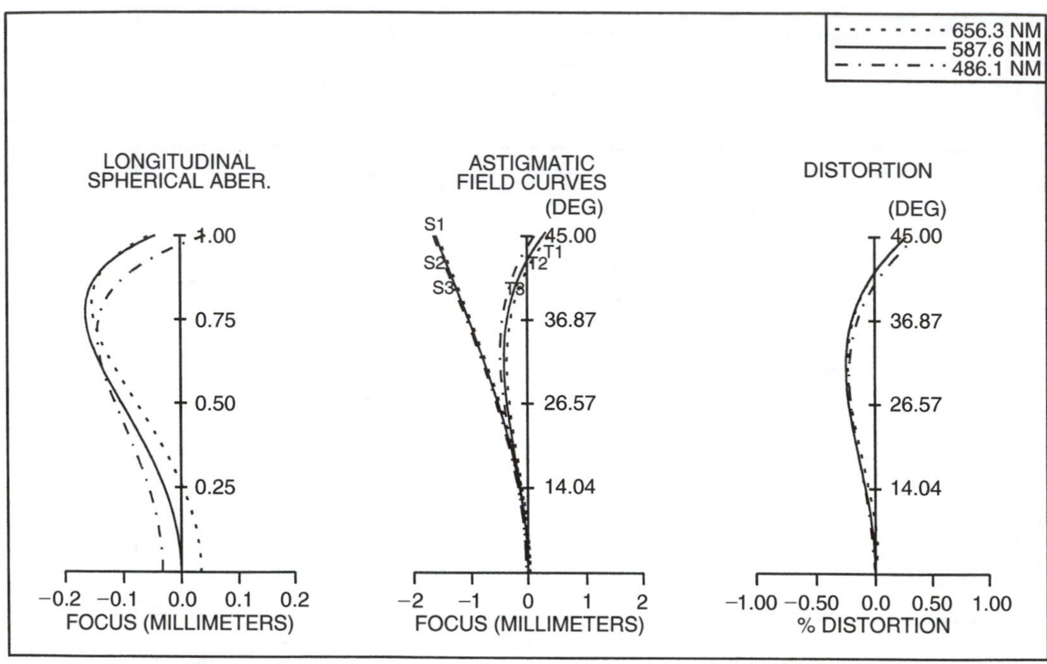

FIGURE 7.101 Field plots for wide-angle lens (CODE V).

width of this ray intercept plot slightly at the edge of the field. The oblique spherical aberration does not seem to be as significant here as the astigmatism correction. The plot of the field and distortion in Figure 7.101 is instructive in understanding the aberration content. The distortion is balanced between large amounts of third- and fifth-order distortion. The amount of spherical aberration is small relative to the field effects. The field is reasonably flat, leading to a best focus located at some distance from the paraxial focus. The MTF values at the best focus are shown in Figure 7.102 and show a relatively low average response. If the MTF versus focus is examined, the reason for this can be seen from the plots in Figure 7.103, which indicates that the optimum focus position is clearly field dependent. This is supported by the MTF versus field plots in Figure 7.104.

An alternate method of imaging across a wide field can be accomplished by using the parameters allowed by modern films. The requirement for exposure outdoors on a bright day with ISO speed 400 film is F/11 at 1/125 s exposure. When this parameter is considered, another approach is to design a lens which exploits this condition. A classical wide-angle lens is the "Hypergon," which is a pair of meniscus single elements surrounding a central stop. Such lenses can give useful imagery over a 120° total field at F/20. Taking this classical design as a starting point (Smith 1992) and changing to a plan for

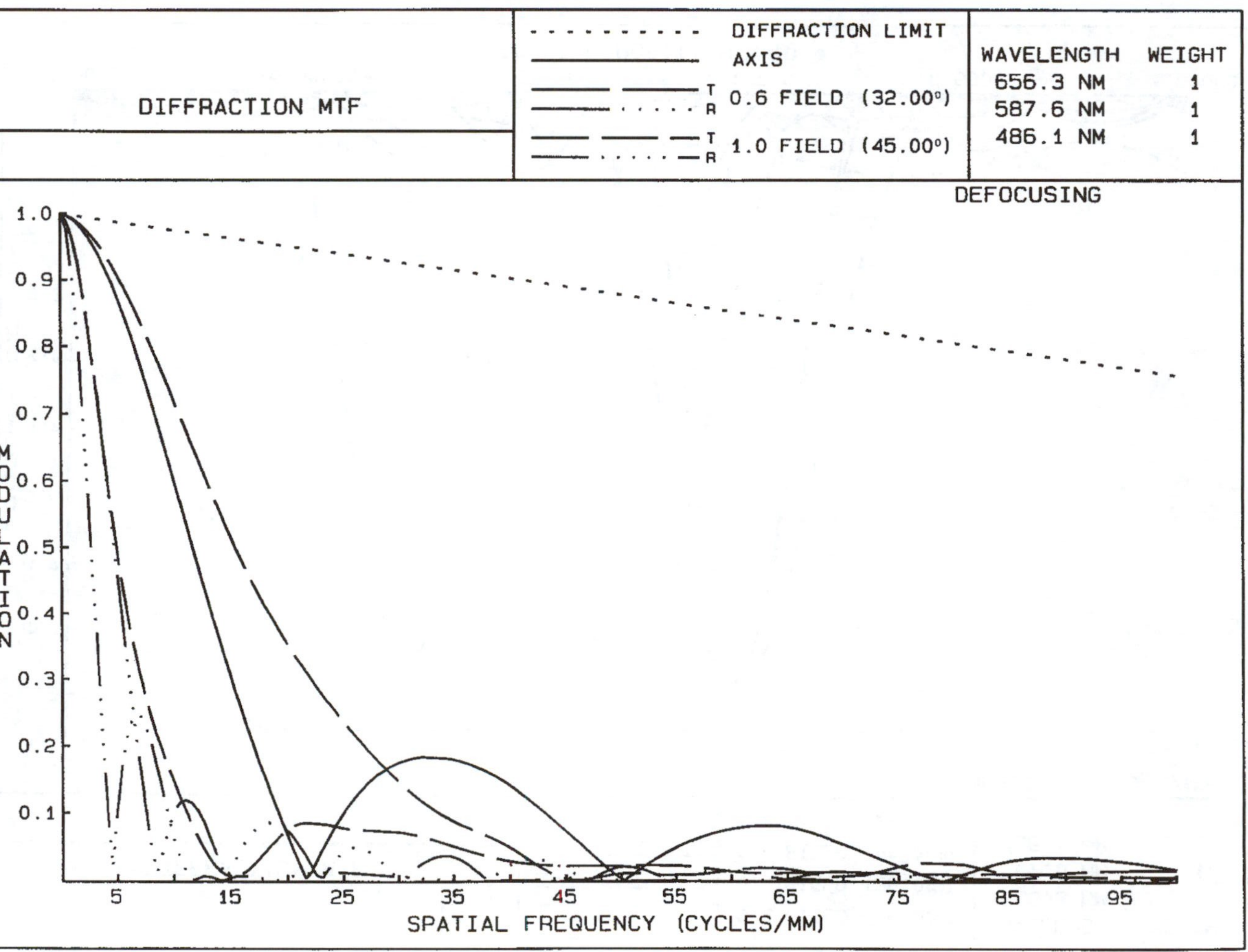

FIGURE 7.102 MTF plots for wide-angle lens (CODE V).

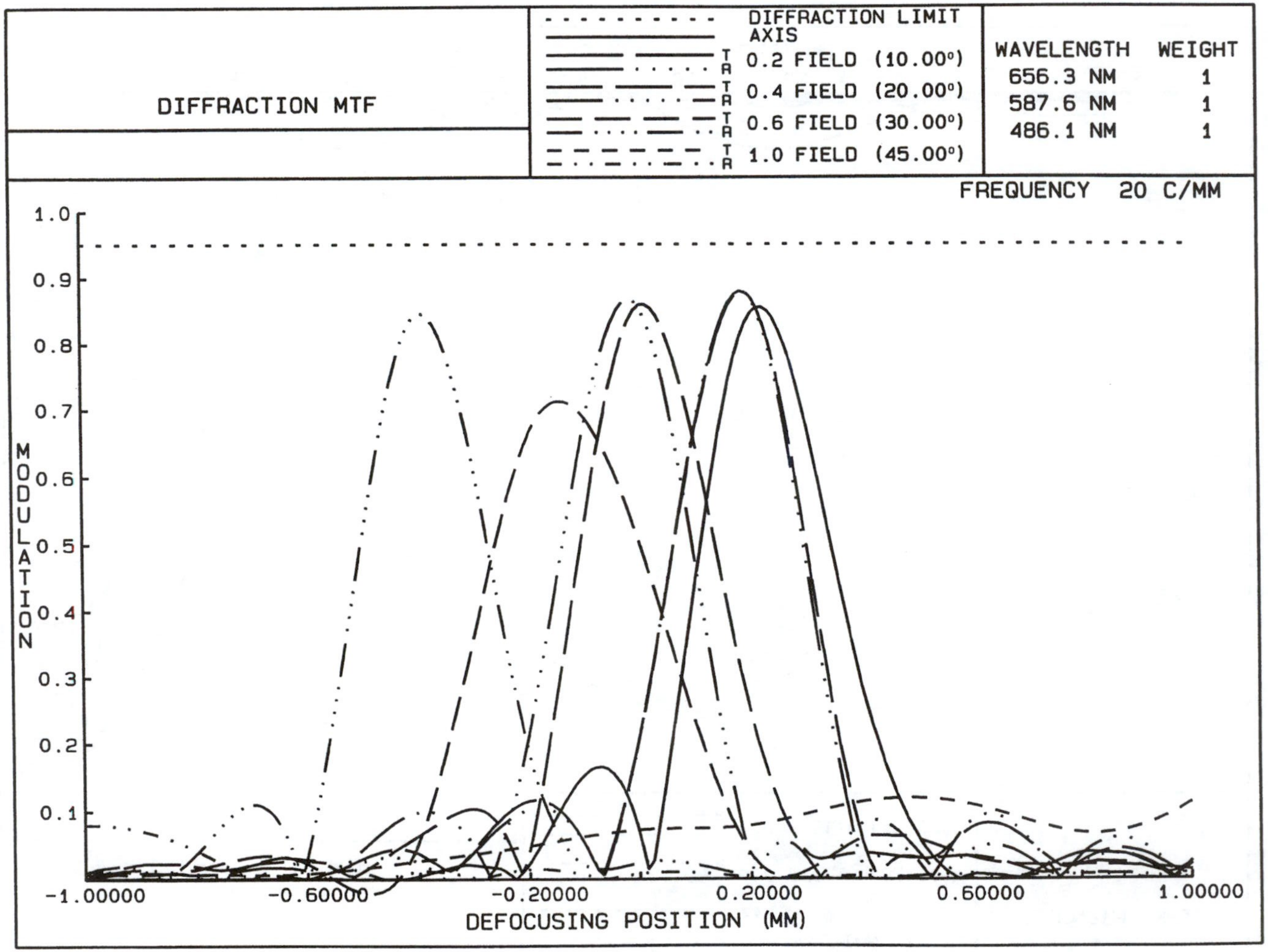

FIGURE 7.103 MTF versus focus for wide-angle lens (CODE V).

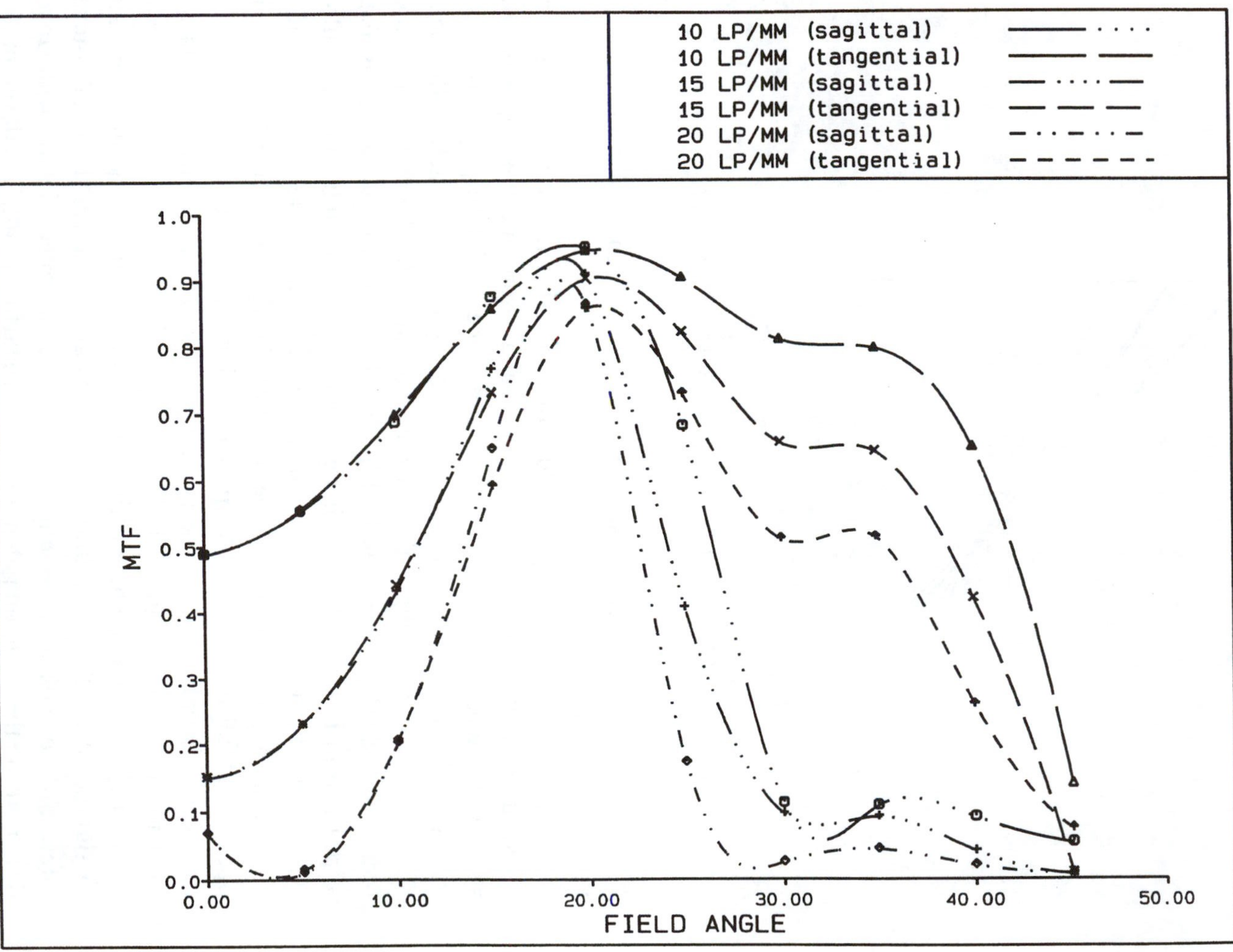

FIGURE 7.104 MTF versus field for wide-angle lens (CODE V).

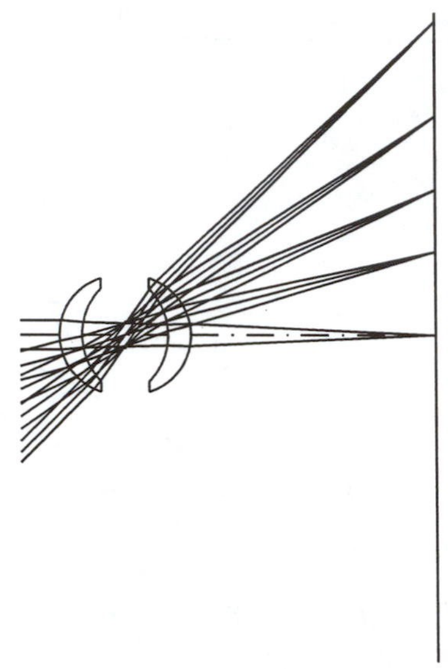

FIGURE 7.105 Hypergon type lens (CODE V).

F/11 at a 45° field can easily be accomplished on a lens design program. In this case CODE V was used.

The result is shown in Figure 7.105. In this design the two elements were constrained to stay identical during the optimization, and the stop was constrained to lie halfway between the elements. The ray intercept plot is shown in Figure 7.106 and illustrates a reasonably good image across the field, limited primarily by astigmatism at the outer field. At the best compromise focus, the MTF plots are as in Figure 7.107, and the MTF versus field appears as in Figure 7.108. The image quality is comparable to the Angulon type of lens, but of course with an aperture diameter 3.1 times smaller, and a relative illumination of about one-tenth of the Angulon when that lens is operated at full aperture. This Hypergon variant is, of course, a much simpler lens, which is tailored for a specific application. The configuration data for the lens are listed in Figure 7.109.

The point to be made here is that the operating conditions can often lead to significantly different solutions to an imaging problem. It is, of course, interesting to look at what improvements can be made in the wide-angle design, using the Angulon patent design as a starting point. The development of a new lens from an old one is best carried out using some small steps at the beginning. Because there are many surfaces, and the design form for the Angulon is close to a desirable one, the best approach is to choose all of the curvatures as variables, and freeze the thicknesses as a start. The default merit function can be used for the starting exploration, but the constraints which keep the design in the region of interest have to be added. In this case

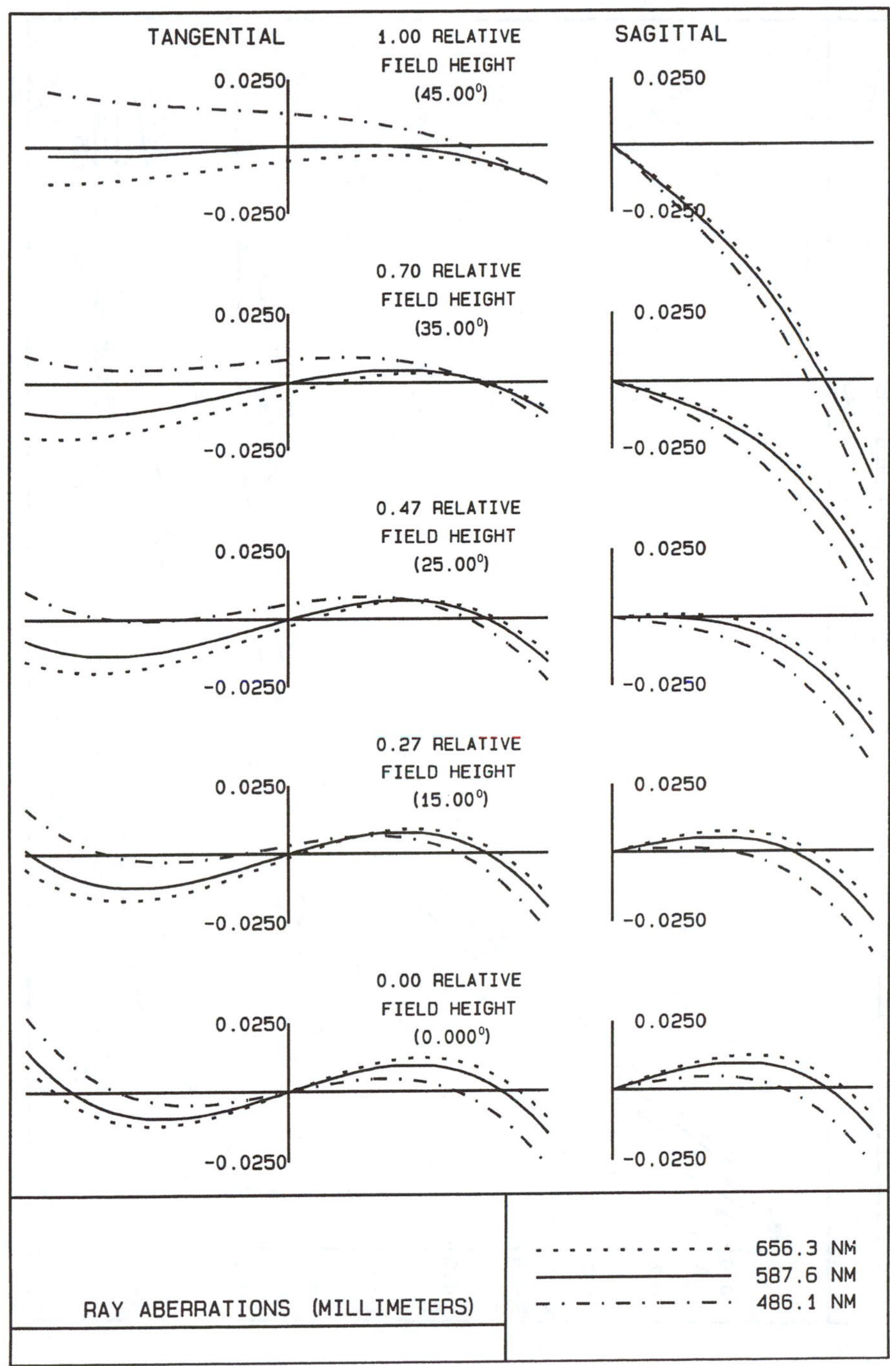

FIGURE 7.106 Ray intercept plot for hypergon lens (CODE V).

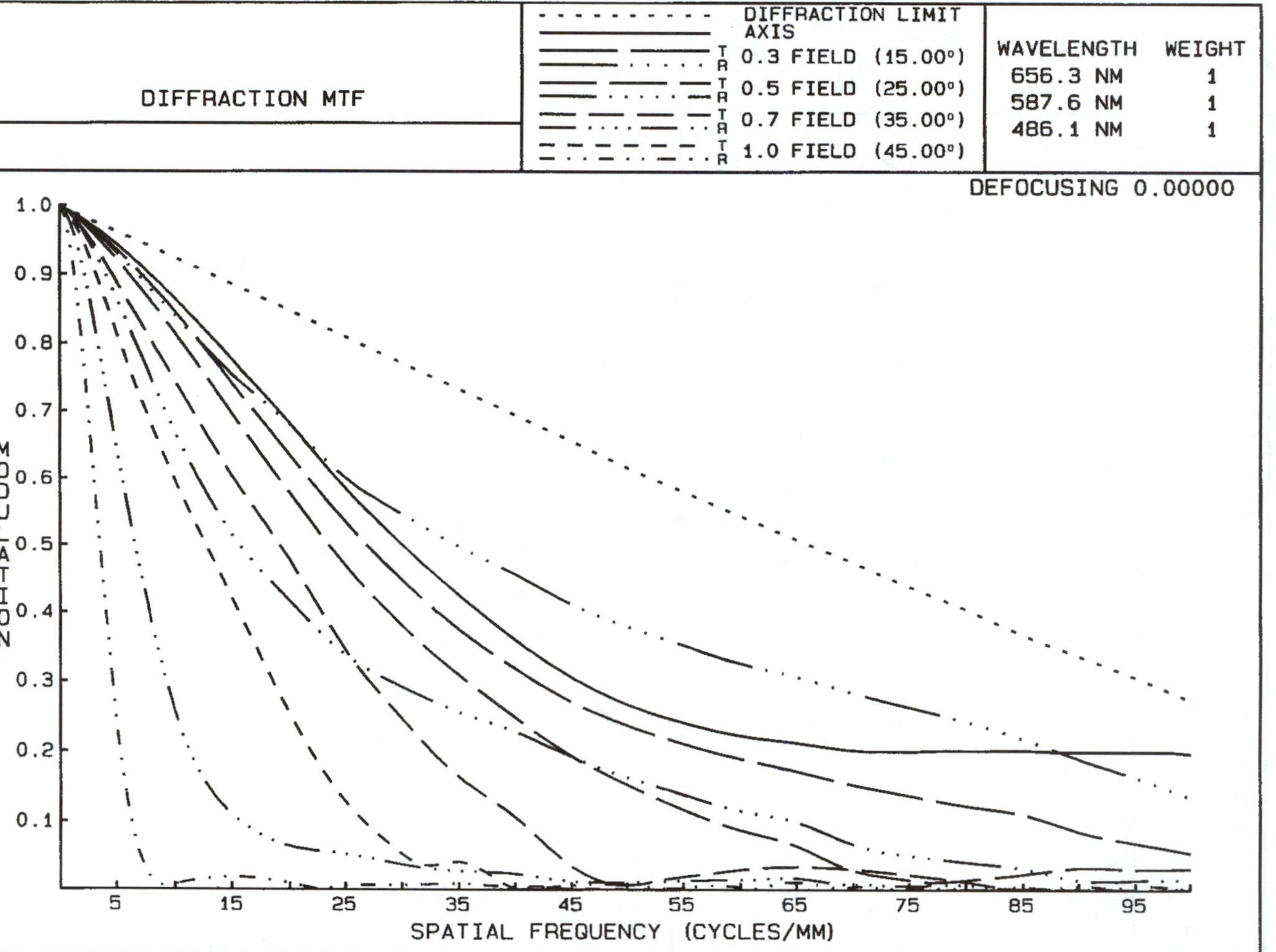

FIGURE 7.107 MTF plots for hypergon design (CODE V).

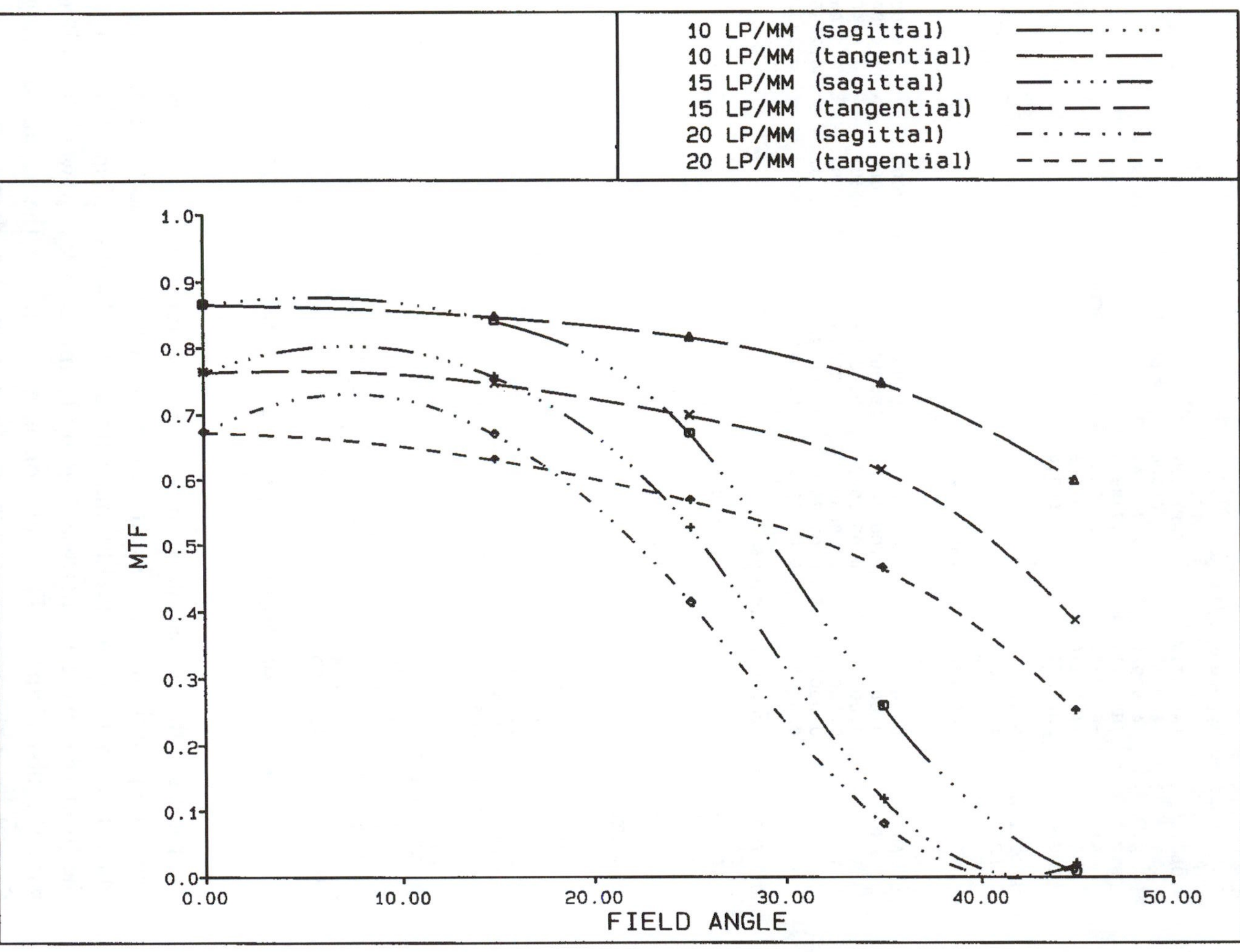

FIGURE 7.108 MTF versus field for hypergon design (CODE V).

```
CODE V> lis
      Modified Hypergon F/11
                    RDY                THI              GLA
 > OBJ:          INFINITY          INFINITY
    1:            4.00170          1.500000         BK1_SCHOTT
    2:            4.53982          2.748544
  STO:           INFINITY          2.748544
    4:           -4.53982          1.500000         BK1_SCHOTT
    5:           -4.00170         16.392956
  IMG:           INFINITY         -0.668916

 SPECIFICATION DATA
    EPD          1.95450
    DIM               MM
    WL            656.30      587.60      486.10
    REF                2
    WTW                1           1           1
    XAN          0.00000      0.00000      0.00000      0.00000      0.00000
    YAN          0.00000     15.00000     25.00000     35.00000     45.00000
    VUX          0.00000      0.00860      0.02375      0.04602      0.07460
    VLX          0.00000      0.00860      0.02375      0.04602      0.07460
    VUY          0.00000      0.01375      0.05726      0.13142      0.23850
    VLY          0.00000      0.03949      0.09272      0.16812      0.33219

 APERTURE DATA/EDGE DEFINITIONS
    CA
    CIR S1                    3.600000
    CIR S2                    3.100000
    CIR S4                    3.100000
    CIR S5                    3.600000

 INFINITE CONJUGATES
    EFL          21.5000
    BFL          16.3930
    FFL         -16.3930
    FNO          11.0003
    IMG DIS      15.7240
    OAL           8.4971
    PARAXIAL IMAGE
     HT          21.5000
    ANG          45.0000
    ENTRANCE PUPIL
     DIA          1.9545
     THI          5.1070
    EXIT PUPIL
     DIA          1.9545
     THI         -5.1070
```

FIGURE 7.109 Program listing of design data for hypergon (CODE V).

the focal length is constrained to be 21.5 mm, and the distortion is to be at least comparable to the present lens. Several iterations led to a lens with much improved image quality, but with about 5% distortion. Also one of the front negative elements was tending to move to less power and collide with the first element. This collision is apparent on the layout drawing. Several manual adjustments were made to keep this element clear of the first element, at least out to the clear aperture. Further adjustments involved constraining the third-order distortion to several different values until a reasonable balance, almost matching the original lens, was obtained. At that point the lens shown in Figure 7.110 was developed.

This modified wide-angle lens appears basically similar to the starting point, with a major difference that the second element has started to

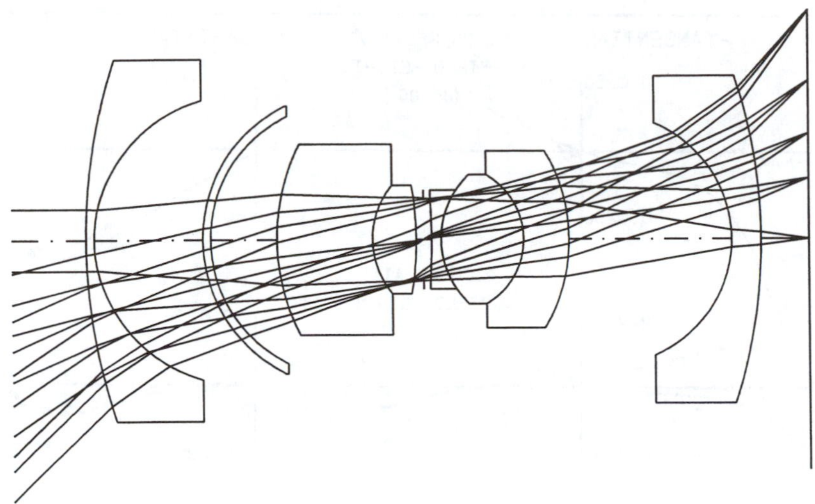

FIGURE 7.110 Modified wide-angle lens (CODE V).

disappear. This element is an almost zero-power concave shell. This is a clue to the designer that an attempt to continue the design with one less element would be a useful direction in which to go. The ray trace plots for the modified lens are shown in Figure 7.111. These are clearly improved over the original lens, at least for the inner portion of the field. There is still a significant increase in image blur at the edges of the field. The field plots in Figure 7.112 show that this is still a member of the basic design class. The major change is a better balancing of the spherical aberration and of the astigmatic fields.

This rebalancing has paid off in image quality. Figures 7.113 and 7.114 show the MTF values for this lens, adjusted to a best focus position. These show distinctly higher contrast at all frequencies. The lens is certainly improved as a result of the rebalancing. The MTF versus field plots indicate that the same "cliff" in the MTF occurs at the edge of the field.

Another example of the "art part" of optical design shows up in any attempt to remove the weak shell element in the design. Simply deleting the element is not very profitable, as this element has been contributing to the aberration balance over the field. Simply erasing the element places the lens immediately in a new design region, which may not be close enough to a good design to be accessible. The approach is to use the design program to remove the element. The lens parameters and thickness can be added as operands in the merit function. The design will proceed and systematically reduce the element to insignificance, and it then can be removed from a lens that is within the aberration targets.

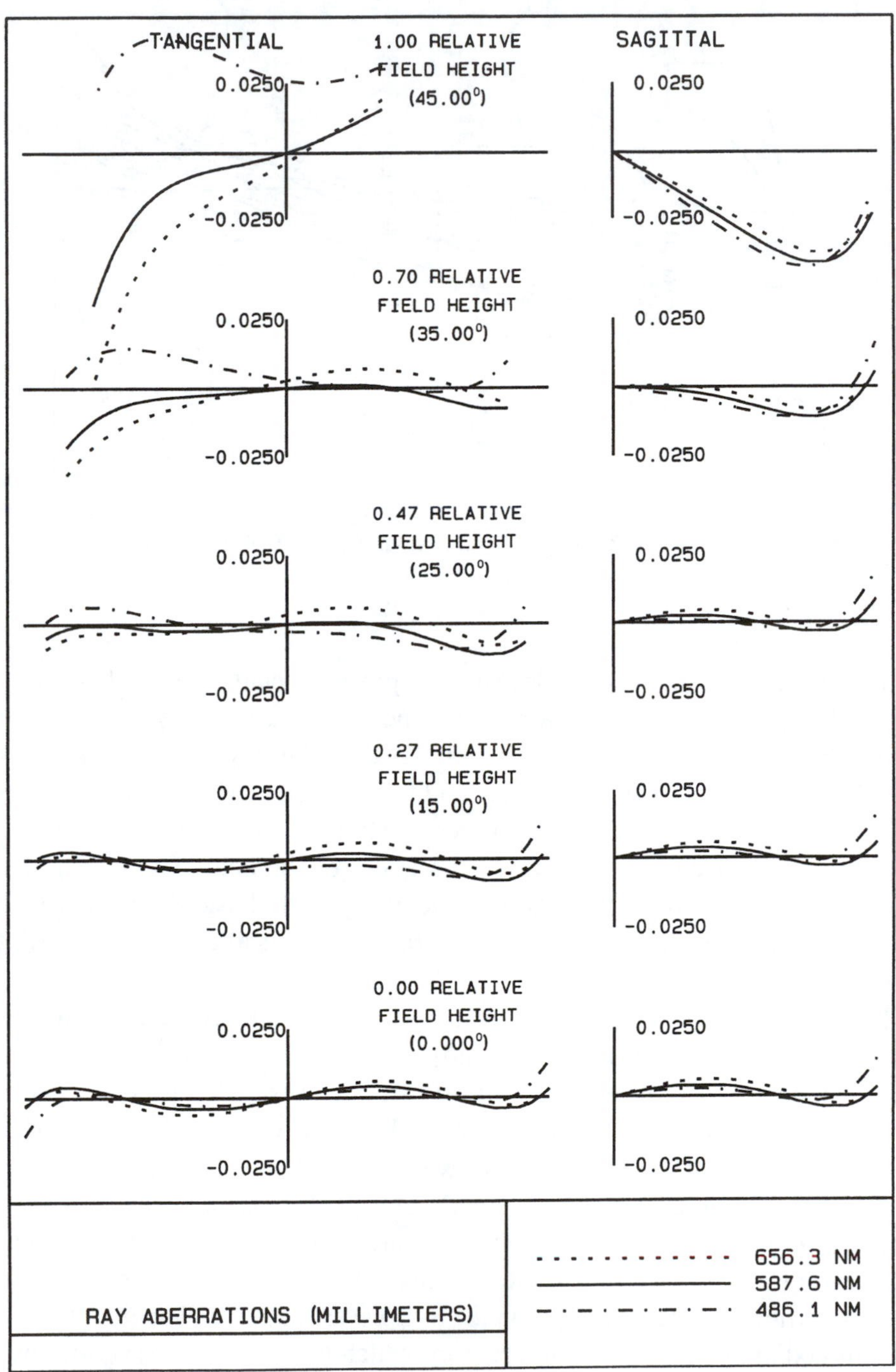

FIGURE 7.111 Ray trace plot for modified wide-angle lens (CODE V).

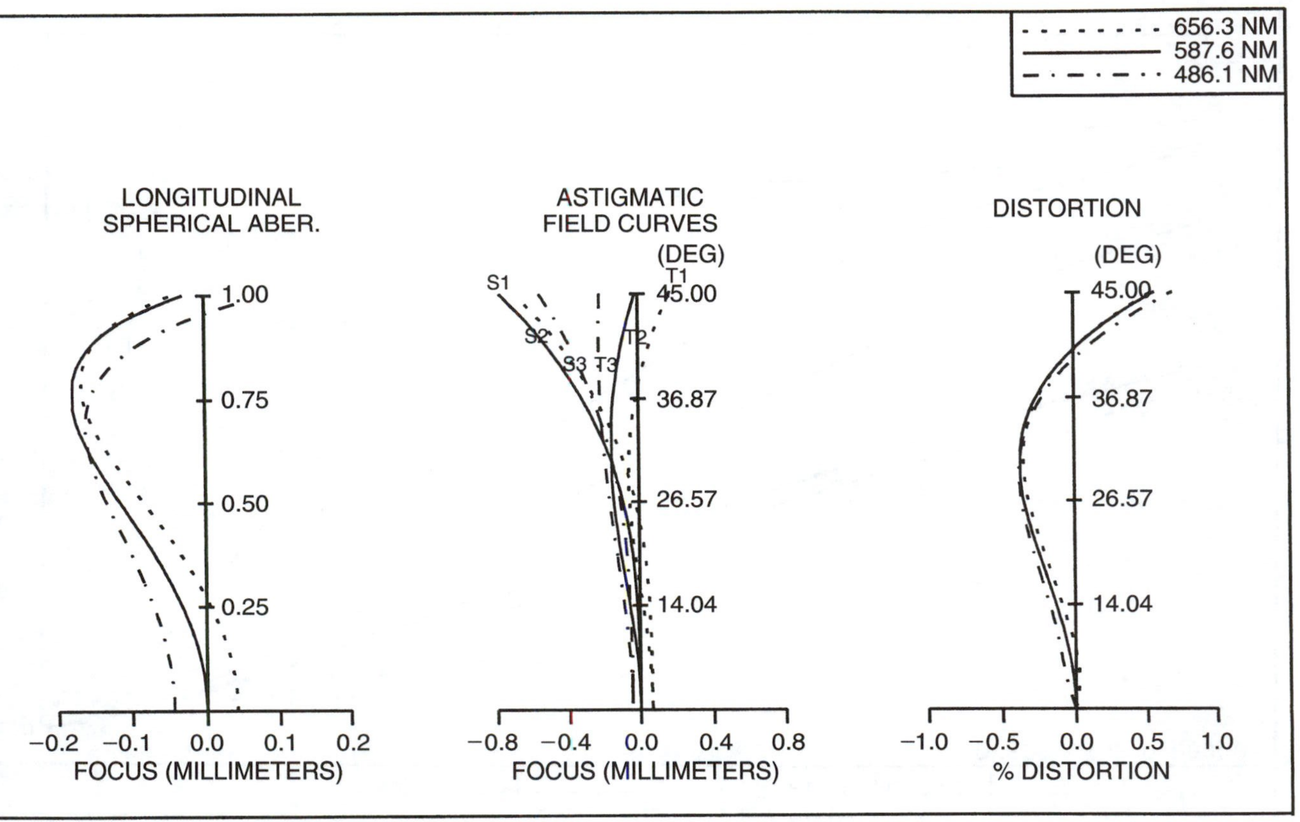

FIGURE 7.112 Field plots of modified wide-angle lens (CODE V).

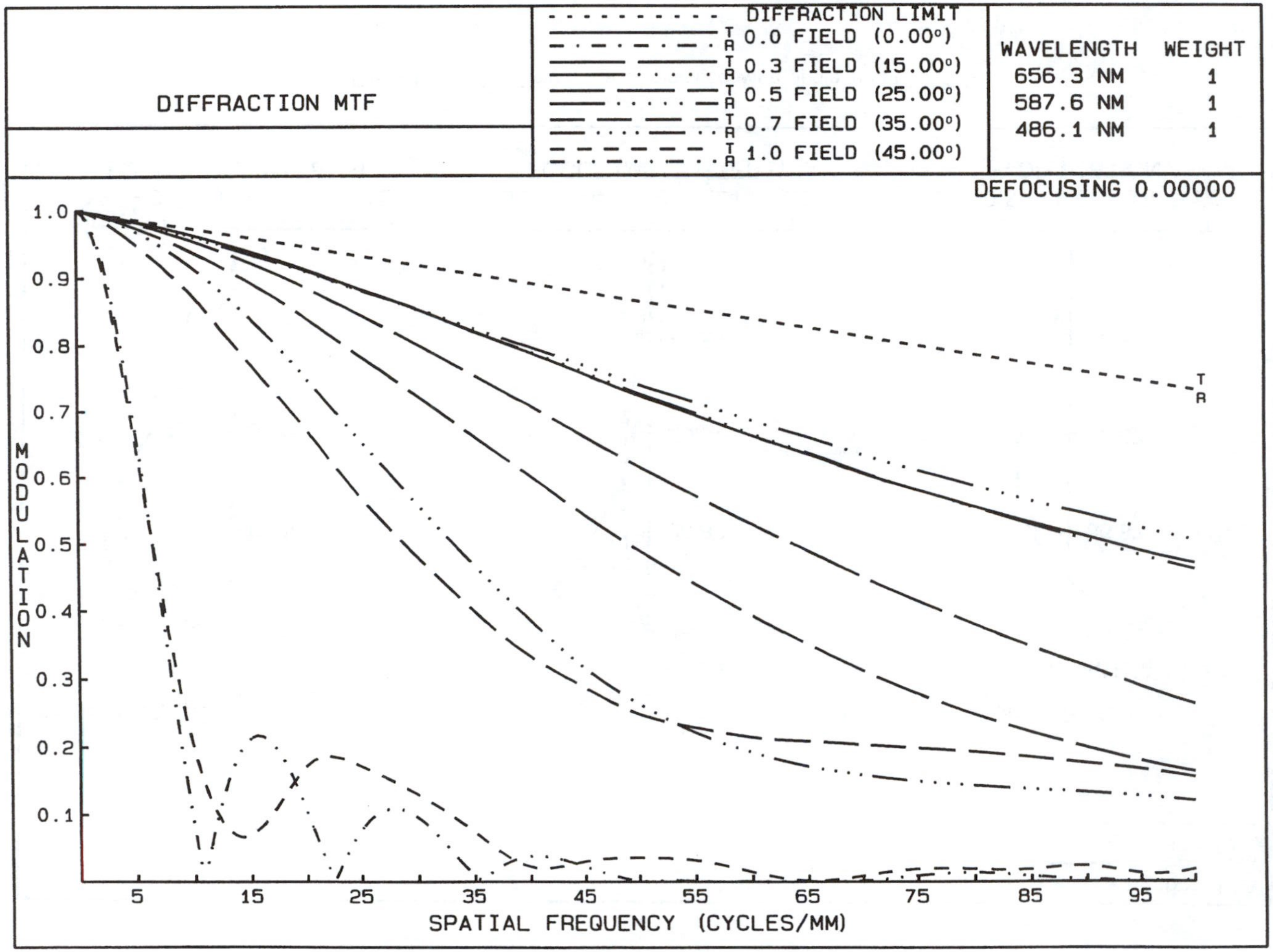

FIGURE 7.113 MTF plots of modified wide-angle lens (CODE V).

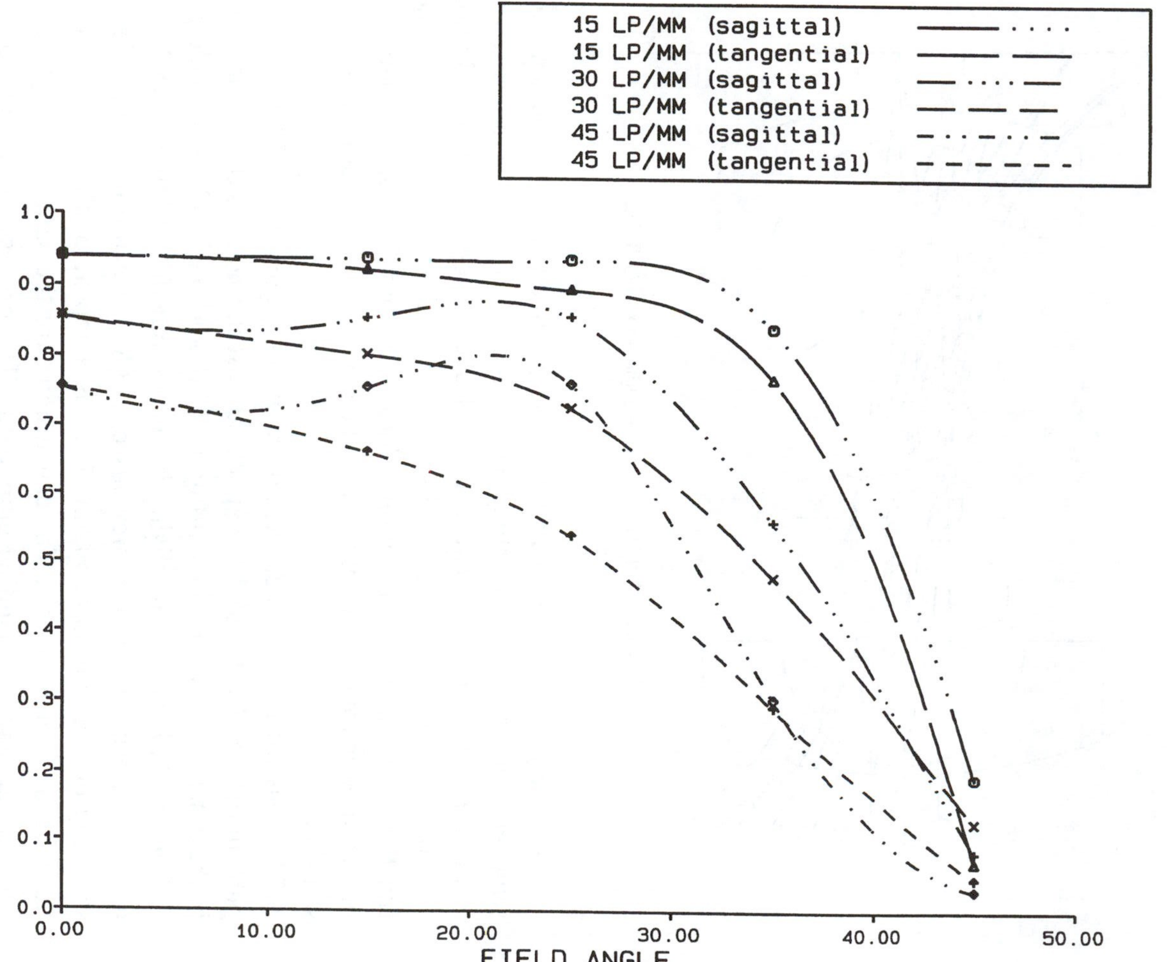

FIGURE 7.114 MTF versus field for modified wide-angle lens (CODE V).

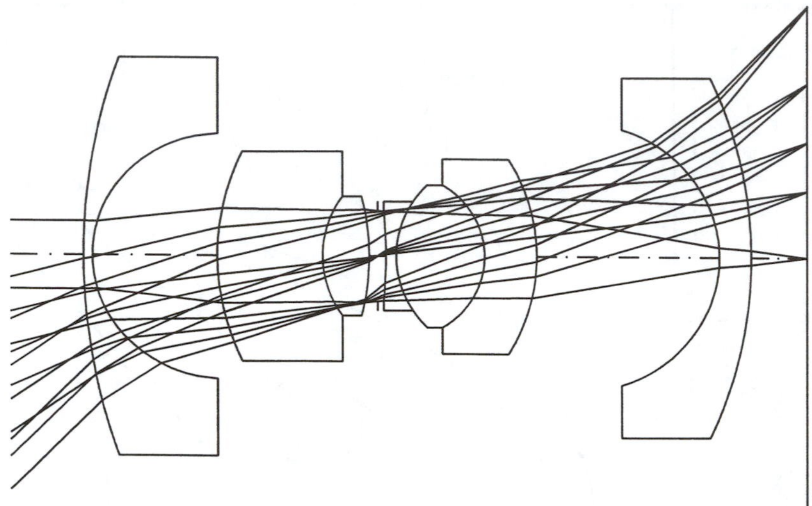

FIGURE 7.115 New wide-angle lens (CODE V).

Alternately, the designer can freeze the parameters of the element to be removed, successively converting the element into a zero-thickness, zero-power element. This is also effective, and leaves the designer some additional control on removing the element. In some cases this is preferable, as the merit function may need some alteration to locate an equivalent design region. This procedure was followed to encourage the lens to move into a good region of solution.

This lens with one element less is shown in Figure 7.115. The aberration plots in Figures 7.116 and 7.117 appear almost the same as the previous stage in the design. The MTF plots in Figure 7.118 and the MTF versus field plots in Figure 7.119 show some slight differences, but the design is probably just about as acceptable as the previous one. (As an example of the effect of sampling on computation, Figure 7.120 is a repeat of the previous figure, but with more field points. The conclusions are the same, but slight differences do appear by including more data.) The process so far has led to an improved design with one fewer element. The lens configuration is shown in Figure 7.121.

This design has been stopped at this point, as it suffices to show how modern lens design programs are capable of optimizing a starting point once the goal is set and appears to be accessible. Any further improvements are up to the reader to attempt. There are a number of things that can be looked at as possible design improvements. The element thicknesses and spacings can be varied. The glass types were not changed, but could be used as variables. It is clear that the goal for improvement needs to be defined. Continuing to iterate without a goal for improvement would be rather fruitless.

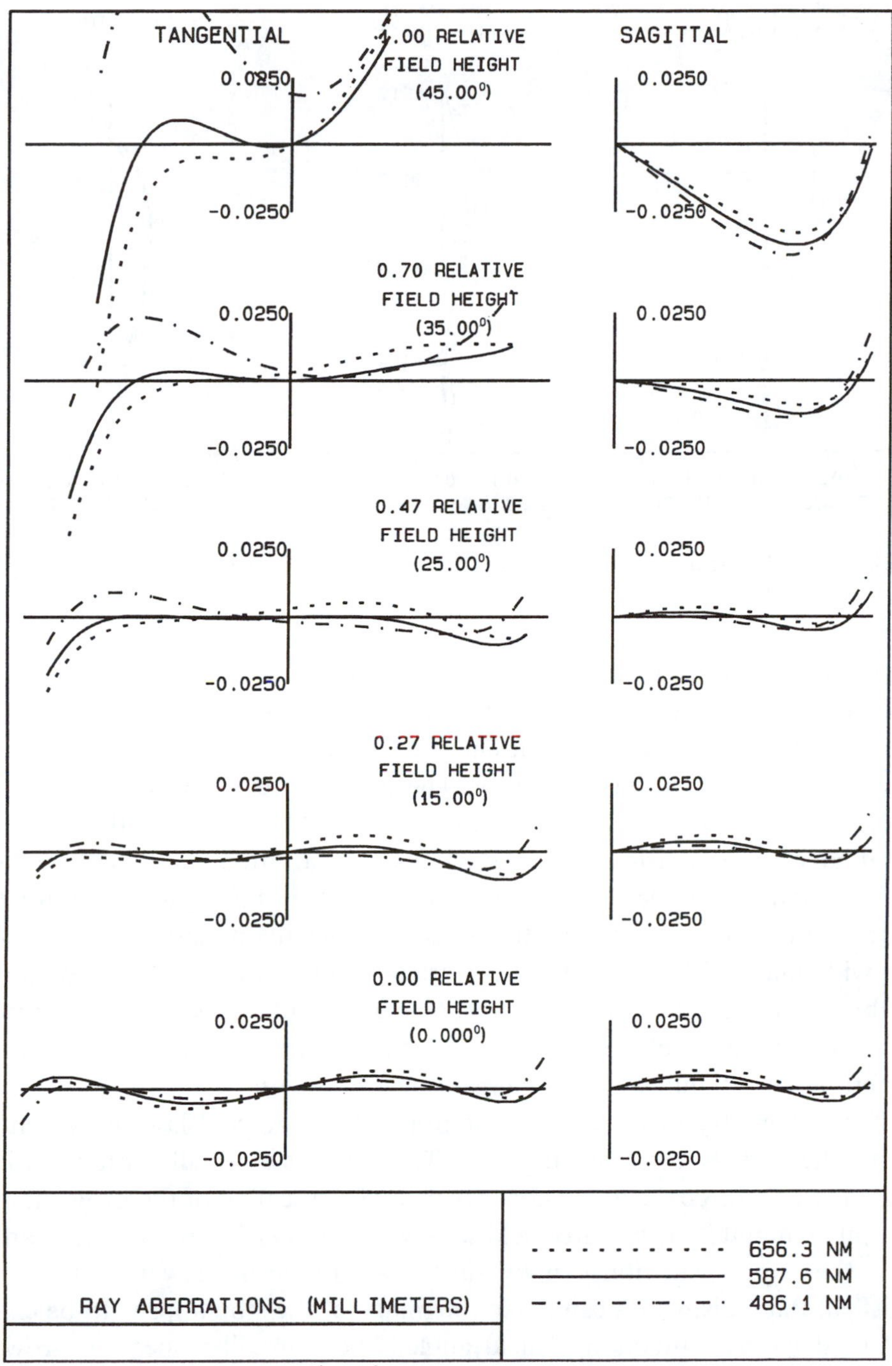

FIGURE 7.116 Ray intercept plots of new wide-angle lens (CODE V).

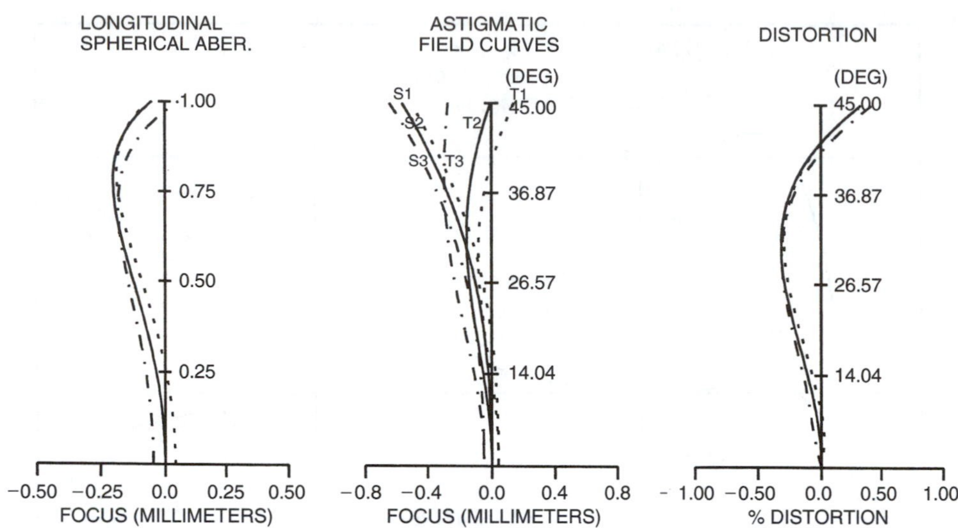

FIGURE 7.117 Field plots for new wide-angle lens: ······, 656.3 nm; ——, 587.3 nm; ·—·—·, 486.1 nm (CODE V).

Improvement of the MTF at the edge of the field is a good goal. Making the elements easier to fabricate and assemble would also be a worthy goal.

It should be obvious to the designer that there are many lenses that will have almost identical image quality. The art is in selecting the most useful design out of this large set of possible designs. The values of the figure of merit have not been listed here. There is a small variation for each of the final designs, and the last lens, with one last element, has a higher value for the figure of merit. Comparison on the basis of image quality indicates that there is a wide range of lenses that are essentially equivalent. The best design will be the one that can be fabricated most efficiently and inexpensively. For most real-world design problems, a few percentage points lost on MTF is usually preferable to a significant cost increase in production.

The radiometry of all lenses is dominated by the $\cos^4$ law for variation across the field, as noted in Chapter 2. This effect is especially noted in wide-angle lenses. The $\cos^4$ law is a consequence of the scaling of the projection of a pupil illuminated by a source field satisfying Lambert's law. In a wide-angle lens, the pupil distortion can be such that the cosine law is not exactly satisfied. Pupil distortion can make the image of the stop appear to be changed in shape as a function of field angle. The stop will appear to grow in diameter with field angle, instead of decreasing in diameter as the anamorphic pupil appearance would suggest. If this is the case, the solid angle subtended by the entrance pupil at the object will be larger than that determined by the $\cos^4$ law, and more energy will be collected. Therefore the amount of irradiance on the focal surface will be greater than that predicted

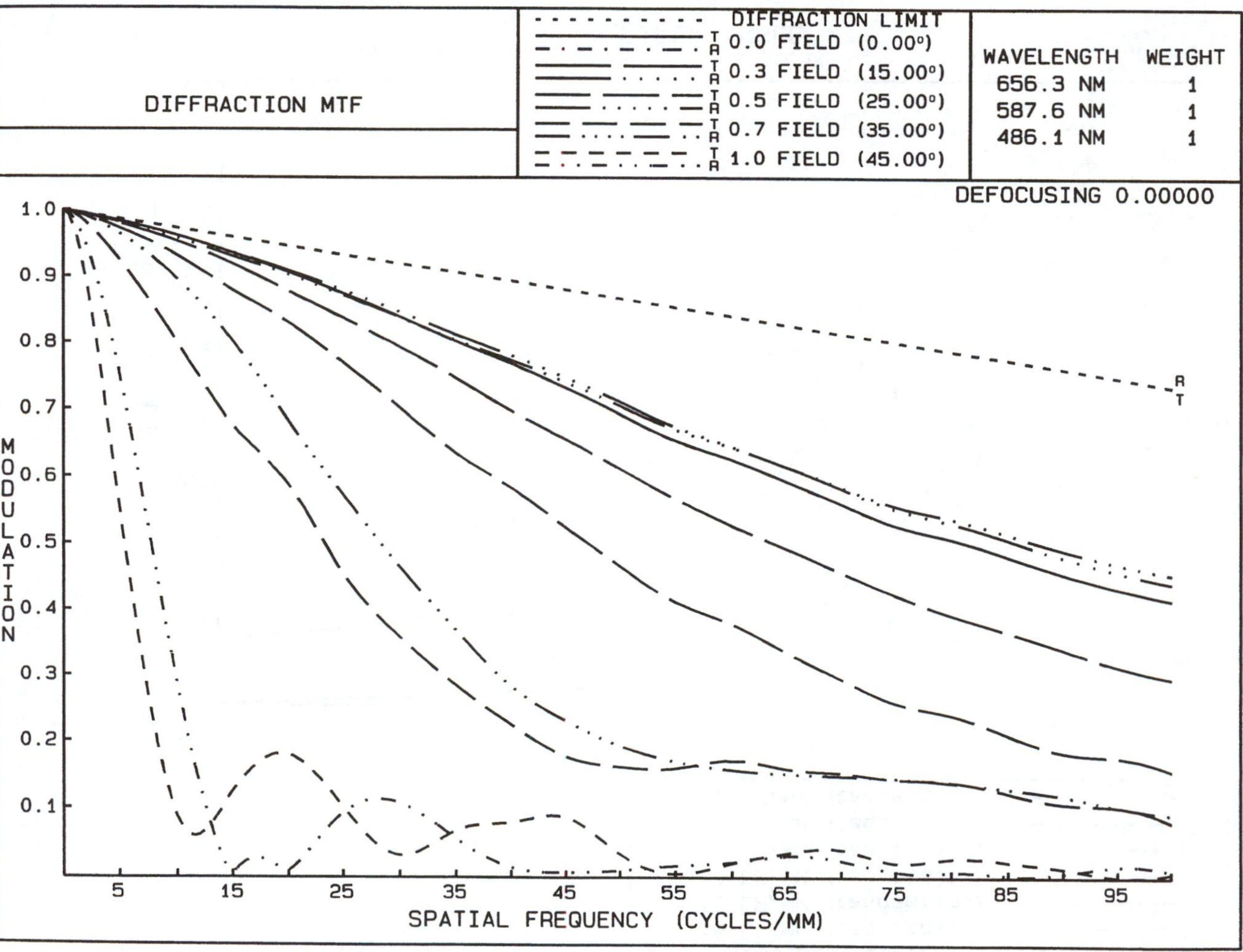

FIGURE 7.118 MTF plots for new wide-angle lens (CODE V).

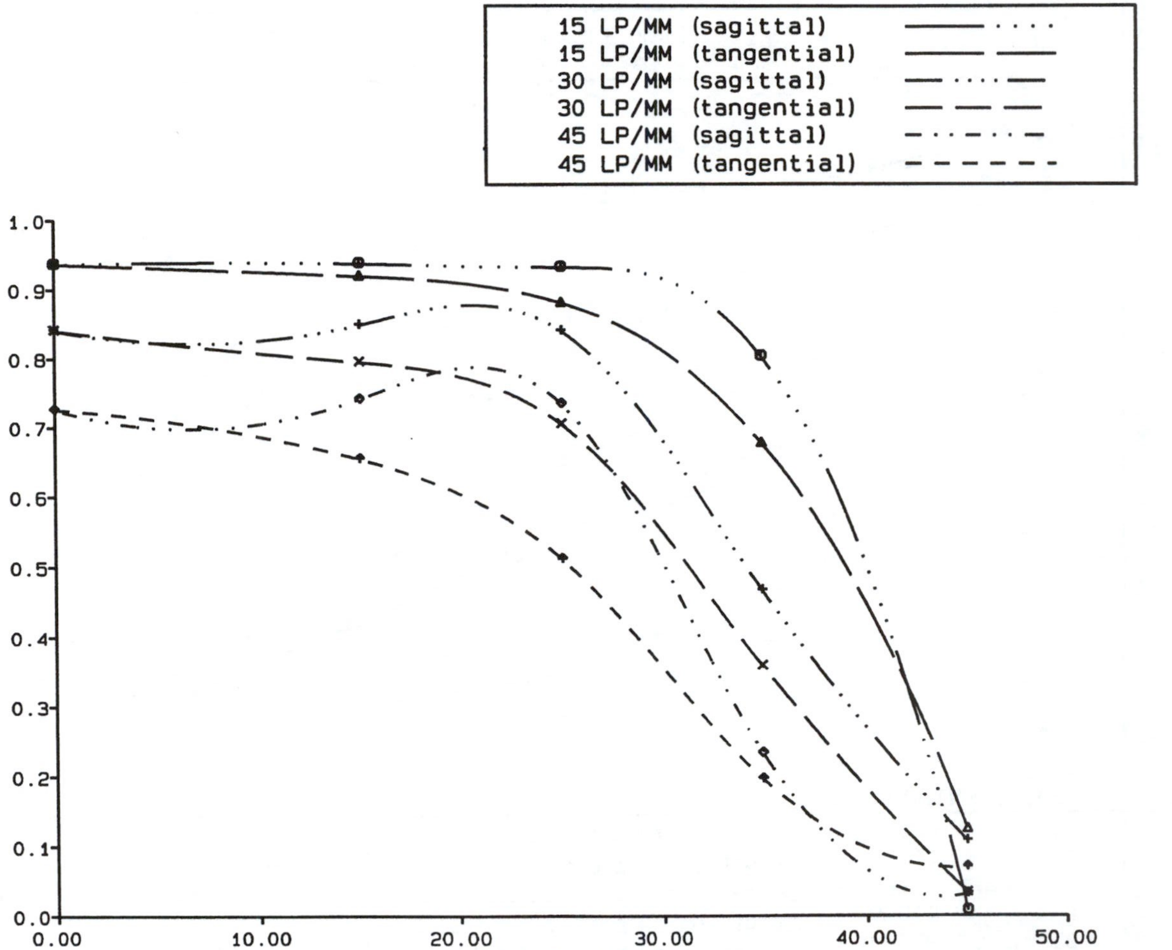

FIGURE 7.119 MTF versus field for new wide-angle lens (CODE V).

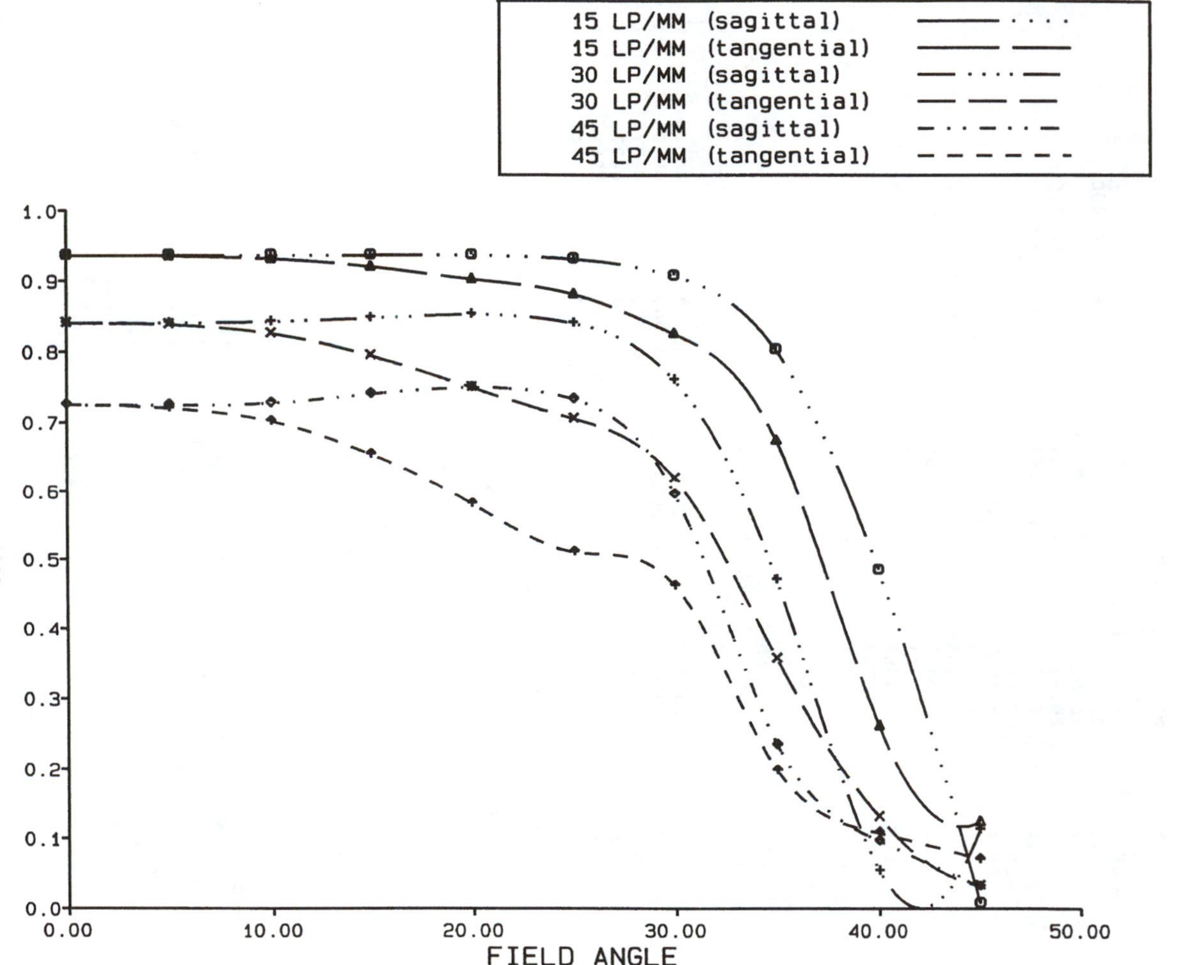

FIGURE 7.120 Replot of previous figure with more data points (CODE V).

```
   Modified Wide Angle Lens
               RDY              THI               GLA
OBJ:         INFINITY         INFINITY
  1:         51.57704         0.735453        503800.667000
  2:         10.50089        10.000000
  3:         20.88957         8.674015        720500.503000
  4:          8.66623         3.936835        607390.510000
  5:        -20.17637         0.692191
STO:         INFINITY         0.692191
  7:        -39.67794         0.843607        569930.575000
  8:          8.49533         7.419419        625000.590000
  9:         -6.98125         4.239668        719660.293000
 10:        -16.07213        15.055149
 11:        -11.40586         2.703870        642000.581000
 12:        -35.63773         4.732472
IMG:         INFINITY        -0.161732

SPECIFICATION DATA
 EPD          6.27298
 DIM               MM
 WL            656.30     587.60     486.10
 REF                2
 WTW                1          1          1
 XAN          0.00000      0.00000      0.00000      0.00000      0.00000
 YAN          0.00000     15.00000     25.00000     35.00000     45.00000
 VUX          0.07416      0.06868      0.05880      0.04374      0.02347
 VLX          0.07416      0.06868      0.05880      0.04374      0.02347
 VUY          0.07416      0.07130      0.09132      0.09515      0.55582
 VLY          0.07416      0.10175      0.09868      0.11241      0.11853

APERTURE DATA/EDGE DEFINITIONS
 CA
 CIR S1                    15.767513
 CIR S2                    13.114876
 CIR S3                     8.105132
 CIR S4                     4.460917
 CIR S5                     3.957661
 CIR S7                     3.852291
 CIR S8                     4.046926
 CIR S9                     5.381007
 CIR S10                    7.487670
 CIR S11                   10.913577
 CIR S12                   14.197146

REFRACTIVE INDICES
  GLASS CODE                      656.30       587.60       486.10
 503800.667000                  1.501468     1.503799     1.509024
 720500.503000                  1.716198     1.720497     1.730528
 607390.510000                  1.603809     1.607388     1.615724
 569930.575000                  1.566918     1.569928     1.576834
 625000.590000                  1.621773     1.624998     1.632370
 719660.293000                  1.712541     1.719655     1.737113
 642000.581000                  1.638639     1.641998     1.649693

INFINITE CONJUGATES
 EFL          21.4997
 BFL           4.7325
 FFL         -10.0407
 FNO           3.4273
 IMG DIS       4.5707
 OAL          54.9924
 PARAXIAL IMAGE
  HT          21.4997
 ANG          45.0000
 ENTRANCE PUPIL
  DIA          6.2730
  THI         12.3194
 EXIT PUPIL
  DIA          6.0316
  THI        -15.9399
```

FIGURE 7.121 Program listing of design data for new wide-angle lens (CODE V).

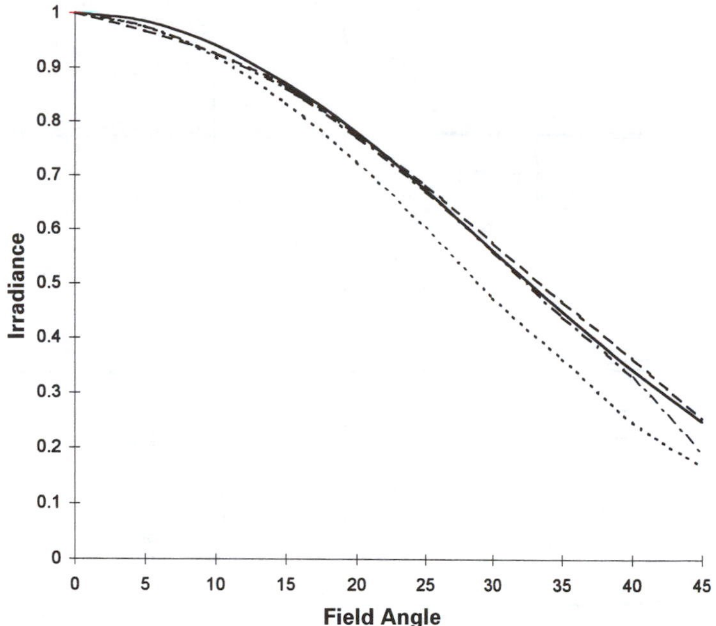

FIGURE 7.122 Plot of image irradiance versus field: ——, $\cos^4$; - - -, Angulon; ······, Hypergon; ·—·—·, modified.

by the $\cos^4$ law. If this pupil distortion exists, it can serve as a compensator for losses due to vignetting as well.

Figure 7.122 shows a plot of the image plane relative irradiance as a function of field angle that was computed during the computation of the MTF. It is evident that the Angulon and modified Angulon lenses show this effect, which offsets, at least, the vignetting losses. The Hypergon design does not show this effect, and shows irradiance below that of the $\cos^4$ law. The effect can be examined by looking at the ray intercept plots made with respect to the entrance pupil coordinates rather than the coordinates on the aperture stop, as in Figure 7.123. The ray plots cover the entire height of the pupil, compared with the vignetting losses shown in the original plots in Figure 7.100. This effect is relatively small for this lens, but is essential in the design of extremely wide-field lenses. The radiometry conditions are an important part of evaluating the design of wide-angle objectives.

The effect of pupil shape on the irradiance is dramatically shown in a design by Roosinov intended to cover a 113° total field of view. The lens is shown in Figure 7.124, in which the extreme pupil shift with field can be observed by following the direction of the incoming and exiting chief rays to their intersection with the axis. Figure 7.125 is a plot of the pupils relative to a fixed circular stop shape. The off-axis bundles actually appear larger due to the shift of the pupils. Figure 7.126 plots the relative irradiance across the

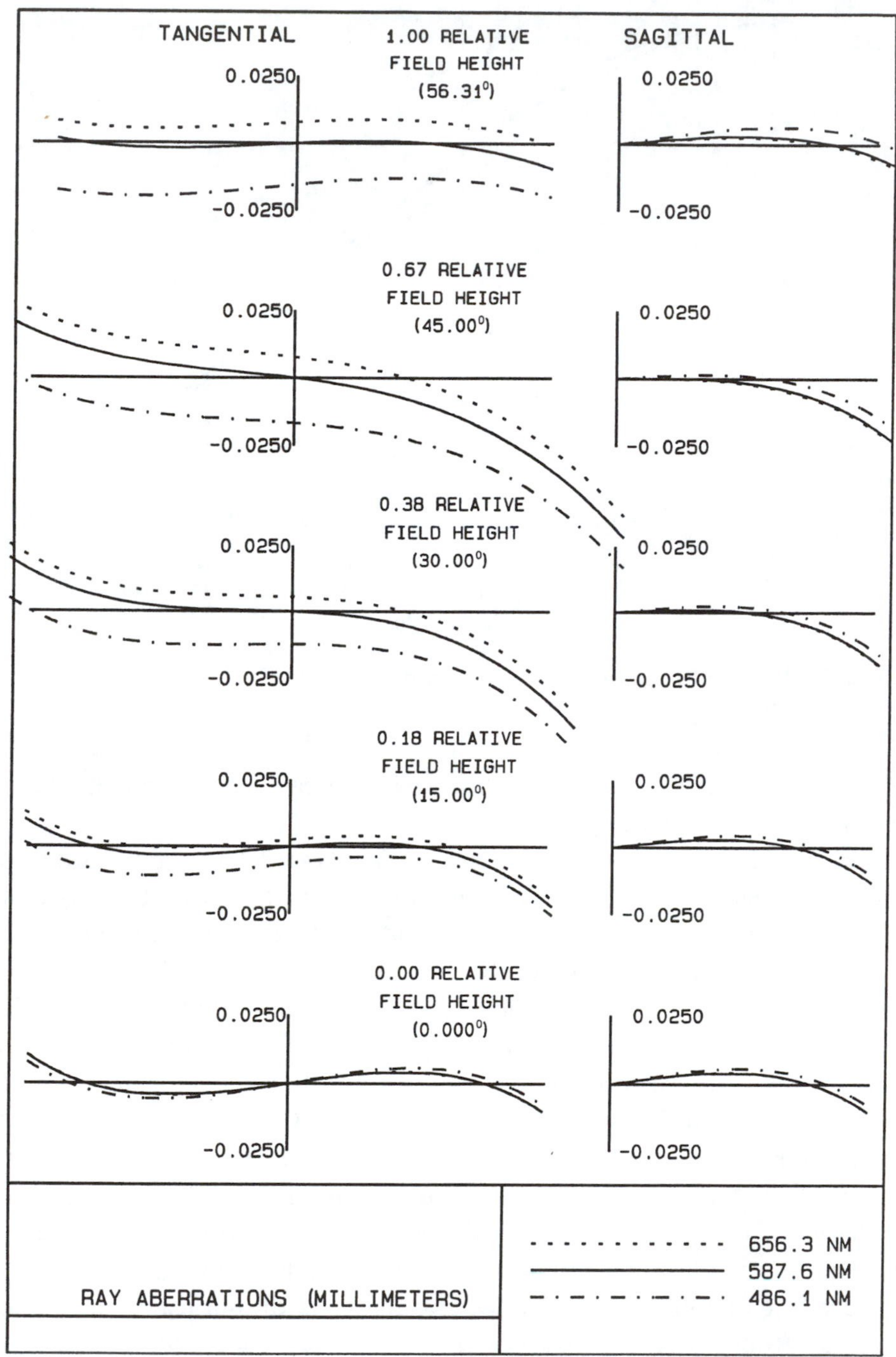

FIGURE 7.123 Ray intercept plot relative to entrance pupil coordinates (CODE V).

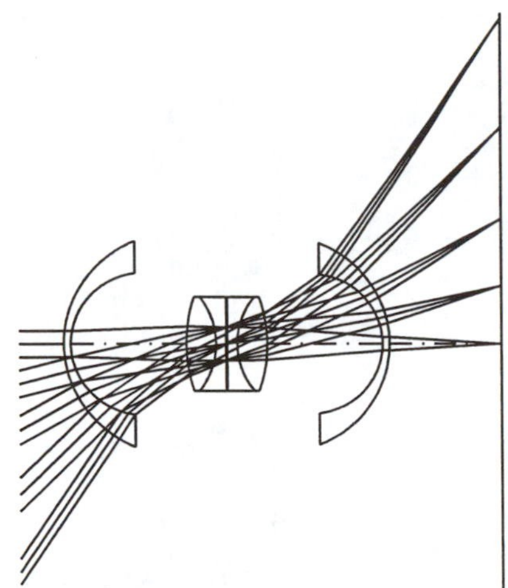

FIGURE 7.124 Example of 113° field lens (CODE V).

field compared with the $\cos^4$ law. It is obvious that vignetting is not a useful tool in controlling aberrations for very-wide-angle systems. Calculation of the illumination in wide-angle lenses is actually a delicate process, and the designer should be sure that the computation is carried out correctly. The calculation shown here is the geometrical optics effect and does not include the effect of reflection losses at coated or uncoated surfaces in the lens. This would require knowledge of the details of the angular reflection and transmission characteristics of the thin-film coatings on each surface of the lens.

7.1.5 ZOOM LENSES

Multiconfiguration optical systems have properties that change with the mode of operation of the lens system. Examples are zoom lenses, spectrometric devices, image splitters, and multiconjugate systems. A zoom lens provides images with different magnifications that vary with the setting of the element separations that serve as zoom parameters. These separations must be chosen so that the focus position of the image remains unchanged during the variation in magnification. A muticonjugate lens is a special case of a "zoom" designed to have specified imaging characteristics at a discrete number of positions, usually two or more different conjugates. The design of such systems is subject to the difficult limitation that the aberrations of any element vary with conjugate position. The art in designing a zoom lens involves the finding of a set of components that minimizes the variation of the aberrations. In addition, the aberrations, particularly the distortion, may be balanced across the zoom range so that, on average, the effects of the aberrations are minimized.

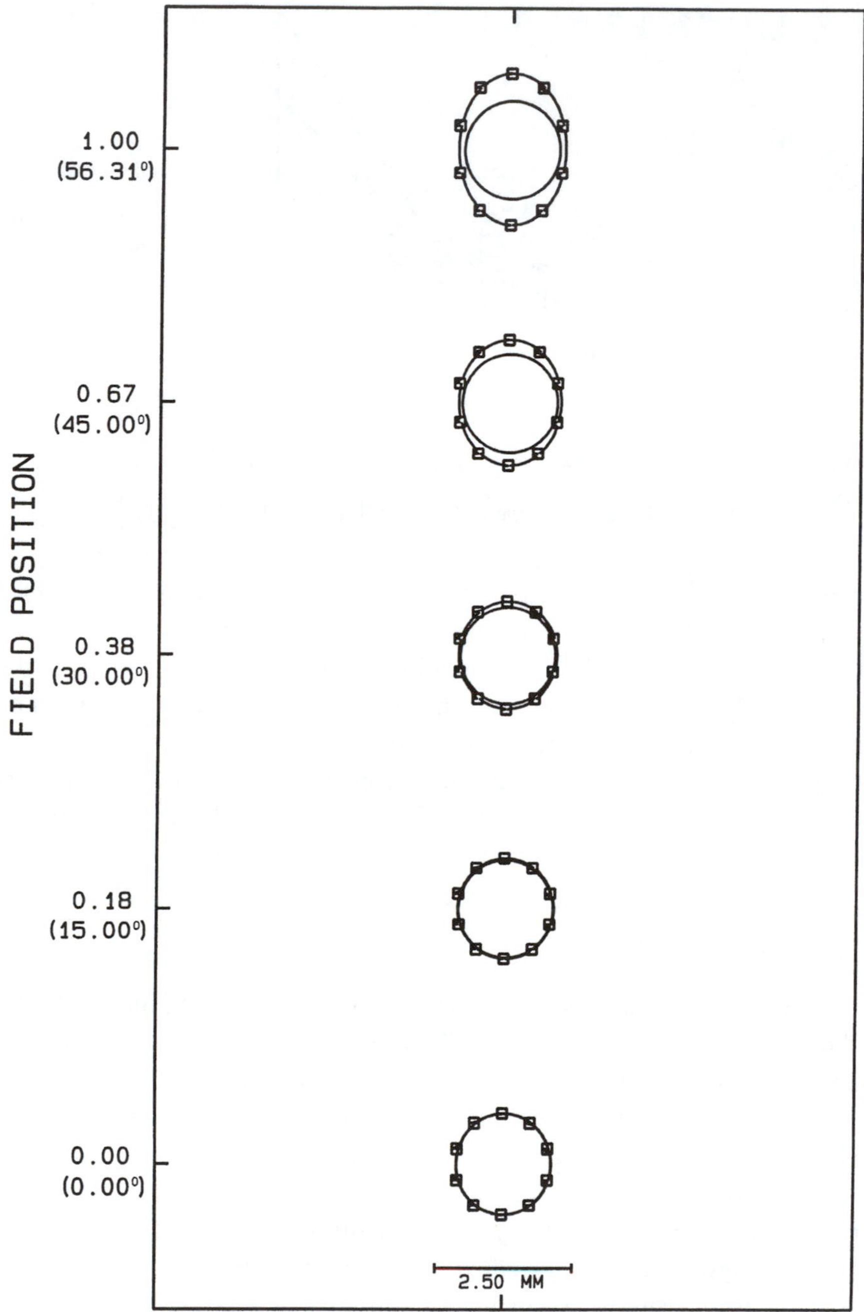

FIGURE 7.125 Aperture footprint plots for lens in previous figure (CODE V).

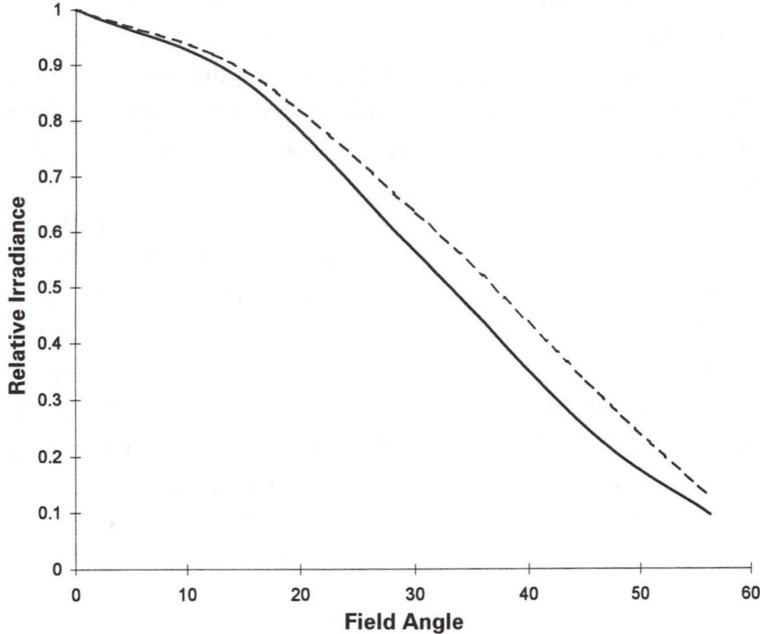

FIGURE 7.126 Plot of image irradiance versus field for Roosinov lens (– – –) compared with $\cos^4$ law (——).

The design of multiconfiguration lenses is accomplished by the simultaneous optimization over several configurations which contain common elements. A multiconfiguration optical system is described for the computer program by stating some specific parameters within the lens that can change under various conditions of use. For a zoom lens this would mean that certain spaces are designated as adjustments for the focus and magnification, but that the curvatures and indices are common to all configurations. In other cases, a lens might be required to operate at several magnifications with equal capability. The lens is specified as usual, but is treated by the design optimization as a set of lenses in which the optical parameters are identical, but the conjugate positions are different for the various configurations. The optimization proceeds by varying the lens parameters while attempting to simultaneously minimize the merit functions and meeting the required magnifications at each configuration.

Zoom lenses are the principal topic of this section. Some of the techniques that would apply to spectroscopic or multiconjugate systems can be inferred from this discussion. The use of zoom lenses has become ubiquitous in camera and video systems. There are many different requirements and configurations of zoom lenses that are required in modern imaging systems. A modest database of zoom lenses attached to the CODE V design program in early 1995 included over 400 different lenses. There are several hundred patents outstanding on zoom lenses. Each zoom design problem is different, with differing goals for image quality, weight, and cost. Therefore, this section can only cover the basic concepts and examine a few samples of such lenses. The design of these lenses once was a specialized task for a few designers. Today most designers are likely to encounter a variable magnification system at some time in their career.

Usually only a small number of configurations, or zoom positions, are chosen to be optimized, and it is presumed that the lens is well behaved for configurations that may lie between the stated configurations. For a zoom lens, the choice of the configurations to represent the entire range of magnifications is critical. Usually three magnifications are chosen, representing the extremes of the desired zoom range and the geometrical average of the magnification extremes of the range. After optimization of the design at the chosen magnifications, it is advisable to verify that no major surprises remain in the lens when the system is zoomed to other intermediate configurations.

As in design of lenses for a single configuration, there are physical limits about what can be accomplished with a given number of components, or with a stated type of lens. Any multiconfiguration optical system is subject to physical laws, and the optimization process must frequently be probed to determine the extent of image quality that is possible. Additional limits upon the allowable changes in the configuration that are imposed by the application must also be considered.

Zoom lenses are special cases of multiconfiguration optical systems that change focal length or scale by motion of some elements in the lens. Usually the motions used are of axial separations of two or more components. (Some zoom systems may involve reflective elements that may also require tilting and decentering in order to accomplish the imaging goals.) The magnification or focal length change is accomplished continuously by coupled motions of lens elements, although in some cases insertion or removal of additional elements is used to change the range of magnification in a discrete manner. The targets in the design are acceptable image quality with a specified magnification and focal position for each configuration of the lens.

The starting point for a zoom lens design is usually the determination of the first-order properties of the lens. There are several approaches that can be used, some of which are analytic, but most are numerical. The basic determination of the powers and motions of the components requires the expression of the components as effective power elements or "thin" lenses, and the successive application of the thin-lens paraxial ray trace. A generic example will be used to illustrate the procedure.

Two moving components are required, at least. A general concept is that the position of one component sets the magnification, and the position of the second element is used to stabilize the focal position. There is usually a very

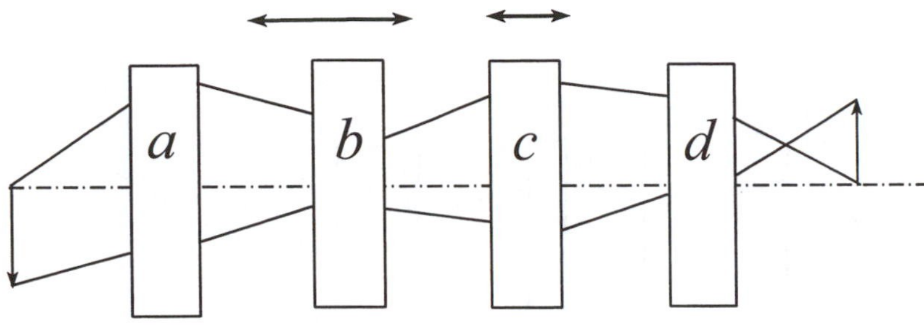

a = Object Fixing Lens
b = Magnification Adjusting Lens
c = Focus Position Adjustment Lens
d = Image Fixing Lens

NOTE: Functions of lenses may be interchanged to meet system requirements

FIGURE 7.127 Concept of mechanically compensated zoom lens: *a*, object-fixing lens; *b*, magnification adjustment lens; *c*, focus position adjustment lens; *d*, image-fixing lens.

nonlinear relationship between the motions of the two components. The starting point for a two-component zoom requires four powers and positions be specified. The outer two of these components set the object and image conjugate positions. The separation between the outer components establishes the space within which the two zooming components will move. The powers of the outer components can be adjusted to place the conjugate positions viewed by the zooming components at convenient locations. In general, the greater the powers of the components in the moving pair, the greater will be the accessible zoom range for the system.

The relative aperture, or F/number, of the system will be determined by the aperture stop and the magnification. It is possible, in fact likely, that the F/number will be permitted to change during the zoom, with electronic control of exposure adjusting the image brightness. In most cases, the issues of weight, space, complexity, and cost are very important boundary conditions.

A generic zoom lens will contain four types of components, as shown in Figure 7.127. In this figure, the outer lenses serve to place the optical location of the object and image at convenient locations for the zoom elements, *b* and *c*, to adjust the focus and magnification. These outer elements also define the physical space available for the motion of the zooming components. Element *b* is moved to change the magnification, and element *c* moves with a nonlinear motion relative to element *b* to adjust the focus position to keep the image at the fixed location. In the design, both lenses must move in a coupled manner, and the assignment of magnification fixer can be attached to either component. If it is desired to keep the image-side F/number fixed, the stop is located at element *d*. If the numerical aperture on the object side needs to be maintained, the stop will be attached to element *a*.

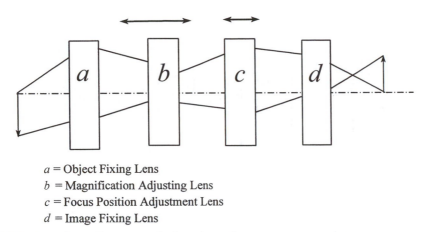

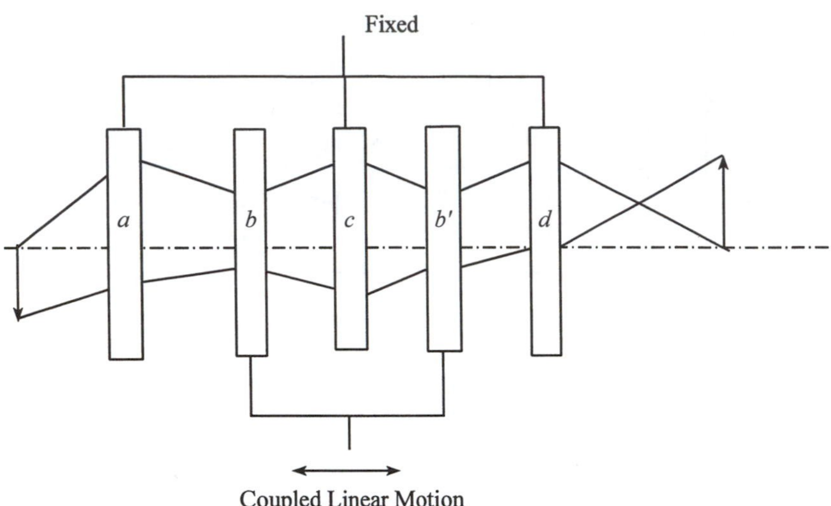

FIGURE 7.128 Concept of optically compensated zoom lens.

Actually, any of these elements could be used for any of the functions. For example, element *c* could be fixed in position with the stop attached, and element *d* could be adjusted to maintain the focus position. In this case the image-side F/number would change with magnification. This could be an advantage, as a progressive reduction in the relative aperture at long focal lengths could reduce the diameter requirements for elements *a* and *b*, and thus reduce the weight of the entire lens. Fixing the aperture stop at element *d* places the exit pupil at a fixed location, with the entrance pupil size and location being determined by the first-order properties of the front three components. An additional zooming component could be used to stabilize the location of the entrance pupil if required.

Additional functions may be included in a zoom lens system, such as a macro-close focusing capability, because any of the components may be used to extend the focusing range. All of these functions are subject to the first-order optics that determines image position and magnification. There are, of course, additional restrictions regarding the aberrations and change of aberrations with zoom functions that will be discussed shortly. The mechanical realization of a zoom lens of this type requires that a nonlinear cam arrangement is required to drive the two components relative to each other. This will generally require a complex mechanical drive to be used. In some cases, a pair of electronically related servo motors can be used.

An important variation of the basic zoom lens approach described in Figure 7.127 is an optically corrected zoom, as illustrated in Figure 7.128. The addition of a third component to the zooming pair permits locating a variable magnification solution in which the two lenses, *b* and *b*′, are linked

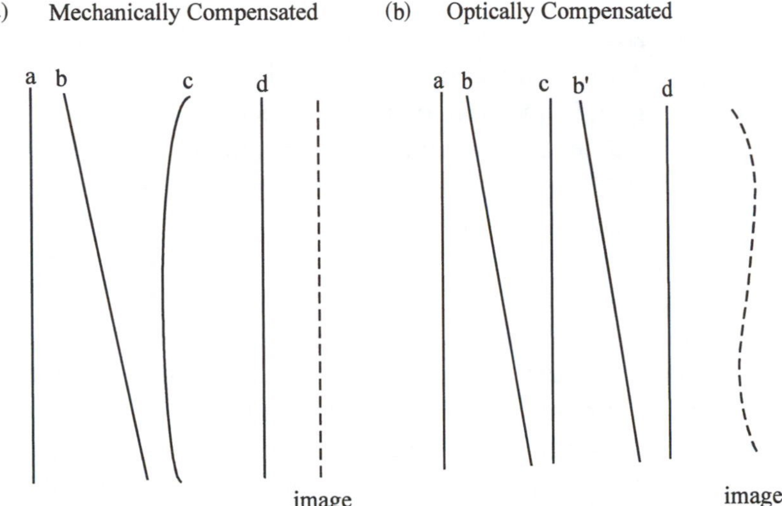

FIGURE 7.129 Lens motions and focal position for (a) mechanically compensated and (b) optically compensated zoom lenses.

to move together as a unit around the central lens which has the opposite power. Solutions can be found in which the image focus position is approximately fixed, while the magnification changes with the joint fixed relation motion of the lenses. This solution is subject to the general condition that the number of crossing points of the image focus at the defined image location are set by the number of components used in the zoom portion of the lens. The use of three zoom components as shown leads to a solution in which the focus position moves about the desired location with three crossing points. Figure 7.129 illustrates the concept for the two approaches. The mechanically or cam-compensated lens diagram of lens positions, on the left, moves lenses b and c to stabilize the image position. The optically compensated situation is shown on the right, in which the components b and b' move together, and the image position is stabilized within some tolerance that is established by the first-order optics of the zoom lens system. The addition of more lenses to the zooming package increases the number of crossing points, and consequently provides better focus compensation over a wider zoom range. There are obvious optical and complexity tradeoffs relative to the choice of mechanically versus optically compensated zoom systems. Modern manufacturing techniques permit the fabrication of quite nonlinear cams with little difficulty, so that the better focus control, smaller number of required elements and potentially smaller packaging of a lens of given zoom ratio of the mechanically compensated zoom lens is generally preferred.

The method of evaluating the properties of a zoom lens can best be demonstrated by carrying out an example. The lens chosen as an example is shown in Figure 7.130 and is taken from the lens library in CODE V. The lens is a 4× zoom, focal length 9–36 mm, F/2.0 for a 14 mm field diameter. The lens is large compared to the focal length, with the overall length approximately 3.5 times the longest focal length. The stop is located at the rear group in order to maintain a fixed F/number for the camera the lens is intended for. In Figure 7.130(b), the moving component group *b* is the fourth through sixth element, and the *c* moving component is the seventh element from the front of the lens. The stop is located more or less in the middle of the fixed rear-elements group.

The MTF plots for this lens are presented in Figure 7.131. They are for three zoom positions with focal positions fixed as stated in the lens specifications. The interpretation of these plots is that the MTF is fairly consistent for the three zoom positions, with a field maximum changing from 21° half angle at the short focal length to 5.8° at the long focal length. Because the final F/number is maintained constant, the entrance pupil diameter increases by a factor of four in zooming across the range. The lens is actually a bit better at the extremes of the zoom range than at the center, but, as noted below, this might be improved by an adjustment in the motions of the elements. In general, this lens would provide from 700 to 1500 resolvable elements across the field of view when operated at the F/2 aperture. This is a good match for most detectors of this size, and although we do not know what precise specifications this lens was designed to meet, the results are reasonable.

The MTF plots are of interest to the system user, but the lens designer can obtain a better appreciation of the correction in this lens by examining ray intercept plots, as presented in Figure 7.132. Examination of the ray plots shows a reasonably consistent balance of aberration across the zoom range. There is some coma, which changes sign across the zoom range. The ray plots of the center focal length indicate that a probable improvement in the focus could be chosen. Since this is a zoom lens, any alteration in the focus position needs to be made by adjusting the motion of the compensating lens in the zooming pair. The designer has used vignetting to limit the growth of the oblique spherical aberration in the lens. The vignetting effect has also limited the effect of the change in astigmatism and coma with zoom setting.

Another view of the aberrations can be seen in the field plots shown in Figure 7.133. The spherical aberration and astigmatic field plots are more or less constant across the zoom range. The distortion, however, makes a consistent change from negative to positive across the zoom. (Note that the plot scale changes in the three distortion plots.) This characteristic is very common in zoom lenses. Attempts to eliminate the distortion change in zoom lenses usually are not successful. The explanation for the distortion change is related to the natural change of spherical aberration with conjugate position of the zoomed elements, which are located well away from the stop position.

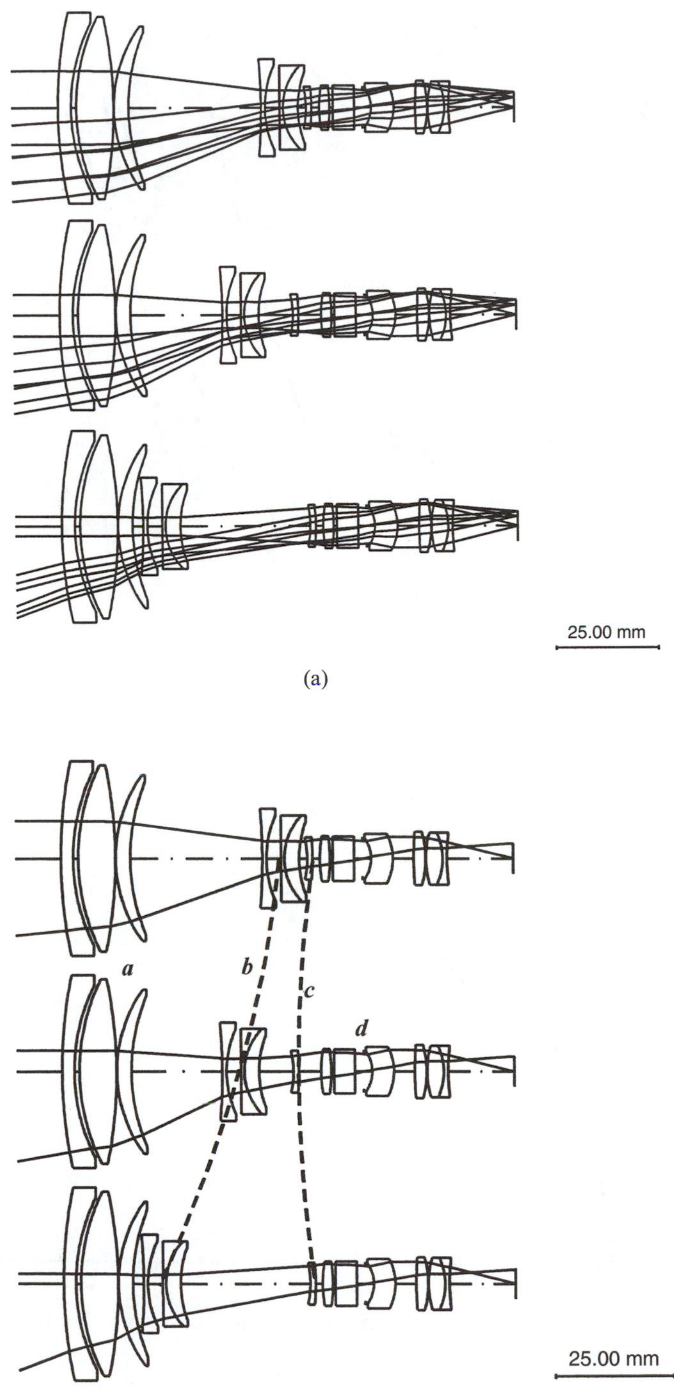

FIGURE 7.130 Sample zoom lens. (a) Layout and ray path; (b) motions of zooming components (CODE V).

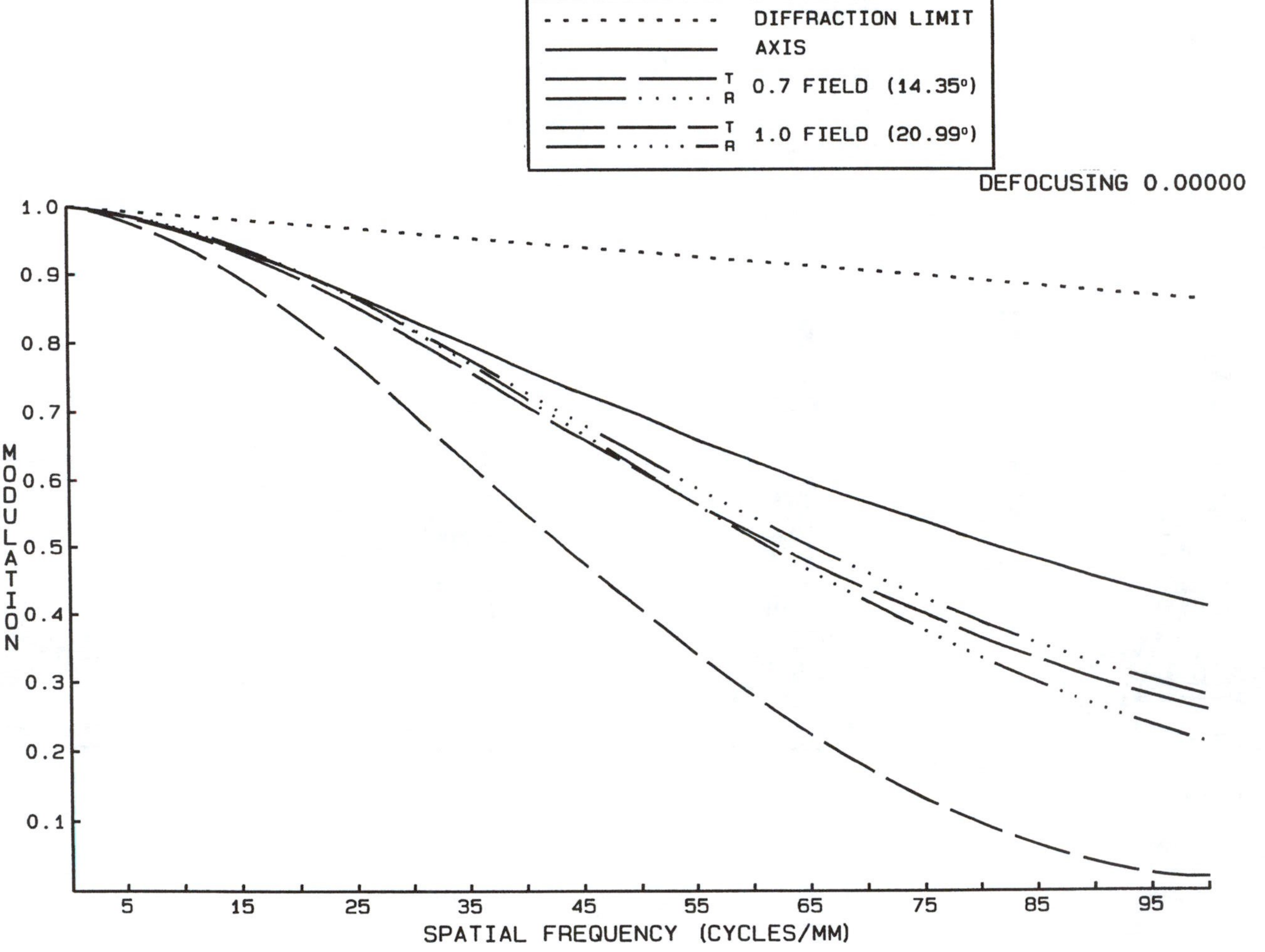
DIFFRACTION LIMIT
AXIS
T
R 0.7 FIELD (14.35°)
T
R 1.0 FIELD (20.99°)
DEFOCUSING 0.00000
MODULATION
1.0
0.9
0.8
0.7
0.6
0.5
0.4
0.3
0.2
0.1
5
15
25
35
45
55
65
75
85
95
SPATIAL FREQUENCY (CYCLES/MM)

(a)

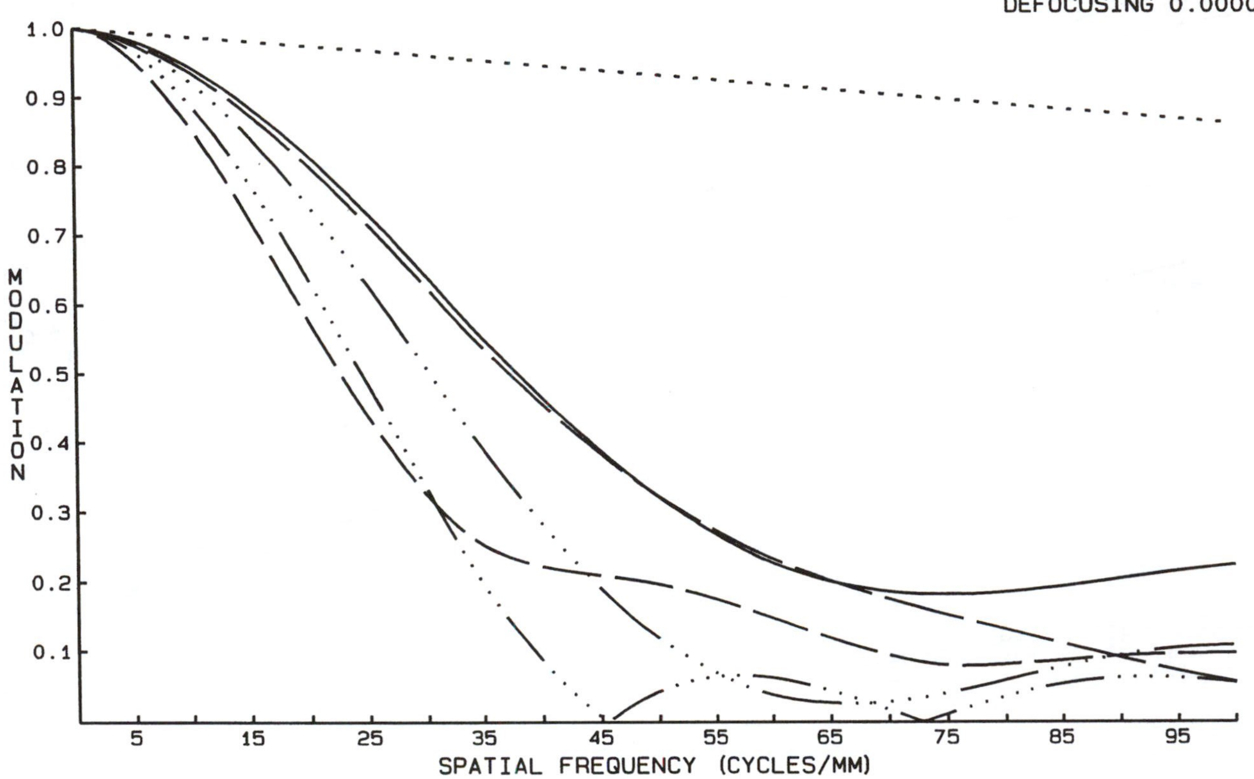
DEFOCUSING 0.00000
MODULATION
1.0
0.9
0.8
0.7
0.6
0.5
0.4
0.3
0.2
0.1
5
15
25
35
45
55
65
75
85
95
SPATIAL FREQUENCY (CYCLES/MM)

(b)

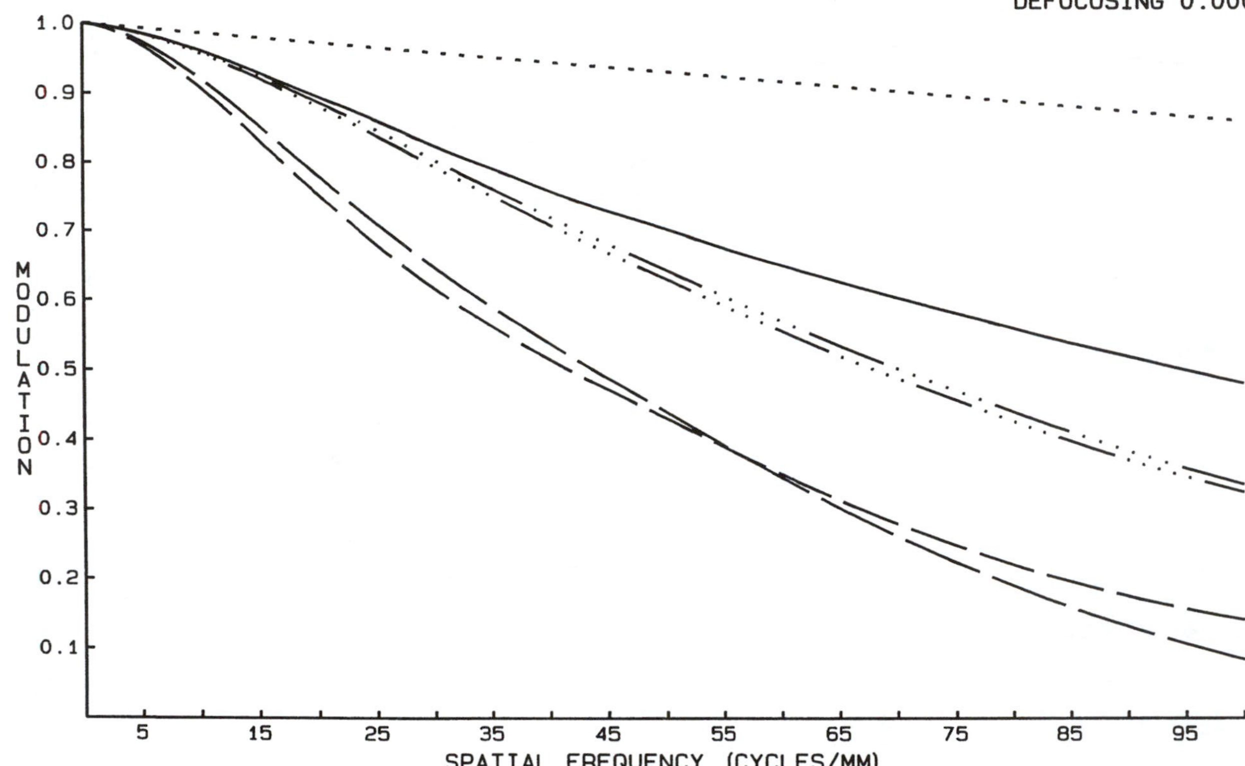

(c)

FIGURE 7.131 MTF plots for sample zoom lens. (a) Short focal length; (b) medium focal length; (c) long focal length (CODE V).

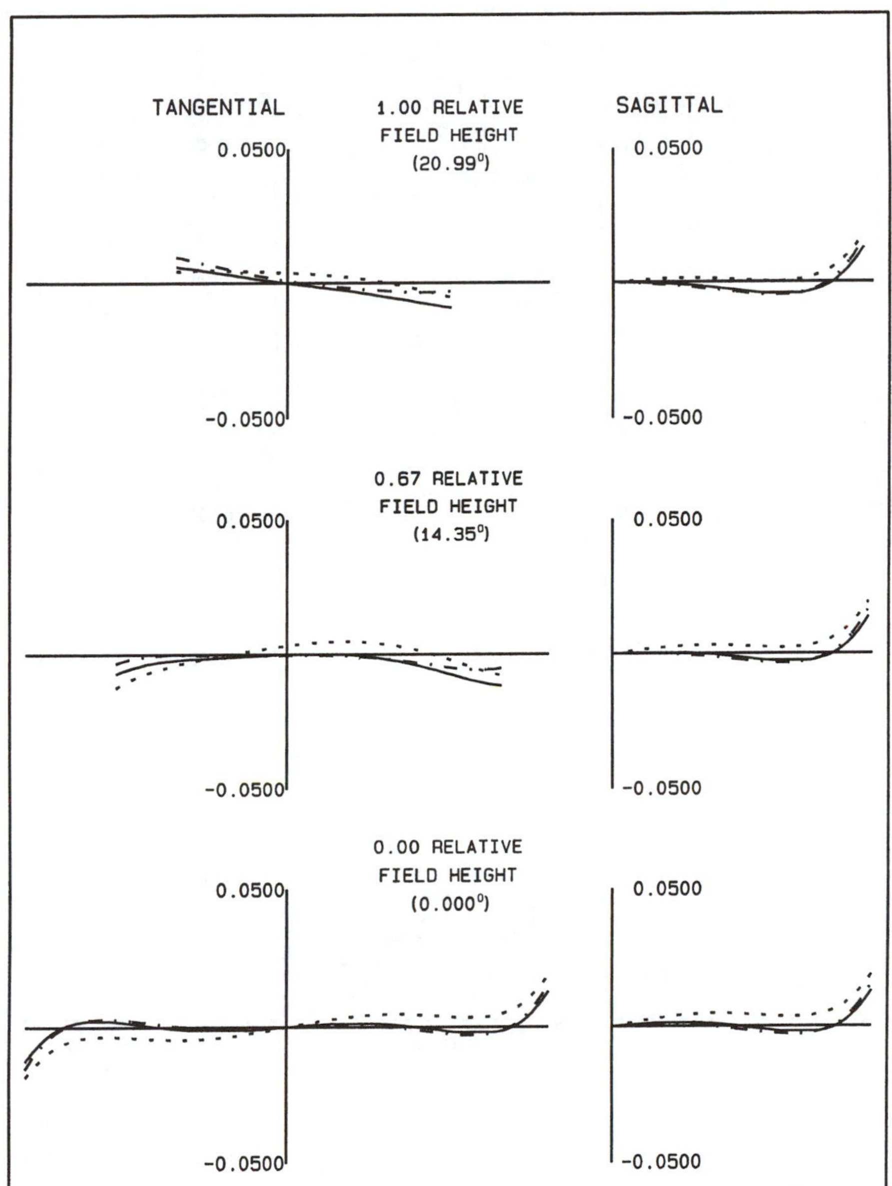
TANGENTIAL
1.00 RELATIVE
FIELD HEIGHT
(20.99°)
SAGITTAL
0.0500
-0.0500
0.0500
-0.0500
0.67 RELATIVE
FIELD HEIGHT
(14.35°)
0.0500
-0.0500
0.0500
-0.0500
0.00 RELATIVE
FIELD HEIGHT
(0.000°)
0.0500
-0.0500
0.0500
-0.0500

(a)

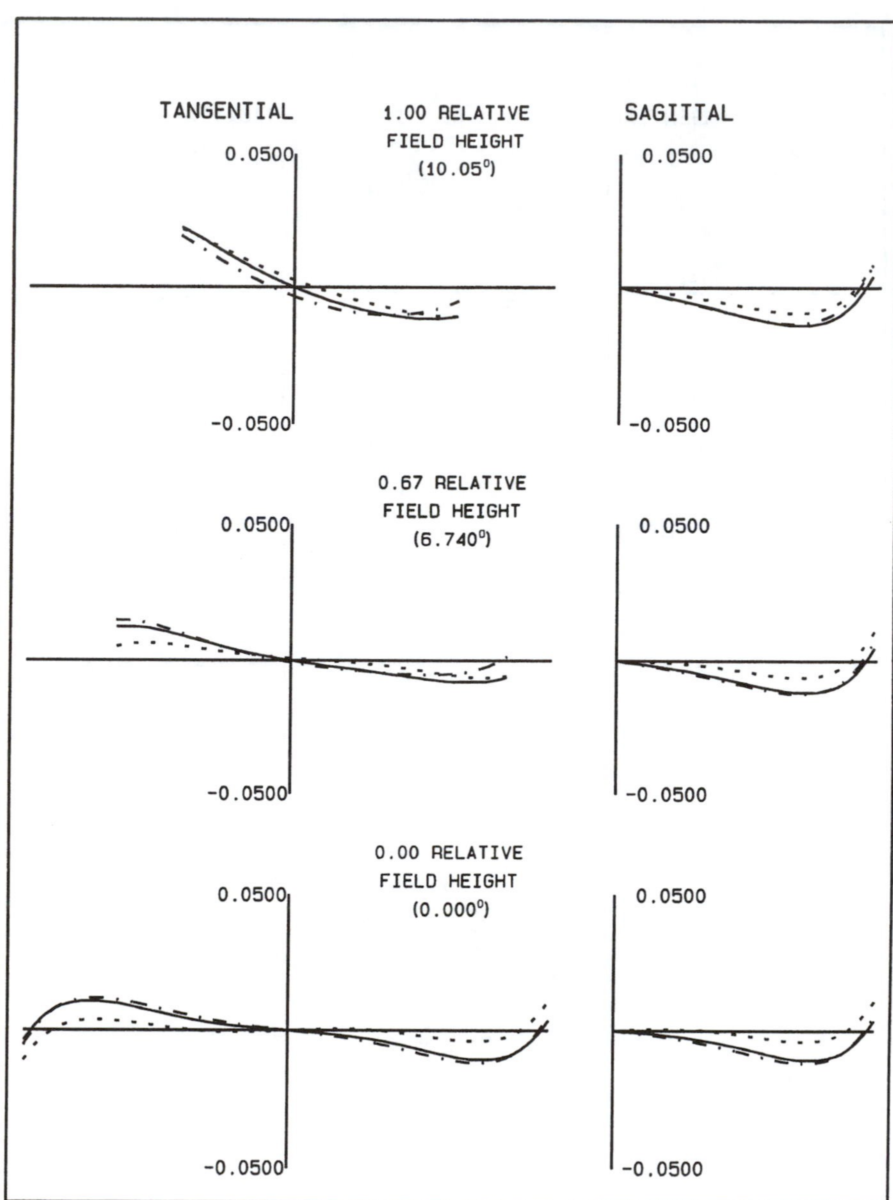

TANGENTIAL
1.00 RELATIVE
FIELD HEIGHT
(10.05°)
SAGITTAL
0.0500
-0.0500
0.0500
-0.0500
0.67 RELATIVE
FIELD HEIGHT
(6.740°)
0.0500
-0.0500
0.0500
-0.0500
0.00 RELATIVE
FIELD HEIGHT
(0.000°)
0.0500
-0.0500
0.0500
-0.0500

(b)

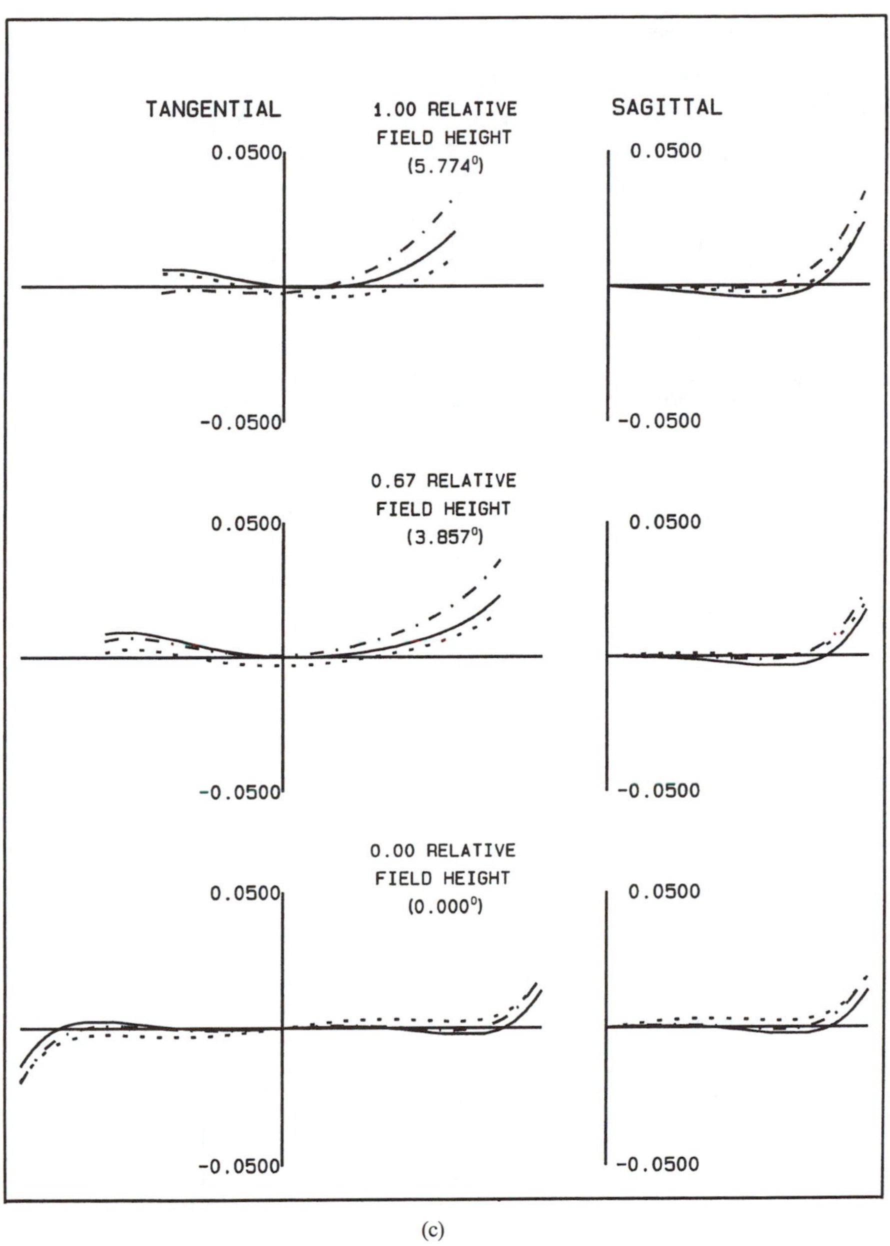

(c)

FIGURE 7.132 Ray intercept plots for sample zoom lens. (a) Short focal length; (b) medium focal length; (c) long focal length (CODE V).

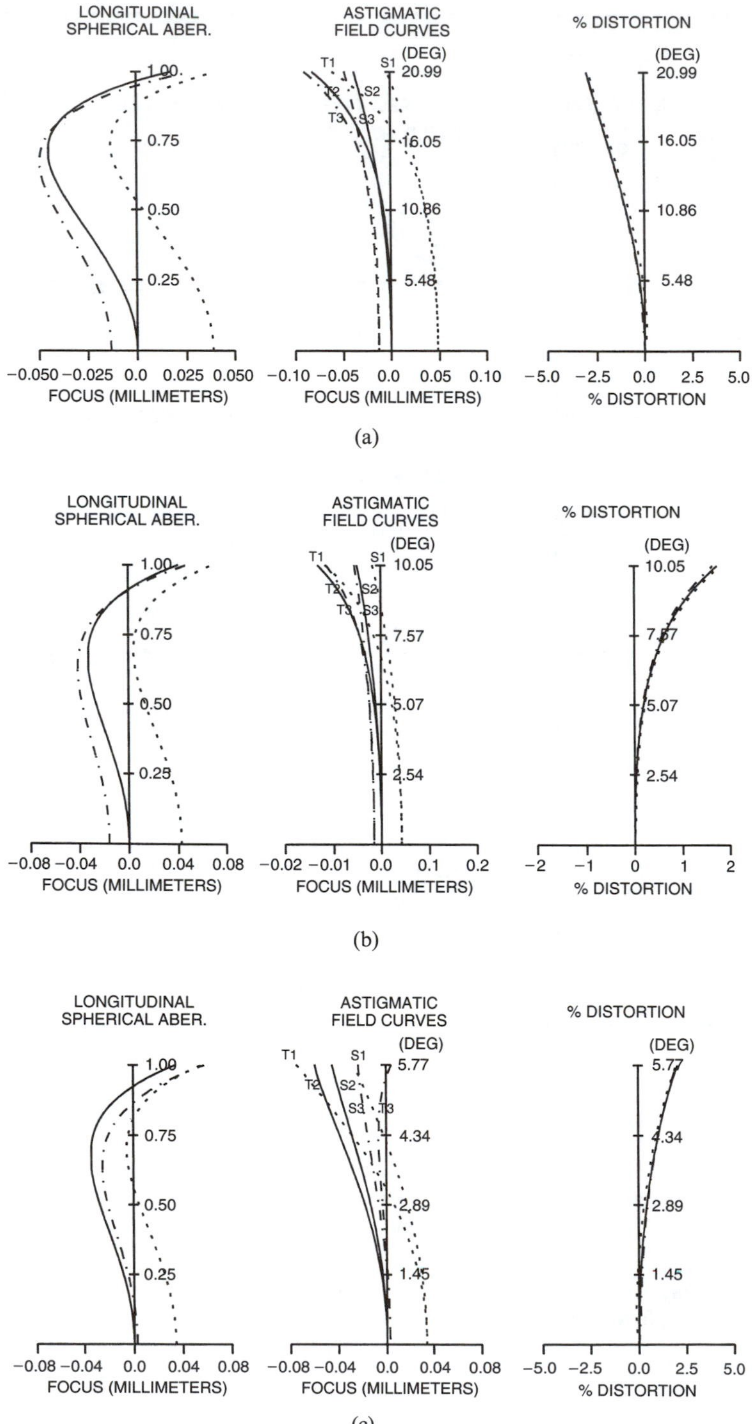

FIGURE 7.133 Field plots for sample zoom lens. (a) Short focal length; (b) medium focal length; (c) long focal length. ······, 650 nm; ——, 550 nm; ·—·—·, 480 nm (CODE V).

The design of a zoom lens generally begins with a first-order layout of the system. In many cases this can be developed analytically. The easy access of optimization features in lens design programs provides an alternate approach that usually offers considerably greater insight into the design process, and makes the investigation of alternate designs quite easy. In the paraxial optics section of Chapter 2 some examples were worked out for establishing the first-order layout of a lens system with varying magnification and established boundary conditions. A similar approach can be used for starting a zoom lens design.

As a sort of generic example, we will look at carrying out the design of a zoom system based upon the choices made in the problem of Chapter 2. The design goal is to provide as much as possible of a zoom range from five to ten times magnification as possible, but while maintaining the conjugates of the original system. Since the zoom system is to be added to the existing lens, we need only consider the object and aperture position as the starting point for the layout. Here the *a* and *d* components of Figure 7.127 are represented by the screen as the object, and the image, and the aperture stop is frozen at the location of the original lens. The spreadsheet analysis of Chapter 2 indicated that this was a very constrained space. The lens design will also indicate some very tight constraints.

A lens design program, in this case ZEMAX, was arbitrarily chosen for the design. The powers and spacings of the lenses obtained in the spreadsheet setup were inserted into the program. The lenses were represented by equivalent paraxial lenses by using an option in the program. Figure 7.134 is taken from the normal output for the configuration of the system. The first-order layout is confirmed by the lens design program. (At this point it is evident that the lens design program could have been used for the lens setup as well as the spreadsheet.) No information about the aberrations of the system is available as the paraxial lens model does not include lens aberrations.

Reading Figure 7.134 may seem a bit complex at first. The problem being worked on consists of the original object as the image formed on the screen by the lens already set into the system. Therefore, the object is the screen at the right end of the layout. The image is formed on this same screen location. The rays are a bit confusing because the upper bundle of rays are actually rays from the virtual object of the screen. The upper set of rays just below the top bundle is the maximum field rays actually being projected toward the screen at the edge of the field at about 0.9 magnification of the full screen height. This is, of course, nine times the original slide dimension.

The next step taken is to replace the paraxial thin lenses with real thin lenses, that is components with surfaces but no thickness. For want of any better choice, Bk7 glass was used for the lenses. It is obvious that multi-element lenses will be required for color and other aberration correction.

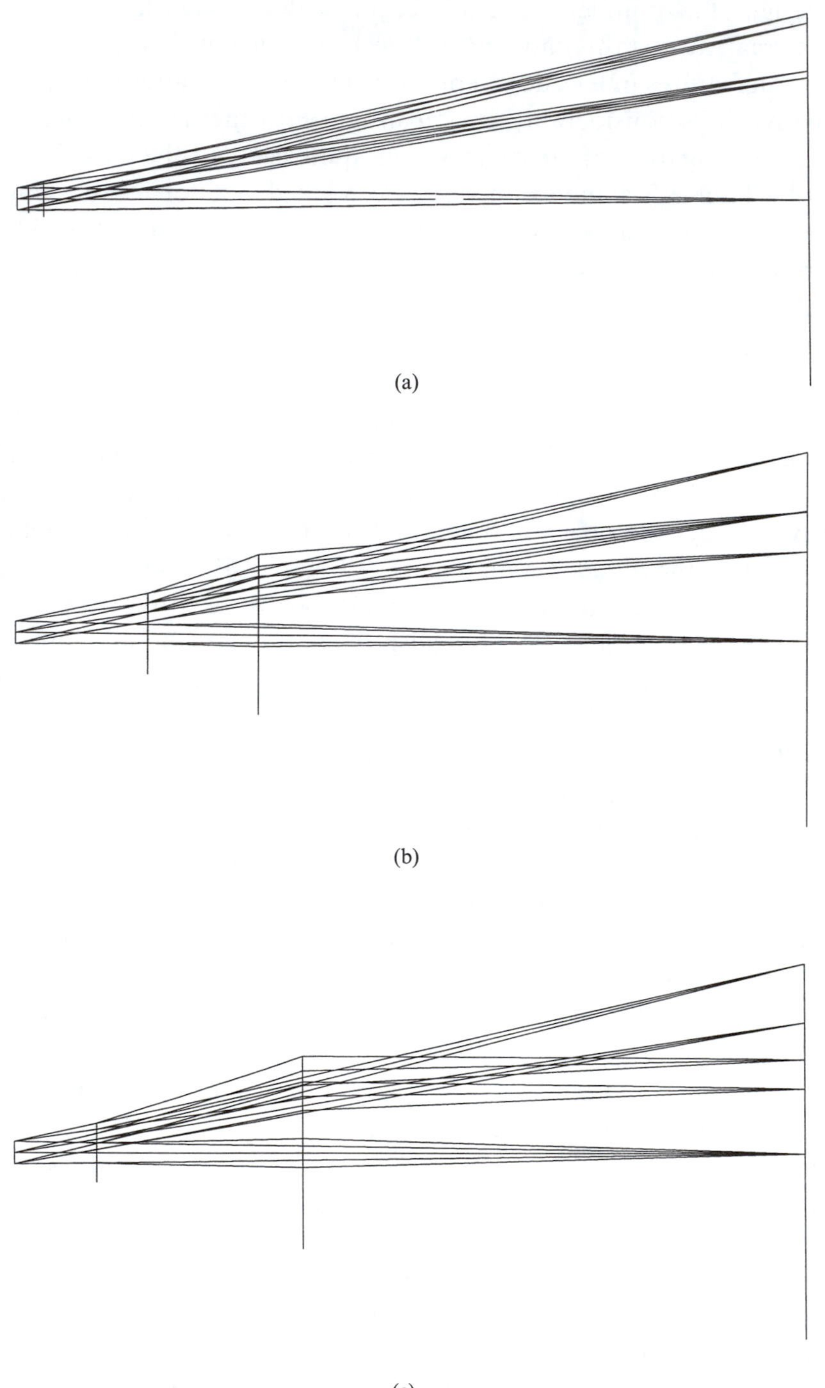

FIGURE 7.134 Ray layout diagrams with power components for zoom system for the viewer problem. (a) 9× magnification; (b) 7× magnification; (c) 5× magnification (ZEMAX).

The layout in Figure 7.135 was obtained by a minor optimization for the first-order parameters of the system. The peculiar-looking zero-thickness lenses show that significant thickness will be required to make the elements feasible. Now that actual simple thin lenses have replaced the paraxial power components, the actual rays can be traced, and the huge amounts of aberration from the lenses are evident.

Some idea of the aberration correction can now be obtained. Examination of the field and distortion plots indicates a huge amount of distortion is present at the five-times magnification end. The fact that the elements come very close together at the 9.5-times end of the range is expected, but is not a problem at this stage of the design. It is obvious that the major correction problems will lie with the distortion, the lateral color, and the variation of the astigmatic field curves. These aberrations, as well as the first-order requirement will have to be added to the merit function.

Obviously more complexity will be required in the individual lenses. As a guess, use of three elements for each lens in the zoom system would seem to be a reasonable starting point. Replacement of the single lenses by three-element achromatic lenses of real thickness can be carried out in many ways. The most direct way is to introduce the elements as a rough achromat, with two crowns surrounding the central flint element. The added thickness is temporarily absorbed by modifying the overall length target in the merit function to accommodate the change. A first solution is sought by use of direct optimization. As various problems appear with the lens, more constraints are added to the merit function. A constraint to ensure that the zoom variable thicknesses are always greater than zero needs to be explicitly stated, otherwise unrealistic mathematical solutions with negative element separation will be obtained. The distortion is controlled at the five-times end of the range, where it is the worst.

Once the distortion and chromatic aberrations are controlled, other aberrations may be added to the merit function. For this program, it is convenient to add a default merit function for the RMS ray error at two or three fields. During an attempt to find the initial solution, the basic nature of the design changes, and wanders away from the first-order design. The change in aberrations and the limits imposed by keeping the lenses from colliding at one end of the zoom range suggest that only about 80% of the zoom range can be accommodated with the constraints imposed by this design. Therefore the magnification range is reduced to cover from five times to about 8.5 times.

The designer must pay attention to the general course of the design, altering targets as required to approach a feasible solution. Modification of requirements, such as the limitations at the end of the zoom range, is necessary to locate a region of solution. Because this is an academic demonstration solution, we will accept this limitation. If the customer was insistent, it would

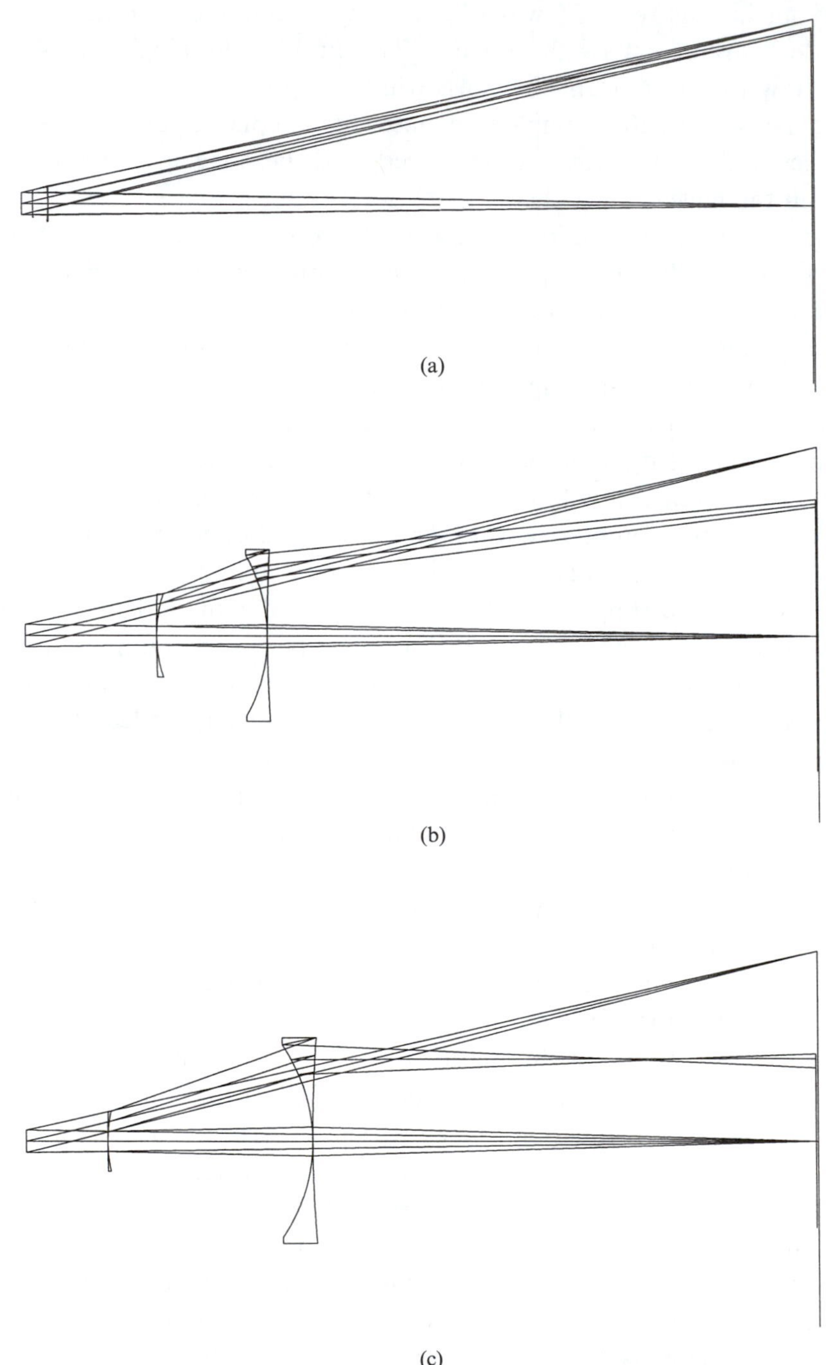

FIGURE 7.135 Ray layout diagrams after initial optimization for first-order parameters. (a) 9.5× magnification; (b) 7× magnification; (c) 5× magnification (ZEMAX).

be necessary to add some stationary or fixed elements adjacent to the initial fixed lens that would alter the magnification and location of the object plane for the zoom lens, and permit adjustment of the range, rather than limiting the solution to two moving components.

Once the ray-based merit function is under control, and the solution is seen to be feasible, the overall length constraint can be reduced, forcing the lenses to fit within the required space. It is also noted that the lenses tend to become large in diameter, as this provides more leverage in finding a solution. A constraint of maximum aperture on the last surface of the moving elements is also imposed to keep the lens within practical reason. The designer may also experiment with changing the zoom range to determine the possible aberration effects.

After several optimization steps, the lens shown in the layout in Figure 7.136 results. The aberration correction is demonstrated by the ray plots and field plots in Figure 7.137. At this point the MTF can be computed, and is shown in Figure 7.138. The MTF for this lens indicates that spatial frequencies between five and ten lines per millimeter will be recorded on the screen. At this point the lens is barely acceptable for the purpose. Improvements in the design would include addition of a component to each moving lens. A better, probably more profitable, approach would be to add a lens group at the location of the original lens to serve as a "group *a*" of Figure 7.127, to place the object for the zooming lenses in a more convenient location.

The data for this lens is provided in Figure 7.139. The reader may wish to follow up on this design by trying to improve the solution following some of the suggestions made above. Eventually, this design should be combined with the lens designed for the original objective purpose in section 7.1.3.

7.1.6 ASPHERIC LENSES

The approach to the design of lenses with aspheric surfaces generally follows the procedures already established, with the exception that the unique nature of the aberration contributions from aspheric additions to the surface need to be considered. This will influence the starting point for the design because aspheric surfaces permit aberration correction with fewer surfaces than are required under a limitation to spherical surfaces. The reason is that the addition of an aspheric to a spherical base surface permits changing the mixture of aberrations generated at that surface. The designer also has to keep in mind the difficulties of producing aspherics, and should consider approaches to the fabrication and testing of such surfaces.

The importance of aspherics in design arises from the ability to modify the shape of a power-producing spherical surface to change the aberrations produced at that surface. The base spherical surface provides the image-

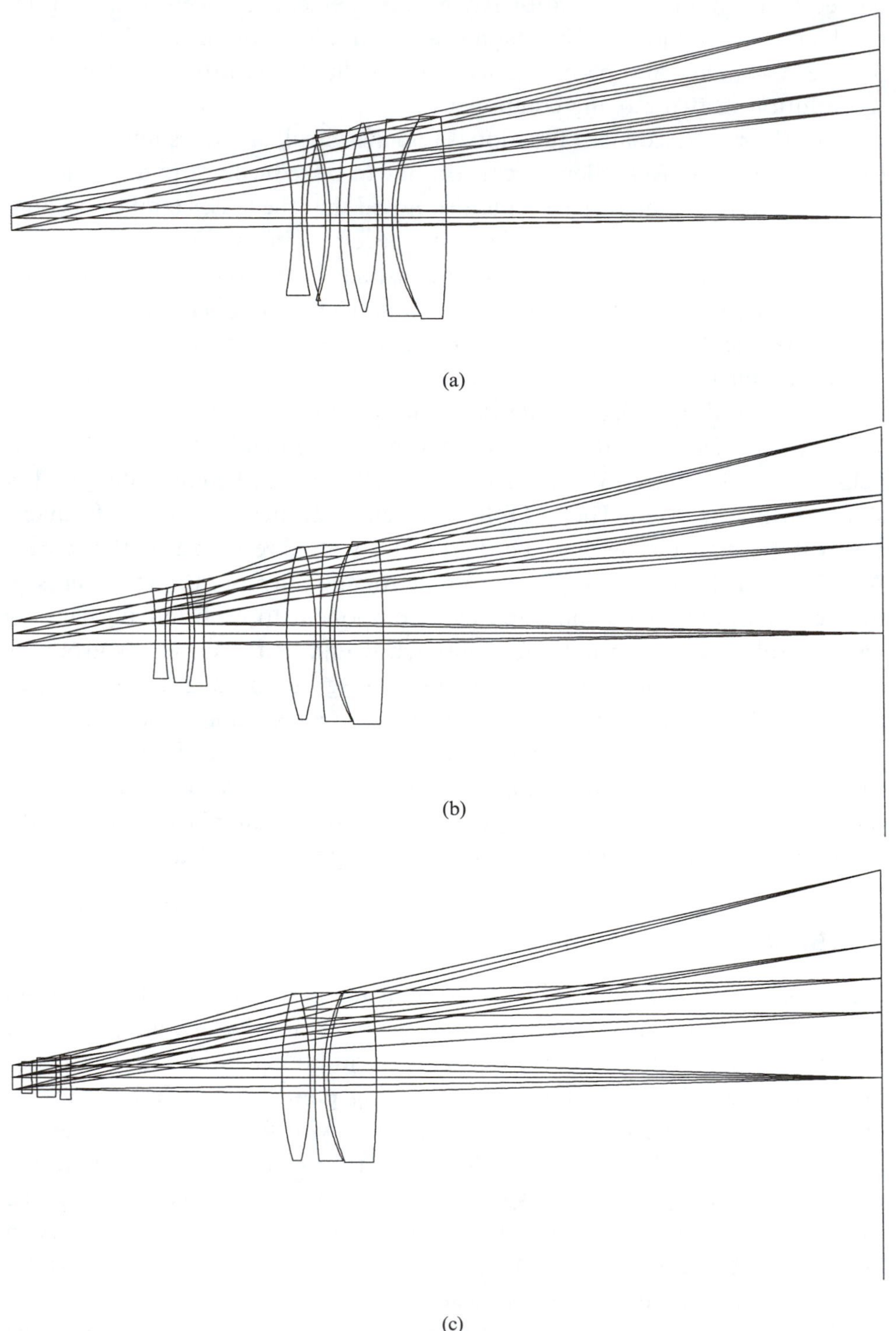

FIGURE 7.136 Ray layout diagram after aberration optimization. (a) 8.5× magnification; (b) 7× magnification; (c) 5× magnification (ZEMAX).

forming power, but produces aberrations, principally spherical aberration, as discussed earlier. Addition of an aspheric to the base sphere does not alter the paraxial image forming properties, but modifies the high-order content of the wavefront following the surface. A familiar example of a single aspheric lens systems is found in the parabolic reflector. The base sphere sets the focal length of the surface, an aspheric figuring, which produces a paraboloidal surface, corrects the spherical aberration produced when imaging an object at infinity. A single aspheric permits complete correction only at a single point in the field so that simply making surfaces aspheric does not ensure that a perfect lens will result at other than a single field point. Multiple aspherics can be used to alter the rate of change of aberrations with field.

Cylindric and toric surfaces form a special case of aspheric surfaces. These surfaces have a different base curvature in mutually perpendicular directions, which results in different paraxial base coordinates in the two directions. This makes it possible to design anamorphic optical systems with different magnifications in the two azimuths. Cylindric surfaces may also have shapes that depart from pure cylinders or toroids. The additions to the base cylinder surface behave in a manner similar to the aspheric surfaces to be discussed in this section. The anamorphic applications of toric and cylindric surfaces will be discussed in section 7.2.2. In this section, nominally rotationally symmetric aspheric departures from a base spherical surface will be considered.

The basic optical surface is the sphere. The sag formula for this surface is the now familiar

$$z_{sphere} = \frac{c(x^2 + y^2)}{1 + \sqrt{1 - c^2(x^2 + y^2)}} \tag{7.4}$$

The spherical power is necessary for forming images and defines the paraxial image coordinates. An aspheric surface is one which deviates from this form, but which leaves the base sphere and the paraxial optics unchanged.

There are several types of aspherics, and several different definitions in various design programs. The most common aspheric form is the conic surface of revolution. To be specific, these surfaces should be called conicoids, as they are actually members of the family of conics, but are rotated about the optical axis. The mathematical form describing such aspherics is

$$z_{sphere} = \frac{c(x^2 + y^2)}{1 + \sqrt{1 - (1 + \kappa)c^2(x^2 + y^2)}} \tag{7.5}$$

where z is the surface sag parallel to the axis of the conic surface and (x, y) are the coordinates perpendicular to the axis. The value of the conic constant, κ, is related to the eccentricity of the conic by

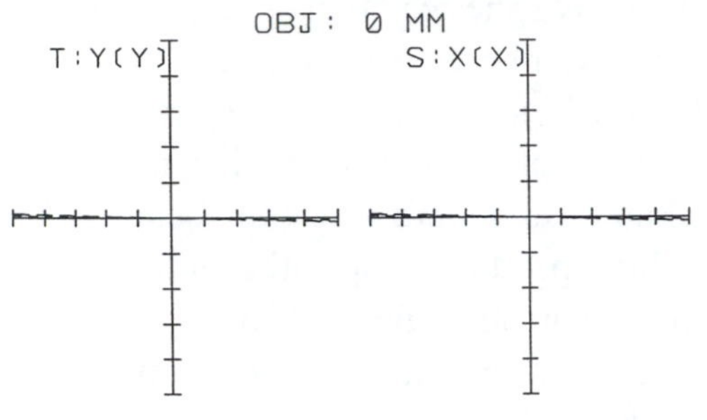

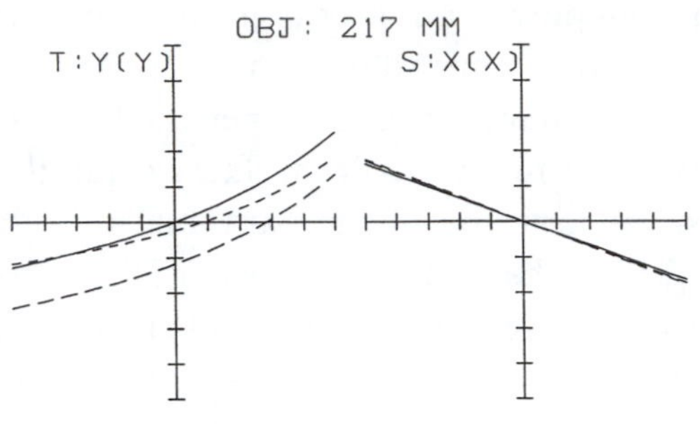

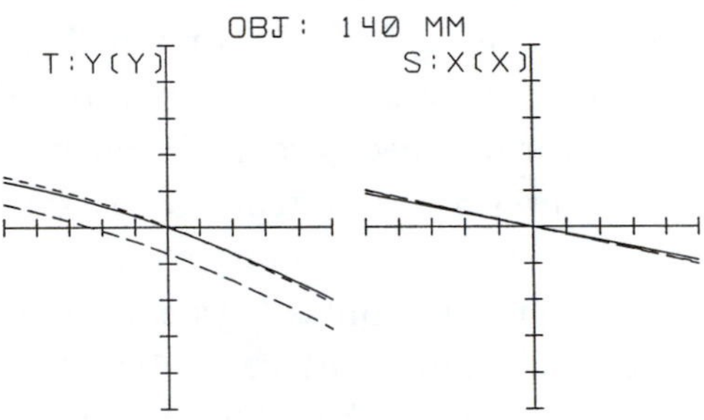

MAXIMUM SCALE: -500.0000 TO 500.0000 MICRONS.
0.550 0.450 0.650

(a)

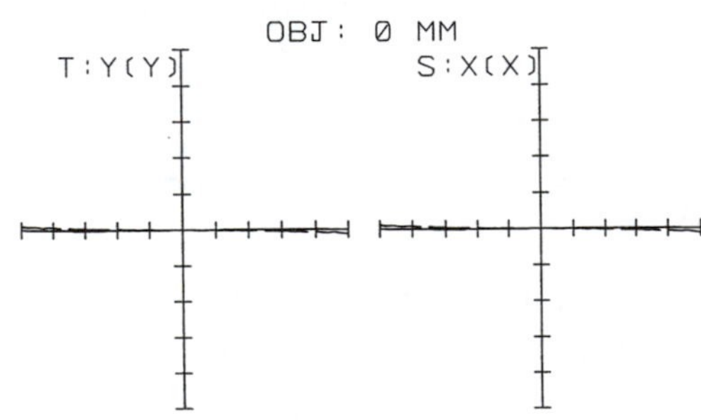

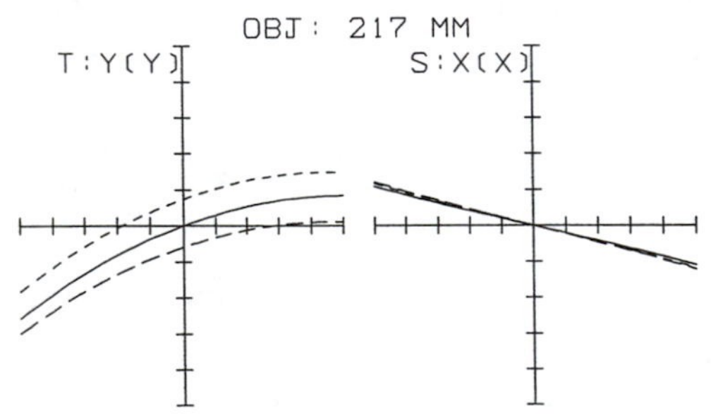

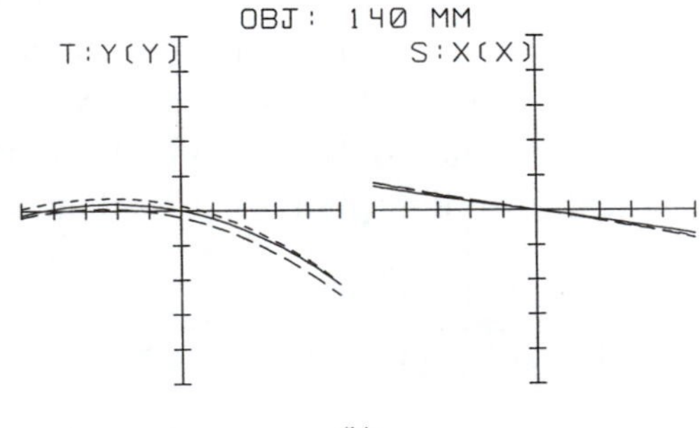

(b)

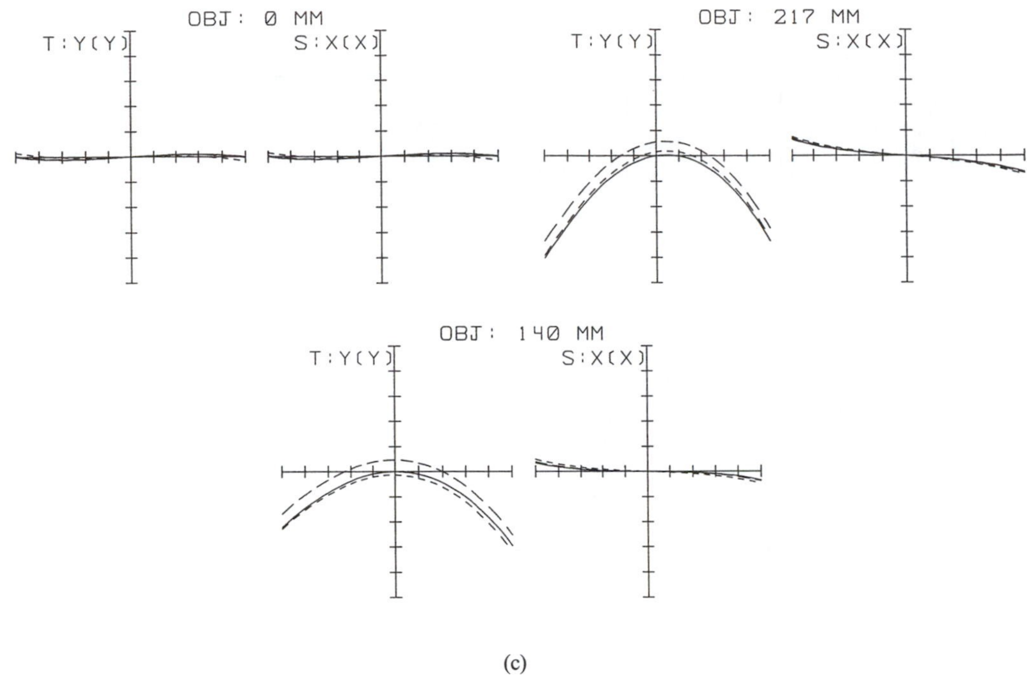

(c)

FIGURE 7.137 Ray intercept plots for zoom system. (a) 8.3× magnification; (b) 6.9× magnification; (c) 5.0× magnification (ZEMAX).

$$\kappa = -\epsilon^2 \tag{7.6}$$

The conic that is generated depends upon the choice of κ. For a sphere $\kappa = 0$, as shown earlier. A spherical surface is defined by a single radius of curvature over the surface. The change in local curvature and direction of each section of the wavefront refracted or reflected from the surface will depend upon the curvature of the surface and the angle of incidence upon the surface. The discussion of close skew rays in section 2.1.5 applies with the astigmatic power being determined by the apparent difference in curvature when the chief ray is incident nonnormally on the surface.

The family of conicoids obviously forms a continuous family of deviations from a sphere. For other values of κ the local curvature of the surface changes across the surface, and will also depend upon the direction along the surface. These surfaces are conicoids, or conic surfaces that are rotated about the major axis of the surface. Since the foci of the conic lie along the major axis, the surface will form perfect aberration-free images between the focus points of the selected conic. The case of $\kappa = -1$ is a paraboloid. For reflecting surfaces, a zero spherical aberration condition exists when one conjugate is at infinity, and the other at the focal point of the surface. When κ lies between -1 and 0, the surface is a section of an ellipsoid with

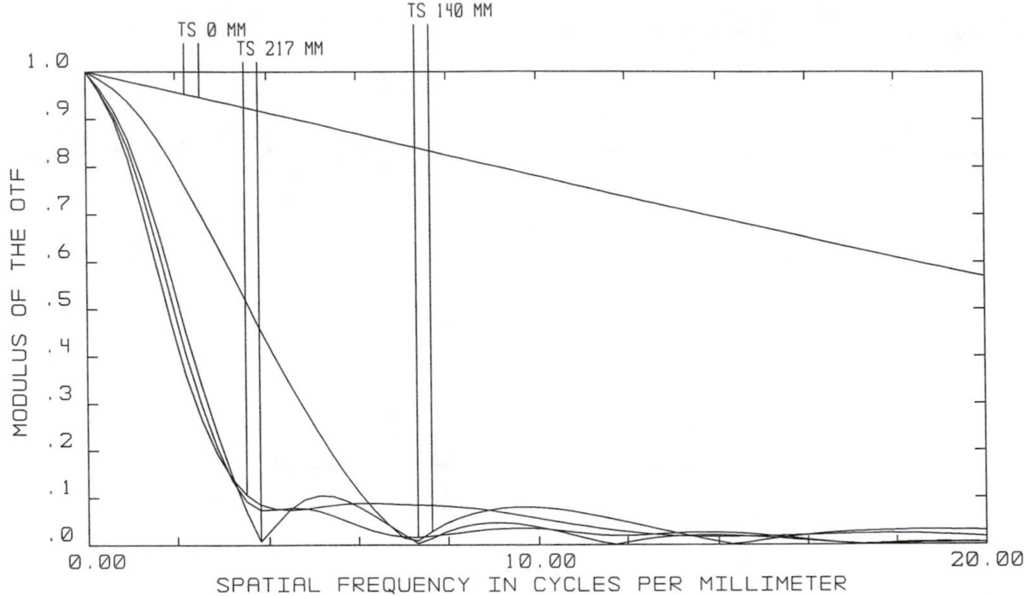
TS 0 MM
TS 217 MM
TS 140 MM
1.0
.9
.8
.7
.6
.5
.4
.3
.2
.1
.0
MODULUS OF THE OTF
0.00
10.00
20.00
SPATIAL FREQUENCY IN CYCLES PER MILLIMETER

(a)

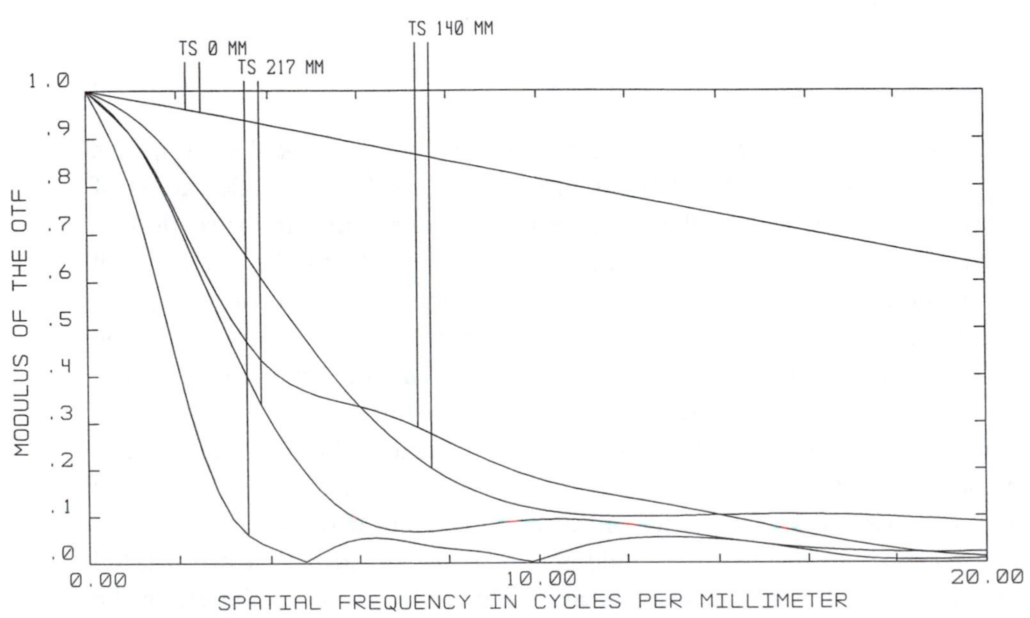
TS 0 MM
TS 217 MM
TS 140 MM
1.0
.9
.8
.7
.6
.5
.4
.3
.2
.1
.0
MODULUS OF THE OTF
0.00
10.00
20.00
SPATIAL FREQUENCY IN CYCLES PER MILLIMETER

(b)

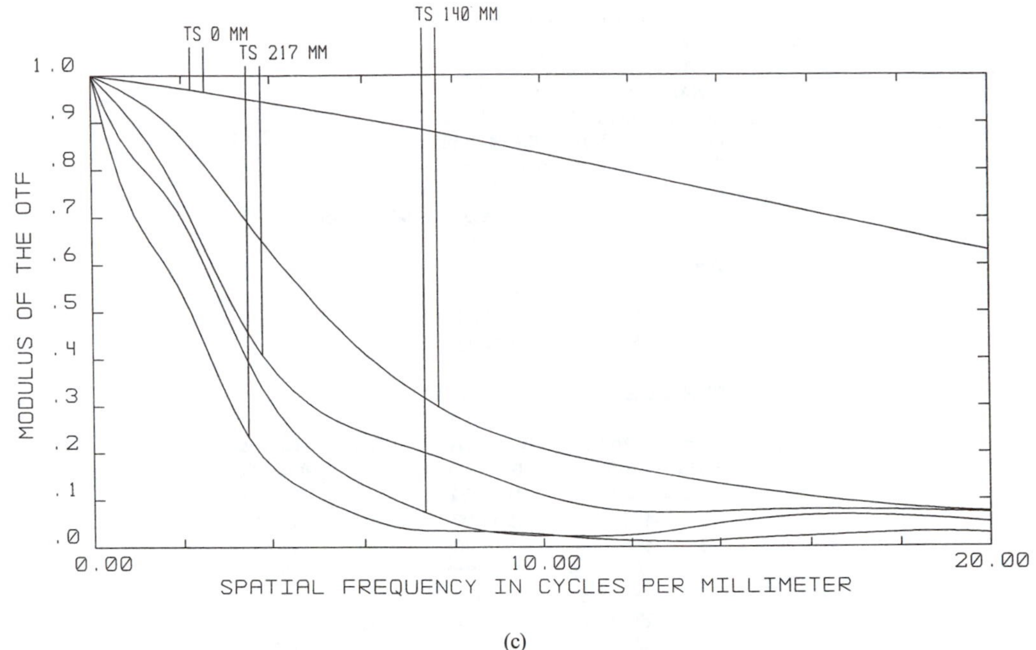

FIGURE 7.138 MTF plots for zoom system. (a) 8.3 × magnification; (b) 6.9× magnification; (c) 5.0× magnification (ZEMAX).

the major axis along the optical axis. Zero spherical aberration imagery exists between the two foci of the ellipsoid. The surfaces for $\kappa < -1$ are sections of a hyperboloid. Spherical aberration free foci exist, but in this case, one conjugate is a real focus, the other is a virtual focus. The surfaces for $\kappa > 0$ form members of the continuous set of conic surfaces, but are not strictly members of the family of conicoids. These are ellipsoids rotated about the minor axis, and are called oblate spheroids.

The difference in curvature across the surface leads to two principal curvatures at each location on the surface. The astigmatic effect for close skew rays will be altered by these curvatures, providing an additional degree of control of the aberrations. There will be no astigmatic effect for ray bundles along a chief ray passing through the foci of the surfaces. This provides an extremely potent aberration control for conic surfaces distributed in space, as may be encountered in some scanning systems.

In addition to a conic surface, additional aspheric terms can be added. The most common is a rotationally symmetric polynomial aspheric. This results in a surface sag of the form

$$z_{aspheric} = z_{conic} + (ad)(x^2 + y^2)^2 + (ae)(x^2 + y^2)^3 + (af)(x^2 + y^2)^4 + (ag)(x^2 + y^2)^5 \tag{7.7}$$

Title: Spherical Surface Design
Surfaces : 14
Stop : 1
Total Track : 909.82
Image Space F/# : 20.1165
Obj. Space N.A. : 0.0142858
Parax. Mag. : 0.499895
Entr. Pup. Dia. : 26
Field Type : Object height in Millimeters

Wavelengths : 3

1	0.550000	1.000000
2	0.450000	1.000000
3	0.650000	1.000000

SURFACE DATA SUMMARY:

Surf	Radius	Thickness	Glass	Diameter
OBJ	Infinity	-909.9000		433.2666
STO	Infinity	10.0062		26.0000
2	-432.6733	10.0000	SK14	30.3594
3	455.3004	5.0000		33.5558
4	293.9639	20.0000	TIF6	36.4187
5	-354.8667	5.0000		41.6500
6	-277.6973	10.0000	BASF1	43.4568
7	486.9299	220.3756		46.7965
8	358.8866	30.0000	FK51	174.6180
9	-389.4308	5.0000		176.3057
10	1023.225	10.0000	LF5	177.2636
11	208.8645	5.0000		176.4342
12	249.9304	50.0000	FK51	177.3721
13	-1144.263	529.4380		180.0177
IMA	Infinity	0.0000		208.5113

MULTI-CONFIGURATION DATA:

Configuration 1:
Thickness 1 : 10.00618 Variable
Thickness 7 : 220.3756 Variable
Configuration 2:
Thickness 1 : 146.9509 Variable
Thickness 7 : 88.22503 Variable
Configuration 3:
Thickness 1 : 293.552 Variable
Thickness 7 : 8.892064 Variable

FIGURE 7.139 Program listing of design data for zoom system. Note data for spacing changes at end of list (ZEMAX).

which is a general rotationally symmetric tenth-order aspheric. In most cases, the polynomial is added to a spherical base surface, that is one in which $\kappa = 0$. There is no basic reason why the surface-defining polynomial is truncated at order ten, and higher orders can be found in most lens design programs.

There are many other forms for the aspheric addition to a surface. The polynomial can be two-dimensional with a large set of aspheric coefficients. The aspheric could also be similar to a rotationally symmetric surface, but with an asymmetry in the coefficients in which the sags of constant depth of departure from the base surface form elliptical zones.

Other general aspheric forms can be generated by defining the asphericity as a set of sag or surface heights which are interpolated using a spline function. The requirement in the fitting is to match the aspheric departure and slope between adjacent sample points on the surface by using a cubic spline-fitting algorithm. There are two generally used forms. One is a radial spline which can be used to fit an arbitrary set of aspheric deformations along a radius. The surface is then represented by the rotational form of the spline-fit surface. In some programs a full two-dimensional spline surface can be generated in which a two-dimensional arbitrary set of aspheric sags is interpolated in two dimensions. The amount of data to be handled in describing the aspheric, and the tactics used in designing, are clearly much more complicated with a spline representation.

The influence of any of these aspherics on image formation is determined by the addition to the wavefront error, or OPD, generated by the base surface. In general, the effect of the aspheric is to add aberration terms to the transmitted or reflected wavefront that generally match the symmetry of the aspheric term. It is most useful to start with an understanding of the effect of conics and asymmetric polynomial aspherics in order to explain the effect of aspherics.

Because the generation of aspheric surfaces is a mathematical phenomenon, it is not surprising that many of the references to aspherics are mathematical rather than pragmatic in content. Some of the important mathematical references on the use of aspherics are in the book by Korsch (1991). This book contains a very extensive survey of the aberrations of aspheric surfaces.

In order to obtain a working knowledge of the use of aspherics, first note that the leading term of the difference between the base sphere and the aspheric (including the conic) is, to fourth order,

$$z_{4th\,order} = z_{sphere} + [ad - \tfrac{1}{8}\kappa c^3 (x^2 + y^2)^2] \tag{7.8}$$

indicating that the effect of a conic through fourth order is the same as an added fourth-order term. The introduction of $-\frac{1}{8}\kappa c^3$ incorporates the leading term in the power series expansion of the conic after subtracting the base curvature. The aberration produced by the base spherical surface will now have a component added due to the surface figuring. This adds a significant parameter to the design of a lens, as surfaces can be "tuned" to introduce desired amounts of spherical aberration.

The development of the aberrations introduced by an aspheric is quite simple through fourth order, at least. Because this wavefront error is in addition to that produced by the base spherical term and contains no second-order terms, the fourth-order wavefront error introduced by an aspheric can simply be added to the aberration produced by the spherical base surface.

The Optical Path Difference introduced by an aspheric of strength α is

$$\begin{aligned} \text{OPD} &= (n' - n)\alpha(x^2 + y^2)^2 \\ &= (n' - n)\alpha y^4 \rho^4 \end{aligned} \tag{7.9}$$

where the latter form places the error in terms of the paraxial marginal ray height on the aspheric surface and the relative height in the pupil. Interpreting the fourth-order aspheric shape as an aberration coefficient leads to

$$S_{\mathrm{I}_a} = 8(n' - n)\left[-\frac{\kappa}{8}c^3 + (ad)\right]y^4 = (n' - n)[-\kappa c^3 + 8(ad)]y^4 \tag{7.10}$$

which is added to the existing spherical aberration contribution from the base surface. Choice of the appropriate value of κ or ad will permit adding a specified amount of spherical aberration contribution to base spherical aberration contribution of the surface. If desired, the net third-order spherical aberration from the surface can be set to zero. Alternately, the aspherization can be used to set the total spherical aberration in the lens to zero, permitting the surface to introduce any necessary amount of spherical aberration.

The effect of aspherics on higher-order aberrations is not quite as simple to obtain. In general, it can be said that addition of an nth-order aspheric term will principally influence the nth-order wavefront aberration terms. Each term will also influence the effect of the contributions from the higher-order aspheric terms because of the alteration of the base surface shape by the lower-order terms. As before, the fourth-order aberration contributions provide a useful guide to the aberration contributions of the aspheric surface figuring.

If the stop is in contact with the surface, there are no additional third-order aberration contributions with field. The only fourth-order aberration arising from an aspheric at the stop will be spherical aberration. If the stop is located away from the aspheric surface, then additional field-dependent aberrations arise which may be determined by applying a stop shift to the wavefront error introduced by the aspheric portion of the surface. Then, for each aspheric surface, the aberration contributions will be:

$$\begin{aligned} S_{\mathrm{I}_a} &= (n' - n)[-\kappa c^3 + (ad)]y^4 \\ S_{\mathrm{II}_a} &= \frac{\bar{y}}{y} S_{\mathrm{I}_a} \\ S_{\mathrm{III}_a} &= \left(\frac{\bar{y}}{y}\right)^2 S_{\mathrm{I}_a} \\ S_{\mathrm{IV}_a} &= 0 \\ S_{\mathrm{V}_a} &= \left(\frac{\bar{y}}{y}\right)^3 S_{\mathrm{I}_a} \end{aligned} \tag{7.11}$$

On the surface, the marginal paraxial height is y and the chief ray height is $\bar{y}$. Note the zero contribution to the Petzval sum because of the lack of a second-order term in the aspheric portion of the surface.

The use of a single aspheric in a lens system, located on a surface close to the stop, has an extremely significant effect upon the design. An aspheric can be used to exactly compensate the spherical aberration, without affecting the field dependent aberrations. Therefore, all of the aberrations in the lens can be optimized essentially without regard to controlling the spherical aberration, and the aspheric strength used to reset the spherical aberration correction. Because the source of field-flattening astigmatism has been shown to arise from the distribution of spherical aberration away from the stop, this is a technique that can be used to explore different regions of solution space, as well as provide a final corrected design.

If the aspheric is on a refractive surface, the spherical aberration contribution of the aspheric will vary with wavelength. The magnitude of this variation can be computed by replacing the indices by the difference in index over the wavelength interval:

$$\begin{aligned} \delta S_{\mathrm{I}_{a_\lambda}} &= (\delta n'_\lambda - \delta n_\lambda)[\kappa c^3 + 8(ad)]y^4 \\ &= \left(\frac{\delta n'_\lambda - \delta n_\lambda}{n' - n}\right) S_{\mathrm{I}_a} \end{aligned} \tag{7.12}$$

For the most usual case, where the aspheric is on an air–glass interface,

$$\delta S_{\mathrm{I}_{a_\lambda}} = \frac{S_{\mathrm{I}_a}}{V} \tag{7.13}$$

where V is the V-number of the glass. The amount of chromatic variation of the spherical aberration introduced by the aspherization is therefore proportional to the aspheric strength. For a crown glass, the variation of spherical aberration over the C–F spectral range will be about one sixtieth of the amount of spherical aberration correction. For an aspheric on a reflecting surface, the spherochromatism is obviously zero.

Chromatic variation of spherical aberration can be compensated by using two aspherics on surfaces close to the stop, with each aspheric placed on a crown–flint pair of glasses. The design procedure is similar to that of correction of chromatic aberration, but compensating the high-order chromatic variation from the aspheric. There is a limit to the correction of spherochromatism by this technique that is similar to the secondary color limitation for first-order color correction in an achromatic doublet.

For all aspherics, the contribution to higher than fourth-order wavefront aberrations will vary with field. If the aspheric is located at a distance from the stop, the correction will change with the field. The higher-order contributions will also vary with the field due to the change in the angle of incidence of the off-axis bundle and because of the anamorphic oblique mapping of the pupil with field. These field-dependent effects are usually somewhat different from other field dependencies in the lens, and can often be put to good use in a design.

The constancy of spherical aberration with field holds only for third-order aberrations. The obliquity of an off-axis beam across an aspheric will induce a contribution of oblique spherical aberration due to the change of mapping of the ray bundle against the aspheric. This will add or subtract from the intrinsic oblique spherical content of the basic lens.

The use of multiple aspherics in a lens is usually less important, except in reflective systems, where the number of available surfaces is small. For any design, placing a number of aspherics on surfaces at different distances from the pupil will permit modification of the aberration contributions from these surfaces to alter the coma and astigmatism and distortion. Because aspherics contain only high-order terms, they do not affect the Petzval sum directly. In many cases the new freedom in setting aberrations resulting from aspherics will permit adjustments in the other variables to allow changes in the Petzval surface.

Aspheric surfaces contain high-order surface contributions, which can directly affect the strategy used in design. There are many forms of aspheric surfaces, as described in Chapter 2. The variables available start with the conic constant for the surface, which modifies the shape of the surface. High-order polynomial terms can add surface shape changes of high order that are most effective in changing the spherical aberration contributions of the surface at the corresponding high order. Just as a second-order surface curvature can introduce spherical aberration of high order, the introduction of an aspheric of fourth order will introduce primarily fourth-order wavefront error, but also some amount of higher-order aberration. The mixture of aberrations depends upon the object–image relations on the surface, as well as the effect of aberrations produced prior to the aspheric surface.

Scaling of a conic requires no change in the conic constant, since it characterizes the form of the surface. Because each of the polynomial coefficients is a factor attached to a different power of the surface aperture, each coefficient will vary differently with scaling. For example, the fourth-order coefficient will vary with the third power of the aperture. The selection of the appropriate amount of change in the coefficient will greatly affect the linearity of the movement in design space. If polynomial coefficients are used as variable in the optimization process, scaling of an aspheric with the dimensions of the design can produce significant changes in the metric of design space. In general, the amount of change produced by the incremental change of an aspheric coefficient used in an optimization process should be established as a fixed number of wavelengths of error at the edge of the aperture.

This will bring significant changes in the size of the aspheric coefficient change with the choice of scale of the lens.

The selection of rays to be used in defining a merit function when aspherics are used is also important. Since the principal effect of aspheric coefficient changes will be for a ray near the edge of the aperture, a systematic approach to choosing the ray for inclusion in generating a merit function is important. Success in design will be enhanced by using a set of rays spaced unequally across the aperture. There are reasonable choices, but not a specific choice, and designers will find that experimentation to locate the optimum aperture coordinate set of rays is necessary.

Consideration of methods for testing the aspheric to be produced is usually necessary. Aspherics are usually tested using an optical null test. In this test special lenses with easily tested and verified spherical surfaces are used to provide an aberrated wavefront that will match or compensate the aberrations produced by the aspheric under specified test conditions. The relative magnitude of orders of spherical aberration that are produced by spherical surfaces generally follow a reduced amount with order.

The leading aberration term for conic aspherics or fourth-order aspherics will be a fourth-order contribution to the wavefront aberration. Sixth- and higher-order coefficients will produce aberration contents generally proportional to the magnitude of the coefficients. The design of an aspheric which has relative amounts of high orders that do not naturally follow the relative ratios of contributions from spherical surfaces will provide some great difficulties in testing. In addition, the normal configuration of a test using refracting components will produce undercorrected spherical aberration, which can be used to null or compensate the overcorrected spherical aberration from an aspheric. If the aspheric has the opposite form, which requires the generation of aberrated compensating wavefronts that are not naturally generated by spherical surfaces, there can be significant difficulties with developing a simple null test for the aspheric. Other tests, such as holographic or binary reference elements could, of course, be used.

Ray tracing through an aspheric surface requires some additional considerations. Conics are easily handled, because the same approach that is used in tracing spherical surfaces may be used, but with an addition to the formulae for transfer to the surface. Similarly, toroidal and cylindric surfaces may be ray traced using a formula ray trace as in the case of spherical surfaces.

Ray tracing through general aspherics is accomplished by an iterative procedure. The transfer of the ray to the tangent plane to the surface is exactly the same as for a spherical or conicoid surface. Transfer of the ray to the aspheric is accomplished by projecting along the ray direction until the ray coordinates and a location on the surface coincide. That point will then be the intersection point with the surface. The gradient of the function describing the surface is then computed at that point to obtain the direction of the normal to the surface. The usual refraction (or reflection) rules can be applied to determine the continuing direction of the ray. Transfer to the succeeding surface normal is then accomplished using standard ray transfer procedures.

The iterative process is carried out until the difference between the projected ray coordinate and the computed coordinate on the surface falls below a specified tolerance. Usually this is set to be either a small fraction of a wavelength (perhaps 0.0001 waves) or the smallest significant figure usable by the computer. This has the added advantage of making the iterative ray trace the most accurate method of tracing rays on a computer, because there are no cumulative ray-tracing round-off errors.

The high-order nature of aspheric deformations provides a useful tool in design because the effect of the aspheric can be localized to sections of the surface. For example, the tenth-order coefficient will have a significant effect upon rays passing through the outer portion of the aperture, but no effect on rays passing through the center of the aperture. Lower orders will have principal effects upon zones of the aperture that are closer to the center of the aperture. Therefore balancing or correction of high-order spherical aberration contributions can be carried out on a term-by-term basis. Obviously the selection of the ray coordinates in the aperture that are used for calculating the aberrations contained in the merit function can be quite critical. If high-order aspheric coefficients are being used, there should be a crowding, or an addition, of rays at coordinates crowded toward the outer portion of the aperture. The exact selection is usually not critical, and may have to be altered as the design proceeds, anyway. A set of useful starting points are at 0.7, 0.85, 0.9, and 0.97 when the fourth-, sixth-, eighth- and tenth-order terms, respectively, are being used in a design. Unless aperture sampling spacings similar to these are used, the merit function may not adequately represent the effect of the aspheric terms.

The preceding discussion of aspherics is generic, referring to any type of aspheric. The following sections will describe the approaches that are useful for particular applications.

7.1.6.1 Reflectors

Reflective systems are extremely attractive for application to many imaging problems. Use of reflecting components does not produce chromatic aberrations. Furthermore, the reflecting surfaces required may be supported by lightweight mirror substrates. Both aspects are desirable for large optical systems.

The unattractive aspect of reflective optical systems is the limitations on forms that are imposed by the requirement that each of the reflecting components must not block light reflected by other reflective components. This leads to a reasonable limitation upon the number of surfaces that may be used. Obscuration and penetrations through surfaces are usually required, leading to great difficulty in producing images over a wide field of view.

Because reflective systems usually contain at most two or three power surfaces, and usually cover a small field of view, it is possible to describe the image-forming properties quite well with only the use of third-order aberrations. There are many possible ways of developing formulae that can be used to obtain the required conic constants and third-order residual aberrations for such systems, given the desired first-order properties of the systems. A good and reasonably practical reference on applications of reflective telescopes is Schroeder's book on telescope optics (Schroeder 1992). A very useful, but somewhat more mathematical reference is the comprehensive book by Dietrich Korsch on reflective optics (Korsch 1991). Some very practical and well worked out examples of the design of various astronomical systems is to be found in the book by Rutten and van Venrooij on telescope optics (Rutten and van Venrooij 1988). A less widely known, but very important, approach is the use of plate diagram analysis in understanding reflective systems, which was developed by C. R. Burch in 1942. This technique can be found in Linfoot's book on advances in optics (Linfoot 1955). The books by Schroeder and Rutten provide many examples and parametric plots of the aberration expected from various telescopes.

The cited examples do not cover all of the possible references on the design of telescope and reflective optics. There is no point in repeating the discussions in these many excellent references on the subject. Therefore, the discussion here will be directed toward the use of the techniques and approaches already discussed in the design of lenses. The designer who is a generalist in design, and not a specialist in reflective systems can easily extend his or her capabilities to understand and design catadioptric systems. The appropriate use of lens design software can produce rapid convergence to a solution, and permits complete analysis of the imaging properties.

The constraints involved in designing a reflective system must first be satisfied by placing a series of reflecting powered surfaces with base spheres determining the power of each element into locations in space such that the magnification and image locations required are obtained. The basic aberrations are computed, and then the aspheric figuring of the surfaces adjusted to obtain the desired aberration correction. If necessary, refracting aspheric components can be used to provide a balance of the imagery across the field of view. Under some conditions, the aspheric components can be tilted

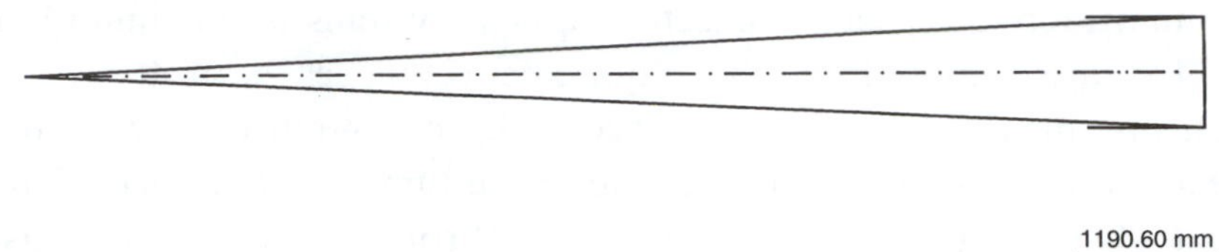

FIGURE 7.140 Layout of single-mirror, F/8, 1 m diameter reflecting system.

and decentered to provide an offset in the aberration correction across the field, in order to obtain an unobscured imaging condition.

The design approach starts with the assumption of a rotationally symmetric imaging system. Setting up the curvatures and locations of the surfaces is almost a "connect the dots" process in which the entrance pupil size and the final focal length and the location of the image is selected. Following this the desired height of the marginal paraxial ray on each component and the locations of the components are selected. The design program can use angle solves and thickness solves to set up the base curvatures of the surfaces.

The simplest reflecting image forming system is a single spherical mirror with the stop located at the mirror. Figure 7.140 shows one example of such a system. The example chosen is a one meter diameter reflector with a final focal F/number of ten. These parameters lead to a final value of $n'u' = -0.05$. Since the index of refraction in the space following the reflective surface is $n' = -1.000$, the paraxial angle is $u' = 0.05$. As no surprise, the radius of curvature of the surface is 20.0 m. At this point the field angle selected (arbitrarily, to determine the useful field coverage) will be set at ten arc minutes, or 0.1666°. The aperture stop is placed at the surface. The ray aberrations and the wavefront aberrations are illustrated in Figure 7.141. The principal limiting aberration is spherical aberration, but coma and astigmatism are apparent from the ray intercept plots.

The obvious correction method for the spherical aberration is to add an aspheric to the reflective surface. It is well known that the aspherization that corrects the spherical aberration when the object is at infinity is when the surface becomes a paraboloid. To demonstrate this, as well as the continuous nature of the family of conics, the third-order spherical aberration contribution for a single reflecting surface can be found by adding the spherical base contribution and the aspheric contribution

$$S_{\mathrm{I}_{total}} = S_{\mathrm{I}} + S_{\mathrm{I}_{aspheric}}$$

$$= -yn^2 i^2 \left(\frac{u'}{n'} - \frac{u}{n} \right) + (n' - n) 8 \left[\frac{-\kappa c^3}{8} + ad \right] y^4 \qquad (7.14)$$

These can be interpreted using the refractive index relationship of $n' = -n = -1$ and the magnification $m = -l'/l$ to be

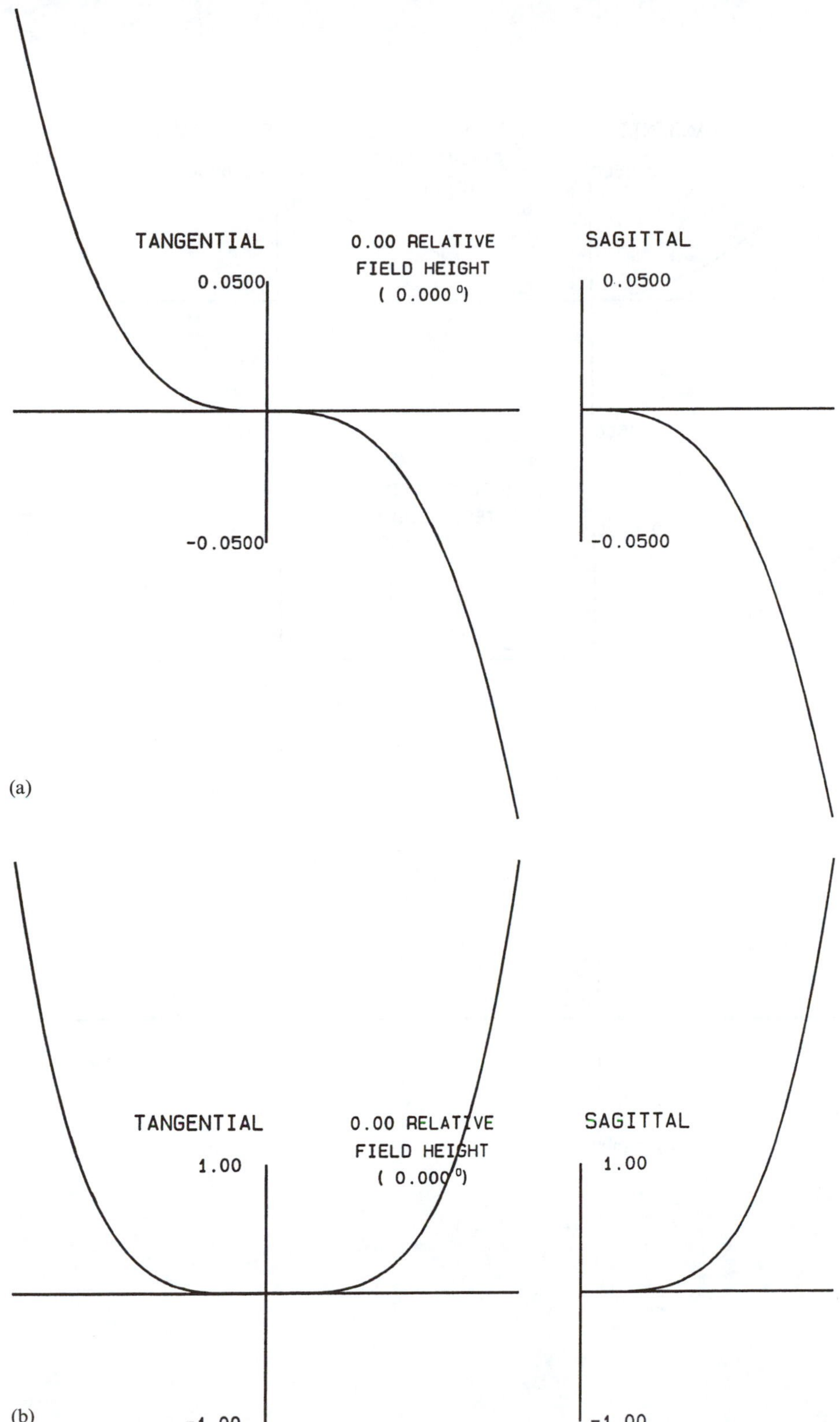

FIGURE 7.141 Aberrations for a single spherical reflecting element. (a) Ray intercept plot; (b) OPD plots (wavelength 587.6 nm).

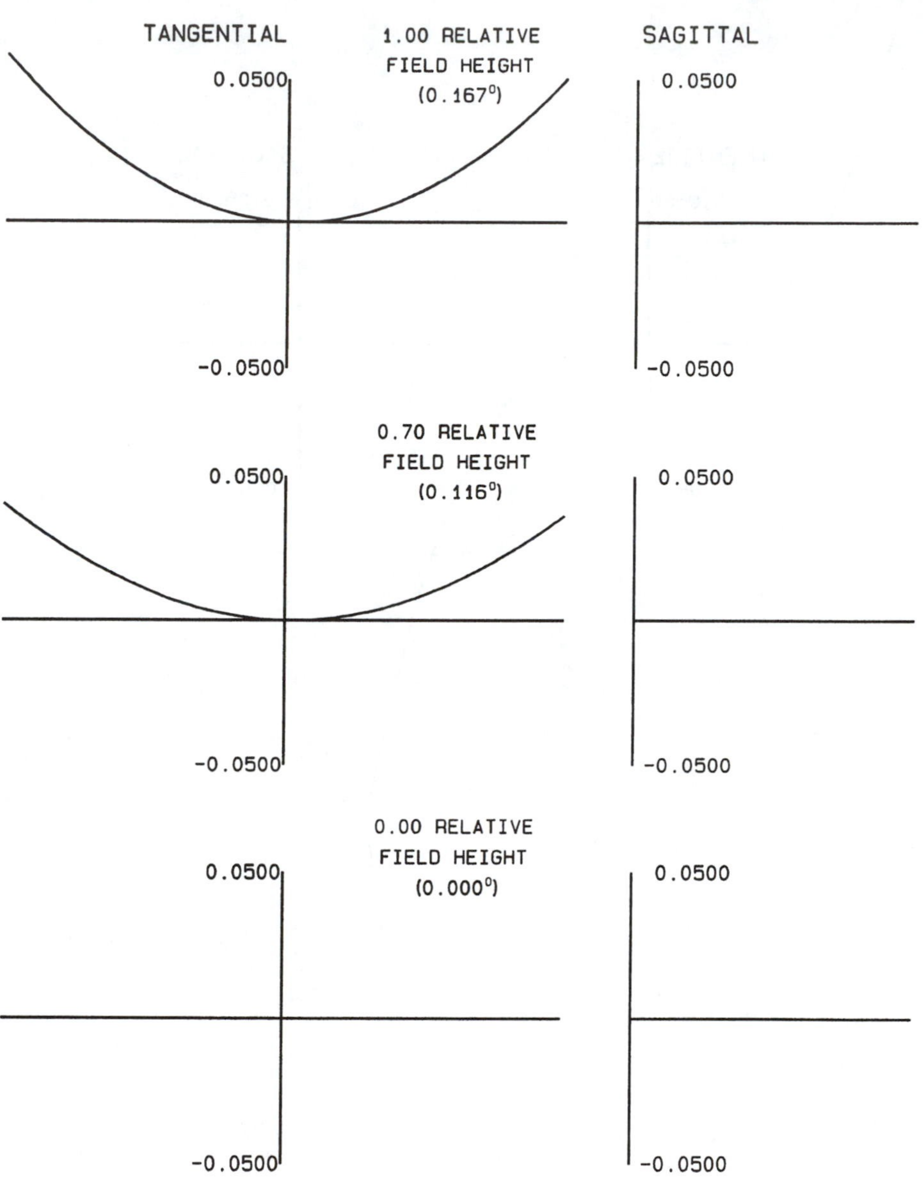
TANGENTIAL
1.00 RELATIVE
FIELD HEIGHT
(0.167°)
SAGITTAL
0.0500
-0.0500
0.0500
-0.0500
0.70 RELATIVE
FIELD HEIGHT
(0.116°)
0.0500
-0.0500
0.0500
-0.0500
0.00 RELATIVE
FIELD HEIGHT
(0.000°)
0.0500
-0.0500
0.0500
-0.0500

(a)

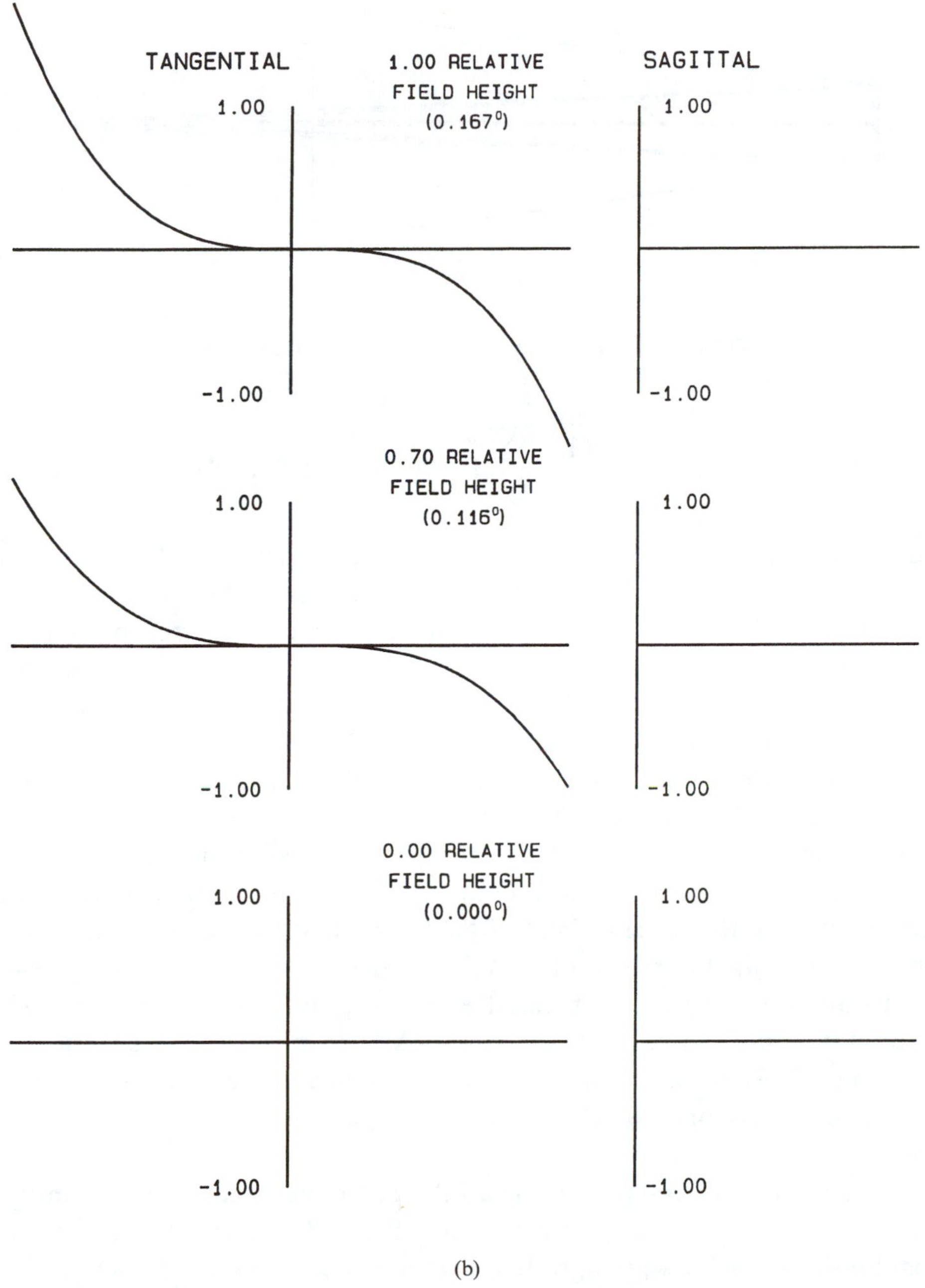

(b)

FIGURE 7.142 Aberrations for a single paraboloidal reflecting element. (a) Ray intercept plot; (b) OPD plots (wavelength 587.6 nm).

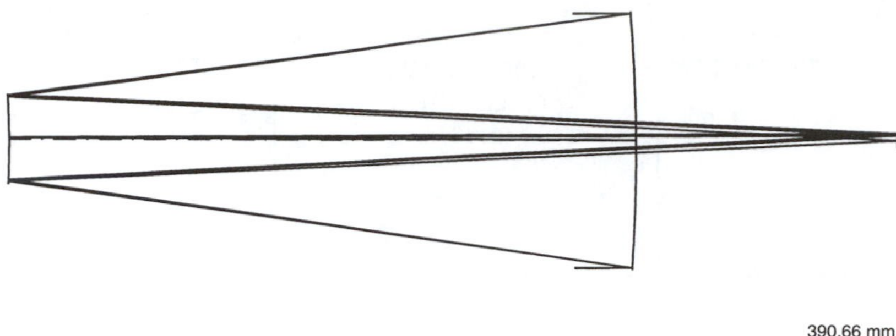

FIGURE 7.143 Layout of an F/2.8 to F/10 Cassegrain.

$$S_{\mathrm{I}_{total}} = y^4\left(\frac{1}{r} - \frac{1}{l}\right)^2\left(\frac{1}{l'}\right)(1 - m) + 2\kappa c^3 y^4 \tag{7.15}$$

For the case of $l = \infty$ then $l' = r/2$ and $m = 0$, so that $\kappa = -1$ in order that the spherical aberration be zero. This is, of course, a paraboloid.

Figure 7.142 shows the ray intercept and wavefront error plots for a paraboloid with the same focal length and diameter as the previous spherical example. The aberration has been completely corrected for the axial image. The residual aberration with field, principally coma, is now evident. The image limitations of a paraboloid can be stated in terms of the angular size of the tangential geometric coma blur in a simple equation.

There are additional limits to the imagery imposed by the obscuration of the paraboloid by the image surface. In most cases, it is desirable to move the image out from the space in front of the paraboloid by use of a fold mirror. The incoming light bundle is obscured or vignetted by the fold mirror, leading to an actual centrally obscured entrance pupil for the telescope. The amount of coma can be expressed as an RMS wavefront error, as described in Chapter 3. Systematic plots of the residual coma versus diameter and field angle can be developed which indicates the region of applicability of a paraboloid.

The next level of complexity for a fully reflective system is a two-mirror system. There are several possible forms of useful systems. The two major possibilities are a Cassegrain, with a diverging secondary mirror, and a Gregorian with a converging secondary. The parameters involving these systems are discussed in an article by Wetherell and Rimmer (1972). The first-order parameters available in the design are the initial F/number from the primary mirror, the final F/number following the secondary, the obscuration, and the location of the image surface. Figure 7.143 shows a possible design for a 1 m aperture Cassegrain with a final F/number of ten. In each case the final image location is (arbitrarily) located 0.1 m behind the primary mirror. A

1° diameter field is used here. There are actually a family of possible designs which vary in obscuration and choice of the primary mirror F/number.

The design in Figure 7.143 provides an image at a desired location and focal length. The aberration content will vary depending upon the choice for the aspherics in the system. Figure 7.144 shows the OPD plots for a "classical" Cassegrain, in which the primary is a paraboloid, and the aspheric hyperboloid on the secondary is selected to provide zero spherical aberration. Observation of these plots indicates that the coma from this system is equal to that of the simple F/10 paraboloid, but that the astigmatism and field curvature is different.

Because there are two parameters available, the aspheric figurings of the primary and secondary, there is additional scope for correction. In the classical Cassegrain, these aspherics are used to independently correct the spherical aberration arising at the primary and the secondary. In the Ritchey–Chretian form of the Cassegrain, these aspherics are set to correct the spherical aberration and the coma in the final image, leaving the spherical aberration not fully corrected to zero by each element individually. The first-order parameters remain unchanged, so that the systems look the same in form.

Figure 7.145 shows the ray intercept plots for this family of systems. The coma is now corrected, but the astigmatism has increased. Since the coma is first order with field, the imagery is good over a widened field of view, until the astigmatism grows enough to dominate the aberrations.

The set of two-mirror systems forms a continuous field of possibilities. Two other named forms of Cassegrains exist. The first is the Dall–Kirkham, which sets the secondary spherical and varies the asphericity of the primary to obtain zero spherical aberration. This has the advantage of a readily tested secondary, but at the cost of a greater amount of coma, and a resulting narrower useful field. The extreme example of aspheric transfer is the inverse Dall–Kirkham, in which the primary is made spherical, and all of the correction is vested in the secondary. The imagery is truly bad for such a design, and almost all such telescopes that have been fabricated have been replaced with either classical or Ritchey–Chretian forms.

If the stop is moved to the center of a spherical reflecting surface, the chief ray angle of incidence on the surface is zero. By symmetry, the coma will be zero, and the imaging properties are not dependent upon field angle. Also by symmetry, it is obvious that the image will fall upon a curved surface having a radius equal to the focal length of the spherical surface. Placing an aspheric surface at this location will permit correction of the spherical aberration. The result is the Schmidt system, which produces no third-order spherical aberration, coma, or astigmatism, but which suffers from the image being located on a curved surface. This basic correction of the spherical aberration permits systems with large numerical apertures and fields.

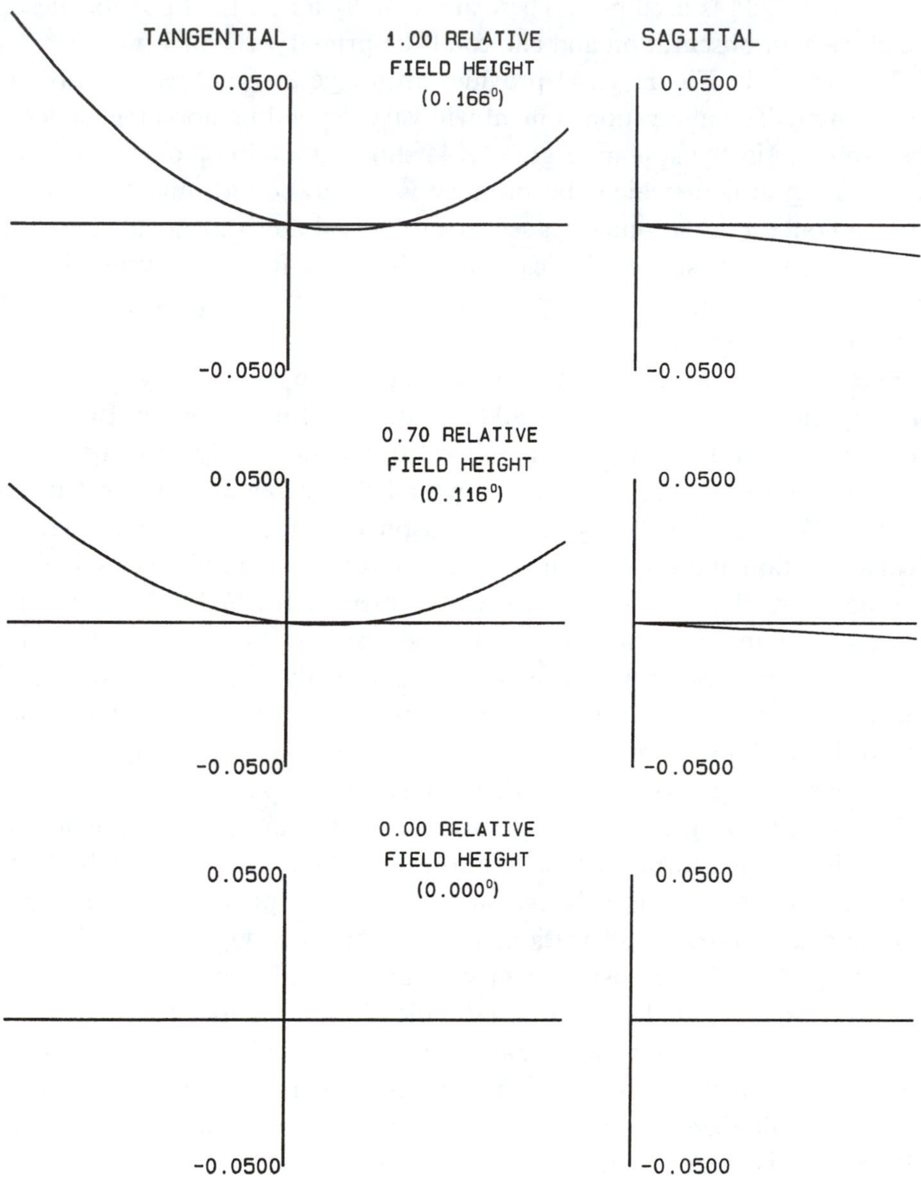
TANGENTIAL
1.00 RELATIVE
FIELD HEIGHT
(0.166°)
SAGITTAL
0.0500
-0.0500
0.0500
-0.0500
0.70 RELATIVE
FIELD HEIGHT
(0.116°)
0.0500
-0.0500
0.0500
-0.0500
0.00 RELATIVE
FIELD HEIGHT
(0.000°)
0.0500
-0.0500
0.0500
-0.0500

(a)

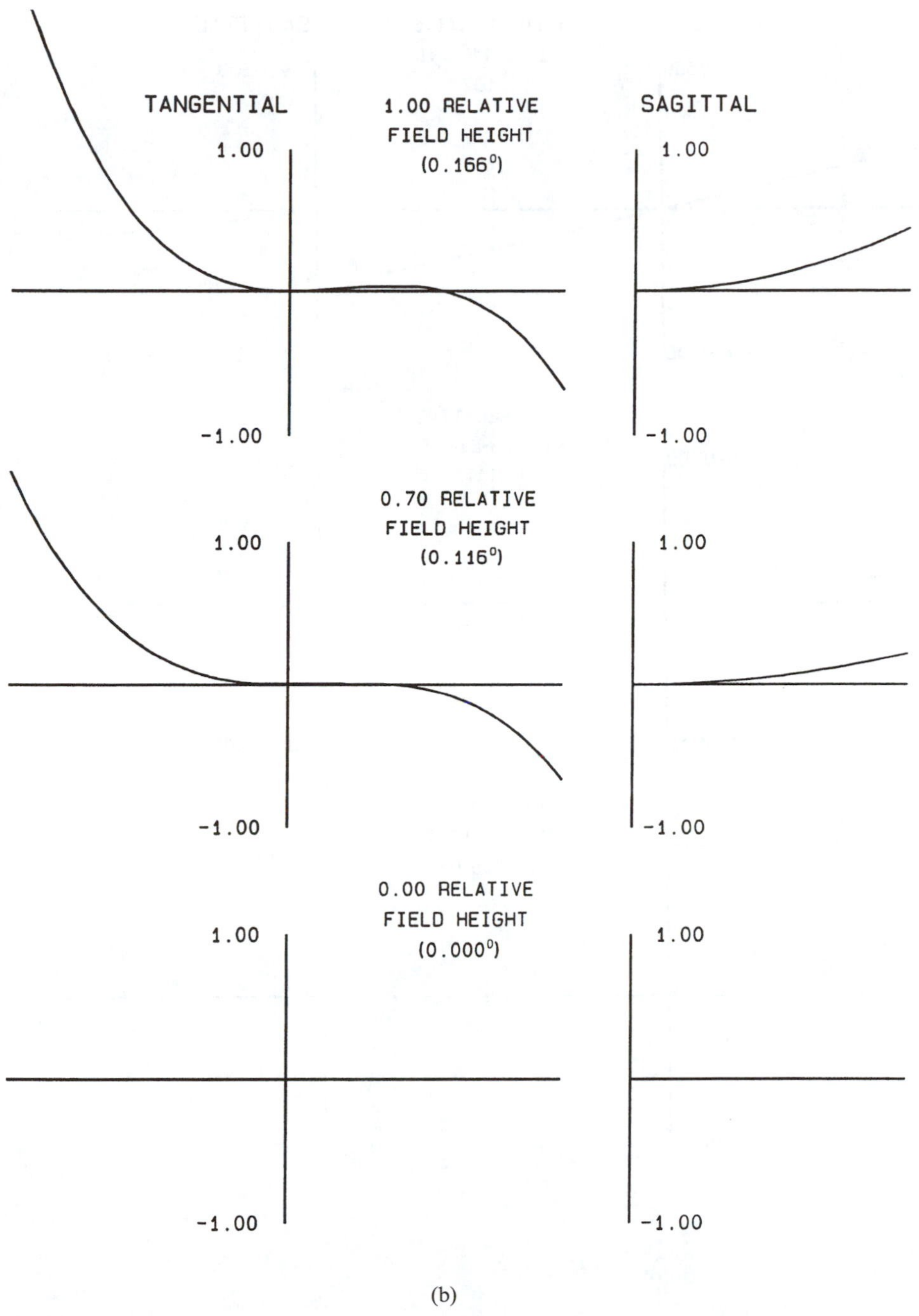

(b)

FIGURE 7.144 Aberrations for a classical Cassegrain system. (a) Ray intercept plot; (b) OPD plots.

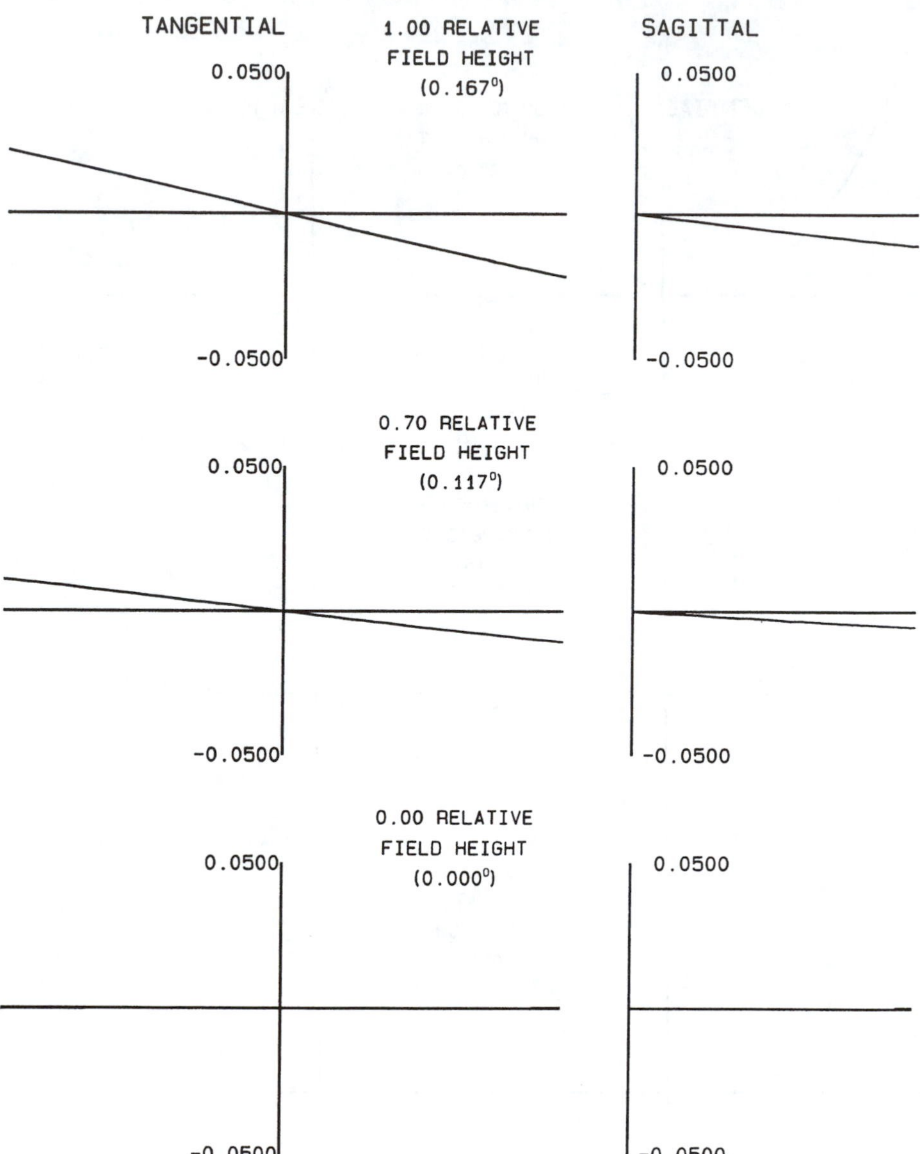
TANGENTIAL
1.00 RELATIVE
FIELD HEIGHT
(0.167°)
SAGITTAL
0.0500
-0.0500
0.0500
-0.0500
0.70 RELATIVE
FIELD HEIGHT
(0.117°)
0.0500
-0.0500
0.0500
-0.0500
0.00 RELATIVE
FIELD HEIGHT
(0.000°)
0.0500
-0.0500
0.0500
-0.0500

(a)

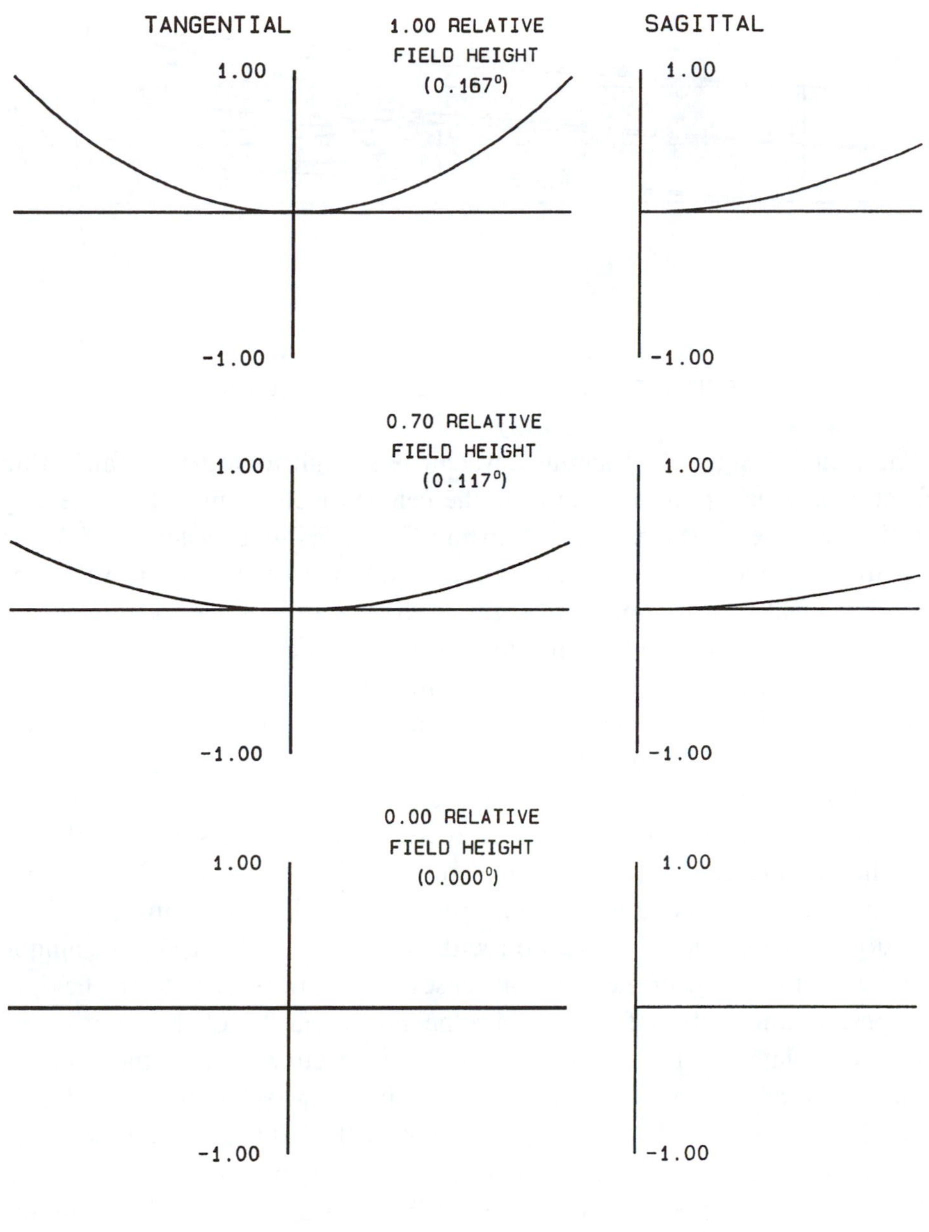

(b)

FIGURE 7.145 Aberrations for a coma-corrected Cassegrain. (a) Ray intercept plot; (b) OPD plots.

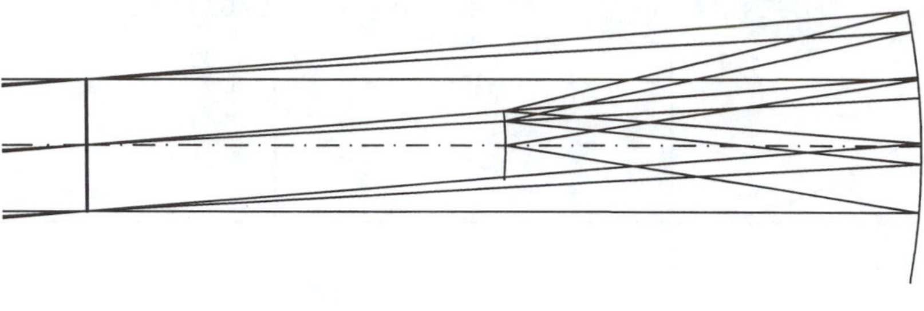

FIGURE 7.146 Layout of an F/3 Schmidt system.

The initial design of a Schmidt system is straightforward. A fairly thin refractive aspheric plate is placed at the center of curvature of a reflecting spherical surface. In the example shown in Figure 7.146, a system of F/3, 1 m aperture, intended for a 10° total circular field is shown. It is known that the final image will be on a spherical surface whose radius is the focal length of the system, therefore an appropriate image surface radius is used in the design. All of the aberrations will be referred to this surface. The design process is quite simple. The aspheric term on the plate is used to correct the spherical aberration to zero.

The design variable initially available is the fourth-order aspheric figuring of the plate. When this is varied to correct the third-order spherical aberration, the ray intercept plot shown in Figure 7.147 is obtained. The ray trace indicates that there is residual high-order spherical aberration, as well as variation of the spherical aberration with wavelength. The design technique is similar to that used in the previous lenses. Rays are traced at zero field, at full aperture and at 0.7 of the aperture height. Variables of the fourth- and sixth-order plate coefficients as well as the base curvature of the aspheric surface are used. The ray intercepts in the primary wavelength are corrected for both rays, and the difference in intercept of the extreme wavelengths are corrected for the 0.7 height rays. A distance solve is placed on the spherical reflecting surface to ensure that the ray aberrations are analyzed in the paraxial image location.

The result is the ray intercept shown in Figure 7.148. The spherical aberration is corrected. The chromatic variation of spherical aberration is compensated by the deliberate introduction of a small amount of longitudinal spherical aberration, resulting from the addition of some small amount of power in the aspheric plate. The image is no longer located precisely at a focal distance from the spherical primary, as there is a small amount of power in the aspheric plate. The ray intercept plots show that the correction holds reasonably well across the field of the lens, except for oblique spherical aberration, and some high-order coma. This uses up all of the available variables.

The basic design now exists. Optimization of the design requires some additional work. Examination of Figure 7.146 shows that the primary mirror must be increased in diameter to avoid vignetting. The usable field of the system also blocks the center of the aperture. For discussion, presume that the only obscuration will be the actual field of the lens. This results in an aperture obstruction. Since the light at the center of the aperture is blocked, it may be that selection of a zone of 0.7 of the aperture as the compensation zone for the spherochromatism may not be optimum. Some experimentation with choice of rays is indicated.

The oblique spherical aberration is the limiting field aberration. The source of this aberration may be understood by reference to Figure 7.146. The circular geometry of the aspheric plate maps onto a circular zone on the primary mirror only at zero field. For oblique angles, a circular zone on the plate maps into an elliptical zone on the primary mirror. Since the spherical aberration correction requires a zonal match on both surfaces, the mismatch leads to a residual and asymmetric oblique spherical aberration.

In previously discussed refractive lenses, the oblique spherical aberration was balanced by the deliberate introduction of astigmatism. This can be done with the Schmidt system to some small extent, by shifting the plate away from the center of curvature of the primary mirror. This will also lead to the introduction of coma, so that the improvement is very limited. The addition of an aspheric to the primary mirror will add a variable that can be used to compensate the coma, and some image quality improvement is possible.

There are several variants of the Schmidt principle. The pate can be used in front of a spherical or aspheric Cassegrain to lead to a Schmidt Cassegrain. This is discussed in a classic paper by Baker (1940). Extreme fields and apertures may be achieved with Schmidt type designs by using multiple corrector plates, or introducing refractive shells into the system that restore the anamorphic projection of the aspheric on the primary mirror.

This brief discussion only skims the surface of the possible reflective and catadioptric possibilities. The design process is quite the same as that of the refractive systems discussed earlier in this book. The obscuration of the aperture by components or image surface needs to be included in the design process, and the merit function altered to include these possibilities. Most frequently, the best approach is to locate a set of spherical components such that the image is in a desired location. Then the aspherics are added to correct the aberrations.

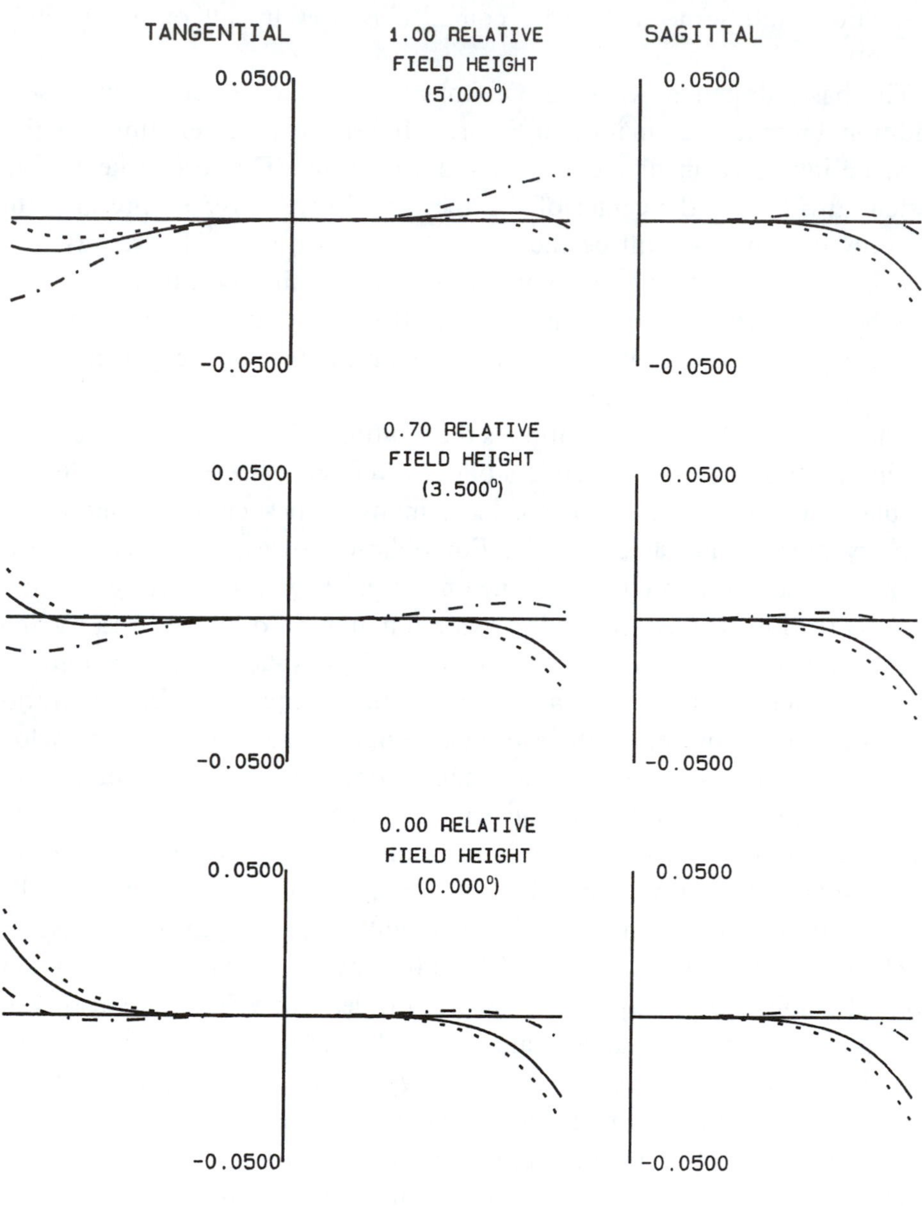
TANGENTIAL
1.00 RELATIVE
FIELD HEIGHT
(5.000°)
SAGITTAL
0.0500
-0.0500
0.0500
-0.0500
0.70 RELATIVE
FIELD HEIGHT
(3.500°)
0.0500
-0.0500
0.0500
-0.0500
0.00 RELATIVE
FIELD HEIGHT
(0.000°)
0.0500
-0.0500
0.0500
-0.0500

(a)

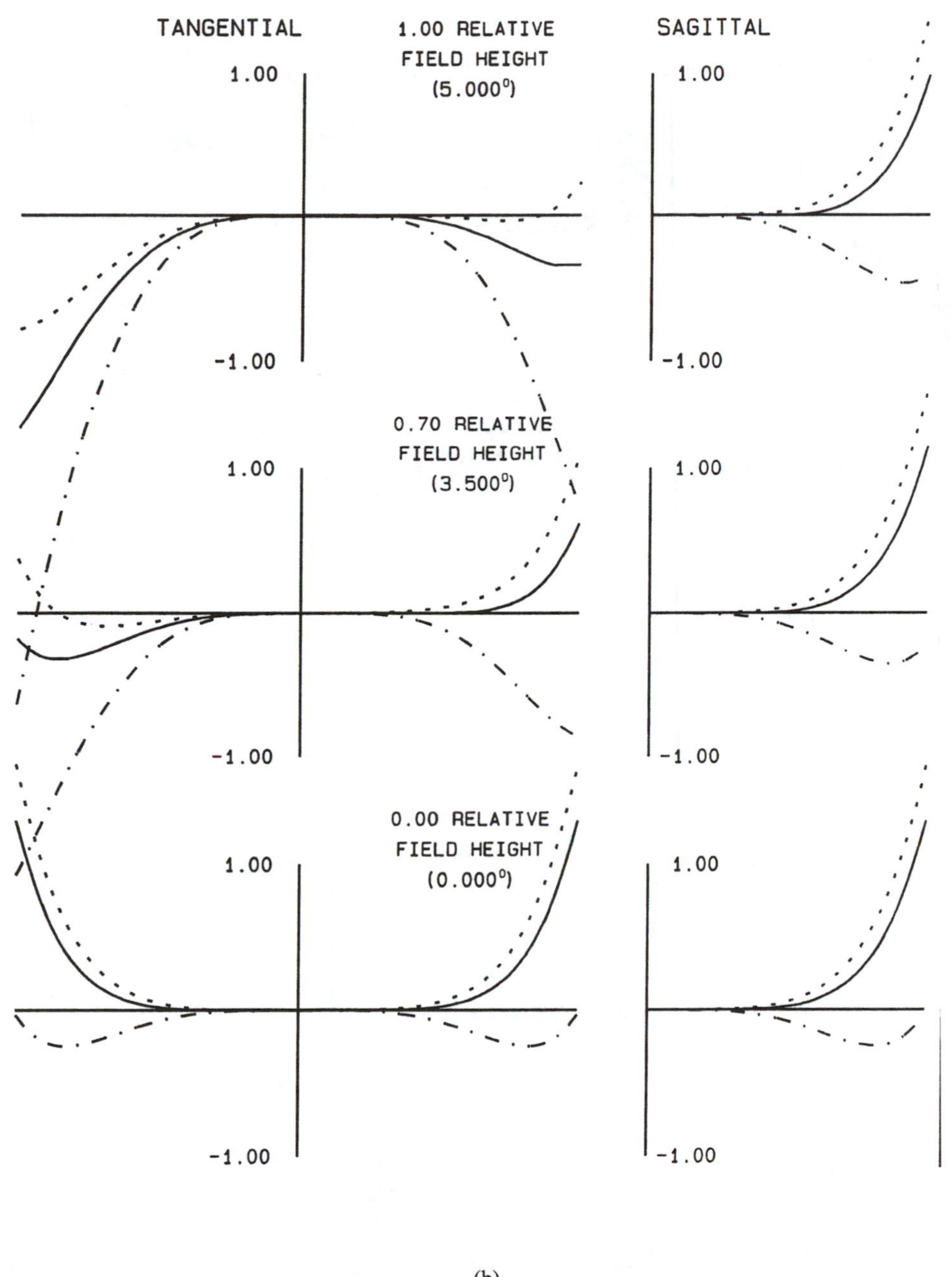

(b)

FIGURE 7.147 Aberrations for a basic Schmidt system. (a) Ray intercept plots; (b) OPD plots. ······, 656.3 nm; ——, 587.6 nm; ·—·—·, 486.1 nm.

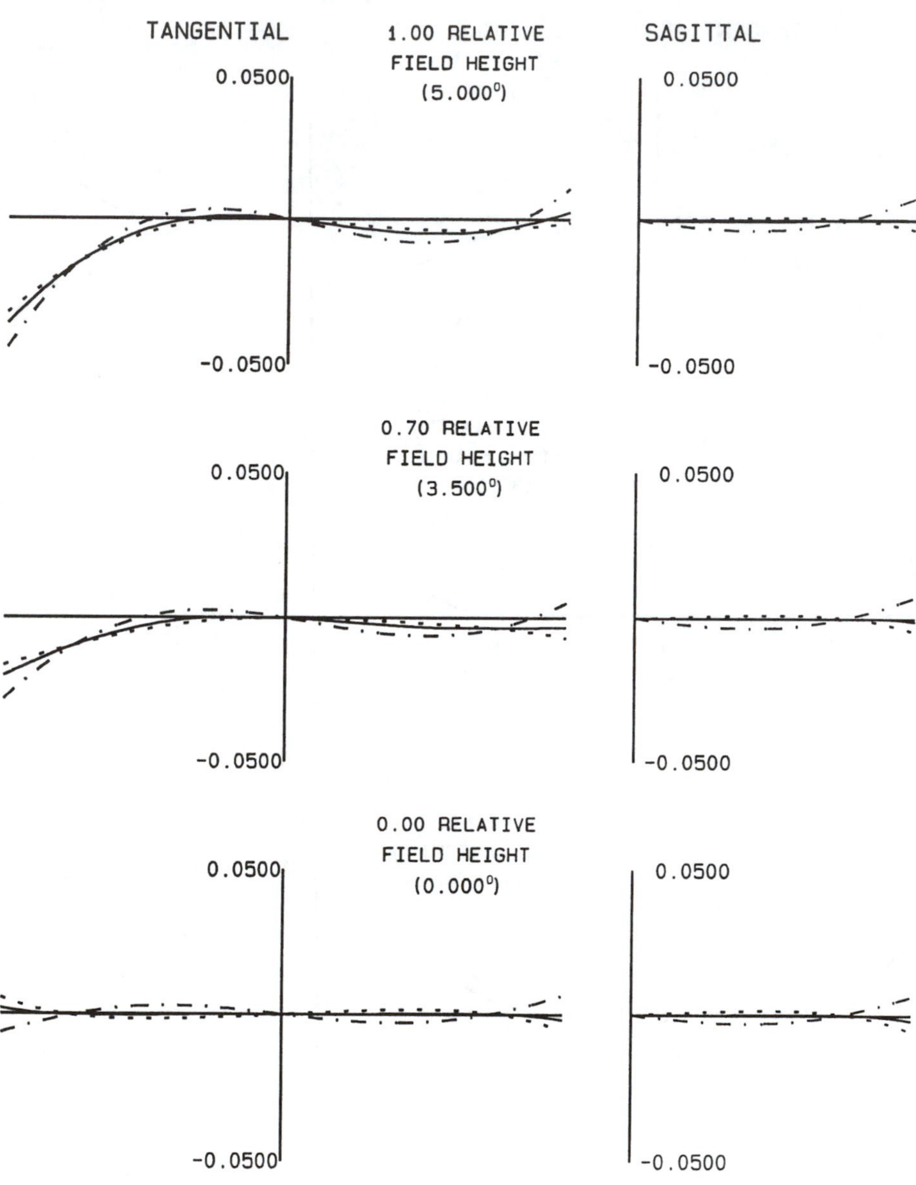

TANGENTIAL
1.00 RELATIVE
FIELD HEIGHT
(5.000°)
SAGITTAL
0.0500
-0.0500
0.0500
-0.0500
0.70 RELATIVE
FIELD HEIGHT
(3.500°)
0.0500
-0.0500
0.0500
-0.0500
0.00 RELATIVE
FIELD HEIGHT
(0.000°)
0.0500
-0.0500
0.0500
-0.0500

(a)

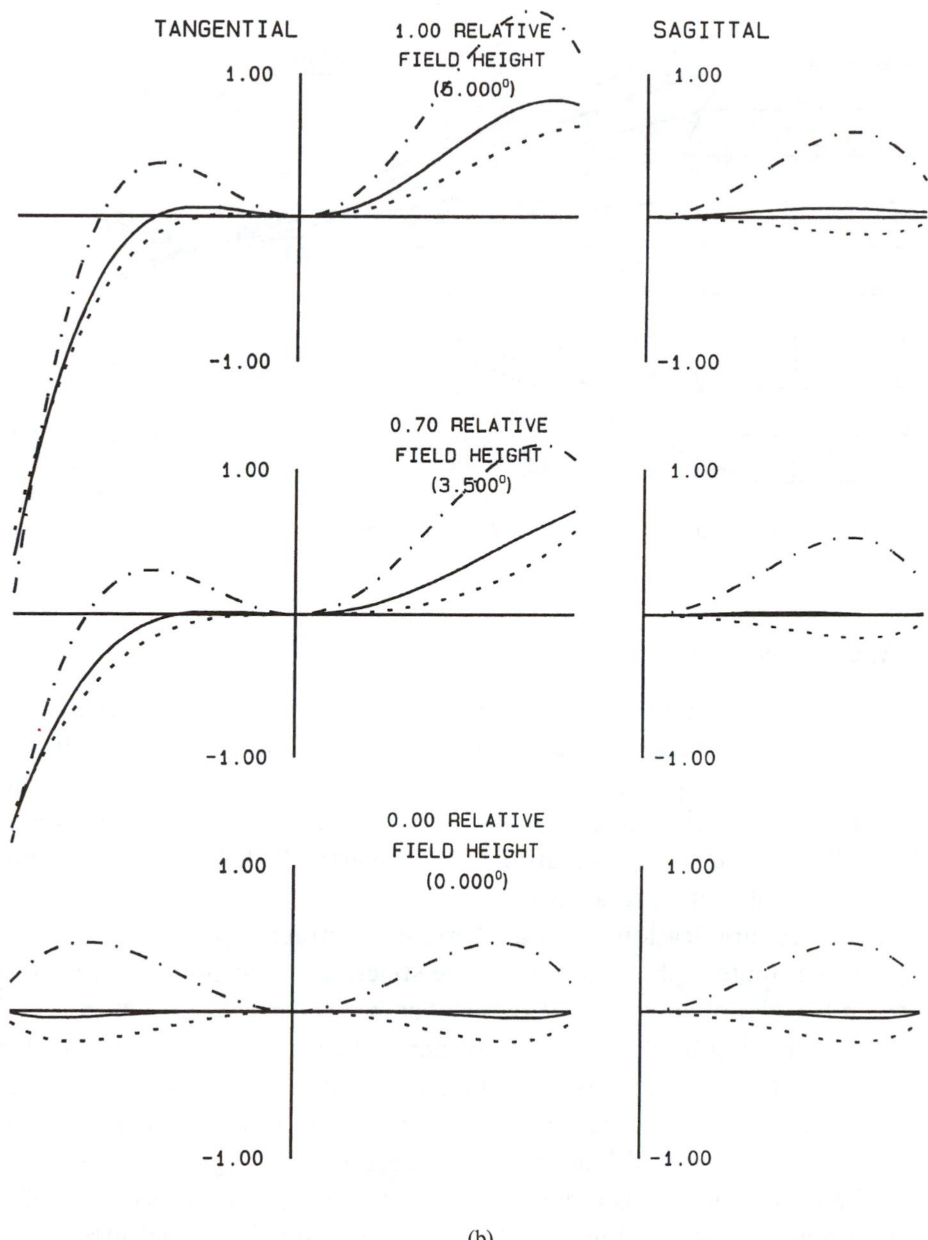

FIGURE 7.148 Aberrations for an optimized Schmidt system. (a) Ray intercept plots; (b) OPD plots. ······, 656.3 nm; ——, 587.6 nm; ·—·—·, 486.1 nm.

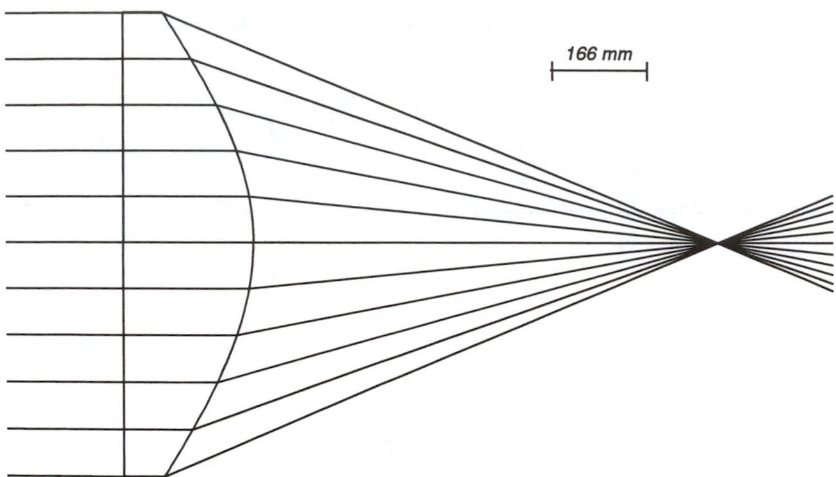

FIGURE 7.149 Aspheric singlet lens. NA = 0.5, focal length = 800 mm (OSLO).

7.1.6.2 Refractors

The use of aspherics in refractive lens design provides the ability to reduce the number of components, or to relax the correction requirements by adding a useful new variable so that a net improvement in the design can result. The addition of an aspheric will change the spherical aberration correction from a surface. The variation of aberration with wavelength, field, and conjugate position will still remain in all cases.

The simplest application of an aspheric in a refracting lens system is the addition of a single aspheric to correct the spherical aberration. Figure 7.149 shows an F/1 plano-convex singlet which has been corrected for all orders of spherical aberration by the use of an aspheric second surface. The particular surface that accomplishes this correction is a hyperboloid, with conic constant equal to minus the square root of the refractive index of the glass. The dispersion of the glass will limit perfect correction to a single wavelength. A perfect image on the axis is obtained, as can be seen from the ray trace plot for the on-axis image in Figure 7.150. Note the small residual error that results from the use of a single conic surface. The irregularities in the plot are due to numerical truncation in the computer, as the scale of the plot (0.5×10^{-13}) indicates. This indicates that the conic surface is a natural surface for zero aberration, but the correction for coma is not influenced by this aspheric surface, as is seen in Figure 7.151. In the plot across the field, the coma grows rapidly. At the edge of the 0.1° field, the coma has grown to almost 2 mm in the tangential direction. This is actually much greater than the paraxial field height.

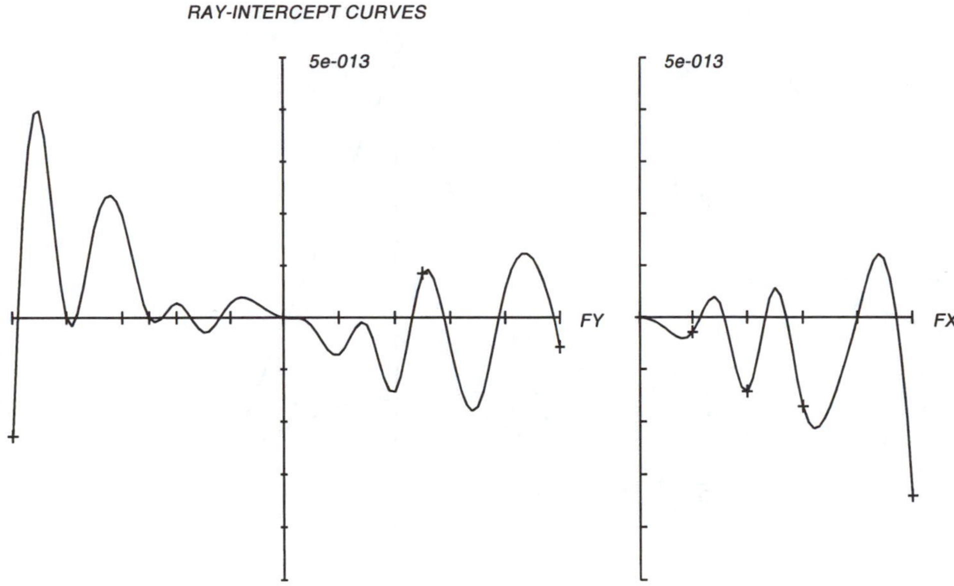

FIGURE 7.150 Ray trace plot for axial image from aspheric lens. Note extremely small scale, which shows effect of finite computer word length in ray trace (OSLO).

Improvement of the imaging in the region of the axis can be obtained by bending the lens to eliminate the third-order coma, as well as the third-order spherical aberration. The asphericity is also added to the first surface, because it is the primary power surface and is the main generator of aberration for the lens. This optimization was done to obtain the lens shown in Figure 7.152. It is now found that the high-order spherical aberration is not automatically zero, so that a sequence of high-order aspheric terms need to be added to correct the aberration. The example shown used the sixth- through the sixteenth-order aspheric terms to correct a set of equally spaced rays in the pupil. The residual error is quite small, of the order of a micron, with the residual OPD of the order of 0.1 waves. The aberration is no longer corrected for all orders by the choice of a conic asphere, since the ray path through the lens including the second surface requires an aspheric that is no longer a conic. This behavior is typical of correction of an aspheric system in which aberration exists between the surfaces of the lens. The use of the high orders produces an irregularity in the ray intercept curves as shown in Figure 7.153. The designer can experiment with successively higher orders to find which surface is adequate for correction. Because a finite number of ray intercepts is used in the optimization, the choice and placement of the rays will drive the aspheric terms that are obtained.

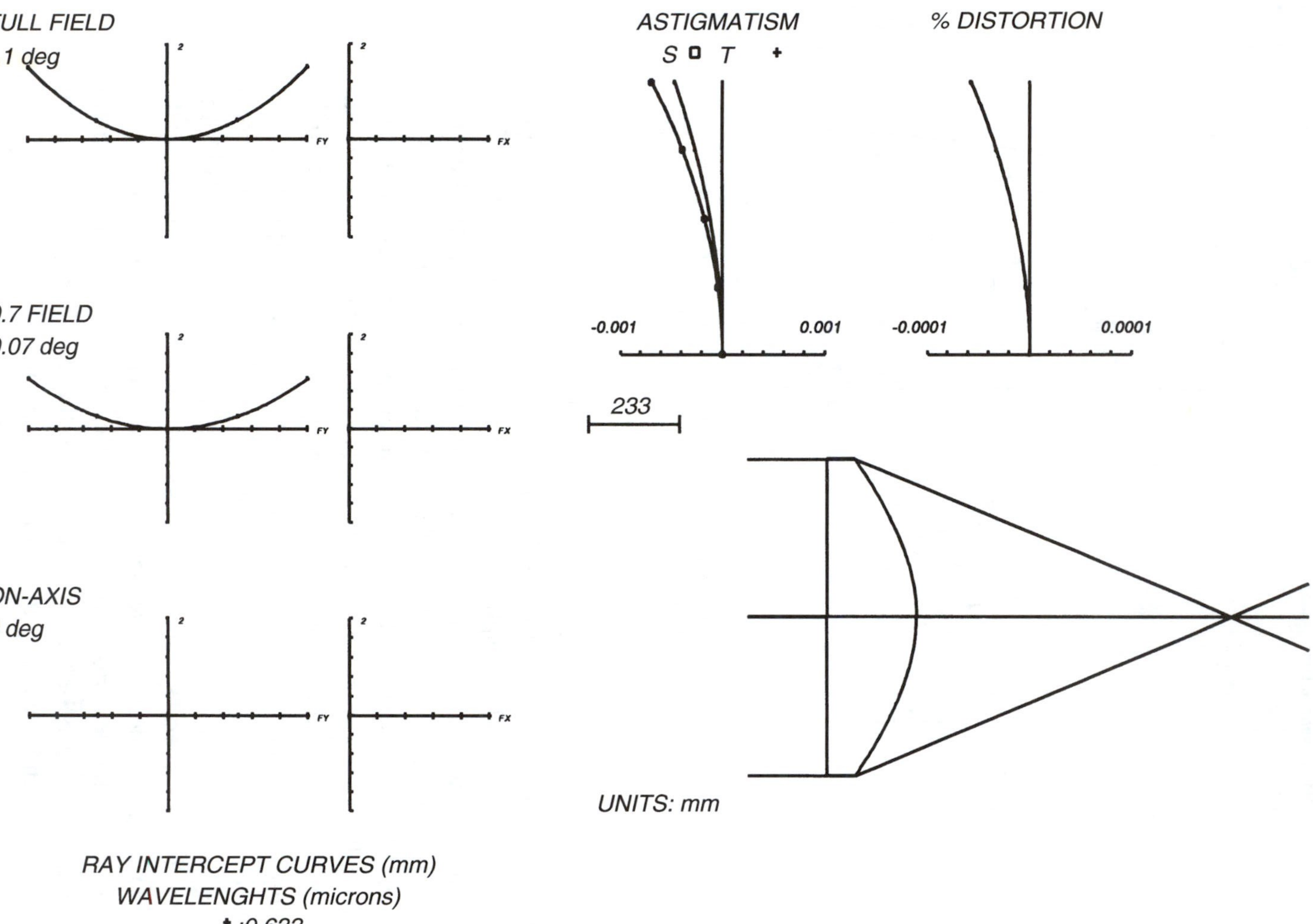

FIGURE 7.151 Monochromatic ray plots for aspheric lens (OSLO).

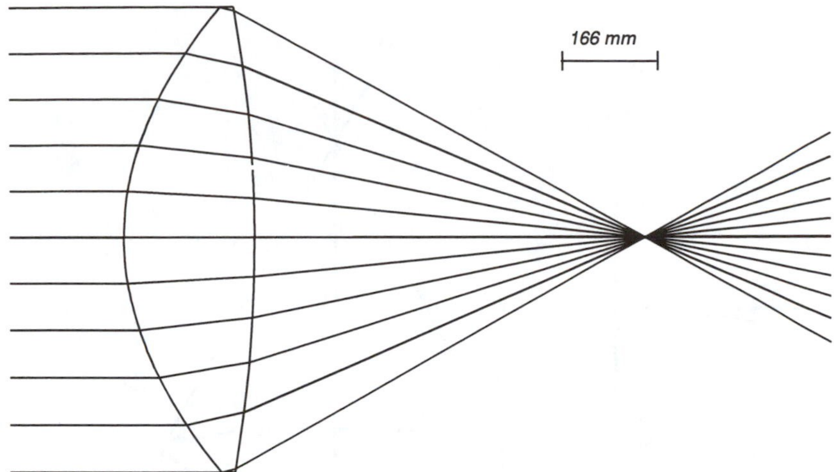

FIGURE 7.152 Aplanatic aspheric lens (OSLO).

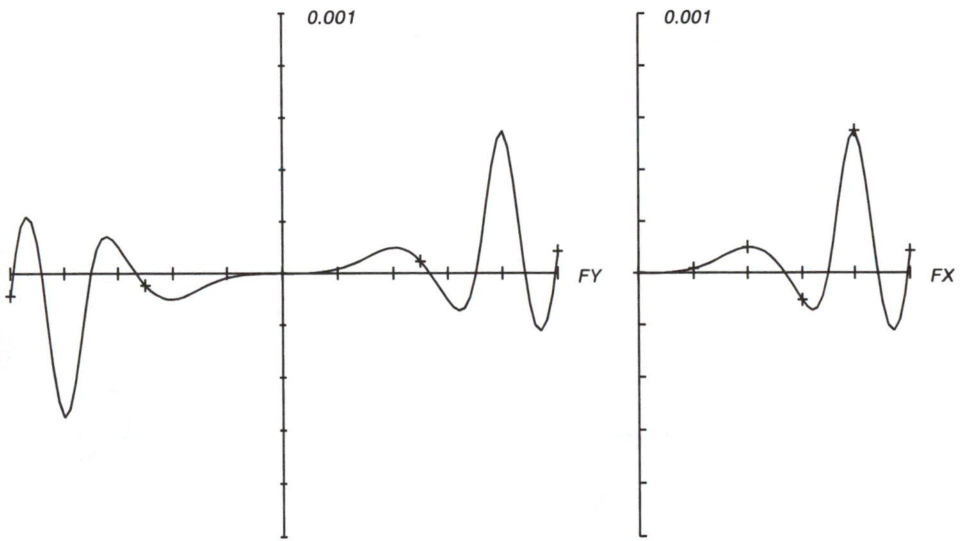

FIGURE 7.153 Axial ray intercept plots for aplanatic aspheric lens (OSLO).

There is some art as well as science in the selection of the representation of the aspherics in problems of this sort. The lower-order aberration balances are based upon the logic discussed earlier in this section. The higher-order terms will be a consequence of the path taken toward the solution of the design. Another interesting aspect of the design is shown in the ray plots across the field, shown in Figure 7.154. The coma is indeed less, about eight times less, as can be seen from the scale of the plots. The shape of the ray intercept plots indicates that the residual coma is fifth-order coma, not the

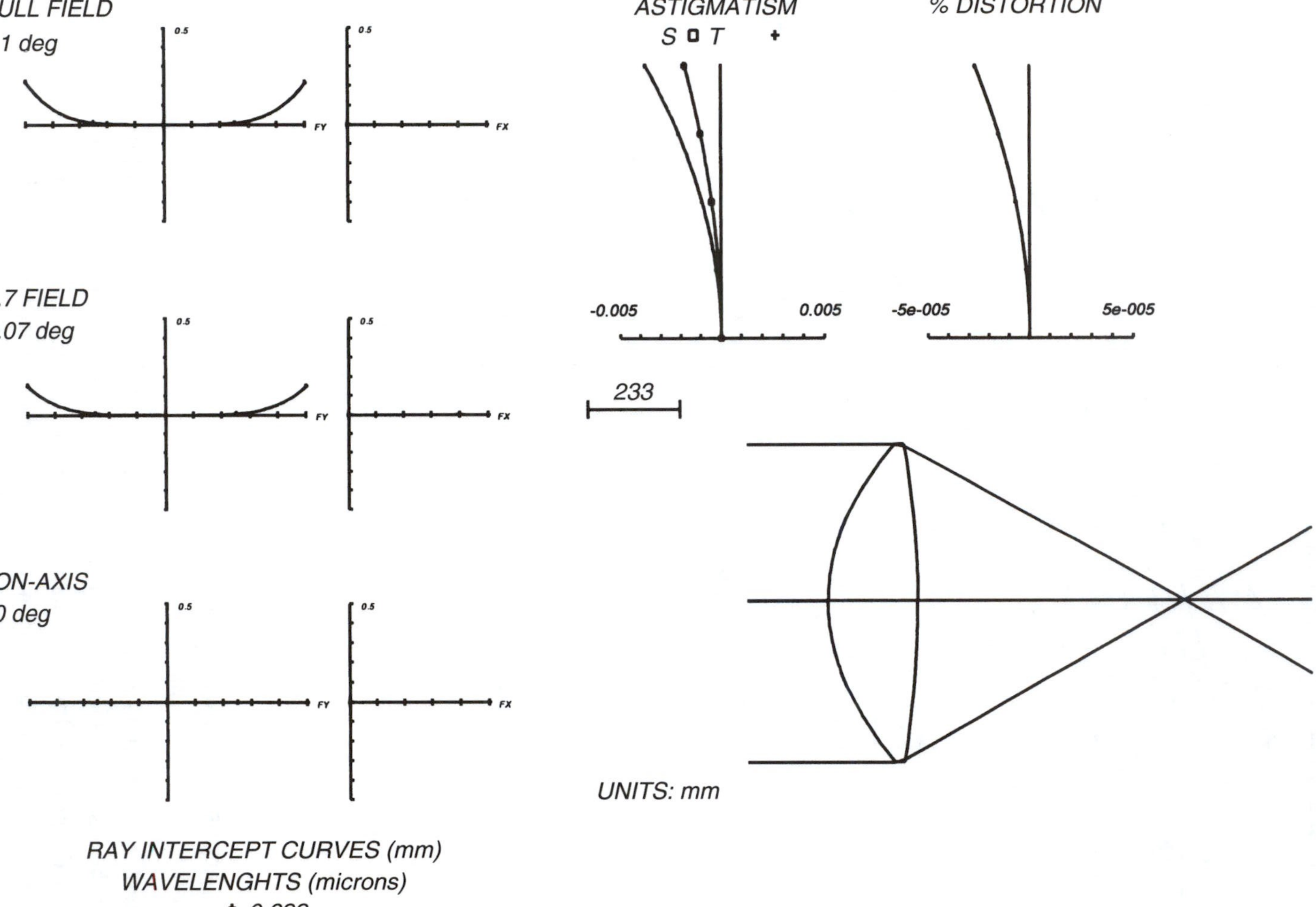

FIGURE 7.154 Ray plots for aplanatic aspheric lens (OSLO).

(a) NA .5 Aspheric Lens

Srf	Radius	Thickness	Aperture Radius	Glass
0	----	1.0000e+20	1.7453e+19	AIR
1	----	225.0000	400.0000	BK7
2	-412.01736	800.0000	400.2592	AIR
3	----	----	1.3962	

Conic and Polynomial Surface Data

Srf	CC	AD	AE	AF	AG
2	-2.295495	----	----	----	----

Wavelengths

1	0.6328

(b) Aplanatic Aspheric Lens

Srf	Radius	Thickness	Aperture Radius	Glass
0	----	1.0000e+20	1.7453e+17	AIR
1	483.066864	225.000000	400.0000	BK7
2	-2.3598e+03	673.319579	400.0000	AIR
3	----	----	1.3963	

Conic and Polynomial Surface Data

Srf	CC	AD	AE	AF	AG
1	-0.809120				

Aspheric Surface Data

Polynomial coefficients of order N

N=	4	6	8	10	12
	0.0	-6.2750e-16	-1.3500e-21	-1.0643e-27	1.1161e-33

14	16
3.9343e-39	-8.883e-44

FIGURE 7.155 Design data for aspheric singlets. (a) Plano-aspheric lens; (b) aplanatic aspheric lens (OSLO).

third-order coma seen in the previous design. This is an indication that the strategy taken in finding the solution should be different, with the fifth-order coma added to the merit function, or that a ray-based coma aberration be incorporated. The difficulty of locating a solution of the high-order aspheric terms still remains, however. The data for the two singlet designs is shown in Figure 7.155.

The first example required complete correction for the spherical aberration. For imaging conditions in which the correction does not need to be perfect, an approximate aspheric can be important. An example is a condenser lens, which is used to image the filament of a projection lamp into the aperture of a projection lens. The numerical aperture required for such a lens can be quite high, as it must equal the field angle of the projection lens. A projector having a 20° half field will require a condenser numerical aperture of 0.34, which corresponds to an F/1.4 lens. The spherical aberration produced by such a lens is extremely large. The image quality that is required is not very high, as the goal is primarily to get all of the light collected from the filament through the aperture of the projection lens, but the spherical aberration usually needs to be minimized. In order to use a minimum number of elements for the condenser, molded glass aspheric surfaces are commonly used. Since the numerical aperture of the condenser is related to the field of view of the projection lens, the distribution of rays across the condenser aperture may differ considerably from a sine condition, or coma-corrected, imaging condition to enhance the uniformity of illumination.

An important use of aspherics in design of objectives is the addition of an aspheric to change the intrinsic amount of third-order aberration that can be expected from an element in a complex lens. Chapter 3 showed that some deliberately introduced spherical aberration is necessary to obtain astigmatism correction. This behavior is not changed by introducing aspherics. The degree of freedom obtained from aspherics can be used to adjust the spherical aberration to a desirable value, while maintaining other characteristics of the lens. Simply sprinkling aspherics throughout a lens will not make perfect imagery possible, except for one location in the field.

The use of a single aspheric to permit improvement of the images in a lens is shown by adding an aspheric to the central element of a triplet. The solution can be permitted to proceed without concern for the spherical aberration, and the asphericity used to correct the final spherical aberration. It is important to use this added variable wisely, by adding some additional conditions to the lens to permit an improvement in the aberration correction across the field.

7.2 Nontraditional Designs

The majority of optical systems are designed using rotational symmetry and homogenous optical materials with smooth continuous surfaces. The reason for this is the normal requirement of image symmetry as well as the availability of materials and fabrication and test methods for producing optics. There are some special applications of optics, which have become important in recent years, that require techniques not available in traditional systems. In some cases these techniques require the incorporation of transfer of radiation by coherent or partially coherent methods, as in beam transfer or illumination systems. In other cases, new techniques for modifying the passage of light through the system, such as gradient index or diffractive surfaces, may be required. In a few (but an increasing number of) problems, surfaces with arbitrary symmetry, such as toric surfaces or general asymmetric aspherics, may be usefully employed.

This section will review some of these topics, and show how the basic knowledge of design procedures discussed earlier in this book may be used or modified to meet the need of such nontraditional designs. The basic approach to design will, in most cases, follow those of the previous section. The fundamental system properties may be described by a geometrical optical ray model for the passage of light through the system. Formation of images with isoplanatic imagery and controlled field curvature will require adherence to the principles discussed in Chapters 2 and 3. The optimization of the lens requires the establishment of a merit function, which will describe the aberration set to be minimized. This merit function must be representative of the desired result. In the case of imaging, this is reasonably well known. In the case of optimization for coherent or incoherent energy transfer, the description of the desired "image" requires some additional care.

7.2.1 BEAM TRANSFER SYSTEMS

Beam transfer systems are intended to convey light from a coherent source, usually a single-mode laser, through a region of space. The divergence and size of the beam must be adjusted to match some required optics or irradiance profile. The operation may be to tailor one-way passage of light through a space, or to permit multiple passes as a part of a laser resonator. The action of the optical components can be modeled using geometrical optics, but the evaluation of the beam transfer requires physical optics. In many cases, design of a beam transfer system can proceed using Gaussian beams in a manner similar to paraxial optical analysis. In other cases, truncation of the beam, or use of optical components of dimensions of a few wavelengths may require detailed modeling using diffraction.

Laser beam transfer may also take place through optoelectronic components such as modulators, nonlinear switches, and wavelength convertors. In each case, the aberrations produced by these devices need to be considered, as well as the focusing properties of the devices. The propagation of a Gaussian beam through an optical system has been dealt with in many articles and textbooks. Gaussian beam propagation is to beam transfer calculations as paraxial optics is to geometrical optics. The use of a lens design program with Gaussian beam modeling associated with the passage along a real ray adds a very important capability to the design process. The usual textbook Gaussian beams describe axially symmetric beam transfer. Tracing a Gaussian beam along a ray includes the change in beam parameters that occurs from the astigmatic errors described in section 2.1.5. The direction and the astigmatic dislocation of the beam waist that is produced by passage nonsymmetrically through a component is modeled, as well as the exact direction of the centroid of the Gaussian beam. Therefore the variation of focus with azimuth in the pupil is obtained. In many cases this is all that is required to significantly accurately optimize the beam.

Incorporation of other aberrations into the analysis requires the modeling of the system as a full-scale wavefront weighted by a Gaussian apodization. The geometrical evaluation of the imaging will provide the aberration information. A diffraction calculation with a Gaussian weighting will provide the actual diffraction image. There are two numerical difficulties that can occur in most design programs, however. The size of the matrix used in the diffraction calculations is chosen to be convenient for the majority of computations of diffraction images. This use of the default array dimensions may lead to poor sampling of the wavefront and thus erroneous results. The second difficulty that may arise is the propagation or diffraction calculation may be based upon a far-field (Frauenhofer) model of the diffraction, which will not permit accurate near-field propagation. Several design programs provide the opportunity for the user to select the mode of diffraction so that good modeling is accomplished. It is good practice for the designer to carry out some calculations using a system for which the result is well known before trusting completely any diffraction or propagation computations on a system with aberrations that will complicate the results.

When this modeling of the results is understood, the designer can proceed with confidence in using the techniques described earlier in this book to pursue optimization of coherent beam transfer systems.

7.2.2 CYLINDRIC AND TORIC OPTICAL SYSTEMS

These systems are aspheric systems, with the added issue that the paraxial power is different in two principal azimuths. Most design programs permit the designation of separate paraxial base sets of the principal azimuths. The paraxial requirements for image position and magnification can be established in the usual manner. The aberrations that result in such systems are subject to new symmetry conditions, and no lens design programs include a good set of aberration coefficients for this case. The principles established for optimization of imaging systems using ray-based merit functions will apply. The designer needs to be sure that the number of fields and number of rays used in setting up the merit function are adequate to describe the image properties. The designer will have to take the responsibility for establishing the appropriate merit function in many cases.

The entrance and exit pupil conditions in these systems can be a bit complicated, as there may be immense pupil aberrations. The designer needs to examine carefully the construction of the wavefront that is used in a diffraction calculation in order to determine if the computation is correct. The usual suggestion that examination of simple cases with stopped-down pupils be used in such calculations in order to develop confidence in the results is strongly recommended.

7.2.3 SCANNERS

Scanners are a special class of optical systems that use active components such as a rotating prism or mirror, or perhaps a Bragg cell, to move a beam across a space in a controlled manner. One type of scanner is used to move a modulated beam across a photographic film to build up an image from a video input line by line. The inverse of this scanner may be used to examine an object element-by-element and convert this to a video data stream, as in a film scanner or a barcode scanner. Other popular scanners are not intended for image formation but for obtaining information about specific targets, such as the product codes on grocery items. In this section we will consider only one type of image-forming scanner, that of mechanical rotating prismatic scanner with an imaging system that produces a reasonably high-resolution line image. The scanning system receives a beam from a modulated laser beam. In order to cover the line on a flat field, the active facet of the spinner must pass through the entrance pupil of the scanning lens, and the lens must convey an unvignetted beam for all scan angles.

In order to reduce the effect of angular errors between facets of the spinner, scanners of this type are usually anamorphic, with significantly different focal lengths and magnifications in the x and y directions. A ten-to-one ratio of magnifications is common. The scanning lens then will contain cylindric components, with one direction of the field covering a wide field and the other covering a very narrow field. In this case, it is only essential that the pupil in the wide-angle direction be well defined.

There are many possible forms for scanners. In most cases the active component is a spinning mirror. Other active components could be rotating wedges, or spinning holograms. There is a growing use of digital and acousto-optic or electrooptic deflectors. One of the major design problems is finding the correct modeling for the action of the active part of the scanning optics. In most cases this is accomplished by using the multiconfiguration option of the design program to provide dynamic elements that closely, or exactly, simulate the behavior of the active component.

7.2.4 ASYMMETRIC OPTICAL SYSTEMS

Most of the previous discussion has been on the use of rotationally symmetric components aligned on a single optical axis of symmetry. In some cases nonrotationally symmetric components or nonaxially symmetric locations of components are required to satisfy particular imaging requirements. Common refractive nonsymmetric systems are anamorphic magnification systems in which the image is compressed in one azimuth. Common reflective systems are unobscured-aperture reflecting telescopes.

There are two general cases. The most usual case requires the assemblage of eccentric components to behave as a rotationally symmetric lens would. The second is one in which a required transformation, such as a magnification difference, is maintained between two principal azimuths. Sometimes a specific mapping is required between the object and image, which can only be accomplished by some unconventional use of imaging components.

There is a wide variety of such systems. The design process follows the procedures discussed earlier in this book. The greatest difficulty is usually in ensuring that the modeling of the system is correct. The designer may have to make trial adjustments of parameters such as tilts and decenters in order to be sure that the desired effect is being produced. In some cases the use of isometric or perspective drawings of the lens will be required to avoid some interference with the imaging beam that occurs in unusual three-dimensional folding of lens systems. This can be especially difficult when a wide-angle imaging system is involved.

The merit function content is based upon the same logic as before, except that the elements of the merit function need to completely cover the aberration possibilities that may exist. As in the case of toroidal systems, lens design programs do not contain natural aberration coefficient descriptions of the aberrations of various symmetry systems. The aberrations must be judged from the appearance of the ray intercept plots and spot diagrams. Adequate sampling of the pupils is required to ensure that the result of convergence of the merit function will ensure convergence of the image quality.

There are numerous problems in the accurate modeling of asymmetric systems. The intrinsic tilts and decenters of powered components will produce linear tilt distortion and other nonuniformity of the pupils. The rules about isoplanatic imagery still apply, however, even though the coordinates are not always obvious. The designer may have to add some pupil aberration control to the design in order to force conditions such as the aplanatic condition to be present. All lens design programs offer this option, but it is up to the designer to understand how to carry it out.

7.2.5 GRADIENT INDEX OPTICS (GRIN MATERIALS)

Gradient index optics is the controlled propagation of light rays through inhomogeneous media. Rays passing through a medium in which the index of refraction is a function of position follow a path defined by the local gradient of the index of refraction. The ray passage is described by the solution of a differential equation. Ray tracing through inhomogeneous media is therefore considerably more complicated and lengthy than for media with constant index of refraction. The general ray-tracing process is to trace a ray by increments, evaluating the ray-tracing differential equation at each step. Using short steps provides more accuracy in the ray tracing at the cost of more computation time. Aberrations produced by passage through inhomogeneous media can be evaluated by fitting optical path errors, as in homogeneous optics. Aberration coefficients can be derived by knowing the functional dependence of the index of refraction in the medium. The determination of the OPD and position of rays in the exit pupil is adequate to fully describe the imagery that occurs.

It is possible to have index gradients that have arbitrary symmetry, as in the index variations in the atmosphere that produce wavefront errors after passing through a turbulent atmosphere. The combinations used in optical design are considerably more limited, and are static with time. There are three basic type of gradient optics that are used. The gradient types are spherical, radial, and axial. As implied, this indicates three-, two-, and one-dimensional gradients. The ray behavior is complicated by the fact that surfaces in a lens may cross through iso-index contours and provide a change in the angle of refraction that depends upon the coordinates at which the ray intersects the surface.

The gradient index material may be described in a number of ways. The usual description involves a base index, and coefficients of linear or higher-order variation of the index from the base value. Usually the index gradient is less than 1% of the nominal index of refraction of the material, although in some cases the amount of the gradient can be as much as 25% to 30% of the base index of refraction. The gradient can interact with the surfaces of the element, causing a linear gradient to have a radial effect upon a beam. In such cases, a gradient index element can have much the same effect upon image formation as an aspheric surface.

Some frequently used formulae for describing the gradient material are a linear gradient through the lens, described by

$$n(z) = n_0 + a_1 z + a_2 z^2 + a_3 z^3 \tag{7.16}$$

Another option is a radial gradient across the aperture described by

$$n(x, y, z) = n_0 + b_2(x^2 + y^2) + b_4(x^2 + y^2)^2 \tag{7.17}$$

where the b coefficients are usually chosen to have even symmetry only, for focusing and aberration reasons discussed in earlier chapters.

In three dimensions, a useful symmetry is spherical symmetry:

$$n(x, y, z) = n_0 + c_2(x^2 + y^2 + (z - z_0)^2) + c_4(x^2 + y^2 + (z - z_0)^2)^2 \tag{7.18}$$

where z_0 is the distance from the front surface to the center of symmetry of the spherical gradient.

The actual equations that are used in various programs generally follow these forms, but will differ in the manner of specifying the coefficients. Sometimes it is quite difficult to visualize the symmetry presented by these forms. The index can, of course, follow any form desired.

The paraxial optics of gradient lenses can be a bit complicated. The location of images is determined by the index gradient as well as the powers of the surfaces. One important application of radial gradient materials is rod lenses, in which the image positions are entirely determined by the index gradient. The important fact to remember is that the conjugate image positions that eliminate the linear and quadratic wavefront errors will establish the appropriate image position.

Modeling these components in a lens design program has several difficulties. The first problem is that the index profile must match the profile that will be obtained in the material. This matching should also include the variation of the index and gradient with wavelength. The satisfactory resolution of modeling of the material may require the participation of the designer in prescribing tests that can verify this relationship. The next step is to define the second-order image formation positions for the lens, which usually has to be obtained by examining the second-order wavefront error. Beyond this, the optimization process can proceed much as usual, as the requirements upon image quality and isoplanatism are common to all optical image-forming systems.

7.2.6 DIFFRACTIVE AND BINARY OPTICS

The passage of a ray across a surface is determined by the indices of refraction and the angle of incidence. If a locally periodic diffraction grating is impressed on the surface, diffraction of the section of wavefront about the incident ray will take place. The direction of the diffracted energy will constitute a new ray direction which will change the apparent refracted or reflected ray direction. The amount of energy that will be diffracted into various orders to define the new ray direction will depend upon the efficiency of the local grating. This efficiency is determined by the amplitude and phase characteristics of the impressed grating. It is obvious from physical optics that a finite area of incident wavefront is required to produce a defined diffracted wavefront, but each order of the diffracted wavefront can be described by the direction of a bundle of rays. In this manner, the aberration and propagation characteristics of the diffracted rays may be used in a conventional manner to describe the aberrations and image forming characteristics of the diffracted wavefronts.

Diffraction of light by a grating follows the well-known grating equation

$$\sin\theta - \sin\theta_0 = \frac{m\lambda}{p} \tag{7.19}$$

where p is the period of the grating, m is the order, λ is the wavelength of the light in the current index of refraction, and θ and θ' are the angles of the ray incident and following the grating. This equation can be used in ray tracing by computing the local grating period and direction, and appropriately applying the equation at each point of incidence of the ray on the surface. It is understood, of course, that this is a geometrical representation of the passage of the rays through the surface, and the efficiency of propagation into each order is not stated by this equation. The local period can be constant, as in the case of a linear ruled grating on a surface, or may vary with position on the surfaces, as in a zone plate or a holographic optical element. The grating thus must be described by a period and a direction for each location on the surface. In some cases these parameters will be obtained for an analytical description in terms of the construction parameters for the holographic setup that will produce the grating. In other cases, a polynomial representation will be used to describe the grating properties.

There are many approaches to the design of optical systems with diffractive components. The earliest suggestions are based upon the use of Fresnel zone plates as imaging elements. The design of spectrometers with linear ruled gratings is well developed. Representation of a zone plate is basically that of a circular grating with a spatial frequency varying linearly with the height in the aperture. In this case, p is inversely proportional to the aperture height. A little thought will show that a plane thin zone plate will produce an image of a distant point source at a distance from the plate inversely proportional to the order of the diffraction and directly proportional to the period of the grating at specific aperture height. The focal length of a Fresnel zone plate is given by

$$f = \frac{Y_m^2}{m\lambda} \qquad d_m = \frac{f\lambda}{Y_m} \tag{7.20}$$

where Y_m is the radius of the aperture at which the mth Fresnel zone is found. This will determine the focal length or, of course, the power of the zone plate, in terms of the construction parameter of the plate and the wavelength. The period at the edge of the aperture of the zone plate is given by d_m, and varies inversely as the height from the axis of the plate. The aberrations of such imaging can be determined by ray tracing.

If the zone plate is placed upon a curved surface, the ray-tracing rules apply, but the aberrations produced will be different. Similarly, if the spatial frequency of the grating varies by a higher power than linear, the focusing grating can introduce a high-order focusing to the wavefront, and can act similarly to an aspheric surface.

The most interesting applications of diffractive optics arise from the use of complex gratings, such as holograms, as optical components. These diffractive components are produced by recording the interference of light from two or more coherent sources. Replacement of the two sources by one source and the recorder hologram will reproduce the second source, as in a normal hologram. Placing any complex object prior to the hologram will produce an image which is related to the process required to regenerate the second source. The aberrations of such holographic optics have been studied extensively. Computer-generated diffractive structures based upon defining the local period in a functional manner, and then quantizing the actual surface phase or intensity structures in some specific manner, lead to many other possible forms. The ray analogy becomes inadequate for dealing with many of these diffraction structures, and a full diffraction or propagation calculation is required to describe the image formation taking place.

Most lens design programs permit the description of diffractive components in several ways. Holographic optics can be generated by describing the coordinates of the constructing sources. A function form may be used to describe a computer-generated diffractive structure. The local period is computed from the chosen description. Ray tracing can proceed, once the diffraction order is specified. The actual efficiency of propagation will, of course, require a detailed diffraction calculation.

There is a simplified approach to design with diffractive optics that can be used to explore regions of possible solution. This model, commonly called the Sweatt model (Sweatt 1978), replaces the complicated grating with an equivalent refractive component of vanishingly small thickness and arbitrarily high index of refraction. Such replacement components can contain aspheric surfaces of any chosen symmetry. In the limit, these effective components match the behavior of the actual diffractive surface for a specified diffraction order.

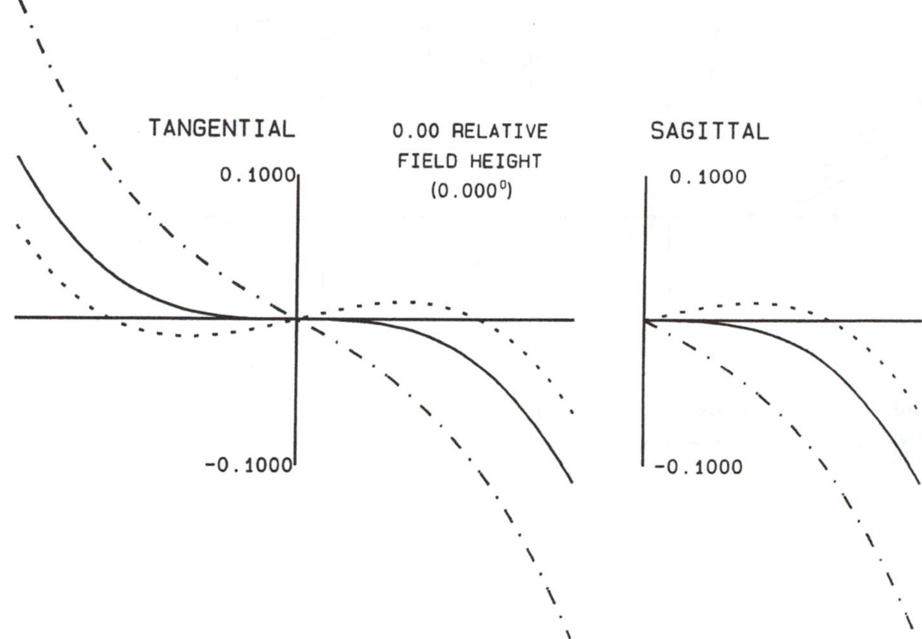

FIGURE 7.156 Ray trace data on a convex-plano singlet: ······, 656.3 nm; ——, 587.6 nm; ·—·—·, 486.1 nm.

The versatility of these diffractive optical components is immense. The limits upon application are also significant. The major limit in application is the strong wavelength dependence of the diffracted light. The dispersive characteristics of a diffractive component can be described in analogy to a refractive component by a *V*-number. Using the zone plate focal length equation, 7.20, the effective *V*-number of a diffractive component is

$$V = \frac{\lambda}{\delta\lambda} = -3.453 \tag{7.21}$$

for the usual *F–C* band. This indicates a very dispersive effective material, and the inverse behavior indicated by the negative *V*-number is significant. It is possible to use a zone plate of the appropriate power to compensate the chromatic characteristics of a refractive component. Compensation for chromatic aberration will be accomplished by an addition of power of the same sign as the basic lens. Note that the actual dispersion behavior is different, however. The partial dispersion for the usual *F–C* spectral region will be 0.4059, which makes the effect of dispersion quite different from that of a normal glass. For example, Bk7 has a partial dispersion of 0.3079.

A simple example will indicate some of the possibilities of a diffractive surface. Start with a single reasonably thin F/5, 100 mm focal length element made of Bk7 glass. The ray intercept curves for this convex-plano element are shown in Figure 7.156. The plots show a significant amount of

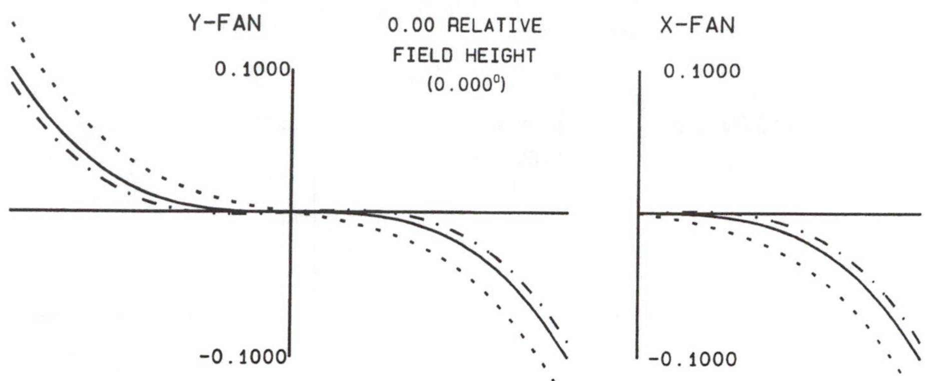

FIGURE 7.157 Ray intercept plot for lens in the previous figure, color-corrected with a diffractive optical element on the plano surface: ······, 656.3 nm; ——, 587.6 nm; ·–·–·, 486.1 nm.

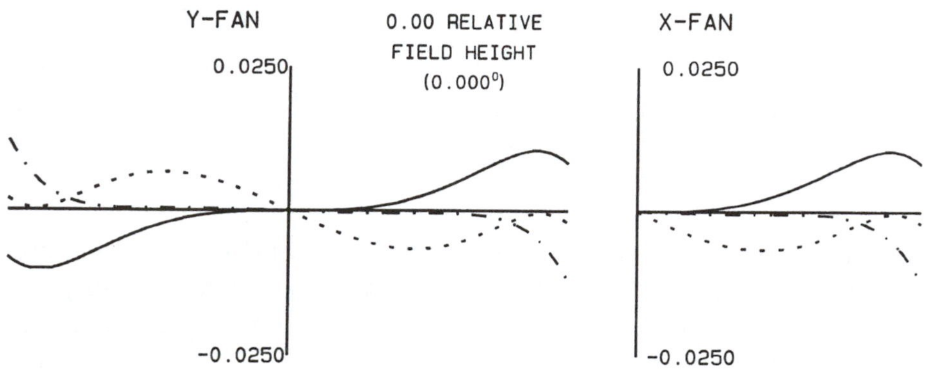

FIGURE 7.158 Lens of the previous figure with an aspherizing grating added to the spherical surface: ······, 656.3 nm; ——, 587.6 nm; ·–·–·, 486.1 nm.

spherical and chromatic aberration as can be expected. A diffractive component modeled by having a quadratic phase behavior with aperture, is added to the plano surface. Figure 7.157 results from correcting the difference in the ray aberration at the *c* and *F* wavelengths to zero at seven-tenths of the aperture to be corrected. This is the normal correction for chromatic aberration, and the plot shows that the chromatic aberration is corrected and the residual error is almost all spherical aberration. Then, a fourth-order "aspherizing" grating is applied to the plano surface to correct the spherical aberration. (It is possible that a better choice for the location of the fourth-order grating would be the first surface, but the problems of fabrication need to be considered as well as theory.) The net correction for this component is shown in Figure 7.158, which is now several times reduced over the original ray intercept plot. The spherical aberration is

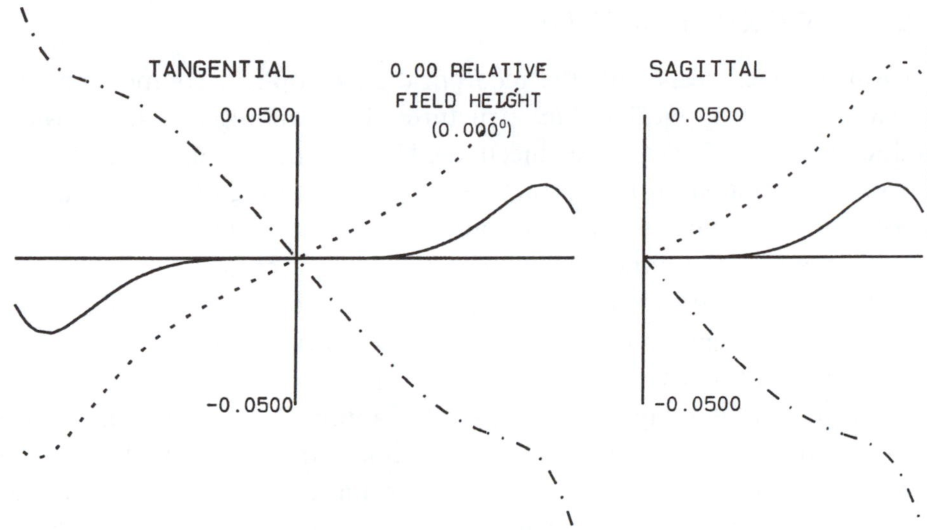

FIGURE 7.159 Aspheric lens, no diffractive optical element: ······, 656.3 nm; ——, 587.6 nm; ·–·–·, 486.1 nm.

corrected, and the major remaining aberration is a secondary chromatic error due to the mismatch of the partial dispersions of the glass and the diffractive component. Just as a comparison, Figure 7.159 is a ray intercept plot for the case of an aspheric and no diffractive surface correction.

The conclusions will change with increased numerical aperture, but it is clear that the diffractive surface provides an important color correction tool that can assist in the minimization of aberrations. Higher orders of diffractive phase correction are possible, and it is clear that compensation for variation of spherical aberration with wavelength could be carried out. It is important in carrying out an exercise of this type that the addition of a diffractive component must be accompanied by additions to the merit function that properly use these capabilities. This example examined only the axial image correction with a single element, but it is evident that permitting the shape of the lens to change will add additional variables that may be useful in correction across a wide field.

The example shown illustrates the potential for aberration correction. Once the parameters for the power distribution for the diffractive component are determined, the efficiency of the element, and the potential of shaping the surface relief that is used to form the diffractive surface needs to be used to maximize the efficiency into the desired diffraction order. The use of diffractive components in multielement systems is somewhat harder to understand. As one example of the possibilities, consider the nominal single meniscus construction.

7.2.7 INTEGRATED OPTICS

Integrated optics deals with the incorporation of optical components into optical and electrooptical surface structures. The building blocks are waveguides, fibers, and free-space connections. Usually sources and detectors will be closely integrated into the structure, and the coupling of sources, lenses, fibers, and waveguides is an important subject. The design of these planar components closely follows the rules of geometrical optics. The image formation is dominated by wave propagation. It is obvious that one of the next growth areas in lens design will be in the design and integration of these components into new types of optical systems.

The process of design follows much of the approach discussed earlier. The definition of the merit function and attaching adequate modeling to the design programs has yet to be carried out. Most integrated component design is presently carried out using programs that are specific to the technology.

7.2.8 ILLUMINATION OPTICS

Optical systems falling into this category are those intended for conveying illumination rather than an image. Similar principles of design apply as in imaging systems, except that the criterion for success is measured in the concentration and uniformity of the irradiance on a specified surface.

There is a special set of illumination devices called nonimaging light concentrators. These succeed in increasing the efficiency of light transfer by permitting drastic violations of the sine condition. The optical principles involved in such devices follow the rules of geometrical optics, but the technique and criteria for design differ considerably from imaging optical systems.

The illumination design problem is that of conveying as much of the available energy from a source through an aperture onto a surface with an irradiance distribution as close to some defined form as possible. The source may be incoherent, such as a light bulb or the sun, or may be coherent, such as a laser. Some level of partial coherence is also likely. The surface being conveyed to may be at an indefinite distance, in which case the angular distribution of the irradiance pattern becomes important. In many cases, the illumination system will operate in conjunction with some imaging system. The figure of merit then becomes that of imaging the source distribution into the pupil of the imaging system but with as uniform as possible a distribution of the irradiance on the image surface.

Some lens design programs include features to carry out an exact calculation of the image irradiance by computation of the fundamental radiometric integral from ray trace data. This will provide a considerably more accurate estimate of appearance of the irradiance on the focal surface. However, it is the responsibility of the user to be sure that the modeling of the optical system is correct, and properly matches the system to be built.

The design process is similar to that described earlier in this book. The variables, constraints, and surface descriptions need to be identified. The merit function is based upon ray tracing, but with a calculation of the illumination function over a desired surface as the measure of "image" quality.

REFERENCES

Baker 1940 Baker, J. G. *A Family of Flat-field Cameras, J. Am. Phil. Soc.* 82, 339

Betensky 1992 Betensky, E. Zoom lens principles and types, *Proc. SPIE* CR41, 88

Cox 1964 Cox, A. *A System of Optical Design*, Focal Press, London

Glatzel 1980 Glatzel, E. New lenses for microlithography, *Proc. SPIE* 237, 310

Hopkins 1962 Hopkins, R. E. *Optical Design, MIL-HDBK-141*, available from Sinclair Optics, Fairport, NY

Kingslake 1946 Kingslake, R. Classification of photographic lens types, *J. Opt. Soc. Am.* 36, 251

Kingslake 1978 Kingslake, R. *Lens Design Fundamentals,* Academic Press, New York

Korsch 1991 Korsch, D. *Reflective Optics*, Academic Press, Boston

Laikin 1995 Laikin, M. *Lens Design*, Marcel Dekker, New York

Linfoot 1955 Linfoot, E. H. *Recent Advances in Optics*, Oxford University Press, London

Mandler 1980 Mandler, W. Design of basic double Gauss lenses, *Proc. SPIE* 237, 222

Ray 1988 Ray, S. *Applied Photographic Optics*, Focal Press, London

Rutten and van Venrooij 1988 Rutten, H. and van Venrooij, M. *Telescope Optics Evaluation and Design*, Willmann-Bell, Richmond, VA

Schroeder 1992 Schroeder, D. J. *Astronomical Optics*, Academic Press, New York

Smith 1992 Smith, W. J. *Modern Lens Design*, McGraw-Hill, New York

Sweatt 1978 Sweatt, W. C. Designing and constructing thick holographic optical elements, *Appl. Opt.* 17, 1220

Wetherell and Rimmer 1972 Wetherell, W. and Rimmer, M. General analysis of aplanatic Cassegrain, Gregorian and Schwarzschild telescopes, *Appl. Opt.* 11, 2817

Wöltche 1980 Wöltche. W. Optical systems design with reference to the evolution of the double Gauss lens, *Proc. SPIE* 237, 202

8 SUMMARY AND CONCLUSIONS

8.1 Putting It All Together

In this book, the process of lens design has been dissected. The combination of art and science necessary to successfully carry out a design have been demonstrated. Each request for a design requires a combination of the skills that are brought to be resident in the mind of a skilled designer. The components for success in optical design are acquired through a combination of study and practice. The same is true of any acquired skill.

By now it is evident that optical design requires access to up-to-date computer programs in order to be competitive. It should also be evident that the computer program alone does not produce the design. The algorithms resident in the program are a consequence of the history and ingenuity that have taken place in the field. Each new design task requires a new path to be generated under the guidance of the designer using the computation tools. The successful designer does not just react to a specification provided by the customer, but is an active participant in developing the solution to the problem.

This book has been directed toward supplying a view into the process of design. The introduction to geometrical and physical optics, aberrations, and image evaluation defines the basis for optical design methods. The examples carried out on a number of types of lenses illustrate how the process of design is carried out using different computer design tools.

A guide to carrying out a design is now useful. Most of the steps that are required have been described somewhere in this book. The usual beginning of a design project is a statement of need from a customer. Usually the customer is not well versed in optical design, but has an appreciation of a need to be fulfilled. The details of how the problem is to be solved are usually secondary to a functional description of the requirement. This request is given to the designer, and is the most important part of the first interaction between the designer and the customer.

Figure 8.1 is a flow diagram for the process of optical design. Each block in the diagram is a step in the process. The arrows and decision points are important to the successful flow of activity leading to a final successful conclusion. As a simplified road-map to this book, Figure 8.1 includes the numbers of the chapters containing the most important information related to each step of the process.

A starting point for the designer is the list of topics that a designer should be aware of when beginning a design. This list is shown in Figure 1.2. The designer should begin reviewing this list as he or she reviews the initial problem statement. The optical requirements should be made as clear as possible. If it is not possible to obtain a unique set of optical specifications, then the range of allowable solutions that are to be evaluated should be determined. (As an aside, the negotiating skills of the designer also must be attuned to the possibility of a stubborn customer who is unwilling to discuss any points of the initial specifications, or to provide any clarification as to vague points. A wise designer will decide then as to whether this is a wise venture to enter into. Psychology and human relations are a very real part of the design business!)

The first step in actually beginning the design is a check for feasibility. The requirements for illumination and image quality must be physically achievable. Verification against the image quality limitations in Chapters 2 and 4 is important. An evaluation against the tolerance limits discussed in Chapter 6 can be used to determine how close the requirements are to the physical limits. A layout of the proposed design must be made to examine the general fit to the customer's required interface.

Next the starting point, the variables, and the merit function must be determined. Frequently the starting point can be a previous design, a patent, or even one of the designs discussed in this book, or one of the other sources listed in the references. (If the basic design is taken from a patent, a check upon the breadth of the patent claims should be considered before carrying on too far.) If a start from first principles is required, a spreadsheet approach can be used to determine the first-order requirements. Alternately, the first-order arrangement may be determined directly using a lens design program. (At the time of writing, some graphical lens setup programs are beginning to become available.)

The decision about usable materials next needs to be made. The long list in Figure 1.2 again becomes important. Availability, weight, cost, spectral region, and ability to fabricate drive this decision as much as the image quality requirements. The task of setting the variables has been discussed in several places in this book. The choice of merit function, and the strategy for applying the aberration goals and boundary conditions is obviously important. Today the choice is most likely to start with one of the default

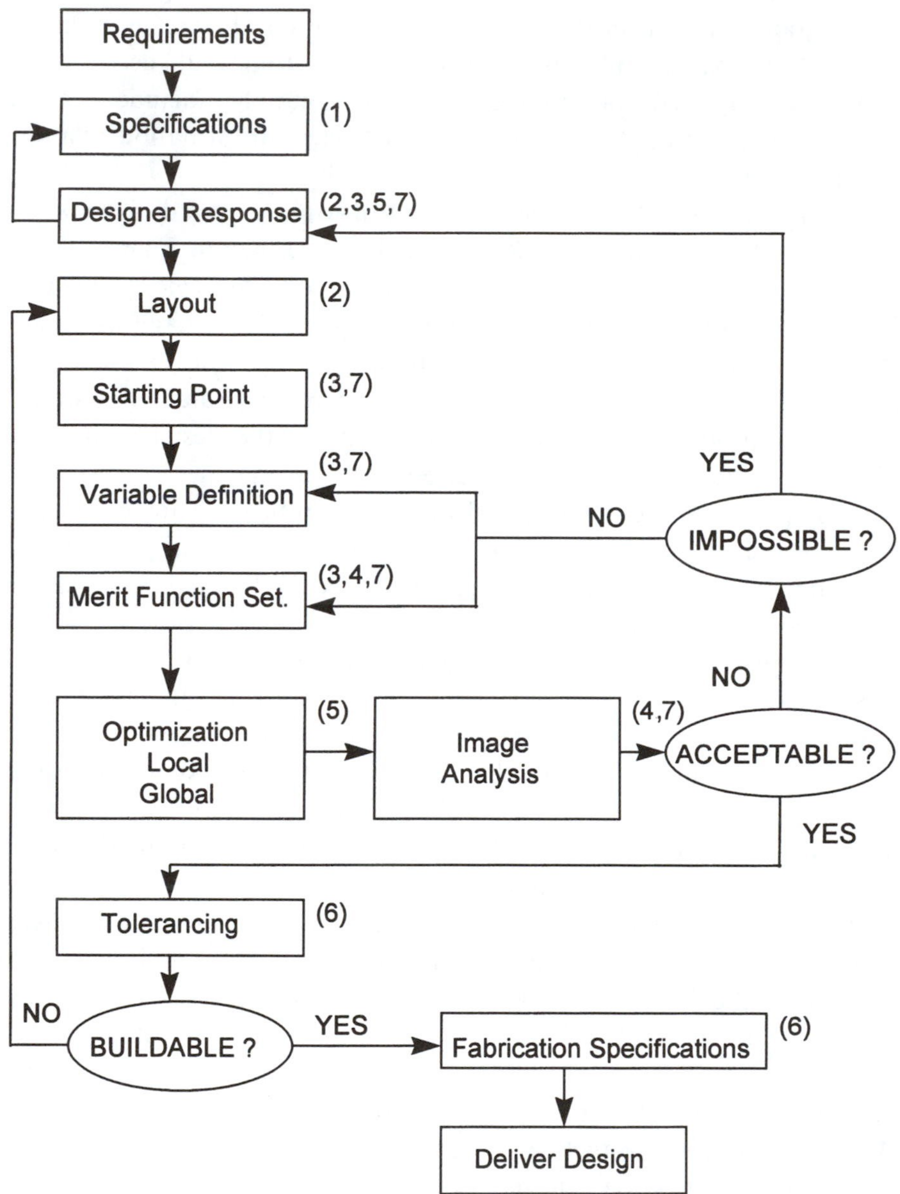

FIGURE 8.1 The important steps involved in carrying out an optical design. Numbers in parentheses identify the chapters relating to the steps shown.

merit functions and add constraints as they appear to be needed. In some cases, the initial merit function will most appropriately be built up as the problem proceeds.

Analysis of the intermediate designs, as well as the final design, needs to be made with the content of Figure 1.2 in mind. The designer will have to decide whether the solution is a step in the right direction, and whether the desired aim will actually be reached. Chapter 4 contains a number of examples of sample image quality evaluations, including the behavior with focus and field, that will be useful in determining whether a lens is actually the best solution for the problem.

Modern global optimization techniques are quite useful in permitting the designer to scan a number of possible regions in variable space. There is a tendency to simply select the lowest merit function values to provide additional starting points of completing the design. However, a scan of the nature of many of the proposed solutions that the global search provides will often provide clues as to the limits on behavior of the lens that can be very useful.

As usual, the job is not done until the tolerancing is completed. In most cases, this will be the least interesting part of the task for most designers, as it does not appear to be as creative as optimization. In reality, this is not true, as the success in fabricating the lens will be directly related to the suggestions that result from the tolerancing operation.

Finally, the designer must be a constant communicator. The complex interaction of optics, mechanics, electronics and environment in most modern instruments has eliminated the possibility of "tossing a design over the transom" to the system engineer and fabrication shop. (Actually, today, few offices have transoms. Most do not even have doors.) The designer needs to communicate efficiently the lessons learned during design that might affect decisions made elsewhere on the project. The designer also has to learn how to express the state of the design not in "designer talk" but in language understandable to the customer.

8.2 Developing a Discipline

The steps just discussed are important parts of the flow of activity in a specific project. A successful designer will develop a solid technical basis for decisions to be made at each step of the design. This is obtained partially by experience, somewhat by training, and significantly by study and awareness of other work being carried out in the field. A very solid understanding of basic geometrical and physical optics is required. Secondarily an appreciation of, and sensitivity to, the issues involved in practical aspects of mechanical engineering, optical fabrication, and optical system integration is required. The reader is encouraged to delve deeper into the subject through some of the references given in this book, as well as following the reference trail included in each of those books and articles.

8.3 Closing the Task

Some of the examples in this book illustrated another subtle point in optical design. The designer must know when to stop, not just to work until the budget is exhausted. The astute designer will sense when the design has either exceeded expectations, or has stagnated, and a new direction is indicated.

In any case, we have reached the end of this book. Many of the important topics in optical design have been addressed in the depth necessary to develop an appreciation and some skill in the subject. The reader is encouraged not to stagnate in his or her pursuance of further knowledge of the art and science of optical design.

INDEX